Microbial Biotechnology

Microbial Biotechnology

Role in Ecological Sustainability and Research

Edited by

Pankaj Chowdhary
Sujata Mani
Preeti Chaturvedi

WILEY

This edition first published 2023

Registered Office
John Wiley & Sons, Inc., 111 River Street, Hoboken, NJ 07030, USA

Editorial Office
Boschstr. 12, 69469 Weinheim, Germany

For details of our global editorial offices, customer services, and more information about Wiley products visit us at www.wiley.com.

Wiley also publishes its books in a variety of electronic formats and by print-on-demand. Some content that appears in standard print versions of this book may not be available in other formats.

Library of Congress Cataloging-in-Publication Data

Names: Chowdhary, Pankaj, editor. | Mani, Sujata, editor. | Chaturvedi, Preeti, 1977– editor.
Title: Microbial biotechnology : role in ecological sustainability and research / edited by Pankaj Chowdhary, Sujata Mani, Preeti Chaturvedi.
Other titles: Microbial biotechnology (John Wiley & Sons)
Description: First edition. | Hoboken, NJ, USA : John Wiley & Sons, Inc., 2023. | Includes bibliographical references and index.
Identifiers: LCCN 2022032480 (print) | LCCN 2022032481 (ebook) | ISBN 9781119834458 (Hardback) | ISBN 9781119834465 (adobe pdf) | ISBN 9781119834472 (epub)
Subjects: LCSH: Microbial biotechnology.
Classification: LCC TP248.27.M53 M5158 2023 (print) | LCC TP248.27.M53 (ebook) | DDC 660.6/2–dc23/eng/20220821
LC record available at https://lccn.loc.gov/2022032480
LC ebook record available at https://lccn.loc.gov/2022032481

Cover Design: Wiley
Cover Image: © Aliaksei Marozau/Shutterstock

Set in 10/12pt TimesNewRomanMTStd by Straive, Pondicherry, India

CONTENTS

Part III: Current Trends and Future Possibilities

CONTRIBUTORS

Nurul H. Adenan
School of Science,
Monash University Malaysia,
Selangor, Malaysia;
School of Biology,
Universiti Teknologi,
MARA (UiTM),
Negeri Sembilan, Malaysia

Gautam Anand
Department of Biotechnology,
Deen Dayal Upadhyaya Gorakhpur University,
Gorakhpur, Uttar Pradesh, India;
Department of Plant Pathology and Weed Research,
Agricultural Research Organization,
The Volcani Center,
Rishon LeZion, Israel

Fuad Ameen
Department of Botany and Microbiology,
College of Science,
King Saud University,
Riyadh, Saudi Arabia

Juliana H-P Américo-Pinheiro
Post-graduate Program in Environmental Sciences,
Brazil University,
São Paulo, Brazil

Arnab Banerjee
Department of Environmental Science,
Sant Gahira Guru Vishwavidyalaya,
Ambikapur, Chhattisgarh, India

Paul O. Bankole
Department of Pure and Applied Botany,
College of Biosciences,
Federal University of Agriculture,
Abeokuta, Ogun, Nigeria

Piotr Barbaś
Potato Agronomy Department,
Plant Breeding, and Acclimatization Institute,
National Research Institute,
Serock, Poland

María-del-Carmen Durán-Domínguez-de-Bazúa
Laboratorios de Ingeniería Química Ambiental y de Química Ambiental,
Departamento de Ingeniería Química,
Facultad de Química,
Universidad Nacional Autónoma de México, UNAM,
Mexico City, Mexico

Enrique R. Bazúa-Rueda
Laboratorios de Ingeniería Química Ambiental y de Química Ambiental,
Departamento de Ingeniería Química,
Facultad de Química,
Universidad Nacional Autónoma de México, UNAM,
Mexico City, Mexico

Akanksha Behl
Research Scholar, Amity Institute of Forensic Sciences,
Amity University,
Noida,Uttar Pradesh, India

Marisela Bernal-González
Laboratorios de Ingeniería Química Ambiental y de Química Ambiental,
Departamento de Ingeniería Química,
Facultad de Química,
Universidad Nacional Autónoma de México, UNAM,
Mexico City, Mexico

Sartaj A. Bhat
River Basin Research Center,
Gifu University,
Gifu, Japan

Muhammad Bilal
School of Life Science and Food Engineering,
Huaiyin Institute of Technology,
Huai'an, China

Andreas Burkovski
Department of Biology,
Friedrich-Alexander-Universität Erlangen-Nürnberg,
Erlangen, Germany

Tania Chalotra
School of Biotechnology,
Shri Mata Vaishno Devi University,
Jammu and Kashmir, India

Preeti Chaturvedi Bharagava
Aquatic Toxicology Laboratory,
Environmental Toxicology Group,
CSIR-Indian Institute of Toxicology Research,
Lucknow, Uttar Pradesh, India

Ángel E. Chávez-Castellanos
Laboratorios de Ingeniería Química Ambiental y de Química Ambiental,
Departamento de Ingeniería Química,
Facultad de Química,
Universidad Nacional Autónoma de México, UNAM,
Mexico City, Mexico

Ram N. Bharagava
Department of Environmental Microbiology,
School for Environmental Sciences,
Babasaheb Bhimrao Ambedkar University (A Central University),
Lucknow, Uttar Pradesh, India

Bernadetta Bienia
Food Production and Safety Plant,
Carpathian State University in Krosno,
Krosno, Poland

Subir K. Bose
Department of Biotechnology & Microbiology,
Meerut Institute of Engineering and Technology,
Meerut, Uttar Pradesh, India

Shraddha Chauhan
Aquatic Toxicology Laboratory,
Environmental Toxicology Group,
CSIR-Indian Institute of Toxicology Research,
Lucknow, Uttar Pradesh, India

Shaohua Chen
State Key Laboratory for Conservation and Utilization of Subtropical Agro-Bioresources,
Guangdong Province Key Laboratory of Microbial Signals and Disease Control,
Integrative Microbiology Research Centre,
South China Agricultural University,
Guangzhou, China;
Guangdong Laboratory for Lingnan Modern Agriculture,
Guangzhou, China

Jyoti Choudhary
Department of Biotechnology,
Chinmaya Degree College,
Haridwar, Uttarakhand, India;
Department of Botany and Microbiology,
Gurukul Kangri Vishwavidyalya,
Haridwar, Uttarakhand, India

Pankaj Chowdhary
Environmental Microbiology Laboratory,
Environmental Toxicology Group,
CSIR-Indian Institute of Toxicology Research (CSIR-IITR),
Lucknow, Uttar Pradesh, India

Guangyu Cui
State Key Laboratory of Pollution Control and Resource Reuse,
Tongji University,
Shanghai, China

Digvijay Dahiya
Department of Biotechnology,
National Institute of Technology,
Tadepalligudem, Andhra Pradesh, India

Ramesh C. Dubey
Department of Botany and Microbiology,
Gurukul Kangri Vishwavidyalya,
Haridwar, Uttarakhand, India

Shruti Dwivedi
Department of Biotechnology,
Deen Dayal Upadhyaya Gorakhpur University,
Gorakhpur, Uttar Pradesh, India

Luiz F. R. Ferreira
Waste and Effluent Treatment Laboratory,
Institute of Technology and Research (ITP),
Tiradentes University,
Aracaju, Sergipe, Brazil

Kuruvalli Gouthami
Department of Biochemistry,
School of Allied Health Sciences,
REVA University,
Bangalore, India

Sanjay Govindwar
Department of Biochemistry,
Shivaji University,
Kolhapur, Maharashtra, India;
Department of Earth Resources and Environmental Engineering,
Hanyang University,
Seoul, South Korea

Supriya Gupta
Department of Biotechnology,
Deen Dayal Upadhyaya Gorakhpur University,
Gorakhpur, Uttar Pradesh, India

Manoj Kumar Jhariya
Department of Farm Forestry,
Sant Gahira Guru Vishwavidyalaya,
Ambikapur, Chhattisgarh, India

Suhas Kadam
Research Institute for Dok-do and Ulleung-do Island,
Department of Biology,
School of Life Sciences,
Kyungpook National University, Daegu, South Korea;
Department of Biochemistry,
Shivaji University,
Kolhapur, Maharashtra, India

Bhargab Kalita
Microbial Biotechnology and Protein Research Laboratory,
Department of Molecular Biology and Biotechnology,
Tezpur University,
Tezpur, Assam, India;
Proteomics Lab,
National Centre for Cell Science,
Pune, Maharashtra, India

Maximilian Koch
Department of Biology,
Friedrich-Alexander-Universität Erlangen-Nürnberg,
Erlangen, Germany

Nahid Khan
Department of Farm Forestry,
Sant Gahira Guru Vishwavidyalaya,
Ambikapur, Chhattisgarh, India

Rahul Khandare
Amity Institute of Biotechnology,
Amity University,
Maharashtra, India

Ashish Kumar
School of Biotechnology,
Shri Mata Vaishno Devi University,
Jammu and Kashmir, India

Abinaya Lakshmi
PG Department of Microbiology,
The American College,
Madurai, Tamil Nadu, India

Iswareya Lakshimi
PG Department of Microbiology,
The American College,
Madurai, Tamil Nadu, India

Lavanya Lakshminarayana
Department of Biochemistry,
School of Allied Health Sciences,
REVA University,
Bangalore, India

Manuel J. Leal-Gutiérrez
Laboratorios de Ingeniería Química
Ambiental y de Química Ambiental,
Departamento de Ingeniería Química,
Facultad de Química,
Universidad Nacional Autónoma de
México, UNAM,
Mexico City, Mexico

Xuyang Lei
Department of Resource and
Environmental Engineering,
Hebei Vocational University of Technology
and Engineering,
Xingtai, China

Fusheng Li
River Basin Research Center,
Gifu University,
Gifu, Japan

Sujata Mani
Department of Biochemistry,
Gramin Science Vocational College,
Nanded, Maharashtra, India

Amarnath Mishra
Amity Institute of Forensic Sciences,
Amity University Uttar Pradesh,
Noida, Uttar Pradesh, India

Sandhya Mishra
Environmental Technologies Division,
CSIR – National Botanical Research
Institute,
Lucknow, Uttar Pradesh, India

Sarad Kumar Mishra
Department of Biotechnology,
Deen Dayal Upadhyaya Gorakhpur
University,
Gorakhpur, Uttar Pradesh, India

Sikandar I. Mulla
Department of Biochemistry,
School of Allied Health Sciences,
REVA University,
Bangalore, India

Anandkumar Naorem
ICAR– Central Arid Zone Research
Institute,
Regional Research Station
Bhuj, Gujarat, India

Abhishek Patel
ICAR– Central Arid Zone Research
Institute,
Regional Research Station
Bhuj, Gujarat, India

Atin K. Pathak
School of Energy Management,
Shri Mata Vaishno Devi University,
Jammu and Kashmir, India

Piotr Pszczółkowski
Experimental Station for Cultivar
Assessment of Central Crop Research
Centre,
Dębowa Kłoda, Poland

Abbas Rahdar
Department of Physics,
University of Zabol,
Zabol, Iran

Neelu Raina
School of Biotechnology,
Shri Mata Vaishno Devi University,
Jammu and Kashmir, India

Abhay Raj
Environmental Microbiology Laboratory,
Environmental Toxicology Group,
CSIR-Indian Institute of Toxicology
Research (CSIR-IITR),
Lucknow, Uttar Pradesh, India

Abhishek Raj
Pt. Deendayal Upadhyay College of
Horticulture and Forestry,
Dr. Rajendra Prasad Central Agriculture
University,
Samastipur, Bihar, India

Zainab Rao
Department of Biotechnology,
Chinmaya Degree College,
Haridwar, Uttarakhand, India

Stefan Rosiwal
Department of Material Sciences,
Friedrich-Alexander-Universität Erlangen-Nürnberg,
Erlangen, Germany

Barbara Sawicka
Department of Plant Production and Commodities Science,
University of Life Sciences in Lublin,
Lublin, Poland

Vikas Sharma
Aquatic Toxicology Laboratory,
Environmental Toxicology Group,
CSIR-Indian Institute of Toxicology Research,
Lucknow, Uttar Pradesh, India

Parul Shukla
Centre for Sustainable Polymers,
Indian Institute of Technology Guwahati,
Guwahati, Assam, India

Swati Shukla
Department of Biotechnology,
Chinmaya Degree College,
Haridwar, Uttarakhand, India

Anuradha Singh
Aquatic Toxicology Laboratory,
Environmental Toxicology Group,
CSIR-Indian Institute of Toxicology Research,
Lucknow, Uttar Pradesh, India

Sourabh Kumar Singh
Research Scholar, Amity Institute of Forensic Sciences,
Amity University,
Noida,Uttar Pradesh, India

Shrutika Singla
Research Scholar, Amity Institute of Forensic Sciences,
Amity University,
Noida,Uttar Pradesh, India

Dominika Skiba
Department of Plant Production and Commodities Science,
University of Life Sciences in Lublin,
Lublin, Poland

Julio A. Solís-Fuentes
Laboratorios de Ingeniería Química Ambiental y de Química Ambiental,
Departamento de Ingeniería Química,
Facultad de Química,
Universidad Nacional Autónoma de México, UNAM,
Mexico City, Mexico

Yashdeep Srivastava
Department of Biotechnology,
RR Institute of Modern Technology,
Lucknow, Uttar Pradesh, India

Immanuel Suresh
PG Department of Microbiology,
The American College,
Madurai, Tamil Nadu, India

Aiman Tanveer
Department of Biotechnology,
Deen Dayal Upadhyaya Gorakhpur University,
Gorakhpur, Uttar Pradesh, India

Adeline S-Y Ting
School of Science,
Monash University Malaysia,
Selangor, Malaysia

Shiva Kumar Udayana
Department of Soil Science and Agricultural Chemistry,
MS Swaminathan School of Agriculture,
Centurion University of Technology and Management,
Gajapati, Odhisha, India

Swati Upadhyay
Department of Biotechnology,
Dr. A.P.J. Abdul Kalam Technical University (APJAKTU),
Lucknow, Uttar Pradesh, India

Vadamalai Veeraraghavan
Department of Biochemistry,
School of Allied Health Sciences,
REVA University,
Bangalore, India

Dinesh Yadav
Department of Biotechnology,
Deen Dayal Upadhyaya Gorakhpur University,
Gorakhpur, Uttar Pradesh, India

Kanchan Yadav
Department of Biotechnology,
Deen Dayal Upadhyaya Gorakhpur University,
Gorakhpur, Uttar Pradesh, India

Sangeeta Yadav
Department of Biotechnology,
Deen Dayal Upadhyaya Gorakhpur University,
Gorakhpur, Uttar Pradesh, India

Naik Yaseera
Department of Environmental Sciences,
Government Degree College Anantnag,
Jammu and Kashmir, India

PREFACE

The book *Microbial Biotechnology: Role in Ecological Sustainability and Research* discusses the potential role of microbial biotechnology in the betterment of our daily lifestyle. Today, microbial biotechnology is a rapidly growing segment of life sciences/biological sciences. It is well known that rapid industrialization and urbanization have resulted in contamination of all components of the environment, increasing public concern over the environmental brunt of wastewater polluted by anthropogenic sources. To mitigate this, numerous traditional wastewater treatment techniques, both physical and chemical, such as sedimentation or filtration techniques, oil–water separators, adsorption, membranes, coagulation, adsorption, activated sludge, trickling filtration, precipitation, and oxidation processes have been applied effectively. Besides, biological strategies through bacteria, fungi, algae, actinomycetes, etc., have also been used to remove environmental pollutants. But these conventional wastewater treatment methods have some limitations, require a noteworthy amount of energy, and most importantly involve pricey paraphernalia and their upkeep in maintaining microorganisms. From an environmental point of view, these recent and classic treatment technologies should be amplified to formulate them in a more viable and feasible manner. Contaminant mitigation or removal by using microbial technology is an attractive and potential alternative. Recent development in the field of biotechnology, molecular biology, ecology, and microbiology has been applied to develop different novel treatment methods involving novel strains of microorganisms and their desirable properties that could be applicable in the process of bioremediation. Various types of beneficial microbes are present in the ecosystem and they can play key roles in mitigating climate concerns, increasing green production technology, improving agriculture productivity, and providing a means of earning a livelihood.

On the other hand, pathogenic microorganisms are a warranted introduction to emergent therapies and disease prevention, and to gradually increasing agricultural profitability using microbial biocontrol agents and bio-fertilizers. Similarly, various potential microbes play critical roles in regulating the environment via their involvement in the production and intake/consumption of greenhouse gases (GHGs) and other air pollutants from the environment. Environmental pollutants such as industrial and pharmaceutical waste have emerged as a global threat, creating widespread antibiotic resistance and giving rise to drug-resistant strains of pathogens. The book details the environmental problems posed by antibiotics, including the various types of toxic environmental pollutants discharged from both natural and anthropogenic activities and their toxicological effects in environments, humans, animals, and plants. This book also highlights the recent advanced and innovative methods for the useful degradation and bioremediation of organic pollutants, heavy metals, dyes, etc., in wastewater. This book covers a wide range of topics: environmental microbiology, biotechnology, nanotechnology, green chemistry, environmental science, and environmental engineering, among others.

It is our hope that this book will also enhance the knowledge base of students, environmental scientists, environmental biotechnologists, microbiologists, biomedical scientists, and policymakers working in environmental microbiology, biotechnology, environmental sciences, and medical microbiology with both basic and more advanced facts about environmental issues and their challenges. Moreover, readers can also get up-to-date information and some background learning about existing environmental problems, their effects on human health, and ways to control or contain these effects by employing various effective approaches.

The editors would like to express their sincere thanks to the contributors for submitting their work in a timely and proper manner. The editors are also thankful to national and international reviewers for evaluation and valuable suggestions and comments to improve the book for readers. Dr. Chowdhary acknowledges the support received from their family, especially their father (Mr. Ram Chandra), and mother (Mrs. Malti Devi). Further, the editors also acknowledge the cooperation received from the Wiley team, and for their guidance to finalize this book.

Editors

Pankaj Chowdhary, PhD
Environmental Microbiology Laboratory, Environmental Toxicology Group
CSIR-Indian Institute of Toxicology Research (CSIR-IITR), Lucknow, India;
Society for Green Environment, New Delhi, India

Sujata Mani, PhD
Department of Biochemistry, Gramin Science Vocational College, Maharashtra, India.

Preeti Chatuervedi, PhD
Aquatic Toxicology Laboratory, Environmental Toxicology Group
CSIR-Indian Institute of Toxicology Research (CSIR-IITR), Lucknow, India

ABOUT THE EDITORS

Dr. Pankaj Chowdhary is Founder and President, Society for Green Environment (SGE) (www.sgeindia.org) at New Delhi India. Currently he is working as a Postdoctoral Fellow in CSIR-Indian Institute of Toxicology Research. He received his post-graduate degree (2011) in Biotechnology from Deen Dayal Upadhyaya Gorakhpur University Uttar Pradesh (UP), India. Afterward, he obtained his PhD (2018) in the area of Microbiology from the Department of Microbiology at Babasaheb Bhimrao Ambedkar University (A Central University), Lucknow, Uttar Pradesh, India. His PhD work has mainly focused on the role of ligninolytic enzyme-producing bacterial strains in the decolorizing and degradation of coloring compounds from distillery wastewater. His main research areas are microbial biotechnology, biodegradation and bioremediation of environmental contaminants in industrial wastewaters, metagenomics, biofuel, and bioenergy production. He has edited three books: *Emerging and Eco-friendly Approaches for Waste Management*; *Contaminants and Clean Technologies*; and *Microorganisms for Sustainable Environment and Health*. He has also authored two international books: *New Technologies for Reclamation of Industrial Wastewater* and *Recent Advances in Distillery Waste Management for Environmental Safety*. He has published more than seventy (70) research publications including research/review papers and two book reviews in national and international peer reviewed journals of high impact published by Springer, Elsevier, Royal Society of Chemistry (RSC), Taylor & Francis Group, and Frontiers. He has also published many national and international book chapters and magazine articles on the biodegradation and bioremediation of industrial pollutants. He has presented many posters/papers relevant to his research areas at national and international conferences. He has also served as a potential reviewer for various national and international journals in his respective areas of research. He is a lifetime member of the Association of Microbiologists of India (AMI); Indian Science Congress Association (ISCA) Kolkata, India; and The Biotech Research Society, India (BRSI).

Dr. Sujata Mani is currently working as an Assistant Professor and is actively engaged in teaching and research activities in the Department of Biochemistry, Gramin Science Vocational College, Vishnupuri, Nanded, Maharashtra, India. She has completed her PhD (2018) degree in Microbiology from Babasaheb Bhimrao Ambedkar Central University, Lucknow, Uttar Pradesh, India. During her PhD, she was awarded the Rajiv Gandhi National Fellowship (Both JRF and SRF). Her doctorate work was focused on the degradation and detoxification mechanisms of crystal violet from textile wastewater for environmental safety. Her major areas of research are biodegradation and bioremediation of synthetic dyes in textile wastewater through bacterial species producing ligninolytic enzymes. She has published review/research papers in reputed national and international journals with high impact factors. She has also published many book chapters in national and international edited books from Springer, Elsevier, and CRC Press. She has also qualified ICAR-NET in microbiology. In addition, she has guided graduation and post-graduation students through their dissertation processes. She is a lifetime member of the Association of Microbiologists of India (AMI) and the Indian Science Congress Association (ISCA), Kolkata, India.

Dr. Preeti Chaturvedi is currently working as a Senior Scientist in Environmental Microbiology Division at the CSIR-Indian Institute of Toxicology Research, Lucknow. Currently, her team is involved in working on the synthesis of native and engineered (surface modifications) biochar using different types of catalysts for enhancing the adsorption capacity for remediation and adsorption of pollutants from wastewater samples. In addition, she is involved in the development of green technology by effluent treatment for sustainable environment and health. The major areas of her research include bioremediation, toxicity assessment, metagenomics, microbial biotechnology, and proteomics. She has published more than 50 research/review papers in peer-reviewed SCI journals published by Springer, Elsevier, and Frontiers. She has also published national and international book chapters and magazine articles. She has been serving as an active reviewer for various reputable national and international journals related to her field. She has successfully completed a major research project funded by State Mission for Clean Ganga & Ground Water. She is a lifetime member of the Biotech Research Society of India (BRSI) and Association of Microbiologists of India (AMI).

Part I
Microorganism: An Introduction

1

Microbes and Environment: Recent Advancement in Environmental Biotechnology

Pankaj Chowdhary[1], Sujata Mani[2], Parul Shukla[3], and Abhay Raj[1]

1.1. INTRODUCTION

Rapid urbanization and industrialization have led to various anthropogenic activities that cause severe environmental impacts including soil pollution (use of fertilizers, pesticides), water pollution (leaching of chemicals, oil spills, etc.), and air pollution (emission of greenhouse gases, CO_2, NOx, SO_2, etc.). To tackle this situation, "environmental biotechnology" has come into consideration, which can offer various opportunities to provide environmental protection [1, 2]:

(i) Biological treatment of agricultural, industrial, hospital, and domestic effluents;
(ii) Bioremediation or biodegradation of contaminants present in soil and water;
(iii) Preservation and conservation of distinct species;
(iv) Monitoring fate of environmental pollutants;
(v) Sustainable production of bioproducts with less waste generation and toxic byproducts;
(vi) Utilization of organic biomass for bio-energy generation;
(vii) Employment of genetic engineering to produce better crop productivity and high yield;
(viii) Synthesis of biofertilizers to be utilized in agriculture/horticulture;
(ix) Forensic and diagnostic practices;
(x) Disease prevention and treatment; and
(xi) Food prevention and nutrition.

[1]*Environmental Microbiology Laboratory, Environmental Toxicology Group, CSIR-Indian Institute of Toxicology Research (CSIR-IITR), Lucknow, Uttar Pradesh, India*
[2]*Department of Biochemistry, Gramin Science Vocational College, Nanded, Maharashtra, India*
[3]*Centre for Sustainable Polymers, Indian Institute of Technology Guwahati, Guwahati, Assam, India*

Microbial Biotechnology: Role in Ecological Sustainability and Research, First Edition.
Edited by Pankaj Chowdhary, Sujata Mani, and Preeti Chaturvedi.

The development of affordable and potent biological techniques accompanying genome alteration, bioinformatics tools, high-resolution imaging and analytical instruments, and single-cell techniques has opened new avenues in various applications. Such applications use microbial communities such as viruses, bacteria (archaea and eubacteria), fungus, algae, protozoa, and other biomass, which have revolutionized the current scenarios. Various organic and inorganic contaminants, including radionuclides, heavy metals, PAHs (polycyclic aromatic hydrocarbons), PCBs (polychlorinated biphenyls), metalloids, and pesticides have been removed from wastewater, contaminated soil, and sediments using microorganisms [3, 4]. Microbes facilitate biodegradation via a series of biochemical reactions that involve enzymes. However, physicochemical characteristics (moisture content, pH, temperature) of the environment, concentration and chemical nature of the contaminant, presence of oxygen, and nutrition availability to microorganisms are the most influential factors affecting bioremediation [5]. For instance, *Bacillus, Methanogens, Nocardia, Azotobacter Pseudomonas, Rhizopus arrhizus, Ganoderma applantus, Rhodococcus, Aspergillus niger, Arthrobacter, Methosinus, Mycobacterium, Stereum hirsutum, Corynebacterium, Pleurotus ostreatus, Flavobacterium alcaligenes, Phormidium valderium, Chlorella sp., Chlamydomonas sp., Parachlorella sp.,* [6, 7] have been utilized for mitigation of toxic pollutants. Moreover, to accelerate and provide efficient removal of pollutants, recombinant DNA (rDNA) technology has been employed. With the advent of rDNA technology, the area of biodegradation/bioremediation has been rejuvenated in terms of creating and developing novel microbial strains known as genetically modified organisms (GMOs) with high removal and degradation efficiency. For this purpose, various strategies have been used: (i) screening and cloning of potent genes; (ii) enhancing the enzyme expression; (iii) expressing degradation genes to construct engineered novel strains; and (iv) fusion of protoplast for enhancing gene functions [8]. In a study, recombinant *Caulobacter crescentus* for removal of soluble heavy metals [9], recombinant *Deinococcus radiodurans* for uranium removal from acidic/neutral aqueous wastes, vector pET21a(+)-merA for mineral mercury [10], and recombinant *Pseudomonas guguanensis* for crude oil-contaminated soil have been used. Besides bioremediation, use of numerous bacterial strains has been demonstrated in vaccine production, enzyme production, alcohol production, biofuel production, healthcare product synthesis, nanotechnology, and in production of transgenic crops. For example, 3-hydroxypropionic acid was produced using recombinant *Klebsiella pneumoniae* L17 in the bioelectrochemical system [11]. Similarly, recombinant *Escherichia coli*, namely phenylalanine ammonia-lyase (PAL), has been used for the synthesis of flavonoids [12].

Nevertheless, it is complicated to use and rely on engineered microbes as it is difficult to assess their effect on indigenous microbes and the ecosystem. In addition, genetic manipulation can cause certain undesirable outcomes such as the creation of herbicide-resistant weeds or adverse impacts on soil microbiota, reduced soil fertility, etc. This requires strict containment protocols and guidelines for the application of microbial communities in diverse fields [13].

This chapter provides comprehensive details of various microbial strains including bacteria, fungus, viruses, protozoa, and algae utilized in various sectors (science and technology, cosmetics, industries, diagnostics, etc.). Furthermore, all possible bioremediation applications such as remediation of environmental pollutants (organic and inorganic) have been covered along with an introduction to commercially utilized microbial strains. The current chapter emphasizes the beneficial impact and use of microorganisms in biofertilizers, fermentation, biopesticides, bioherbicides, and bioinsecticides, as well as a few other applications. Furthermore, the current state and extent of microbial biotechnology, as well as the benefits and disadvantages of employing microbial communities in many regions for numerous purposes, are explained.

1.2. MICROBES AND ENVIRONMENT

Microorganisms are omnipresent; they can be found everywhere, such as in air, water, food, soil, animal intestines, and also in extreme environments such as hot springs, glaciers, deep-sea vents, and rocks. Microbes cannot be seen with the naked eye since our eye's resolution is limited to 100–200 μm and microbial size ranges from about 0.2–200 μm; therefore, a microscope is needed to see them. There are a few exceptions, like fruiting bodies of some fungi, which are found to be larger. The extensive array of microorganism habitats reflects their vast diversity in metabolic as well as in biochemical traits that might have occurred through natural selection or genetic variations in their populations. Microorganisms provide plentiful substantiation of their presence, both favorably such as through bread production, fermenting sugar to wine and beer, flavoring cheese, and producing valuable products like antibiotics and insulin, and unfavorably through decaying matter or in spreading disease. The value of microbes in the earth's ecosystem is immeasurable. Microorganisms play a very important role in balancing our ecosystem as disintegrating animals and plants remain by converting them into simpler substances that can be easily used by other organisms [4, 14, 15]. This group of microorganisms includes algae, bacteria, fungi, protozoa, and viruses, which play a principal role in numerous natural processes that have been well summarized in the following subsections.

1.2.1. Bacteria (Archaea and Eubacteria)

Microbiology subjects came into the limelight through studies on bacteria and their importance to humans with experiments performed by scientists like Louis Pasteur, Robert Koch, and others during the late 1800s. The technique of culturing and isolating pure culture from the mixed culture in laboratories was usually performed for bacteria, but later was modified and applied for all microorganisms. The microbial world has been characterized into either prokaryotic or eukaryotic, among which bacteria have been put under the prokaryotic category. Until the late 1970s, all bacterial groups were thought to be closely related, but the later discovery of ribosomal RNA distinguished it into distinct groups, mainly eubacteria, archaea, and eukarya. Presently, eubacteria are known as true bacteria and form the main bacterial domain [16, 17].

Bacteria are the smallest free-living prokaryotic organisms whose cell size varies from 0.2 to 1.0 μm in length and exists in various cell shapes such as cocci, rods, and spirals. Although there are unicellular cells, the majority appears either in pairs, tetrads, chains, or clusters. They lack true nuclear envelopes and the genome is composed of single double-stranded DNA, which is found free in the cytoplasm. The bacterial genome codes around 3000–4000 genes and their lengths are approximately 4–6 million nucleotides. The cell envelope of bacteria is composed of two layers, of which the inner layer is composed of phospholipids and the outer cell wall layer is made up of lipids, proteins, and carbohydrates. The bacterial cell moves through flagella and file filaments, i.e. pili, which enables them to attach with other cells or soil particles. These pili also help in the process of conjugation, where one cell transfers its genetic material to other cells with their help [16, 17]. The bacterial cells reproduce asexually, a process known as binary fission, where one bacterial cell divides into two genetically identical bacterial cells. One of the most important characteristics of bacteria is their reaction during gram staining. On this basis, bacteria get divided into two types; Gram-positive and Gram-negative, and both have different physiology and cell structures.

Bacteria and archaea look alike when viewed through a microscope but have different biochemical activity, chemical composition, and even environments. Carbon is the building block for carrying out energy-driven activities like cell biosynthesis and metabolism. Oxygen is utilized by most of the bacterial cells but only a few, as well as archaea, grow anaerobically, i.e. in absence of oxygen by utilizing alternate electron acceptors such as sulfates and nitrates. All true bacterial cell wall consists of peptidoglycan, which is lacking in archaeans. Earlier, archaeans were noted to survive only under harsh conditions such as high acid or salt levels and in high temperatures, which is why they were known as extremophiles. But now we have learned that they can survive in normal environments and can be found in ecosystems such as soil. Based on morphology, it becomes very difficult to distinguish between bacteria and archaea. Later 16S rRNA phylogenetic sequencing classifications differentiated three domains of life, eukaryotes, bacteria, and archaea, which revealed archaea to be closer to eukaryotes than bacteria [18].

1.2.2. Fungus

Fungi are eukaryotic organisms and thus are more closely related to plants and animals than bacteria and archaea. Their cell wall is rigid, composed of chitin and glucans, and may be either unicellular or multicellular. Some fungi are so small that they can be viewed only through a microscope, while some are very large in structure, such as bracket fungi and mushrooms, which can be easily grown in soil or on damp logs. Fungi are heterotrophic organisms, e.g. saprophytic fungi, which feed on dead and decaying organic materials for energy [17]. Unicellular fungi such as yeast grow in a cylindrical thread-like structure, commonly known as hyphae, which might be either septate or non-septate. Hypha is a major group of fungi that comprises mycelium occupying the largest surface area in the soil, producing an array of enzymes. These enzymes start acting on the organic materials present in the soil to produce sufficient energy and nutrients that are required for fungal growth. In fungi, the reproduction process occurs through both asexual budding or binary fission and sexually through spore formation. Fungus is diverse and plays an important role in decomposition, as well as acting as predators, pathogens, mutualists, and endophytes of plants [16].

1.2.3. Algae

In contrast to bacteria, algae are eukaryotic organisms consisting of chlorophyll, which carry out the process of photosynthesis like plants, and also have a rigid cell wall. Algae are usually found in aquatic environments and soggy soil conditions. Algae may exist as unicellular or multicellular eukaryotes and can be found in various sizes from microscopic to approximately 120 m in length. They are also motile and exist in various shapes such as spindle, spherical, rod, or club-shaped. Multicellular algae are found in various forms as well as degrees of convolution. Some are found in colonies, some as filaments in which cells are attached from end to end, and some even aggregate to form single cells [4, 16].

1.2.4. Protozoa

These are single-celled eukaryotic microorganisms found in different shapes such as spherical, oval, or elongated but some have been reported with different shapes at different life cycle stages. Like algae, these also exist in a range of sizes from 1 μm in diameter to 2000 μm. Some protozoa are like animal cells in lacking the cell wall and ingesting food particles, but some are like plants, known as phytoflagellate protozoa, which perform the process of photosynthesis to obtain their energy needs [4, 16]. Some animal-like protozoan cells swim in water with cilia or flagella, or through beating actions which can be easily seen through a microscope in a drop of pond water. Some protozoa, such as amoebas, do not swim but creep by extending their cell portion as a pseudopod; this form of locomotion is known as an amoeboid movement.

1.2.5. Viruses

Viruses are extensively present in nature, causing infection to plants, animals, and microbes after coming in contact with them. Viruses lack their metabolic machinery to synthesize protein and energy generation, thus depending on host cells for carrying out their all-important functions, and are therefore known as obligate parasites. Once they enter any host cell, they take over the energy generating and protein-synthesizing system of cells for their purposes. Viruses have an extracellular form for carrying viral nucleic acid form only surrounded by a protein coat called a capsid (protects genes in the environment or outside the host cell) from one host cell to another cell [16]. The infectious virus particle with mature structure is called a virion, which ranges from 20–300 nm in size. Since most of the viruses are less than 150 nm in size, they are visible only through an electron microscope.

1.3. MICROBIAL INVOLVEMENT IN ECOLOGICAL/ ENVIRONMENTAL SUSTAINABILITY

Microorganisms are present in large numbers everywhere, and they play very important roles in biogeochemical cycles. Generally, most of the microbes present in our surroundings are beneficial to plants, animals, and humans, except for a few. More than half of the breathable oxygen in the air is generated by microbes. Microbes have the capability to keep their ecosystem clean if it's not overloaded with pollutants (Figure 1.1). The involvement of the microbes in cleaning up pollutants from the ecosystem and their benefits for humans and the environment has been summarized in detail in the following subsections.

1.3.1. Remediation of Environmental Pollutants

Water and soil resources worldwide have been heavily polluted by different types of organic as well as inorganic pollutants. These pollutants persist for a long time in the environment, causing severe threats to life. These pollutants reach to freshwater bodies through leaching as well as runoff methods from contaminated soil. The organic and inorganic pollutants reported in the soil as well as in water, such as PAHs, PCBs, pesticides, explosives, heavy metals, metalloids, and radionuclides, are discussed in detail in the following subsections.

1.3.1.1. Organic Pollutants

The widespread application of organic compounds over the last few decades has consequently led to insidious and persisting environmental threats. The long-term application of organic compounds in public health sectors and agricultural fields resulted in irrigational soils, surface and water, and foodstuff contaminations. The residues of the compounds in the environment have been reported even several years after their application [19]. Some examples of these organic compounds are pesticides, BTEX (benzene, toluene, ethylbenzene, and xylene), PAHs, PCBs, and explosives [19]. However, the major hurdle associated with the cleanup solutions of these organic compounds is the high cost of their remediation processes. But in recent years, various microbes have been reported in degrading these toxic

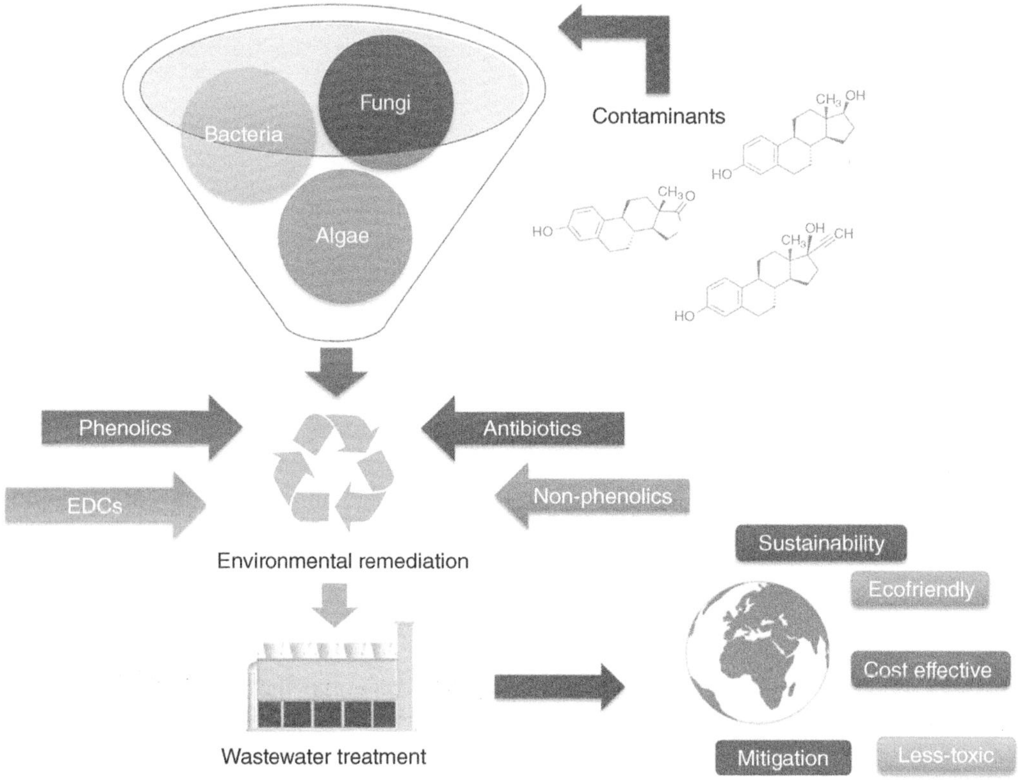

Figure 1.1 Contaminants and microbial treatments for environmental sustainability.

compounds in laboratory conditions [20–22]. Microbial bioremediation of some of the toxic organic compounds has been summarized here, and also their degradation through plant–bacterial association is summarized in Table 1.1.

1.3.1.1.1. Polycyclic Aromatic Hydrocarbons Groups of hydrophobic compounds whose molecules consist of two or more aromatic benzene rings are known as PAHs. PAHs are measured as the most recalcitrant compound among all included here. For the biodegradation of organic compounds of the PAHs group, several bacterial organisms were isolated but were able to metabolize compounds having low or intermediate molecular weight. For biodegradation of high molecular weight molecules, bacterial strains require voluntarily high carbon sources [19]. This is the what hastens the microbial biodegradation of PAHs in collaboration with some other sources like plants, to reduce its toxic effects more effectively. Daane [41] isolated a bacterial strain from rhizospheric soil of salt marsh plants that was capable of biodegrading PAHs from the contaminated sites. Similarly, Kuiper [42] found bacterial strain *Pseudomonas putida* PCL1444, which showed potential biodegradation of PAHs by utilizing the root exudates. In recent years, Xun [43] discovered that arbuscular mycorrhizal fungus improved the quality of soil with the application of PGPR (plant growth-promoting rhizobacteria) by increasing their dehydrogenase, sucrase, and urease activities along with the degradation of petroleum hydrocarbons. Scientists have also found the potential remediation of some PAHs such as pyrene, phenanthrene, and naphthalene by endophytes [44].

Table 1.1 Plant–bacteria partnership in biodegradation of organic pollutants.

Class	Target pollutants	Plants utilized	Bacterial strains	References
Pesticides	Paraquat	Cowpea	*B. aryabhattai*	[23]
	Aldrin	—	*Pseudomonas fluorescens* and *Bacillus polymyxa*	[24]
	Lindane	*Zea mays*	*Streptomyces* sp.	[25]
	Phenanthrene	*Salix* sp.	*Pseudomonas putida*	[26]
	Crude Oil	*Tectona grandis, Azadirachta indica*	*Pseudomonas aeruginosa*	[27]
	Hexachlorocyclohexane	*Cytisus striatus*	*Rhodococcus erythropolis*	[28]
	Chlorpyrifos	*Lolium perenne*	*Bacillus pumilus*	[29]
PCBs	PCB	*Phalaris arundinacea, Arabidopsis thaliana*	*Rhodococcus*	[30]
	3,3′,4,4′-TCB	*Astragalus sinicus*	*Mesorhizobium* sp.	[31]
	PCB	*Zea mays*	Plant growth promoting bacterium	[32]
	Weathered PCBs	*Medicago sativa*	*Rhizobium meliloti*	[33]
PAHs	Dibenzothiophene, naphthalene, Fluorene	*Populus deltoides* × *Populus nigra*	*Burkholderia fungorum*	[34]
	Different PAH compounds	*Medicago sativa*	*Rhizobium meliloti*	[35]
	Phenanthrene and Pyrene	*Lolium multiforum*	*Acinetobacter* sp.	[36]
	Fluoranthene, Pyrene, and Benzo[a] pyrene	*Medicago sativa*	*Flavobacterium* sp., *Bacillus* sp.	[37]
Explosives	TNT, RDX, HMX	*Populus deltoidesnigra*	*Methylobacterium populi*	[38]
	Toluene	*Lupinus luteus*	*Burkholderia cepacia*	[39]
	TNT	*Brous erectus*	*Pseudomonas* sp.	[40]

1.3.1.1.2. Polychlorinated Biphenyls (PCBs) These are synthetic organic compounds with a high boiling point, chemical stability, non-flammability, and electric insulating properties. These compounds are involved in various industrial and commercial products but proved to be very toxic in the environment, thus requiring solid biodegradation mechanisms. Some bacterial strains such as *Agrobacterium tumefaciens* have been reported to enable the plants in phytoremediation for absorbing a greater amount of PCBs and other toxic pollutants from the soil as well as from groundwater [19]. Some plants such as *Morusrubra* (red mulberry) have been reported to increase the activity as well as the growth of bacterial communities that have the potential of biodegrading PAHs and PCBs [45]. A similar report has also been reported with the bacterium *Burkholderia* sp. LB400. In another study conducted on artificially contaminated soils, *Ouillayasaponins* showed increased biodegradation of PCBs in soils [19].

1.3.1.1.3. Pesticides Most of the pesticides residing in the environment can be easily degraded through the metabolic pathways of both plants and microbes [19, 46]. But long-term persisting compounds have demonstrated the limitations regarding the use of microbes for their biodegradation. Therefore, microbial transformation has helped in more uptake and biodegradation of these compounds. A noteworthy reduction was reported in the concentration of a p′-DDE pesticide, which is found in the rhizospheres of pumpkin, lucerne, spinach, zucchini, and ryegrass. More degradation was reported in the rhizospheres or near root zones than that present in the bulk of soil [19, 47]. In a study, increased biodegradation of chloroacetamide herbicides was reported by the combined use of chemical benoxacor with herbicide-detoxifying bacterial strain *Pseudomonas fluorescens* [48]. In another study, wheatgrass inoculated with microbial consortium at the contaminated site was able to tolerate PCP (pentachlorophenol) [49]. A prompt of dibenzofuran by recombinant microbial strain *Rhizobium tropici* (PBK3-IS) and PCP from *Sphingobium chlorophenolicum* has also been reported [50, 51]. Some scientists have also reported the remediation of various pesticides with the involvement of endophytes. Biodegradation of 2,4-D was later reported by three endophytic Pseudomonas strains found in endophytes of hybrid cottonwood [52]. In another study, the enhanced removal of 2,4-D by pea plant was found by the colonization of its roots with endophyte *P. putida* [53]. These studies suggest the enhanced bioremediation of toxic pesticides with endophytes in combination with microbial strains.

1.3.1.1.4. Explosives Several scientists have also studied the potentiality of endophytes in biodegrading several types of explosives like RDX and TNT; for example, an endophyte *Methylobacterium populum* sp. Nov., strain BJ001 showed great potential in degrading these explosives (Khan and Doty, 2011). Several species of grasses such as *Anthoxanthum odoratum*, *Bromus erectus*, and *Loliumperenne* have been inoculated with bacterial strain *Pseudomonas* sp. strain I4, which was TNT transforming, and this combination shows very reduced levels when applied to soil [40]. A study by Rylott et al. [54, 55] reported the increased removal of RDX in military areas when bacterial expressions flavodoxinreductase (*xplb*) and flavodoxin cytochrome P450 were fused in transgenic plants. But remediation of explosives by combining with plants needs further improvement by using chemical amendments.

1.3.1.2. Inorganic Pollutants

Numerous types of synthetic chemicals are released in the water ecosystem daily, either intentionally or unintentionally from various sources. Most of them persist in the environment for a long time and their biodegradation is a big challenge. The inorganic pollutants group is mainly comprised of heavy metals, radionuclides, and metalloids released from industrial effluents, mining, application of phosphate fertilizers, burning of fossil fuels, and from municipal solid wastes, and though these processes pollutants get accumulated in the food chain, thus causing serious health hazards to life in the environment. Here, the biodegradation of these pollutants is discussed. Also, plant–bacteria partnership in biodegradation of these pollutants has been compiled in Table 1.2.

1.3.1.2.1. Heavy Metals For the remediation of metal-contaminated soils, various bacterial species have been deployed either alone or in partnership with plant species. In this respect, various heavy metal accumulating plant species have been planted for bioremediation of heavy metal contaminated soils and works with PGP (plant growth-promoting) bacteria. PGP bacteria have the capability of increasing plant growth by producing various secretions such as biosurfactants, auxin, phosphorus solubilization, siderophores, exopolysaccharides, and organic acid [58, 76]. Bacterial species show metal adsorption property, which plays a very important role in metal uptake. Nanda et al. [77] reported in their study that since being omnipresent, bacteria and fungi have developed metal-resistant properties. An experiment performed by Rahman et al. [78] showed the degradation of lead and cadmium by the lead-resistant bacterium *Staphylococcus hominis* strain AMB-2. In another

Table 1.2 Plant–bacteria partnership in biodegradation of inorganic pollutants.

Class	Target pollutants	Plants utilized	Bacterial strains	References
Heavy metals	Pd, Zn, Cu, Cd	*H. cannabinus*	*Enterobacteriaceae, Pseudomonadaceae*, and *Comamonadaceae*	[56]
	As, Zn, Cu	*Oryza satvia*	*Bacillus* sp.	[57]
	Cd	*Zea mays*	*Bacillus* sp. *Enterobacter* sp.	[58]
	Ni, Zn, Fe	*Brassica juncea, Ricinus communis*	*Psychrobacter* sp. *Pseudomonas* sp.	[59]
	Cu	*Elsholtzia splendens*	*Pseudomonas putida*	[60]
	Cd, Zn, Pb	*Salix viminalis*	*Rahnella* sp.	[61]
	Cd	*Triticum aestivum, Zea mays*	*Klebsiella* sp. *Bacillus* sp.	[62]
	Cd	*Zea mays*	*Ralstonia eutropha, Chryseobacterium humi*	[63]
	Ni, Pb, Cr	*Cicer arietinum, Zea mays*	*Pseudomonas* sp. *Ralstonia* sp.	[64, 65]
Metalloids	As	*Pterisvittata*	*Delftia* sp. *Bacillus* sp. *Variovorax* sp. *Pseudomonas* sp. *Pseudoxanthomonas* sp.	[66]
	As	*Zea mays*	*Pseudomonas koressnsis Rhodococcus aetherivorans*	[67]
	Se	*Astragalus sp. Stanleya sp.*	*Endophytic bacteria*	[68]
	As	*Pterisvittata*	*Arsenic reducing bacteria*	[69]
	As	*Populus deltoides*	*Agrobacterium radiobacter*	[70]
Radionuclides	U, Sr	*Agrostis capillaris, Deschampsia flexuosa, Festucarubra, Helianthus annus*	*Microbial consortia*	[71]
	Cs, Sr, Pu, Am	Agricultural plants	–	[72]
	U, Cs, Sr	*Helianthus annus*	—	[73]
	U, Ra	*Helianthus annus*	—	[74]
	U	—	*Halommonas* sp. *Clostridium* sp.	[75]

study, Say [79] demonstrated degradation of lead, copper, and cadmium by the filamentous fungi *Phanerochaeta chrysosporium*. De et al. [80] utilized Hg-resistant bacteria like *Bacillus pulilus, P. aeruginosa, Alcaligenes faecalis,* and *Brevibacterium iodinium* for the potential removal of lead and cadmium. In another study, they reported a 78% reduction in chromium through a bacterial consortium (*Acinetobacter* sp. and *Arthrobacter* sp.). In a study conducted on lead, chromium, and cadmium reduction by various microbes, *Bacillus megaterium* was reported to give the highest reduction percentage followed by *B. subtilis* [81].

For years, fungi have been widely used in the degradation of toxic heavy metals, and most of the studies conducted showed that lifeless and active fungi have played a very important role in the biodegradation of inorganic chemicals [82, 83]. A study by Srivastava and

Thakur [84] reported the efficiency of *Aspergillus* sp. in degrading chromium from tannery wastewater. Park et al. [85] showed that the biomass from dead fungi of *Rhizopus oryzae, Saccharomyces cerevisiae, Penicillium chrysogenum,* and *Aspergillus niger* could be utilized for reducing toxic Cr(VI) into less toxic Cr(III). The different moieties such as amide, phosphate, hydroxyl, and carboxyl present on the surface of algae act as metal-binding sites [86]. Goher et al. [87] studied the degradation of copper, cadmium, and lead by using dead cells of *Chlorella vulgaris*.

During recent years, the studies on microbial bioremediation of heavy metals by microbial-fuel-based techniques, microbial gene transfer, and biofilm-mediated techniques have shown strong challenges. These techniques would further help in developing more improved methods for the bioremediation of heavy metals from the environment.

1.3.1.2.2. Metalloids Metalloids in the environment get accumulated not only because of geological processes but also through some human activities. For carrying out some biological systems, metalloids are essential, but their high accumulation might pose a serious threat to living organisms. To handle this toxic problem, nature has created some plants as well as microorganisms. The most important method of bioremediation of mobilized and immobilized arsenic is by bacteria [88]. Bacteria found in the roots of *Cirsium arvense* showed a reduction in As toxicity by producing IAA (indole acetic acid), ACC (1-aminocyc lopropane-1-carboxylate) deaminase, and siderophore [89]. Shagol et al. [67] identified six species such as *Pseudomonas, Paenibacillus, Brevibacterium, Rhodococcus, Rahnella,* and *Microbacterium* that were capable of tolerating high arsenite as well as arsenate concentrations. In microbial detoxification of As, As(V) gets reduced into As(III) by enzyme arsenate reductase, which is based on *Ars* operon expression. When any bacteria residing in the rhizosphere converts some metalloids such as Hg, Te, Se, Tn, Pb, Sb, As into its volatile form then it is known as its detoxification mechanism. In some cases, bacteria in the presence of arsenite oxidase utilize arsenite as an electron donor for oxidizing it into arsenate [90].

1.3.1.2.3. Radionuclides Radionuclides are life-threatening hazardous substances being released into the environment from nuclear power plants and their waste disposal sites, accidental release of Pu, U, etc., and as fission products such as Cs, Sr. But fortunately, there are some plant and microbial species capable of accumulating, transforming, and solubilizing these radionuclides [91]. Francis [75] showed accumulation of U in *Halomonas* sp., which bonded to its phosphate group, while in other studies they found a reduction of U(VI) into U(IV) with bacterial species *Clostridium* [91].

Like microbes, plants have also shown potential in accumulating radionuclides, which takes place in a two-step process. In the first step, pollutants are taken up by the plant roots, and in the second step, the pollutants are accumulated into the protoplast through cell membranes by utilizing energy [92]. Scientists have reported the accumulation of Cs and Sr in sunflower and rhizofiltration of Cs, Sr, and U [73, 93]. Rodriguez et al. [74] studied the translocation factor of U and Ra from soil to plant (sunflower) and found Ra translocation in the shoot while U accumulation in roots of sunflower but in crops, the radionuclides Cs, Pu, Sr, Am found to be accumulated in its vegetative parts [72] (Figure 1.2).

1.3.2. Beneficial Microbes in Agriculture

The productivity of agricultural soil is always improved with the presence of microorganisms. For controlling pests, weeds, and plant diseases, and enhancing plant growth, human has always developed bio-fertilizers as well as bio-pesticides from naturally occurring

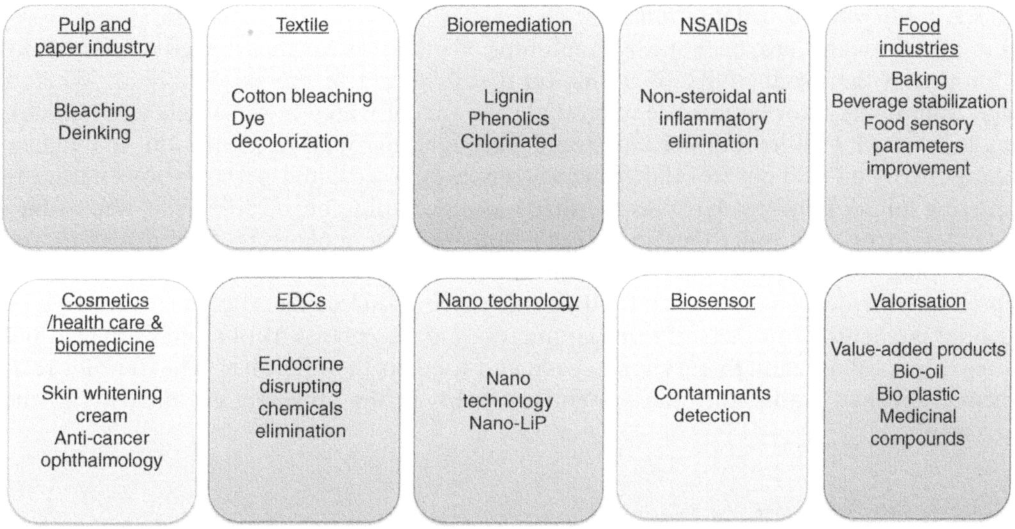

Figure 1.2 An overview of processing, pollutants, and valorization.

microbes. Plants along with these helpful microbes also help in balancing nutrient and gaseous cycles. Some of the uses of microorganisms in the agricultural field have been summarized in the following subsections.

1.3.2.1. Biofertilizers

The most important nutrients for plant growth are nitrogen and phosphate occurring naturally in the ecosystem, but plants are unable to extract them in high enough quantities. Nitrogen is found free in the air while phosphate is found in a bound state abundantly in soil. Phosphate helps crops with stress tolerance, quality, maturity, and also in fixing nitrogen either directly or indirectly. But phosphate gets sheltered in the soil through chemical reactions if not utilized by plants hastily, making it less available to plants. Plants themselves cannot unbolt the phosphates for which a fungus, *Penicillium bilaii,* is a key to unlock it from the soil. *Penicillium bilaii* releases an organic acid for dissolving phosphates in the soil and makes it available for roots to utilize. Therefore, bio-fertilizers are made from these fungi either by coating them on seeds or by directly applying them onto the ground near plant roots, where they get attached around it, preventing the attachment of other non-beneficial organisms.

Another microbe utilized for making bio-fertilizers is *Rhizobium,* which lives forming nodules on plants' roots. These nodules convert free nitrogen present in the environment into organic forms that are further utilized by plants. The bacterium lives in the roots, hence, it's transferring the nutrients directly to plants. On the other hand, legumes utilize naturally occurring nitrogen instead of nitrogen fertilizers because of the presence of friendly microbes on their roots in large populations. The production of bio-fertilizers has limited the use of chemical fertilizers which helps in preserving the environment for upcoming generations. Some other examples of bio-fertilizers are Bio-N, BIO-Fix, NitroPlus, Mykovam, and Mycogroe. Mykovam and Mycogroe help in absorbing phosphorus as well as water from the soil, whereas mycorrhizal fungus helps in preventing infections caused by pathogens and making plants heavy metal- and drought-tolerant [94].

1.3.2.2. Microbial Fermentation

For ages, humans have been utilizing millions of microbes for their benefits, specifically in improving the productivity of agricultural soils found in the same habitat. When a single microbial cell or large-scale microbial cells are cultivated in the absence of air (O_2), then it is known as microbial fermentation. Through microbial fermentation, to enhance plant growth and also control the growth of diseases, weeds, and pests, various naturally occurring microorganisms have been utilized for developing bio-fertilizers as well as bio-pesticides. Most of the microbes found in the soil are helping plants in their growth as well in absorbing essential nutrients from the soil and in return, plants produce wastes as byproducts for microbes as their food for an energy source. The higher absorbance of essential nutrients from the soil through microbes makes plants' root systems bigger and stronger, providing enough surviving space and food to the microbes. This mutual relationship between plants and microorganisms helps in the recycling of nutrients in the ecosystem.

1.3.2.3. Bio-Pesticides

Apart from plant-friendly microorganisms, other non-friendly microbes are also present in the soil that might cause any damage or tremendous disease to the plants. Scientists have developed biological tools from friendly microbes that use harmful microbes in controlling pests and weeds naturally.

1.3.2.4. Bio-Herbicides

The most constant dilemma for farmers is weeds, which continuously battle with crops for all aspects such as nutrients, water, space, and sunlight. Apart from this, weeds also harbor disease-causing pests and insects, obstruct drainage and irrigation systems, deposit weed seeds on harvested crops, and damage the quality of crops, which may result in reduced crop yields. Farmers tackle weeds through hand weeding, tillage, applying synthetic herbicides, or applying all these techniques together. But the major drawback with tillage is the exposure of valuable topsoil to wind and water erosion, while the use of synthetic herbicides contaminates groundwater, which can cause various human and animal diseases and also fatality to various wildlife species. Therefore, the application of bio-herbicides began for controlling weeds without causing any environmental hazards created by synthetic herbicides.

Bio-herbicides are prepared both from microorganisms such as fungi, bacteria, virus, etc., and firm insects such as painted lady butterfly, parasitic wasps, etc., which target only specific types of weeds. The microorganisms that hold invasive genes are selected for killing weeds by attacking their defense genes. Generally, scientists isolated those microbes whose genes match weeds and cause fatal disease in them. These microbes are used only when weeds become most vulnerable to infection. The benefit of using bio-herbicides is that they can target one weed at a time, leaving the rest of the ecosystem unharmed. They can also survive for even the next growing season where chances of weed infecting crops will be higher. These are less expensive, can be easily managed, don't harm non-targeted organisms, and most importantly, are safe for the environment.

1.3.2.5. Bio-insecticides

Earlier farmers used synthetic pesticides to fight against insects but these pesticides affected productivity, were expensive, polluted the water system by accumulating in the environment, and were killing non-targeted microorganisms. On the other hand, bio-insecticides have a shorter shelf life so they persist for a very short time in the environment. Only small quantities are very effective against insects and are comparatively very safe for humans and animals. These specifically affect only one species, and their mode of action is also very specific. But the achievement of bio-insecticides is mostly affected by several environmental conditions such as pH, UV, moisture, temperature, condition of the soil, and presence of other competitive microbes.

1.3.2.5.1. Fungi-Based Bio-insecticides Fungi can create diseases in approximately 200 different types of insects and are thus gaining importance in making bio-insecticides. One of the greatest achievements in the early 1880s was the discovery of fungus, *Beauveria bassiana* (Bb), which originates in plants and soils worldwide. In China, over 2 million hectares of land are sprayed annually with Bb for controlling forestry pests. The major advantage of using Bb is that it does not grow on warm-blooded organisms and it has a short life in rivers or water basins. The fungal-based bio-insecticides make it very difficult for insects to develop any kind of resistance since they attack in various ways once applied.

1.3.2.5.2. Virus-Based Bio-insecticides The virus researchers have been testing for creating virus-based insecticides are of rod-shaped baculoviruses, which affect insect pests akin to potato beetles, flea beetles, corn borers, and aphids. Bertha armyworms attack flax, canola, and vegetable crops while in their larval stages, for which one particular strain is used as a controlling agent. Traditional insecticides do not attack at this stage when most of the damage is done, but virus-based insecticides stop the damage from the early stages.

1.3.3. Miscellaneous Microbial Applications

1.3.3.1. Pesticidal Intermediates

This is a type of living microorganism that produces pesticides constituting only inactive microorganisms. These are applied as an agricultural function. The live microorganisms used for producing pesticides are reviewed under the TSCA (Toxic Substances Control Act) as an intermediate while the final pesticide consisting of dead microorganisms are reviewed by the EPA's office of pesticide program under FIFRA (Federal Insecticide Fungicide and Rodenticide Act). Final submission of pesticide intermediates has remained.

1.3.3.2. Biosensors

The use of microbes showing an array of correspondent molecules indicating the presence of target molecules is known as microbial sensors. The reported genes used for microbial sensors should produce a signal that can be seen with the naked eye such as the production of bioluminescence, e.g. *luc* or *lux* genes, color, e.g. blue color produced by the breakdown of X-galactopyranoside by β-galactosidase or fluorescence, e.g. *gfp* or DsRed. The first microorganism to be field tested was Monsanto's *Pseudomonas chlororaphis,* which is a genetically engineered microorganism into which the β-galactosidase gene was inserted to enable their detection property of microbes in the environment. Another microorganism manipulated with a pesticidal gene containing DsRed protein and reviewed under TSCA was *A. xylosoxidans*. For detection of PAHs including methyl salicylate and naphthalene, another biosensor bacterial strain *P. fluorescens* Hk44 was inserted with gene *lux* bioluminescence. And for detection of unexploded ordinance, namely trinitrotoluene (TNT), trichloroethylene (TCE), and BTEX compounds like benzene, ethylbenzene, toluene, and xylene, another strain *P. putida* was biosensor with *lux* genes. Several other biosensor microorganisms were also developed in-situ generally for detection of heavy metals such as arsenic, copper, lead, zinc, cadmium, nickel, cobalt, and different forms of mercury [95].

In the future, biosensor microorganisms by using intergeneric microbes can also be developed for detecting bioterrorist agents, pathogenic strains of *E. coli* as well as *Salmonella*, chemical fertilizers used in agricultural fields, and some other toxic environmental pollutants.

1.3.3.3. Modification of Weather

The intergeneric microorganisms earlier manipulated were involved in weather modification like for prevention of frost damage on strawberries by ice-minus *Pseudomonas syringae*. For increasing snow volume, a commercial product, Snomax, was prepared from a strain of *P. syringae,* which increases the nucleation temperature of the water. But in the 1980s under the Plant Protection Act, the USDA (United States Department of Agriculture) led to review these two products since *Pseudomonas syringe* is a plant pathogen.

1.3.3.4. Use of Algal Biomass

Presently, algae are extensively used in the production of biofuel feedstocks by R&Ds. Microalgae is very advantageous in the production of high yield biomass, utilizing waste, saline, or produced water, recycling of carbon from CO_2 produced from industrial gases, etc. These advantages are achieved since microalgal culture has a rapid growth rate, high oil content, and high cell density. Algae can produce various fuels together with gaseous compounds such as methane, hydrogen, and sort of usual liquid hydrocarbons. But currently, much of the research focus is given to the production of liquid transportation fuels like diesel, gasoline, and jet fuels from algal biofuels. Several types of algae can be easily cultured for recovering oil from them. They can be photosynthetically grown either naturally or through artificial lighting. Biofuel production from algae has begun in recent years but earlier it was well used in producing pharmaceutical as well as nutraceutical commercial products such as vitamins, steroids, fatty acids, carotenoids, polysaccharides, phycobilins, etc., for human and animal use [96].

1.4. MODERNIZATION IN ENVIRONMENTAL/MICROBIAL BIOTECHNOLOGY

Microbial biotechnology is usually defined as the "utilization of microorganisms or their products for profitable purposes." Since the beginning of history, the use of traditional biotechnology has been reported in baking bread, brewing beverages, and breeding food crops as well as animals [97]. Presently, more importance is given to the establishment of new hybrid genes through biotechnology [98]. Earlier agriculturalists practiced a primitive form of biotechnology for recognizing the better qualities of plant and animal species through cross-pollination or cross-breeding methods [99]. The history of both establishment of microbiology as a science and the use of biotechnological techniques are exacerbated by each other. With the acceleration in biotechnology, the microbiology field also gets stepped up in developing different beneficiary products for humanity. Biotechnology and microbiology are mixed up together in this modern era for selective breeding and training of animals, cultivation of crops with more productivity and nutrition, and also utilizing microbes in producing large varieties of products such as bread, cheese, yogurt, wine, and beer [100].

Previously, microbial biotechnology was focused on producing foods and on industrial applications but in the modern era, it has been accepted as a tool for developing genetically modified crops as well as strains with amplified accurateness concerning precise functions. Through the standard applications of microbial biotechnology, new species, and their selection has been discovered, and improvement in strains with certain functional straits has been improvised.

1.4.1. Current Status and Scope of Microbial Biotechnology

The beginning of the modern era of microbial biotechnology began in the year 1972 with the evolution of new technology known as rDNA technology, which advanced new heights and approaches in this field of science and has now been around for 50 years [101]. The discovery of rDNA technology propelled biotechnology to new heights. In recent years, advancement in some other areas such as fermentation technology, nanobiotechnology, microbial physiology, screening of novel metabolites and their strain improvement, cell immobilization, cell fusion, etc., has also helped in the modernization of microbial biotechnology [102].

1.4.1.1. Recombinant Microorganisms

From the 1970s to the twenty-first century, industrial microbiology has played a major role in the exploitation and establishment of microbial genetic discoveries for which various microbes such as *Aspergillus niger*, *Bacillus subtilis*, *E. coli*, *Pichiapastoris*, etc., were used as a host for producing glycosylated recombinant protein [103]. But for its successful application in higher organisms, entrenched systems and understanding is required. One of the outstanding results of growth in microbial biotechnology is the development of human insulin and human growth hormone within just four years of discovery of rDNA technology [101].The second result is the establishment of a new company in various areas such as downstream processing and biochemical engineering, which were completely dedicated to innovations through genetic approaches and their explorations. The major push was in the area of production of rare mammalian peptides such as antibodies, enzymes, growth factors, hormones, and biological response modifiers [103].

1.4.1.2. Vaccine Production

Another main aspect of modern technology is the production of vaccines, of which the first vaccine produced was from yeast for hepatitis B virus surface antigen. The development of recombinant vaccines helped in removing the problems related to conventional vaccines where persons receiving them failed to be defended or were reinfected with the disease.

1.4.1.3. Enzyme Production

Earlier production of enzymes was done through the fermentation process by commercial industries before the modernization of microbial biotechnology techniques. But after the discovery of rDNA technology, companies immediately involved themselves in manufacturing enzymes with an improvised method. The most common example of enzyme production through rDNA technology is recombinant chymosin for cheese manufacture, carbohydrases, lipases, proteases, and recombinant lipase for detergents [104–107]. Later, recombinant therapeutic enzymes including cerozyme, human DNAase, and tissue plasminogen activator were produced, which were used for the treatment of gastrointestinal and rheumatic disorders, metabolic diseases, cancer, and thromboses.

1.4.1.4. Alcohol Production

Alcohol is a primary metabolite of fermented hexose sugar by microbe *S. cerevisiae* while lactose and pentose sugar are used as a substratum for fermentation by *Kluyveromyces fragilis* and *Candida* species, respectively [108]. But after the discovery of rDNA technology, certain recombinant microbes such as *E. coli* were used for obtaining excellent ethanol products.

1.4.1.5. Healthcare Products

The first hybridoma technology was introduced in the year 1975, but it became widespread several years after its introduction only after some improvement in its technology [109]. Further, recombinant proteins were utilized for improving health care products; one example is DNA vaccines, which usually consist of properly engineered plasmid DNA prepared from *E. coli* on a large industrial scale [110]. In the twenty-first century, the most challenging human health issue was cancer, as a large population was suffering from different types. But now, various standard methods are practiced for curing cancer, among which immunotherapy is considered to be the most promising one and uses specific types of microorganisms [111]. Microorganisms have exceptionally antitumor properties, which after clinical trials can be used for cancer treatments. Presently, anticancer products obtained from the bacterial community are helping to complete recovery and are considered a complementary standard treatment method [112].

1.4.1.6. Nanobiotechnology

In nature, various microbes are capable of producing various nanoparticles, either through intracellular or extracellular routes, of metal oxides such as Ag, Au, Fe, and Zn. Some of these microbial species are actinomycetes, *Aspergillus niger*, *Bacillus cereus*, *Candida utilis*, *E. coli*, *Trichoderm aviride*, etc. [113]. In the modern era, researchers have shown a massive interest in forming nanoparticles from alloys and elemental oxides because of their widespread use in coating, catalysis, optical materials, biomedical uses, and electronics [114].

1.4.1.7. Transgenic Crops and Genetically Modified Organisms

To obtain certain specific traits as well as enhanced or improved nutritional content in animals, plants, or microbes through in-vivo techniques is known as GMOs. The modification is done by transferring the desired gene from chromosomes of any particular animal, plant, or microorganism into the host cell species [115]. Through GMO, various disease resistance, herbicide resistance, insect-resistant (*Bacillus thuringiensis*), enhanced crops such as corn, cotton, soybean, rice, brinjal, etc., either in yield, durability, or size have been successfully developed [116, 117].

In the modern era, microbial technology has been accepted and applied as a more complex science that has led to the development of various enhanced and improved microbial strains capable of managing several issues like xenobiotic degradation, rapid drug manufacturing, producing primary as well as secondary metabolites, enhanced crop yields, etc. (Table 1.3).

1.5. PROS AND CONS OF MICROBIAL BIOTECHNOLOGY

Biotechnology is the branch of science that applies biochemistry, engineering, and microbiology to achieve technological applications of the capabilities of microorganisms and tissue culture cells. The pros and cons of this technology depend on the range of genes altered in living cells for their applications in different fields like medical, agriculture, marine, etc. The improvisation in technology from time to time has aroused various pros and cons. When the technology is applied in a very appropriate manner it creates beneficial effects for humans, but when it is mishandled, it can cause various downsides [130]. To understand this technology well, we need to go through all pros and cons of this field.

1.5.1. Pros of Microbial Biotechnology

1.5.1.1. Improvement in Medical Treatments and Lowering Infection Rate

With the study and applications of technology, it became possible to understand some genetic diseases, their reasons, and effective ways of managing them, for example certain

Table 1.3 List of genetically engineered microorganisms and their different applications.

	Engineered microorganism	Modified gene expression	Function/applications	References
Dye degradation	*Phanerochaete chrysosporium*	Co-expression of *lacIIIb*, *vpl2*, *mnp1*, *lipH8*	2,6-dimethoxyphenol (DMP) and guaiacol (GUA) -*N*,*N*-dimethyl-*p*-phenylenediamine sulfate (DMPPDA) as azo dye, and Remazol Brilliant Blue R (RBBR) as an anthraquinone dye were removed using co-expression of laccases and peroxidase enzymes	[118]
	Saccharomyces cerevisiae	pVT-100U-*MtL* or pVT-100U-*TtL* (LCC1 isoform of *TtL*)	*TtL* completely decolorized Bromophenol Blue, Coomassie Brilliant Blue, Remazol Brilliant Blue R	[119]
	Streptomyces lividans strain	Homologous expression	Indigo carmine removal due to enhanced thermostability	[120]
Heavy Metal Removal	*Pseudomonas aeruginosa* (Pse-w)	Metallothioneins	Metallothioneins promoted immobilization of cadmium (Cd^{2+}) inducing high Cd^{2+} removal efficiency	[121]
	E. coli BL 21	*merA*	merA gene was cloned into pET21a(+) for mercury removal	[17]
	Deinococcus radiodurans (radiation-resistant)	*merH*, a new ion transporter gene incorporation from *M. marinum* strain	Degrades ionic mercury	[120]
Organic xenobiotic/ Poly aromatic hydrocarbons (PAHs)	*Pseudomonas putida* S16	Glutamate decarboxylase (GAD)	Nicotine degradation by acid stress tolerant bacteria created by GAD system or IrrE regulator	[122]
	Pseudomonas putida	*NahH* encoded dioxygenase production	Phenanthrene and pyrene were removed via C230 enzyme catalysis	[123]
	Escherichia coli-Pseudomonas aeruginosa	Catechol 2,3 Dioxygenase	Phenanthrene dioxygenase exhibited significant activity for Phenanthrene bioremoval	[124]
Pesticide removal	*Pseudomonas putida* KT2440	Carbofuran hydrolase genes and green fluorescent protein (*gfp*)	Chlorpyrifos and carbofuran present in soil were removed.	[125]
	Sphingobiumwenxiniae JZ-1	*pbaA1A2BC*	3-Phenoxybenzoate 1′,2′-dioxygenase Sphingobiumwenxiniae JZ-1 enzyme degraded 3-Phenoxybenzoate (PBA)	[126]
	Sphingomonas sp. DC-6	*CndA*	Acetochlor degradation via Oxygenase activity	[8]
Remediation of Oil spills	*Pseudomonas guguanensis*	Catechol 2,3 Dioxygenase	Removal of crude oil contaminated soil	[127]
	Pseudomonas putida	Catechol 2,3 Dioxygenase encoded gene *nahH* cloned into pUC18	Removal of hydrocarbons via C230 enzyme activity	[128]
	Pseudomonas aeruginosa strain DAB	Amplification of rhamnosyltransferase 1 complex *RhlAB* with the native operon promoter	Removal of crude oil by releasing rhamnolipid biosurfactants	[129]

inherited disabilities linked with folic acid lacking in the mother's body. These generic health conditions can be avoided by developing precise methods. Such advancements have made it possible to treat some diseases arising from genetic makeup. Experts were also able to understand the correct cause for alteration in genes for causing any type of disease like cancer and develop correct or appropriate therapy for targeting them. The application of biotechnology has also helped in dealing with diseases caused by microbes that might affect a large population. Researchers study the causes of its spread and develop effective strategies on how to stop it from spreading and becoming pandemic to protect the population vulnerable to it [131].

1.5.1.2. Improvement of Quality in Crops

The addition or change in the nutritional values of crops has turned to be a great way of preserving resources. A lot of nourishment can be supplied to people by eating very little food. Apart from this, the application of biotechnology helped in growing some crops in places that were unable to grow. Biotechnology applications have also helped in developing various types of crops that had been affected by pests, drought conditions into pest-resistant and drought-resistant crops. Various seed varieties with improved qualities have also been developed, which have contributed a lot toward food production. Other processes along with gene alteration have enabled the production of various highly resistant traits, making crops available to grow in poor rainfall areas with improved yields. Hence, biotechnology science is very helpful in the security of food crops.

1.5.1.3. Improvement in Crop Yields and Reduced Usage of Pesticides

The crops depending on regular climatic conditions now can be grown in any part of the world; this is made possible because of the advancement in technology. Researchers have also developed drought-resistant and other climate-resistant crops to counter those hindrances that affected the growth of the crop and in return this has resulted in increased crop yields. Worldwide farmers now can get better, high-yielding, and resistant seeds for cultivation.

Scientists have also developed pest-resistant along with drought-resistant crops, enabling them to survive pests as well as parasite attacks, which ultimately limits the use of pesticides. Use of too many or too many varieties of pesticides has led to environmental pollution and also has hindered the fertility of the soil, thus pest-resistant crops have proven to be highly effective. Farmers who limited the use of pesticides started using pest-resistant seeds, which are easy to grow with high yields.

1.5.2. Cons of Microbial Biotechnology

To produce human beneficial products, biotechnology science controls and manipulates the biological systems of microorganisms, which was proven to be very advantageous. Along with these benefits, however, technology has also shown some drawbacks as described here.

1.5.2.1. Awful Impact on Agriculture

The genetically modified plants are prepared after manipulation in their genetic elements of unmodified crops. Sometimes, some of the traits get transferred to weeds, creating herbicide-resistant plants as well, which is a major concern to technologists.

1.5.2.2. Impact on Biodiversity

Biotechnology utilizes microorganisms such as bacteria, fungi, etc., for its genetic modification and to develop new strains. This genetically modified microbe, if it escapes to the ecosystem, can become wild and disturb the entire ecosystem.

1.5.2.3. Impact on Soil Fertility

Biotechnology helped in developing crops with high nutrients that they usually receive from the soil itself, making the soil less fertile. If the soil is put into cultivation with the same crop, then it might take many years to recover in its nutrient rate and puts the soil at the risk of losing its viability. To maintain the fertility rate, crop rotation and some other treatment processes should be practiced. Therefore soil viability and fertility should be checked from time to time.

1.5.2.4. Can Cause Destruction

The major drawback of biotechnology is that it can be easily converted into a biological weapon, which could become a reason for massive destruction. For example, new medicines developed with the help of biotechnology can at the same time be used for weaponizing diseases. On the other hand, through genetic modification, we can transform a simple microbe into a destructive one that can threaten human life.

1.5.2.5. Biotechnology Field with Many Unknowns

There are plenty of things that are still unknown about the biotechnology field. Several consequences might arise if anything goes wrong during the process of genetic modification. Several questions are still not answered.

There are numerous pros and cons with technology; therefore, it becomes essential to weigh properly each benefit and drawback before getting started with any technology.

1.6. CHALLENGES AND FUTURE PROSPECTIVE

Recent scientific discoveries have allowed environmental biotechnologists to use microorganisms to benefit society. Microbial genomic research has resulted in breakthroughs such as improved microbes for biologically controlling plant and animal pests, modified plant and animal pathogens for reduced virulence, the development of new microbial agents for cleaning up contaminated water and soil from agricultural overspill, tools for better disease diagnosis and the preparation of improved vaccines, and advancements in fermentation organisms as well as the development of new industrial catalysts.

The challenging fact is not collecting information about microbial communities but it's about how to and what to do with these known data. But even more challenging will be handling the new purposes of those that are not yet used on a large scale, including biodegrading of trace organic pollutants affecting mammals' hormonal systems, conversion of organic wastes to renewable energy in place of treatment demanding energy-consumption, and bioremediating the pollutants created from agricultural overspill or drainage. To overcome these big challenges, we need to use the following steps for achieving success: (i) utilization of powerful tools; (ii) aiming for immense remuneration; (iii) application of contemporary substances; (iv) application of good engineering; and (v) biomass preservation and quantification. The assessment of consequences, challenges, and opportunities of modern biotechnology is essential for industries as well as for policymakers.

CONFLICTS OF INTERESTS

The authors declare that they have no conflict of interest.

ACKNOWLEDGMENTS

Pankaj Chowdhary acknowledges the Council of Scientific and Industrial Research (CSIR), New Delhi, India for Research Associate (RA) work. Further, the authors are highly thankful to Director CSIR-Indian Institute of Toxicology Research (IITR), Lucknow (India) for his valuable encouragement and support to complete this manuscript. This is IITR MS No. IITR/SEC/2022-2023/17.

REFERENCES

1. Singh, S. and Jaiswal, K. (2016). Biotechnological applications of microbes for the remediation of environmental pollution. In: *Biotechnology: Trends and Applications* (ed. R. Singh and M. Trivedi), 179–214. USA: Studium Press LLC.
2. Timmis, K., Ramos, J.L., de Vos, W. et al. (2016). *Microbial Biotechnology 9* (5): 529.
3. Chowdhary, P. and Raj, A. (ed.) (2020). *Contaminants and Clean Technologies*. CRC Press.
4. Chowdhary, P., Raj, A., Verma, D., and Akhter, Y. (ed.) (2020). *Microorganisms for Sustainable Environment and Health*. Elsevier.
5. Abatenh, E., Gizaw, B., Tsegaye, Z., and Wassie, M. (2017). The role of microorganisms in bioremediation-a review. *Open Journal of Environmental Biology 2* (1): 38–46. https://doi.org/10.17352/ojeb.000007.
6. Verma, S. and Kuila, A. (2019). Bioremediation of heavy metals by microbial process. *Environmental Technology and Innovation 14*: 100369.
7. Leong, Y.K. and Chang, J.S. (2020). Bioremediation of heavy metals using microalgae: recent advances and mechanisms. *Bioresource Technology 303*: 122886.
8. Liu, B. and Duan, I. (2019). Mitigation of environmental pollution by genetically engineered bacteria - current challenges and future perspectives. *Science of the Total Environment 667*: 444–454.
9. Xu, Z., Lei, Y., and Patel, J. (2010). Bioremediation of soluble heavy metals with recombinant *Caulobacter crescentus*. *Bioengineered Bugs 1*: 3.
10. Sefidan, T. (2017). Bioremediation of mineral mercury via construction of recombinant vector pET21a(+)-merA. *Journal of Microbial World 10* (1): 6–17.
11. Kim, C., Kim, M.Y., Michie, I. et al. (2017). Anodic electro-fermentation of 3-hydroxypropionic acid from glycerol by recombinant *Klebsiella pneumoniae* L17 in a bioelectrochemical system. *Biotechnology for Biofuels 10*: 199.
12. Chouhan, S., Sharma, K., Zha, J. et al. (2017). Recent advances in recombinant biosynthesis of polyphenols. *Frontiers in Microbiology 8*: 2259.
13. Borchert, E., Hammerschmidt, K., Hentschel, U. et al. (2021). Enhancing microbial pollutant degradation by integrating eco-evolutionary principles with environmental biotechnology. *Trends in Microbiology 29* (10): 908–918.
14. Chaturvedi, P., Singh, A., Chowdhary, P. et al. (2020). Occurrence of emerging sulfonamide resistance (sul1 and sul2) associated with mobile integrons-integrase (intI1 and intI2) in riverine systems. *Science of the Total Environment 751*: 142217. https://doi.org/10.1016/j.scitotenv.2020.142217.
15. Chaturvedi, P., Chowdhary, P., Singh, A. et al. (2021). Dissemination of antibiotic resistance genes, mobile genetic elements, and efflux genes in anthropogenically impacted riverine environments. *Chemosphere* https://doi.org/10.1016/j.chemosphere.2021.129693.
16. Pelczar, R.M. and Pelczar, M.J. (2020). Microbiology. Encyclopedia Britannica. https://www.britannica.com/science/microbiology. Accessed 23 June 2021.
17. Gupta, A., Gupta, R., and Singh, R.L. (2017). Principles and applications of environmental biotechnology for a sustainable future. In: *Applied Environmental Science and Engineering for a Sustainable Future* (ed. R.L. Singh). Switzerland AG: Springer Nature https://doi.org/10.1007/978-981-10-1866-4_3.
18. Woese, C., Kandler, O., and Wheelis, M. (1990). Towards a natural system of organisms: proposal for the domains Archaea, Bacteria, and Eucarya. *Proceedings of the National Academy of Sciences of the United States of America 87*: 4576–4579.

19. Chaudhry, Q., Blom-Zandstra, M., Gupta, S. et al. (2005). Utilizing the synergy between plants and rhizosphere microorganisms to enhance breakdown of organic pollutants in the environment. *Environmental Science and Pollution Research 12*: 34–48.
20. Imran, M., Arshad, M., Khalid, A. et al. (2015). Decolorization of reactive black-5 by *Shewanella* sp. in the presence of metal ions and salts. *Water Environment Research 87*: 579–586.
21. Hussain, S., Maqbool, Z., Ali, S. et al. (2013). Biodecolorization of Reactive Black-5 by a metal and salt tolerant bacterial strain *Pseudomonas* sp. RA20 isolated from Paharang Drain effluents in Pakistan. *Ecotoxicology and Environmental Safety 98*: 331–338.
22. Chowdhary, P., Sammi, S.R., Pandey, R. et al. (2020). Bacterial degradation of distillery wastewater pollutants and their metabolites characterization and its toxicity evaluation by using Caenorhabditiselegans as terrestrial test models. *Chemosphere 261*: 127689. https://doi.org/10.1016/j.chemosphere.2020.127689.
23. Inthama, P., Pumas, P., Pekkoh, J. et al. (2021). Plant growth and drought tolerance-promoting bacterium for bioremediation of Paraquat pesticide residues in agriculture soils. *Frontiers in Microbiology 12*: 604662. https://doi.org/10.3389/fmicb.2021.604662.
24. Wang, B., Wang, Q., Liu, W. et al. (2017). Biosurfactant-producing microorganism Pseudomonas sp. SB assists the phytoremediation of DDT-contaminated soil by two grass species. *Chemosphere 182*: 137–142.
25. Alvarez, A., Benimeli, C.S., Saez, J.M. et al. (2015). Lindane removal using Streptomyces strains and maize plants: a biological system for reducing pesticides in soils. *Plant and Soil 395*: 401–413.
26. Khan, Z., Roman, D., Kintz, T. et al. (2014). Degradation, phytoprotection and phytoremediation of phenanthrene by endophyte*Pseudomonas putida*, PD1. *Environmental Science and Technology 48*: 12221–12228.
27. Yenn, R., Borah, M., Boruah, H.D. et al. (2014). Phytoremediation of abandoned crude oil contaminated drill sites of Assam with the aid of a hydrocarbon-degrading bacterial formulation. *International Journal of Phytoremediation 16*: 909–925.
28. Becerra-Castro, C., Kidd, P.S., Rodríguez-Garrido, B. et al. (2013). Phytoremediation of hexachlorocyclohexane (HCH)-contaminated soils using *Cytisusstriatus*and bacterial inoculants in soils with distinct organic matter content. *Environmental Pollution 178*: 202–210.
29. Ahmad, F., Iqbal, S., Anwar, S. et al. (2012). Enhanced remediation of chlorpyrifos from soil using ryegrass (*Lolliummultiflorum*) and chlorpyrifos-degrading bacterium *Bacillus pumilus* C2A1. *Journal of Hazardous Materials 237*: 110–115.
30. Vergani, L., Mapelli, F., Suman, J. et al. (2019). Novel PCB-degrading *Rhodococcus* strains able to promote plant growth for assisted rhizoremediationof historically polluted soils. *PLoS One 14* (8): e0221253. https://doi.org/10.1371/journal.pone.0221253.
31. Teng, Y., Li, X., Chen, T. et al. (2015). Isolation of the PCB-degrading bacteria *Mesorhizobium* sp. ZY1 and its combined remediation with *Astragalussinicus* L. for contaminated soil. *International Journal of Phytoremediation 18* (2): 141–149. https://doi.org/10.1080/15226514.2015.1073667.
32. Luo, C., Wang, S., Wang, Y. et al. (2015). Effects of EDDS and plant-growth-promoting bacteria on plant uptake of trace metals and PCBs from e-waste–contaminated soil. *Journal of Hazardous Materials 286*: 379–385.
33. Teng, Y., Luo, Y., Sun, X. et al. (2010). Influence of arbuscularmycorrhiza and rhizobium on phytoremediation by alfalfa of an agricultural soil contaminated with weathered PCBs: a field study. *International Journal of Phytoremediation 12*: 516–533.
34. Andreolli, M., Lampis, S., Poli, M. et al. (2013). Endophytic*Burkholderiafungorum*DBT1 can improve phytoremediation efficiency of polycyclic aromatic hydrocarbons. *Chemosphere 92*: 688–694.
35. Teng, Y., Shen, Y., Luo, Y. et al. (2011). Influence of *Rhizobium meliloti* on phytoremediation of polycyclic aromatic hydrocarbons by alfalfa in an aged contaminated soil. *Journal of Hazardous Materials 186* (2–3): 1271–1276. https://doi.org/10.1016/j.jhazmat.2010.11.126.
36. Yu, X.Z., Wu, S.C., Wu, F.Y. et al. (2011). Enhanced dissipation of PAHs from soil using mycorrhizal ryegrass and PAH-degrading bacteria. *Journal of Hazardous Materials 186*: 1206–1217.

37. Zhang, J., Yin, R., Lin, X. et al. (2010). Interactive effect of biosurfactant and microorganism to enhance phytoremediation for removal of aged polycyclic aromatic hydrocarbons from contaminated soils. *Journal of Health Science 56*: 257–266.
38. Van Aken, B., Peres, C.M., Doty, S.L. et al. (2004). *Methylobacteriumpopuli* sp. nov., a novel aerobic, pink-pigmented, facultativelymethylotrophic, methane-utilizing bacterium isolated from poplar trees (*Populusdeltoides* x *nigra* DN34). *International Journal of Systematic and Evolutionary Microbiology 54*: 1191–1196.
39. Barac, T., Taghavi, S., Borremans, B. et al. (2004). Engineered endophytic bacteria improve phytoremediation of water soluble, volatile, organic pollutants. *Nature Biotechnology 22*: 583–588.
40. Siciliano, S.D. and Greer, C.W. (2000). Plant-bacterial combinations to phytoremediate soil contaminated with high concentrations of 2,4,6-trinitrotoluene. *Journal of Environmental Quality 29*: 311–316.
41. Daane, L.L., Harjono, I., Zylstra, G.J. et al. (2001). Isolation and characterization of polycyclic aromatic hydrocarbon degrading bacteria associated with the rhizosphere of salt marsh plants. *Applied and Environmental Microbiology 67*: 2683–2691.
42. Kuiper, I., Kravchenko, L.V., Bloemberg, G.V. et al. (2002). *Pseudomonas putida* strain PCL 1444, selected for efficient root colonization and naphthalene degradation, effectively utilizes root exudates components. *Molecular Plant-Microbe Interactions 15*: 734–741.
43. Xun, F., Xie, B., Liu, S. et al. (2015). Effect of plant growth-promoting bacteria (PGPR) and arbuscularmycorrhizal fungi (AMF) inoculation on oats in saline-alkali soil contaminated by petroleum to enhance phytoremediation. *Environmental Science and Pollution Research 22*: 598–608.
44. Khan, Z. and Doty, S. (2011). Endophyte assisted phytoremediation. *Current Topics in Plant Biology 12*: 97–104.
45. Hegde, R.S. and Fletcher, J.S. (1996). Influence of plant growth stage and season on the release of root phenolics by mulberry as related to development of phytoremediation technology. *Chemosphere 32*: 2471–2479.
46. Hussain, S., Devers-Lamrani, M., El-Azahari, N. et al. (2011). Isolation and characterization of an isoproturon mineralizing *Sphingomonas*sp. strain SH from a French agricultural soil. *Biodegradation 22*: 637–650.
47. White, J.C. (2011). Plant-facilitated mobilization and translocation of weathered 2,2-bis(p-chlor ophenyl):-1,1-dichloroethylene (p,p'-DDE) from an agricultural soil. *Environmental Toxicology and Chemistry 20*: 2047–2052.
48. Hoagland, R.E., Zablotowicz, R.M., and Locke, M.A. (1997). An integrated phytoremediation strategy for chloroacetamide herbicides in soil. In: *Phytoremediation of Soil and Water Contaminants* (ed. R.E. Hoagland), 92–105. Washington USA: American Chemical Society.
49. Miller, E.K., Boydston, J.A., and Dyer, W.E. (1998). Mechanisms of pentachlorophenol phytoremediation in soil. In: *Pacific Northwest Regional Meeting of the ASAE*, Fairmont Hot Springs, Montana, USA. (Sep 10-12), 7.
50. Saiki, Y., Habe, H., Yuuki, T. et al. (2003). Rhizoremediation of dioxin-like compounds by a recombinant *Rhizobium tropici* strain expressing carbazole 1,9a-dioxygenase constitutively. *Bioscience, Biotechnology, and Biochemistry 67*: 1144–1148.
51. Dams, R.I., Paton, G.I., and Killham, K. (2007). Rhizoremediation of pentachlorophenol by *Sphingobium chlorophenolicum* ATCC 39723. *Chemosphere 68*: 864–870.
52. Germaine, K., Keogh, E., Garcia-Cabellos, G. et al. (2004). Colonization of poplar trees by gfp expressing bacterial endophytes. *FEMS Microbiology Ecology 48*: 109–118.
53. Germaine, K.J., Liu, X., Cabellos, G.G. et al. (2006). Bacterial endophyte enhanced phytoremediation of the organochlorine herbicide 2,4-dichlorophenoxyacetic acid. *FEMS Microbiology Ecology 57*: 302–310.
54. Rylott, E.L., Jackson, R.G., Edwards, J. et al. (2006). An explosive-degrading cytochrome P450 activity and its targeted application for the phytoremediation of RDX. *Nature Biotechnology 24*: 216–219.

55. Makris, K.C., Sarkar, D., and Datta, R. (2009). Coupling indigenous biostimulation and phytoremediation for the restoration of 2,4,6-trinitrotoluene-contaminated sites. *Journal of Environmental Monitoring 12*: 399–403.
56. Chen, Y., Ding, Q., Chao, Y. et al. (2018). Structural development and assembly patterns of the root-associated microbiomes during phytoremediation. *Science of the Total Environment 644*: 1591–1601.
57. Hare, V. and Chowdhary, P. (2019). Changes in growth responses in rice plants grown in the arsenic affected area: implication of As resistant microbes in mineral content and translocation. *SN Applied Sciences 1*: 882. https://doi.org/10.1007/s42452-019-0945-y.
58. Ahmad, I., Akhtar, M.J., Asghar, H.N. et al. (2016). Differential effects of plant growth-promoting rhizobacteria on maize growth and cadmium uptake. *Journal of Plant Growth Regulation 35*: 303–315.
59. Ma, Y., Rajkumar, M., Rocha, I. et al. (2015). Serpentine bacteria influence metal translocation and bioconcentration of *Brassica juncea* and *Ricinuscommunis* grown in multi-metal polluted soils. *Frontiers in Plant Science 5*: 757. https://doi.org/10.3389/fpls.2014.00757.
60. Xu, C., Chen, X., Duan, D. et al. (2015). Effect of heavy-metal-resistant bacteria on enhanced metal uptake and translocation of the Cu-tolerant plant, *Elsholtziasplendens*. *Environmental Science and Pollution Research International 22*: 5070–5081.
61. Janssen, J., Weyens, N., Croes, S. et al. (2015). Phytoremediation of metal contaminated soil using willow: exploiting plant-associated bacteria to improve biomass production and metal uptake. *International Journal of Phytoremediation 17*: 1123–1136.
62. Ahmad, I., Akhtar, M.J., Zahir, Z.A. et al. (2014). Cadmium-tolerant bacteria induce metal stress tolerance in cereals. *Environmental Science and Pollution Research 21*: 11054–11065.
63. Moreira, H., Marques, A.P.G.C., Franco, A.R. et al. (2014). Phytomanagement of Cd-contaminated soils using maize (*Zea mays* L.) assisted by plant growth-promoting rhizobacteria. *Environmental Science and Pollution Research 21*: 9742–9753.
64. Tank, N. and Saraf, M. (2009). Enhancement of plant growth and decontamination of nickel-spiked soil using PGPR. *Journal of Basic Microbiology 49*: 195–204.
65. Braud, A., Jezequel, K., Bazot, S., and Lebeau, T. (2009). Enhanced phytoextraction of an agricultural Cr, Hg- and Pb-contaminated soil by bioaugmentation with siderophore producing bacteria. *Chemosphere 74*: 280–286.
66. Lampis, S., Santi, C., Ciurli, A. et al. (2015). Promotion of arsenic phytoextraction efficiency in the fern *Pterisvittata*by the inoculation of As-resistant bacteria: a soil bioremediation perspective. *Frontiers in Plant Science 6*: 80. https://doi.org/10.3389/fpls.2015.00080.
67. Shagol, C.C., Krishnamoorthy, R., Kim, K. et al. (2014). Arsenic-tolerant plant-growth-promoting bacteria isolated from arsenic-polluted soils in South Korea. *Environmental Science and Pollution Research 21*: 9356–9365.
68. Lindblom, S.D., Valdez-Barillas, J.R., Fakra, S.C. et al. (2013). Influence of microbial associations on selenium localization and speciation in roots of *Astragalus* and *Stanleya* hyperaccumulators. *Environmental and Experimental Botany 88*: 33–42.
69. Yang, C., Zhou, Q., Wei, S. et al. (2012). Chemical-assisted phytoremediation of CD-PAHs contaminated soils using *Solanumnigrum*L. *International Journal of Phytoremediation 13*: 818–833.
70. Wang, Q., Xiong, D., Zhao, P. et al. (2011). Effect of applying an arsenic-resistant and plant growth-promoting rhizobacterium to enhance soil arsenic phytoremediation by *Populusdeltoides* LH05-17. *Journal of Applied Microbiology 111*: 1065–1074.
71. Langella, F., Grawunder, A., Stark, R. et al. (2014). Microbially assisted phytoremediation approaches for two multi-element contaminated sites. *Environmental Science and Pollution Research 21*: 6845–6858.
72. Kozhakhanov, T.E., Lukashenko, S.N., and Larionova, N.V. (2014). Accumulation of artificial radionuclides in agricultural plants in the area used for surface nuclear tests. *Journal of Environmental Radiation 137*: 217–226.
73. Prasad, M.N.V. (2007). Sunflower (*Helinathusannuus* L.) - a potential crop for environmental industry. *Helia 30*: 167–174.

74. Rodriguez, P.B., Tome, F.V., Fernandez, M.P. et al. (2006). Linearity assumption in soil-to-plant transfer factors of natural uranium and radium in *Helianthus annuus*L. *The Science of the Total Environment 361*: 1–7.
75. Francis, A.J., Gillow, J.B., Dodge, C.J. et al. (2004). Association of uranium with halophilic and nonhalophilic bacteria and archaea. *Radiochemica Acta 92*: 481–488.
76. Glick, B.R. (2010). Using soil bacteria to facilitate phytoremediation. *Biotechnology Advances 28*: 367–374.
77. Nanda, M., Kumar, V., and Singh, D.K. (2019). Multimetal tolerance mechanisms in bacteria: the resistance strategies acquired by bacteria that can be exploited to 'clean-up' heavy metal contaminants from water. *Aquatic Toxicology 212*: 1–10.
78. Rahman, Z., Thomas, L., and Singh, V.P. (2019). Biosorption of heavy metals by a lead (Pb) resistant bacterium, Staphylococcus hominis strain AMB-2. *Journal of Basic Microbiology 59*: 477–486.
79. Say, R., Denizli, A., and Arica, M.Y. (2000). Biosorption of cadmium(II), lead(II) and copper(II) with the filamentous fungus *Phanerochaete chrysosporium*. *Bioresource Technology 76*: 67–70.
80. De, J., Ramaiah, N., and Vardanyan, L. (2008). Detoxification of toxic heavy metals by marine bacteria highly resistant to mercury. *Marine Biotechnology 10* (4): 471–477.
81. Abioye, O.P., Oyewole, O.A., Oyeleke, S.B. et al. (2018). Biosorption of lead, chromium and cadmium in tannery effluent using indigenous microorganisms. *Brazilian Journal of Biological Sciences. 5* (9): 25–32.
82. Tiwari, S., Singh, S.N., and Garg, S.K. (2013). Microbially enhanced phytoextraction of heavy-metal fly-ash amended soil. *Communications in Soil Science and Plant Analysis 44* (21): 3161–3176.
83. Fu, Q.Y., Li, S., and Zhu, Y.H. (2012). Biosorption of copper (II) from aqueous solution by mycelial pellets of *Rhizopusoryzae*. *African Journal of Biotechnology 11* (6): 1403–1411.
84. Srivastava, S. and Thakur, I.S. (2006). Isolation and process parameter optimization of *Aspergillus* sp. for removal of chromium from tannery effluent. *Bioresource Technology 97* (10): 1167–1173.
85. Park, D., Yun, Y.S., Jo, J.H. et al. (2005). Mechanism of hexavalent chromium removal by dead fungal biomass of *Aspergillusniger*. *Water Research 39* (4): 533–540.
86. He, J. and Chen, J.P. (2014). A comprehensive review on biosorption of heavy metals by algal biomass: materials, performances, chemistry, and modeling simulation tools. *Bioresource Technology 160*: 67–78.
87. Goher, M.E., El-Monem, A.M.A., Abdel-Satar, A.M. et al. (2016). Biosorption of some toxic metals from aqueous solution using nonliving algal cells of *Chlorella vulgaris*. *Journal of Elementology 21* (3): 703–714.
88. Wang, S. and Zhao, X. (2009). On the potential of biological treatment for arsenic contaminated soils and groundwater. *Journal of Environmental Management 90*: 2367–2376.
89. Cavalca, L., Zanchi, R., Corsini, A. et al. (2010). Arsenic-resistant bacteria associated with roots of the wild *Cirsiumarvense* (L.) plant from an arsenic polluted soil, and screening of potential plant growth-promoting characteristics. Systems of. *Applied Microbiology 33*: 154–164.
90. Gadd, G.M. (2010). Metals: minerals and microbes: geomicrobiology and bioremediation. *Microbiology 156*: 609–643.
91. Francis, A.J. (2006) Microbial transformations of radionuclides released from nuclear fuel reprocessing plants. International Symposium on Environmental Modeling and Radioecology, Rakkasho, Aomori, Japan (Oct. 18–20).
92. Cecal, A., Popa, K., Caraus, I. et al. (2002). Uranium and thorium uptake on the hydrophilic plants. In: *Uranium in the Aquatic Environment III* (ed. J.B. Merkel, B. Planer-Friederich and C. Wolkersdorfer). Freiberg: Springer.
93. Soudek, P., Tykva, R., Vankova, R. et al. (2006). Accumulation of radioiodine from aqueous-solution by sunflower (*Helianthus annuus* L.) at the laboratory scale. *Environmental and Experimental Botany 57*: 220–225.
94. Sánchez-Esteva, S., Gómez-Muñoz, B., Jensen, L.S. et al. (2016). The effect of *Penicilliumbilaii* on wheat growth and phosphorus uptake as affected by soil pH, soil P and application of sewage sludge. *Chemical and Biological Technologies in Agriculture 3*: 21. https://doi.org/10.1186/s40538-016-0075-3.

95. Shin, H.J. (2011). Genetically engineered microbial biosensors for in situ monitoring of environmental pollution. *Applied Microbiology and Biotechnology 89* On-line publication https://doi.org/10.1007/s00253-010-2990-8.
96. Oilgae Newsletter. http://www.oilgae.com/blog/2010/09/algae-as-a-sourceof-pharmaceuticals-nutraceuticals.html. 2010; Sept 15.
97. Nath, A., Gupta, A., Neopany, B. et al. (2016). Biotechnology and traditional fermented foods. In: *Indigenous Fermented Foods of South East Asia*, 773–828. Boca Raton, FL: CRC Press ISBN 9780367377076.
98. Rodrigues, P.J. and Lira, C.F. (2018). The biological Anthropocene: Rethinking novelty organisms, interactions, and evolution. *arXiv Populations and Evolution* 1812.05662.
99. Lammerts van Bueren, E.T., Struik, P.C. et al. (2018). Towards resilience through systems-based plant breeding. A review. *Agronomy for Sustainable Development 38* (42): 1–21.
100. Gupta, V.K., Treichel, H., Shapaval, V. et al. (ed.) (2018). *Microbial Functional Foods and Nutraceuticals*. Wiley ISBN:9781119048961.
101. Khan, S., Ullah, M.W., Siddique, R. et al. (2016). Role of recombinant DNA technology to improve life. *International Journal of Genomics 20*: 16.
102. Fiedurek, J., Trytek, M., and Szczodrak, J. (2017). Strain improvement of industrially important microorganisms based on resistance to toxic metabolites and abiotic stress. *Journal of Basic Microbiology 57* (6): 445–459.
103. Vieira Gomes, A., Souza Carmo, T., Silva Carvalho, L. et al. (2018). Comparison of yeasts as hosts for recombinant protein production. *Microorganisms 6* (2): 38.
104. Chandra, R. and Chowdhary, P. (2015). Properties of bacterial laccases and their application in bioremediation of industrial wastes. *Environmental Science: Processes & Impacts 17*: 326–342.
105. Chowdhary, P., Shukla, G., Raj, G. et al. (2019). Microbial manganese peroxidase: a ligninolytic enzyme and its ample opportunities in research. *SN Applied Sciences 1*: 45. https://doi.org/10.1007/s42452-018-0046-3.
106. Belenkaya, S.V., Rudometov, A.P., Shcherbakov, D.N. et al. (2018). Biochemical properties of recombinant chymosin in Alpaca (*Vicugnapacos* L.). *Applied Biochemistry and Microbiology 54* (6): 569–576.
107. Grunwald, P. (2018). Immobilized biocatalysts. *Catalysts 8* (9): 386.
108. Donova, M.V. (2017). Steroid bioconversions. In: *Microbial Steroids, 1645* (ed. J.L. Barredo and I. Herráiz), 1–13. New York: Humana Press.
109. Tulchinsky, T.H. and Varavikova, E.A. (2014). *The New Public Health*, 3e. Elsevier, Academic Press, ISBN: 978-0-12-415766-8.
110. Singh, A.K., Narayanan, K.G.S., and Sharma, P. (2017). Effect of climatic factors on cash crop farming in India: an application of Cobb-Douglas production function model. *International Journal of Agricultural Resources, Governance and Ecology 13* (2): 175–210.
111. Mohammed, S., Bakshi, N., Chaudri, N. et al. (2016). Cancer vaccines: past, present, and future. *Advances in Anatomic Pathology 23* (3): 180–191.
112. Lukasiewicz, K. and Fol, M. (2018). Microorganisms in the treatment of cancer: advantages and limitations. *Journal of Immunology Research 20*: 18.
113. Li, X., Xu, H., Chen, Z.S. et al. (2011). Biosynthesis of nanoparticles by microorganisms and their applications. *Journal of Nanomaterials 2011*: 270974. https://doi.org/10.1155/2011/270974.
114. Lee, S.H. and Jun, B.H. (2019). Silver nanoparticles: synthesis and application for nanomedicine. *International Journal of Molecular Sciences 20* (4): 865.
115. Arya, D. (2015). Genetically modified foods: benefits and risks. Massachusetts Medical Society Committee on nutrition and physical activity. http://www.massmed.org/gmo.
116. Kumar, A., Agarwal, D.K., Kumar, S. et al. (2019). Nutraceuticals derived from seed storage proteins: implications for health wellness. *Biocatalysis and Agricultural Biotechnology 17*: 710–717.
117. Zilberman, D., Holland, T., and Trilnick, I. (2018). Agricultural GMOs-what we know and where scientists disagree. *Sustainability 10* (5): 1514.
118. Coconi-Linares, N., Ortiz-Vazquez, E., Fernandez, F. et al. (2015). Recombinant expression of four oxidoreductases in *Phanerochaete chrysosporium* improves degradation of phenolic and non-phenolic substrates. *Journal of Biotechnology 209*: 76–84.

119. Herkommerova, K., Dostal, J., and Pichova, I. (2018). Decolorization and detoxification of textile wastewaters by recombinant Myceliophthorathermophila and Trametestrogii laccases. *3 Biotech. 8*: 505.
120. Pant, G., Garlapati, D., Agrawal, U. et al. (2021). Biological approaches practiced using genetically engineered microbes for a sustainable environment: a review. *Journal of Hazardous Materials 405*: 124631.
121. Huang, J., Liu, Z., Li, S. et al. (2016). Isolation and engineering of plant growth promoting rhizobacteria *Pseudomonas aeruginosa* for enhanced cadmium bioremediation. *The Journal of General and Applied Microbiology 62* (5): 258–265.
122. Zhou, Z., Liu, Y., Zanaroli, G. et al. (2019). Enhancing bioremediation potential of Pseudomonas putida by developing its acid stress tolerance with glutamate decarboxylase dependent system and global regulator of extreme radiation resistance. *Frontiers in Microbiology 10*: 2033.
123. Mardani, G., Mahvi, A., Hashemzadeh-Chaleshtori, M. et al. (2016). Degradation of phenanthrene and pyrene using genetically engineered dioxygenase producing Pseudomonas putida in soil. *Genetika 48* (3): 837–858.
124. Jia, X., He, Y., Jiang, D. et al. (2019). Construction and analysis of an engineered Escherichia coli- *Pseudomonasaeruginosa* co-culture consortium for phenanthrene bioremoval. *Biochemical Engineering Journal 148*: 214–223.
125. Gong, T., Liu, R., Che, Y. et al. (2016). Engineering Pseudomonas putida KT2440 for simultaneous degradation of carbofuran and chlorpyrifos. *Microbial Biotechnology 9*: 6.
126. Cheng, M., Chen, K., Guo, S. et al. (2015). PbaR, an IclR family transcriptional activator for the regulation of the 3-Phenoxybenzoate 1′,2′-dioxygenase gene cluster in Sphingobiumwenxiniae JZ-1 T. *Applied and Environmental Microbiology 81*: 8084–8092.
127. Ajona, M. and Vasanthi, P. (2021). Bio-remediation of crude oil contaminated soil using recombinant native microbial strain. *Environmental Technology and Innovation 23*: 101635.
128. Mardani, G., Mahvi, A.H., Hashemzadeh-Chaleshtori, M. et al. (2017). Application of genetically engineered dioxygenase producing Pseudomonas putida on decomposition of oil from spiked soil. *Judishapur Journal of Natural Pharmaceutical Products 12* (3).
129. He, D., Li, L., and Huang, M. (2017). Characterization of rhamnolipidbiosurfactants produced by recombinant Pseudomonas aeruginosa strain DAB with removal of crude oil. *Biotechnology Letters 39*: 1381–1388.
130. Ivanov, V. and Hung, Y.T. (2010). Application of environmental biotechnology. *Environmental Biotechnology Handbook of Environmental Engineering 10*: 1–17.
131. Hanlon, G. and Hodges, N. (2013). *Essential Microbiology for Pharmacy and Pharmaceutical Sciences*. UK: Wiley.

2

Environmental and Industrial Applications of Biosurfactants

Sandhya Mishra[1], Shaohua Chen[2,3], and Ram Naresh Bharagava[4]

2.1. INTRODUCTION

Environmental contamination is becoming a serious threat to the environment because of contaminants' extensive use in anthropogenic sectors of mining, ore processing, leather tanning, paints and pigments, preservatives, combustion of fossil fuels, electroplating, fertilizer, chemical industry, refineries, pharmaceuticals, and cosmetics [1, 2]. The wastewaters/sludge released from such industrial activities contain a wide range of organic and inorganic pollutants such as chlorinated compounds, aromatic hydrocarbons, endocrine-disrupting pollutants, phenolics, heavy metals, xenobiotic substances, pesticides, phthalates, etc., directly into the environment, producing harmful effect on living beings [3, 4]. These compounds are highly toxic and less biodegradable in nature and thus pose a major challenge for environmental safety and human health protection [5, 6]. Therefore there is an urgent need for appropriate treatment technology for the treatment of contaminated sites and removal of these hazardous pollutants. Although several physicochemical treatment technologies such as reverse osmosis, chemical adsorption and precipitation, electro-dialysis, membrane filtration, oxidation, and reduction reactions have been developed and applied for the degradation and removal of toxic environmental contaminants, these treatment approaches are not suitable for the pollution-free remediation of environmental contaminants [7].

However, one of the most environmentally friendly advanced remediation processes, "bioremediation," is greatly considered for the decontamination and restoration of polluted

[1]*Environmental Technologies Division, CSIR – National Botanical Research Institute, Lucknow, Uttar Pradesh, India*

[2]*State Key Laboratory for Conservation and Utilization of Subtropical Agro-Bioresources, Guangdong Province Key Laboratory of Microbial Signals and Disease Control, Integrative Microbiology Research Centre, South China Agricultural University, Guangzhou, China*

[3]*Guangdong Laboratory for Lingnan Modern Agriculture, Guangzhou, China*

[4]*Department of Environmental Microbiology, School for Environmental Sciences, Babasaheb Bhimrao Ambedkar University (A Central University), Lucknow, Uttar Pradesh, India*

Microbial Biotechnology: Role in Ecological Sustainability and Research, First Edition.
Edited by Pankaj Chowdhary, Sujata Mani, and Preeti Chaturvedi.

sites [8]. It is the widely accepted clean-up strategy for the degradation of toxicants and pollutants from contaminated environments without producing harmful products [9, 10]. Bioremediation involves the metabolic capabilities of microorganisms in the removal of pollutants and thus, it is a most suitable and promising technology these days [11, 12].

Microbial surfactants have interestingly emerged as a new bioremediation strategy used in industrial as well as environmental applications for the degradation and detoxification of hazardous contaminants due to their high economic effectivity and biodegradation potential [13, 14]. Biosurfactants are extensively used in the pharmaceutical, petroleum, cosmetic, detergent, paint, food, and wastewater treatment industries [15, 16]. Biosurfactants can enhance the bioremediation process by increasing the bioavailability and solubility of toxic pollutants. Biosurfactants are becoming a potential candidate for bioremediation technology due to their ionic nature, low toxicity, multi-functionality, surface activities, and environmental potentiality [13, 17].

2.2. BIOSURFACTANTS AND THEIR PROPERTIES

Biosurfactants are amphiphilic, surface-active compounds that are obtained from a variety of microbial sources produced either extracellularly or from their cell surface. Biosurfactant basic chemical structure mainly contains two moieties: hydrophobic head and hydrophilic tail [18, 19]. The hydrophilic moiety includes anions, cations, peptides, polysaccharides, and amino acids, and the hydrophobic moiety contains unsaturated and saturated fatty acids [20]. Biosurfactants can be classified based on the chemical structure of their hydrophobic component into several categories like (i) glycolipids, (ii) lipopeptides, (iii) fatty acids, and (iv) polymer type, according to their chemical structure (Figure 2.1) [21]. Most of these compounds are anionic or neutral, but a few of them are cationic. However, based on molecular mass of the surfactant, biosurfactants are generally categorized into two main classes: (i) *Low molecular mass biosurfactant:* includes glycolipids, lipopetide, and phospholipids; and (ii) *High molecular mass biosurfactants:* includes particulate and polymeric surfactants. High molecular mass biosurfactants are mostly stabilizing agents while low molecular mass biosurfactants lower the interfacial surface tension.

Biosurfactants have greater advantages over synthetic surfactants due to their ability to synthesize from renewable feed stock and can have potent biodegradability, low surface and interfacial activity, greater specificity at high temperature, and alkaline or acidic pH, low toxicity, and potent biocompatibility [22, 23]. A variety of microorganisms such as *Bacillus, Pseudomonas, Candida, Burkholderia*, and *Acinetobacter* have been characterized as potential biosurfactant producers and have been used in bioremediation studies for the degradation of environmental pollutants [24, 25]. Because of their amphiphilic structure, biosurfactants increase the surface area of hydrophobic water-insoluble substances, increase water bioavailability of these substances, and transform bacterial cell surface properties. Surface activity makes excellent emulsifiers, foaming, and dispersing agents.

Biosurfactants have immense ability to reduce the interfacial and surface tension between two surfaces. It has been reported that biosurfactants can decrease the surface tension value of water from 70 to 25–28 mN/m [26]. Biosurfactant activities of surface-active compounds depend on their specific concentration known as critical micelle concentration (CMC). It is the minimum concentration of biosurfactant required to give maximum surface tension reduction of water and initial micelle formation. Biosurfactant molecules initiate micelle formation, bilayers, and vesicles at concentrations above the CMC. Micelle formation facilitates biosurfactants to reduce the interfacial and surface tension and increase bioavailability and solubility of hydrophobic organic compounds and metals [27, 28].

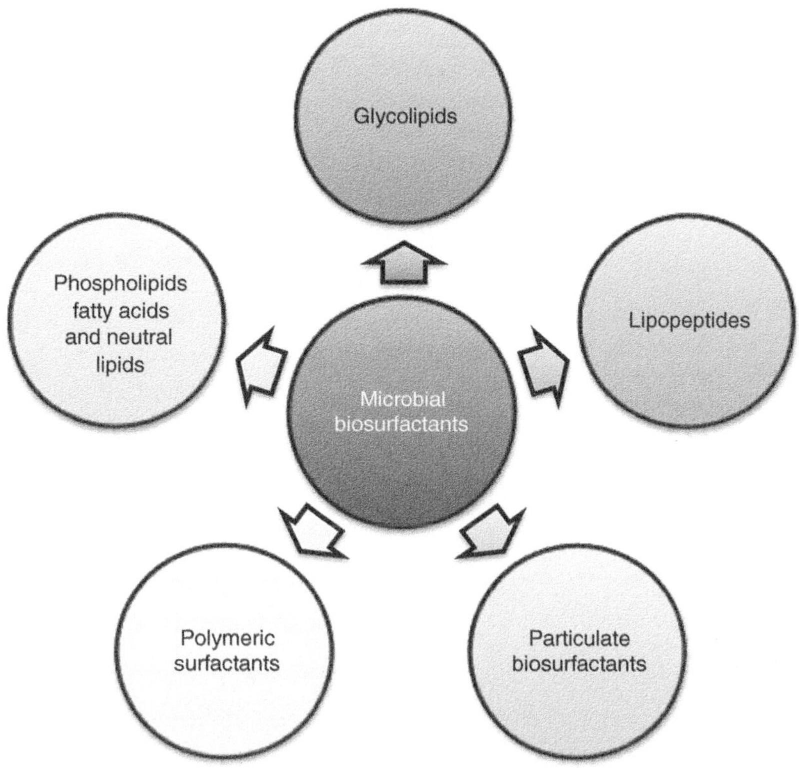

Figure 2.1 Types of microbial biosurfactants.

Biosurfactants are less toxic and more readily biodegradable and safer than synthetic surfactants as they are principally obtained from microorganisms by using low-cost and sustainable feed stocks such as soyabean oil, agricultural residues, molasses, palm oil, and waste cooking oil instead of using glucose, hydrocarbons, and glycerols [29]. Because of their low toxicity profile, biosurfactants have great advantages in pharmaceutical, industrial, food and cosmetic, and bioremediation applications [30, 31]. Furthermore, biosurfactant molecules have been reported to exhibit greater stability and activity at wider pH range (acidic and basic) and temperature in extreme environments, and can also retain their activity even after sterilization. They can also have excellent salt-tolerant capability and are reported to tolerate salt concentrations up to 10%, whereas most of the chemical surfactant inactivate at a salt concentration of >2% [32]. Several biosurfactants such as polymyxin, lichenysin, fengycin, and pumilacidin showed a strong antimicrobial activity and therefore can be used as effective and safe therapeutic agents [33, 34].

2.3. TYPES OF BIOSURFACTANTS

2.3.1. Glycolipids

Glycolipids are the most commonly derived microbial biosurfactants that mainly contain saccharide (including sugar moiety like glucose, galactose, and rhamnose, etc.) and lipid moiety (saturated or unsaturated fatty acids, or hydroxyl aliphatic acids). Generally

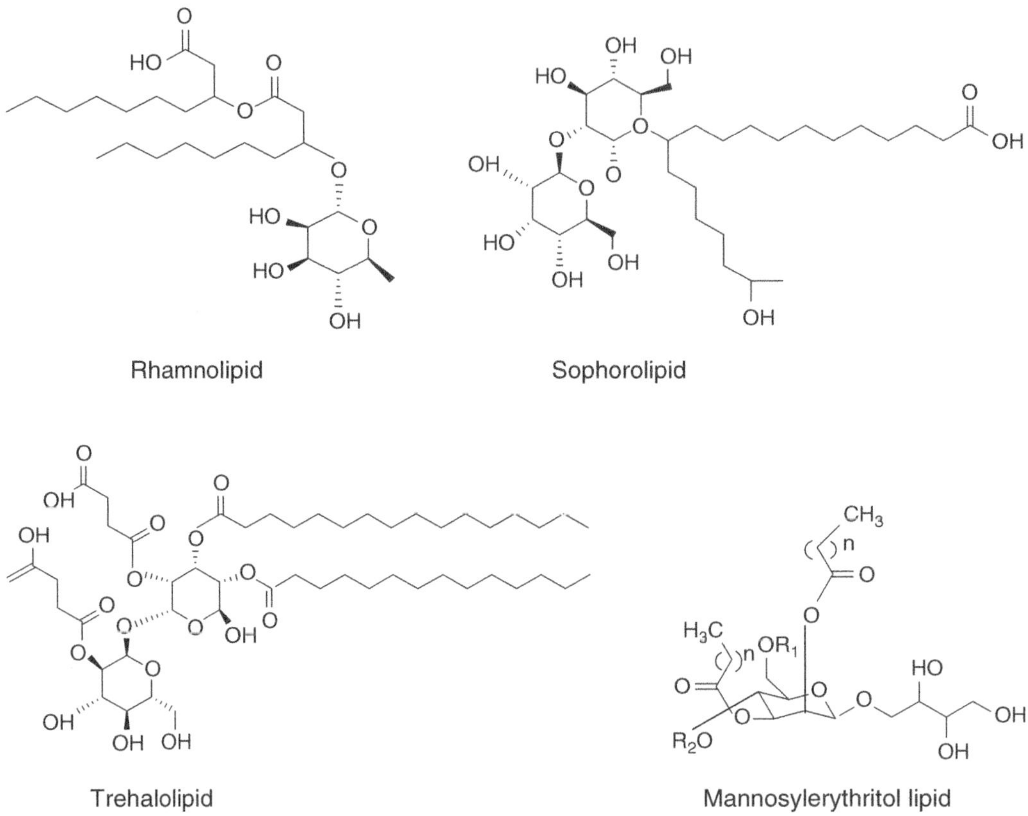

Figure 2.2 Chemical structures of glycolipid biosurfactants produced by microorganisms.

rhamnolipids, sophorolipids, trehalolipids, and mannosylerythritol lipids are included in the group of glycolipids (Figure 2.2) [35].

2.3.1.1. Rhamnolipids

Rhamnolipids contain a hydrophilic group, consisting of either one or two rhamose molecules, with a glycosidic linkage to the hydrophobic group made up of one or two β hydroxy fatty acids chains with 8–14 carbon atoms. *Pseudomonas aeruginosa* is the most primary producer of rhamnolipid with mixture different congeners. It produces two forms of rhamnolipids, i.e. mono and di-rhamnolipids [36]. These are the most popular glycolipid biosurfactants, used in various remediation technologies such as metal removal, oil recovery, hydrocarbon degradation, and pharmaceuticals due to their great structural variety, unique metal binding capacities, and selectivity over the synthetic surfactants or chelates [37]. Rhamnolipids have the ability to weaken bacterial adsorption and enhance bacterial transport to or throughout the remediation sites for successful bioaugumentation process [38].

2.3.1.2. Sophorolipids

Sophorolipids are amphiphilic molecules composed of the disaccharide sophorose (2′-O-β-D-glucopyranosyl-β-D-glycopyranose) linked with glycosidic bond to a long chain of fatty acid containing 16–18 carbon atoms with one or more saturation. Sophorolipids can be classified as anionic (acidic) or non-anionic (lactonic) biosurfactants [39]. Sophorolipids have higher water solubility and better foaming agents, thus they are extensively used in environmental applications as well as in the cosmetics and food industries. They have also been reported to have excellent antimicrobial, biocidal, cytotoxic, anticancer, and pro-inflammatory activities and thus are often used in the pharmaceutical industries for antibiotics and drugs [40].

2.3.1.3. Trehalolipids

Trehalolipids are mainly obtained from several actinomycetes and bacterial strains such as *Rhodococcus* sp. and *Mycobacterium* sp. These glycolipids are mainly composed of disaccharide trehalose linked with two β-hydroxyl branched fatty acids. It has been reported that *Rhodococcus* sp. can synthesize three forms of trehalolipids, namely monomycolates, dimycolates, and trimycolates [41]. Trehalolipids have excellent emulsifying properties and thus are utilized in oil spill and oil recovery treatment process.

2.3.1.4. Mannosylerythritol Lipids

Mannosylerythritol lipids are amphipathic molecules containing 4-O-β-D-mannopyranso sylerythritol or 1-O-β-D-mannopyranosyl-erythritol as a hydrophilic head group and a fatty acyl group as the hydrophobic unit [42]. Mannosylerythritol lipids usually have one or two acetyl groups at C-40 and C-60 of mannose moiety and due to different number and location of mannose moiety. Mannosylerythritol lipids are classified as MEL-A, MEL-B, MEL-C, and MEL-D. These glycolipids are mainly derived from fungal strains; *Ustilago* and *Pseudozyma*. Mannosylerythritol lipids have been widely used in environmental industries due to their excellent surface activity, biocompatibility, and restorative function [43, 44].

2.3.2. Lipopeptides

Lipopeptides are cyclic peptides, acylated with fatty acids that are differ in length and structure. It comprises d-amino acids or iso- or anteiso-hydroxy fatty acids. Lipopeptides shows high structural diversity in their compounds due to the variation in peptide composition, as well as different degrees of branching and oxidation. Surfactin, polymixin, viscosin, fengycin, and iturin are the best examples of antimicrobial lipopeptides (Figure 2.3) [45, 46].

2.3.2.1. Surfactins

Surfactins are amphipathic cyclic lipopeptides of seven amino acids and different 3-β-hydroxy fatty acids (12–16 carbon) with the main component being 3-hydroxy-13-methyl-myristic acid. It mainly synthesized from the members belonging to the genus Bacillus, such as *Bacillus pumilus, Bacillus subtilis, Bacillus amyloliquefaciens*, and *Bacillus licheniformis* [47]. Surfactins exhibit a wide range of biological activities such as interaction with target cell membrane, inhibiting clot formation, lysing bacterial spheroplasts and protoplas, and possessing potential for various medical applications due their strong antimicrobial and antiviral properties against pathogenic bacteria, fungi, and viruses. Surfactins are effectively applied in the removal of heavy metals, hydrocarbons, and oil recovery treatment plants [48, 49].

2.3.2.2. Polymixins

Polymyxins are cationic and branched cyclic decapeptides. Polymyxin B, which is derived from *Bacillus polymyxa*, is the best example of polymyxin, which occurs as decapeptide with eight amino acids forming a ring, linked to a branched fatty acid. Due to the properties of

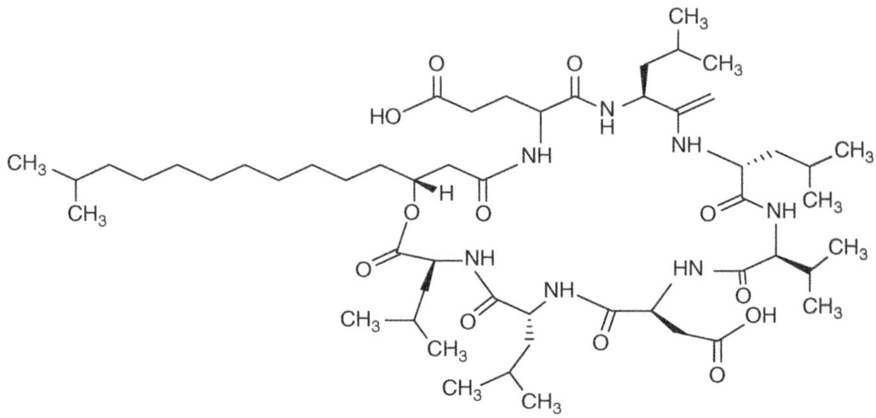

Surfactin

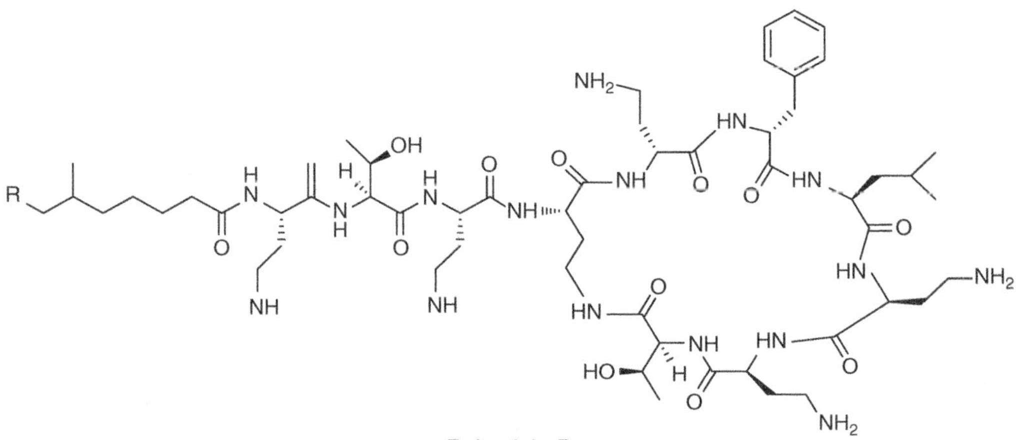

Polymixin B

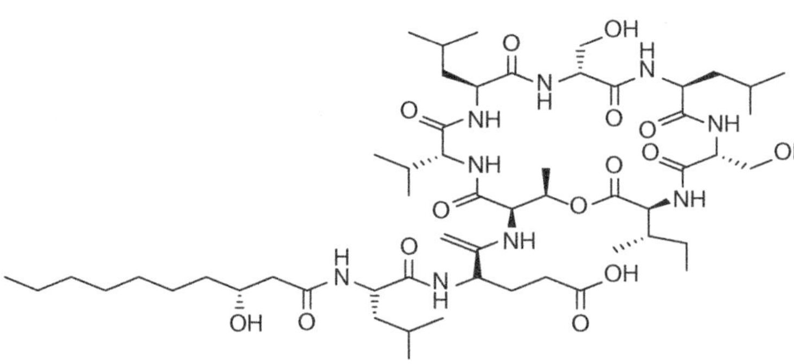

Viscosin

Figure 2.3 Chemical structures of lipopeptide biosurfactants produced by microorganisms.

being integrated into bacterial cell walls of Gram-negative bacteria thereby influencing permeability, polymyxins are known as antibiotics. Besides polymyxin B, polymyxin E, which is also known as colistin, is used for application as an antibiotic [33].

2.3.2.3. Viscosin

Viscosin is a cyclic lipopeptide linked to a fatty acid with a pronounced pattern of hydrophobic amino acid residues. Viscosin is synthesized from Gram-negative bacterial strains such as *Pseudomonas fluorescence, Pseudomonas libanensis*, and *Pseudomonas viscosa*. It has been reported that viscosin favors surface spreading on plant roots in plant growth-promoting rhizobacteria and displays antiviral properties, and thus can act as a potential plant growth-promoting and phyto-protective agent [50].

2.3.3. Phospholipids and Fatty Acids

A few microorganisms and a yeast have the ability to produce unsaturated fatty acids and phospholipids during the cultivation *n*-alkanes. *Acinetobacter* sp. HO1-N and *Acinetobacter* sp. 1-N have phosphotidyl ethanolamine-rich vesicles that consist of protein, phospholipids, and lipopolysaccharides. These vesicles play a crucial role in the uptake of alkane by microbial cells. These biosurfactants are essential for therapeutic applications.

2.3.4. Polymeric Biosurfactants

Biosurfactants constituted from polysaccharides bound to protein are called polymeric biosurfactants. Alasan, emulsan, liposan, biodispersan, and some other polysaccharide-protein complexes are all well-studied examples of polymeric biosurfactants. RAG-1 and BD4 are the most prominent examples of microbial emulsans. RAG-1 is a complex mixture of exoploysaccharide and lipopolysaccharide, whereas BD4 is a protein polysaccharide mixture with a polysaccharide composed of heptasaccharides [51, 52].

2.4. APPLICATIONS OF BIOSURFACTANTS

2.4.1. In the Pharmaceutical and Therapeutic Industries

Biosurfactants are used extensively in pharmaceutical fields. Biosurfactants demonstrate antimicrobial action against pathogenic microorganisms, parasites, and infections. Several biosurfactants have strong antimicrobial activity and therefore can be used as effective and safe therapeutic agents [53, 54]. The lipopeptide class of biosurfactants is mainly reported for antimicrobial activities. Surfactin and polymixin B produced from *Bacillus* sp. are the best known lipopeptides with strong antimicrobial activities [34]. Other lipopeptides such as lichenysin, fengycin, pumilacidin, and bacillomycins showed antimicrobial properties [29, 33]. A rhamnolipid surfactant produced from soyabean oil showed positive antimicrobial activity against *Micrococcus leuteus, Bacillus cereus, Staphylococcus aureus*, and *Mucor miehei* as investigated by Nitschke [55]. Sophorolipids are also shown to have antiviral activity against human immunodeficiency virus [40].

Moreover, several biosurfactant antifungal activities of cellibiose, rhamnolipids, fengycin, and surfactin have also been reported against phyto-pathogenic fungi [56–58]. Biosurfactant-based liposomes have potential applications in gene transfaction compared to commercially available cationic liposomes [59].

2.4.2. In the Cosmetic Industry

Biosurfactants have large-scale application in the cosmetic industry as they incorporate to help healthy skin physiology through their use as beautifying agents. Compared to chemical surfactants, biosurfactant-based products are much safer and have great potential to be used in the formulation of cosmetic products due to their ecofriendly nature and biodegradability [60, 61]. Sophorolipids and mannosylerythritol lipids are well reported for their exceptional hygroscopic properties to boost skin hydration and enhance radiance of dermal fibroblast metabolism. Sophorolipids are mainly used in facial make-up creams, shampoo, perfumes, and formulations used for the treatment of acne and infections [62]. Sophorolipids are also incorporated in cellulite treatments as they help to reduce the subcutaneous fat overload by stimulating leptin synthesis in adipocytes. Mannosylerythritol lipids, mainly derived from *Pseudozyma* spp. of yeasts, are widely used in sprays, powders, lipsticks, nail care, body massage oils, and other items. Moreover, mannosylerythritol lipids exhibit great potential to enhance water retention in the stratum corneum and to repair damaged hair. They are also used as skin-whitening agents to reduce hyperpigmentation skin problems by suppressing the melanocyte production and improving skin fairness [63].

Nowadays, biosurfactant-based Japanese cosmetics brand using surfactin-derived lipopeptides are highly popular in the market [64]. Surfactin biosurfactants have superb foaming properties and low CMC properties, and thus are important in the formulation of dermatological products, anti-wrinkle cosmetics, cleansing products, and water and oil emulsions [65, 66]. Moreover, some studies have reported that rhamnolipids can also be applied in personal skincare and cosmetics due to their biocompatibility and excellent emulsification properties. Biosurfactant-based beauty products and moisturizers can be ideal for an effective skin care routine as they have great potential to stimulate collagen renewal and control other factors that cause harmful effects on skin [31].

2.4.3. In Bioremediation of Heavy Metals

Heavy metal contamination is becoming a serious threat to the environment because of their extensive use and bioaccumulation in soil and water systems. Several heavy metals, such as arsenic (As) mercury (Hg), lead (Pb), chromium (Cr), and cadmium (Cd), are highly toxic in nature even at low concentration because of their chemical complexity, persistence, and non-biodegradability [7, 67]. Biosurfactants are one of the outstanding microbial products that have been successfully employed in the remediation of metal-contaminated environments [13, 16]. The efficacy of biosurfactants for the removal of heavy metals has been investigated and reported by several authors who have described that biosurfactants can perform heavy metal removal either by associating with heavy-metal ions resulting in a biosurfactant–metal complex or via accumulation at a solid solution interface, which leads to the direct interaction of the metal [68, 69]. Biosurfactants are competent to form complexes with free, non-ionic forms of metals in solution. Following Le-Chatelier's principle, this complexation reduces the solution phase activity of metal and enhances the metal desorption. Biosurfactants can directly interact with absorbed metal at the solid solution interface under conditions of reduced interfacial tension, which allows biosurfactants to accumulate at solid solution interface [70].

Mulligan [19, 28, 68] investigated the effectiveness of biosurfactants on the removal of heavy metals from soils. New approaches and attempts for metal stabilization using biosurfactants have also been addressed by Gnanamani [71] who showed bioremediation of Cr(VI) by a biosurfactant producing marine isolate *Bacillus* sp. MTCC 5514. Huang and Liu [72] investigated the removal of Cd and Pb from industrial wastewater by using a biosurfactant produced from the *Pseudomonas* sp. strain LKS06. Ahuekwe [73] reported the metal removal activities for Cr (23.11%), Cd (15.71%), Pb (9.93%), Zn (7.29%), and Cu (4.96%) at 70 mg/l concentration of respective metal solution by a sophorolipid biosurfactant producing yeast *Candida bombicola*.

Moreover, in another study, a sophorolipid producing yeast *Staemerella bombicola* CGMCC 1576 was used to study the effects of Cd and Pb removal in batch soil washing from artificially contaminated soil. Crude total sophorolipids showed better removal efficiency, 83.6% and 44.8% for Cd and Pd, respectively, at higher biosurfactant concentration of 8% than synthetic surfactants. Furthermore, the fermentation broth of *S. bambicola* removed 95 and 52% Cd and Pb, respectively [74]. Similarly, Arab and Mulligan [75] utilized a microbial biosurfactant produced by *Candida lipolytica* as an ecofriendly tool for the removal of heavy metals from mine tailing.

2.5. CONCLUSIONS

Increased industrial and anthropogenic activities have resulted in the continuous introduction of toxic pollutants and heavy metals into the environment, which pose severe and hazardous threats to living beings. Considering the severity of environmental pollutants, the application of biosurfactants has aroused great interest and attention for the valuable remediation of contaminated sites. Biosurfactants and biosurfactant-producing microorganisms seem to be an encouraging approach that improves the efficiency of bioremediation technology because of their versatility, biodegradability, economical effectivity, and environmental safety. Also, biosurfactants are potentially less toxic than chemical surfactants and thus have great applications in the cosmetic and pharmaceutical industries. However, there are certain limitations regarding the high production cost, and the selection of potent biosurfactant-producing microbial strains confines the success of biosurfactant technology to large-scale applications. Optimized culture conditions, use of cheap renewable substrates, and novel, efficient techniques to find potent biosurfactant-producing microbes and purification of biosurfactants from them via downstream processing could make the production process more economically practical.

REFERENCES

1. Mishra, S., Lin, Z., Pang, S. et al. (2021). Recent advanced technologies for the characterization of xenobiotic-degrading microorganisms and microbial communities. *Frontiers in Bioengineering and Biotechnology 9*: 632059.
2. Bhatt, P., Bhatt, K., Huang, Y. et al. (2020). Esterase is a powerful tool for the biodegradation of pyrethroids insecticides. *Chemosphere 244*: 125507.
3. Embrandiri, A., Kiyasudeen, S.K., Rupani, P.F. et al. (2016). Environmental xenobiotics and its effect on natural ecosystem. In: *Plant Responses to Xenobiotics Singh A* (ed. S. Prasad and R. Singh), 1–18. Singapore: Springer.
4. Dhakal, K., Gadupadi, G.S., and Robertson, L.W. (2017). Sources and toxicities of phenolic polychlorinated biphenyls (OH-PCBs). *PCBs Risk Evaluation and Environmental Protection 25*: 16277–16290.
5. Dinka, D.D. (2018). Environmental xenobiotics and their adverse health impacts-a general review. *Journal of Environment Pollution and Human Health 6*: 77–88.
6. Maurya, P.K. (2016). Bioaccumulation of xenobiotics compound of pesticides in riverine system and its control technique: a critical review. *Journal of Industrial Pollution Control 32*: 580–594.

7. Mishra, S., Bharagava, R.N., More, N. et al. (2019). Heavy metal contamination: an alarming threat to environment and human health. In: *Environmental Biotechnology for Sustainable Future*, 103–125. Singapore: Springer.
8. Paul, D., Pandey, G., Pandey, J. et al. (2005). Accessing microbial diversity for bioremediation and environmental pollutant restoration. *Trends in Biotechnology 23* (3): 135–142.
9. Bhatt, P., Gangola, S., Bhandari, G. et al. (2021). New insights into the degradation of synthetic pollutants in contaminated environments. *Chemosphere 268*: 128827.
10. Malla, M.A., Dubey, A., Yadav, S. et al. (2018). Understanding and designing the strategies for the microbe-mediated remediation of environmental contaminants using omics approaches. *Frontiers in Microbiology 9*: 1132.
11. Azubuike, C.C., Chikere, C.B., and Okpokwasili, G.C. (2016). Bioremediation techniques-classification based on site of application: principle, advantages, limitations, and prospects. *World Journal of Microbiology and Biotechnology 32* (11): 180–180.
12. Zhang, W., Lin, Z., Pang, S. et al. (2020). Insights into the biodegradation of lindane (γ-hexacholocyclohexane) using a microbial system. *Frontiers in Microbiology 11*: 522.
13. Mishra, S., Lin, Z., Pang, S. et al. (2021). Biosurfactant is a powerful tool for the bioremediation of heavy metals from contaminated soils. *Journal of Hazardous Materials 418*: 126253.
14. Jimoh, A. and Lin, J. (2019). Biosurfactant: a new frontier for greener technology and environmental sustainability. *Ecotoxicology and Environmental Safety 184*: 109607.
15. Akbari, S., Abdurahman, N.H., Yunus, R.M. et al. (2018). Biosurfactants-a new frontier for social and environmental safety; a mini review. *Biotechnology Research and Innovations 2*: 81–90.
16. Das, A.J., Lal, S., Kumar, R. et al. (2017). Bacterial biosurfactants can be an ecofriendly and advanced technology for remediation of heavy metal and co-contaminated soil. *International Journal of Environmental Science and Technology 14*: 1343–1354.
17. Mao, X., Jiang, R., Xiao, W. et al. (2015). Use of surfactants for the remediation of contaminated soils: a review. *Journal of Hazardous Materials 285*: 419–435.
18. Fracchia, L., Cavallo, M., and Giovanna, M.M.I. (2012). Biosurfactants and emulsifiers biomedical and related applications- present status and future potentials. In: *Biomedical Science Engineering and Technology* (ed. N. Ghista), 325–370. InTech.
19. Mulligan, C.N., Sharma, S.K., and Mudhoo, A. (2014). *Biosurfactants: Research Trends and Applications*. CRC Press.
20. Basak, G. and Das, N. (2014). Characterization of sophorolipid biosurfactants produced by *Crytococcus* sp. VITGBN2 and its application on Zn(II) removal from electroplating wastewater. *Journal of Environmental Biology 35*: 1087–1094.
21. Roy, A. (2017). Review on the biosurfactants: properties, types and its applications. *Journal of Fundamentals of Renewable Energy and Applications 8*: 248.
22. Usman, M.M., Dadrasnia, K.T., Lim, A.F. et al. (2016). Application of biosurfactants in environmental biotechnology, remediation of oil and heavy metal. *AIMS Bioengineering 3*: 289–304.
23. Karlapudi, A.P., Venkateshwarulu, T.C., Tammineedi, J. et al. (2018). Role of biosurfactant in bioremediation of oil pollution-a review. *Petroleum 4* (3): 241–249.
24. Rastogi, S. and Kumar, R. (2020). Remediation of heavy metals using non-conventional adsorbents and biosurfactants-producing bacteria. In: *Environmental Degradation: Causes and Remediation Strategies* (ed. V. Kumar, J. Singh and P. Kumar). India: AGRO Environ Media, Agriculture and Environmental Science Academy (AESA) https://doi.org/10.26832/aesa-2020-edcrs-010.
25. Varjani, S.J. (2017). Microbial degradation of petroleum hydrocarbons. *Bioresource Technology 223*: 277–286.
26. Sarubbo, L.A., Rocha Junior, R.B., Luna, J.M. et al. (2015). Some aspect of heavy metals contamination remediation and role of biosurfactants. *Chemistry and Ecology 31*: 707–723.
27. Pacwa-Plociniczak, M., Plaza, G.A., Piotrowska-Seget, Z. et al. (2011). Environmental applications of biosurfactants: recent advances. *International Journal of Molecular Sciences 12*: 633–654.

28. Mulligan, C.N. (2005). Environmental applications for biosurfactants. *Environmental Pollution 133*: 183–198.
29. Carolin, F., Kumar, S.P., and Ngueagni, P.T. (2021). A review on new aspect of lipopeptide biosurfactant: types, production, properties and its application in the bioremediation. *Journal of Hazardous Materials 407*: 124827.
30. Santos, D.K.F., Rufino, R.D., Luna, J.M. et al. (2016). Biosurfactants: multifunctional biomolecules of the 21st century. *International Journal of Molecular Sciences 17*: 401.
31. Adu, S.A., Naughton, P.J., Marchant, R. et al. (2020). Microbial biosurfactants in cosmetics and personal skincare pharmaceutical formulations. *Pharmaceuticals 12*: 1099.
32. Cheng, Y., He, H., Yang, C. et al. (2016). Challenges and solutions for bofiltration of hydrophobic voltage organic compounds. *Biotechnology Advances 34* (6): 1091–1102.
33. Landman, D., Georgescu, C., Martin, D.A. et al. (2008). Polymyxins revisited. *Clinical Microbiology Reviews 21*: 449–465.
34. Das, P., Mukherjee, S., and Sen, R. (2008). Antimicrobial potential of a lipopeptide biosurfactant derived from a marine *Bacillus circulans*. *Journal of Applied Microbiology 104*: 1675–1684.
35. Zhao, F., Zhou, J.-D., Ma, F. et al. (2016). Simultaneous inhibition of sulfate-reducing bacteria, removal of H_2S and production of rhamnolipid by recombinant *Pseudomonas stutzeri* Rhl: applications for microbial enhanced oil recovery. *Bioresource Technology 207*: 24–30.
36. Amani, H. (2015). Study of enhanced oil recovery by rhamnolipids in a homogeneous 2D micromodel. *Journal of Petroleum Science and Technology 128*: 212–219.
37. Henkel, M., Geissler, M., Weggenmann, F. et al. (2017). Production of microbial biosurfactants: status quo of rhamnolipid and surfactin towards large scale production. *Biotechnology Journal 12*: 1600561.
38. Zhong, H., Liu, G., Jiang, Y. et al. (2017). Transport of bacteria in porous and its enhancement by surfactants for bioaugmentation: a review. *Biotechnology Advances 35*: 490–505.
39. Van Bogaert, I.A. and Soetaert, W. (2011). Sophorolipids. In: *Biosurfactant* (ed. G. Soberon Chavez), 179–210. Germany: Springer.
40. Shah, V., Doncel, G.F., Seyoum, T. et al. (2005). Sophorolipids, microbial glycolipids with antihuman immunodeficiency virus and sperm immobilizing activities. *Antimicrobial Agents and Chemotherapy 49*: 4093–4100.
41. White, D.A., Hird, L.C., and Ali, S.T. (2013). Production and characterization of trehalolipids biosurfactant produced by the novel marine bacterium *Rhodococcus* sp., strain PMLo26. *Journal of Applied Microbiology 115*: 744–755.
42. Bakur, A., Niu, Y., Kuang, H. et al. (2019). Synthesis of gold nanoparticles derived from mannosylerythritol lipid and evaluation of their bioactivities. *AMB Express 9*: 62.
43. Fan, L., Li, H., Niu, Y. et al. (2016). Characterization and inducing melanoma cell apoptosis activity of mannosylerythritol lipids-a produced from *Pseudozyma aphidis*. *PLoS One 11* (2): e0148198.
44. Morita, T., Konishi, M., Fukuoka, T. et al. (2006). Characterization of the genus *Pseudozyma* by the formation of glycolipid biosurfactants, mannosylerythritol lipids. *FEMS Yeast Research 7* (2): 286–292.
45. Mnif, I. and Ghribi, D. (2015). Review lipopeptides biosurfactants: mean classes and new insights for industrial, biomedical and environmental application. *Peptide Science 104*: 129–147.
46. Baltz, R.H., Miao, V., and Wrigley, S.K. (2005). Natural products to drugs: daptomycin and related lipopeptide antibiotics. *Natural Product Reports 22* (6): 717–741.
47. Shaligram, N.S. and Singhal, R.S. (2010). Surfactin- a review on biosynthesis fermentation, purification and applications. *Food Technology and Biotechnology 48* (2): 119–134.
48. Phulpoto, I.A., Yu, Z., Hu, B. et al. (2020). Production and characterization of surfactin-like biosurfactant produced by novel strain *Bacillus nealsonii* S2MT and its potential for oil contaminated soil remediation. *Microbial Cell Factories 19*: 145.
49. Felix, A.K.N., Martins, J.J., and Almeida, J.G.L. (2019). Purification and characterization of a biosurfactant produced by *Bacillus subtilis* in cashew apple juice and its application in the remediation of oil contaminated soil. *Colloids and Surfaces, B: Biointerfaces 175*: 256–263.

50. Alsohim, A.S., Taylor, T.B., Barrett, G.A. et al. (2014). The biosurfactant viscosin produced by *Pseudomonas fluorescens* SBW25 aids spreading motility and plant growth promotion. *Environmental Microbiology 16* (7): 2267–2281.
51. Shekhar, S., Sundaramanickam, A., and Balasubramaniam, T. (2014). Biosurfactant producing microbes and its potential applications: a review. *Critical Reviews of Environmental Science and Technology 45*: https://doi.org/10.1080/10643389.2014.955631.
52. Kaplan, N., Zosim, Z., and Rosenberg, E. (1987). Reconstitution of emulsifying activity of *Acinetobacter calcoaceticus* Bd4 emulsan by using pure polysaccharide and protein. *Applied and Environmental Microbiology 53* (2): 440–446.
53. Singh, P. and Cameotra, S.S. (2004). Enhancement of metal bioremediation by use of microbial surfactant. *Biochemical and Biophysical Research Communications 319*: 291–297.
54. Marchant, R. and Banat, I.M. (2012). Microbial biosurfactants: challenges and opportunities for future exploitation. *Trends in Biotechnology 30*: 558–565.
55. Nitschke, M., Costa, S.G., and Contiero, J. (2009). Structure and application of rhamnolipid surfactant produced in soybean oil waste. *Applied Biochemistry and Biotechnology 160*: 2066–2074.
56. Borah, S.N., Goswami, D., Sarma, H.K. et al. (2016). Rhamnolipid biosurfactant against *Fusarium verticillioides* to control stalk and ear rot disease of Maize. *Frontiers in Microbiology 7*: 1505.
57. Velho, R.V., Medina, L.F.C., Segalin, J. et al. (2011). Production of lipopeptides among Bacillus strains showing growth inhibition of phytopathogenic fungi. *Folia Microbiologica 56* (4): 297.
58. Grover, M., Nain, L., Singh, S.B. et al. (2010). Molecular and biochemical approaches for characterization of antifungal trait of a potent biocontrol agent *Bacillus subtilis* RP24. *Current Microbiology 60* (2): 99–106.
59. Kitamato, D. (2008). Naturally engineered glycolipids biosurfactants leading to distinctive self-assembling properties. *Yakugaki Zasshi 128*: 695–706.
60. Varvaresou, A. and Iakovou, K. (2015). Biosurfactants in cosmetics and biopharmaceuticals. *Letters in Applied Microbiology 61*: 214–223.
61. Vecino, X., Cruz, J.M., Moldes, A.A. et al. (2017). Biosurfactants in cosmetic formulations: trends and challenges. *Critical Reviews in Biotechnology 37*: 911–923.
62. Lourith, N. and Kanlayavattanakul, M. (2009). Natural surfactants used in cosmetics: glycolipids. *International Journal of Cosmetic Science 31*: 255–261.
63. Yoo, J.W., Hwang, Y.K., Bin, S. et al. (2019). Skin whitening composition containing mannosylerythritol lipid. U.S. 20190231668A1 (Patent).
64. Lu, J.K., Wang H.M., and Xu, X.R. (2016). U.S. 20160030322A1: Applications of Surfactin in Cosmetic Products (Patent).
65. Gupta, P.L., Rajput, M., Oza, T. et al. (2019). Eminence of microbial products in cosmetic industry. *Natural Products and Bioprospecting 9*: 267–278.
66. Carrillo, C., Teruel, J.A., Aranda, F.J. et al. (2013). *Biochemistry and Biophysics Acta 1611*: 91.
67. Mishra, S. and Bharagava, R.N. (2016). Toxic and genotoxic effect of hexavalent chromium in environment and its bioremediation strategies. *Journal of Environmental Science and Health Part C: Environmental Carcinogenesis and Ecotoxicology Reviews 34*: 1–32.
68. Mulligan, C.N., Yong, R.N., and Gibbs, B.F. (2001). Heavy metal removal from sediments by biosurfactants. *Journal of Hazardous Materials 85*: 111–125.
69. Lal, S., Ratna, S., Said, O.B. et al. (2018). Biosurfactant and exopolysaccharide-assisted rhizobacterial technique for the remediation of heavy metal contaminated soil: an advancement in metal phytoremediation technology. *Environmental Technology and Innovation 10*: 243–263.
70. Miller, R.M. (1995). Biosurfactant-facilitated remediation of metal contaminated soils. *Environmental Health Perspectives 103*: 59–62.
71. Gnanmani, A., Kavitha, V., Radhakrishnan, N. et al. (2010). Microbial products (biosurfactants and extracellular chromate reductase) of marine microorganism are the potential agent reduce the oxidative stress induced by toxic heavy metals. *Colloids and Surfaces B, Biointerfaces 79*: 334–339.

72. Huang, W. and Lin, Z.N. (2013). Biosorption of Cd(II)/Pb(II) from aqueous solution by biosurfactant-producing bacteria: isotherm kinetic characteristic and mechanism studies. *Colloids and Surfaces. B, Biointerfaces 105*: 113–119.
73. Ahuekwe, E.F., Okoli, B.E., Stanley, H.O., and Kinigoma, B. (2016). Evaluation of hydrocarbon emulsification and heavy metal detoxification potentials of sophorolipid biosurfactants produced from waste substrates using yeast and mushroom. SPE African Health Safety, Security, Environment and Social Responsibility Conference and Exhibition. Accra, Ghana. http://10.0.8.70/183578-MS.
74. Qi, X., Xu, X., Zhong, C. et al. (2018). Removal of cadmium and lead from contaminated soils using sophorolipids from fermentation culture of *Starmerella bombicola* CGMCC1576 fermentation. *International Journal of Environmental Research and Public Health 15*: 2334.
75. Arab, F. and Mulligan, C.N. (2018). An eco-friendly method for heavy metal removal from mine tailings. *Environmental Science and Pollution Research 25*: 16202–16216.

3

Synbiotic Effects of Human Milk on Neonatal Health: Probiotic Early Microflora and Prebiotic Oligosaccharides in Symphony

Swati Shukla[1], Ramesh C. Dubey[2], and Jyoti Choudhary[1,2]

3.1. INTRODUCTION

The human fetus resides in the intrauterine environment and encounters the external microbial world for the first time post-delivery. An adult has a strong immune system, due to the acquired immunity achieved during their lifetime encounters with various pathogens, but neonates are born with immature innate immunity and are inexperienced in developing acquired immunity. Breastfeeding is usually the only source of nutrition and a supplement to new antigenic substances in the surrounding environment. This is contestable that neonates have immunosuppressed predicaments and function differently from adults [1]. Breastmilk introduces a wide range of microbial populations to the immature immune system of infants. Breastmilk is often the only source of nutrition for infants during the first six months. This is a crucial period for development of the intestinal immune system as infants don't receive additional diet besides breastmilk, which could expose infants to new immunogens [2]. Human milk serves as a complete diet for a newborn because it contains 3–5% fat, 0.8–0.9% proteins, 6.9–7.2% carbohydrates, and 0.2% minerals, as well as human milk oligosaccharides (HMOs) and microflora [3]. These HMOs enhance the colonization of healthy microflora in an infant's intestines and help to smooth the gastrointestinal and physiological system [4]. Bode reported that HMOs also work as a "Bifidus factor" [5]. Bifidus factors are the compounds that could enhance the growth of bifidobacteria in any product or on natural sites such as the intestine, urinary tract, etc., where they tend to provide various health benefits by stabilizing the microflora. These Bifidus factors are almost negligible in cow's milk. During vaginal birth, the microflora are established in neonates at the time of delivery, when neonates come in contact with the mother's vaginal walls. But this is not the case in a cesarean delivery because neonates only receive the first microflora in a later stage: breastmilk.

[1]*Department of Biotechnology, Chinmaya Degree College, Haridwar, Uttarakhand, India*

[2]*Department of Botany and Microbiology, Gurukul Kangri Vishwavidyalya, Haridwar, Uttarakhand, India*

Microbial Biotechnology: Role in Ecological Sustainability and Research, First Edition.
Edited by Pankaj Chowdhary, Sujata Mani, and Preeti Chaturvedi.

Epidemiological studies also show that breastfeeding reduces the chances of gastrointestinal tract (GIT) infection, respiratory infection, diarrhea, and colitis in both developed and developing countries [2]. Oligosaccharides and commensal microflora of human milk not only stimulate the secretory antibody system but also awaken the oral tolerance that is the suppressive mechanism of the body against antigens [6]. Interaction among HMOs, intestinal microflora, and the glycans present on the mucosal surface of the intestine incline the ontogeny of mucosal immunity and the genesis of the enteric infection [7]. Colonization pattern and population of commensal bacteria on the surface of the intestine have a significant impact for the development of innate and adaptive immune systems [8].

3.2. SOURCES OF NUTRITION FOR INFANTS

3.2.1. Breastfeeding

International agencies such as the FDA (Food and Drug Administration) and WHO (World Health Organization) recommend breastfeeding as the main mode of infant feeding for the entire first year of life, and thereafter as long as it is pleasant to a given mother–infant pair [9]. From a nutritional perspective, infancy is a critical and vulnerable period when a single food is adequate as the sole source of nutrition.

3.2.2. Formula Milk for Infants

In situations when breastmilk is not available, formula milk supplements can provide nutritional requirements. The development of the infant intestinal flora is very much dependent on the mode of feeding. Breastfeeding favors early bonding between mothers and infants, as well as providing the best nutritional source to infants [10]. However, there are some situations when breastmilk needs to be replaced with alternative options. An alternative mode for feeding infants that is used widely throughout the world is bottle feeding. Here, the infants are being fed with formula milk, cow's milk, or other animal milk, as well as human milk from a human milk bank. Formula milk fulfills the nutritional requirements of infants, but it does not provide the bioactive entities which dictate the immunity of infants. Most commercial formula milk for infants are cow's milk, which at times may cause gastrointestinal blood loss and various allergies [11]. There are cases where formula milk has been reported to cause various other problems such as obesity, allergies, and higher crying rate [12, 13]. Commercial infant formula lacks the microflora prebiotic substances and other bioactive components of human milk [14]. Many previous studies have suggested a positive relationship between obesity or sudden excess weight gained during infancy with formula milk feeding [15]. The composition of fecal and intestinal microflora found in infants that feed on formula milk versus those that feed on breastmilk is significantly different as well [16].

Manufacturers of commercial formula milk tried to imitate the composition of human milk as closely as possible but so far, they have yet to achieve it. A human milk bank or a breastmilk bank acts as a substitute instead of formula, but it is not a sustainable source and lacking in the desirable microbiota. This is because the human milk bank pasteurizes the milk to increase its shelf life [17].

3.3. ROLE OF INTESTINAL MICROFLORA IN NEONATAL HEALTH

There are approximately 100 trillion microbes that reside in close harmony in the human body [18]. These commensal microbes play a key role in human health. The community of microbe that lives in or on the human body is collectively called microbiota and the genes that are regulated by this microbiota are called the microbiome. Some of the microbes from this microbiota can be cultured in laboratory conditions.

Recently culture-independent high-throughput sequencing has embellished and expanded our knowledge about the available microbial population in the human body [18]. It is interesting to note that the human intestinal microbiome contains 150 times more genes than the total human genome [19]. This microbiota plays a crucial role in deciding the immune-modulation and metabolic phenotype [20, 21]. Intestinal microflora serves as a physical barrier and protection against pathogens through competitive exclusion as well as through the production of the bacteriocin. The microbiota also helps in the development of intestinal mucosa [22]. The bacterial population in the GI (gastrointestinal) tract has been reported to be able to produce butyrate, which helps to fight food allergies, and a sufficient amount of butyrate can also be produced by human milk microflora [23]. Probiotic strains, for example *Lactobacilli shirota* and *Lactobacillus casei* are capable of induce IL12 and IFN expression and modulation in the systemic immune system, which helps in reducing allergic reactions [24].

Some of these studies indicated that the immune cells of the GI tract can also track the intestinal microflora [25, 26]. The bacterial component transmits the STAT-mediated signals over pattern-recognition receptors and directs the release of pro-inflammatory cytokines (such as Tumor necrosis factor-α, interleukin-12) and anti-inflammatory cytokines (such as transforming growth factor-β and IL-10) [27, 28]. As a result of changes in physiological condition, gut microbiota composition and population vary from the stomach to colon and also from infancy to adulthood (Figure 3.1).

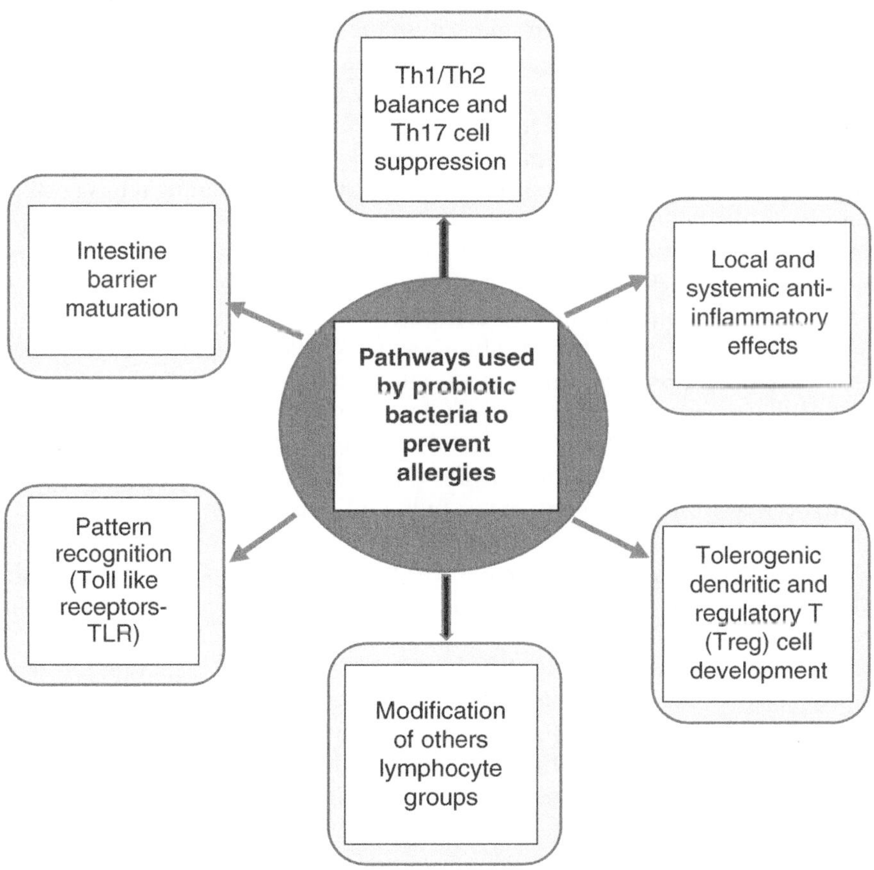

Figure 3.1 Mechanism of probiotics to overcome allergies. *Source:* Based on reference [105].

Gut microbes such as *Lactobacillus, Bifidobacteria, Bacteroids*, and *Enterobacteria* digest the carbohydrates and produce the short-chain fatty acids that can fulfill the high energy requirements of growing children. The gut microflora produces the antimicrobial proteins that evoke IL-10 and other cytokines and protect from inflammatory bowel disease (IBD) and other serious diseases [29]. Gut flora can modulate the T-helper cell response, thus infants who have a low number of microbes in their gut have a high risk of contracting allergies or developing asthma, celiac disease, and diabetes. Feeding methods are the only avenue to introduce healthy bacterial populations and Bifidus factors to neonates. Here, they will get colonized in the GI tract, which allows infants to fight against various infections and other serious diseases. It has been observed that breastfed infants are more likely to have diverse microflora and approximately twofold higher bacterial population in the gut compared to those infants that are formula-fed. A breastfed infant showed a high abundance of *Bifidobacteria* and *lactobacillus* species, and this plays a key role in the infant's health [30, 31].

3.4. FACTORS THAT INFLUENCE THE NEONATAL MICROBIOME

The microflora of GI tract is a complex and progressive ecosystem that contains 10^{11}-10^{12} bacteria per gram of colonic content and 60% of total fecal mass [19, 32]. Gut microflora composition and progression are influenced by various internal and external factors [33]. Gut microbiota influences the differentiation of gut epithelial cells and immune cells, and plays a crucial role in the nutritive, metabolic, immunological, and protective functions of infants [19]. Dysbiosis of the gut flora may cause the pathogenesis of immunological, cardiovascular, and metabolic diseases, as well as other intestinal disorders [34]. Numerous intrinsic and extrinsic factors may be responsible for the imbalance of human commensal microflora. The establishment of gut microflora is a complex and continuous process that is incepted at the time of birth and continued for several years after birth with the influence of the feeding method and formula diet of infants [35, 36]. The neonatal gut has an abundance of oxygen due to the facultative aerobes, especially the *Enterobacteria*, *Enterococcus*, and *Streptococcus* species, that start to colonize first. *Escherichia coli* and *Enterococcus fecali* are the most abundant species in the GI tract followed by *Klebsiella* and *Enterobacter* [37], which makes them susceptible to various diseases.

After the completion of the entire gestation age in a sterile intrauterine environment, neonates usually come in contact with the vagina, and bacterial colonization is initiated during delivery [38, 39]. Common prenatal interventions such as C-sections and the use of antibiotics or any other medication given to the mother during the gestation period are the major components that shape the newborn microflora [33]. Bacteria that colonized the newborn intestine are derived from the vagina, breastmilk, and intestine of the mother [31]. Indeed, there is clinical evidence that the fecal microbial composition of neonates has a close similarity with the vagina, milk, and intestinal microbial composition, and also depends on the diet of the mother [40, 41]. Similarly, infants delivered through cesarean section harbor a microbial population similar to the mother's skin microbes, such as *Staphylococcus* and *Propioni bacterium spp,* that are distant from the mother's vaginal flora [39]. Interestingly, the umbilical cord has also been shown to harbor commensal bacteria in the case of neonates born by cesarean section [42]. Studies have shown that delivery mode influences the immunological reaction during the first two years of a neonate's life, due to the transition in the gut microbiota development line. So expectedly, babies delivered by cesarean section constitute lower bacterial count of bifidobacteria in fecal matter compared to vaginally delivered newborns [43].

Other than the mode of delivery, feeding methods also influence the microbial composition of the neonate's gut, and discursively affect the physiology and morphology of intestinal mucosa as well as the pancreatic function [44]. Breastfed neonates have a more stable microbial population and possess twice the number of microbes in the GI tract than formula-fed neonates [45]. Similarly, Bergstrom et al. reported a dramatic change in bacterial abundance after the introduction of formula milk or other solid food in the infant's diet [46]. The antibiotic treatment for neonates and/or mothers during the gestational period is also a major disruptor for microbiota composition and shown to increase the population of antibiotic-resistant pathogens [47, 48]. For infants undergoing antibiotic treatment, breastmilk is the only singular source of probiotics to improve the beneficial microbial population. Hospitalization, gestational age, and mother's diet during the gestational period are some other possible factors that may influence the microbial composition of a neonatal gut.

3.5. NUTRITIVE ASPECTS OF BREASTMILK

3.5.1. Pre- and Probiotic Aspects of Human Milk

Human milk is often referred to as the chief source of nutrition for the first year of an infant's life. During the first year, newborns are most susceptible to enteric disease, which is reported to be lower in the case of breastfed infants. Breastmilk contains all the nutrients and essentials for the growth of infants as well as bioactive ingredients that are required for the growth of beneficial bacteria, which in turn enhance the immune system [49] (Figure 3.2).

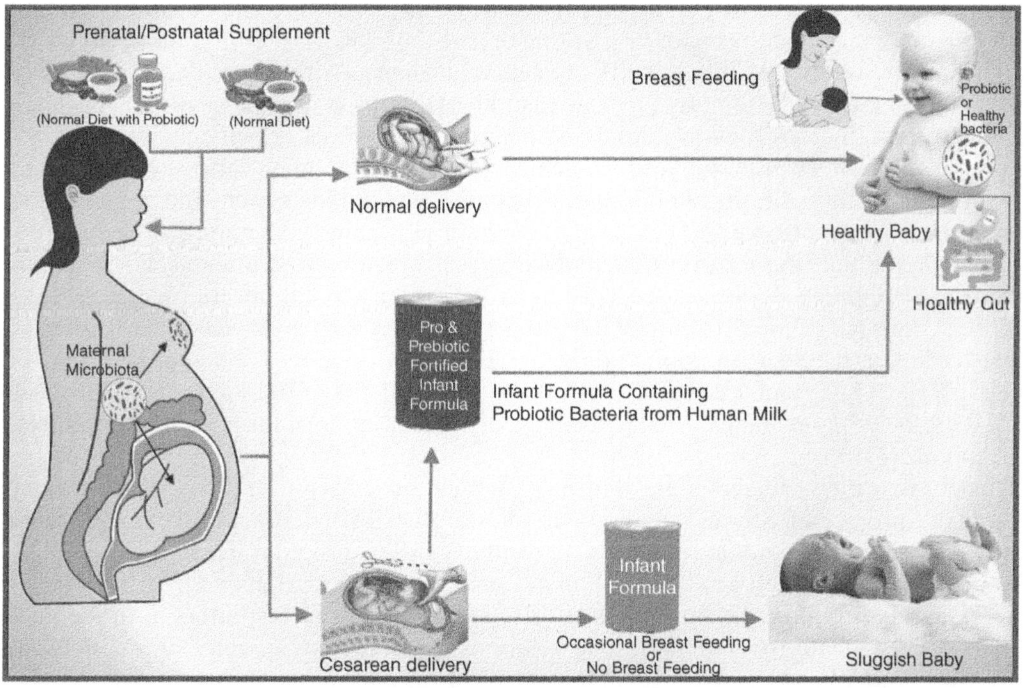

Figure 3.2 Schematic representation of microbiota exchange between mother and infant. The potential nutraceutical interventions for neonatal health are also depicted schematically in the center. *Source:* Lanak/Adobe Stock.

HMOs are the unconjugated complex carbohydrates that boost infant health by modulating the intestinal microflora and by enhancing nutritional absorption [5]. Recently, HMOs concentrations in human milk has been shown to differ across geographies [50]. One liter of human milk contains 5–10 g of HMOs [51]. Human milk contains unique glycans and the composition of these glycans largely depends on the genotype of the host mother [49]. HMOs serve as prebiotic and nourish host health by modulating the microflora such as *Bifidobacterium bifidum* and other *Lactobacilli* species in the intestine [52]. HMOs also exhibit significant antimicrobial activity, by competitive displacement of common pathogens such as *E. coli, Salmonella, Listeria, Streptococcus agalactiae,* and *Campylobacter*, from the infant gut [53, 54]. Milk other than breastmilk generally does not provide these important prebiotic oligosaccharides, which leads to various gastric disorders at times [55]. Soluble glycans and oligosaccharides present in human milk form the biofilm on the intestinal mucosa, and thereby restrict the pathogen binding with mucosal receptors of the intestine [56]. Oligosaccharides are the third major component of solid constituents in human milk. The dominant oligosaccharide in 80% of women is 2′-fucosyllactose, present at a concentration of approximately 0.088 oz/l [5]. Newburg reported that the signature of oligosaccharides in the urine and feces of the breastfed infants is similar to breastmilk HMOs, which have lactose at the reducing end, with a carbohydrate core and sialic acid at the non-reducing end [57]. A high risk of morbidity ratio has been recorded in prematurely delivered infants [58]. The combined effect of HMOs, glycans, and microflora, in principle, could stabilize premature and high-risk infant life [51].

3.5.2. Role of Human Milk Flora in Neonatal Diseases

Breastmilk plays a vital role in the first two years of an infant's life and beyond. An infant who feeds on breastmilk has a lower crying rate and stays healthier from infancy to adulthood. Immune profiles of human milk are generally aligned with geographical and socioeconomic human living standards [59]. Breastmilk develops a suitable environment in the infant GI tract, by providing probiotic and prebiotic substances to make infants comfortable with the new extrauterine environment, and helps to combat various infections by enhancing immunity. Infants are born with a suppressed immune system and are unable to produce sufficient antibodies. They have a low level of serum IgA compared to an adult, which makes infants more susceptible to a variety of infections and diseases. For example, infants in Scotland were more susceptible to Crohn's disease (CD), ulcerative colitis (UC), and IBD, however, this distribution has been reduced rapidly by the popularization of breastfeeding in the Scottish population [60]. CD and UC can be detected at any stage of one's life, but 15–25% of cases have been diagnosed in 0–5 year-old babies. This problem may be resolved by enhancing breastfeeding, or feeding of formula milk that resembles human milk [61].

Obesity is a physiological state that has emerged as a major health concern among infants in Europe [62]. Studies on animal models of obesity have concluded that gut flora population could modulate obesity [20]. Interestingly, a similar trend of microbial population shift has been observed in humans on a weight-reduction diet [63]. The role of *Bifidobacterium spp.* in the improvement of the inflammatory responses in obese mice has also been demonstrated [64]. The relationship between gut microflora, immune responses, and obesity is well documented. This may induce obesity during infancy, which could continue throughout life [65–67]. Gut microflora dysbiosis plays a crucial role in obesity-related disorders, such as diabetes, atherosclerosis, and non-alcoholic fatty liver disease [68]. The World Health Organization has also endorsed that modification of intestinal microflora through breastfeeding reduces the risk of obesity, type-2 diabetes, and dyslipidemia, as well as maintaining blood pressure in infants [69]. Breastfeeding has also been associated with increased cholesterol levels during infancy, and lower levels during adulthood [70].

Neonatal necrotizing enterocolitis (NEC) is a severe inflammatory intestinal disorder in newborns. Although the pathogenic pathway is not completely understood, inappropriate innate immune response and excessive inflammatory reaction in an immature intestine may be associated with NEC. Successful colonization of probiotic bacteria assert the manipulation of enteric microbiota and reduce the abundance bacterial population, causing NEC in premature babies [71].

Several studies have reported that there are higher incidences of NEC in formula-fed infants compared to breastfed ones [72]. One plausible explanation for the reduced incidence of NEC in breastfed infants could well be the composition of the microflora developed in the intestine due to breastmilk, which could protect pathogenic organisms [73].

3.6. MICROBIOME EXCHANGE BETWEEN MOTHERS AND INFANTS IN THE INTRAUTERINE AND EXTRAUTERINE ENVIRONMENT

The fetus needs to grow appropriately in utero since this is essential for offspring development and is also critical for long-term health, post birth. Jenmalm and co-workers reported gestational age as an important factor for fetal immunity development [74]. The amniotic cavity is a well-defined pathway for the invasion of microbes [75]. In fact, bacteria have been successfully cultured from chorioamniotic issue [76]. Various other studies also indicated that the amniotic cavity contains culturable and non-culturable microbial populations [77]. The gestational age placenta and the delivery birth canal are some other common routes of microbial exposures to neonates [78]. Funkhouser and Bordenstein have shown that the placental and birth canal bacterial population are the first to colonize in the infant gut [79]. These bacteria function significantly during the early stage of infancy and continue for rest of life. The number and diversity of the microbial population in early infancy is estimated to be very low, but it expands quickly and gets more diverse through breastfeeding. Moles et al. has observed that the bacterial population in fecal samples of preterm and postnatal infants was significantly different, suggesting a shift in the equilibrium of the bacterial population post birth [80]. Probiotic feeding of the pregnant person during pregnancy and breastfeeding can prove to be a preventive and therapeutic tool to improve the dysbiosis condition of gut microbiota during the c-section delivery, perinatal, and infant period [81].

3.6.1. Microbial Population in the Intrauterine Environment

The most common bacterial culprits detected in amniotic fluid of preterm births are, based on culture-dependent detection, *Ureaplasma urealyticum*, *Mycoplasma hominis*, *Fusobacterium nucleatum*, *Gardnerella vaginalis,* and *Bacteroides* spp. [82], while Group B *Streptococcus*, *E. coli,* and *Listeria monocytogenes* are less frequently detected [83]. Post birth, during the postnatal development of an infant, breastfeeding is a principal source of microbial exchange between the mother and neonates. It is already established that during breastfeeding, a mother delivers a large population of bacteria to neonates [84]. Human breastmilk is largely dominated by a few species such as *Staphylococcus, Streptococcus, Pseudomonas, Cornybacterium, Propionibacterium Ralstoniaetc, Bifidobacterium*, and *Lactobacillus* bacteria [85]. These species transfer from the mother's gut to the neonatal gut. It has been demonstrated that bacteria from neonate fecal matter and those from breastmilk are of similar types [86]. This underpins that breastfeeding is the major postnatal route of mother–infant microbial exchange, and it takes over from the initial colonization seeded by mother's gut. Although historically it was assumed that the intrauterine environment for infants is sterile until delivery, some studies revealed that infants come in contact with bacteria in the intrauterine environment and suggest that these bacteria may influence the microbial colonization pattern of the infant before birth [87]. But evidence is also available that shows any kind of treatment or antibiotic therapy of the mother can affect the infant's gut flora indirectly [88]. Mothers under antibiotic treatment during pregnancy showed higher counts of *Enterobacteria* and *Enterococci,* and lower proportions of *bifidobacteria* [89]. Fallani et al. also reported that antibiotic therapy of mothers during breastfeeding may alter the bacterial population of the infant's intestine [31]. These studies support the possible transfer of microbes from mother to fetus so the idea that the intrauterine environment is sterile is challenged [90]. As bacteria can get transferred from mothers to neonates, probiotic intake by a mother could influence the infant gut microflora. The consumption of probiotic strain and prebiotics by Finnish mothers, during the prenatal and breastfeeding phases, altered the microbial development pattern in infants. Other studies also revealed that probiotic consumption *(Lactobacillus rhamnosus, Bifidobacterium longum* or *L. paracasei, and B. longum*) during the last two months of the gestational period as well as during breastfeeding increased similar bacterial colony counts in the fecal sample of infants, and a similar pattern of microflora transfer was observed between mother and infants [91]. In a similar study, it was reported that infants who fed on their mother's milk, without any probiotic and prebiotic prescription, had lower *Bifidobacterium* counts in the GI tract. Probiotic treatment of infants can maintain long-term healthy intestinal microflora stability.

The genetic makeup of the mother, mode of delivery, maternal nutrition, and lactation stage influence human milk composition. Microflora of milk also varies with the individual and by geographical region [92]. The lactation period and gestational age are important aspects that influences the mother's milk microflora [86]. Mother's skin, mother–infant oral cavity, vagina, and intestinal flora of the mother are other factors that influence the microflora of breastmilk. Microbial diversity of human milk has also been shown to vary with different lactation stages [85]. Human milk microflora is affected by the mode of delivery, diet of the mother, the treatment the mother received during pregnancy and also by the intake of probiotics during gestational age and post-delivery [93].

3.7. HUMAN MILK PROBIOTIC-BASED MILK FORMULA FOR INFANTS

Probiotics and prebiotics are beneficial to infant's health, ease their digestion by modulating the environment of the GI tract, and eventually decrease the complexity of the first year of life. They have shown beneficial effects in infants for protection against infections such as diarrhea [94], necrotizing enterocolitis, eczema [95], and atopic dermatitis [96]. Genetically engineered microorganisms producing HMOs are applied to formula milk of infants to match the nutritional composition of mother's milk as much as possible [97]. Therefore, it appears safe to recommend the use of extensively tested probiotics and prebiotics for neonates. Breastmilk contains probiotic microflora and prebiotics like

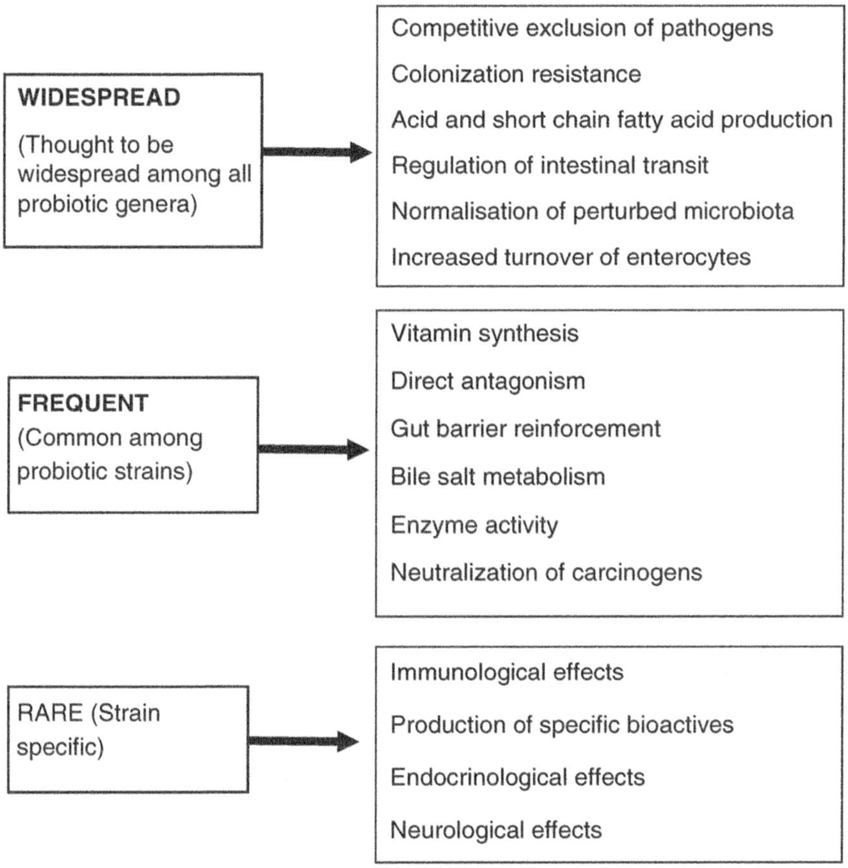

Figure 3.3 Possible pathways of action of probiotics. *Source:* Based on reference [106].

oligosaccharides, but such bioactive and immunomodulatory substances are not available in infant formula milk that is available in the market. Breastmilk is generally dominated by the *Bifidobacterium* species or varies according to individual and geographical conditions, so the bacteria could be isolated from human milk and collected from different geographies for further evaluation based on *in vitro* probiotic attributes as per the WHO guidelines. Bacteria isolated from milk should fulfill the probiotic criteria such as acid, bile tolerance, and attachment to the intestinal surface and should exert antimicrobial activity [98] (Figure 3.3). Collado et al. reported that some physiological, genetic, and environmental factors like weight gain, dietary supplements, and sanitization habits during pregnancy can also determine the diversity of probiotic bacteria in the intestine [99] Thus, these are most crucial and this sensitive point should be focused on to isolate the probiotic bacteria from the healthy females. The bacteria characterized as probiotic should be more promising for strengthening the immune system of infants and is generally recognized as safe. The high concentration and structural diversity of human's milk oligosaccharides are unique to humans. Research explains the presence of potential probiotic bacteria and HMOs, which are unique to the breastmilk [100] and could be used as tool for the protection of infants from various diseases (Table 3.1).

Table 3.1 Effects of probiotics in children.

Condition	Effect of probiotic as group	Examples of probiotics with documented or promising efficiacy
Acute gastroenteritis (treatment)	Approximately 1-day reduction in the duration of diarrhea	**Strong recommendation** Lactobacillus GG S. boulardii **Weak recommendation** ▶ *L. reuteri* DSM 17938
Antibiotic-associated diarrhea (prevention)	Reduce risk	▶ *Lactobacillus* GG ▶ *S. boulardii*
Nosocomial diarrhea (prevention)	Reduce risk	▶ *Lactobacillus* GG
Infections in children attending day care centres (prevention)	Reduce risk	▶ *Lactobacillus* GG ▶ *L. reuteri* DSM 17938
Allergy (prevention)	Reduce risk of eczema	WAO suggests the use of probiotics in select high-risk populations to reduce the risk of eczema; however, there is no clear indication regarding which probiotic(s) to use
NEC (prevention)	Reduced risk of NEC and mortality in infants who were born	No clear indications from scientific societies regarding which probiotic strain(s) should be recommended
H. pylori infection	Reduced risk of side effects and increased eradication rate	No clear indications which strain(s) to use Promising ▶ *S. boulardii*
Infantile colic (management)	Reduced crying time	▶ *L. reuteri* DSM 17938 (documented in breastfed infants)
Abdominal pain- related functional gastrointestinal disorders	Certain probiotics reduced intensity of pain (especially in patients with irritable bowel syndrome)	More studies are needed to identify beneficial strains. Promising ▶ *Lactobacillus GG* ▶ *VSL#3*
Induction of remission in ulcerative colitis	Limited evidence suggests that probiotics added to standard therapy may provide modest benefits	▶ *E. coli Nissle* 1917 ▶ *VSL#3*
Induction of remission in Crohn's disease	Insufficient evidence	Not recommended
Functional constipation	Until more data are available, the use of probiotics should be considered investigational	Not recommended

Source: Based on reference [101].

3.8. FORTIFICATION OF INFANT FORMULA WITH SYNBIOTICS

Microbial dysbiosis in the infant gut is responsible for the incidence and severity of diarrhea [102]. In recent years, probiotics supplementation has increased to modulate and diversify the gut microbiota in cesarean-born neonates [103]. Adding probiotics and other breastmilk-based bioactive substances to infant formula may be a feasible strategy to combat these etiologies. As discussed earlier, prebiotics or probiotics supplementation for pregnant women – which produce changes in the maternal gut microbiota – could get further transferred to the infant, at the time of delivery or during breastfeeding [104]. However, this approach may have its limitations in cases of C-section deliveries and with mothers who are not able to breastfed their infants. Hence there is an unmet need to develop infant milk formula that can fulfill the infant need for breastmilk.

The approach should be to first isolate the complex HMOs and probiotic bacteria naturally present in the breastmilk. Introduction of these crucial symbiotic components into infant formulas would hopefully generate a gut microbiota that is comparable to that of a breastfed infant. Neonates deprived of breastmilk would feed on the human milk-based probiotic bacteria and prebiotics like oligosaccharides in supplemented formula milk that can reduce the morbidity and mortality among the infants. The infants can stay healthy and be able to attain a stable microflora in their GI tract, through modulating the gut environment by including these components of human's milk in infant formula. Therefore, fortified infant milk formula with HMOs and probiotic bacteria originating from breastmilk can provide mother's benefits to these infants.

3.9. CONCLUSION

Breastmilk is the master source of nutrition for human infants. While it is ideal, and recommended, for most new mothers to breastfeed their newborns, not all of them can do so due to several reasons. Fortified infant milk formula with HMOs as prebiotic and probiotic bacteria isolated from breastmilk, could effectively provide mother's secrets to these bottle-fed infants. These infants could well be as healthy as their breastfed peers and can attain a stable microflora in their GI tract, by optimally including near-natural symbiotic components of human milk in infant formula. In addition, probiotics and human milk-based prebiotics isolated from human milk could also strengthen the general probiotic pool used for the population at large.

ACKNOWLEDGEMENTS

The authors acknowledge the contributions of Mr. Ajeet Raina for his help with artwork. We extend our gratitude to Professor Ramesh Chand Dubey for their swift administrative support.

REFERENCES

1. Dietert, R.R. and Piepenbrink, M.S. (2008). The managed immune system: protecting the womb to delay the tomb. *Human & Experimental Toxicology 27* (2): 129–134.
2. Martin, R., Nauta, A., Ben Amor, K. et al. (2010). Early life: gut microbiota and immune development in infancy. *Beneficial Microbes 1* (4): 367–382.
3. Jenness, R. (1979). The composition of human milk. *Seminars in Perinatology 3* (3): 225–239.

4. Perez-Escalante, E., Alatorre-Santamaria, S., Castaneda-Ovando, A. et al. (2020). Human milk oligosaccharides as bioactive compounds in infant formula: recent advances and trends in synthetic methods. *Critical Reviews in Food Science and Nutrition 10*: 1–34.
5. Bode, L. (2012). Human milk oligosaccharides: every baby needs a sugar mama. *Glycobiology 22* (9): 1147–1162.
6. Brandtzaeg, P. (1996). History of oral tolerance and mucosal immunity. *Annals of the New York Academy of Sciences 778* (1): 1–27.
7. Newburg, D.S. and Walker, W.A. (2007). Protection of the neonate by the innate immune system of developing gut and of human milk. *Pediatric Research 61* (1): 2–8.
8. Round, J.L. and Mazmanian, S.K. (2009). The gut microbiota shapes intestinal immune responses during health and disease. *Nature Reviews Immunology 9* (5): 313–323.
9. Verduci, E., Giannì, M.L., and Di Benedetto, A. (2019). Human milk feeding in preterm infants: what has been done and what is to be done. *Nutrients 12* (1): 44.
10. Lessen, R. and Kavanagh, K. (2015). Position of the academy of nutrition and dietetics: promoting and supporting breastfeeding. *Journal of the Academy of Nutrition and Dietetics 115* (3): 444–449.
11. Castillo, E.F., Dekonenko, A., Arko-Mensah, J. et al. (2012). Autophagy protects against active tuberculosis by suppressing bacterial burden and inflammation. *Proceedings of the National Academy of Sciences 109* (46): E3168–E3176.
12. Baird, J., Fisher, D., Lucas, P. et al. (2005). Being big or growing fast: a systematic review of size and growth in infancy and later obesity. *BMJ 331* (7522): 929.
13. Barreto, S.M., Pinheiro, A.R., Sichieri, R. et al. (2005). Análise da estratégia global para alimentação, atividadefísica e saúde, da Organização Mundial da Saúde. *Epidemiologia e serviços de saúde 14* (1): 41–68.
14. Sanchez-Pozo, A., Lopez, J., Pita, M.L. et al. (1986). Changes in the protein fractions of human milk during lactation. *Annals of Nutrition and Metabolism 30* (1): 15–20.
15. Weber, M., Grote, V., Closa-Monasterolo, R. et al. (2014). Lower protein content in infant formula reduces BMI and obesity risk at school age: follow-up of a randomized trial. *The American Journal of Clinical Nutrition 99* (5): 1041–1051.
16. Roberts, A.K., Chierici, R., Sawatzki, G. et al. (1992). Supplementation of an adapted formula with bovine lactoferrin: 1. Effect on the infant faecal flora. *Acta Paediatrica 81* (2): 119–24124.
17. Lucas, A. and Cole, T.J. (1990). Breast milk and neonatal necrotising enterocolitis. *The Lancet 336* (8730–8731): 1519–1523.
18. Whitman, W.B., Coleman, D.C., and Wiebe, W.J. (1998). Prokaryotes: the unseen majority. *Proceedings of the National Academy of Sciences 95* (12): 6578–6583.
19. O'Hara, A.M. and Shanahan, F. (2006). The gut flora as a forgotten organ. *EMBO Reports 7* (7): 688–693.
20. Ley, R.E., Turnbaugh, P.J., Klein, S. et al. (2006). Human gut microbes associated with obesity. *Nature 444* (7122): 1022–1023.
21. Wang, B. and Li, L. (2015). Who determines the outcomes of HBV exposure? *Trends in Microbiology 23* (6): 328–329.
22. Hooper, L.V., Stappenbeck, T.S., Hong, C.V. et al. (2003). Angiogenins: a new class of microbicidal proteins involved in innate immunity. *Nature Immunology 4* (3): 269–273.
23. Paparo, L., Nocerino, R., Ciaglia, E. et al. (2021). Butyrate as a bioactive human milk protective component against food allergy. *Allergy 76* (5): 1398–1415.
24. Cross, M.L. (2002). Microbes versus microbes: immune signals generated by probiotic lactobacilli and their role in protection against microbial pathogens. *FEMS Immunology and Medical Microbiology 34* (4): 245–253.
25. Smythies, L.E., Sellers, M., Clements, R.H. et al. (2005). Human intestinal macrophages display profound inflammatory anergy despite avid phagocytic and bacteriocidal activity. *The Journal of Clinical Investigation 115* (1): 66–75.
26. Wu, H.J. and Wu, E. (2012). The role of gut microbiota in immune homeostasis and autoimmunity. *Gut Microbes 3* (1): 4–14.

27. Miettinen, M., Sobin, L.H., and Sarlomo-Rikala, M. (2000). Immunohistochemical spectrum of GISTs at different sites and their differential diagnosis with a reference to CD117 (KIT). *Modern Pathology 13* (10): 1134–1142.
28. Iwasaki, A. and Medzhitov, R. (2015). Control of adaptive immunity by the innate immune system. *Nature Immunology 16* (4): 343–353.
29. Sartor, R.B. (2008). Microbial influences in inflammatory bowel diseases. *Gastroenterology 134* (2): 577–594.
30. Bezirtzoglou, E., Tsiotsias, A., and Welling, G.W. (2011). Microbiota profile in feces of breast- and formula-fed newborns by using fluorescence in situ hybridization (FISH). *Anaerobe 17* (6): 478–482.
31. Fallani, M., Young, D., Scott, J. et al. (2010). Intestinal microbiota of 6-week-old infants across Europe: geographic influence beyond delivery mode, breastfeeding, and antibiotics. *Journal of Paediatric Gastroenterology and Nutrition 51* (1): 77–84.
32. Eckburg, P.B., Bik, E.M., Bernstein, C.N. et al. (2005). Diversity of the human intestinal microbial flora. *Science 308* (5728): 1635–1638.
33. Mackie, R.I., Sghir, A., and Gaskins, H.R. (1999). Developmental microbial ecology of the neonatal gastrointestinal tract. *The American Journal of Clinical Nutrition 69* (5): 1035s–1045s.
34. Macia, L., Thorburn, A.N., Binge, L.C. et al. (2012). Microbial influences on epithelial integrity and immune function as a basis for inflammatory diseases. *Immunological Reviews 245* (1): 164–176.
35. Fanaro, S., Chierici, R., Guerrini, P. et al. (2003). Intestinal microflora in early infancy: composition and development. *Acta Paediatrica 92*: 48–55.
36. Arumugam, M., Raes, J., Pelletier, E. et al. (2011). Enterotypes of the human gut microbiome. *Nature 473* (7346): 174.
37. Penders, J., Thijs, C., Vink, C. et al. (2006). Factors influencing the composition of the intestinal microbiota in early infancy. *Pediatrics 118* (2): 511–521.
38. Palmer, C., Bik, E.M., DiGiulio, D.B. et al. (2007). Development of the human infant intestinal microbiota. *PLoS Biology 5* (7): e177.
39. Dominguez-Bello, M.G., Costello, E.K., Contreras, M. et al. (2010). Delivery mode shapes the acquisition and structure of the initial microbiota across multiple body habitats in newborns. *Proceedings of the National Academy of Sciences 107* (26): 11971–11975.
40. Prince, A.L., Chu, D.M., Seferovic, M.D. et al. (2015). The perinatal microbiome and pregnancy: moving beyond the vaginal microbiome. *Cold Spring Harbor Perspectives in Medicine 5* (6): a023051.
41. Chu, D.M., Meyer, K.M., Prince, A.L. et al. (2016). Impact of maternal nutrition in pregnancy and lactation on offspring gut microbial composition and function. *Gut Microbes 7* (6): 459–470.
42. Jimenez, E., Fernandez, L., Marin, M.L. et al. (2005). Isolation of commensal bacteria from umbilical cord blood of healthy neonates born by cesarean section. *Current Microbiology 51* (4): 270–274.
43. Huurre, A., Kalliomäki, M., Rautava, S. et al. (2008). Mode of delivery–effects on gut microbiota and humoral immunity. *Neonatology 93* (4): 236–240.
44. Le Huerou-Luron, I., Blat, S., and Boudry, G. (2010). Breast-v. formula-feeding: impacts on the digestive tract and immediate and long-term health effects. *Nutrition Research Reviews 23* (1): 23–36.
45. Granger, C.L., Embleton, N.D., Palmer, J.M. et al. (2021). Maternal breastmilk, infant gut microbiome and the impact on preterm infant health. *Acta Paediatrica 110* (2): 450–457.
46. Bergstrom, A., Skov, T.H., Bahl, M.I. et al. (2014). Establishment of intestinal microbiota during early life: a longitudinal, explorative study of a large cohort of Danish infants. *Applied and Environmental Microbiology 80* (9): 2889–2900.
47. Dethlefsen, L., Huse, S., Sogin, M.L. et al. (2008). The pervasive effects of an antibiotic on the human gut microbiota, as revealed by deep 16S rRNA sequencing. *PLoS Biology 6* (11): e280.
48. Jernberg, C., Löfmark, S., Edlund, C. et al. (2010). Long-term impacts of antibiotic exposure on the human intestinal microbiota. *Microbiology 156* (11): 3216–3223.

49. Chaturvedi, P., Warren, C.D., Altaye, M. et al. (2001). Fucosylated human milk oligosaccharides vary between individuals and over the course of lactation. *Glycobiology 11* (5): 365–372.
50. McGuire, M.K., Meehan, C.L., McGuire, M.A. et al. (2017). What's normal? Oligosaccharide concentrations and profiles in milk produced by healthy women vary geographically. *The American Journal of Clinical Nutrition 105* (5): 1086–1100.
51. Bode, L. (2009). Human milk oligosaccharides: prebiotics and beyond. *Nutrition Reviews 67* (suppl_2): S183–S191.
52. Gyorgy, P., Norris, R.F., and Rose, C.S. (1954). Bifidus factor. I. a variant of lactobacillus bifidus requiring a special growth factor. *Archives of Biochemistry and Biophysics 48* (1): 193–201.
53. Gura, T. (2014). Nature's first functional food. *Science 345*: 747–749.
54. Ackerman, D.L., Doster, R.S., Weitkamp, J.H. et al. (2017). Human milk oligosaccharides exhibit antimicrobial and antibiofilm properties against group B streptococcus. *ACS Infectious Diseases 3* (8): 595–605.
55. Kolida, S., Tuohy, K., and Gibson, G.R. (2002). Prebiotic effects of inulin and oligofructose. *British Journal of Nutrition 87* (S2): S193–S197.
56. Dai, D., Nanthkumar, N.N., Newburg, D.S. et al. (2000). Role of oligosaccharides and glycoconjugates in intestinal host defense. *Journal of Pediatric Gastroenterology and Nutrition 30*: S23–S33.
57. Newburg, D.S. (1999). Human milk glycoconjugates that inhibit pathogens. *Current Medicinal Chemistry 6* (2): 117–128.
58. Garcia-Basteiro, A.L., Quinto, L., Macete, E. et al. (2017). Infant mortality and morbidity associated with preterm and small-for-gestational-age births in southern Mozambique: a retrospective cohort study. *PLoS One 12* (2): e0172533.
59. Ruiz, L., Espinosa-Martos, I., García-Carral, C. et al. (2017). What's normal? Immune profiling of human milk from healthy women living in different geographical and socioeconomic settings. *Frontiers in Immunology 30* (8): 696.
60. Armitage, E., Drummond, H.E., Wilson, D.C. et al. (2001). Increasing incidence of both juvenile-onset Crohn's disease and ulcerative colitis in Scotland. *European Journal of Gastroenterology & Hepatology 13* (12): 1439–1447.
61. Kim, S.C. and Ferry, G.D. (2004). Inflammatory bowel diseases in pediatric and adolescent patients: clinical, therapeutic, and psychosocial considerations. *Gastroenterology 126* (6): 1550–1560.
62. Bammann, K., Peplies, J., De Henauw, S. et al. (2014). Early life course risk factors for childhood obesity: the IDEFICS case-control study. *PLoS One* (2): 9, e86914.
63. Ley, R.E. (2010). Obesity and the human microbiome. *Current Opinion in Gastroenterology 26* (1): 5–11.
64. Bo, T.B., Wen, J., Zhao, Y.C. et al. (2020). Bifidobacterium pseudolongum reduces triglycerides by modulating gut microbiota in mice fed high-fat food. *The Journal of Steroid Biochemistry and Molecular Biology 198*: 105602.
65. Lillycrop, K.A. and Burdge, G.C. (2011). Epigenetic changes in early life and future risk of obesity. *International Journal of Obesity 35* (1): 72–83.
66. Heijmans, B.T., Tobi, E.W., Stein, A.D. et al. (2008). Persistent epigenetic differences associated with prenatal exposure to famine in humans. *Proceedings of the National Academy of Sciences 105* (44): 17046–17049.
67. Oddy, W.H. (2012). Infant feeding and obesity risk in the child. *Breastfeeding Review 20* (2): 7–12.
68. Fukuda, S. and Ohno, H. (2014). Gut microbiome and metabolic diseases. In: *Seminars in Immunopathology*, vol. *36*, 1, 103–114. Berlin Heidelberg: Springer.
69. Reichelt, A.C., Loughman, A., Bernard, A. et al. (2020). An intermittent hypercaloric diet alters gut microbiota, prefrontal cortical gene expression and social behaviours in rats. *Nutritional Neuroscience 23* (8): 613–627.
70. Owen, C.G., Whincup, P.H., Kaye, S.J. et al. (2008). Does initial breastfeeding lead to lower blood cholesterol in adult life? A quantitative review of the evidence. *The American Journal of Clinical Nutrition 88* (2): 305–314.
71. van Best, N., Trepels-Kottek, S., Savelkoul, P. et al. (2020). Influence of probiotic supplementation on the developing microbiota in human preterm neonates. *Gut Microbes 12* (1): 1826747.

72. Quigley, M., Embleton, N.D., and McGuire, W. (2019). Formula versus donor breast milk for feeding preterm or low birth weight infants. *Cochrane Database of Systematic Reviews 7*: 1–63.
73. Greenwood, C., Morrow, A.L., Lagomarcino, A.J. et al. (2014). Early empiric antibiotic use in preterm infants is associated with lower bacterial diversity and higher relative abundance of Enterobacter. *The Journal of Pediatrics 165* (1): 23–29.
74. Svensson, J., Jenmalm, M.C., Matussek, A. et al. (2011). Macrophages at the fetal–maternal interface express markers of alternative activation and are induced by M-CSF and IL-10. *The Journal of Immunology 187* (7): 3671–3682.
75. DiGiulio, D.B., Gervasi, M.T., Romero, R. et al. (2010). Microbial invasion of the amniotic cavity in pregnancies with small-for-gestational-age fetuses. *Journal of Perinatal Medicine 38*: 495–502.
76. Goldenberg, R.L., Hauth, J.C., and Andrews, W.W. (2000). Intrauterine infection and preterm delivery. *New England Journal of Medicine 342* (20): 1500–1507.
77. Jacobsson, B., Mattsby-Baltzer, I., Andersch, B. et al. (2003). Microbial invasion and cytokine response in amniotic fluid in a Swedish population of women with preterm prelabor rupture of membranes. *Acta Obstetricia et Gynecologica Scandinavica 82* (5): 423–431.
78. Viscardi, R.M. (2010). Ureaplasma species: role in diseases of prematurity. *Clinics in Perinatology 37* (2): 393–409.
79. Funkhouser, L.J. and Bordenstein, S.R. (2013). Mom knows best: the universality of maternal microbial transmission. *PLoS Biology 11* (8): e1001631.
80. Moles, L., Gomez, M., Heilig, H. et al. (2013). Bacterial diversity in meconium of preterm neonates and evolution of their fecal microbiota during the first month of life. *PLoS One 8* (6): e66986.
81. Navarro-Tapia, E., Sebastiani, G., Sailer, S. et al. (2020). Probiotic supplementation during the perinatal and infant period: effects on gut dysbiosis and disease. *Nutrients 12* (8): 2243.
82. Han, Y.W., Shen, T., Chung, P. et al. (2009). Uncultivated bacteria as etiologic agents of intra-amniotic inflammation leading to preterm birth. *Journal of Clinical Microbiology 47* (1): 38–47.
83. Wang, X., Buhimschi, C.S., Temoin, S. et al. (2013). Comparative microbial analysis of paired amniotic fluid and cord blood from pregnancies complicated by preterm birth and early-onset neonatal sepsis. *PLoS One 8* (2): e56131.
84. Koenig, J.E., Spor, A., Scalfone, N. et al. (2011). Succession of microbial consortia in the developing infant gut microbiome. *Proceedings of the National Academy of Sciences 108* (Supplement 1): 4578–4585.
85. Hunt, K.M., Foster, J.A., Forney, L.J. et al. (2011). Characterization of the diversity and temporal stability of bacterial communities in human milk. *PLoS One 6* (6): e21313.
86. Khodayar-Pardo, P., Mira-Pascual, L., Collado, M.C. et al. (2014). Impact of lactation stage, gestational age and mode of delivery on breast milk microbiota. *Journal of Perinatology 34* (8): 599–605.
87. Jimenez, E., Marín, M.L., Martín, R. et al. (2008). Is meconium from healthy newborns actually sterile? *Research in Microbiology 159* (3): 187–193.
88. Reck, C., Tietz, A., Müller, M. et al. (2018). The impact of maternal anxiety disorder on mother-infant interaction in the postpartum period. *PLoS One 13* (5): e0194763.
89. Tanaka, S., Kobayashi, T., Songjinda, P. et al. (2009). Influence of antibiotic exposure in the early postnatal period on the development of intestinal microbiota. *FEMS Immunology and Medical Microbiology 56* (1): 80–87.
90. Zhu, L., Luo, F., Hu, W. et al. (2018). Bacterial communities in the womb during healthy pregnancy. *Frontiers in Microbiology 9*: 2163.
91. Grzeskowiak, Ł., Collado, M.C., Mangani, C. et al. (2012). Distinct gut microbiota in southeastern African and northern European infants. *Journal of Pediatric Gastroenterology and Nutrition 54* (6): 812–816.
92. Quinn, E.A. (2014). Too much of a good thing: evolutionary perspectives on infant formula fortification in the United States and its effects on infant health. *American Journal of Human Biology 26* (1): 10–17.
93. Armstrong, J. and Reilly, J.J. (2002). Breastfeeding and lowering the risk of childhood obesity. *The Lancet 359*: 2003–2004.

94. Sazawal, S., Black, R.E., Ramsan, M. et al. (2006). Effects of routine prophylactic supplementation with iron and folic acid on admission to hospital and mortality in preschool children in a high malaria transmission setting: community-based, randomised, placebo-controlled trial. *The Lancet 367* (9505): 133–143.
95. Rautava, S., Kainonen, E., Salminen, S. et al. (2012). Maternal probiotic supplementation during pregnancy and breast-feeding reduces the risk of eczema in the infant. *Journal of Allergy and Clinical Immunology 130* (6): 1355–1360.
96. Lee, J., Seto, D., and Bielory, L. (2008). Meta-analysis of clinical trials of probiotics for prevention and treatment of pediatric atopic dermatitis. *Journal of Allergy and Clinical Immunology 121* (1): 116–121.
97. Cheng, L., Akkerman, R., Kong, C. et al. (2021). More than sugar in the milk: human milk oligosaccharides as essential bioactive molecules in breast milk and current insight in beneficial effects. *Critical Reviews in Food Science and Nutrition 61* (7): 1184–1200.
98. Choudhary, J., Dubey, R.C., Sengar, G. et al. (2019). Evaluation of probiotic potential and safety assessment of lactobacillus pentosus MMP4 isolated from mare's lactation. *Probiotics and Antimicrobial Proteins 11* (2): 403–412.
99. Collado, M.C., Isolauri, E., Laitinen, K. et al. (2008). Distinct composition of gut microbiota during pregnancy in overweight and normal-weight women. *The American Journal of Clinical Nutrition 88* (4): 894–899.
100. Oozeer, R., Van Limpt, K., Ludwig, T. et al. (2013). Intestinal microbiology in early life: specific prebiotics can have similar functionalities as human-milk oligosaccharides. *The American Journal of Clinical Nutrition 98* (2): 561S–571S.
101. Szajewska, H. (2016). What are the indications for using probiotics in children? *Archives of Disease in Childhood 101* (4): 398–403.
102. Chassard, C., de Wouters, T., and Lacroix, C. (2014). Probiotics tailored to the infant: a window of opportunity. *Current Opinion in Biotechnology 26*: 141–147.
103. Yang, W., Tian, L., Luo, J. et al. (2021). Ongoing supplementation of probiotics to cesarean-born neonates during the first month of life may impact the gut microbial. *American Journal of Perinatology 38* (11): 1181–1191.
104. Gueimonde, M., Sakata, S., Kalliomäki, M. et al. (2006). Effect of maternal consumption of lactobacillus GG on transfer and establishment of fecal bifidobacterial microbiota in neonates. *Journal of Pediatric Gastroenterology and Nutrition 42* (2): 166–170.
105. Nazir, Y., Hussain, S.A., Abdul Hamid, A. et al. (2018). Probiotics and their potential preventive and therapeutic role for cancer, high serum cholesterol, and allergic and HIV diseases. *BioMed Research International 29*: 2018.
106. Hill, C., Guarner, F., Reid, G. et al. (2014). Expert consensus document: the international scientific Association for Probiotics and Prebiotics consensus statement on the scope and appropriate use of the term probiotic. *Nature Reviews Gastroenterology & Hepatology 11*: 506–514.

4

Metabolic Engineering of Microbes for the Production of Plant-Based Compounds

Subir K. Bose[1], Swati Upadhyay[2], and Yashdeep Srivastava[3]

4.1. INTRODUCTION

More than 200 000 distinct chemical compounds are formed from the plant kingdom, most of which are derived from advanced metabolism from different tissues of the plants [1]. In addition to growth, development, defense, and protection, plant-based metabolites are also a good source of nutrients, foods, and medicines for humans. Plants have the capability to synthesize various secondary metabolites that have medicinal properties, for example *Withania somnifera, Mentha arvensis,* and *Cinchona officinalis*. In addition to medicinal uses, these metabolites are also utilized for obtaining flavors, fragrances (*Cymbopogon citratus*), repellents (Pyrethrum from *Chrysanthemum coccineum*), and nutrients (*Moringa oleifera*) [2–5]. These value-added compounds in plants are produced due to a variety of interlinked metabolic pathways (Figure 4.1). In the current scenario, scientists are looking to increase yield and production of plant-based metabolites, mainly plant secondary metabolites, which have high cost and low yield [6–9].

For the plant metabolite production, the microorganisms that are used for engineering have high growth rates, and cultivation media for enhancing yield and biomass concentrations are also cheap and sustainable [10]. The metabolic abilities of microorganisms, including yeast and bacterial systems, are widely exploited for the production of novel/desired compounds in chemical and pharmaceutical industries. In addition, because of their toxic side effects, there is an increasing trend to substitute chemical drugs. Scientists are looking to produce natural compounds via different biotechnological approaches, including microbial fermentation.

[1]Department of Biotechnology & Microbiology, Meerut Institute of Engineering and Technology, Meerut, Uttar Pradesh, India

[2]Department of Biotechnology, Dr. A.P.J. Abdul Kalam Technical University (APJAKTU), Lucknow, Uttar Pradesh, India

[3]Department of Biotechnology, RR Institute of Modern Technology, Lucknow, Uttar Pradesh, India

Microbial Biotechnology: Role in Ecological Sustainability and Research, First Edition.
Edited by Pankaj Chowdhary, Sujata Mani, and Preeti Chaturvedi.

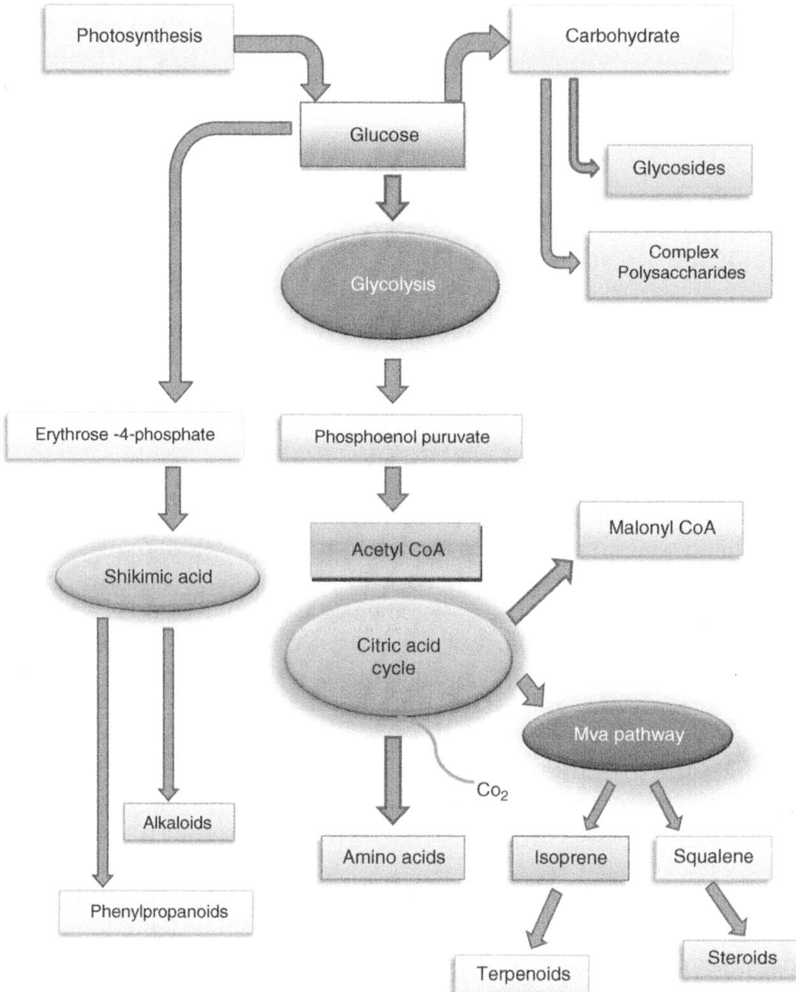

Figure 4.1 Inter-relationship of biosynthetic pathways leading to secondary constituents to plants.

Using direct manipulation of genetic constituent using pathway engineering (r-DNA technology), it becomes feasible to change the metabolic abilities of a variety of microorganisms. An approach known as metabolic engineering resulted in the increased production and yield of metabolites in a short period. Metabolic engineering is based on analyzing the problems associated with the metabolic design that can be preceded by building of a mathematical or in- silico model with the help of bioinformatics approaches [11]. These models are used to "draw" a better metabolic network by proposing an altered genotype of the model/host microorganisms. By using modern molecular biology approaches and in-silico modeling platforms, it is possible to perform such a hypothesis at the root level. Considering the successful use of mathematical models in metabolic engineering, such models are suitable to be implemented in the fields outside industrial biotechnology [12]. Mathematical modeling-based metabolic engineering provides a proposal for in-silico biology, with a proven role in the field of life science and medical science [13].

The complexity of cellular metabolism and regulation often creates challenges in metabolic engineering (e.g. alteration of a metabolic pathway to increase flux of certain metabolite), and therefore the elucidation of whole metabolome is needed. Complete genome sequences of the desired microorganism are open information that provide a platform to build a metabolic model on a genomic scale. Due to the advancement in experimental biology, modern biology has changed its emphasis from a traditional approach and has shifted its focus to a cellular-level, "global," holistic perspective [14]. This led to the creation of significant experimental databases, also known as "omics databases." Plenty of data on numerous microorganisms that help to alter their metabolome is also available on such platforms. Reconstruction of the frame of the metabolic network and its conversion into a mathematical format is the first step of this approach, followed by statistical analysis and computational treatment [15–18].

In this chapter, our focus is to explore new plant-based products that are produced via genome-scale alterations and modifications by using microbial models/systems with the help of metabolic engineering and r-DNA technology. Genome-scale information that can be helpful for metabolic engineering and further for the improvement of yield and production of peculiar primary, as well as secondary, metabolites (primarily plant-based) having medicinal properties are also summarized.

4.2. METABOLIC ENGINEERING OF MICROBES

According to the type and nature of metabolites, they can be divided into two classes: primary and secondary metabolites. Primary metabolites are those organic substances that are directly involved in the growth and development of an organism; an organism is not able to survive with deficiency of these substances [19].

4.3. METABOLIC ENGINEERING OF MICROBES FOR PRIMARY METABOLITE PRODUCTION

The culture of microalgae is the source of several products like carbohydrates, lipids, and pigments [20–22]. In addition to aquatic food and obtaining nutrition, microalgae are used widely for the production of metabolites such as polysaccharides, pigments, and polyunsaturated fatty acids [23]. The single-celled red algae Porphyridium species has considerable commercial importance as it accumulates sulfated polysaccharides, polyunsaturated fatty acids, phycobilins, and other bioactive substances [24].

In *Streptomyces coelicolor* (a Gram-positive soil bacteria), increased production of an antibiotic actinorhodin (Act) is achieved by manipulation of a central carbohydrate metabolism for shifting the carbon flux to Act production [25]. Similarly, industrial bacitracin production by *Bacillus licheniformis* can be enhanced by engineering main transcription factors involved in carbon, nitrogen, and phosphorus metabolism [26]. Enniatin is a cyclodepsipeptide that is used for the treatment of topical bacterial and fungal infections by using an engineered bacterial strain of *Bacillus subtilis*, by overexpression of *esyn* gene (eukaryotic non-ribosomal peptide synthetase gene) under the control of an acetoin inducible promoter system [27].

Engineered *Escherichia coli* is efficiently utilized in disaccharide trehalose (C12 disaccharide) production from xylose (C5 sugar) [28]. In addition to this, by using simple sugars as substrate, engineered *E. coli* strain was also reported to successfully produce tailored fatty esters, fatty alcohols, and waxes [29].

Microorganisms such as bacteria, yeasts, and fungi are also used for the production of biosurfactants such as glycolipids, phospholipids, lipo-polysaccharides, and lipopeptides.

One biosurfactant, i.e. Rhamnolipids, was successfully synthesized from the bacterium *Pseudomonas aeruginosa* and widely used as a biopesticide and in therapeutics [30]. The culture of mixed microbial biofilm from wastewater sludge can be used for rhamnose production by providing them with volatile fatty acids and glucose as carbon sources. In this microbial biofilm, there is an abundance of Xanthobacter species that is known to produce rhamnose as *zeaxanthin rhamnoside* [31].

B. subtilis is a popularly known strain for industrial riboflavin (intermediate of primary metabolism) production. It was observed that during fermentation, riboflavin production decreases along with the decrease in dissolved oxygen content. By engineering genes involved in purine, metabolism can be a better target to enhance the riboflavin production at later stages of fermentation [32].

Since ubiquinone is stored in the lipid bilayer, it was reported that the human lipid binding/transfer protein Saposin B (hSapB) mediated metabolic sink system can enhance ubiquinone (CoQ_8) production in an engineered *E. coli* [33].

Corynebacterium glutamicum was discovered by Japanese researchers as a source of glutamic acid production. During this process, intracellular metabolism is regulated in such a way to increase 2-oxoglutarate (TCA cycle intermediate) production so that that flux can be shifted toward glutamic acid biosynthesis [34].

Saccharomyces cerevisiae proved to be a suitable host system for engineering and production of aromatic compounds as well due to their capability to heterologously express long metabolic pathways, including CyP450. High levels of p-coumaric acid, which has high commercial value, as well as a precursor molecule for secondary metabolite biosynthesis, was produced by metabolic engineering of *S. cerevisiae* [35]. Similarly, *Ralstonia eutropha* is metabolically engineered to produce polyhydroxyalkanoate (PHA) by using intermediates of fatty acid metabolism [36]. Engineered *S. cerevisiae* can produce a tremendous amount of lipase that helps in the degradation of triglycerides and releases fatty acid components.

Fatty acids are regarded as an essential platform compound synthesized by the sustainable use of engineered microbial systems. Expression of genes involved in the fatty acid generation and its modification in engineered strains permit commercial utilization of yeast for bioconversion of low-cost materials into value-added lipid products [37]. In addition to this, ricinoleic acid (12-hydroxyoctadec- cis-9-enoic acid), an essential fatty acid utilized widely in the synthesis of synthetic fibers and cosmetic products, is obtained from the castor bean. It was reported that this acid is produced in large amounts by engineering another important fungal species, *Pichia pastoris* [38]. Punicic acid, a conjugated fatty acid, is used as a medicine in several health disorders and is an essential component of seed oil from *Punica granatum* and *Trichosanthes kirilowii*. *Schizosaccharomyces pombe* was engineered successfully for commercial production of punicic acid for nutraceutical applications [39].

4.4. METABOLIC ENGINEERING OF MICROBES FOR SECONDARY METABOLITES

Plants secondary metabolites are a group of organic compounds, biosynthesized naturally from plants, bacteria, or fungi, which do not participate directly in the growth process, development, or reproductive mechanism of the organism [40, 41]. They may be the end product of primary metabolism, e.g. bacterial antibiotics, fungal alkaloids, steroids, essential oils, phenolics, nucleosides, phenazines, peptide, and growth factors. Many plant secondary metabolites have medicinal properties, including terpenoids, alkaloids, and phenylpropanoids [42, 43]. Endophytes are recognized as prospective alternatives for terpenoid bioproduction, either directly or indirectly, due to the long-term harmony and coevolution in plant-endophyte symbioses [44]. The limitations of supply are the main hurdle in the large-scale use of these molecules. The use of microbes provided an opportunity for rapid and cost-effective biosynthesis of these compounds; therefore microbes are targeted by the scientific community as a source for the manufacture of plant-based compounds [45].

4.4.1. Terpenoids Biosynthesis

Terpenoids are the largest group of plant secondary metabolites and the most diverse class, including more than 50000 compounds [1, 46–48]. Terpenoids are biosynthetically derived from the basic five-carbon (C-5) units isopentenyl diphosphate (IPP) and dimethylallyl diphosphate (DMAPP) via the cytosolic mevalonate (MVA) from Acetyl-CoA and plastidial MEP pathway (2-*C*-methyl-D-erythritol-4-phosphate/Rohmer pathway) from pyruvate and glyceraldehyde-3-phosphate [49]. Monoterpenes are volatile 10 carbon (C-10) containing compounds [50]. Terpenoids play essential roles in plant primary metabolism and provide the building blocks of isoprenoids for pigments biosynthesis involved in photosynthesis and electron carriers in respiration (quinone). Other pigments such as carotenoids, ubiquinones, plastoquinones, sterols, and dolichols also have a terpenoid backbone. They are also involved in the synthesis of phytohormones such as abscisic acid (sesquiterpene, 15-C), gibberellins (diterpene 20-C), cytokinins, strigolactones, and the brassinosteroids [10]. A variety of microorganisms are used as hosts to produce plant based terpenoids (Table 4.1).

4.4.1.1. Monoterpenoids

Monoterpenoids are secondary metabolites having 10 carbon-containing compounds (10 C) [62]. Some volatile monoterpenoids are produced through microorganisms using an *E. coli* and yeast model system with the help of metabolic engineering, including geraniol, pinene, and limonene [51, 52, 54]. Eucalyptol was synthesized artificially by Ignea et al. [55] and Mendez-Perez et al. [56]. Similarly, linalool was synthesized by the Mendez-Perez et al. in 2017 and Camesasca et al. in 2018 [8, 56].

4.4.1.2. Sesquiterpenoids

Sesquiterpenes (C15) are secondary metabolites based on the 15-Carbon skeleton, containing compounds, derived from FPP (farnesyl pyrophosphate). They are biosynthesized by the condensation of GPP (geranyl pyrophosphate) and isopentenyl pyrophosphate (IPP) in *S. cerevisiae* and naturally from IPP and dimethylallyl pyrophosphate (DMAP) moiety in plants. Artemisinin and amorphadiene are potent new-generation antimalarial drugs, synthesized from a sesquiterpene precursor artemisinic acid. Artemisinic acid is the primary precursor of artemisinin discovered by Paddon et al. and expressed artificially in strains of *S. cerevisiae*, whereas amorphadiene precursors were also produced through an engineered *E. coli* model [63, 64].

It was observed that the yield of artemisinic acid had reached $25\,g\,l^{-1}$ by using engineered *S. cerevisiae* [67]. There are several other biologically active sesquiterpenes, produced through genetic engineering using yeast as a model, for example, β-farnesene [57, 58], bisabolene [59], and α-humulene (anti-inflammatory, antibacterial, and appetite suppressant) [60]. On the other hand, β-caryophyllene [61, 74], santalene [65], and patchoulol [66] were biosynthesized by using engineered *E. coli*.

4.4.1.3. Diterpenoids

Diterpenoids are plant secondary metabolites having 20 carbons (20 C) in their backbone. Some volatile diterpenoids are produced through microorganisms using *E. coli* as a model system with the help of metabolic engineering for, e.g. jolkinol C [68]. Taxadiene biosynthesis was established in yeast systems through metabolic engineering, and is an essential step toward taxol (Paclitaxel) production [11, 69, 70].

Table 4.1 Genetically modified (GM) microbes engineered for the biosynthesis of plant-based terpenoids.

S. No.	Engineered microbes stain	Type of secondary metabolites	Secondary metabolites	Application	References
				TERPENOIDS	
1	*E. coli, & S. cerevisiae*	**Monoterpenoids (10C)**	Geraniol	Perfumery industry, scent product and soap.	[51, 52]
2	*E. coli*		Pinene	α-pinene is mainly used for insects repellent. Also used in production of many perfumes and deodorants.	[53]
3	*E. coli,* and *S. cerevisiae*		Limonene	It is used to promote weight loss, prevent and treatments of cancer, and bronchitis. Used in shampoo, perfume, cosmetics, chewing gum, and in foods, beverages, and also used as a flavoring agent.	[6, 54]
4	*S. cerevisiae* and *E. coli*		Eucalyptol	It is an active ingredient in mouthwash and used as a cough suppressant. Used in the treatments of asthma via anti-inflammatory cytokine inhibition and also to control airway mucus hypersecretion.	[55, 56]
5	*E. coli* and *S. cerevisiae*		Linalool	It is an active ingredient of all scent in perfumed and hygiene products (60–80% of) and used in shampoos, soaps, lotions, and detergents.	[8, 56]
1	*S. cerevisiae* and *E. coli*	**Sesquiterpenone (15C)**	β-farnesene	Aphids used natural trans-β compound as a death signal to warn away other individuals; in potatoes its act as a natural insect repellent.	[57, 58]
2	*E. coli* and *S. cerevisiae*		Bisabolene	Used in artificial oil of bergamot, myrrh, and lemon, and also used as fixative for neroli bases.	[59, 60]
3	*S. cerevisiae*		α-humulene	Effective anti-inflammatory, antibacterial, and appetite suppressant.	[61]
4	*E. coli*		β-caryophyllene	Non-steroidal anti-inflammatory drug, a fragrance, and a metabolite	[62]
5	*E. coli* and *S. cerevisiae*		Amorphadiene	Artemisinin biosynthesis pathway amorpha-4, 11-diene synthase (ADS) is the first committed as well as limiting step, ADS levels are correlated with artemisinin yield; when it increases artemisinin also increased. Amyris company has developed a new method for high production of artemisinin.	[63, 64]
6	*S. cerevisiae*		Santalene	Santalene is found in allspice and is a constituent of sandalwood oil. α-Santalene is a good flavoring ingredient.	[65]

7	*S. cerevisiae*		Patchoulol	Patchoulol is also used in the synthesis of the chemotherapy drug Taxol.	[66]
8	*S. cerevisiae* (baker's yeast)		Artemisinic acid	It has uses in many pharmacological activities, such as antimalarial activity, antipyretic effect, anti-tumor activity, antibacterial activity, etc.	[67]
1	*S. cerevisiae*	**Diterpenoids (20C)**	Jolkinol C	It is of diterpenoid nature	[68]
2	*E. coli*, *S. Cerevisiae and E. coli*		Taxadiene	*Taxol* is a potent chemotherapy drug for breast cancer; taxadiene is the first committed steps in the biosynthesis of taxol,	[11, 69, 70]
	E.coli and *S. cerevisiae*	**Triterpenoids (30C)**	α-amyrin	All three amyrins occur in the surface wax of tomato fruit.	[40, 71]
1	*E. coli* and *S. cerevisiae*	**Tetraterpenoids(40-C)**	β-carotene	All carotenoids are antioxidants. An antioxidant is a substance that inhibits the oxidation of other molecules. It protects the body from free radicals.	[53, 72]
2	*E.coli* and *S. cerevisiae*		lycopene	It is an antioxidant found in tomatoes, and helps reduce exercise-related asthma attacks.	[7, 73]

4.4.1.4. Triterpenoids

Triterpenoids are the compounds associated with human health, therapeutics, and societal needs. They perform many functions such as anticancer, anti-inflammatory, antiparasitic, antimicrobial, antiallergenic, antispasmodic, antihyperglycemic, and immunomodulatory properties [75]. Triterpenoids are compounds with a backbone based on six isoprene units, which are derived biosynthetically from squalene (the acyclic C30 hydrocarbon). Their cyclic structures are relatively complex; most of them belong to alcohols, carboxylic acids, and aldehydes. Triterpenoids are 30 carbons (30 C) containing compounds and are mainly composed of three terpene units with the molecular formula $C_{30}H_{48}$ [70]. In plants, squalene (the precursor of all steroids) gives rise to many pharmacologically active triterpenoidal compounds. These compounds are also obtained by microbial engineering approaches for, e.g. lupeol, β-amyrin and α-amyrin [40, 71].

4.4.1.5. Tetraterpenoids

Tetraterpenes are 40 carbon (40-C) containing compounds, and formed by the condensation of two molecules of GGPP (geranylgeranyl pyrophosphate), and produced phytoene, which is the precursor of carotenoids (such as β-carotene and lycopene). Using genetic engineering tools, artificial synthesis of β-carotene and lycopene was established in laboratories by different research groups [7, 72, 73].

4.4.1.6. Alkaloids

Alkaloids are secondary metabolites having low molecular weight with heterocyclic nitrogen or sometime exocyclic nitrogen atoms. More than 20000 natural alkaloids exist, which mostly exhibit analgesic (morphine), psychotropic (mescaline and cocaine), stimulant (caffeine and ephedrine), anti-inflammatory (berberine), antibacterial (sanguinarine), antitussive (codeine), antispasmodic (papaverine), antimalarial (quinine obtained from cinchona), and anticancerous activity (vinblastine and vincristine) [76, 77].

4.4.2. Benzylisoquinoline Alkaloids (BIAs)

Benzylisoquinoline alkaloids are a large, diverse group of (~2500 compound) specialized plants secondary metabolites [78]. Although the ecophysiological functions of most BIAs are not known to date, the maximum compound belongs to this group have medicinal properties. Some of the BIAs exhibits potent pharmacological activities, including morphine and codeine (narcotic analgesics); papaverine, a potent vasodilatory compound; noscapine, a potent anticancerous drug; antimicrobial agents (berberine and sanguinarine); and the muscle relaxant (+)-tubocurarine. Pyne et al. in 2019 reconstituted BIA biosynthetic pathways in the yeast *S. Cerevisiae* [1]. Several other BIAs are also artificially (using yeast and *E. coli*) synthesized like (S)-reticuline [9, 79] and thebaine [80], while hydrocodone and noscapine were artificially synthesized [24, 81].

4.4.2.1. Monoterpene Indole Alkaloids

Indole alkaloids are a class of alkaloids that possess a structural moiety of indole. Many indole alkaloids include isoprene groups and are thus popularly known as terpene indole (secologanin tryptamine) alkaloids. It is also one of the largest class of alkaloids [82]. Tryptophan is the primary aromatic amino acid, which is the biochemical precursor molecule for the synthesis of indole alkaloids. Artificial biosynthesis of the plant-derived alkaloid strictosidine was achieved in the yeast model [83].

4.4.2.2. Phenylpropanoids

The phenylpropanoids are organic compounds that are naturally biosynthesized by plants when two aromatic amino acids are added; one is phenylalanine, and another is tyrosine. The name of phenylpropanoids is derived from the six-carbon (aromatic phenyl group) and the three-carbon propene tail of coumaric acid, which play a key role in the biosynthesis of phenylpropanoid. 4-coumaroyl-CoA is the precursor compound in biosynthesis of thousands of phenylpropanoids, including lignols (precursors of lignin and lignocellulose), flavonoids, isoflavonoids, coumarins, a-urones, catechin, stilbenes, and phenylpropanoids [84]. Cinnamic acid is produced from the coumaroyl moiety. It is found everywhere in the plant kingdom, where it acts as an essential and structural component derived from phenylalanine ammonia-lyase and phenylalanine, having a peculiar function including protection from ultraviolet light (UV light), defending against pathogens and herbivores, and mediating plant–pollinator interactions as scent and pigment compounds in flowers [85].

4.4.2.3. Stilbenoids

Stilbenoids are obtained from the hydroxylation of stilbene. Resveratrol is an important member of this group with anti-infective, antioxidant, and cardioprotective property that was produced by microbial engineering of yeast and has reached a production level at 5.0 g l^{-1} [86, 87]. Groups of stilbenoid derivatives that were produced artificially by microbe models are pinostilbene, pterostilbene, pinosylvin, and piceatannol (Table 4.1) [88–91].

4.4.2.4. Flavonoids

Like stilbenoids, a central branch point flavonoid naringenin has been synthesized from glucose above 200 mg l^{-1} using microbial technology [74, 92]. Flavonoid pathways are often modified with the help of metabolic engineering (pathway engineered) into two-model system yeast and *E. coli* to obtain the dihydrochalcone phloretin and its derivatives [93] as well as the flavanone and flavonols including liquiritigenin, kaempferol, dihydrokaempferol [94, 95], quercetin [35], and fisetin (Table 4.1). The flavonoid pathway was targeted to produce anthocyanin pigments in *E. coli* [96] and a single engineered strain of yeast by the research group of Eichenberger and his colleague [95, 97].

Another flavonoids precursor is pinocembrin, a novel natural compound that is very useful in the pharmaceutical industry for its anti-inflammatory, antimicrobial, anticancer, and antioxidant activities. Due to two separate fermentation steps involved in its biosynthesis, the quantity of the desired product is low, so it is not suitable for industrial bulk production. With the help of stepwise modification via genetic engineering in the metabolic pathway involved in this process, the compound was produced efficiently. Adapting fed-batch cultivation of modified engineered strains in the fermenter resulted in the production of (2S)-pinocembrin (432.4 mg l^{-1}) [93, 98]. Other flavonoids are also produced using different microbial strains with the genetic engineering technique, including scutellarin [99]. Genistein is an isoflavone that is an angiogenesis inhibitor [100] and the artificial biosynthesis of pelargonidin-3-O-glucoside in the microbial model was demonstrated [96, 97].

4.4.3. Coumarins

Coumarins are an organic chemical compound that is chemically 1-benzopyran-2-one (with formula $C_9H_6O_2$) of benzopyrone class and naturally found in many plants. Moreover, the primary application of coumarins is that they are used as a blood thinner to keep blood flowing smoothly and prevent the formation of blood clots. They have other biological properties, including antiviral, antimicrobial, anti-inflammatory, antioxidant, antidiabetic, and enzyme inhibitory activity. Umbelliferone (7-hydroxycoumarin), 4-hydroxycoumarin, and esculetin [101, 102] are derivatives of coumarin that are very useful and are compounds in high demand. Due to its wide range of applications, scientists are looking for the natural synthesis of this compound acquiring microbial r-DNA technique for more yields [101, 102].

4.5. FUTURE PROSPECTS AND CONCLUSION

Plant-based metabolites play important roles in medicine, agriculture, and industry. The use of plant secondary metabolites for treatment has been observed since time immemorial. Due to increasing market demand, over the past few decades, the tremendous development in biosynthesis of plant-based metabolites has been seen through the metabolic engineering of microbial platforms. Metabolic engineering of microbes has made it possible to generate complex natural products viz. carotenoids from simple carbon sources, on both laboratory as well as industrial scale. The continuous demand of the plant-based primary and secondary metabolites, especially for medicinal purposes, makes it challenging to fulfill the demands of medicinal availability. Elucidation of biosynthetic pathways, synthetic biology, and in-silico modeling methods helped to produce complex and valuable compounds. Searching for the alternate solution, the use of r-DNA technology using genetically modified organisms (GMOs), helped humankind through increased yield and productivity of natural medicinal constituents/compounds. However, the successful release of GM microbes has been opened in many countries, in particular by the willingness of a particular remaining country to allow the execution of GM microbes under a large scale.

REFERENCES

1. Pyne, M.E., Narcross, L., and Martin, V.J. (2019). Engineering plant secondary metabolism in microbial systems. *Plant Physiology 179* (3): 844.
2. Sangwan, N.S., Tripathi, S., Srivastava, Y. et al. (2017). Phytochemical genomics of Ashwagandha. In: *Science of Ashwagandha: Preventive and Therapeutic Potentials*, 3–36. Cham: Springer.
3. Srivastava, Y. and Sangwan, N.S. (2020). Improving medicinal crops through phytochemical perspective: Withania somnifera (Ashwagandha). In: *Advancement in Crop Improvement Techniques* (ed. N. Tuteja), 275–295. Woodhead Publishing.
4. Kumar, S., Kumar, A., Gupta, P., and Srivastava, Y. (2020). Plants based insecticides as promising source of biopesticides – A review. *Indian Journals 11* (1): 51–57.
5. Srivastava, Y., Upadhyay, S., Ahmad, M. et al. (2021). Role of specific spices in fortifying the human immune system. In: *Traditional Herbal Therapy for the Human Immune System* (ed. A. Husen), 397–410. CRC Press.
6. Willrodt, C., David, C., Cornelissen, S. et al. (2014). Engineering the productivity of recombinant Escherichia coli for limonene formation from glycerol in minimal media. *Biotechnology Journal 9* (8): 1000–1012.
7. Chen, Y., Xiao, W., Wang, Y. et al. (2016). Lycopene overproduction in Saccharomyces cerevisiae through combining pathway engineering with host engineering. *Microbial Cell Factories 15* (1): 1–3.
8. Camesasca, L., Minteguiaga, M., Farina, L. et al. (2018). Overproduction of isoprenoids by Saccharomyces cerevisiae in a synthetic grape juice medium in the absence of plant genes. *International Journal of Food Microbiology 3* (282): 42–48.
9. DeLoache, W.C., Russ, Z.N., Narcross, L. et al. (2015). An enzyme-coupled biosensor enables (S)-reticuline production in yeast from glucose. *Nature Chemical Biology 11* (7): 465–471.
10. Dudareva, N., Negre, F., Nagegowda, D.A. et al. (2006). Plant volatiles: recent advances and future perspectives. *Critical Reviews in Plant Sciences 25* (5): 417–440.

11. Ding, M.Z., Yan, H.F., Li, L.F. et al. (2014). Biosynthesis of taxadiene in Saccharomyces cerevisiae: selection of geranylgeranyl diphosphate synthase directed by a computer-aided docking strategy. *PLoS One 9* (10): e109348.
12. Perez-Garcia, O., Lear, G., and Singhal, N. (2016). Metabolic network modeling of microbial interactions in natural and engineered environmental systems. *Frontiers in Microbiology 7*: 673.
13. Garcia-Albornoz, M.A. and Nielsen, J. (2013). Application of genome-scale metabolic models in metabolic engineering. *Industrial Biotechnology 9* (4): 203–214.
14. Patil, K.R., Åkesson, M., and Nielsen, J. (2004). Use of genome-scale microbial models for metabolic engineering. *Current Opinion in Biotechnology 15* (1): 64–69.
15. Likić, V.A., McConville, M.J., Lithgow, T. et al. (2010). Systems biology: the next frontier for bioinformatics. *Advances in Bioinformatics 2010*: 268925.
16. Zhang, S., Ye, B.C., Chu, J. et al. (2006). From multi-scale methodology to systems biology: to integrate strain improvement and fermentation optimization. *Journal of Chemical Technology and Biotechnology 81* (5): 734–745.
17. Wishart, D.S. (2007). Current progress in computational metabolomics. *Briefings in Bioinformatics 8* (5): 279–293.
18. Weckwerth, W. (2011). Green systems biology—from single genomes, proteomes and metabolomes to ecosystems research and biotechnology. *Journal of Proteomics 75* (1): 284–305.
19. Williams, D.H., Stone, M.J., Hauck, P.R. et al. (1989). Why are secondary metabolites (natural products) biosynthesized? *Journal of Natural Products 52* (6): 1189–1208.
20. Khan, M.I., Shin, J.H., and Kim, J.D. (2018). The promising future of microalgae: current status, challenges, and optimization of a sustainable and renewable industry for biofuels, feed, and other products. *Microbial Cell Factories 17* (1): 1–21.
21. Dolganyuk, V., Andreeva, A., Budenkova, E. et al. (2020). Study of morphological features and determination of the fatty acid composition of the microalgae lipid complex. *Biomolecules 10* (11): 1571.
22. Castillo, T., Ramos, D., García, T. et al. (2021). Mixotrophic cultivation of microalgae: an alternative to produce high-value metabolites. *Biochemical Engineering Journal 12*: 108183.
23. Wu, J., Gu, X., Yang, D. et al. (2021). Bioactive substances and potentiality of marine microalgae. *Food Science & Nutrition 9* (9): 5279–5292.
24. Li, S., Ji, L., Shi, Q. et al. (2019). Advances in the production of bioactive substances from marine unicellular microalgae Porphyridium spp. *Bioresource Technology 292*: 122048.
25. Ryu, Y.G., Butler, M.J., Chater, K.F. et al. (2006). Engineering of primary carbohydrate metabolism for increased production of actinorhodin in Streptomyces coelicolor. *Applied and Environmental Microbiology 72* (11): 7132–7139. Epub 1 September 2006.
26. Cai, D., Zhu, J., Zhu, S. et al. (2019). Metabolic engineering of main transcription factors in carbon, nitrogen, and phosphorus metabolisms for enhanced production of bacitracin in Bacillus licheniformis. *ACS Synthetic Biology 8* (4): 866–875.
27. Zobel, S., Kumpfmüller, J., Süssmuth, R.D. et al. (2015). Bacillus subtilis as heterologous host for the secretory production of the non-ribosomal cyclodepsipeptide enniatin. *Applied Microbiology and Biotechnology 99* (2): 681–691.
28. Wu, Y., Wang, J., Shen, X. et al. (2018). Investigating the strategies for microbial production of trehalose from lignocellulosic sugars. *Biotechnology and Bioengineering 115* (3): 785–790.
29. Steen, E.J., Kang, Y., Bokinsky, G. et al. (2010). Microbial production of fatty-acid-derived fuels and chemicals from plant biomass. *Nature 463* (7280): 559–562.
30. Tan, Y.N. and Li, Q. (2018). Microbial production of rhamnolipids using sugars as carbon sources. *Microbial Cell Factories 17* (1): 1–3.
31. RedCorn, R.M., Hillman, E.T., Solomon, K.V. et al. (2019). Xanthobacter-dominated biofilm as a novel source for high-value rhamnose. *Applied Microbiology and Biotechnology 103* (11): 4525–4538.
32. You, J., Pan, X., Yang, C. et al. (2021). Microbial production of riboflavin: biotechnology advances. *Metabolic Engineering 68*: 46–58.
33. Xu, W., Yuan, J., Yang, S. et al. (2016). Programming Saposin-mediated compensatory metabolic sinks for enhanced ubiquinone production. *ACS Synthetic Biology 5* (12): 1404–1411.

34. Kataoka, N., Vangnai, A.S., Pongtharangkul, T. et al. (2019). Engineering of Corynebacterium glutamicum as a prototrophic pyruvate-producing strain: characterization of a ramA-deficient mutant and its application for metabolic engineering. *Bioscience, Biotechnology, and Biochemistry 83* (2): 372–380.
35. Rodriguez, A., Chen, Y., Khoomrung, S. et al. (2017). Comparison of the metabolic response to over-production of p-coumaric acid in two yeast strains. *Metabolic Engineering 44*: 265–272.
36. Riedel, S.L., Lu, J., Stahl, U. et al. (2014). Lipid and fatty acid metabolism in Ralstonia eutropha: relevance for the biotechnological production of value-added products. *Applied Microbiology and Biotechnology 98* (4): 1469–1483.
37 Dyer, J.M., Chapital, D.C., Kuan, J.W. et al. (2002). Metabolic engineering of Saccharomyces cerevisiae for production of novel lipid compounds. *Applied Microbiology and Biotechnology 59* (2–3): 224–230.
38. Meesapyodsuk, D., Chen, Y., Ng, S.H. et al. (2015). Metabolic engineering of Pichia pastoris to produce ricinoleic acid, a hydroxy fatty acid of industrial importance. *Journal of Lipid Research 56* (11): 2102–2109.
39. Garaiova, M., Mietkiewska, E., Weselake, R.J. et al. (2017). Metabolic engineering of Schizosaccharomyces pombe to produce punicic acid, a conjugated fatty acid with nutraceutic properties. *Applied Microbiology and Biotechnology 101* (21): 7913–7922.
40. Takemura, M., Tanaka, R., and Misawa, N. (2017). Pathway engineering for the production of β-amyrin and cycloartenol in Escherichia coli—a method to biosynthesize plant-derived triterpene skeletons in E. coli. *Applied Microbiology and Biotechnology 101* (17): 6615–6625.
41. Anulika, N.P., Ignatius, E.O., Raymond, E.S. et al. (2016). The chemistry of natural product: plant secondary metabolites. *International Journal of Technology Enhancements and Emerging Engineering Research 4* (8): 1–9.
42. Jimenez-Garcia, S.N., Vazquez-Cruz, M.A., Guevara-Gonzalez, R.G. et al. (2013). Current approaches for enhanced expression of secondary metabolites as bioactive compounds in plants for agronomic and human health purposes-a review. *Polish Journal of Food and Nutrition Sciences 63* (2): 1–10.
43. Cheynier, V., Comte, G., Davies, K.M. et al. (2013). Plant phenolics: recent advances on their biosynthesis, genetics, and ecophysiology. *Plant Physiology and Biochemistry 72*: 1–20.
44. Tiwari, P., Srivastava, Y., and Bae, H. (2021). Trends of pharmaceutical design of Endophytes as anti-infective. *Current Topics in Medicinal Chemistry 21* (17): 1572–1586.
45. Gorlenko, C.L., Kiselev, H.Y., Budanova, E.V. et al. (2020). Plant secondary metabolites in the battle of drugs and drug-resistant bacteria: new heroes or worse clones of antibiotics? *Antibiotics 9* (4): 170.
46. Connolly, J.D. and Hill, R.A. (1991). *Journal: Dictionary of Terpenoids*, 153–653. Chapman and Hall/CRC.
47. Croteau, R.B., Davis, E.M., Ringer, K.L. et al. (2005). (−)-Menthol biosynthesis and molecular genetics. *Naturwissenschaften 92* (12): 562–577.
48. Bose, S.K., Yadav, R.K., Mishra, S. et al. (2013). Effect of gibberellic acid and calliterpenone on plant growth attributes, trichomes, essential oil biosynthesis and pathway gene expression in differential manner in Mentha arvensis L. *Plant Physiology and Biochemistry 66*: 150–158.
49. Sharkey, T.D., Wiberley, A.E., and Donohue, A.R. (2008). Isoprene emission from plants: why and how. *Annals of Botany 101* (1): 5–18.
50. Alonso, W.R., Rajaonarivony, J.I., Gershenzon, J. et al. (1992). Purification of 4S-limonene synthase, a monoterpene cyclase from the glandular trichomes of peppermint (Mentha x piperita) and spearmint (Mentha spicata). *The Journal of Biological Chemistry 267* (11): 7582–7587.
51. Liu, W., Xu, X., Zhang, R. et al. (2016). Engineering Escherichia coli for high-yield geraniol production with biotransformation of geranyl acetate to geraniol under fed-batch culture. *Biotechnology for Biofuels 9* (1): 1–8.
52. Jiang, G.Z., Yao, M.D., Wang, Y. et al. (2017). Manipulation of GES and ERG20 for geraniol overproduction in Saccharomyces cerevisiae. *Metabolic Engineering 41*: 57–66.

53. Yang, J. and Guo, L. (2014). Biosynthesis of β-carotene in engineered E. coli using the MEP and MVA pathways. *Microbial Cell Factories 13* (1): 1–1.
54. Jongedijk, E., Cankar, K., Ranzijn, J. et al. (2015). Capturing of the monoterpene olefin limonene produced in Saccharomyces cerevisiae. *Yeast 32* (1): 159–171.
55. Ignea, C., Cvetkovic, I., Loupassaki, S. et al. (2011). Improving yeast strains using recyclable integration cassettes, for the production of plant terpenoids. *Microbial Cell Factories 10* (1): 1–8.
56. Mendez-Perez, D., Alonso-Gutierrez, J., Hu, Q. et al. (2017). Production of jet fuel precursor monoterpenoids from engineered Escherichia coli. *Biotechnology and Bioengineering 114* (8): 1703–1712.
57. Meadows, A.L., Hawkins, K.M., Tsegaye, Y. et al. (2016). Rewriting yeast central carbon metabolism for industrial isoprenoid production. *Nature 537* (7622): 694–697.
58. You, S., Yin, Q., Zhang, J. et al. (2017). Utilization of biodiesel by-product as substrate for high-production of β-farnesene via relatively balanced mevalonate pathway in Escherichia coli. *Bioresource Technology 243*: 228–236.
59. Peralta-Yahya, P.P., Ouellet, M., Chan, R. et al. (2011). Identification and microbial production of a terpene-based advanced biofuel. *Nature Communications 2* (1): 1–8.
60. Özaydın, B., Burd, H., Lee, T.S. et al. (2013). Carotenoid-based phenotypic screen of the yeast deletion collection reveals new genes with roles in isoprenoid production. *Metabolic Engineering 15*: 174–183.
61. Zhang, C., Liu, J., Zhao, F. et al. (2018). Production of sesquiterpenoid zerumbone from metabolic engineered Saccharomyces cerevisiae. *Metabolic Engineering 49*: 28–35.
62. Yang, J., Li, Z., Guo, L. et al. (2016). Biosynthesis of β-caryophyllene, a novel terpene-based high-density biofuel precursor, using engineered Escherichia coli. *Renewable Energy 99*: 216–223.
63. Tsuruta, H., Paddon, C.J., Eng, D. et al. (2009). High-level production of amorpha-4, 11-diene, a precursor of the antimalarial agent artemisinin, in Escherichia coli. *PLoS One 4* (2): e4489.
64. Westfall, P.J., Pitera, D.J., Lenihan, J.R. et al. (2012). Production of amorphadiene in yeast, and its conversion to dihydroartemisinic acid, precursor to the antimalarial agent artemisinin. *Proceedings of the National Academy of Sciences of the United States of America 109* (3): E111–E118.
65. Tippmann, S., Scalcinati, G., Siewers, V. et al. (2016). Production of farnesene and santalene by Saccharomyces cerevisiae using fed-batch cultivations with RQ-controlled feed. *Biotechnology and Bioengineering 3* (1): 72–81.
66. Albertsen, L., Chen, Y., Bach, L.S. et al. (2011). Diversion of flux toward sesquiterpene production in Saccharomyces cerevisiae by fusion of host and heterologous enzymesAppl. *Environmental Microbiology 77* (3): 1033–1040.
67. Paddon, C.J., Westfall, P.J., Pitera, D.J. et al. (2013). High-level semi-synthetic production of the potent antimalarial artemisinin. *Nature 496* (7446): 528–532.
68. Wong, J., de Rond, T., d'Espaux, L. et al. (2018). High-titer production of lathyrane diterpenoids from sugar by engineered Saccharomyces cerevisiae. *Metabolic Engineering* (45): 142–148.
69. Engels, B., Dahm, P., and Jennewein, S. (2008). Metabolic engineering of taxadiene biosynthesis in yeast as a first step towards Taxol (Paclitaxel) production. *Metabolic Engineering 10* (3–4): 201–206.
70. Ajikumar, P.K., Xiao, W.H., Tyo, K.E. et al. (2010). Isoprenoid pathway optimization for Taxol precursor overproduction in *Escherichia coli. Science 330* (6000): 70–74.
71. Zhu, M., Wang, C., Sun, W. et al. (2018). Boosting 11-oxo-β-amyrin and glycyrrhetinic acid synthesis in Saccharomyces cerevisiae via pairing novel oxidation and reduction system from legume plants. *Metabolic Engineering 45*: 43–50.
72. Wang, R.Z., Pan, C.H., Wang, Y. et al. (2016). Design and construction of high β-carotene producing Saccharomyces cerevisiae. *China Biotechnology 36* (7): 83–91.
73. Sun, T., Miao, L., Li, Q. et al. (2014). Production of lycopene by metabolically-engineered Escherichia coli. *Biotechnology Letters 36* (7): 1515–1522.

74. Yang, D., Kim, W.J., Yoo, S.M. et al. (2018). Repurposing type III polyketide synthase as a malonyl-CoA biosensor for metabolic engineering in bacteria. *Proceedings of the National Academy of Sciences of the United States of America 115* (40): 9835–9844.
75. Ajikumar, P.K., Tyo, K., Carlsen, S. et al. (2008). Terpenoids: opportunities for biosynthesis of natural product drugs using engineered microorganisms. *Molecular Pharmaceutics 5* (2): 167–190.
76. Kaur, R.A. and Arora, S.A. (2015). Alkaloids-important therapeutic secondary metabolites of plant origin. *Journal of Critical Review 2* (3): 1–8.
77. Habli, Z., Toumieh, G., Fatfat, M. et al. (2017). Emerging cytotoxic alkaloids in the battle against cancer: overview of molecular mechanisms. *Molecules 22* (2): 250.
78. Dastmalchi, M., Park, M.R., Morris, J.S. et al. (2018). Family portraits: the enzymes behind benzylisoquinoline alkaloid diversity. *Phytochemistry Reviews 17* (2): 249–277.
79. Matsumura, E., Nakagawa, A., Tomabechi, Y. et al. (2018). Microbial production of novel sulphated alkaloids for drug discovery. *Scientific Reports 8* (1): 1-0.
80. Galanie, S., Thodey, K., Trenchard, I.J. et al. (2015). Complete biosynthesis of opioids in yeast. *Science 349* (6252): 1095–1100.
81. Nakagawa, A., Matsumura, E., Koyanagi, T. et al. (2016). Total biosynthesis of opiates by stepwise fermentation using engineered Escherichia coli. *Nature Communications 7* (1): 1–8.
82. Saiman, M.Z. (2014). *Terpenoids and Terpenoid Indole Alkaloids in Catharanthus roseus Cell Suspension Cultures*. Leiden University.
83. Brown, S., Clastre, M., Courdavault, V. et al. (2015). De novo production of the plant-derived alkaloid strictosidine in yeast. *Proceedings of the National Academy of Sciences of the United States of America 112* (11): 3205–3210.
84. Brey, L.F., Włodarczyk, A.J., Thøfner, J.F. et al. (2020). Metabolic engineering of Synechocystis sp. PCC 6803 for the production of aromatic amino acids and derived phenylpropanoids. *Metabolic Engineering 57*: 129–139.
85. Thakur, S., Pandit, K., Kumar, A. et al. (2021). Phenylpropanoid biosynthesis and its protective effects against plants stress. In: *Environmental Stress Physiology of Plants and Crop Productivity*, vol. *6* (ed. T. Kaur and S. Arora), 144–157. Sharjah, UAE: Bentham Science Publishers.
86. Kumar, N. and Maheshwari, K.K. (2014). 20 Nutrigenomics: nurturing of genotype and role in human health. In: *Phytochemicals of Nutraceutical Importance* (ed. D. Prakash and G. Sharma), 324–349. CABI.
87. Li, M., Schneider, K., Kristensen, M. et al. (2016). Engineering yeast for high-level production of stilbenoid antioxidants. *Scientific Reports 6* (1): 1–8.
88. Kang, S.Y., Lee, J.K., Choi, O. et al. (2014). Biosynthesis of methylated resveratrol analogs through the construction of an artificial biosynthetic pathway in E. coli. *BMC Biotechnology 14* (1): 1–1.
89. Heo, K.T., Kang, S.Y., and Hong, Y.S. (2017). De novo biosynthesis of pterostilbene in an Escherichia coli strain using a new resveratrol O-methyltransferase from Arabidopsis. *Microbial Cell Factories 16* (1): 1–8.
90. Wu, J., Zhang, X., Zhu, Y. et al. (2017). Rational modular design of metabolic network for efficient production of plant polyphenol pinosylvin. *Scientific Reports 7* (1): 1–5.
91. Jeandet, P., Sobarzo-Sánchez, E., Clément, C. et al. (2018). Engineering stilbene metabolic pathways in microbial cells. *Biotechnology Advances 36* (8): 2264–2283.
92. Lehka, B.J., Eichenberger, M., Bjørn-Yoshimoto, W.E. et al. (2017). Improving heterologous production of phenylpropanoids in Saccharomyces cerevisiae by tackling an unwanted side reaction of Tsc13, an endogenous double-bond reductase. *FEMS Yeast Research 17* (1): fox004.
93. Eichenberger, M., Lehka, B.J., Folly, C. et al. (2017). Metabolic engineering of Saccharomyces cerevisiae for de novo production of dihydrochalcones with known antioxidant, antidiabetic, and sweet tasting properties. *Metabolic Engineering 39*: 80–89.
94. Duan, L., Ding, W., Liu, X. et al. (2017). Biosynthesis and engineering of kaempferol in Saccharomyces cerevisiae. *Microbial Cell Factories 16* (1): 1-0.

95. Levisson, M., Patinios, C., Hein, S. et al. (2018). Engineering de novo anthocyanin production in Saccharomyces cerevisiae. *Microbial Cell Factories 17* (1): 1–6.
96. Jones, J.A., Vernacchio, V.R., Collins, S.M. et al. (2017). Complete biosynthesis of anthocyanins using E. coli polycultures. *MBio 8* (3): e00621-17.
97. Eichenberger, M., Hansson, A., Fischer, D. et al. (2018). De novo biosynthesis of anthocyanins in Saccharomyces cerevisiae. *FEMS Yeast Research 18* (4): foy046.
98. Wu, J., Zhang, X., Zhou, J. et al. (2016). Efficient biosynthesis of (2S)-pinocembrin from d-glucose by integrating engineering central metabolic pathways with a pH-shift control strategy. *Bioresource Technology 218*: 999–1007.
99. Liu, X., Cheng, J., Zhang, G. et al. (2018). Engineering yeast for the production of breviscapine by genomic analysis and synthetic biology approaches. *Nature Communications 9* (1): 1-0.
100. Trantas, E., Panopoulos, N., and Ververidis, F. (2009). Metabolic engineering of the complete pathway leading to heterologous biosynthesis of various flavonoids and stilbenoids in Saccharomyces cerevisiae. *Metabolic Engineering 11* (6): 355–366.
101. Lin, Y., Shen, X., Yuan, Q. et al. (2013). Microbial biosynthesis of the anticoagulant precursor 4-hydroxycoumarin. *Nature Communications 4* (1): 1–8.
102. Yang, S.M., Shim, G.Y., Kim, B.G. et al. (2015). Biological synthesis of coumarins in *Escherichia coli*. *Microbial Cell Factories 14* (1): 1–2.

5

Quorum Sensing and Environmental Sustainability

Arnab Banerjee[1], Manoj Kumar Jhariya[2], Abhishek Raj[3], and Nahid Khan[2]

5.1. QUORUM SENSING – A BRIEF INTRODUCTION

Quorum sensing (QS) is a term that defines cell-to-cell communication among living bacterial cells, as well as an important process to run physiological progress in living beings. It is a very useful process for the bacterial community to communicate with each other through a chemical signaling process. Some important functions of microbes are mediated through QS systems; these include biomolecule formation, virulence development, and genetic recombination, as well as formation of biofilm [1]. There are some QS mediated problems, such as occurrence of diseases viz. dental plague and cancer, agricultural production, and yield reduction, as well as various aquaculture practices along with wastewater treatment [2–6].

In simple terms, the process of QS is a mode of communication between bacterial colonies through a cell signaling mechanism under the regulation of gene expression. The process is very much dependent upon the production of auto-inducer molecules, which act as the communication channel in the form of signaling molecules [7]. Thus, the authors have discussed QS as gene expression and its pattern on a global scale from a microorganism perspective [8]. From an ecological perspective, the QS system is very much important for both Gram-positive and negative bacteria in order to maintain their ecological niche. It is also beneficial for the pathogenic microorganisms as an energy-intensive process during host–cell infection [9]. From a pathological perspective, such a mechanism is very important for humanity in terms of spreading of diseases and pathological invasion of pathogens [9, 10].

A living organism's environment has a big impact on its existence. Now in the era of climate change, a continuous change is taking place both in external and internal environments of living beings [11, 12]. Therefore, adaptating to this environmental change is very

[1]*Department of Environmental Science, Sant Gahira Guru Vishwavidyalaya, Ambikapur, Chhattisgarh, India*

[2]*Department of Farm Forestry, Sant Gahira Guru Vishwavidyalaya, Ambikapur, Chhattisgarh, India*

[3]*Pt. Deendayal Upadhyay College of Horticulture and Forestry, Dr. Rajendra Prasad Central Agriculture University, Samastipur, Bihar, India*

Microbial Biotechnology: Role in Ecological Sustainability and Research, First Edition.
Edited by Pankaj Chowdhary, Sujata Mani, and Preeti Chaturvedi.

much essential in order to maintain the macro and micro niches that exist within the ecosystem through food chains and food webs [13–16]. Microorganisms have thus evolved the mechanism of QS for detection of these environmental changes to cope up with such climatic extremes. QS bacterial communities tend to alter their behavioral physiology in order to increase their survivability rate. In this regard, auto-inducing molecules such as AIs and their concentrations change along with the bacterial population count [10]. The bacterial community tends to maintain their life within the host cells of plant and animal, and thus this cell signaling mechanism becomes very important for them to survive within the host [17]. However, significant variation exists in the gene regulation process and accordingly the behavior of the microorganisms change. It was observed that microorganisms can exhibit specific characteristics due to operation of the QS system [9, 10].

The present chapter deals with proper illustration of the concept of QS along with its diverse application in the field of environment and sustainability. It also addresses various challenges and issues followed by facts that need to be explored by the future academic world. Moreover, it will lead to exploration of diverse applications of QS that can be used for the wellbeing and welfare of society.

5.2. CONCEPT AND MECHANISM IN QUORUM SENSING

The basic mechanism of the mode of operation of the QS system within the bacterial cells involves the production of AIs molecules, which are sensed through the bioreceptors present within the bacterial cells (Figure 5.1). Further, this leads to proper gene expression to express the specific behavior for the bacterial cells. In this way the harmonization and cohesiveness between the bacterial cells within a bacterial colony exists in a sustainable manner.

The building block of QS includes the auto-inducing molecules, which are usually formed within the bacterial cells during their growth. Accordingly, the concentration of auto-inducing molecules depends upon bacterial cell density [18]. The basic mechanism behind the operation of the QS system includes the increase in concentration of auto-inducing molecules up to a threshold level that stimulates the bacterial cell at the molecular level by triggering specific gene expression, resulting in physiological changes that benefit the bacterial community [18, 19]. There is significant level of variation among the cell signaling process of different microbial species. Usually, Gram-negative bacteria tend to use smaller molecules

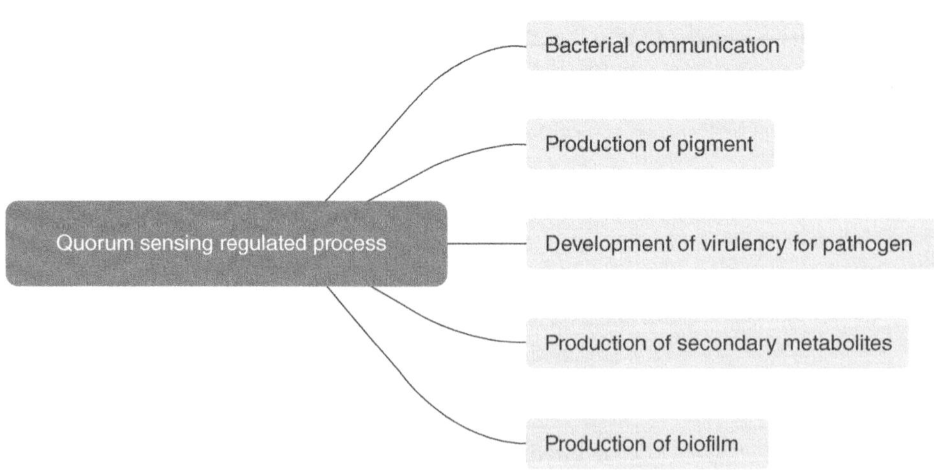

Figure 5.1 Quorum sensing-regulated molecular process.

in the form of auto-inducers. AHLs (N-acylated-L-homoserine lactones) are the common auto-inducer molecules that are used by Gram-negative microbes [18]. In the case of Gram-positive bacteria, the major auto-inducer molecules appear to be cyclical peptides [20]. Recent research has revealed the role of AI-2 molecules as the QS mechanism between Gram-positive and Gram-negative bacteria [18].

5.3. QUORUM SENSING AND THE ENVIRONMENT

Environment has a big influence over the QS process of microorganisms. The nature and attributes of the physical environment have been reported to influence the cell signaling process [21]. As it is a known fact that biosynthesis of AI molecules is a crucial step to initiate the QS system, the environmental conditions regulate the propagation of signals and signal concentration that will mediate the process. Under an alkaline environment, rapid degeneration of the signaling process was recorded by different workers due to unstable nature of the AI molecules [22]. This is very problematic aspect as it may not initiate the receptor cells present within the bacterial cells. Further, exogenous materials may enhance or decrease the sensing process of the auto-inducing molecules due to inhibition of gene expression for biosynthesis of auto-inducing molecules.

Under an ex-situ environment, i.e. beyond the laboratory condition, the signaling molecules required for operation of QS may become vulnerable. Alteration of the natural environment may impose significant problems for the signaling molecules to operate smoothly in a specific direction. Such alterations in the external environment as well as in the signal molecules may provide necessary information about the ambient environment to the living cell. The chemical configuration of the signaling molecules makes them more susceptible to environmental changes. As a consequence of such a process, the activity of signal molecules is hindered and therefore cell–cell communication becomes inactive [23].

Research reports revealed that there is a higher level of diversity with respect to habitat condition as well as bacterial species that undergo specific changes, which is ecologically known as species interaction (Table 5.1). This would lead to sensing of different chemical signals for different chemical processes. A specific environment is required for a specific microorganism, which is dependent upon the QS system to generate the response. Microbial metabolism may also influence the sensing process, which responds accordingly to the prevailing external environment [34]. Research reports revealed that microbial metabolism of different microbial communities can be enhanced through the QS system [35].

5.4. APPLICATIONS OF QUORUM SENSING

From an environmental perspective, QS seems to provide significant potentiality in its various spheres (Figure 5.2, Table 5.2) [5]. It has other wider dimensions in the field of health and safety as well as in agriculture production [5, 45]. Considering the application of QS, it plays a major role in the field of wastewater treatment [46]. Wastewater treatment is a critical issue, in which use of microbial consortium in the form of biological treatment provides a suitable, ecofriendly alternative. These beneficial microbes degrade contaminants and inhibit the pathogenetic process, as well as having wider application in disease diagnosis and formulation of antimicrobial agents [46, 47]. Within the biological treatment process, culture of microbial cells tends to degrade the pollutants into less-toxic forms that helps in bacterial metabolism and growth. Such a process, therefore, has become the key step in wastewater treatment in the form of activated sludge process, rotating biological contractor,

Table 5.1 Quorum sensing in various environments.

Nature of environment	Type of microbes	Mechanism and mode of action	References
Salty environment	Halophiles	AHLs production by *Halomonas* isolates	[24]
		H. halophilus- production of AI-2 signals	[25]
		Micro-algae *Dunaliella salina* – production of antagonistic molecules for QS	[26]
Acidic environment	Acidophiles	*F. acidarmanus* – Genes present in the genome are regulating the process of biofilm formation	[27]
		Acidithiobacillus ferrooxidans – Genes *afeI* and *afeR* code for specific proteins that develop resistance against heavy metal toxicity	[28]
		V. cholera – CqsA regulates the production of excess gelatinous biofilm leading to its protection against the acidic environment of the alimentary tract of the human body	[29]
High temperature	Thermophiles	*Thermus* sp. GH5 – undergoes AHL signaling by producing AHL precursors under cold stress	[30]
		P. furiosus in combination with *T. maritime* produces AI-2 type signal with no phenotypic expression	[31]
Cold Environment	Psychrophiles	*Pseudoaltermonas haloplanktis* contains *mtnN* gene produces AI-2 type signal	[32]
High pressure Environment	*Piezophiles*	*P. profundum* contains AI-2 signaling systems that regulates the metabolism of the microbe.	[33]

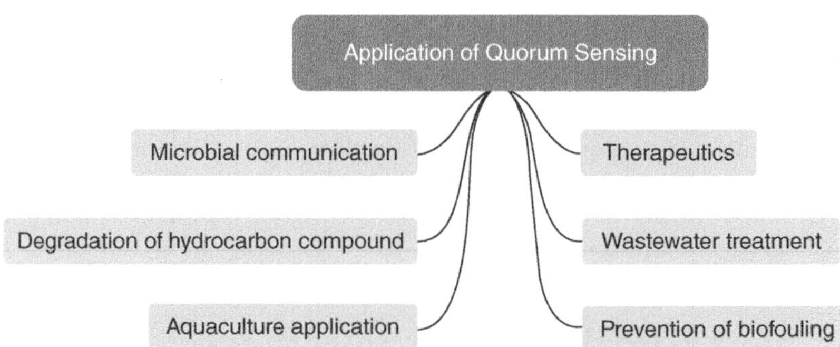

Figure 5.2 Diverse applications of quorum sensing.

fixed films system, etc. In all these systems, proper cell–cell signaling of the bacterial cells helps to degrade the pollutants and thus QS plays a significant role in pollution remediation [48].

Biofilm is an important structure that comprises a bacterial consortium surrounded by a gelatinous matrix of various biomolecules, which help in the bioremediation process [49]. Various bacterial cells such as *Acinetobacter*, *Pseudomonas*, etc., participate in the formation of biofilms that help in the bioremediation process [50, 51]. In the case of *Pseudomonas*, the QS system tends to help in biofilm formation through operon regulation of various biomolecules that frame the matrix of biofilm [50]. According to research studies, organic molecules such

Table 5.2 Modes of operation of Quorum sensing: Molecular approaches.

Operational sectors	Mode of application	References
Pathology	*D. pulchra* produces furanones compound that inhibits QS system of various gram-negative microbes	[36]
	QS regulates pathogenesis of *P. aeruginosa* through production of various multiple virulence factors as well as production of biofilm	[37, 38]
Environment	Gram-negative bacteria in the marine environment help to prevent biofouling through inhibition of QS through kojic acid production.	[39]
Agriculture	Pathogens of plant origin (*Pectobacterium carotovora, Pantoea stewartii*) and other aquatic pathogens regulate their virulency through various QS systems	[40]
	TraR-based QS found in *A. tumefaciens* helps to develop pathogenesis in plants	[41]
	P. chlororaphis strain 30–84 utilizes AHL-based QS to regulate antibiotics production for protection of wheat from fungal disease	[42]
	QS systems seem to regulate biological nitrogen fixation process within the symbiotic relationship between symbiotic nitrogen fixing relationship.	[43]
Aquaculture	*Aeromonas salmonicida* and *Aeromonas hydrophila* utilize AHL-based QS systems to regulate their virulence.	[3]
	QS systems of *V. harveyi* have been shown to express virulency under in vivo conditions.	[44]

as the exogenous 3-oxo-octanoyl-L-homoserinelactone (C8-oxo-HSL) influence the growth of *Pseudomonas*, which ultimately leads to formation of extracellular polymeric compounds in the biofilm system [52]. The QS system tends to influence the growth of biofilm by increasing biomass under constant nutrient supply and hence outfitting all other microorganisms [53].

Under natural conditions, formation of biofilm can have bidirectional effects for the human population. Biofilms are mainly used for degradation of various forms of pollutants, which includes organic and inorganic compounds. Further, biofilms tend to maintain a suitable environment for natural processes to takes place effectively [54]. Various studies have revealed that biofilm formation in sequential batch reactors used in wastewater treatment was proven to be highly effective in decontaminating the contaminants. The aerobic granules in the biofilm tend to perform the process effectively and hence reduce the cost of the process [55]. QS play a key role in such a wastewater treatment process through production of specific molecules such as AHLs by different microorganisms [56]. The QS mechanism in this process involves induction of the AHL molecules, which helps in bioremediation [57]. Such molecules help the bacterial colony to absorb the pollutants present in wastewater on its cellular surface, leading to formation of biofilm [58]. AHL molecules help in cell–cell signaling leading to degradation of the contaminants [59]. Similarly, the signaling molecules under the QS system, along with EPS production, help to maintain the structural configuration of the biofilm as reported in various earlier works [60]. AHL molecules also lead to biosurfactant production along with proteins of extracellular origin [61].

Biofouling is a negative consequence in the field of biofilm formation as the microbial community gets adsorbed over the solid matrix. This therefore reduces the efficiency of bioreactors for waste treatment and increases the operational cost [62]. The generation of AHL molecules was confirmed during a biofouling process in the reactor environment [63]. Further, it was reported that the activity of AHL increases the pressure of the trans-membrane of a microbial biological reactor [64]. Research studies have mentioned that inhibition of the QS system based on auto-inducing molecules tends to reduce the biofouling event. Various organic hydrocarbon compounds are reported to regulate biofilm formation under the QS signaling process [65]. Considering the toxicity of the hydrocarbon compounds, less-toxic or non-toxic compounds are usually selected as a suitable strategy to combat biofouling in the microbial bio-reactor. This in turn does not hamper the growth of the bacteria.

Research reports revealed that the QS system found in bacteria helps to degrade organic hydrocarbon pollutants as well as other inorganic and organic compounds [66]. The AHL molecules in the QS system have been reported to be involved in the biodegradation process of phenol through an activated sludge system. The process has been reported to take place for a longer duration. Another research report based on *Pseudomonas aeruginosa* revealed that the RhlI/RQS system tends to facilitate the biodegradation process in the wastewater from various sources. The process involves biodegradation of the pollutants followed by nitrification [67]. Short chain AHL molecules were reported as active degraders of nicotine whereas the long chain AHL molecules offer resistance against the biodegradation process of nicotine [68]. Major enzymes involved in the biodegradation process seem to be regulated by the auto-inducer molecules followed by the existing QS system. For example, biodegradation of antranilate and phenol through *P. aeruginosa* involves active participation of rhlI/R QS system [56]. Further, in the process of denitrification rhlI/R, the QS system tends to regulate the activity of various genes coding for various enzymes or protein molecules [69].

Aquaculture practice generates a significant amount of revenue for both developed and developing nations. In aquaculture, pathological infestation and disease outbreaks are major issues that often hamper the net economic gain and result in a loss of productivity. The QS system has been reported to be actively involved in regulation of disease-causing pathogens that negatively impact the aquaculture practice. Various pathogenic molecules secreted by various organisms are usually regulated by the QS system [70]. For example, enzyme activity in *Vibrio* spp. is often regulated by QS [71]. Research reports revealed that attenuation of the QS system in the pathogenic organism increases the survivability rate of various shrimp and prawn species [72]. A consequence of close correlation between the virulence of pathogens and the QS system, ecological interventions seem to be the most suitable alternative to arrest pathogenic diseases and their treatment [73]. Various compounds performing the cell signaling process under the QS system have been reported by various workers to arrest bacterial virulence through such a system [72].

It was observed that chemicals used in aquaculture practice tend to show higher mortality in the aged organism and therefore new alternatives in the form of natural compounds have been discovered. For example, Cinnamaldehyde tend to protect freshwater prawn species from pathogenic infection through disruption of QS system [74]. Application of probiotic microbes to regulate aquaculture diseases has shown significant promise through inhibition of the QS system of the pathogens. For instance, AHL molecules generated through auto-inducing activity may lead to inhibition of growth of pathogenic microbes [3, 75]. Microalgal species applied in aquaculture reflects inhibition of pathogens through production of QS-generated antagonistic metabolites [26].

QS is a process that tends to regulate the growth of bacterial population through auto-inducers and influences various bacterial physiological processes. Auto-inducing molecules seem to influence the virulence and mode of action of bacterial species as observed in *P. aeruginosa*. It also influences biofilm formation. Various enzyme systems have been reported from the genus *Bacillus* that tend to biodegrade AHLs signaling molecules.

However, all these enzymes are sensitive to abiotic factors such as pH and temperature, and hence can be easily regulated by elevating the temperature of water and thus have no effect in the treatment of drinking water.

5.5. QUORUM SENSING AND ENVIRONMENTAL SUSTAINABILITY

The sustainability of the earth's ecosystem is under the severe threat of pollution, different forms of environmental degradation, climate change, and other mega events [76–78]. It was observed that different forms of pollution involve the presence of unwanted materials (some of them recalcitrant compounds) that poses a challenge to degradation [79–82]. In this prospect, microorganisms as decomposers act as a boon for humankind, undergoing metabolic disintegration and making the environment contamination and pollution free. Wastewater pollution has now become a problem due to unprecedented growth of the industry and urbanization. Currently, people are investigating biodegradation through the use of microbes. On a long-term basis, it was observed that the efficiency of bioreactors for treating wastewater gradually slows down, which is not good for the industrial health. Within the reactor system, it was observed that the food to microbe ratio becomes unbalanced with the lowering of bacterial growth. Such problems can be addressed through biofilm formation under the aegis of the QS system of the microbes. The QS system has the capacity to restrain the bacteria within the bioreactor system for effective degradation of pollutants in the wastewater. Further, this novel approach has been studied by several workers using different microorganisms targeting specific pollutant molecules. Even in the natural system, bacterial groups have been reported to effectively degrade organic hydrocarbon compounds through the triggering of auto-inducer molecules present in the bacterial cells. Inventory of the role of the QS system in the bioremediation process has made it much more effective in treating wastewater pollutants [67, 83].

5.6. OPPORTUNITIES AND CHALLENGES OF QUORUM SENSING

QS is the latest molecular development in the field of molecular technology and its application. Further, it was observed that such systems prevalent in microorganisms help to regulate the gene expression, which in turn regulates the level of auto-inducers. These auto-inducers often promote the expression of QS-dependent genes in order to develop adaptability among the living organisms. Studies have reported that such auto-inducer mediated QS systems tend to regulate various important functions of microorganisms. However, the future of such QS-based systems tends to be promising in the area of ecological application and biofilm formation.

Bioremediation is the latest ecofriendly technology that includes various techniques in the form of zoo-remediation, phytoremediation, and micro-remediation [84, 85]. The aim of these processes includes biodegradation of pollutants to make a green and clean environment. Studies have revealed that QS has a most significant role to play in the process of biodegradation at a molecular level. However, the challenge appears to explore the potentiality, effectivity, and applicability of the QS system toward biodegradation and environmentally clean processes. QS in biodegradation may have a positive impact over the pollutant degradation rate by making the environment contaminant free. It seems that more study and exploration on the QS system and its potential role in biodegradation is required for the wellbeing of humankind. Further, degradation pathways involving aromatic compounds highlight new molecules involved in QS. Therefore, proper regulation of the aromatic compounds through the cell-signaling process are one of the new challenges that needs to be overcome in the future.

In the field of medical science, QS can be effectively used for screening of antimicrobial agents that help to ameliorate bacterial infections. However, it is yet to be used in real-life applications. The QS system and its gene regulation process involve a networking system and therefore, isolating a particular pathway of QS would not hamper the functionality of the entire network system. Further, development in these areas in relation to nature would be a good option and would generate new scientific knowledge and advancement.

5.7. RESEARCH AND DEVELOPMENT TOWARD ACHIEVING ENVIRONMENTAL SUSTAINABILITY

Smooth functioning of QS is very much dependent upon the auto-inducer molecules that help in the recognition of cell–cell signaling. Recent research and development in QS have been aimed toward identifying various molecules that inhibit the sensing process [18]. Most of the studies related to QS have been done under mixed laboratory conditions in order to assess and understand the functionality of the QS system. This, therefore, provided an insight into the biochemical and evolutionary aspects of QS, as well as developed the conceptual framework in the form of how QS regulates the molecular biology of cell signaling in the bacterial community. This is a very important aspect in research and development to explore the future possibilities of QS in environmental sustainability. Microbes have an inherent role in nature in the form of decomposers. They also play a significant role in nutrient cycling. All these processes are the result of bacterial metabolism, which needs to be studied at its molecular level. However, further research needs to be implemented under natural conditions, considering the complexities of the environmental factors that may influence the QS process. The major challenge behind the study of QS lies in understanding bacterial communication in natural habitats [86].

Forests are an important area to explore in terms of their multifaceted ecological services. In the era of climate change, it was observed that forests or vegetation seem to acts as major carbon sink, and thus are an important tool for combating the changing climate. From this perspective, ecological function within the forest ecosystem such as the food chain and nutrient cycling becomes the fundamental process to maintain forest productivity and well-being [87–90]. Litter decomposition and forest floor soil chemistry happen to be regulated by the soil microbes [91–93]. Therefore, cell–cell communication in the form of QS is a very important aspect for study. Studies have revealed that auto-inducer molecules present in soil and litter help to perform specific function through proper gene expression. QS may tend to regulate the communication between bacterial communities and hence the ecosystem processes. QS may be highly functional in the forest soil communities due to nutrient-poor conditions in the absence of proper ground vegetation and herbaceous layer [94]. QS has been reported to be a significant factor in decomposition of recalcitrant litter as well as litter accumulated over a long time. However, the rate of litter decomposition increases under stress conditions, which may be due to the smooth operation of QS [94, 95].

5.8. FUTURE PROSPECTS FOR QUORUM SENSING

QS is a molecular approach that involves intraspecific and interspecific communication that regulates community structure and function. Recently, the QS system and its associated mechanisms have received worldwide attention in various environmental cleaning operations. The QS process has mediated the easy treatment of wastewater and degradation of toxic contaminants and pollutants, including biofouling control. However, such approaches seem to have limited application and have yet to be explored properly by the scientific community. Further, the QS system is very much monospecific and, therefore, multi-specific studies need to be done in the near future. Moreover, the ecological function of the QS system appears to be complex in nature through the involvement of more than one species under in-situ conditions. Scientific experimentation is also yet to occur properly in these aspects. From a future perspective, it is a big challenge to evaluate the potential of the QS system in wastewater treatment. There is a huge knowledge gap, and more studies are required in both lab and land conditions for effective implementation of the QS system on a broader scale. This also needs to be explored at gene and community levels in order to reveal the ecological complexity for the benefit of human beings [96].

The mode of operations of QS systems under natural conditions is yet to explored properly. Research reports reveal the availability of a limited number of signal molecules that are under operative mode [97]. It has been reported that QS signaling is altered through changes within the natural environment. However, this requires a significant level of exploration. The genetic mechanisms behind the mode of operation of the QS system are not well explored scientifically, and are present under a learning phase through metagenomic analysis. Such studies help to realize the diverse pattern of the QS system under variable environments [98]. The influence of the environment in the process of gene regulation is very important in order to facilitate the smooth functioning of the QS system and cell–cell communication among the bacterial cells. It would thus regulate the gene expression process.

Forestry is becoming an important field from a future perspective to combat mega events such as climatic extremes. The ecological service, in the form of microclimate regulation, prevention of natural hazards, and carbon sink, has made it a valuable resource for humans [99, 100]. Further, ecological processes such as nutrient cycling, litter decomposition, and carbon flux tend to be regulated through molecular processes in the form of QS systems. QS has been found to play potential role in the litter decomposition process but due to a lack of information such issues require adequate future research in order to maintain sustainable forest management. The rate of litter decomposition influenced by QS is a big question under in-situ conditions. Proper exploration in these aspects is the future directive of forestry research in the field of QS.

5.9. CONCLUSION

QS is an innovative mechanism of microorganisms that helps to regulate the coordination between bacterial cells within a colony. The system helps the bacteria to occupy specific niches, which hamper human wellbeing due to pathogenic activities of the microbes. However, QS seems to have diverse applications in the field of pharmacy, agriculture, environment, biotechnology, and other allied fields. From a future perspective, more exploration is required in this field in order to identify different QS systems and their potential roles for sustainability. Understanding the mechanism of QS is very important from a therapeutic perspective in order to treat infectious diseases. This would help in environmental cleanup operations and maintain overall environmental health across the globe.

REFERENCES

1. Lade, H., Paul, D., and Kweon, J.H. (2014). Quorum quenching mediated approaches for control of membrane biofouling. *International Journal of Biological Sciences 10*: 550–565.
2. Li, L., Hooi, D., Chhabra, S.R. et al. (2004). Bacterial N-acylhomoserine lactone-induced apoptosis in breast carcinoma cells correlated with down-modulation of STAT3. *Oncogene 23*: 4894–4902.

3. Defoirdt, T., Boon, N., Bossier, P. et al. (2004). Disruption of bacterial quorum sensing: an unexplored strategy to fight infections in aquaculture. *Aquaculture 240*: 69–88.
4. Wang, W.Z., Morohoshi, T., Ikenoya, M. et al. (2010). AiiM, a novel class of *N*-acylhomoserine lactonase from the leaf-associated bacterium *Microbacterium testaceum*. *Applied and Environmental Microbiology 76*: 2524.
5. Galloway, W.R., Hodgkinson, J.T., Bowden, S. et al. (2012). Applications of small molecule activators and inhibitors of quorum sensing in Gram-negative bacteria. *Trends in Microbiology 20*: 449–458.
6. Basavaraju, M., Sisnity, V.S., Palaparthy, R. et al. (2016). Quorum quenching: signal jamming in dental plaque biofilms. *Journal of Dental Sciences 11*: 349–352.
7. Syal, K. (2017). Novel method for quantitative estimation of biofilms. *Current Microbiology 74* (10): 1194–1199. https://doi.org/10.1007/s00284-017-1304-0.
8. Turovskiy, Y., Kashtanov, D., Paskhovar, B. et al. (2007). Quorum sensing: fact, fiction, and everything in between. *Advances in Applied Microbiology 62*: 191–234.
9. LaSarre, B. and Federle, M.J. (2013). Exploiting quorum sensing to confuse bacterial pathogens. *Microbiology and Molecular Biology Reviews 77* (1): 73–111. https://doi.org/10.1128/MMBR.00046-12.
10. Papenfort, K. and Bassler, B.L. (2016). Quorum sensing signal-response systems in Gram-negative bacteria. *Nature Reviews. Microbiology 14* (9): 576–588. https://doi.org/10.1038/nrmicro.2016.89.
11. Raj, A., Jhariya, M.K., Yadav, D.K. et al. (2020). *Climate Change and Agroforestry Systems: Adaptation and Mitigation Strategies*, 383. US & Canada: Apple Academic Press Inc., CRC Press- a Tayler and Francis Group https://doi.org/10.1201/9780429286759.
12. Banerjee, A., Jhariya, M.K., Yadav, D.K. et al. (2020). *Environmental and Sustainable Development Through Forestry and Other Resources*, 400. US & Canada: Apple Academic Press Inc., CRC Press- a Tayler and Francis Group https://doi.org/10.1201/9780429276026.
13. Jhariya, M.K., Banerjee, A., Meena, R.S. et al. (2019). *Sustainable Agriculture, Forest and Environmental Management*, 606. Singapore: Springer https://doi.org/10.1007/978-981-13-6830-1.
14. Jhariya, M.K., Yadav, D.K., and Banerjee, A. (2019). *Agroforestry and Climate Change: Issues and Challenges*, 335. US & Canada: Apple Academic Press Inc., CRC Press- a Tayler and Francis Group https://doi.org/10.1201/9780429057274.
15. Khan, N., Jhariya, M.K., Yadav, D.K. et al. (2020). Herbaceous dynamics and CO_2 mitigation in an urban setup- a case study from Chhattisgarh, India. *Environmental Science and Pollution Research 27* (3): 2881–2897. https://doi.org/10.1007/s11356-019-07182-8.
16. Khan, N., Jhariya, M.K., Yadav, D.K. et al. (2020). Structure, diversity and ecological function of shrub species in an urban setup of Sarguja, Chhattisgarh, India. *Environmental Science and Pollution Research 27* (5): 5418–5432. https://doi.org/10.1007/s11356-019-07172-w.
17. Rutherford, S.T. and Bassler, B.L. (2012). Bacterial quorum sensing: its role in virulence and possibilities for its control. *Cold Spring Harbor Perspectives in Medicine 2*: a012427, 1–25. https://doi.org/10.1101/cshperspect.a012427.
18. Galloway, W.R.J.D., Hodgkinson, J.T., Bowden, S.D. et al. (2011). Quorum sensing in Gram-negative bacteria: small-molecule modulation of AHL and Al-2 quorum sensing pathways. *Chemical Reviews 111*: 28–67.
19. McInnis, C.E. and Blackwell, H.E. (2011). Design, synthesis, and biological evaluation of abiotic, non-lactone modulators of LuxR-type quorum sensing. *Bioorganic & Medicinal Chemistry 19*: 4812–4819.
20. Thoendel, M., Kavanaugh, J.S., Flack, C.E. et al. (2011). Peptide signaling in the Staphylococci. *Chemical Reviews 111*: 117–151.
21. Decho, A.W., Norman, R.S., and Visscher, P.T. (2010). Quorum sensing in natural environments: emerging views from microbial mats. *Trends in Microbiology 18* (2): 73–80.
22. Fuqua, C., Parsek, M.R., and Greenberg, E.P. (2001). Regulation of gene expression by cell-to-cell communication: acyl-homoserine lactone quorum sensing. *Annual Review of Genetics 35*: 439–468.
23. Mashburn, L.M. and Whiteley, M. (2005). Membrane vesicles traffic signals and facilitate group activities in a prokaryote. *Nature 437*: 422–425.

24. Llamas, I., Quesada, E., Martínez-Cánovas, M.J. et al. (2005). Quorum sensing in halophilic bacteria: detection of N-acyl-homoserine lactones in the exopolysaccharide-producing species of *Halomonas. Extremophiles 9*: 333341.
25. Averhoff, B. and Muller, V. (2010). Exploring research frontiers in microbiology- recent advances in halophilic and thermophilic extremophiles. *Research in Microbiology 161*: 506514.
26. Natrah, F.M.I., Kenmegne, M.M., Wiyoto, W. et al. (2011). Effects of micro-algae commonly used in aquaculture on acyl-homoserine lactone quorum sensing. *Aquaculture 317*: 53–57.
27. Baker-Austin, C., Potrykus, J., Wexler, M. et al. (2010). Biofilm development in the extremely acidophilic archaeon *Ferroplasma. acidarmanus* Fer1. *Extremophiles 14*: 485491.
28. Wenbin, N., Dejuan, Z., Feifan, L. et al. (2011). Quorum-sensing system in *Acidithiobacillus. ferrooxidans* involved in its resistance to Cu2+. *Letters in Applied Microbiology 53*: 8491.
29. March, J. and Bentley, W. (2004). Quorum sensing and bacterial cross-talk in biotechnology. *Current Opinion in Biotechnology 15*: 495502.
30. Johnson, M., Montero, C., Connors, S. et al. (2005). Population density-dependent regulation of exopolysaccharide formation in the hyperthermophilic bacterium *Thermotoga. maritima. Molecular Microbiology 55*: 664674.
31. Nichols, J., Johnson, M., Chou, C. et al. (2009). Temperature, not LuxS, mediates AI-2 formation in hydrothermal habitats. *FEMS Microbiology Ecology 68*: 173181.
32. Medigue, C., Krin, E., Pascal, G. et al. (2005). Coping with cold: the genome of the versatile marine Antarctica bacterium *Pseudoalteromonas. haloplanktis* TAC125. *Genome Research 15*: 1325–1335. https://doi.org/10.1101/gr.4126905.
33. Rezzonico, F. and Duffy, B. (2008). Lack of genomic evidence of AI-2 receptors suggests a non-quorum sensing role for luxS in most bacteria. *BMC Microbiology 8*: 119.
34. Hense, B.A., Kuttler, C., Muller, J. et al. (2007). Does efficiency sensing unify diffusion and quorum sensing? *Nature Reviews. Microbiology 5*: 230–239.
35. Baumgartner, L.K., Reid, R.P., Dupraz, C. et al. (2006). Sulfate reducing bacteria in microbial mats: changing paradigms, new discoveries. *Sedimentary Geology 185*: 131–145. https://doi.org/10.1016/j.sedgeo.2005.12.008.
36. Thiel, V., Vilchez, R., Sztajer, H. et al. (2009). Identification, quantification, and determination of the absolute configuration of the bacterial quorum-sensing signal autoinducer-2 by gas chromatography-mass spectrometry. *ChemBiochem 10*: 479–485.
37. Antunes, L.C.M., Ferreira, R.B.R., Buckner, M.M.C. et al. (2010). Quorum sensing in bacterial virulence. *Microbiology 156*: 2271–2282. https://doi.org/10.1099/mic.0.038794-0.
38. Brint, J.M. and Ohman, D.E. (1995). Synthesis of multiple exoproducts in Pseudomonas aeruginosa is under the control of RhlR-RhlI, another set of regulators in strain PAO1 with homology to the autoinducer responsive LuxR-LuxI family. *Journal of Bacteriology 177*: 7155–7163.
39. Dobretsov, S., Teplitski, M., Bayer, M. et al. (2011). Inhibition of marine biofouling by bacterial quorum sensing inhibitors. *Biofouling 27*: 893–905.
40. von Bodman, S.B., Bauer, W.D., and Coplin, D.L. (2003). Quorum sensing in plant-pathogenic bacteria. *Annual Review of Phytopathology 41*: 455–482.
41. Haudecoeur, E. and Faure, D. (2010). A fine control of quorum-sensing communication in Agrobacterium tumefaciens. *Communicative & Integrative Biology 3*: 84–88.
42. Maddula, V.S.R.K., Zhang, Z., Pierson, E.A. et al. (2006). Quorum sensing and phenazines are involved in biofilm formation by *Pseudomonas chlororaphis* (aureofaciens) strain 30–84. *Microbial Ecology 52*: 289–301.
43. Cao, H., Yang, M., Zheng, H. et al. (2009). Complex quorum-sensing regulatory systems regulate bacterial growth and symbiotic nodulation in Mesorhizobium tianshanense. *Archives of Microbiology 191*: 283–289.
44. Defoirdt, T., Boon, N., Sorgeloos, P. et al. (2008). Quorum sensing and quorum quenching in Vibrio harveyi: lessons learned from in vivo work. *The ISME Journal 2*: 19–26.
45. Yong, Y.C. and Zhong, J. (2013). Impacts of quorum sensing on microbial metabolism and human health. *Advances in Biochemical Engineering/Biotechnology 131*: 25–61.

46. Yong, Y.C., Wu, X.Y., Sun, J.Z. et al. (2015). Engineering quorum sensing signaling of *Pseudomonas* for enhanced wastewater treatment and electricity harvest: a review. *Chemosphere 140*: 18–25.
47. Bhardwaj, A.K., Vinothkumar, K., and Rajpara, N. (2013). Bacterial quorum sensing inhibitors: attractive alternatives for control of infectious pathogens showing multiple drug resistance. *Recent Patents on Anti-Infective Drug Discovery 8*: 68–83.
48. Feng, L., Wu, Z.Y., and Yu, X. (2013). Quorum sensing in water and wastewater treatment biofilms. *Journal of Environmental Biology 43*: 437–444.
49. Parsek, M.R. and Greenberg, E.P. (2005). Sociomicrobiology: the connections between quorum sensing and biofilms. *Trends in Microbiology 13*: 27–33.
50. Sakuragi, Y. and Kolter, R. (2007). Quorum-sensing regulation of the biofilm matrix genes (*pel*) of *Pseudomonas aeruginosa*. *Journal of Bacteriology 189*: 5383–5386.
51. Kang, Y.S. and Park, W. (2010). Contribution of quorum-sensing system to hexadecane degradation and biofilm formation in *Acinetobacter* sp. strain DR1. *Journal of Applied Microbiology 109*: 1650–1659.
52. Xia,S.,Zhou,L.,Zhang,Z.etal.(2012).InfluenceandmechanismofN-(3-oxooxtanoyl)-L-homoserine lactone (C8-oxo-HSL) on biofilm behaviors at the early stage. *Journal of Environmental Sciences 24*: 2035–2040.
53. Frederick, M.R., Kuttler, C., Hense, B.A. et al. (2011). A mathematical model of quorum sensing regulated EPS production in biofilm communities. *Theoretical Biology & Medical Modelling 8*: 8–36.
54. Singh, R., Paul, D., and Jain, R.K. (2006). Biofilms: implications in bioremediation. *Trends in Microbiology 14*: 389–397.
55. Wagner, J., Guimarães, L.B., Akaboci, T.R. et al. (2015). Aerobic granular sludge technology and nitrogen removal for domestic wastewater treatment. *Water Science and Technology 71*: 1040–1046.
56. Yong, Y.C. and Zhong, J.J. (2013). Regulation of aromatics biodegradation by *rhl* quorum sensing system through induction of catechol meta-cleavage pathway. *Bioresource Technology 136*: 761–765.
57. Wang, M.Z., He, H.Z., Zheng, X. et al. (2014). Effect of *Pseudomonas* sp. HF-1 inoculum on construction of a bioaugmented system for tobacco wastewater treatment: analysis from quorum sensing. *Environmental Science and Pollution Research 21*: 7945–7955.
58. Ren, T.T., Li, X.Y., and Yu, H.Q. (2013). Effect of N-acy-l-homoserine lactones- like molecules from aerobic granules on biofilm formation by *Escherichia coli* K12. *Bioresource Technology 129*: 655–658.
59. Li, Y., Lv, J., Zhong, C. et al. (2014). Performance and role of N-acyl-homoserine lactone (AHL)-based quorum sensing (QS) in aerobic granules. *Journal of Environmental Sciences 26*: 1615–1621.
60. Tan, C.H., Koh, K.S., Xie, C. et al. (2014). The role of quorum sensing signaling in EPS production and the assembly of a sludge community into aerobic granules. *ISME Journal 8*: 1186–1197.
61. Lv, J., Wang, Y., Zhong, C. et al. (2014). The effect of quorum sensing and extracellular proteins on the microbial attachment of aerobic granular activated sludge. *Bioresource Technology 152*: 53–58.
62. Drews, A. (2010). Membrane fouling in membrane bioreactors- characterisation, contradictions, cause and cures. *Journal of Membrane Science 363*: 1–28.
63. Cuadrado-Silva, C.T., Castellanos, L., Arévalo-Ferro, C. et al. (2013). Detection of quorum sensing systems of bacteria isolated from fouled marine organisms. *Biochemical Systematics and Ecology 46*: 101–107.
64. Yeon, K.M., Cheong, W.S., Oh, H.S. et al. (2009). Quorum sensing: a new biofouling control paradigm in a membrane bioreactor for advanced wastewater treatment. *Environmental Science & Technology 43*: 380–385.
65. Song, X.N., Cheng, Y.Y., Li, W.W. et al. (2014). Quorum quenching is responsible for the underestimated quorum sensing effects in biological wastewater treatment reactors. *Bioresource Technology 171*: 472–476.

66. Zhang, K., Zheng, X., Shen, D.S. et al. (2015). Evidence for existence of quorum sensing in a bioaugmented system by acylated homoserine lactone-dependent quorum quenching. *Environmental Science and Pollution Research International 22*: 6050–6056.
67. Valle, A., Bailey, M.J., Whiteley, A.S. et al. (2004). N-Acyl- L-homoserine lactones (AHLs) affect microbial community composition and function in activated sludge. *Environmental Microbiology 6*: 424–433.
68. Montgomery, K., Charlesworth, J.C., LeBard, R. et al. (2013). Quorum sensing in extreme environments. *Life 3*: 131–148.
69. Toyofuku, M., Nomura, N., Fujii, T. et al. (2007). Quorum sensing regulates denitrification in *Pseudomonas aeruginosa* PAO1. *Journal of Bacteriology 189*: 4969–4972. https://doi.org/10.1128/JB.00289-07.
70. Natrah, F.M.I., Ruwandeepika, H.A.D., Pawar, S. et al. (2011). Regulation of virulence factors by quorum sensing in *Vibrio harveyi*. *Veterinary Microbiology 154*: 124–129.
71. Ha, C., Kim, S.K., Lee, M.N. et al. (2014). Quorum sensing-dependent metalloprotease VvpE is important in the virulence of *Vibrio vulnificus* to invertebrates. *Microbial Pathogenesis 71–72*: 8–14.
72. Pande, G.S.J., Scheie, A.A., Benneche, T. et al. (2013). Quorum sensing-disrupting compounds protect larvae of the giant freshwater prawn *Macrobrachium rosenbergii* from *Vibrio harveyi* infection. *Aquaculture 406–407*: 121–124. https://doi.org/10.1016/j.aquaculture.2013.05.015.
73. Singh, R.P. (2015). Attenuation of quorum sensing-mediated virulence in Gram- negative pathogenic bacteria: implications for the post-antibiotic era. *Medicinal Chemistry Communications 6*: 259–272.
74. Brackman, G., AlQuntar, A.A., Enk, C.D. et al. (2013). Synthesis and evaluation of thiazolidinedione and dioxazaborocane analogues as inhibitors of AI-2 quorum sensing in *Vibrio harveyi*. *Bioorganic & Medicinal Chemistry 21*: 660–667.
75. Gutiérrez-Barranquero, J.A., Reen, F.J., McCarthy, R.R. et al. (2015). Deciphering the role of coumarin as a novel quorum sensing inhibitor suppressing virulence phenotypes in bacterial pathogens. *Applied Microbiology and Biotechnology 99*: 3303–3316.
76. Jhariya, M.K., Meena, R.S., and Banerjee, A. (2021). *Ecological Intensification of Natural Resources for Sustainable Agriculture*, 655. Singapore: Springer https://doi.org/10.1007/978-981-33-4203-3. Pp 655.
77. Jhariya, M.K., Meena, R.S., Banerjee, A. et al. (2021). *Natural Resources Conservation and Advances for Sustainability*, 623. Elsevier, Academic Press ISBN: 9780128229767.
78. Jhariya, M.K., Banerjee, A., Meena, R.S. et al. (2021). *Sustainable Intensification for Agroecosystem Services and Management*, 870. Singapore: Springer https://doi.org/10.1007/978-981-16-3207-5.
79. Banerjee, A., Meena, R.S., Jhariya, M.K. et al. (2021). *Agroecological Footprints Management for Sustainable Food System*, 514. Singapore: Springer https://doi.org/10.1007/978-981-15-9496-0.
80. Banerjee, A., Jhariya, M.K., Meena, R.S. et al. (2021). Ecological footprints in agroecosystem-an overview. In: *Agroecological Footprints Management for Sustainable Food System* (ed. A. Banerjee, R.S. Meena, M.K. Jhariya and D.K. Yadav), 1–23. Singapore: Springer https://doi.org/10.1007/978-981-15-9496-0_1.
81. Banerjee, A., Jhariya, M.K., Raj, A. et al. (2021). Land footprint management and policies. In: *Agroecological Footprints Management for Sustainable Food System* (ed. A. Banerjee, R.S. Meena, M.K. Jhariya and D.K. Yadav), 221–246. Singapore: Springer https://doi.org/10.1007/978-981-15-9496-0_7.
82. Banerjee, A., Jhariya, M.K., Raj, A. et al. (2021). Energy and climate footprint towards the environmental sustainability. In: *Agroecological Footprints Management for Sustainable Food System* (ed. A. Banerjee, R.S. Meena, M.K. Jhariya and D.K. Yadav), 415–443. Singapore: Springer https://doi.org/10.1007/978-981-15-9496-0_14.
83. Huang, Y., Zeng, Y., Yu, Z. et al. (2013). *Insilico* and experimental methods revealed highly diverse bacteria with quorum sensing and aromatics biodegradation systems–a potential broad application on bioremediation. *Bioresource Technology 148*: 311–316.

84. Jhariya, M.K., Yadav, D.K., and Banerjee, A. (2018). Plant mediated transformation and habitat restoration: phytoremediation an eco-friendly approach. In: *Metallic Contamination and its Toxicity* (ed. A. Gautam and C. Pathak), 231–247. New Delhi: Daya Publishing House, A Division of Astral International Pvt. Ltd.
85. Banerjee, A., Jhariya, M.K., Yadav, D.K. et al. (2018). Micro-remediation of metals: a new frontier in bioremediation. In: *Handbook of Environmental Materials Management* (ed. C. Hussain), 1–36. Cham: Springer https://doi.org/10.1007/978-3-319-58538-3_10-1.
86. Whiteley, M., Diggle, S.P., and Greenberg, E.P. (2017). Bacterial quorum sensing: the progress and promise of an emerging research area. *Nature 551* (7680): 313–320.
87. Raj, A. and Jhariya, M.K. (2021). Site quality and vegetation biomass in the tropical Sal mixed deciduous forest of Central India. *Landscape and Ecological Engineering 17*: https://doi.org/10.1007/s11355-021-00450-1.
88. Raj, A. and Jhariya, M.K. (2021). Carbon storage, flux and mitigation potential of tropical Sal mixed deciduous forest ecosystem in Chhattisgarh, India. *Journal of Environmental Management 293* (1): 112829. https://doi.org/10.1016/j.jenvman.2021.112829.
89. Khan, N., Jhariya, M.K., Raj, A. et al. (2021). Soil carbon stock and sequestration: implications for climate change adaptation and mitigation. In: *Ecological Intensification of Natural Resources for Sustainable Agriculture* (ed. M.K. Jhariya, R.S. Meena and A. Banerjee), 461–489. Singapore: Springer https://doi.org/10.1007/978-981-33-4203-3_13.
90. Khan, N., Jhariya, M.K., Raj, A. et al. (2021). Eco-designing for sustainability. In: *Ecological Intensification of Natural Resources for Sustainable Agriculture* (ed. M.K. Jhariya, R.S. Meena and A. Banerjee), 565–595. Singapore: Springer https://doi.org/10.1007/978-981-33-4203-3_16.
91. Jhariya, M.K. and Singh, L. (2021). Herbaceous diversity and biomass under different fire regimes in a seasonally dry forest ecosystem. *Environment, Development and Sustainability 23* (5): 6800–6818. https://doi.org/10.1007/s10668-020-00892-x.
92. Jhariya, M.K. and Singh, L. (2021). Effect of fire severity on soil properties in a seasonally dry forest ecosystem of Central India. *International journal of Environmental Science and Technology 18*: 3967–3978. https://doi.org/10.1007/s13762-020-03062-8.
93. Jhariya, M.K. and Singh, L. (2021). Productivity and carbon accumulation in tropical dry deciduous forest ecosystem. In: *Tropical Dry Forests: Emerging Features and Ecological Perspectives* (ed. R.K. Chaturvedi, R. Singh and R. Bhadouria), 269–287. NY USA: Nova Science.
94. Michalet, R., Brooker, R.W., Cavieres, L.A. et al. (2006). Do biotic interactions shape both sides of the hump-back model of species richness in plant communities? *Ecology Letters 9*: 767–773.
95. Strickland, M.S., McCulley, R.L., and Bradford, M.A. (2013). The effect of a quorum-quenching enzyme on leaf litter decomposition. *Soil Biology and Biochemistry 64*: 65–67. https://doi.org/10.1016/j.soilbio.2013.04.008.
96. Maddela, N.R., Sheng, B., Yuan, S. et al. (2019). Roles of quorum sensing in biological wastewater treatment: a critical review. *Chemosphere 221*: 616–629. https://doi.org/10.1016/j.chemosphere.2019.01.064.
97. Hmelo, L. and van Mooy, B.A.S. (2009). Kinetic constraints on acylated homoserine lactone-based quorum sensing in marine environments. *Aquatic Microbial Ecology 54*: 127–133.
98. Handelsman, J. (2008). Metagenomics is not enough. *DNA and Cell Biology 27*: 219–221.
99. Raj, A., Jhariya, M.K., Khan, N. et al. (2021). Ecological intensification for sustainable development. In: *Ecological Intensification of Natural Resources for Sustainable Agriculture*, 137–170. Singapore: Springer https://doi.org/10.1007/978-981-33-4203-3_5.
100. Raj, A., Jhariya, M.K., and Khan, N. (2021). The importance of forest for soil, food and climate security in Asia. In: *Biodiversity, Conservation and Sustainability in Asia*, South and Middle Asia, vol. *2* (ed. M. Ozturk, S.M. Khan, V. Altay, et al.). Switzerland: Springer https://doi.org/10.1007/978-3-030-73943-0_3.

6

Endophytic Microbes: Potential Source of Antibiotic Production

Jyoti Choudhary[1,2], Sarad Kumar Mishra[3], Zainab Rao[1], and Swati Shukla[1]

6.1. INTRODUCTION

Over the last several decades, emphasis on drug discovery and developmental processes from natural sources has increased. About 28% of new chemical elements and about 42% of anticancer drugs that have been identified worldwide from 1981 to 2006 were natural products and their derivatives [1]. Along with plants, microorganisms have also been found to be rich sources of bioactive metabolites. As of 2002, approximately 22 000 bioactive compounds have been reported from microorganisms. Due to the rise of life-threatening viruses and drug-resistant bacteria, complications from disease after organ transplantation, and excessive rise in fungal infections in overall the population, scientists continued to work in this field to deal with these medical issues. Endophytic microbes found in the tissues of living plants remain relatively unexplored and unresearched. Reportedly, these microorganisms could potentially be rich sources of natural bioactive products for medicinal use, industry, and agriculture. There are approximately 300 000 species of plants on Earth, one or more types of endophytes reside in inter- or intracellular spaces of these plants, creating an immense diversity [2], of which only some of the plant-associated endophytes have been studied. In future, there is a high possibility of finding new endophytic species and their related natural products. These plant-associated endophytes play important roles in enhancing nutrient availability, producing bioactive compounds, and directly or indirectly influencing plant growth and survival against infection by phytopathogens, etc. [2–4].

Endophytes grow inside plants and have been found to be useful sources of biologically active secondary metabolites, for example alkaloids, phenolic acids, tannins, and terpenoids. These secondary metabolites produce antimicrobial, cytotoxic, anti-inflammatory, and antioxidant activities. Endophytes and host plants have special relationships with each other ranging from symbiotic to pathogenic, which influence the production of metabolic products

[1]*Department of Biotechnology, Chinmaya Degree College, Haridwar, Uttarakhand, India*

[2]*Department of Botany and Microbiology, Gurukul Kangri Vishwavidyalya, Haridwar, Uttarakhand, India*

[3]*Department of Biotechnology, Deen Dayal Upadhyaya Gorakhpur University, Gorakhpur, Uttar Pradesh, India*

Microbial Biotechnology: Role in Ecological Sustainability and Research, First Edition.
Edited by Pankaj Chowdhary, Sujata Mani, and Preeti Chaturvedi.

in plants. Endophytes' applications in medicine and human health can be helpful in developing new antibiotics against pathogens that have developed resistance to many currently available antibiotics. It has been found that from the 300 000 plant species, one or more endophytes are found in most plants. The plant hosts of endophytic microbes do not show symptoms of disease, at least during the endophytic phase of their life cycle [5]. Due to the emergence of antibiotic resistance in microbes against several drugs, the production of more effective antibiotics is necessary. Endophytes have been found useful in developing effective antibiotics against most pathogens, as endophytes produce a variety of bioactive compounds that can be helpful in the elimination of various diseases. This chapter intends to provide a list of various endophytes, the secondary metabolites isolated from them, and various antibiotics that have been produced to serve human needs.

6.1.1. Definition

The term endophytes has been defined in various ways by different scientists. Debarry came up with endophytes as a term to describe fungi that live in the tissue of epiphytes [6]. Carroll used it for those microorganisms that cause asymptomatic infection within plant tissues, excluding pathogenic and mycorrhizal fungi [7]. Petrini modified Carroll's definition to include all microorganisms that reside in plant tissues without causing apparent harm to hosts throughout their entire life cycles [8]. Wilson further expanded it and included both fungi and bacteria, which for all or part of their life cycle invade living plants and cause asymptomatic infections within plant tissues, but cause no disease [5]. Later, Azevado defined all microorganisms residing inside plants without producing visible symptoms as "endophytes" [9]. According to Stone, the term "endophyte" is an all-encompassing topographical term that includes all organisms that, during a variable period of their life, occupy the living tissues of their hosts without any symptoms [10]. However, Bacon and White more precisely defined that an endophytic fungus lives in a mycelial form in biological association with the living plant, at least for some period [11].

Although by definition endophytes do not cause disease in their hosts, they are obligatorily heterotrophic and are often congeneric with a taxonomically diverse range of pathogenic microorganisms.

6.1.2. Origin and Evolution of Endophytes

It has known for over a hundred years that bacterial endophytes exist: fossilized tissues of stems and leaves indicate the presence of microorganisms associated with plants, which proved that associations of plant and microbe may have evolved around the time when higher plants first appeared on earth [12]. According to Rodriguez and Redmann, the symbiotic association of plants and microbes most probably belongs to an earlier time, since vascular plants emerged [13]. Under certain environmental conditions, various microbes penetrate the plant tissues and actively invade the cell wall by hydrolyzing, using hydrolytic enzymes such as pectinase and cellulase through the wound [14–16]. It is believed that some microbial endophytes have originated from the rhizospheric plants that entered through root hairs and colonized the root tissues [17].

During long-term simultaneous evolution of endophytic microbes with the plant host, the endophytes developed adaptation toward the microenvironment within the plant tissues, which may involve symbiotic association between plant host and endophytes or exchange of genes between host and residing microbes [18, 19]. Such exchange of properties has been observed in the production of taxol from the endophytic fungi that were obtained from the Yew (*Taxus*) tree species [18].

6.1.2.1. Endophytes and their Biodiversity

Endophytes are ubiquitously present in plants from tropical, temperate, and boreal forests with the host plants being anything from herbaceous to woody trees in different habitats, ranging from extreme arctic alpine and xeric environments to mesic temperate and tropical forests. The most biologically diverse terrestrial ecosystems are tropical and temperate rainforests. Even though the most endangered of these locations only span 1.44% of the land's surface, they are home to more than 60% of the world's terrestrial biodiversity [20]. As a result, one may expect regions with significant plant endemicity to contain particular endophytes that evolved with the endemic plant species. Various endophytic bacterial strains have been reported among several healthy plant species, for instance, in crop plants such as rice, wheat, tomato, carrot, potato, citrus, different woody trees, ferns, and club mosses. Roots are the host to more bacterial populations in comparison to leaves and stems. More than 40 new taxa have been found to date by polyphasic taxonomic approaches, including four new genera: *Plantactinospora; Actinophytocola; Phytohabitans;* and *Jhisengella*. However, the list is no longer complete (Table 6.1) because there is a lot of interest in this topic and new endophytes are being discovered all the time. Based on DNA, the majority of these published species shared more than 97% of their DNA with their closest known relatives on the basis of 16S rRNA gene sequence analysis.

Table 6.1 Reported endophytic bacteria and their hosts.

Endophytes	Plant Species	References
Actinophytocola oryzae	*Oryza sativa* L. çv. RD6	[21]
Azoarcus sp.	Kallar grass, rice	[22, 23]
Azorhizobium caulinodans	Rice	[22]
Azospirillum brasilense	Banana	[24]
Burkholderia cepacian	Yellow lupine, citrus plants	[25, 26]
Burkholderia pickettii	Maize	[27]
Burkholderia sp.	Banana, pineapple, rice	[22, 24]
Jiangella alba J.	Maytenus austroyunnanensis	[28]
Methylobacterium mesophilicum	Citrus plants	[29]
Micromonospora lupine	*Lupinus angustifolius*	[30]
Micromonospora pisi	*Pisum sativum*	[31]
Micromonospora saelicesensis	*Lupinus angustifolius*	[30]
Pseudonocardia adelaidensis	*Eucalyptus microcarpa*	[32]
Pseudonocardia artemisiae	*Artemisia annua* L.	[33]
Pseudonocardia eucalypti	*Eucalyptus microcarpa*	[32]
Rhizobium (Agrobacterium) radiobacter	Carrot, rice	[34]
Streptomyces artemisiae	*Artemisia annua* L.	[33]
Chromobacterium violaceum	Rice	[35]
Herbaspirillum seropedicae	Banana, rice, sorghum, sugarcane, maize	[22, 36]
Herbaspirillum rubrisulbalbicans	Sugarcane	[36]
Klebsiella pneumoniae	Soybean	[37]
Klebsiella variicola	Banana, rice, maize, sugarcane	[38]
Klebsiella terrigena	Carrot	[34]
Klebsiella Oxytoca	Soybean	[37]
Antoe as.p	Rice, soybean	[37, 39]
Bacillus spp.	Citrus plants	[25, 29]
Bacillus megatarium	Maize, carrot, citrus plants	[25, 27, 34]
Clostridium	Grass *Miscanthus sinensis*	[40]
Paenibacillus odorifer	Sweet potato	[41]

Endophytes includes both bacteria and fungi. More than 200 genera from 16 phyla are found to be endophytes ranging from Gram-positive to Gram-negative bacteria, for example *Agrobacterium, Acinetobacter, Achromobacter, Bacillus, Brevibacterium, Chromobacterium, Enterobacter, Methylobacterium, Pseudomonas, Streptomyces*, etc. Most of the bacterial endophytes are classified as phyla *Actinobacteria, proteobacteria,* and *firmicutes* [42]. It has been found that *Streptomyces* sp. produce approximately 76% of the total metabolites in the form of antibiotics. Two types of endophytic fungi have been isolated to date; the first is the clavicipitaceous endophytes found in grass, and the second is the non-clavicipitaceous endophytes that reside in vascular plants, ferns, and angiosperms. The major endophytic fungi isolated belong to the group ascomycota and basidiomycota.

6.2. ENDOPHYTIC MICROBES AND SECONDARY METABOLITES

New antibiotics, chemotherapeutic medicines, and agrochemicals that are extremely effective, have minimal toxicity, and have a low environmental impact are in high demand because of the development of resistance in pathogenic organisms (*Staphylococcus*, *Mycobacterium*, and *Streptococcus*) to known drugs, as well as the threat of naturally resistant organisms. Natural products and the organisms that produce these products provide prospects for drug development innovation. Therefore, wild unexplored territories of endophytes need to be explored for their potential usefulness.

Endophytes offer great opportunities for discovery of novel antimicrobial compounds effective against antibiotic-resistant bacteria (Methicillin-resistant *Staphylococcus aureus,* Vancomycin-resistant *Enterococcus,* and multidrug-resistant *Mycobacterium tuberculosis*). The novel antimicrobial compounds can also replace the existing important antibiotics, which are costly and known to have nephrotoxicity and ototoxicity. They possess a variety of biological activities such as immunosuppressive, inhibitors of enzyme, antimicrobial agents, etc. A plant-derived natural medicine is usually not up to an appropriate level since it is produced at a given developmental stage or under specific environmental conditions, stress, or nutrient availability. Furthermore, it can take a long time for plants to reach a proper development phase for product accumulation and extraction. In light of these limitations, endophytes may be a viable alternative source of bioactive natural chemicals. *Pseudomonas*, *Burkholderia*, and *Bacillus* sp. are only a few of the typical soil bacterial genera that have endophytes. [43]. Antibiotics, anticancer, volatile chemical compounds, antifungal, antiviral, insecticidal, and immunosuppressive substances are among the secondary metabolites produced by these species. Endophytic organisms have yielded a diverse spectrum of biologically useful chemicals. Many endophytic bacteria and fungi specially isolated from medicinal plants have been reported as having potential to kill or inhibit pathogenic bacteria, fungi, and viruses are among the dangerous microorganisms. As a result, there is a good probability that endophytic bacteria might be used to manufacture antimicrobial products. Various antimicrobial compounds reported from endophytic microbes are shown in Tables 6.2 and 6.3. Various metabolites have been used as antibiotics, such as bleomycin, adriamycin, mithramycin, and daunomycin, which were used as antitumor compounds. Endophytic fungi are important sources of antibiotics like penicillin, tetracycline, cephalosporin, and immunosuppressive agents. The antitumor agent taxol was produced from the fungus *Taxomyces andreanae*. Fungi imperfecti and ascomycetes, the most common makers of secondary metabolites, have been discovered producing about 6400 compounds. *Penicillium* species produce 900 compounds, *Aspargillus* species produce 950 compounds, and *Fusarium* species produce 350 compounds. Basidiomyctes or mushrooms produce 2000 active compounds; 140 metabolites from yeast and 60 from Myxomycetes

Table 6.2 List of reported antimicrobial compounds isolated from endophytic fungi.

Host Plant	Endophytic Fungi	Compound	Activity	References
Tripterygium wilfordii	*Rhinocladiella* sp.	Cytochalasin	Antibiotic	[44]
Eucryphia cordifolia	*Muscodor albus* & *Gliocladium* sp.	Volatile organic compounds	Antibiotic	[45]
Scapania verrucosa	*Chaetomium fusiforme*	Acetic acid, valeric acid, 3methyl-, methyl ester, butane-2,3-diol	Antifungal & antitumor	[46]
Quercus sp.	*Cytonaema* sp.	Cytonic acid A&B	Antiviral	[47]
Terminalia morobensis	*Pestalotiopsis. Microspora*	Pestacin & isopestacin	Antimicrobial and Antioxidant	[48, 49]
Cinnamomum zeylanicum	*Muscodor albus*	Volatile organic compound	Antifungal & antibacterial	[50]
Taxus cuspidata	*Periconia* sp.	Periconicins A &B	Antibacterial	[51]
Citrus limon	*Penicillium digitatum, P. citrinum*	Tryptoquialanine A, Tryptoquialanine C, 15-dimethyl-2-epi-fumiquinazoline A, deoxytryptoquialanone, Citrinadin A, Deoxycitrinadin A, Chrysogenamide A	Antifungal	[52]
Phyllanthus reticulatus	*Geotrichum candidum*	–	Antibacterial and Antifungal	[53]

have been isolated [65]. Some culturable endophytic bacteria and their novel natural products with biotechnological applications (Figure 6.1) and their diverse biological activities have been studied.

6.2.1. Endophytes as a Source of Antibiotics

6.2.1.1. Endophytic Fungi as a Source of Antibiotics

Endophytic fungus has been found to produce a diverse range of bioactive chemicals. Endophytic fungi, *Taxomyces andreanae*, of the yew plant *Taxus brevifolia* Nutt, developed Taxol, a strong anticancer medicine. [18]. Though the amount of taxol produced by endophytic fungi is minimal in comparison to that produced by plants, the fungi's short generation period and fast growth rate make them ideal candidates for drug manufacture. In India, taxol was isolated from two endophytes, *Colletotrichum gloeosporioides* and *Bartalinia robillardoides*, both of which have potent cytotoxic properties [67].

The extract from the culture broth of endophytic fungi *Trichoderma* sp. isolated from Vinca plants, found in Iran, exhibited strong activity against the pathogenic fungi *Pyricularia oryzae*, *Aspergillus fumigatus,* and *Botrytis cinera*. The compound was identified as trichodermin by using the ^{1}H-NMR and 13C-NMR spectroscopic technique [68]. Some important antimicrobial compounds isolated from various endophytic fungi, including their activity, are shown in Table 6.2.

The antibacterial activity of javanicin, a naphthaquinone produced by the endophytic *Chloridium* sp. of *Azadirachta indica* under liquid and solid media culture, was reported

Table 6.3 List of bioactive metabolites with potent activity from endophytic bacteria.

Endophyte	Host	Chemical nature	Compound	Activity	References
Streptomyces sp. strain NRRL 30562	*Kennedia nigriscans*	Peptide	Munumbicin A, B, C and D	Antibiotic	[54]
Streptomyces sp. GT 2002/1503	*Bruguiera gymnorrhiza*	Pentacyclic indolosesquiterpine	Xiamycin A	Anti HIV	[55]
Pseudomonas viridiflava EB 273	*Lactuca saliva*	Lipopeptides	Ecomycine B, C	Antifungal	[56]
Streptomyces sp. HK 10595	*Kennedelia candel*	Pentacyclic indolosesquiterpine	Xiamycin, Indosespine, and Sespenine	Antibacterial	[57]
Pseudomonas syringae	*Nicotiana benthamiana*	Lipopeptides	Pseudomycins A, B, C and D	Antifungal	[58]
Streptomyces sp. strain NRRL 30566	*Greevillea pteridifolia*	Peptide	Kakadumycin	Antibiotic	[59]
Streptomyces sp. MSU-2110	*Monstera* sp.	Peptide	Coronamycins	Antibiotic	[50]
Streptomyces hygroscopicus TP-A0451	*Pteridium aquilinum*	δ-lactone	Pterocidin	Antitumor	[60]
Streptomyces sp. NRRL 30562	*Kennedia nigriscans*	Peptide	Munumbicins E-4 and E-5	Antibiotic	[54]
Streptomyces sp. TP-A0456	*Cryptomeria japonica*	Butyrolactones	Cedarmycins A and B	Antifungal	[61]
Streptomyces sp. MaB-QuH-8	*Maytenus aquifolia* Mart.	Heterocyclic compounds	Celastramycins A and B	Antimicrobial	[62]
Micromonospora lupini	*Lupinus angustifolius*	Anthraquinones	Lupinacidins A and B	Antitumor	[63]
Streptomyces sp. SUCI	*Ficus benjamina*	Phenols	Lansai B and C	Weakly anticancer and anti-inflammatory	[64]
Streptomyces sp. TP-A0595	*Allium tuberosum*	Alkaloids	6-Prenylindole	Antifungal	[61]

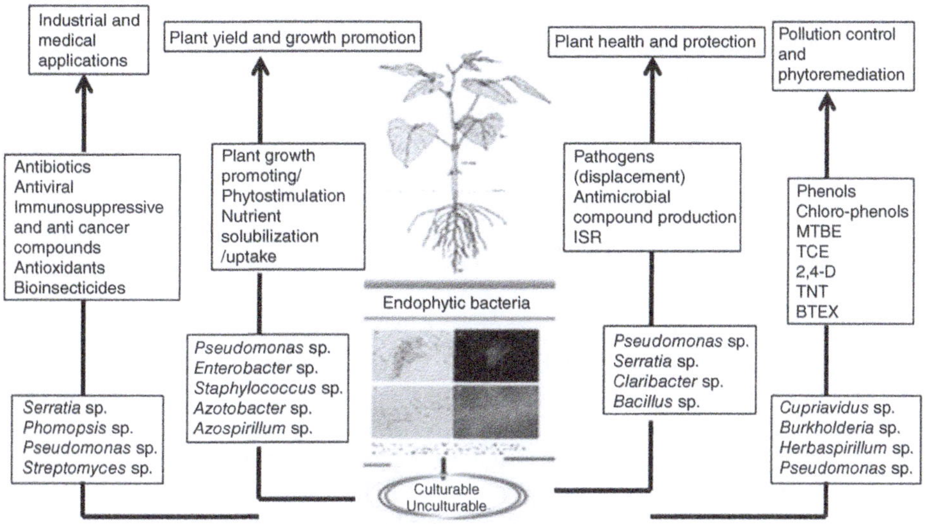

Figure 6.1 Schematic diagram of the different plant–bacterial endophyte interactions that have been studied and their applications [66].

to be robust against *Pseudomonas* sp., which exhibit both human and plant pathogenic properties [69].

The imperfect stage of the fungus *Pezicula cinnamomea* linked with hardwood species in Europe is *Cryptosporiopsis quercina*. It was discovered as an endophyte on *Tripterigeum wilfordii*, a medicinal plant. *Cecidomyia quercina* (produces cryptocandin, an antimycotic) had antifungal activity against *Candida albicans* and *Trichophyton* sp., i.e. two human fungal pathogens [2]. This cryptocandin molecule contains a number of unusual hydroxylated amino acids as well as 3-hydroxy-4-hydroxymethylproline, a new amino acid. Antimycotics, echinocandins, and pneumocandins are all connected to this bioactive molecule. Many bioactive compounds have been produced from *C. quercina*. Cryptocandin also acts against various plant pathogenic fungi such as *Sclerotinia sclerotiorum* and *Botrytis cinerea*. Cryptocandin is also used against a number of fungal diseases of the skin and nails. A unique tetrameric acid, cryptocin was also isolated from *C. quercina*, which possesses potent activity against *Pyricularia oryzae* and is also active against various human pathogenic fungi. [70, 71].

Pestaliotiopsis microspora, a rainforest endophyte, is the source of many secondary metabolites. Ambuic acid, an antifungal agent, is one of the metabolites identified from *P. Microspora*. *P. Microspora* has also been isolated from *Torreya taxifolia*, an endangered tree that contains various antifungal chemicals including pestaloside, an aromatic glucoside, and two pyrones (pestalopyrone and hydroxypestalopyrone). Phytotoxic activity is also present in these products. Pestalotiopsins A and B, two novel caryophyllene sesquiterpenes, were isolated from *P. microspora*, which is an endophyte of *T. brevifolia*. The type of products separated from this endophyte vary, depending on the cultural circumstances and the origin of the plant [72–74].

A non-peptidal fungal metabolite (L-783, 281) was isolated from an endophytic fungus *Pseudomassaria* sp. collected from an African rainforest near Kinshasa in the Democratic Republic of the Congo. It acts as an insulin mimetic but, unlike insulin, does not get destroyed in the digestive tract and is given orally. In two mice models of diabetes, oral delivery of this chemical resulted in a considerable reduction in blood glucose levels. [75].

Immunosuppressive medications are currently used to avoid allograft rejection in transplant patients, but they may also be used to treat autoimmune disorders, including insulin-dependent diabetes and rheumatoid arthritis in the future. *Fusarium subglutinans*, an endophytic fungus identified from *T. Wilfordii*, produces the immunosuppressive compounds diterpene pyrones subglutinols A and B. In the mixed lymphocyte reaction (MLR) and thymocyte proliferation (TP) assays, subglutinols A and B are equipotent with an IC_{50} of 0.1 μM. In the same assay systems, cyclosporin, which is also an immunosuppressant and a fungal metabolite, was roughly as potent in the MLR assay and 10^4 more potent in the TP assay [76].

An endophyte, *P. microspora*, obtained from a combretaceaous plant, *Terminalia morobensis,* grows in the Sepik River basin in Papua New Guinea and is found to be the source of two compounds, pestacin and isopestacin. Both these compounds exhibit antimicrobial and antioxidant activity as they show structural similarity with flavonoids [48, 49].

The endophytic fungus *Nigrospora sphaerica* (URM-6060) and *Pestalotiopsis maculans* (URM-6061) were identified from healthy leaves of the medicinal plant *Indigofera suffruticosa* Miller, which is used as traditional medicine in Brazil to cure a variety of diseases. Among all endophytic fungi isolated from *I. suffruticosa*, *N. sphaerica* and *P. maculaus* demonstrated the best antibacterial activity against both Gram-negative and Gram-positive bacteria [77]. In addition, antimicrobial compounds such as clavatol, chaetomugilin D, guignardic acid, colletotric acid, virdicatol, enfumafugin, jesterone, pestacin, javanicin, ecomycins, pseudomycins, pestalopyrone, etc., producing endophytic fungi has also been reported.

6.2.1.1.1. Muscodor Albus *as a Source of Volatile Antibiotics*

Muscodor albus is an endophytic fungus that grows on the small limbs of the *Cinnamomum zeylanicum* (Cinnamon tree). To isolate such endophytic fungus, all the plant tissues were collected and placed in a plastic box and the endophytic fungus growth was studied post incubation. It was concluded that only one of the fungus isolates grew and this fungus thus inhibited the growth of other fungus by producing volatile compounds. Thus, endophytic fungus can make volatile compounds with biological activity. Gas chromatography-mass spectrometry (GC-MS) was used to identify the volatile chemicals released by the fungus. The presence of at least 28 volatile compounds was observed, belonging to five general classes of organic substances: alcohols, lipids, esters, ketones, and acids. All these five classes possess some inhibitory effect against the test fungi and bacteria. The esters were the most potent inhibitory chemical class. Commercially, mycofumigation of the fungus *M. albus* has been used to manage human and plant infections. After *M. albus*, other endophytic fungi were also isolated that produces volatile antibiotics. *Muscodor roseus*, an endophytic fungus that produces volatile chemicals, was discovered from a tree species in Australia's Northern Territory. This fungus was shown to be just as effective at inhibiting pathogen growth as *M. albus* [78–80].

6.2.1.2. Endophytic Bacteria as a Source of Antibiotic Production

As many endophytic bacteria possess antimicrobial activity, several antibiotics were developed from actinobacteria like munumbicins A-D, celastramycins A-B, kakadumycins, and demethylnovobiocins [54, 59, 61, 62].

6-Prenylindole, a simple chemical discovered from the endophytic bacterium *Streptomyces* sp. TPA0595, has antifungal action against the plant disease *Fusarium oxysporum.*

Two chemicals, cedarmycins A and B, were identified from the strain *Streptomyces* sp. TP-A0456, which was collected from a cedar twig. Cedarmycins A had antifungal activity in vitro against *Candida glabrata*, with a MIC of 0.4 μg/ml [61].

Munumbicins E-4 and E-5, two chromophoric peptide antibiotics, were identified from endophytic *Streptomyces* NRRL 30562, which also produced the broad range antibiotics munumbicins A-D. Both compounds were found to have wide antibacterial action against both Gram-positive and Gram-negative microorganisms [81].

From the endophytic *streptomyces* sp. Hedaya48, an antimycotic compound saadamycin was isolated, which showed antimycotic activity against dermatophytes and other clinical fungi [82].

The peptide antibiotics kakadumycins were discovered in the endophytic bacterium *Streptomyces* NRRL30566, found in the fern-leaved Grevillea tree. *Grevillea pteridifolia* is native to Australia's northern area. Kakadumycin A is efficient against *P. Falciparum* and has antibacterial action similar to Munumbicins. Echinomycin, a quinoxaline antibiotic and possible anticancer medicine discovered from *Streptomyces* sp., is chemically related to kakadumycin [59, 83, 84].

The indolo sesquiterpenes identified from prokaryotes are known as xiamycins. Strain GT2002/1503 of the endophyte *Streptomyces* sp. was isolated from the mangrove plant *Bruguiera gymnorrhiza*. Xiamycin-A, a novel pentacyclicindolosesquiterpene, and its methyl ester-2 were discovered. Xiamycin-A has anti-HIV action that is selective. Three novel indolosesquiterpenes, xiamycin B, indosespene, and sespenine, were discovered in *Streptomyces* sp. strain HKI0595, a bacterial endophyte of the ubiquitous mangrove tree *Kandelia candel*. Xiamycins have antibacterial properties against a variety of microorganisms, including methicillin-resistant bacteria *S. aureus* and Vancomycin-resistant *Enterococcus faecalis* [55, 57].

Streptomyces sp. (MSU-2110) was isolated from *Monstera* sp., an epiphytic vine. Coronamycin, a new peptide molecule, was isolated from *Streptomyces* sp. (MSU-2110). With an IC50 of 9.0 ng ml^{-1}, this chemical is active against the malaria pathogen *Plasmodium falciparum*. Coronamycin is also effective against pythiaceous fungi and *Cryptococcus neoformans*, a human fungal pathogen [50]

Maytansinoids are anticancer agents identified from the plant-associated actinomycetes *Actinosynnema pretiosum* (19-membered macrocyclic lactams linked to ansamycin antibiotics) [85].

Another maytansinoid, ansacarbamitocins, was discovered in the soil actinomycete *Amycolatopsis* CP2808 of the *Pseudonocardiaceae* family. *A. Pretiosum,* which produces ansamitocin, also belongs to the family *Pseudonocardiaceae* [86].

Various species of *Pseudomonas* have been reported that produce phytotoxic compounds and antibiotics. *Pseudomonas viridiflava* is a bacterium that is found in the tissues of several grass species and has been isolated to make ecomycins. The ecomycins belong to lipopeptides, which possess molecular weights of 1153 and 1181 Da. The structure of ecomycins contain various amino acids such as serine, glycine, alanine, threonine, homoserine, and β-hydroxyaspartic acid. Ecomycin is a chemical that fights human pathogenic fungus like *C. neoformans* and *C. albicans* [56].

Pseudomycins are peptides with antifungal properties produced by plant-associated *Pseudomonad*. Pseudomycins are cyclic depsipeptides made by acylation of the OH group of the N-terminal serine with the carboxyl group of L-terminal chlorothreonine's carboxyl group. Pseudomycin works against the plant and human pathogenic fungus, i.e. *C. albicans, C. neoformans, Ceratocystis ulmi,* and *Mycosphaerella fijiensis*. L-chlorothreonine, L-hydroxyaspartic acid, and D- and L-diaminobutryic acid are all found in pseudomycins [58, 87].

Bacillus cereus, an endophytic microbe isolated from *Adhatoda beddomei*, was used to make silver nanoparticles (AgNps). After a three-day incubation period at room temperature, the endophytic bacterium reduced AgNO3 solution to produce these nanoparticles. Antibacterial activity of AgNps was found to be adequate against a few harmful bacteria, including *E. coli, Pseudomonas aeruginosa,* and *S. aureus* [88].

Plectranthus tenuiflorus root, stem, and leaves were used to isolate endophytic bacteria. From 28 endophytic bacteria, 8 endophytic strains exhibited antimicrobial properties against human pathogenic microorganisms like *E. coli*, *S. aureus*, *Klebsiella pneumonia*, *Streptococcus agalactiae*, *Proteus mirabilis*, and *C. albicans*. These eight endophytic isolates were identified as *Bacillus* sp., *Bacillus megaterium*, *Bacillus pumilus*, *Bacillus licheniformis*, *Micrococcus luteus*, *Paenibacillus* sp., *Pseudomonas* sp., and *Acinetobacter calcoaceticus* [89].

Flacourtia jangomas bacterial endophytes demonstrated excellent broad range antibacterial efficacy against both Gram-positive and Gram-negative bacteria. The bacterial endophytes isolated from the fruit of *F. jangomas* (FjF_1 and FjF_2) showed best results, which was further identified as *Bacillus subtilis* and *B. cereus* after molecular characterization [90].

Endophytic bacteria have yielded a vast number of bioactive substances that have found applications in medicine. Furthermore, endophytic bacteria found in medicinal plants are more likely to have antibiotic-like bioactive chemicals. In recent years, researchers have focused on the potential of endophytic bacteria to colonize plants and their antimicrobial activity. Endophytic bacteria's interaction with plants and generation of bioactive natural compounds provides a possibility to extract useful medications for use in pharmaceuticals. Table 6.3 lists some key bioactive substances and their activity extracted from endophytic bacteria.

6.3. RATIONALE FOR PLANT SELECTION TO ISOLATE NOVEL ENDOPHYTIC MICROORGANISMS

To provide the best opportunity to identify unique and novel endophytic bacteria, as well as those that create novel bioactive chemicals, it is vital to understand the methods and rationale used. There are several species of plants with great biodiversity. In order to find endophytes, inventive and imaginative tactics must be applied [20]. For the collection of each plant for endophyte isolation, some specific rationales are applied. There are several hypotheses for plant selection strategy [2]:

- Plants from a typical environmental situation, particularly those with distinctive biology and novel survival strategies, can be considered for endophyte isolation.
- Plants with an ethnobotanical history that are used by indigenous peoples for specific purposes and applications.
- Endophytic microbes with novel active natural products are more likely to lodge in plants that are indigenous and have extraordinary longevity, or that colonized a certain ancient land mass such as Gondwanaland, than other plants.
- Endophytic strains with high biodiversity within plants can be examined in plants growing in places with high biodiversity.

6.3.1. Isolation and Cultivation of Endophytes

Following plant selection, a specific protocol for endophyte isolation must be followed, as isolation is the most important step in obtaining pure cultures, and host species, sampling strategy, host-endophyte and interendophyte interactions, tissue type and ages, geographic and habitat distribution, culture conditions, surface sterilant, and selective media all influence endophyte detection and enumeration [16]. Coombs and Franco have examined and introduced endophyte isolation processes in detail, including plant sampling, surface sterilization, and particular medium [91, 92]. Surface sterilization, which kills all surface microorganisms, is the first step in endophytic isolation. It is commonly done by soaking plant tissues in an oxidant or general sterilant for a length of time, then rinsing with sterile water. 70–90% ethanol, sodium hypochlorite (3–10%), and hydrogen peroxide are common surface disinfectants. Surfactants such as Tween 20, Tween 80, and Triton X-100 can be used to improve the efficiency of surface sterilization [91].

The common protocol involves three steps as defined by Coombs and Franco, but a five-step procedure with added sodium thiosulfate solution after being treated with sodium hypochlorite is also recommended to improve cultivation efficiency on media plates due to the fact that thiosulfate can suppress the effect of residual sodium hypochlorite on plant material surfaces, which may kill the endophytes and make them unable to form colonies on the plates. [28]. A 10% sodium bicarbonate solution is also used by some scientists after the treatment of plant tissues with sodium hypochlorite [93]. To prevent fungus from overgrowing the samples and masking the actinobacteria, they bathed the plant tissues in 10% sodium bicarbonate. Plant tissues must be pretreated before endophytes may be isolated. Drying plant components on surface-sterilized plates at 80 °C or 100 °C for 15–30 minutes to kill bacteria is common. The materials were chopped into small shards of around 0.2 × 1.0 cm using aseptic techniques. [39, 92, 94–96]. Plant fragments are transferred to isolation media and cultured for several days after surface sterilization. Soft and sensitive tissues can be pestled and homogenized in a mortar with extraction solution or buffer, followed by bacterial isolation using the gradient dilution method. Gram-positive organisms have been isolated using maceration, vacuum, and pressure bomb procedures [97, 98]. Other scientists have also advised using new combination enzymatic hydrolysis and differential centrifugation bacterial cell enrichment methods [28, 99, 100].

6.3.2. Screening for Antimicrobial Activity

Many of the methods listed in the following subsections have been used to detect antibacterial activity of some endophytes, and are among the many approaches available for evaluating antimicrobial activity.

6.3.2.1. Agar Diffusion Method

This approach has been used to test antimicrobial activity for many years. Fleming employed the agar diffusion method for the first time in 1924 [101]. This approach is commonly used for antimicrobial activity screening, particularly for biologically produced chemicals in which the agar well diffusion assay and the disc assay are included. In this experiment, an antimicrobial chemical was administered to an agar plate using a paper disc or a well. As the drug diffuses into the agar, it creates a concentration gradient that is inversely proportional to the distance between the disc and the well. The degree of inhibition is measured by the size of the inhibition zone surrounding the disc or well (Figure 6.2). This test is generally qualitative and this method requires uniform, rapid, and aerobic growth

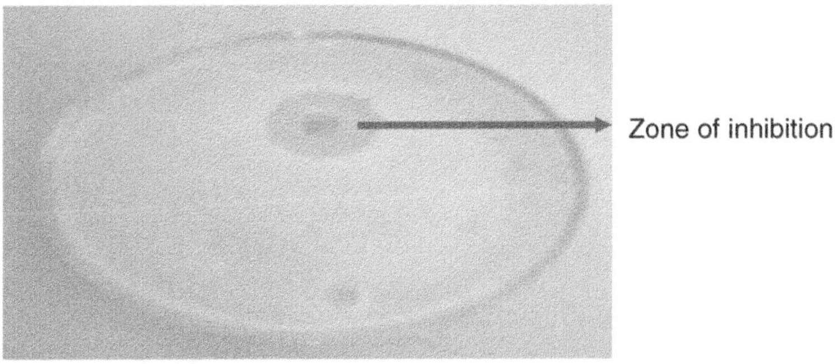

Figure 6.2 Screening of solvent extract of antimicrobial compound produced by endophytes by the agar well diffusion method against pathogenic bacteria.

of the indicator microorganism on agar plate. Piddock reported that highly hydrophobic antimicrobial compound cannot diffuse in agar, hence this method is not suitable to screen hydrophobic antimicrobial compound.

The antimicrobial activity of *Bacillus amyloliquefaciens* (BAC03) was checked through disc diffusion method. A spore suspension of 100 µl of various *Streptomyces* sp. was spread on yeast malt extract agar media, 15 µl of 10^7 CFU ml^{-1} of BAC03 was placed as a drop on a sterile filter paper disc (5 mm diameter), and this disc was placed on agar media. After incubation, the zone of inhibition was measured with the ruler [102].

Sunkar and Nachiyar used *B. cereus*, an endophytic bacterium isolated from the *Adhatoda beddomei*, to make silver nanoparticles (AgNPs). Antibacterial activity of AgNPs against harmful bacteria such as *E. coli, P. aeruginosa,* and *S. aureus* was shown to be the best activity. The antibacterial property of silver nanoparticles was checked by agar well diffusion method [103] in which 100 µl of the test microorganism was spread on the nutrient agar plate with the help of spreader. With the use of a sterile cork borer, wells were created, into which 100 µl of AgNP solution was poured and incubated at 37 °C. For assessing the antibiotic activity of certain microorganisms, several modified approaches based on the agar diffusion method have been used. The agar spot test, postponed antagonism assay, and spot-on grass assay are examples of these processes [104].

Phoma medicaginis (GenBank accession number MK517550) and *Fusarium equiseti* (GenBank accession number MK517551) were isolated from *Mikania cordata* (a perennial vine that is well established in Sri Lanka). Both endophytes were found to have wide antibacterial action against Gram-positive and Gram-negative bacteria. The Kirby-Bauer disc diffusion method was used to test the antimicrobial activity of these fungus [105].

6.3.2.2. Agar and Broth Dilution Method

Quantitative approaches such as agar and broth dilution are used for microorganisms with varying growth rates, as well as anaerobic and microaerophilic bacteria [106]. The results are reported as the minimum inhibitory concentration (MIC), which is the lowest concentration of antimicrobial agent that stops microorganism growth after a specific incubation period. In this test, an antimicrobial drug is serially diluted and then introduced in a single concentration to nonselective broth or melting agar media in a culture tube or plate, which is then inoculated with the test organism and incubated. The inhibitory effects on Gram-positive and Gram-negative bacterial growth were observed to be relatively modest concentrations of MIC values ranging from 0.156 to 0.625 mg ml^{-1} for the endophytic bacterium *B. amyloliquefaciens* [107].

A total of 14 endophytic fungi were obtained from plant samples located in Egypt (Wadi EL-Natrun). Some fungal extracts had the best antibacterial activity against *S. aureus* with MIC values ranging from 12.3 to 31.25 µg ml^{-1}, antibacterial activity against *P. aeruginosa* with MIC values ranging from 13.7 to 31.25 µg ml^{-1}, and antifungal activity against *C. albicans* with MIC values ranging from 35.0 to 125 µg ml^{-1} among these isolated endophytes [108].

6.3.3. Automated Turbidometric Assay

The influence of a drug on the growth or death kinetics of a microbe is determined using a turbidometric test based on an automated system. At concentrations below the MIC, it offers information on the effect of an antibiotic that may produce a delayed lag phase or a lowered growth rate. The method necessitates a highly sensitive instrument since bacterial growth is tracked by measuring the turbidity of the broth medium. It is possible that growth at log 5.0 CFU ml^{-1} will not be detectable [109].

6.4. CONCLUSION AND FUTURE PERSPECTIVE

Endophytes have been discovered to be a significant source of new physiologically active chemicals with a broad range of antibacterial action and considerable structural diversity. Researches in endophytes isolated from plants are gaining industrial as well as biotechnological relevance. Medical plants and its associated endophytes are currently attracting the scientists in their assessment of their potential in pharmaceuticals, agriculture, and medicine. Drug resistance in bacteria, the prevalence of life-threatening viruses, issues with disease in organ donation, and a massive increase in fungal infections among the world's population have all compelled scientists to look for novel antibiotics to address these medical issues. These plant-associated endophytes play an important role in enhancing nutrient availability, producing bioactive compounds, directly or indirectly influencing plant growth, and surviving against infection by phytopathogens. Endophytic bacteria have received a lot of interest in recent years for their ability to produce new and improved antibacterial compounds. It could be due to their close and beneficial relationship with plants, which leads to the creation of a variety of antibacterial chemicals. Antibiotic-producing endophytic bacteria are being exploited in plant pathology as a source for novel antifungal chemicals for biological control of pre- and post-harvest illnesses.

So, researchers who want to enter into this unexplored territory of endophytes have the opportunity to find interesting species having potential for novel natural products. Thus, screening endophytes associated with medicinal plants for bioactive substances would undoubtedly aid researchers in the development of pharmacological and antimicrobial compounds that will assure justifiable human health and action against antibiotic resistance.

REFERENCES

1. Newman, D.J. and Cragg, G.M. (2007). Natural products as sources of new drugs over the last 25 years. *Journal of Natural Products 70* (3): 461–477.
2. Strobel, G. and Daisy, B. (2003). Bioprospecting for microbial endophytes and their natural products. *Microbiology and Molecular Biology Reviews 67* (4): 491–502.
3. Schulz, B., Römmert, A.K., Dammann, U. et al. (1999). The endophyte-host interaction: a balanced antagonism? *Mycological Research 103* (10): 1275–1283.
4. Kloepper, J.W., Lifshitz, R., and Zablotowicz, R.M. (1989). Free-living bacterial inocula for enhancing crop productivity. *Trends in Biotechnology 7* (2): 39–44.
5. Wilson, D. (1995). Endophyte: the evolution of a term, and clarification of its use and definition. *Oikos 1*: 274–276.
6. De Bary, A. (1866). Morphologie und Physiologie der Pilze, Flechten, und Myxomyceten. In: *Hofmeister's Handbook of Physiological Botany*, vol. *Vol. II*. Leipzig, Germany.
7. Carroll, G.C. (1986). The biology of endophytism in plants with particular reference to woody perennials. *Microbiology of the Phyllosphere* 203–222.
8. Petrini, O. (1991). Fungal endophytes of tree leaves. In: *Microbial Ecology of Leaves*, 179–197. New York, NY: Springer.
9. Stone, J.K., Bacon, C.W., and White, J.F. Jr. (2000). An overview of endophytic microbes: endophytism defined. *Microbial Endophytes 25*: 199–236.
10. Azevedo, J.L., Maccheroni, W. Jr., Pereira, J.O. et al. (2000). Endophytic microorganisms: a review on insect control and recent advances on tropical plants. *Electronic Journal of Biotechnology 3* (1): 15–16.
11. Bacon, C.W. and White, J.F. Jr. (2000). Physiological adaptations in the evolution of endophytism in the Clavicipitaceae. In: *Microbial Endophytes*, 251–276. CRC Press.
12. Strobel, G.A. (2003). Endophytes as sources of bioactive products. *Microbes and Infection 5* (6): 535–544.

13. Rodriguez, R.J. and Redman, R.S. (1997). Fungal life-styles and ecosystem dynamics: biological aspects of plant pathogens, plant endophytes and saprophytes. *Advances in Botanical Research 24*: 169–193.
14. Suto, M., Takebayashi, M., Saito, K. et al. (2002). Endophytes as producers of xylanase. *Journal of Bioscience and Bioengineering 93* (1): 88–90.
15. Carrim, A.J., Barbosa, E.C., and Vieira, J.D. (2006). Enzymatic activity of endophytic bacterial isolates of Jacaranda decurrens Cham (Carobinha-do-campo). *Brazilian Archives of Biology and Technology 49*: 353–359.
16. Zhang, H.W., Song, Y.C., and Tan, R.X. (2006). Biology and chemistry of endophytes. *Natural Product Reports 23* (5): 753–771.
17. Misko, A.L. and Germida, J.J. (2002). Taxonomic and functional diversity of pseudomonads isolated from the roots of field-grown canola. *FEMS Microbiology Ecology 42* (3): 399–407.
18. Stierle, A., Strobel, G., and Stierle, D. (1993). Taxol and taxane production by Taxomyces andreanae, an endophytic fungus of Pacific yew. *Science 260* (5105): 214–216.
19. Germaine, K., Keogh, E., Garcia-Cabellos, G. et al. (2004). Colonisation of poplar trees by gfp expressing bacterial endophytes. *FEMS Microbiology Ecology 48* (1): 109–118.
20. Mittermeier, R.A., Myers, N., Mittermeier, C.G. et al. (1999). *Hotspots: Earth's Biologically Richest and most Endangered Terrestrial Ecoregions*. SA, Agrupación Sierra Madre, SC: CEMEX.
21. Indananda, C., Matsumoto, A., Inahashi, Y. et al. (2010). Actinophytocola oryzae gen. Nov., sp. Nov., isolated from the roots of Thai glutinous rice plants, a new member of the family Pseudonocardiaceae. *International Journal of Systematic and Evolutionary Microbiology 60* (5): 1141–1146.
22. Engelhard, M., Hurek, T., and Reinhold-Hurek, B. (2000). Preferential occurrence of diazotrophic endophytes, Azoarcus spp., in wild rice species and land races of Oryza sativa in comparison with modern races. *Environmental Microbiology 2* (2): 131–141.
23. Reinhold-Hurek, B., Hurek, T., Gillis, M. et al. (1993). Azoarcus gen. Nov., nitrogen-fixing proteobacteria associated with roots of Kallar grass (Leptochloa fusca [L.] Kunth), and description of two species, Azoarcus indigens sp. Nov. and Azoarcus communis sp. Nov. *International Journal of Systematic and Evolutionary Microbiology 43* (3): 574–584.
24. Weber, O.B., Baldani, V.L., Teixeira, K.D. et al. (1999). Isolation and characterization of diazotrophic bacteria from banana and pineapple plants. *Plant and Soil 210* (1): 103–113.
25. Araújo, W.L., Maccheroni, W. Jr., Aguilar-Vildoso, C.I. et al. (2001). Variability and interactions between endophytic bacteria and fungi isolated from leaf tissues of citrus rootstocks. *Canadian Journal of Microbiology 47* (3): 229–236.
26. Barac, T., Taghavi, S., Borremans, B. et al. (2004). Engineered endophytic bacteria improve phytoremediation of water-soluble, volatile, organic pollutants. *Nature Biotechnology 22* (5): 583–588.
27. McInroy, J.A. and Kloepper, J.W. (1995). Survey of indigenous bacterial endophytes from cotton and sweet corn. *Plant and Soil 173* (2): 337–342.
28. Qin, S., Zhao, G.Z., Li, J. et al. (2009). Jiangella alba sp. Nov., an endophytic actinomycete isolated from the stem of Maytenus austroyunnanensis. *International Journal of Systematic and Evolutionary Microbiology 59* (9): 2162–2165.
29. Araújo, W.L., Marcon, J., Maccheroni, W. Jr. et al. (2002). Diversity of endophytic bacterial populations and their interaction with Xylella fastidiosa in citrus plants. *Applied and Environmental Microbiology 68* (10): 4906–4914.
30. Trujillo, M.E., Kroppenstedt, R.M., Fernandez-Molinero, C. et al. (2007). Micromonospora lupini sp. Nov. and Micromonospora saelicesensis sp. Nov., isolated from root nodules of Lupinus angustifolius. *International Journal of Systematic and Evolutionary Microbiology 57* (12): 2799–2804.
31. Garcia, L.C., Martínez-Molina, E., and Trujillo, M.E. (2010). Micromonospora pisi sp. Nov., isolated from root nodules of Pisum sativum. *International Journal of Systematic and Evolutionary Microbiology 60* (2): 331–337.
32. Kaewkla, O. and Franco, C.M. (2011). Pseudonocardia eucalypti sp. Nov., an endophytic actinobacterium with a unique knobby spore surface, isolated from roots of a native Australian eucalyptus tree. *International Journal of Systematic and Evolutionary Microbiology 61* (4): 742–746.

33. Zhao, G.Z., Li, J., Qin, S. et al. (2010). Streptomyces artemisiae sp. Nov., isolated from surface-sterilized tissue of Artemisia annua L. *International Journal of Systematic and Evolutionary Microbiology 60* (1): 27–32.
34. Surette, M.A., Sturz, A.V., Lada, R.R. et al. (2003). Bacterial endophytes in processing carrots (*Daucus carota* L. var. sativus): their localization, population density, biodiversity and their effects on plant growth. *Plant and Soil 253*: 381–390.
35. Phillips, D.A., Martinez-Romero, E., Yang, G.P. et al. (2000). Release of nitrogen: a key trait in selecting bacterial endophytes for agronomically useful nitrogen fixation. In: *The Quest for Nitrogen Fixation in Rice* (ed. J.K. Ladha and P.M. Reddy), 205–217. Manila, The Philippines: International Rice Research Institute.
36. Olivares, F.L., Baldani, V.L., Reis, V.M. et al. (1996). Occurrence of the endophytic diazotrophs Herbaspirillum spp. in roots, stems, and leaves, predominantly of Gramineae. *Biology and Fertility of Soils 21* (3): 197–200.
37. Kuklinsky-Sobral, J., Araújo, W.L., Mendes, R. et al. (2004). Isolation and characterization of soybean-associated bacteria and their potential for plant growth promotion. *Environmental Microbiology 6* (12): 1244–1251.
38. Rosenblueth, M., Martínez, L., Silva, J. et al. (2004). Klebsiella variicola, a novel species with clinical and plant-associated isolates. *Systematic and Applied Microbiology 27* (1): 27–35.
39. Verma, V.C., Kharwar, R.N., and Strobel, G.A. (2009). Chemical and functional diversity of natural products from plant associated endophytic fungi. *Natural Product Communications 4* (11): 1934578X0900401114.
40. Miyamoto, T., Kawahara, M., and Minamisawa, K. (2004). Novel endophytic nitrogen-fixing clostridia from the grass Miscanthus sinensis as revealed by terminal restriction fragment length polymorphism analysis. *Applied and Environmental Microbiology 70*: 6580–6586.
41. Reiter, B., Bürgmann, H., Burg, K. et al. (2003). Endophytic nifH gene diversity in African sweet potato. *Canadian Journal of Microbiology 49* (9): 549–555.
42. Golinska, P., Wypij, M., Agarkar, G. et al. (2015). Endophytic actinobacteria of medicinal plants: diversity and bioactivity. *Antonie Van Leeuwenhoek 108* (2): 267–289.
43. Lodewyckx, C., Vangronsveld, J., Porteous, F. et al. (2002). Endophytic bacteria and their potential applications. *Critical Reviews in Plant Sciences 21* (6): 583–606.
44. Wagenaar, M.M., Corwin, J., Strobel, G. et al. (2000). Three new cytochalasins produced by an endophytic fungus in the genus Rhinocladiella. *Journal of Natural Products 63* (12): 1692–1695.
45. Stinson, M., Ezra, D., Hess, W.M. et al. (2003). An endophytic Gliocladium sp. of Eucryphia cordifolia producing selective volatile antimicrobial compounds. *Plant Science 165* (4): 913–922.
46. Guo, L., Wu, J.Z., Han, T. et al. (2008). Chemical composition, antifungal and antitumor properties of ether extracts of Scapania _antoe as Heeg. And its endophytic fungus Chaetomium fusiforme. *Molecules 13* (9): 2114–2125.
47. Guo, B., Dai, J.R., Ng, S. et al. (2000). Cytonic acids a and B: novel tridepside inhibitors of hCMV protease from the endophytic fungus Cytonaema species. *Journal of Natural Products 63* (5): 602–604.
48. Harper, J.K., Ford, E.J., Arif, A.M. et al. (2003). Pestacin: a 1,3-dihydro isobenzofuran from Pestalotiopsis _antoe asp possessing antioxidant and antimycotic activities. *Tetrahedron 59*: 2471–2476.
49. Strobel, G.A., Ford, E., Worapong, J. et al. (2002). Ispoestacin, an isobenzofuranone from *Pestalotiopsis _antoe asp*, possessing antifungal and antioxidant activities. *Phytochemistry 60*: 179–183.
50. Ezra, D., Castillo, U.F., Strobel, G.A. et al. (2004). Coronamycins, peptide antibiotics produced by a verticillate Streptomyces sp.(MSU-2110) endophytic on Monstera sp. *Microbiology 150* (4): 785–793.
51. Kim, S., Shin, D.S., Lee, T. et al. (2004). Periconicins, two new fusicoccane diterpenes produced by an endophytic fungus Periconia sp. with antibacterial activity. *Journal of Natural Products 67* (3): 448–450.

52. Costa, J.H., Wassano, C.I., Angolini, C.F.F. et al. (2019). Antifungal potential of secondary metabolites involved in the interaction between citrus pathogens. *Scientific Reports 9*: 18647.
53. Pai, G. and Chandra, M. (2018). Antimicrobial activity of endophytic fungi isolated from ethnomedicinal plant Phyllanthus reticulatus poir. *The International Journal of Science and Engineering Invention 7*: 40–46.
54. Castillo, U.F., Strobel, G.A., Ford, E.J. et al. (2002). Munumbicins, wide-spectrum antibiotics produced by Streptomyces NRRL 30562, endophytic on Kennedia nigriscansaa. *Microbiology 148* (9): 2675–2685.
55. Ding, L., Münch, J., Goerls, H. et al. (2010). Xiamycin, a pentacyclic indolosesquiterpene with selective anti-HIV activity from a bacterial mangrove endophyte. *Bioorganic and Medicinal Chemistry Letters 20* (22): 6685–6687.
56. Miller, C.M., Miller, R.V., Garton-Kenny, D. et al. (1998). Ecomycins, unique antimycotics from Pseudomonas viridiflava. *Journal of Applied Microbiology 84* (6): 937–944.
57. Ding, L., Maier, A., Fiebig, H.H. et al. (2011). A family of multicyclic indolosesquiterpenes from a bacterial endophyte. *Organic and Biomolecular Chemistry 9* (11): 4029–4031.
58. Harrison, L., Teplow, D.B., Rinaldi, M. et al. (1991). Pseudomycins, a family of novel peptides from Pseudomonas syringae possessing broad-spectrum antifungal activity. *Microbiology 137* (12): 2857–2865.
59. Castillo, U., Harper, J.K., Strobel, G.A. et al. (2003). Kakadumycins, novel antibiotics from Streptomyces sp. NRRL 30566, an endophyte of Grevillea pteridifolia. *FEMS Microbiology Letters 224* (2): 183–190.
60. Igarashi, Y., Miura, S.S., Fujita, T. et al. (2006). Pterocidin, a cytotoxic compound from the endophytic Streptomyces hygroscopicus. *The Journal of Antibiotics 59* (3): 193–195.
61. Igarashi, Y. (2004). Screening of novel bioactive compounds from plant-associated actinomycetes. *Actinomycetologica. 18* (2): 63–66.
62. Pullen, C., Schmitz, P., Meurer, K. et al. (2002). New and bioactive compounds from Streptomyces strains residing in the wood of Celastraceae. *Planta 216* (1): 162–167.
63. Igarashi, Y., Trujillo, M.E., Martínez-Molina, E. et al. (2007). Antitumor anthraquinones from an endophytic actinomycete Micromonospora lupini sp. nov. *Bioorganic and Medicinal Chemistry Letters 17* (13): 3702–3705.
64. Taechowisan, T., Wanbanjob, A., Tuntiwachwuttikul, P. et al. (2009). Anti-inflammatory activity of lansais from endophytic Streptomyces sp. SUC1 in LPS-induced RAW 264.7 cells. *Food and Agricultural Immunology 20* (1): 67–77.
65. Soković, M.D., Glamočlija, J.M., and Ćirić, A.D. (2013). Natural products from plants and fungi as fungicides. In: *Fungicides–showcases of integrated plant disease management from around the world* (ed. M. Nita). InTechOpen.
66. Ryan, R.P., Germaine, K., Franks, A. et al. (2008). Bacterial endophytes: recent developments and applications. *FEMS Microbiology Letters 278* (1): 1–9.
67. Gangadevi, V. and Muthumary, J. (2008). Taxol, an anticancer drug produced by an endophytic fungus Bartalinia robillardoides Tassi, isolated from a medicinal plant, Aegle marmelos Correa ex Roxb. *World Journal of Microbiology and Biotechnology 24* (5): 717–724.
68. Leylaie, S. and Zafari, D. (2018). Antiproliferative and antimicrobial activities of secondary metabolites and phylogenetic study of endophytic Trichoderma species from Vinca plants. *Frontiers in Microbiology 11* (9): 1484.
69. Kharwar, R.N., Verma, V.C., Kumar, A. et al. (2009). Javanicin, an antibacterial naphthaquinone from an endophytic fungus of neem, Chloridium sp. *Current Microbiology 58* (3): 233–238.
70. Walsh, T.A. (1992). *Emerging Targets in Antibacterial and Antifungal Chemothraphy* (ed. J.A. Sutcliffe and N.H. Georgopapadakou), 349–373. London: Chapman & Hall.
71. Li, J.Y., Strobel, G.A., Harper, J.K. et al. (2000). *Organic Letters 2*: 767–770.
72. Li, J.Y., Harper, J.K., Grant, D.M. et al. (2001). *Phytochemistry 56*: 463–468.
73. Lee, J.C., Yang, X., Schwartz, M. et al. (1995). *Chemistry & Biology 2*: 721–727.
74. Pulici, M., Sugawara, F., Koshino, H. et al. (1996). *Organic Chemistry 61*: 2122–2124.
75. Zhang, B., Salituro, G., Szalkowski, D. et al. (1999). *Science 284*: 974–981.

76. Lee, J., Lobkovsky, E., Pliam, N.B. et al. (1995). *Organic Chemistry 60*: 7076–7077.
77. Santos, I.P., Silva, L.C., Silva, M.V. et al. (2015). Antibacterial activity of endophytic fungi from leaves of Indigofera suffruticosa Miller (Fabaceae). *Frontiers in Microbiology 7* (6): 350.
78. Worapong, J., Strobel, G.A., Ford, E.J. et al. (2001). *Muscodor albus* gen. et sp. Nov., an endophyte from Cinnamomum zeylanicum. *Mycotaxon 79*: 67–79.
79. Worapong, J., Strobel, G.A., Daisy, B. et al. (2002). *Muscodor roseus* anna. Nov. an endophyte from Grevillea pteridifolia. *Mycotaxon 81*: 463–475.
80. Strobel, G.A., Dirkse, E., Sears, J. et al. (2001). Volatile antimicrobials from Muscodor albus, a novel endophytic fungus. *Microbiology 147* (11): 2943–2950.
81. Castillo, U.F., Strobel, G.A., Mullenberg, K. et al. (2006). Munumbicins E-4 and E-5: novel broad-spectrum antibiotics from Streptomyces NRRL 3052. *FEMS Microbiology Letters 255* (2): 296–300.
82. El-Gendy, M.M. and El-Bondkly, A.M. (2010). Production and genetic improvement of a novel antimycotic agent, saadamycin, against dermatophytes and other clinical fungi from endophytic Streptomyces sp. Hedaya48. *Journal of Industrial Microbiology and Biotechnology 37* (8): 831–841.
83. Katagiri, K., Yoshida, T., and Sato, K. (1975). Quinoxaline antibiotics. In: *Antibiotics Mechanism of Action of Antimicrobial and Antitumour Agents* (ed. J.W. Corcoran and F.E. Hahn), 234–251. Heidelberg: Springer Verlag.
84. Waring, M.J. and Wakelin, L.P. (1974). Echinomycin: a bifunctional intercalating antibiotic. *Nature 252* (5485): 653–657.
85. Higashide, E., Asai, M., Ootsu, K. et al. (1977). Ansamitocin, a group of novel maytansinoid antibiotics with antitumour properties from Nocardia. *Nature 270* (5639): 721–722.
86. Snipes, C.E., Duebelbeis, D.O., Olson, M. et al. (2007). The ansacarbamitocins: polar ansamitocin derivatives. *Journal of Natural Products 70* (10): 1578–1581.
87. Ballio, A., Bossa, F., Di Giorgio, D. et al. (1994). Novel bioactive lipodepsipeptides from Pseudomonas syringae: the pseudomycins. *FEBS Letters 355* (1): 96–100.
88. Sunkar, S. and Nachiyar, C.V. (2012). Microbial synthesis and characterization of silver nanoparticles using the endophytic bacterium Bacillus cereus: a novel source in the benign synthesis. *Global Journal of Medical Research 12* (2): 43–50.
89. El-Deeb, B., Fayez, K., and Gherbawy, Y. (2013). Isolation and characterization of endophytic bacteria from Plectranthus tenuiflorus medicinal plant in Saudi Arabia desert and their antimicrobial activities. *Journal of Plant Interactions 8* (1): 56–64.
90. Shukla, S., Naik, G., and Mishra, S.K. (2021). Potential antimicrobial activity of bacterial endophytes isolated from Flacourtia jangomas (Lour.) Raeusch, a less explored medicinal plant. *Journal of Microbiology, Biotechnology and Food Sciences 6* (2021): 473–477.
91. Hallmann, J., Berg, G., and Schulz, B. (2006). Isolation procedures for endophytic microorganisms. In: *Microbial Root Endophytes*, 299–319. Berlin, Heidelberg: Springer.
92. Coombs, J.T. and Franco, C.M. (2003). Isolation and identification of actinobacteria from surface-sterilized wheat roots. *Applied and Environmental Microbiology 69* (9): 5603–5608.
93. Nimnoi, P., Pongsilp, N., and Lumyong, S. (2010). Endophytic actinomycetes isolated from Aquilaria crassna Pierre ex Lec and screening of plant growth promoters production. *World Journal of Microbiology and Biotechnology 26* (2): 193–203.
94. Cao, L., Qiu, Z., You, J. et al. (2004). Isolation and characterization of endophytic Streptomyces strains from surface-sterilized tomato (Lycopersicon esculentum) roots. *Letters in Applied Microbiology 39* (5): 425–430.
95. Verma, V.C., Gond, S.K., Kumar, A. et al. (2009). Endophytic actinomycetes from Azadirachta indica a. Juss.: isolation, diversity, and anti-microbial activity. *Microbial Ecology 57* (4): 749–756.
96. de Oliveira, M.F., da Silva, M.G., and Van Der Sand, S.T. (2010). Anti-phytopathogen potential of endophytic actinobacteria isolated from tomato plants (Lycopersicon esculentum) in southern Brazil, and characterization of Streptomyces sp. R18 (6), a potential biocontrol agent. *Research in Microbiology 161* (7): 565–572.

97. Hallmann, J., Kloepper, J.W., and Rodriguez-Kabana, R. (1997). Application of the Scholander pressure bomb to studies on endophytic bacteria of plants. *Canadian Journal of Microbiology 43* (5): 411–416.
98. Hallmann, J., Quadt-Hallmann, A., Mahaffee, W.F. et al. (1997). Bacterial endophytes in agricultural crops. *Canadian Journal of Microbiology 43* (10): 895–914.
99. Jiao, J.Y., Wang, H.X., Zeng, Y. et al. (2006). Enrichment for microbes living in association with plant tissues. *Journal of Applied Microbiology 100* (4): 830–837.
100. Ikeda, S., Kaneko, T., Okubo, T. et al. (2009). Development of a bacterial cell enrichment method and its application to the community analysis in soybean stems. *Microbial Ecology 58* (4): 703–714.
101. Piddock, L.J. (1990). Techniques used for the determination of antimicrobial resistance and sensitivity in bacteria. *Journal of Applied Bacteriology 68* (4): 307–318.
102. Meng, Q.X., Jiang, H.H., Hanson, L.E. et al. (2012). Characterizing a novel strain of B acillus amyloliquefaciens BAC 03 for potential biological control application. *Journal of Applied Microbiology 113* (5): 1165–1175.
103. Perez, C. (1990). Antibiotic assay by agar-well diffusion method. *Acta Biologiae et Medicinae Experimentalis 15*: 113–115.
104. Hastings, J.W. and Stiles, M.E. (1991). Antibiosis of Leuconostoc gelidum isolated from meat. *Journal of Applied Bacteriology 70* (2): 127–134.
105. Jayatilake, P.L. and Munasinghe, H. (2020). Antimicrobial activity of cultivable endophytic and rhizosphere fungi associated with "mile-a-minute," Mikania cordata (Asteraceae). *BioMed Research International 15*: 2020.
106. Cintas, L.M., Rodriguez, J.M., Fernandez, M.F. et al. (1995). Isolation and characterization of pediocin L50, a new bacteriocin from Pediococcus acidilactici with a broad inhibitory spectrum. *Applied and Environmental Microbiology 61* (7): 2643–2648.
107. Bhoonobtong, A., Sawadsitang, S., Sodngam, S. et al. (2012). Characterization of endophytic bacteria, Bacillus amyloliquefaciens for antimicrobial agents production. *International Conference on Biological and Life Sciences 40*: 6–11.
108. Hamed, A.A., Soldatou, S., Qader, M.M. et al. (2020). Screening fungal endophytes derived from under-explored Egyptian marine habitats for antimicrobial and antioxidant properties in factionalised textiles. *Microorganisms 8* (10): 1617.
109. Davidson, P.M. (1989). Methods for testing the efficacy of food antimicrobials. *Food Technology 43*: 148–155.

7

The Role and Importance of Microorganisms in Environmental Sustainability

Barbara Sawicka[1], Piotr Pszczółkowski[2], Piotr Barbaś[3], Dominika Skiba[1], and Bernadetta Bienia[4]

7.1. INTRODUCTION

The world of microbes is a great treasure if used and exploited wisely, and can in turn contribute to sustainable development. For many years, scientific research has been developing innovative technologies in agriculture and municipal management, e.g. agricultural waste, etc. Naturally occurring microorganisms are consortia of microorganisms that inhabit the soil and surfaces of all living organisms, both inside and outside, which have a large potential for bioregulation, biodegradation, bioleaching, and biocomposting, but also nitrogen fixation, improvement of soil fertility, and the ability to produce plant growth hormones. Without these microorganisms, life on our planet would be difficult [1, 2]. Therefore, restoring the natural environment and preserving life on Earth thanks to indigenous, local microbes converting useless waste into productive bioresources is a fundamental goal of sustainable agriculture. Soil microorganisms increase agricultural productivity. Organisms found in nature can be used to develop biofertilizers, bioregulators, bioherbicides, bioinsecticides, and biopesticides supporting the growth and development of crops and limiting the development of weeds, diseases, and pests. In addition, soil microorganisms help plants absorb more nutrients [1, 3, 4].

Effective microorganisms (EM) are involved in "nutrient recycling," helping plants "absorb" the necessary energy sources. This chapter attempts to emphasize the role of native microorganisms in the sustainability of the agricultural environment through their various

[1]*Department of Plant Production and Commodities Science, University of Life Sciences in Lublin, Lublin, Poland*

[2]*Experimental Station for Cultivar Assessment of Central Crop Research Centre, Dębowa Kłoda, Poland*

[3]*Potato Agronomy Department, Plant Breeding, and Acclimatization Institute, National Research Institute, Serock, Poland*

[4]*Food Production and Safety Plant, Carpathian State University in Krosno, Krosno, Poland*

Microbial Biotechnology: Role in Ecological Sustainability and Research, First Edition.
Edited by Pankaj Chowdhary, Sujata Mani, and Preeti Chaturvedi.

applications in agriculture, such as growth bioregulators, biofertilizers, bioherbicides, bioinsecticides, biopesticides, denitrifiers, etc. [5–14]. The motivation for writing this chapter is to make the public aware of the role of microorganisms in the sustainable development of ecological and pro-environmental research, as well as to motivate the contribution to achieving the goals of food demand in a sustainable and environmentally friendly manner.

7.2. DEFINITIONS

A biomarker is defined as an indicator (measurement) reflecting the interaction between an organism and an environmental factor (physical, chemical, biological). According to the World Health Organization, a biomarker is any substance, structure, or process that can be measured in a plant or animal organism or in their products. They allow the study of changes related to, for example, the physiological state, and forecast the occurrence or detection of diseases, pollutants, and poisoning, and thus affect the health of plants, animals, and humans. Based on the analysis of biomarkers, it is possible to obtain information on whether the organism was exposed to a harmful factor (exposure biomarkers); whether exposure to, e.g. xenobiotic, causes biological effects and health risk (so-called effect biomarkers); whether the organism is sensitive to the toxic effects of a given factor (so-called sensitivity biomarkers) [2, 15–17].

The biomarker of xenobiotic exposure should be easy to extract from biological material and be quantitatively dependent on the absorbed dose of xenobiotic. Biological monitoring consists of examining the content of a toxic substance or metabolites in the tested biological material over time. Knowing the speed and direction of changes that individual substances are subjected to in the body, it is possible to assess the amount of xenobiotic dose used on the basis of biomarker analysis, and on this basis to determine the risk to health. Important xenobiotics are herbicides, insecticides, and fungicides as important agents in the plant production process. Thanks to them, it is possible to effectively fight weed infestation, diseases, and pests affecting crops. However, the long-term use of the same plant protection substances in the same areas may induce resistance of weeds, diseases, and pests to active substances of pesticides, as well as the presence of their residues in soil, and groundwater and in the food [3, 16–21]. The way they work is important. Accordingly, there are: acetyl CoA carboxylase inhibitors, acetolactate synthesis, photosynthesis, and protoporphyrin oxidases, which interfere with the synthesis of dyes, e.g. carotenoids; EPSP synthesis inhibitors; glutamine synthetase; DHP inhibitors by interfering with microtubule formation; inhibitors of the mitosis process; inhibitors of long fatty acid chains; inhibitors of cell wall formation/cellulose synthesis as destabilizers of cell membranes; fat synthesis inhibitors; and auxin transport inhibitors. A classification of herbicides has been developed depending on the mechanism of their action and weed resistance to them [16]. The risk of spreading resistant weed biotopes to herbicides depends on the technology of plant cultivation and the application of herbicides, as well as the biology of weeds and the degree of weed infestation of the crop plants [17].

The residues of active substances of pesticides used in plant cultivation are important risk factors for disorders of lipid metabolism (increased levels of triglycerides, total cholesterol), and to low-density lipoprotein (LDL) fractions and lowering the concentration of high-density lipoprotein (HDL) fractions. This leads to the development of diseases of the cardiovascular system and an increase in atherosclerotic lesions of the aorta [15]. Pesticides used in plant cultivation get into surface waters, which is a big problem [3, 22–24]. Their use may lead to the resistance of weeds, diseases, or insects to active substances of pesticides. The conditions of use of the plant protection products (PPP) can be changed by the product user, while the risk is caused by the interaction between the specific characteristics of the target pest and the product itself [22, 25, 26]. If the risk of practical resistance and the risk of resistance when using restrictions are too high, modifiers can be introduced to reduce the risk to an acceptable level [2, 17, 25]. The residues of PPP are determined using multi-component methods. Many methods are indicated as methods for determining residues of PPP, including high-performance liquid chromatography (HPLC) with a photodiode as a detector, HPLC with UV detection, amperometry, coulometry, gas chromatography, mass spectrometry, and others [2, 26].

7.3. BIOFERTILIZERS

Many approaches have been used to solve the problem of hidden hunger in the world and new methods have been developed to improve basic crops. Biofortification is one such method that scientists have used widely with success. Various biofortification tactics were considered. One method is the use of biofertilizers in many basic food crops. Alam et al. [27] considered their use and limitations at a global level.

The basic nutrients N, P, K, and Mg are naturally present in the environment, but plants have little ability to benefit from them. Alam et al. [27] argue that in order to not face a food shortage for a growing population in 2005–2050, farmers worldwide will have to harvest the same surplus of food each year as they did in the last 2000 years, and estimated that they would need around 250 million tons of nitrogen fertilizer in 2050, more than twice in 2005. The most common economically friendly approach in developed countries is agronomic biofortification, but it tends to ignore the ecological aspects [28, 29].

Biofertilizers are substances that use microorganisms to fertilize the soil. These fertilizers are not harmful to crops. They are collected from animal waste along with microbial mixtures. Microorganisms are used to increase the level of nutrients in plants. The use of biofertilizers in the soil protects plants against diseases. A biofertilizer is a substance that contains living microorganisms. These, when applied to seeds, or the plant surface or soil, colonize the rhizosphere or interior of the plant and promote plant growth by increasing the availability of nutrients in the host plant. Biofertilizers enrich soil and plants with nutrients thanks to the natural processes of nitrogen fixation, dissolving phosphorus and stimulating plant growth by synthesizing and stimulating plant growth [28, 30, 31]. The use of biofertilizers is expected to reduce the use of chemical fertilizers and pesticides.

Biological fertilization is one strategy that can be used to reduce the effects of drought stress and increase food production [32]. The use of biological fertilizers such as arbuscular mycorrhizal fungi (AMF) can effectively improve plant growth and yield under drought stress [33]. For example, inoculating AMF seeds or fungal spores during cultivation is a very effective strategy to enhance plant symbiosis with AMF in the soil [34–36]. The symbiosis of AMF allows for a better transport of water and nutrients from the soil to the host plants. This allows the transfer of carbohydrates from the photosynthetic pathway to AMF [37]. Arbuscular mycorrhiza symbiosis makes it easier for plants to survive the stress of drought thanks to various mechanisms [33], among other things, through greater water uptake by the external hyphae system [38]. As a result, the absorption of nutrients from the soil by plants, such as, in particular, phosphorus, iron, and zinc, is improved [39]. This may stimulate the biosynthesis of chlorophyll and the chlorophyll-protein complex [40]. After establishing AMF–plant symbiosis, AMF hyphae increase the rate of H_2O_2 outflow from the host plants to the rhizosphere, which significantly reduces oxidative stress in the host plants [36]. Mathur and Jajoo [34] and Paravar et al. [36] recently found that drought stress-induced damage to PSII and PSI structure and function was alleviated by AMF colonization.

Ghanbarzadeh et al. [40] found that the colonization of the roots of *Dracocephalum moldavica* L. thanks to AMF enhanced plant growth through an improved antioxidant system and reduction of ROS (reactive oxygen species) formation. Information on the influence of AMF on the growth, yielding, chemical composition, and physiological processes of crops, especially under the influence of drought stress, is limited, so extensive research on the use of AMF in plant cultivation is advisable. Phosphorus biofertilizers are used to determine the level of phosphorus in the soil.

The phosphorus requirement for plant growth is also limited. Phosphorus biocarbons help the soil to obtain the required amount of phosphorus. Phosphates play an important role in plant cultivation as they are responsible for maturity, quality, drought tolerance, and nitrogen transformation, directly or indirectly. *Penicillium bilaii* helps to release phosphate from the soil. It creates an organic acid, which then dissolves phosphates in the soil and makes them available to plants. Biofertilizers together with these organisms are applied by coating the seeds with fungus as inoculation or by introducing it into the soil [41].

Rhizobium is a bacterium that lives on the roots of *Fabaceae* in a collection of cells called "nodules." They can take nitrogen from the air and convert it into organic forms, which can then be used by the plant [28]. These interactions between plants and bacteria occur between *rhizobium* and legumes, but also pseudomonads are frequent and very active participants in the rhizosphere. For example, the *Pseudomonas putida* population adapts its genetic program to adapt when interacting with maize roots. Differently expressed genes were identified in this heterogeneous environment comparing rhizosphere populations with three control situations for different nutrients and lifestyle (planktonic and sedentary). About 100 genes have been found to be differentially expressed from all three controls; the number of upregulated rhizosphere genes is nine times greater than the number of repressed genes, and changes in their expression are more pronounced. The presence of individual nutrients in the root exudates translates into the preferred expression of genes that are involved in the uptake of amino acids and in the metabolism of aromatic substances in the rhizosphere. However, the induction of efflux pumps and enzymes of glutathione metabolism proves that adaptation to unfavorable conditions and response to oxidative stress are essential for the life of bacteria in this environment [41, 42]. This is consistent with two distinct selection forces important for rhizosphere colonization: stress and the availability of specific nutrients. The identification of the domain protein (PilZ) and the domain response regulator (GGDEF/EAL, a gene containing a degenerate phosphodiesterase domain) among the genes supporting the regulatory role suggests the implication of signaling by the secondary messenger c-di-GMP in the plant–Pseudomonas association. This transcriptomic approach to the rhizosphere revealed previously unknown characteristics of bacteria in terms of ecological efficiency [27, 28, 43]. This helps to protect the environment for future generations.

7.4. BIOPESTICIDES

Not all soil microorganisms are plant-friendly. Pathogenic microorganisms can be used for natural weed and pest control. Biopesticides are a special group of active substances used to protect plants. They occur naturally or are synthetically produced but are identical to natural substances. Biopesticides usually break down fairly quickly and some semi-chemical ones are used in very low doses. They also concern many living organisms (so-called "biological protection organisms") [44–47]. The mechanisms of pesticide action are best known and divided into several groups [21, 48, 49]. Biopesticides use genetics and molecular physiology as tools to overcome resistance against ever newer generations of chemicals and antimicrobials [44]. Human exposure to pesticide residues concerns many substances from various sources, such as the environment and food. The negative effects of such combinations can manifest themselves in an unpredictable way and lead to a change in the assessment of risk to human health [3, 46, 50–52]. In the light of the latest research and European and international regulatory changes, the effect of mixtures of pesticide residues used in PPP is aimed at risk analysis and setting maximum pesticide residue limits, as well as planning and enforcing their control [8, 44, 49, 50].

Food safety is seen through inter alia pesticide residues in food. There are strong associations between exposure to pesticides and the incidence of diseases such as prostate cancer, leukemia, lung cancer, and other diseases. Hence the possibility of applying biopesticides in agriculture and horticulture, which are a special group of active substances in plant protection. Their use can show many positive changes, including the reduction of pesticide residues in food, improving the natural environment, and thus reducing the risks for consumers [3, 20, 21, 53]. There are various alternative options to deviate from synthetic pesticides, such as, e.g. biopesticides or natural remedies [51, 53, 54]. They also include many living organisms. Research on biological preparations will allow for better selection of innovative weeding practices and combining traditional weed-control treatments with biological preparations in commercial crops. The use of biodynamic preparations in the cultivation of agricultural and horticultural plants enables the improvement of agrochemical and biological properties of the soil, increases plant resistance to pathogens, and allows for the targeted selection of biologically active compounds [55, 56]. For example, entomopathogenic fungi are recognized as effective means of protection against certain crop pests. Entomopathogenic fungi have recently been used to control the whitefly (*Bemisia tabaci*) and the diamond moth (*Plutella xylostella*). *Beauveria bassiana* GHA (genetic health analysis) strain is used as an insecticide to combat sucking insects, especially those that feed on greenhouse and ornamental plants. *Bacillus thuringiensis* subsp. *tenebrionis* strain NB-176, which destroys *Leptinotarsa decemlineata* larvae, is a bacterial insecticide. On the other hand, *B. bassiana* and *Isaria fumosoroseus* are effective in combating soil pests as well as damaging ground parts of plants. The use of biological preparations in combination with non-chemical weed control methods improves the structure of the soil, increases the number of earthworms in the soil, and reduces CO_2 emissions from the soil, but also reduces the enzymatic activity of the soil [42]. Slow-release biodegradable spheres consist of sodium alginate and skim milk as carriers in bacterial inoculation of plants. The granular preparation of *Alternaria macrospora* is, e.g. used in the control of *Anoda cristata*. The implementation of integrated weed, disease, and pest control in commercial farming and horticulture has led to an increased interest in biopesticides, which in turn will reduce the risk in terms of food safety.

7.5. BIOHERBICIDES

Weeds compete with crop species not only for water, nutrients, light, and space, but they also harbor diseases and pests. Moreover, they clog irrigation and drainage systems and contribute to the deterioration of the quality of the crop, and leave weed seeds in the crops and the soil. Bioherbicides are a way to control weeds without synthetic herbicides endangering the environment [54, 57].

Organic and biodynamic agriculture lacks effective methods of infestation of control (CPB; cardiopulmonary bypass) [55, 57–61]. The biodynamic preparation BD 500, used by Vaitkeviciene [55] and Levickienė [61], increases the microbial biomass and its conversion and activity of soil enzymes, including dehydrogenase. In the soil sprayed with the BD 500 biodynamic preparation, there was a significant increase in microbial biomass during the vegetation period of plants by 11%, compared to the object not sprayed with the biodynamic preparation, both after 14 and 115 days from the application of the preparation. The dehydrogenase activity in the soil turned out to be over 20% higher than in the soil not sprayed with this preparation. Moreover, in such soil the content of assimilable phosphorus, potassium, and nitrogen in the soil increased, which increased the resistance of plants to soil pathogens [42, 60]. According to Keidan [57], weeding by non-chemical methods, i.e. thermal and mechanical methods, and with the use of biological preparations, reduced the amount of winter oilseed rape in the fall compared to self-regulating methods of weed management, and the arable crops were characterized by better wintering, better frost resistance, and higher productivity. Non-chemical weed management methods (thermal, mechanical, biodynamic) in combination with biological preparations improve the soil structure, improve the abundance of earthworms in the soil, and reduce CO_2 emissions from the soil. However, all these practices result in a decline in the activity of soil enzymes [42, 57, 61, 62].

Werle et al. [63] highlighted the identification of crops with a natural ability to suppress weeds and compensate for weed interference. This ability refers to the potential of a crop to reduce or inhibit the emergence or growth of weeds. Some cultivars show different weed suppression abilities when grown together with weeds, compared to monocultures. Tomato or potato cultivars may reduce the production of weed seeds, but the amount of this reduction depends on the density of weeds in the crop and their speed of growth [64]. In response to weed competition, differences between different tomato cultivars have been documented [65]. It has also been proven that some tomato cultivars have a significant tolerance to the parasitic weed lespedeza (*Cuscuta* spp.) [66]. The utility of such varieties with particular weed tolerance abilities can be of value in low input farming systems or in situations where chemical weed control is not possible, such as in organic farming. Weed-inhibiting crop varieties involve the manifestation of joint activity and the interaction of multiple traits instead of one. Plant architecture, growth pattern, and overall morphological performance, such as leaf surface area, are the primary characteristics responsible for increased competition between crop and weed species [67]. Weed-inhibiting plants often exhibit allelopathic properties. Allelopathic capacity was found, among others, in cereal crops such as rye, sorghum, rice, wheat, and in *Cruciferae* plants such as mustards and canola [68]. Phytochemicals involved in weed control include simple phenols, flavonoids, and alkaloids [63, 69]. The allelopathic properties of some plants are potentially valuable for integrated farming systems as soil additives by incorporation of crop residues and inhibition of weed emergence [70]. The discovery of varieties with weed suppression potential could be a helpful resource for plant breeding programs and could provide farmers with an alternative to chemical weed control, thus contributing to a more sustainable farming system. Research on biological preparations will allow for a better selection of weed management practices and combinations with biological preparations in the cultivation of winter crops (such as rape, turnip, wheat, alfalfa, red clover, rye, triticale), especially in a temperate climate. This will also ensure the selection of appropriate biological PPP that will improve the survival of wintering plants and increase soil fertility [62]. The use of biodynamic preparations in the cultivation of agricultural and horticultural plants makes it possible to introduce them in order to improve the agrochemical properties and biological properties of the soil, and thus increase the resistance of crops to drought stress, low temperature stress, pathogens, and the possibility of a targeted selection of PPP. Soil microorganisms are characterized by the presence of invasive genes that can interfere with the defense genes of weeds, thus rendering them harmless. The advantage of using bioherbicides in plant cultivation is that they can survive in the soil environment for a long time until the next growing season, when there will be more weeds to control. This method of weed management is much cheaper than chemical pesticides and can significantly reduce the cost of cultivation. Moreover, it is not harmful to the natural environment compared to herbicides and does not affect non-target organisms [63, 69, 71].

7.6. BIOINSECTICIDES

A huge role in plant protection is assigned to microbiological antibiotics, and an alternative method of pest control is the use of insecticides produced by biological microorganisms [72–74]. The combination and use of many positive functions of microorganisms have become possible thanks to the introduction of mixed cultures containing various types of microorganisms, both aerobic and anaerobic.

Biotechnology is also helping to develop alternative control measures for synthetic insecticides in the control of insect pests. Soil microorganisms that attack bacteria, fungi, and viruses also cause root diseases [47, 61, 75, 76]. The results of the research proved that the effectiveness of *B. bassiana* in protecting potatoes against the Colorado potato beetle is medium to good [47, 49, 75, 77]. Natural substances are starting to play an increasingly important role in plant protection. Most often they are plant secondary metabolites. These include, among others: farnesyls, saponins, and Indian honey extract [78]. Entomopathogenic fungi are considered to be promising means of protection against many pests of arable crops not only in Europe, but also in the rest of the world [47, 70, 79, 80]. However, the use of entomopathogenic fungi in plant protection practice is limited. Entomopathogenic fungi, such as *B. bassiana* or *I. fomosoroseus*, are effective in combating pests not only in soil, but also against aerial parts of plants [77]. Long et al. [81], Laznik et al. [82], and Ropek and Kołodziejczyk [47] proved that the effectiveness of natural preparations in the control of *L. decemlineata* larvae is, however, significantly lower than that of chemical pesticides. Chemical agents protected potatoes more effectively against *L. decemlineata* larvae, as the reduction of this pest 10 days after their single application (spraying with imidacloprid or lambda-cyhalothrin) was about 90%. The most effective protection against this pest is obtained by treating the tubers with thiamethoxam. The lower effectiveness of *B. bassiana* or *I. fomosoroseus* results from their slower action than that of chemical pesticides [82, 83]. Despite the favorable research results, the preparation based on entomopathogenic fungi is not yet registered. New formulas for coating seeds (so-called modifiers) with beneficial organisms can be developed to protect plants at the seedling stage. Bioinsecticides do not stay long in the environment, and they are safer for humans and animals compared to synthetic insecticides. They work in a strictly defined way, often affecting only one species of insect. They are usually slow, and their lifetime is short [5, 14].

7.7. EFFECTIVE MICROORGANISMS

Recently, in Europe and the world, as the interest in healthy, safe food has grown, so has the need for substances that can replace or partially eliminate plant protection chemicals. This type of research is important for the end-user, consumer, and processing industry. Numerous studies [56, 84–98] are aimed at limiting the chemicalization of agriculture, including through the introduction of biologically effective preparations that are an alternative to synthetic fertilizers and conventional PPP. Their goal is to influence growth and development of plants and their protection against pathogens [84, 99, 100].

Kowalska et al. [101] assessed effective microorganisms (EM) for suitability for organic farming. These authors stated that preparations based on EM contain a mixture of various coexisting microorganisms, while they do not contain genetically modified microorganisms.

In the composition of preparations of this type, they confirmed the presence of lactic acid bacteria, photosynthetic bacteria, and yeast [53]. According to Kowalska et al. [101] the EM preparation contains lactic acid bacteria as *Lactobacillus casei* and *Streptococcus lactis*, photosynthetic bacteria such as *Rhodopseudomonas palustrus* and *Rhodobacter spae*, yeasts such as *Saccharomyces album* and *Candida utilis*, Actinomycetes as *Streptomyces album* and *Streptomyces griseus*, and molds such as *Aspergillus orysae* and *Mucor hiemalis*.

7.7.1. Microorganisms in Biological Plant Protection

FAOSTAT [43] estimates that worldwide production increases by 1% per hectare of arable land and is associated with an increase in pesticide use of over 1.8% per hectare. Therefore, the use of pesticides per hectare of arable land increases more than proportionally with the increasing intensity of land use. Thus, increasing the world's food supply by intensifying agriculture may lead to the continued use of pesticides in the future. A good way out of this problem is the use of biological plant protection. There are 220 biological protection products (so-called biopesticides) registered in OECD member countries [46, 54]. The registration of these agents is carried out on similar principles to that of synthetic preparations [54]. The difficult and costly registration procedure for biological plant protection agents means that some producers give up the production of biopreparations, which reduces their range and thus reduces the scope of biological protection in agriculture, but ensures safety for people and the environment and makes them good quality products.

An element of EM research by Zydlik and Zydlik [71] was strawberry resistance to disease. Strawberry leaves were collected after two EM treatments were assessed. These studies were started because after the contact of plant cell membranes with the pathogen, there was a rapid increase in the number of active oxygen species, e.g. O_2^- and H_2O_2, and as a result, an oxygen "burst" occurred, which was cytotoxic to infected plant cells. As a consequence, the cytoskeleton was rearranged, and the thickness of the cell walls increased. In such a situation, according to Kowalska et al. [101], there was so-called systemic acquired immunity, which is the activation of the defense system in uninfected parts of the plant. It is triggered by the production of proteins related to pathogenesis. The signal for this response was salicylic acid. Currently, there are 14 classes of plant proteins involved in this process. These are, among others: glucanases, chitinases, osmotines, protease inhibitors, proteinases, lysozymes, peroxidases (including β-glucosidase), and proteins that act bactericidal and participate in apoptosis. The expression of the genes coding for them depends on the presence of the stimulus coming from the pathogen – the so-called elicitor, picked up only by specific receptors, which, via kinases, transmit a signal for their induction. As a result, it leads to the accumulation of phenols, an increase in the number of flavonoids, the synthesis of numerous proteins, and the formation of several secondary metabolites (phytoalaxins), which may in turn have a toxic effect on pathogens [101].

In the case of some biopreparations, e.g. containing *Bacillus thuringienis*, the production scale is large and is carried out with the use of fermenters with a volume of 100000 cubic liters [46, 102]. The recently increasing demand for biopreparations is related to the development of pro-ecological methods of plant cultivation, including biodynamic and ecological agriculture [95, 103].

In Japan, there is a program aimed at comprehensive environmental protection using EM. In cities, for example, EM technology is used in home composting plants, municipal water intakes, landfills, and sewage treatment plants, etc. Rivers, lakes, ponds, and seas are also cleaned with the help of EM technology. A national revitalization program for the largest inner sea was also adopted there. The direct effect of using EM in agriculture is, however, an increase in crop yields and a reduction in weed infestation [104–106]. In Europe, this technology is becoming more and more common at the level of individual farms, especially organic ones [86, 105, 106].

7.7.1.1. EM in Plant Cultivation

Microorganisms metabolize organic matter, accumulating humus substances, which ultimately lead to the mineralization of organic compounds, thanks to which they recirculate elements necessary for plant production [87, 88, 92, 94, 97, 105, 107]. Among the biopreparations, the best known and used in agricultural practice are vaccines containing bacteria (the so-called rhizobia), which bind atmospheric nitrogen in symbiosis with the roots of legumes [102, 108–111]. *Rhizobium* vaccines as well as other microbiological preparations are only released to the market after meeting the detailed requirements of the registration procedure in the EU. Due to this, vaccines containing symbiotic bacteria of legumes plants are biopreparations with proven effectiveness and good microbiological quality [110, 112].

EM technology was developed in the 1970s at Ryukyus University, Okinawa, Japan [104, 105]. EMs are in liquid form, which consist of naturally occurring beneficial microorganisms, such as: actinomycetes, lactic acid bacteria, photosynthetic bacteria, and yeast [2, 104, 105, 113–115]. Multicultural technology of beneficial microorganisms is gaining popularity nowadays due to its environmentally friendly nature. Its products contain photosynthetic lactic acid, bacteria, ferments, and products resulting from their metabolism [42, 73, 116, 117]. Microorganisms metabolize organic matter, contribute to the accumulation of humic substances, which lead to the mineralization of organic compounds, thanks to which the elements necessary for plant production are recirculated [112]. EM technology uses naturally occurring microorganisms that are able to cleanse and enliven nature. It is used, inter alia, in agriculture, horticulture, forestry, food processing, and other industries [90, 97, 98, 109, 116, 117]. Although microbiological preparations (so-called modifiers) are widely promoted as an ecological strategy to improve the quality of compost, there is no consensus in the literature on their effectiveness. Abebe Nigussie et al. [115] performed a quantitative meta-analysis to determine the overall effect of microbial inoculants on nutrient content as well as on lignocellulose humification and degradation. These analyses indicate a beneficial effect of microbiological modifiers on the content of total nitrogen (+30%), total phosphorus (+ 46%), as well as the compost maturity index (C:N ratio) (−31%) on humification (+60%) and the germination rate (+28%). The average effect of the applied modifiers was −46, −65, and −40%, respectively, for cellulose, hemicellulose, and lignin. However, the magnitude of this effect turned out to be marginal for the bioavailable nutrient concentrations of phosphate, nitrate, and ammonium. The authors [115] conclude that the effectiveness of microbiological modifiers also depends on the form of the modifier, inoculation time, composting method, and the duration of the experiment. These recent discoveries suggest that microbial modifiers are important in accelerating the degradation of lignocellulose. Therefore, researchers should be encouraged to publish also non-significant results to provide a more reliable estimate of the effect size and to clarify doubts about the benefits of microbial inoculation in composting.

The application of EM products using a known formula as an effective microbial activated solution (EMAS) has been tested in Japan, China, Malaysia, Russia, Denmark, Holland, Czech Republic, Poland, Romania, etc., depending on scale, location, and geographic conditions. The main goal of these technologies is to improve the condition of soil, water, and food quality [73, 87, 88, 90, 92–97, 107, 113, 114, 116, 118, 119]. The most important role in this technology is played by atmospheric nitrogen-fixing bacteria, which play a huge role in this process, converting indigestible forms of nitrogen into forms accessible to plants. EM significantly contributes to the improvement of soil structure. The mucous substances released by EM stick to humus and mineral particles, creating clumps of the soil, thus improving its structure [102, 109, 110]. Microorganisms also produce bioactive material useful for plants, such as hormones and growth stimulants that trigger cell division. In the group of biopreparations, the best known and most frequently used in agricultural practice are vaccines containing bacteria (rhizobia) that bind atmospheric nitrogen in symbiosis with the roots of legumes [102, 110–112]. The technology of producing these vaccines consists of collecting various strains of microorganisms and controlling their purity and quality (symbiotic efficiency), which is a multi-stage process.

EM, which are a mixture of naturally occurring, beneficial microorganisms, are used to broadly understand agriculture to improve soil quality, the growth of crops, and improve their quality, especially in biodynamic and organic farming systems [55, 56, 84, 95, 100, 120]. Ecological, biodynamic, and soilless production plant cultivation systems allow us to limit the use of synthetic products in plant and animal production. As a result, they have a significant impact on the size and quality of crops and their suitability as a raw material for processing in the food industry. In the research of Pszczółkowski et al. [96] and Sawicka et al. [97], cultivation methods with EM were used in various combinations compared to the standard technology and success was achieved in the form of an increase in the potato yield, its quality, and health. In other studies, with the use of fungicides and EM, no significant differences were found in the total and commercial yield of potato tubers compared to the control object, but an increase in the multiplication factor and an improvement in the health of the seed material were observed [95]. Van Vliet [109] explain this lack of yield-forming effect by coal mineralization. Xu [84] claims that EM applied to organic fertilizers increases the growth and activity of roots, as well as increases their efficiency and photosynthetic ability, which increases the yield of plants. This is largely due to the higher level of nutrient availability that facilitates the application of EM over time. Despite these successes, the exact mechanisms of EM are not fully understood, requiring further multifaceted research.

There are also mycorrhizal preparations available on the market, which have a beneficial effect on the development of plants and may have a protective effect on them. Alternative preparations are also vaccines based on the fungus *Trichoderma* sp. These vaccines, thanks to the fact that they produce antibiotics and enzymes that degrade the cell walls of pathogens, contribute to the protection of plants against pathogens. Microbiological preparations that can be used in the cultivation of plants are, among others, *Nitrobacter* and *Phosphobacter*. They include microorganisms supplying the soil with hardly digestible forms of phosphorus or nitrogen [74, 91]. In ecological agriculture, interest is also enjoyed by entomopathogenic preparations, which include microbial insecticides limiting the number of harmful insects [72, 89, 91, 92, 121]. In the revitalized soil, the number of fungi increases, which in symbiosis with plant roots (mycorrhiza) accelerate the absorption of nutrients. Thanks to the mycorrhiza, on average 90% more nitrogen, 75% potassium, and 200% phosphorus are used compared to plants without mycorrhiza [72, 91]. Baranowska et al. [108] claim that inoculation of soil with EM, before planting potato tubers, protects them against pox (*Rhizoctonia solani*) and common scab (*Streptomyces scabies*).

In the opinion of many authors [90, 122–124] the use of EM-Farming™ biotechnology in the cultivation of tomato, wheat, and rape accelerates the decomposition of crop residues, eliminates the use of seed dressing, increases the pH without the use of lime, significantly improves the level of humus in the soil, and increases the availability phosphorus, which in turn leads to a higher yield of seeds and straw. EM preparations complement other agrotechnical measures aimed at reducing the presence of pests. They do not directly reduce diseases, but by applying them to the soil, they increase the microbial activity of the soil [108, 121].

According to Marczakiewicz [125], in the cultivation of maize for grain and silage, an important phenomenon that occurs after the use of EM is the conversion of putrefactive processes that cause the development of diseases and pests, auto-oxidation, and regeneration. Therefore, post-harvest residues treated with EM preparations cease to be a source of pathogens development and become food for beneficial microorganisms that transform them into complementary nutrients for plants.

Grabowska [126] states that in horticultural production under covers, the use of EM-Farming™ significantly reduces the process of plant rot, blocks the development of diseases and pests, and stimulates the humification process, as a result of which the plant receives full-value fertilization. This is of great importance in the cultivation of cut flowers susceptible to diseases of the genus Fusarium, such as cloves, lilies, and gerberas. In the case of vegetable cultivation, treatments performed with EM-Farming™ during the harvest period protect plants against fungal diseases such as gray mold, downy mildew, potato blight, and cabbage syphilis. In the case of plants grown under cover, their health and vigor are also improved, the quality and quantity of the crop increases, and the taste and color of the fruit is improved [126].

In the opinion of Janas [73], EM-Farming preparations have a high potential for interaction through specialized enzymes that are capable of transforming harmful chemical compounds into beneficial forms. The use of active EM cultures in potato cultivation, according to Marczakiewicz [125], significantly accelerates the recovery of the soil, restoration of the humus layer, and, consequently, increases the yield. After the restoration of permanent humus, the resistance of potato plants to pests is significantly increased, and thus their need for fertilization is reduced. These preparations can, inter alia, be used for treating seed potatoes by soaking, for spraying or irrigating crops, for inoculating plant seedlings before planting, to reduce or eliminate PPP against pathogens and diseases; for controlling insect pests of crops; for storing vegetables and fruits in cold rooms and warehouses to improve their quality and shelf life; and to increase soil biological activity without affecting plant, human and animal health [51, 116, 117, 121, 127].

On the other hand, biopreparations consisting of appropriately selected strains of microorganisms are considered controversial due to their wide spectrum of activity [102, 110, 111]. According to other authors [107, 128], however, they have a positive effect on the morphological features of plants, their physiological functions, and the properties of the substrate. The use of agents containing EM is of great importance, among others in hop cultivation, in protection against downy mildew and powdery mildew, as well as against pests such as plum-hop aphid and hop spider mite [129].

7.7.2. EM in Food Processing

Currently, biotechnology solves many problems in agriculture and broadly understood food processing [97, 100]. One of them is the EM technology, which is now mainly used in organic farming [2, 92–95, 114–116]. Allahverdiyev et al. [117] report that there are microbial materials behind EM technology (e.g. "Baikal EM1"). They contain lactic acid bacteria, as well as photosynthetic and nitrogen-fixing bacteria, yeasts, and molds. Lactic acid bacteria are found mainly in fermented foods and in the digestive tract of healthy humans and animals. These are *Lactobacillus plantarum, L. casei, Lactobacillus delbrueckii, Lactobacillus fermentum, L. lactobacillus salivarius*, and yeast (*Saccharomyces cerevisiae*), which have a beneficial effect on the fermentation process that distinguishes them from other microbial organisms. These microorganisms play a very significant role in processing technologies for foods [59, 94, 96, 97]. The results obtained by Sawicka et al. [97] prove, inter alia, that the technology with the use of EM-Farming affects the quality of fried potato products (French fries [FF], chips) by optimizing physiological processes (respiration, transpiration, and enzymatic systems), thus ensuring the resistance of tubers to pathogenic organisms and stressors in their environment [112, 121].

In the opinion of Szembowski [90] the application of EM-Farming™ in the biotechnology of agri-food processing allows for higher quality standards, and at the same time limiting the workload, reduces production costs. Some of them are used in industry and households, for example yeast, used in the production of alcohol (wine, beer) and baking bread. Other known microorganisms are lactic acid bacteria, used in dairy processing for the production of, among other things, cheese, as well as acetic acid bacteria used for the production of vinegar [88]. In addition to microorganisms, enzymes produced by microorganisms are also used, including pectinases, applied to clarify fruit juices [60, 100].

According to Boligłowa and Gleń [121] and Allahverdiyev et al. [117], the EM not only enriches food with essential amino acids, but also contributes to the elimination of gastrointestinal diseases in humans and monogastric animals. The EM cultivation system also contributed to a significant reduction in fat absorption in the processing of FF and potato chips (PC) and to a significant reduction of the greatest defect of FF, which are "dark ends." In the case of PC production, the EM cultivation technology contributed to lower absorption of fat by PC, but did not affect its taste, smell, or overall appearance. This suggests that potatoes produced using EM technology are the better raw material for the production of FF and PC than potatoes grown without the use of these microorganisms. The use of the new technology of cultivation with the use of EM will therefore contribute to obtaining a better quality raw material for food processing and thus to the production of noble potato products enriched with exogenous amino acids and vitamin C [97].

According to Tommonaro et al. [123], EM preparations have great biological potential due to specialized enzymes capable of converting chemicals into useful forms. EM are helpful, for example, in the production of trehalose, which shows the ability to protect plants when these are exposed to stressful conditions, e.g. drought or high salinity. This disaccharide prevents the formation of water crystals because it forms hydrogen bonds. After drying, aqueous solutions of trehalose create a glaze, which is an ideal substance for protecting plants against drought and high temperatures [94, 95, 123].

EM are on the lists of preparations used in organic farming around the world, although their use is constantly criticized [102, 110, 111]. Their application methods are various, ranging from spraying the soil during stubble cultivation, harrowing, plowing, or by direct spraying of plants throughout the growing season, by soaking seed potatoes or seedlings before planting, through seed treatment, etc., of seeds including those from the genus *Alternaria*, *Botrytis*, *Fusarium* or *Septoria*.

The future direction of research in this area should be toward an in-depth study of the impact of EM technology on the quality of raw materials and processed food products, as it is directly related to human health.

7.7.3. EM in Composting

Composting, or organic recycling, is a natural way of utilizing and managing waste, which consists of the decomposition of organic matter by microorganisms such as aerobic bacteria, fungi, nematodes, etc. [117, 123, 124]. Henry et al. [130] revealed increased biodiversity and microbial population followed thanks to the increase in the composting efficiency of lignocellulosic waste with the use of effective microbial thermoacids (tEM). Composting with tEM significantly increases the microbial population by an average of 12.0%, compared to composting without tEM. Similarly, the biodiversity of microorganisms in this group increases to 35–44% compared to the control. The greatest increase in population and diversity (91–92%) was observed 30 days after the application of this preparation. During composting, *Bacteroidetes* turned out to be the most dominant group. The degree of gradation of the large compost pellets (<2 mm) increased by 25 and 36%, respectively, after treatment with the tEMA and tEMB microbial preparations. Major differences in microbial structure, including a higher abundance of beneficial microorganisms such as *Bacillus*, were found in the composts treated with tEM. Henry et al. [130] found that tEM can increase biodiversity and microbial population, mainly in the thermophilic phase (above 45 °C). The results of their research are very important in the field of environmental protection and organic farming. In China, effective local microorganisms (LEMs) have been tested as inoculants produced from litter from forest substrates [130, 131]. Assessment of the effects of combining LEM with broiler litter composted to enrich *Glycine max* L. with nitrogen available to plants, nematode trophic group communities, and soybean productivity turned out to be positive. The combination of LEM with composted broiler litter resulted in a rapid increase in nitrogen mineralization in the early growing season and allowed for a stable abundance of many trophic groups of nematodes during droughts. By contrast, this bio-inoculum had no effect on Edamame's nipple. The research of Ney et al. [131] indicates a significant, high potential of LEM to strengthen soil resistance to drought stress – thus ensuring greater safety of the functional agro-ecosystem in changing climatic conditions.

7.7.4. Simultaneous Nitrification and Denitrification and Aerobic Denitrifying Microorganisms (*HNADM*)

Traditional nitrogen removal from the ecological system relies on autotrophic nitrification and an anaerobic denitrification process. This system is based on both autotrophic nitrification and an anaerobic denitrification process [130, 132]. On the other hand, in the ecological system, autotrophic microorganisms nitrify under aerobic conditions, and heterotrophic microorganisms carry out the denitrification process under anaerobic conditions. Both types of microorganisms are characterized by variable tolerance to oxygen concentration, and nitrification and denitrification are mainly in two compartments for nitrogen removal. There is a special type of microbes called heterotrophic nitrification and oxygen denitrification microorganisms (HNADM) that can oxidize ammonium nitrogen and denitrify in the presence of oxygen [132]. HNADM has been reported in many environments. It has been found that HNADM can lead to nitrification and denitrification simultaneously. Some HNADMs are also able to remove phosphorus, suggesting that these organisms have great potential to remove contaminants from wastewater. A comprehensive assessment and summary of HNADM could provide a better picture of their potential capability to target their use. Song et al. [132] summarized and evaluated the nitrogen metabolism pathway in HNADM. They documented the effects of pH, carbon, and C/N ratio on the metabolism of HNADM. They also demonstrated the production of extracellular polymeric substance (EPS), secretion of quorum sensing (QS), and removal of P by HNADM.

7.7.5. EM and Germanium in Sustainable Agriculture

Organic Germanium (Ge) compounds, which are now quite widely used as medical applications and dietary supplements, have a significant positive impact on human health. The inorganic Ge element can be absorbed by plants, accumulated, and converted into organic Ge compounds in various plant tissues, which can then be consumed as Ge-rich foods.

The Ge content may improve the physiological functions of plants, such as salt tolerance or increasing the peroxidase activity in leaves [26]. Hence, fertilizing with exogenous, inorganic Ge is an essential, effective way to obtain more organic germanium. Li et al. [26] determined the effect of exogenous Ge on its accumulation and nutritional properties of garlic using GO_{24} and GeO_2 variant as exogenous Ge. In addition, they introduced an effective microorganism inoculation (EM) factor to assess the combined effect of exogenous application of Ge and EM to garlic seedlings. It turned out that foliar spraying with GeO_2 solutions with a concentration of 6–72 mg l^{-1} had a significant effect on both the accumulation and distribution of Ge in garlic plants. For the production of garlic cloves and bulbs rich in Ge, the foliar application of the plants at a concentration of 9–12 mg l^{-1} GeO_2 during the stimulation period turned out to be the most optimal. A study by Li et al. [26] provided very important information on field production of Ge rich garlic. The foliar application of GeO_2 very significantly influenced the accumulation and distribution of Ge in garlic.

7.7.6. EM in Livestock Production

In 1 cm^3 of EM-Farming™, there are approximately one billion microorganisms that improve the welfare of farm animals in closed facilities [133]. The physiological function of the digestive system of herbivores depends on microbial activities in the gastrointestinal tract, according to Gacka and Kolbusz [134] and Faturrahman et al. [133] to improve digestive processes and thus better feed conversion. In the case of calf rearing, the use of EM improves digestion, improves the health condition of animals, and causes faster weight gain in animals. Preparations containing EM improve the physical condition and physiological functions of cattle, eliminate the occurrence of mycoses, reduce the appearance of diarrhea, reduce the occurrence of respiratory diseases, and reduce the population of insects in livestock buildings [134]. The incidence of inflammation in animals is also reduced. Widuch [135] proved that the use of the preparation EM-Farming™ for grassland significantly reduces the susceptibility of plants to water shortage in these areas, and thus reduces the negative effects of drought. EM preparations are used in animal husbandry, as an additive to fodder and drinking water, cleaning of pig houses and removal of unpleasant odors, and for the production of full-fledged natural fertilizer and the production of silage [133, 134]. The use of EM-Farming™ in livestock buildings indirectly stabilizes the life processes of livestock by improving sanitary conditions (ensuring animal welfare, improving the health status of the herd, creating appropriate zoohygienic conditions in livestock housing). By means of EM contained in EM-Farming™ preparations, it is possible to minimize the excessive concentration of troublesome gases or effectively eliminate the inconvenience affecting the animals' discomfort. Preparations containing EMs can also be used in the case of skin diseases, as well as wounds and abrasions in animals [133, 136–138]. Moreover, the role of agarolytic bacteria as producers of exogenous agarose is also known in animal nutrition [133, 138]. Probionts can colonize the gastrointestinal tract. For example, the activity of agaroses in the gastrointestinal tract of female *Haliotis asinina* was significantly higher than that of those fed with probiotics, compared to *H. asinina* fed without agar, and the gross energy content turned out to be lower than that of *H. asinina* fed a diet supplemented with a mixed culture of algae strains (Alg3.1- Abn1.2), which showed a significantly higher growth rate compared to *H. asinina*, fed with a standard diet under laboratory conditions. Communities of certain groups of marine bacteria produce extracellular agarose enzymes that degrade agar into agar polysaccharide and galactose. The introduction of bacteria into *H. asinina* should therefore increase the proportion of digestive enzymes and improve the digestibility of agar, which is the main component of the feed [139].

In dairy cattle, the effect of using EM is a decrease in the number of somatic cells below 100 000, a visible decrease in inflammation, which makes it possible to eliminate the use of antibiotics [122]. In the opinion of Dylewski [136], the application of EM-Farming™ in livestock buildings causes low-temperature fermentation, which displaces rotting. In an environment dominated by beneficial microorganisms, unpleasant odors, gases, and oxidizing compounds are not produced. Useful microorganisms eliminate the decay process and inhibit the release of gases such as ammonia, hydrogen sulfides, and mercaptans, while nitrogen is bound in manure. EM reduce or eliminate pathogenic microorganisms (*Salmonella, enterococci, Escherichia coli bacteria*), mycoses, which leads to the improvement of animal health, e.g. by improving the digestive process, improving physical condition, improving physiological functions, and increasing weight gain [133].

7.8. MICROBIOLOGICAL ORGANISMS IN SUSTAINABLE AGRICULTURE

Currently, more and more attention is being paid to the sustainable development of agriculture, which uses natural adaptive potential with minimal interference in the environment [41, 71, 95, 102, 103, 110, 118]. The most optimal strategy for achieving this goal is to replace agrochemicals, such as fertilizers, synthetic growth regulators, and chemical PPP, with symbiotic microbial preparations that can improve the nutrition of crops and livestock, as well as provide protection against abiotic (heavy metal contamination, aflatoxins, nitrates and nitrites, pesticide residues) and biotic (pests, diseases) stresses [110, 118]. Zydlik and Zydlik [71] tried increasing the biological activity of the soil thanks to the use of three microbiological preparations: Bacto Fill 10B, EM-5, and Humobak PG. Soil biological activity was determined using the following indicators: soil enzymatic activity, dehydrogenase, protease, and respiratory activity. The application of microbiological preparations contributed to a significant increase in the activity of dehydrogenase in the soil. Two of the three preparations used had a lesser effect on protease activity. Only the use of Humobak PG significantly improved the activity of the protease in the soil. Moreover, they proved that the growing season significantly influences the soil and its enzymatic and respiratory activity. In autumn, the activity of enzymes in the soil was significantly higher than in spring, and the opposite was true in spring – respiratory activity was higher [95, 102]. The tested preparations, however, did not have a significant effect on the pH value. The effect of using these preparations, however, turned out to be more significant in the case of soil organic carbon. The use of microbiological preparations in the soil caused changes in the growth and fruiting of plants, as well as an increase in the average leaf area, fruit weight, and juice extract content in strawberry fruit [41, 71].

The use of microorganisms in sustainable agriculture depends on the genetic correlation of plants with the function of symbiotic cohabitants. Thanks to the use of microorganisms, it is possible to recreate the lumpy structure of the soil, provide macro- and microelements inaccessible to plants, accelerate the humification of organic mass in the soil, strengthen the natural resistance of plants, displace pathogens and pests, neutralize the effects of drought, and optimize the carbon to nitrogen ratio [101]. The agricultural potential of plant-microbial symbiosis stems from their ecological effects, especially those well-studied for nitrogen fixation. This knowledge is based on co-evolutionary ecological and molecular research on the mechanisms of mutual adaptation and speciation of plants and microorganisms [2, 140].

Chowdhary et al. [141] characterized microorganisms that work for sustainable environment and health, as well as eliminate hazardous pollutants released from natural and anthropogenic activities, and have a beneficial effect on the environment and human health as well as clean technologies and pollution abnormalities in the context of microorganisms discharged from various sources, their toxicological effects in the environment on animals and plants, and different approaches to biodegradation and bioremediation.

Kumar et al. [142] show the latest advances in the use of microbes in environmental management. They summarize existing practical applications and provide the latest information that will help develop new practices and applications for EM. This research supports the implementation of microorganisms for the benefit of biotechnology, agriculture, environmental management, soil microbiology and waste management.

In-depth knowledge of microbial-based plant symbiosis can provide effective opportunities for the development of sustainable agriculture, so as to ensure an appropriate amount of food production for humans and animals, while respecting the principles of environmental protection. Managing symbiotic microbial communities is possible thanks to the molecular approach based on the continuity of the microbial pool that regularly circulates in nature all the time among soil and plant niches and in animal natural and agricultural ecosystems [101, 112, 142].

The greatest potential for the use of beneficial microorganisms is popular all over the world in terms of small agroecosystems, in order to promote the maintenance of sustainable plant and soil health and to increase resistance of the agroecosystem to unpredictable climatic disturbances [98]. The evaluation of this natural circulation could allow the creation of a highly sustainable farming system that is microorganism-based, obviously taking into account the ecological and genetic consequences of the extensive use of microorganisms in agricultural practice [142].

Today, EM technology is used by more than 140 countries on most continents, and Brazil is now the largest user of EM technology, where EM has been found to be the most appropriate way to replace traditional management methods with natural methods. This reduces the burning and cutting of the Amazon jungle. Thanks to EM technology, farmers around the world will be able to increase the quantity and quality of crops without affecting the environment [55, 98, 101, 112, 143, 144]. In the opinion of Alam et al. [27], this technology will help to control world hunger. While EM should be considered a promising and sustainable agriculture-friendly technology, more research is necessary to better understand the mechanisms by which EM works in a changing climate.

7.9. PERSPECTIVES

Recently, a very dynamic development of the biotechnology industry has occurred, mainly related to the production of biological PPP. One achievement is that OECD member countries have already registered more than 250 biological PPP based on microorganisms, and more than 1000 different biopesticides are produced worldwide. These are not only agents based on living organisms, but also products containing plant extracts, bioregulators, or plant growth biomodifiers, as well as various types of pheromone traps. The results of the research on microbiological biopreparations and the implementation of preparations containing various microorganisms, such as bacteria (*Bacillus, Pseudomonas*), fungi (*Baeuveria, Matharhizium, Pythium, Trichoderma*) and microscopic nematodes (*Heterorahabditis, Steinernema*), and viruses, which reduce (fight) pathogens and pests of crops in various ways and indicate a great success in biotechnology sciences. Estimates of the sale of biopesticides over the next decade of the twenty-first century are predicted to exceed $1 billion. In the last decade, the production and sale of products containing macroorganisms (i.e. parasitic, and predatory insects, predatory mites). Biopreparations used in strictly controlled conditions, such as in vitro laboratories, aeroponic and hydroponic cultivation, etc., have increased. Their participation in the total sale is currently 55–60% of microbiological preparations and other biological PPP – approx. 2000 (28%) [43]. The registration procedures for biological

PPP based on microorganisms are very similar to the registration of chemical protection products. This means that the process is not only extremely stringent, but also very cost-intensive. The difficult and costly procedure of registering biopesticides causes some producers to depart from the registration procedures and even the production of these preparations, which leads to a limited product range and scope of their application. The number of registrations of biopreparations in various countries varies greatly. Most of this research is carried out in the USA.

However, biological PPP are not used as often in agricultural practice as chemical agents, which is mainly due to the fact that no biopreparations have been developed so far to effectively reduce the most important diseases (powdery mildew, rust, fusariosis, etc.) of the main crops, and above all cereals. Moreover, the effectiveness of most biopesticides is lower than that of chemical PPP [1, 143]. This applies, in particular, to biopreparations based on microorganisms (bacteria and fungi), which are usually used for soil under field conditions, in an environment with a high complexity of interaction between soil microorganisms and other soil co-inhabitants. Usually, preparations that can be applied directly to the protected organ of the plant are more effective. However, they are much more expensive, and their use is more burdensome, especially in large-scale field crops [1, 144].

The future of biological plant protection depends on scientific progress in solving the most important problems, such as: increasing the reliability and effectiveness of biopreparations in field conditions and the development of appropriate biopesticide formulations based on living organisms, adapted to large-area technologies of growing the most important agricultural plants. Many plant protection practices are constantly used in the European Union and have the potential to improve plant protection in the future. These changes will only be effective if they are implemented by farmers and accepted by other stakeholders in society [1, 144].

Biotechnology, especially the technique of genetic modification, would be of great help in increasing the yield and reliability of crops. This technique could provide a rapid increase in the ability of microbes to colonize individual plant organs, especially roots, and the production of antibiotic substances that inhibit the growth of pathogens. Due to strong opposition from social organizations and the release of GMOs (genetically modified organisms) into the environment, one cannot count on rapid progress in this area.

In the case of biopreparations that protect the roots, pre-sowing application would be optimal. However, in the case of preparations containing microorganisms, this method of application is very difficult to carry out, because drying the seeds after treatment with biopreparations kills most of the microorganisms.

It is possible to exclude the use of chemical and biological mortars together because the former is harmful to living organisms contained in biopesticides. This is why we are looking for microorganisms resistant to drying out, with a high ability to intensively colonize the roots, or with a broad spectrum of antifungal activity. However, biological PPP containing microorganisms are successfully registered in OECD countries.

Microorganisms approved for use in the EU are of low risk, as they decompose very quickly and leave no residues that could affect living organisms, either in the environment or in food. Therefore, they meet all the requirements to be classified as biologically active low-risk factors. Nevertheless, when assessing microorganisms, it should be carefully checked whether they can interact with other microorganisms in the environment and produce secondary metabolites with negative effects on humans and animals. Another problem is the issues of interspecific transfer of genetic material and its impact on the environment. For example, the assessment of *B. thuringiensis* subsp. *tenebrionis* strain NB-176.26 raises such doubts [14]. When using plant extracts in plant protection, problems also arise with the proper identification of mixtures of many active substances [145–152]. However, most of the currently approved biopesticides meet the criteria for low-risk active substances. According to the official Annex VI to Regulations (EC) No. 1095/2007 and No. 283/2013 [153–156], biopesticides should be characterized by the absence of any harmful effects on animal health and the quality of groundwater, and not exert any other unacceptable impact on the natural environment. The definitions of biopesticides are also included in the relevant regulations (EC) of the European Parliament and of the Council of Europe, as active substances for which MRLs (maximum residue levels) are not required due to insufficient risk [156, 157]. The entire branch of agricultural sciences, including plant protection, also actively participates in the promotion of biopesticides [158, 159]. The addition of EM in a sustainable agriculture will increase the availability of mineral resources to improve plant growth and their yield. Regardless of the exact mechanism or mechanisms of action of EM, more research is needed to establish the true efficacy of EM in field conditions where plants are exposed to many coexisting biotic (weeds, diseases, pests) and abiotic stresses (changing soil, climatic, meteorological, and graphical conditions).

With regard to precision farming, it can be expected that improved EM application and spraying technologies will improve the efficiency of EM application in plant cultivation and reduce the use of pesticides and environmental pollution [159].

7.10. CONCLUSIONS

Society strives to improve the state of the environment and restore its natural balance. Thanks to the latest advances in biology, genetics, soil science, breeding, engineering, microbiology, and information technology, environmental biotechnologists are now better able to harness microbial communities to provide many services to society. These activities include, but are not limited to: detoxification of water, wastewater, sediment, and soil, as well as renewable energy from biomass, and the detection of heavy metal contamination, aflatoxins, nitrates, or pathogens. They help protect society from exposure to all types of pollutants and from all kinds of pathogens. The effect of microorganisms on composting, assessed using a metagenomic approach, increases the biodiversity of the compost and the population in the thermophilic phase. This is because effective microbes improve the composting rate, compost quality, and their mineralization, and they ultimately make the abundance of beneficial microbes better. Thermoacidic microorganisms have the potential to transform composting. Due to the unique position of EMs and their unpredictable microbiological features and their biosynthetic abilities, under certain environmental conditions, they can contribute to solving various problems in agriculture sciences as well as in other fields of science.

The proper use of microorganisms for social, environmental, and economic benefits is very attractive and contributes to a significant advance in research and its application in practice. The microorganisms of HNADM were also discussed. The influence of cultivation conditions on microorganisms was demonstrated. The mechanism of microbial nitrogen and phosphorus metabolism has been proposed.

REFERENCES

1. Kucharski, J. and Jastrzębska, E. (2005). The role of effective microorganisms in shaping the microbiological properties of soil. *Ecological Engineering 12*: 295–296. (in Polish).
2. Li, X., Guo, Q., Wang, Y. et al. (2020). Enhancing nitrogen and phosphorus removal by applying effective microorganisms to constructed wetlands. *Water 12*: 2443.

3. Mann, R.S. and Kaufman, P.E. (2012). Natural product pesticides: their development, delivery and use against insect vectors. *Mini-Reviews in Organic Chemistry 9*: 185–202.
4. Coskun, M., Coskun, M., and Cayir, A. (2011). Frequencies of micronuclei (MNi), nucleoplasm bridges (NPBs), and nuclear buds (NBUDs) in farmers exposed to pesticides in Çanakkale, Turkey. *Environment International 1*: 93–96.
5. Wong, R.H., Chang, S.Y., and Ho, S.W. (2008). Polymorphisms in metabolic GSTP1 and DNA-repair XRCC1 genes with an increased risk of DNA damage in pesticide-exposed fruit growers. *Mutation Research 654*: 168–175.
6. Biswas, S.K., Shankar, U., Kumar, S. et al. (2015). Effect of bio-fertilizers on the treatment of spot disease and parameters influencing the growth and yield of wheat. *Journal of Pure and Applied Microbiology 9* (4): 3025–3030.
7. Commission Implementing Regulation (EU), No. 540/2011, OJ L 153 of 11.6.2011: 1–186.
8. (2012). European food safety authority. *EFSA Journal 10* (10): sf101. https://www.efsa.europa.eu/en/efsajournal/pub/sf101 (accessed 17 December 2021).
9. European Food Safety Authority (2012). Conclusion on the peer review of the pesticide risk assessment of the active substance garlic extract. *EFSA Journal 10* (2): 2520–2559.
10. European Food Safety Authority (2012). Conclusion on the peer review of the pesticide risk assessment of the active substance gibberellins. *EFSA Journal 10* (1): 2502–2551.
11. European Food Safety Authority (2013). Conclusion on the peer review of the pesticide risk assessment of the active substance Bacillus thuringiensis ssp. tenebrionis strain NB-176. *EFSA Journal 11* (1): 3024–3059.
12. European Food Safety Authority (2013). Conclusion on the peer review of the pesticide risk assessment of the active substance carbon dioxide. *EFSA Journal 11* (5): 3053–3153.
13. European Food Safety Authority (2013). Conclusion on the peer review of the pesticide risk assessment of the active substance orange oil. *EFSA Journal 11* (2): 3090–3144.
14. European Union Pesticides Database (2019). (Online). Available: http://ec.europa.eu/food/plant/protection/evaluation/database_act_subs_en.html (accessed 02.10.2021).
15. Farahat, F.M., Ellison, C.A., and Bonner, M.R. (2011). Biomarkers of Chlorpyrifos exposure and effect in Egyptian cotton field workers. *Environmental Heath Perspectives 119* (6): 801–806. Published online 11 January 2011, https://doi.org/10.1289/ehp.1002873.
16. Kapka-Skrzypczak, L., Cyranka, M., Kruszewski, M. et al. (2011). Plant protection products and farmers' health - biomarkers and the possibility of using them to assess exposure to pesticides. *MedycynaOgólna i Nauki o Zdrowiu 17* (1): 28–32. (in Polish).
17. Skiba, D., Sawicka, B., Pszczółkowski, P. et al. (2021). Impact of crop management and weed control systems of very early potatoes on weed infestation, biodiversity, and tuber health safety. *Life 11* (826): 33. https://doi.org/10.3390/life11080826.
18. Bolognesi, C. (2003). Genotoxicity of pesticides: a review of human biomonitoring studies. *Mutation Research 3*: 251–272.
19. Barbaś, P. and Sawicka, B. (2017). Comparison of the profitability of various methods of weed infestation in edible potato cultivation. *Problems of Agricultural Engineering 2* (96): 5–15. ISSN 1231-0093. (in Polish).
20. Nowacka, A., Gnusowski, B., and Dąbrowski, J. (2006). Remains of protection measures in agricultural crops. *Progress in Plant Protection/Postępy w Ochronie Roślin 47* (4): 79–90. (in Polish).
21. Martínez-Valenzuela, C., Gómez-Arroyo, S., and Villalobos-Pietrini, R. Genotoxic biomonitoring of agricultural workers exposed to pesticides in the north of Sinaloa State, Mexico. *Environment International 8*: 1155–1159.
22. Bortoli, G.M., Azevedo, M.B., and Silva, L.B. (2009). Cytogenetic biomonitoring of Brazilian workers exposed to pesticides. Micronucleus analysis in buccal epithelial cells of soybean growers. *Mutation Research 1–2*: 1–4.
23. Crestani, M., Menezes, C., Glusczak, L. et al. (2007). Effects of Clomazone Herbicide on biochemical and histological aspects of silver catfish (*Rhamdia quelen*) and recovery pattern. *Chemosphere 67*: 2305–2311.

24. Santos, M.D., Crestani, M., Shettinger, M.R. et al. (2005). Effects of the herbicides clomazone, quinclorac, and metsulfuron methyl on acetylcholinesterase activity in the silver catfish (*Rhamadia quelen*). *Ecotoxicology and Environmental Safety 6*: 398–403.
25. Anonymous. Assessment of the effectiveness of plant protection products. Resistance risk analysis. Organisation Européenne et Méditerranéennepour la Protection des Plantes, 2002, PP 1/213(2).
26. Li, X., Pan, Y., Xi, Q. et al. (2021). Effects of exogenous germanium and effective microorganisms on germanium accumulation and nutritional qualities of garlic (*Allium sativum* L.), Scientia. *Horticulture 283*: 110114. https://doi.org/10.1016/j.scienta.2021.110114.
27. Alam, A., Bibi, F., Deshwal, K. et al. (2022). Biofortification of primary crops to eliminate latent hunger: an overview. *Natural Resources for Human Health 2* (1): 91–99. https://doi.org/10.53365/nrfhh/142883.
28. Gupta, K. (2021). Profiles of the bio-fertilizer industry on the market. In: *Biofertilizers: Research and Impact* (ed. I. Inamuddin, M.I. Ahamed, R. Boddula and M. Rezakazemi). Wiley https://doi.org/10.1002/9781119724995.ch18.
29. Gupta, P.K., Balyan, H.S., Sharma, S. et al. (2021). Biofortification and bioavailability of Zn, Fe and Se in wheat: present status and future prospects. *Theoretical and Applied Genetics 134* (1): 1–35. https://doi.org/10.1007/s00122-020-03709-7.
30. Gupta, A.K., Sharma, M.L., Khan, M.A. et al. (2015). Problems faced by tribes in collection and marketing of Non-Timber Forest Products (NTFPs) in Chhattisgarh, India. *Plant Archives 15* (2): 789–793.
31. Singh, M., Biswas, S.K., Nagar, D. et al. (2017). Effect of biofertilizer on growth parameters and potato yield. *Journal of Current Microbiology and Regret Science 6* (5): 1717–1724.
32. De Assis, R.M.A., Carneiro, J.J., Medeiros, A.P.R. et al. (2020). Arbuscular mycorrhizal fungi and organic manure enhance growth and accumulation of citral, total phenols, and flavonoids in Melissa officinalis L. *Industrial Crops and Products 158*: 112981. https://doi.org/10.1016/j.indcrop.2020.112981.
33. Mathur, S., Tomar, R.S., and Jajoo, A. (2019). Arbuscular mycorrhizal fungi (AMF) protect photosynthetic apparatus of wheat under drought stress. *Photosynthesis Research 139*: 227–238. https://doi.org/10.1007/s11120-018-0538-4.
34. Mathur, S. and Jajoo, A. (2020). Arbuscular mycorrhizal fungi protect maize plants from high temperature stress by regulating photosystem II heterogeneity. *Industrial Crops and Products 143*: 111934. https://doi.org/10.1016/j.indcrop.2019.111934.
35. Igiehon, N.O., Babalola, O.O., Cheseto, X. et al. (2021). Effects of rhizobia and arbuscular mycorrhizal fungi on yield, size distribution and fatty acid of soybean seeds grown under drought stress. *Microbiological Research 242*: 126640. https://doi.org/10.1016/j.micres.2020.126640.
36. Paravar, A., Farahani, S.M., and Rezazadeh, A. (2021). L*allemantia* response to drought stress and the use of Arbuscular mycorrhizal fungi. *Crops and Industrial Products 172*: 114002.
37. Plouznikoff, K., Asins, M.J., de Boulois, H.D. et al. (2019). Genetic analysis of tomato root colonization by arbuscular mycorrhizal fungi. *Annals of Botany 124*: 933–946. https://doi.org/10.1093/aob/mcy240.
38. Liu, R., Abdelraouf, E., Bicego, B. et al. Deficit irrigation: a viable option for sustainable confection sunflower (*Helianthus annuus* L.) production in the semi-arid US. *Irrigation Science 36*: 319–328. https://doi.org/10.1007/s00271-018-0588-6.
39. Liu, J., Liu, X., Zhang, Q. et al. (2020). Response of alfalfa growth to arbuscular mycorrhizal fungi and phosphate-solubilizing bacteria under different phosphorus application levels. *AMB Express 10*: 1–13. https://doi.org/10.1186/s13568-020-01137-w.
40. Ghanbarzadeh, Z., Mohsenzadeh, S., Rowshan, V. et al. (2020). Mitigation of water deficit stress in *Dracocephalum moldavica* L. by symbiotic association with soil microorganisms. *Scientia Horticulturae 272*: 109549. https://doi.org/10.1016/j.scienta.2020.109549.
41. Ramos-González, M.I., Matilla, M.A., Quesada, J.M. et al. (2013). Using genomics to discover bacterial lifestyle determinants in the rhizosphere. In: *Molecular Microbiological Ecology of the Rhizosphere* (ed. F.J. de Bruijn) Issue 1. ISBN: 9781118296172. https://doi.org/10.1002/9781118297674.

42. Lee, D.H. and Jacobs, J. (2009). Serum gamma-glutamyl transferase: new insights about an old enzyme. *Journal of Epidemiology and Community Health 63*: 884–886.
43. FAOSTAT. Faostat data base 2019. http://www.fao.org/faostat/en (accessed: 28.09.2021).
44. Council Directive 91/414/EEC of 15 July 1991 concerning the placing of plant protection products on the market. OJ L 230, 19.8.1991: p. 1–32.
45. Regulation (EC) No 1107/2009 of the European Parliament and of the Council on the placing of plant protection products on the market and repealing Council Directives 79/117/EEC and 91/414/EEC. 2009. https://www.fao.org/faolex/results/details/en/c/LEX-FAOC103352 (accessed 14.12.2021)
46. Villaverde, J.J., Sevilla-Morán, B., Sandín-España, P. et al. Biopesticides in the framework of the European Pesticide Regulation (EC) No. 1107/2009. *Pest Management Science 70*: 2–5.
47. Ropek, D. and Kołodziejczyk, M. (2018). Effectiveness of selected insecticides and natural preparations against *L. decemlineata*. *Potato Research 62*: 85–95. https://doi.org/10.1007/s11540-018-9398-8.
48. Rotter, S., Beronius, A., Boobis, A.R. et al. (2019). Overview on legislation and scientific approaches for risk assessment of combined exposure to multiple chemicals: the potential Euro Mix contribution. *Critical Reviews in Toxicology 48* (9): 796–814. https://doi.org/10.1080/10408444.2018.1541964.
49. Defarge, N., Takács, E., Lozano, V.L. et al. (2016). Co-formulants in glyphosate-based herbicides disrupt aromatase activity in human cells below toxic levels. *International Journal of Environmental Research and Public Health 13*: 264–280.
50. Roth N, Wilks MF. Combination ("cocktail") effects of pesticide residues in food SCAHT report for FSVO. Technical Report, 2018, 06 November 2018, https://doi.org/10.13140/RG.2.2.30899.07208.
51. Hameed, T.S. and Sawicka, B. The level consciousness of farmers about a negative influence of pesticides on the environment. Scientific Conference: Agricultural systems in crop production. Production - quality - environment. 25 years of research at IUNG-PIB. Puławy, 06-07.06.2019. In: *Environment* (ed. G. Hołubowicz-Kliza, M. Zarychta and E. Markowska-Strzemska), 116–117. Puławy, Poland, ISBN: 987-83-7562-302-4.
52. Sawicka, B., Egbuna, C., Nayak, A.K. et al. (2020). Chapter 2. Plant diseases, pathogens, and diagnosis. PART I. Green approach to pest and disease control. In: *Natural Remedies for Pest, Disease and Weed Control* (ed. C. Egbuna and B. Sawicka), 1–16. London: Elsevier, Academic Press ISBN: 978-0-12-819304-4, https://doi.org/10.1016/B978-0-12-819304-4.00001-4.
53. Gałązka, A. (2015). Soil contamination with oil derivatives and biological methods of purification. *Kosmos, Problemy Nauk Biologicznych 64* (1) (306): 145–164. (in Polish).
54. Egbuna, C., Sawicka, B., Tijjani, H. et al. Chapter 4. Biopesticides, safety issues and marked trends, PART I. Green approach to pest and disease control. In: *Natural Remedies for Pest, Disease and Weed Control* (ed. C. Egbuna and B. Sawicka), 43–53. London: Elsevier, Academic Press ISBN: 978-0-12-819304-4, https://doi.org/10.1016/B978-0-12-819304-4.00001-4.
55. Vaitkeviciene, N. The Effect of Biodynamic Preparations on the Accumulation of Biologically Active Compounds in the Tubers of ˙Different Genotypes of Ware Potatoes. Ph.D. Thesis, Agricultural Sciences, Agronomy (01A), ASU, Akademija, Kaunas, Lithuania, 2016, pp. 212.
56. Vaitkeviciene, N., Jariene, E., Danilcenko, H. et al. (2016). Effect of biodynamic preparations on the content of some mineral elements and starch in tubers of three colored potato cultivars. *Journal of Elementology 21*: 927–935.
57. Keidan M. Optimization of winter oilseed rape technological parameters in the organic farming system. PhD thesis, Aleksandras Stulginskis University, Kaunas, pp. 221.
58. Kulig, B., Lepiarczyk, A., Oleksy, A. et al. (2010). The effect of tillage system and fore crop on yield and values of LAI and SPAD indicators of spring wheat. *European Journal of Agronomy 2010* (33): 43–51.
59. Sawicka, B. The possibilities of uses of biomarkers to the estimation of exposure of potato to herbicides. 9th International Conference on Biomarkers Expert Speakers. Evidence based. Health Care Professionals. October 26–28: 2017, Osaka, Japan. http://www.omicsgroup.com/conferences/ACS/conference/pdfs/molecular-cancer-biomarkers2017_workshop-program.pdf

60. Levickienė, D., Jarienė, E., Gajewski, M. et al. (2017). Influence of harvest time on biologically active compounds and antioxidant activity in mulberry leaves in Lithuania. *Notulae Botanicae Horti Agrobotanici Cluj-Napoca 45* (2): 431–436.
61. Levickienė D. The Influence of the Biodynamic Preparations on the Soil Properties and Accumulation of Bioactive Compounds in the Leaves of White Mulberry (*Morus alba* L.). PhD thesis, Aleksandras Stulginskis University, Kaunas 2018: pp. 212.
62. NFUP. Nationales Fremd Stoffuntersuchungs Programm, Jahresbericht, Bern, Schweiz, 2017.
63. Werle, R., Lowell, D., Sandeln, L.D. et al. (2014). Predicting the emergence of 23-year-old annual weed species. *Weed Science 62* (2): 267–279. https://doi.org/10.1614/WS-D-13-00116.1.
64. Norris, R.F., Elmore, C.L., Remine, M. et al. (2001). Spatial arrangement, density, and competition between *Barnyardgrass* and *Tomato:* I. Crop growth and yield. *Weed Science 49* (61–68): 378.
65. Ngouajio, M., McGiffen, M.E., and Hembree, K.J. (2009). Tolerance of tomato cultivars to velvetleaf interference. *Weed Science 49* (379): 91–98. https://doi.org/10.1614/0043-1745(2001)049(0091:TOTCTV)2.0.CO, 2 380.
66. Goldwasser, Y., Lanini, W., and Wrobel, R. (2009). Tolerance of tomato varieties to lespedeza dodder. *Weed Science 49*: 520–523. https://doi.org/10.1614/0043-1745.
67. Andrew, I.K.S., Storkey, J., and Sparke, D.L. (2015). A review of the potential for competitive cereal cultivars as a tool in integrated weed 383 management. *Weed Research 55*: 239–248. https://doi.org/10.1111/wre.12137 384.
68. Olofsdotter, M., Navarez, D., Rebulanan, M. et al. (1999). Weed suppressing rice cultivars - does allelopathy play a role? *Weed Research 39*: 441–454. https://doi.org/10.1046/j.1365-3180.1999.00159.x 386.
69. Macías, F.A., Mejpias, F.J.R., and Molinillo, J.M.G. (2019). Recent advances in allelopathy for weed control: from knowledge to applications. *Pest Management Science 75*: 2413–2436. https://doi.org/10.1002/ps.5355 388.
70. Farooq, M., Bajwa, A.A., Cheema, S.A. et al. (2013). Application of allelopathy in crop production. *International Journal of Agriculture and Biology 15* (389): 1367–1378. https://doi.org/10.1007/978-3-642-30595-5_6.
71. Zydlik, Z. and Zydlik, P. (2013). Influence of microbiological products on soil properties under conditions of replantation disease. *Zemdirbyste-Agriculture 100* (1): 19–24. https://doi.org/10.13080/za 2013.100.003.
72. Boligłowa, E. (2005). Potato protection against diseases and pests using Effective Microorganisms (EM) with herbs. In: *Selected Ecological Issues in Modern Agriculture* (ed. Z. Zbytek), 165–170. Poznań: PIMR (in Polish).
73. Janas, R. (2009). Possibilities of using effective microorganisms in ecological systems of crop production. *Problems of Agriculture Engineering 3*: 111–119. (in Polish).
74. Ji, B., Hu, H., Zhao, Y. et al. (2014). Effects of deep tillage and straw returning on soil microorganism and enzyme activities. *Scientific World Journal 1–12*: https://doi.org/10.1155/2014/451493.
75. Kołodziejczyk, M., Szmigiel, A., and Ropek, D. (2009). Effectiveness of potato protection production using selected insecticides to combat potato beetles (*Leptinotarsa decemlineata* Say). *Acta Scientiarum Polonorums. Agricultura 8* (4): 5–14.
76. Regulation EC 1107/2009 of the European Parliament and of the Council of 21 October 2009 concerning the placing of plant protection products. 2009. https://eur-lex.europa.eu/legal-content/PL/ALL/?uri=celex%3A32009R1107
77. Daniel, C. and Wyssa, E. (2010). Field applications of *Beauveria bassiana* to control the Rhizolites cerasi fruit fly. *Journal of Applied Entomology 134* (9–10): 675–681. https://doi.org/10.1111/j.1439-0418.2009.01486.x.
78. Macías, F.A., Rodríguez, M.J., and Molinillo, M.G. Recent advances in weed control allelopathy. From knowledge to application. *Pest Control Science 75* (9): https://doi.org/10.1002/ps.5355.
79. Rudeen, M.L., Jaronski, S.T., Petzold-Maxwell, J.L. et al. (2013). Entomopathogenic fungi in cornfields and their potential to fight the larvae of western corn rootworm *Diabroticavirgiferavirgifera. Journal of Invertebrate Pathology 114* (3): 329–332. https://doi.org/10.1016/j.jip.2013.09.009.

80. Reddy, G.V.P., Tangtrakulwanich, K., Wua, S. et al. (2014). Evaluation of the effectiveness of entomopaths in the control of wireworms (*Coleoptera: Elateridae*) on spring wheat. *Journal of Invertebrate Pathology 120*: 43–49. https://doi.org/10.1016/j.jip.2014.05.005.
81. Long, D.W., Drummond, F.A., and Groden, E. (1998). Sensitivity of Colorado potato beetle (*Leptinotarsa decemlineata*) *to Beauveria bassiana*. *Journal of Invertebrate Pathology 71*: 182–183. https://doi.org/10.1006/jipa.1997.4714.
82. Laznik, Ž., Tóth, T., Lakatos, T. et al. (2010). Inspection of Colorado potato beetle (*L. decemlineata* Say) on potato under field conditions: comparison of the effectiveness of the two strains *Steinernemafeltiae* (Filipjev) and spraying with thiamethoxam. *Journal of Plant Diseases and Protection 117* (3): 129–135.
83. Lepiarczyk, A., Kulig, B., and Stępnik, K. (2005). The effect of simplified soil cultivation and forecrops on LAI development of selected winter wheat varieties for crop rotation. *Fragmenta Agronomica 2* (86): 98–105.
84. Xu, H.L. (2000). Effects of a microbial inoculant and organic fertilizer on the growth, photosynthesis, and yield of sweet corn. *Journal of Crop Production 3*: 183–214.
85. Kalitkiewicz, A. and Kępińska, E. (2008). The use of rhizobacteria to stimulate plant growth. *Biotechnology 2*: 102–114.
86. Mayer, J., Scheid, S., and Oberholzer, H.R. (2008). How effective are 'Effective Microorganisms'? Results from an organic farming field experiment. In: *16th IFOAM Organic World Congress*, 18–20 June, 168–171. Modena, Italy.
87. Kaczmarek, Z., Owczarzak, W., Mrugalska, L. et al. (2007). Effect of effective microorganisms on selected physical and water properties of arable-humus levels of mineral soils. *Journal of Research and Applications in Agricultural Engineering 52*: 73–77. (In Polish).
88. Kaczmarek, Z., Jakubas, M., Grzelak, M. et al. (2008). Impact of the addition of various doses of Effective Microorganisms to arable-humus horizons of mineral soils on their physical and water properties. *Journal of Research and Applications in Agricultural Engineering 53*: 118–121.
89. Kaczmarek, Z., Wolna-Murawska, A., and Jakubas, M. (2008). Change in the number of selected groups of soil microorganisms and enzymatic activity in soil inoculated with effective microorganisms (EM). *Journal of Research and Applications in Agricultural Engineering 53*: 122–128.
90. Szembowski, B. (2009). *Experiences of a Farm in Trankwice with the EM-FarmingTM Biotechnology. Natural Probiotic Microorganisms*, 56–58. Lichen, Poland: Associate Ecosystem (In Polish).
91. Kosicka, D., Wolna-Murawka, A., and Trzeciak, M. (2015). Influence of microbiological preparations on soil and plant growth and development. *Kosmos 64* (2): 327–335. (in Polish).
92. Kołodziejczyk, M. (2014). Effectiveness of nitrogen fertilization and application of microbial preparations in potato cultivation. *Turkish Journal of Agriculture and Forestry 38*: 299–310.
93. Kołodziejczyk, M. (2014). Effect of nitrogen fertilization and microbial populations on potato yielding. *Plant, Soil and Environment 60* (8): 379–386.
94. Pszczółkowski, P. and Sawicka, B. (2018). The impact of the application of biopreparations and fungicides on the yield and selected parameters of the seed value of seed potatoes. *Acta Agrophysica 25* (2): 239–255.
95. Pszczółkowski, P. and Sawicka, B. (2018). The impact of the use of fungicides, microbiological preparations, and herbal extracts on the shaping of the potato yield. *Fragmenta Agronomica 35* (1): 81–93.
96. Pszczółkowski, P., Krochmal-Marczak, B., Sawicka, B. et al. (2021). Effect of the use of effective microorganisms on the color of raw potato tuber flesh for food processing. *Applied Sciences 11* (19): 8959. https://doi.org/10.3390/app11198959.
97. Sawicka, B., Pszczółkowski, P., Kiełtyka-Dadasiewicz, A. et al. (2021). Effect of effective microorganisms on the quality of potatoes in food processing. *Applied Sciences 11*: 1415. https://doi.org/10.3390/app11041415.
98. Koskey, G., Mburu, S.W., Awino, R. et al. (2021). Potential use of beneficial microorganisms for soil amelioration, phytopathogen biocontrol, and sustainable crop production in smallholder agroecosystems. *Frontiers in Sustainable Food Systems 5*: 130. https://doi.org/10.3389/fsufs.2021.606308.

99. Galarreta, J.I.R., Ezpelata, B., Pascualena, J. et al. (2006). Combining ability in early generations of potato breeding. *Plant Breeding 125*: 183–186.
100. Kolasa-Wiącek, A. (2010). Will effective microorganisms revolutionize the world? *Postharvest Technology Convert Food 1*: 66–69. (In Polish).
101. Kowalska, J., Sosnowska, D., Remlein-Starosta, D. et al. (2021). *Effective Microorganisms in Organic Farming*. Poznań: IOR-PIB https://www.ior.poznan.pl/plik,3661,sprawozdanie-mikroorganizmy-w-rol-eko-2011-pdf.pdf.
102. Martyniuk, S. (2011). Effective and ineffective microbiological preparations used in the protection and cultivation of plants as well as reliable and unreliable methods of their evaluation. *Postepy Microbiology 50* (4): 321–328.
103. Okorski, M. and Majchrzak, B. (2008). Fungi colonizing pea seeds after applying the EM 1 microbiological preparation. *Progress in Plant Protection/Postępy w Ochronie Roślin 48* (4): 1314–1318.
104. Higa, T. (1998). Effective microorganisms, concept, and the latest technological achievements. Materials from the Effective Microorganisms Conference for Sustainable Agriculture and Environment. In: *4th Kyusei International Wildlife Agriculture Conference*, 247–248. Bellingham-Washington, USA.
105. Hoshino, Y., Satou, N., and Higa, T. (2002). Remediation and protection of natural ecosystems through the use of effective microorganisms (EM). In: *Proceedings of the 7th International Conference on Kyusei Natural Farming*, 129–137. New Zealand.
106. Higa, T. (2005). Effective microorganisms – technology of the 21st century. In: *Conference: "Effective Microorganisms in the World"*, 20–24. London, UK.
107. Zarzecka, K. and Gugała, M. (2013). Effect of UGmax soil fertilizer on the potato yield and its structure. *Bulletin IHAR 267*: 107–112. (In Polish).
108. Baranowska, A., Zarzecka, K., Gugała, M. et al. (2018). The effect of fertilizer on UGmax soil on the presence of *Streptomyces scabies* on edible potato tubers. *Journal of Ecological Engineering 3*: 68–73. https://doi.org/10.12911/22998993/85743.
109. Van Vliet, P.C.J., Bloem, J., and de Goede, R.G.M. (2006). Microbial diversity, nitrogen loss and grass production after addition of Effective Microorganisms® (EM) to slurry manure. *Applied Soil Ecology 32*: 188–198.
110. Martyniuk, S. (2010). Production of microbiological preparations on the example of symbiotic bacteria in legumes. *Journal of Research and Applications in Agricultural Engineering 55*: 20–23. (In Polish).
111. Martyniuk, S. and Księżak, J. (2011). Evaluation of pseudo-microbial biopreparations used in plant production. *Polish Agronomic Journal 6*: 27–33.
112. Paśmionka, I. and Kotarba, K. (2015). Possibilities of using effective microorganisms in environmental protection. *Cosmos, Problems of Biological Sciences 64* (1): 173–184.
113. Zakaria, Z., Gairola, S., and Shariff, N.M. (2010). Effective microorganisms (EM) technology for water quality restoration and potential for sustainable water resources and management. In: *Proceedings of the International Congress on Environmental Modelling and Software Modelling for Environment's Sake, Fifth Biennial Meeting*, Ottawa, ON, Canada, 5–8 July, 54.
114. Okumoto, S., Shintani, M., and Higa, T. (2019). The use of effective microorganisms and biochar inhibits the transfer of radioactive caesium from soil to plant during continuous Komatsu a cultivation. In: *Proceedings of the International Scientific Conference "Radiobiology: Present", Institute of Radiobiology of the NAS of Belarus*, Gomel, Belarus, 26–27 September, 22.
115. Nigussie, A., Dume, B., Ahmed, M. et al. (2021). Effect of microbial inoculation on nutrient turnover and lignocellulose degradation during composting: a meta-analysis. *Waste Management 125*: 220–234. ISSN 0956-053X https://doi.org/10.1016/j.wasman.2021.02.043.
116. Sawicka, B., Pszczółkowski, P., Noaema, A.H. et al. (2019). Effective microorganisms in agriculture and food processing. In: *Modern Research on the State of the Environment and the Therapeutic Use of Plants* (ed. M. Chwil and M.M. Skoczylas), 45–64. Lublin, Poland: Publishing House of the University of Life Sciences in Lublin.

117. Allahverdiyev, S.R., Kırdar, E., Gunduz, G. et al. (2011). Effective Microorganisms (EM). *Technology in Plants. Technology 14*: 103–106.
118. Pszczółkowski, P., Sawicka, B., Danilcenko, H., and Jariene, E. (2017). The role of microbiological preparations in improving the quality of potato tubers. In: *Proceedings of the International Scientific Conference: 'New trends in food safety and quality'*, 22–23. Akademija, Lithuania, 5–7 October: Aleksandras Stulginskis University.
119. Sawicka, B. and Egbuna, C. (2020). Chapter 1. Pest of agricultural crops and control measures. Part I. Green approach to pest and disease control. In: *Natural Remedies for Pest, Disease and Weed Control* (ed. C. Egbuna and B. Sawicka), 17–28. Elsevier, Academic Press. ISBN: ISBN: 978-0-12-819304-4 https://doi.org/10.1016/B978-0-12-819304-4.00001-4.
120. Sawicka, B. (2008). Rate of spread of fungal diseases on potato plants as affected by application of a biostimulator and foliar fertilizer. In: *Biostimulators in Modern Agriculture. Solanaceous Crops* (ed. Z. Dąbrowski), 68–76. Warsaw. Plantpress: Editorial House: WieśJutra, Limited. ISBN: ISBN 83-89503-55-7.
121. Boligłowa, E. and Gleń, K. (2018). Assessment of effective microorganism activity (EM) in winter wheat protection against fungal diseases. *Ecological Chemistry and Engineering A 15*: 23–27.
122. Gacka, S. (2009). *Economic Aspects of Implementing EM-FarmingTM Biotechnology. Natural Probiotic Microorganisms*, 28–33. Licheń: Ecosystems Association (In Polish).
123. Tommonaro, G., Abbamondi, G.R., Mikołaja, B. et al. (2021). Productivity and nutritional trait improvements with various tomatoes grown effective microorganisms technology. *Agriculture 11* (2): 112. https://doi.org/10.3390/agriculture11020112.
124. Quiroz, M. and Céspedes, C. (2019). Bokashi as a fix and nitrogen source for sustainable farming systems: an overview. *Journal of Soil Science and Plant Nutrition 19* (1): 237–248. https://doi.org/10.1007/s42729-019-0009-99.
125. Marczakiewicz, J. (2009). *Another Year with EM Biotechnology at RZD SGGW Chylice. Natural Probiotic Microorganisms*, vol. *65*. Licheń: Ecosystem Association Publishing House (in Polish).
126. Grabowska, E. (2009). *EM-FarmingTM in Horticultural Production. Natural Probiotic Microorganisms*, 41–42. Licheń: Ecosystem Association.
127. Al-Taweil, H.I., Bin Osman, M., Hamid, A.A. et al. (2009). Development of microbial inoculants and the impact of soil application on rice seedlings growth. *American Journal of Agricultural and Biological Sciences 4*: 79–82.
128. Szewczuk, C., Sugier, D., Baran, S. et al. (2016). The impact of fertilizing agents and different doses of fertilizers on selected soil chemical properties as well as the yield and quality traits of potato tubers. *Annales Universitatis Mariae Curie-Skłodowska, Sectio E Agricultura LXXI* (2): 65–79.
129. Solarska, E. (2010). Development of ecological hop production technology. In: *1st Lublin Scientific and Technical Conference – Microorganisms in Environmental Revitalization - Science and Practice*. Lublin, 23–24 March, 56–64. (in Polish).
130. Henry, A.B., Maung, C.E.H., and Kim, K.Y. (2020). Metagenomic analysis reveals enhanced biodiversity and composting efficiency of lignocellulosic waste by thermoacidophile effective microorganism (tEM). *Journal of Environmental Management 276*: 111252. ISSN 0301-4797, https://doi.org/10.1016/j.jenvman.2020.111252.
131. Ney, L., Franklin, D., Mahmud, K. et al. (2020). Impact of inoculation with local effective microorganisms on soil nitrogen cycling and legume productivity using composted broiler litter. *Applied Soil Ecology 154*: 103567. https://doi.org/10.1016/j.apsoil.2020.103567.
132. Song, T., Zhang, X., Li, J. et al. (2021). A review of research progress of heterotrophic nitrification and aerobic denitrification microorganisms (HNADMs). *Science of the Total Environment 80*: 149319. https://doi.org/10.1016/j.scitotenv.2021.149319.
133. Faturrahman, L., Meryandini, A., Junior, M.Z. et al. (2015). The role of agarolytic bacteria in enhancing physiological function for digestive system of abalone (*Haliotis asinine*). *Journal of Applied Environmental and Biological Sciences 5* (5): 49–56.
134. Gacka, S. and Kolbusz, S. (2009). *Biotechnology EM-FarmingTM – Comprehensive, Natural, Solution in Animal Production Ensuring Animal Welfare. Natural Probiotic Microorganisms*, 102–103. Licheń: Ekosystem Association (In Polish).

135. Widuch, E. (2009). Niewidzialni dobroczyńcy. Naturalne probiotyczne mikroorganizmy. *Wydawnictwo Stowarzyszenia "Ekosystem" Licheń*: 27–28. In Polish.
136. Dylewski, T. (2009). *EM-FarmingTM in Horticultural Production. Natural Probiotic Microorganisms*, 77–78. Licheń: Ecosystems Association (in Polish).
137. Daly, M.J. and Stewart, D.P.C. (1999). Influence of "effective microorganisms" (EM) on vegetative production and carbon mineralization a preliminary investigation. *Journal of Sustainable Agriculture 4*: 15–25.
138. Fatunbi, O. and Ncube, L. (2009). Activities of effective microorganism (EM) on the nutrient dynamics of different organic materials applied to soil. *American Eurasian Journal of Agronomy 2* (1): 26–35.
139. Burd, G.I., Dixon, D.G., and Glick, B.R. (1998). A plant growth-promoting bacterium that decreases nickel toxicity in seedlings. *Applied and Environmental Microbiology 64*: 3663–3668.
140. Gopalasamy, R. (2019). The role of microorganisms in sustainable development. *International Journal of Scientific Research 6* (7): 413.
141. Chowdhary, P., Raj, A., Verma, D., and Akhte, Y. (ed.) (2020). *Microorganisms for Sustainable Environment and Health.*, 1e; ISSN: Paper book: 978-0-12-819001-2, eBook, 978-0-12-819004-3.
142. Kumar, A., Singh, V.K., Singh, P. et al. (2020). *Purification of Environmental Pollutants Via Microbes*, 1e (ed. A.K. Vipin, K. Singh, P. Singh and V.K. Mishra), 474. Woodhead Publishing, eBook ISBN: 9780128232071, Paperback ISBN: 978012821199.
143. Iriti, M., Scarafoni, A., Pierce, S. et al. (2019). Soil application of effective microorganisms (EM) maintains leaf photosynthetic efficiency, increases seed yield and quality traits of bean (*Phaseolus vulgaris* L.) plants grow non different substrates. *International Journal of Molecular Sciences 20*: 2327. https://doi.org/10.3390/ijms20092327.
144. Olle, M. and Williams, I.H. (2013). Effective microorganisms and their impact on vegetable production - a review. *The Journal of Horticultural Science and Biotechnology 88* (4): 380–386. https://doi.org/10.1080/14620316.2013.11512979.
145. SANCO/D3/,rep. Monitoring of Pesticide Residues in Products of Plant Origin in the European Union, Norway, Ireland, and Liechtenstein 2003 Report – Summary, European Commission Health & Consumer Protection Directorate-General Directorate F – Food and Veterinary Office, Unit 4 - Food of plant origin, plant health, processing, and distribution. In cooperation with JRC IRMM, Food Safety and Quality Unit and DG SANCO D3, 2003 http://ec.europa.eu/food/fvo/specialreports/pesticide_residues/summary_2003_en.pdf
146. SANCO/2609/08 (2008a), rev. 1.
147. SANCO/2621/08, (2008b), rev. 2.
148. SANCO/11470/2012 (2012a), rev. 5.
149. SANCO/10363/2012 (2012b), rev. 7.
150. SANCO/10069/2013 (2013a), rev. 1.
151. SANCO/11188/2013 (2013b). rev. 2.
152. Anonymous Guidelines on Active Substances and Plant Protection Products. Technical guidance. 2021. https://ec.europa.eu/food/plants/pesticides/approval-active-substances/guidelines-active-substances-and-plant-protection-products_pl (accessed 19.12/2021)
153. Journal of Laws of The Republic of Poland. Warsaw, 9 October 2015, Item 1569.
154. Commission Regulation (EC) No 1095/2007 of 20 September 2007 amending Regulation (EC) No 1490/2002 laying down further detailed rules for the implementation of the third phase of the work program as referred to in Art. 8 sec. 2 of Council Directive 91/414/EEC and Regulation (EC) No 2229/2004 laying down further detailed rules for the implementation of the fourth phase of the work program referred to in Art. 8 sec. 2 of Council Directive 91/414/EEC (Text with EEA relevance). Journal of Laws OJ L 246, 21.9. 2007, pp. 1. 19–28
155. Commission Regulation [EU] No 284/2013 of 1 March 2013 laying down data requirements for plant protection products. 2013.
156. Commission Implementing Regulation (EU) 2018/679. 2018.
157. Commission Implementing Regulation (EU) 2018/679 of 3 May 2018 renewing the approval of the active substance forchlorfenuron. 2018. https://euroalert.net/oj/70957/commission-implementing-

regulation-eu-2018-679-of-3-may-2018-renewing-the-approval-of-the-active-substance-forchlorfenuron-in-accordance-with-regulation-ec-no-1107-2009-of-the-european-parliament-and-of-the-council-concerning-the-placing-of-plant-protection-products-on-the-market-and-amending-the-annex-to-commission-implementing-regulation-eu-no-540-2011-1

158. Commission Regulation (EC) No 2229/2004 of 3 December 2004 laying down further detailed rules for the implementation of the fourth stage of the Programme of work referred to in Article 8(2) of Council Directive 91/414/EEC. EN Official Journal of the European Union L 379/13. 2004. https://eur-lex.europa.eu/LexUriServ/LexUriServ.do?uri=OJ:L:2004:379:0013:0063:EN:PDF

159. Bremmer J, Riemens M, Reinders M. The future of crop protection in Europe. Panel for the Future of Science and Technology. European Parliamentary Research Service Scientific Foresight Unit (STOA) PE 656.330 – February 2021 EN. https://www.europarl.europa.eu/RegData/etudes/STUD/2021/656330/EPRS_STU(2021)656330_EN.pdf

Part II
Environmental Management: Bioremediation through Nexus Approach

8

Application of Green Remediation Technology in Field of Dye Effluent Management

Suhas Kadam[1,2], Rahul Khandare[3], and Sanjay Govindwar[2,4]

8.1. INTRODUCTION

At the beginning of the twentieth century, scientists developed new methods and technologies to meet a growing population. The consequence of these new techniques is a generation of waste, which has an adverse effect on many lifeforms. Moreover, due to modern lifestyle and urbanization, the environment has been extensively damaged by excessive inorganic and organic waste. These wastes degrade gradually, as well as accumulate in the environment for long decades and become recalcitrant pollutants. Unselective and uncontrolled dumping of factory processing, as well as municipal waste, into water or land has become a critical issue of pollution worldwide.

Industrial textile dyes are one of the most hazardous classes of pollutants. These dyes and stains are used as coloring compounds in the paper, pulp, pharmaceutical, textile, food, paint, plastics, photographic, ink, and cosmetics industries. Textile industries contribute a major share to the economies of developing countries. On the other hand, the dye manufacturing and processing firms at small and large scales are condemned as one of the foulest polluters of water and soil. It is determined that more than 10 000 various pigments are used in industry and above 735×10^3 kg of man-made dyes are manufactured yearly worldwide [1]. Around 10–15% of synthetic textile dyes having carcinogenic and other toxic effects are released during the dyeing and finishing of clothing processes, which ultimately causes a threat to all lifeforms [2]. The toxic and recalcitrant nature of textile wastewater is due to complex compositions like dyestuffs, salts, humectants, surfactants, detergents, dispersants, acids, bases, oxidants, and chlorine residues. Because of the complex and toxic nature of textile effluent, it needs to be treated before discharge.

[1]*Research Institute for Dok-do and Ulleung-do Island, Department of Biology, School of Life Sciences, Kyungpook National University, Daegu, South Korea*

[2]*Department of Biochemistry, Shivaji University, Kolhapur, Maharashtra, India*

[3]*Amity Institute of Biotechnology, Amity University, Maharashtra, India*

[4]*Department of Earth Resources and Environmental Engineering, Hanyang University, Seoul, South Korea*

Microbial Biotechnology: Role in Ecological Sustainability and Research, First Edition.
Edited by Pankaj Chowdhary, Sujata Mani, and Preeti Chaturvedi.

Many different chemical, physical, and biological methods are available for the treatment of textile effluents like charcoal precipitation, adsorption, sedimentation, filtration, flocculation, oxidation, ozonation, electrolysis by gamma radiation, and photochemical degradation. However, these methods have the disadvantage of secondary waste generation, leachates, cost, and other technical difficulties while managing at in situ wetlands.

India is a developing country with lots of small-scale industries; they do not have enough economic support to treat the wastewater. Conventional wastewater treatment requires a large investment, which directly affects the final cost of the product, hence to avoid high costs industries release wastewater in its original form [3]. As a result, rivers, as well as other freshwater resources, are being polluted, and that causes a serious threat to aquatic flora and fauna. For example, a huge number of fish die every year in the river because of the release of toxic dyes. In India, there are many treatment plants present for wastewater, but the wastewater generated is in very large quantities compared to the capacities of the treatment plant, hence they do not have enough time to treat this waste and finally have to discharge the effluent in the untreated form [4]. Many old government industries are unable to treat effluent because they require maintenance; some have improper design and others have insufficient electricity supply. Therefore, these industries release the effluent in toxic form, by which ground and surface water are contaminated [5]. After considering all aspects, there is a need to develop an economic technology with efficient remediation potential.

In the last decade, the use of plants has appeared as a promising green and clean tool for the treatment of textile dyes. Phytoremediation, the use of potential plants for environmental clean-up, is becoming a true "green" technology [6]. Plants and their rhizospheric microbes can efficiently remove pollutants via rhizodegradation, biostimulation, biostabilization, bioaccumulation, phytoextraction, and phytovolatization [7]. In situ phytoremediation is highly feasible for public authorization because of it is easy to run, economical, requires low nutrient input, and is esthetically acceptable, although it is still in experimental stages and requires a lot of attention [8]. Many times, phytotechnology has been found available at less than half the price of alternative physicochemical and biological methods. The elementary step in phytoremediation is a selection of plant species for treatment. Each plant has a different remediation potential, which varies with the type of contaminant. The root system of the plant plays an important role in treatment. Non-edible plants with massive root systems as well as fast-growing roots were preferably used in phyto-treatment [9]. Plant cellular enzymes such as peroxidases, transferase, dehalogenases, hydrolases, oxidoreductase, nitrilases, laccases, and nitroreductases were found to be involved in dye degradation [10, 11]. These enzymes were secreted by plant cells after sensing the oxidative stress of pollutants. Microorganisms growing in the root vicinity also play a role in the dye treatment and support the overall remediation process [12]. Phytoremediation with plant–plant and plant–microbes consortia showed efficient treatment due to a synergistic effect. Plants provide favorable conditions to microbial colonization of the rhizosphere for symbiotic degradation and detoxification of pollutants [13]. Plants and microbes possess different enzymatic cascades, which trigger more mineralization of textile waste than the individual system. Transgenic and genetically engineered plants are one of the new areas of research in phytoremediation.

Heavy metal removal using plants is the most successful, engineered, and accepted approach where phytoremediation technology is concerned. Phytoremediation of textile dyes, however, has remained an overlooked area of research [14]. Although many studies have reported dye removal with plants, most of them have remained at laboratory scales. Pilot-scale demonstrations of textile wastewater treatment have revealed the potential of this technology. The hydroponic phyto-tunnel system has also been utilized for the treatment of textile effluent [15]. Lab-scale horizontal and vertical subsurface flow bioreactors based on plant bacterial synergistic approach have been developed to treat real textile effluent [3, 15]. Some pilot-scale operational systems using macrophytes are on record. For instance, *Phragmites australis, Typha domingensis,* and *Alternanthera philoxeroides* were proposed in independently constructed wetlands studies on removal of textile dyes from wastewater [16–18]. Large-scale constructed wetlands treatment of textile dye effluents involves the cultivation of plants in situ or ex situ with low to moderate levels of contamination for a required period of stabilization and growth, to remediate contaminants from the polluted water and soil as well as to enhance biotransformation (detoxification) of the xenobiotic compounds. Many reports are now available on a variety of constructed phytoreactors to treat real textile wastewater, e.g., *Portulaca grandiflora* has been reported for degradation of Navy Blue HE2R, real textile effluent, and simulated dye mixture [8, 15], whereas a static hydroponic bioreactor of *Pogonatherum crinitum* plants along with immobilized *Bacillus pumilus* was applied for the treatment of textile wastewater [19]. Macrophytes *A. philoxeroides* Griseb and *Ipomoea aquatica* were found to degrade sulfonated dyes Remazol Red, Brown 5R, respectively. Further, they were utilized for decolorization of textile effluent at pilot-scale rhizofiltration unit separately and in combinatorial hybrid reactors as later planted on field constructed lagoons [18, 20]. Large-scale textile effluent treatment by *Salvinia molesta* has been reported to reduce values of COD, BOD, and ADMI by 76, 82, and 81%, respectively [21]. Field treatment of textile industry effluent was also performed efficiently in constructed drenches planted independently with *T. angustifolia, Paspalum scrobiculatum,* and their co-plantation [22]. A 2001 volume of wastewater from the tie and dye industry was shown to be treated with the use of cattail and cocoyam plants in independently engineered wetland systems [23]. Though phytoremediation has many positive points, it is a little bit slower in treatment and needs improvisation to apply at the industrial level. Phytoremediation with pilot- or large-scale hydroponic systems as well as phytoreactors along with other more conventional remedial methods can be used as a final/polishing step of treatment. This includes in situ or ex situ cultivation of plants within a defined range of contamination for the requisite period of stabilization and growth. Additionally, rhizospheric bioremediation of xenobiotic compounds such as textile dyes from polluted water and soil can also be accomplished.

On the other side, the world is suffering from an energy crisis, and we must look for some alternative techniques to addressing the same. Constructed wetlands-microbial fuel cells (CW-MFC) could prove to be one low-cost technique for electricity production. In CW-MFC, plants help in electricity production by enhancing cathode potential when placed at the cathode, as well as electrodes encourage dye decolorization efficiency at anode [24]. Hence, CW-MFC is helping to resolve both problems by enhancing the treatment efficiency of phytoremediation with power generation. It reduces the time required for the treatment of effluent, therefore CW-MFC can be sustained at large-scale effluent treatment plants. A few attempts have been made to achieve electricity generation using CW-MFC with Reactive Brilliant Red X-3B degradation in a constructed wetland with *I. aquatica* plantation. The CW-MFC simultaneously generated the bioelectricity production of $0.117\,W\,m^{-4}$ and achieved a decolorization up to 92.7% [25]. In another experiment, ABRX3 was treated in CW-MFC using *I. aquatica* and they found the highest decolorization and electricity generation of 95.6% and $0.852\,W\ m^3$, respectively, at three days of hydraulic retention time [26]. The industrial effluent soil was also found to produce 0.93 V of electricity continuously for 650 hours in a built MFC [27]. The activated sludge process for the treatment of Acid Navy Blue R with MFC was also found to show a power density of $2.236\,mW\ m^2$ [28].

8.2. DYES AND THEIR HISTORY

Dyes are a coloring agent with a tendency to bind on the surface of applied material. These dyes are segregated into different types according to their application and chemical nature. They are obtainable in numerous colors as well as in shades. Textile dyes are unsaturated aromatic complex chemicals which are absolute soluble in liquids.

Visualize the world devoid of color. Before the Neolithic period, the world was in only black and white shades. The primitive people from the New Stone Age or Neolithic period started to express themselves colorfully. There is evidence of the use of color since 10 200 BCE. Initially, dyes were obtained from animal and plant origins to paint the body, stones, feathers, shells, and stories in ancient caves. Scientists and archeologists studied history and found evidence that white, yellow, black, red, and other several pigments were produced from ochre. Researchers from all over the world failed to state when the first addition of dye to textile fiber came into practice. There are reports found from 3000 BCE in Chinese for preparation of yellow, black, and red color on silk cloth. Several prehistoric Indian texts are found to define the procedure of red color preparation from certain tree bark and wood, use of indigo for preparation of blue color, and diverse yellow shades. In certain areas of America, purple and red shades were obtained from cochineal bugs and used to dye plant fibers. Several of these natural dyes played important roles in changing the lifestyle of primitive society.

Most cultures have used a similar approach but with different techniques for dyeing fabrics. These techniques mainly contain the following steps: soaking a dye-producing material in water and immersing the fabric into that solution. The resultant water solution is left to simmer for up to a week. After this procedure, dye will release from the source material and become attached to the surface of the fabric. But problems remain: how much longer should the simmering period of water keep? In addition to this, many people faced the following difficulties with using natural dyes: inability to replicate similar color shade; quick fading of colored fabrics; and non-availability of raw material throughout the season. Therefore, they had to wait for 8–10 months to prepare a specific color. Moreover, dye content present in the raw material varies from species to species, so again the question arises of the quantity of raw material. These processes of dyeing the fabrics with natural dyes were still in the experimental stage.

In the nineteenth century, the scientists' interest in dye production and treatment gradually increased. This interest facilitated the production of synthetic dye. The first synthetic dye was accidentally produced by W. Perkin in 1856 during the synthesis of quinine (antimalarial drug). Perkin's experimental setup included extraction of black precipitate aniline from coal tar; further alcoholic extract of this showed purple color. The silk dyed by the parent dye stayed purple. He acquired a patent for the first synthetic dye in August 1856 and started a large-scale industry in the United Kingdom. Initially, this dye was named aniline purple; later, it was renamed Mauveine. After the discovery of Mauveine, several experiments were carried out in structural chemistry to produce different synthetic dyes. The different chemicals from coal tar were used for the production of several different synthetic dyes. Then, research turned in the direction of natural dye structures that could be produced in the lab as a synthetic dye.

8.2.1. Dye Classification

Dyes can be classified by many methods. Besides these classifications, every class of dye has a specific bonding pattern, structure, and chemistry. Every dye has a Color Index (CI) based on its characteristics, applications, and color. The classification is based on three methods, viz. source of material, method of application (Table 8.1), and chemical structure (Table 8.2).

Table 8.1 Classes of dyes depending on method of application.

	Dye class	Method	Description	Pollutant associated with dye	Application
1	Direct	Continuous/ Beck / Exhaust	Anionic water-soluble compounds, cellulosics can be dyed directly devoid of mordants.	Unfixed dye, color, fixing agents, salt, defoamer, surfactant, finish.	Cotton, silk, rayon, leather, nylon, other cellulosic.
2	Sulfur	Continuous	Sulfur or sodium sulphide present organic compounds.	Alkali, color, reducing agent, oxidizing agent.	Cotton, other cellulosic
3	Basic	Exhaust/Beck	Water-soluble, very bright in color, require weakly acidic conditions during dyeing.	N/A	Acrylic, some silk, wool and polyesters.
4	Acid	Beck/ Continuous/ Exhaust (carpet)	Anionic water-soluble compounds.	Unfixed dyes, organic acids, color.	Wool, nylon, silk.
5	Disperse	Exhaust/ Continuous High temperature	Water-insoluble.	Organic acids, color, leveling agents, carriers, defoamers, phosphates, lubricants, delustrants, dispersants, diluents.	Polyester, acetate, polyamide, other synthetics.
6	Reactive	Direct/ continuous	Covalent bonding formed with some groups of fiber like -OH, -NH, or –SH.	Unfixed dye, salt, color, surfactant, cationic fixing agents, Defoamer.	Cotton, wool and other synthetic fibers.
7	Mordant	Exhaust	Trairylmethane or oxazine, azo needs mordant.	Color, leveling agents, carriers, phosphates, dispersants, defoamers, delustrants.	Silk, paper wool, leather and cellulose fibers.
8	Pigment	Continuous/ exhaust	Non-ionic nature, particulate or crystalline insoluble salts, azo metal complex phthalocyanin, quinacridone or anthraquinone.	Colorants, dispersants, thickeners, delustrants.	Paper, cellulosic, Cotton, leather, all.
9	Vat	Continuous/ exhaust/ Package	Firstborn dyes, complex chemical structure, insoluble in water.	Alkali, color, reducing agents, oxidizing agents.	Wool, cotton, rayon, other cellulosic

(*Continued*)

Table 8.1 (Continued)

	Dye class	Method	Description	Pollutant associated with dye	Application
10	Anionic and ingrain	Direct	Insoluble products of a reaction between naphthols, phenols or acetoacetylamides and a diazotised aromatic amine.	Alkali, defoamers, coupling component.	Silk, cotton, synthetic fibers.
11	Metal complex	Direct/ Exhaust	16% of the C. I., chromium, nickel or cobalt, copper used in metal complex.	Unfixed dye, alkali, salt. Fixing agents, surfactant, metals.	Cotton, nylon, cellulosic, synthetic fibers.
12	Solvent dye	Direct	Non-ionic phthalocyanine, anthraquinone, diazo, and triarylmethane dyes.	Alkali, neutral fixing agents, surfactant.	Varnish, plastics, ink, fats, and waxes.

Table 8.2 Classes of dye depending on chromophore group.

	Class	Chromophore group
1.	Azo	—N=N—
2.	Anthraquinone	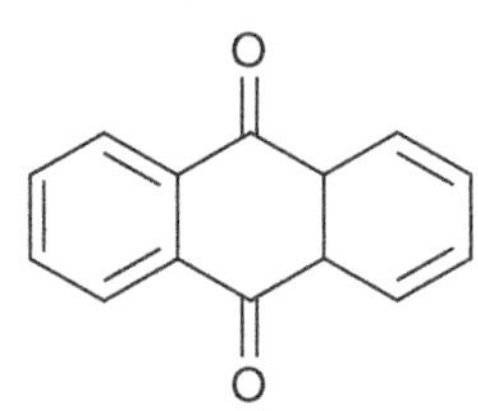
3.	Nitro	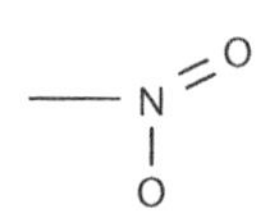
4.	Nitroso	—N=O
5.	Tri phenyl methane	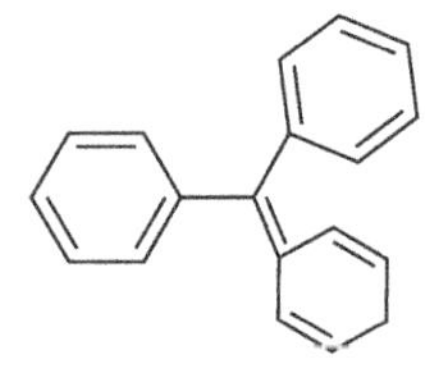
6.	Anthraquinone	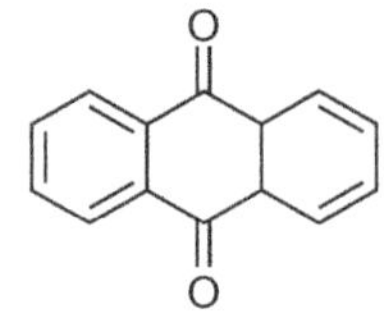
7.	Phthalein	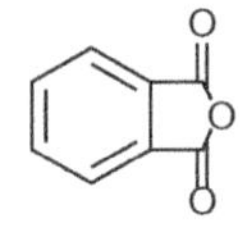

Table 8.2 (Continued)

	Class	Chromophore group
8.	Rhodamine	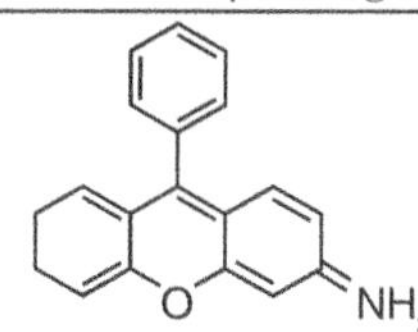
9.	Thiazine	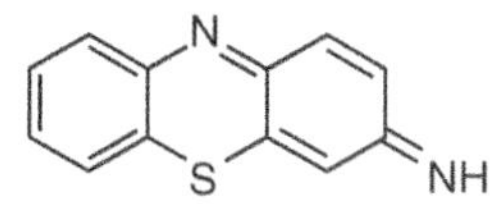
10.	Thiazole	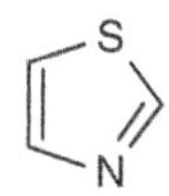
11.	Oxazin	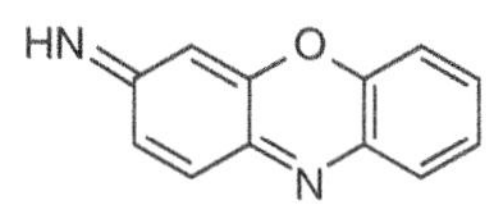
12.	Diaryl methane	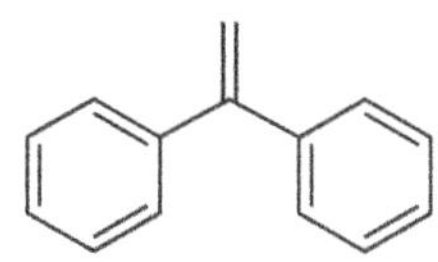
13.	Fluorone	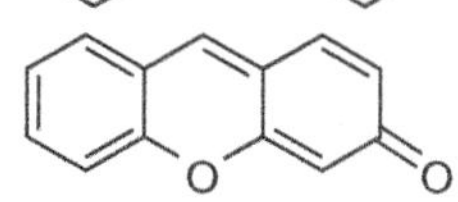
14.	Acridine	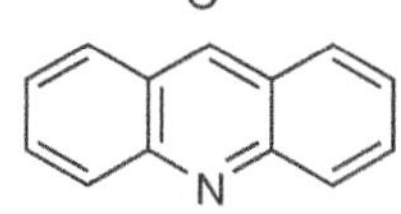
15.	Triaryl methane	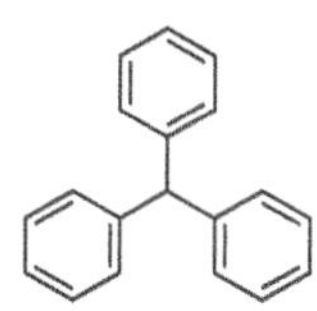

8.2.2. Class of Dyes Depending on the Source of Raw Material

8.2.2.1. Natural Dyes

The compounds used to color fabrics, leather, food, etc., and produced from natural resources like plants and animals are called natural dyes. Most natural dyes are vegetable dyes, which are produced from a plant root, bark, berries, leaves, or wood. Mordants like salt and vinegar are required for the application of these dyes. Indian yellow was the most-used natural dye, produced from cow urine. Kamala tree and Lac insect were used to produce red dyes.

8.2.2.2. Synthetic Dyes

These dyes are synthesized chemically in the laboratory. The structure of natural dyes that have been mimicked unnaturally for large productions is called synthetic dye. These dyes are made from petroleum byproducts like coal tar. The problems that arose during the use of natural dyes, such as the requirement of huge quantities and quick color fading were overcome by the application of synthetic dyes.

8.2.3. Class of Dyes Depending upon the Method of Application

A single dye is unable to color all types of fibers. In addition to this, a specific class of dyes binds or reacts with only a particular type of fibers. Hence dyes are also classified according to their mode of application into direct, sulfur, basic, acid, mordant, vat, reactive, disperse, anionic, metal complexes, solvent dye, and pigment. A detailed description of these classes is given in Table 8.1.

8.2.4. Class of Dyes Depending upon Chromophore Group

The color of dyes is due to the presence of chromophore groups as well as acid and basic groups like NR_2, NH_2, OH, SO_3H, and H. The polar auxochromes present in the dye structure are responsible for making a dye water-soluble. The interaction between oppositely charged groups present on dye and fabric results in colored fabric. The class of dyes depending upon the chromophore group is detailed in Table 8.2.

8.3. EFFECT OF TEXTILE DYES ON LIFEFORMS

In a revolution of European industry, the textile business arose as one of large scale and high-yielding trade. This led to deterioration of freshwater resources; as a result, natural flora and fauna of water bodies were distorted and, moreover, complete ecosystems were destroyed [29].

Many scientists tested the toxic nature of most commonly used dyes, particularly the dyes used in drugs, food, textiles, and cosmetics enlisted by Combes and Haveland-Smith [30]. A few years ago, Puvneshwari et al. [31] reviewed the carcinogenicity of azo dyes. The complexity of these dyes and their metabolites may become malignant carcinogens [32]. The IARC (International Agency for Research on Cancer) [33] announced that benzidine dyes act as an effective mutagen on most mammalians. In Algeria and Spain, Ardystil Syndrome was observed due to dyes that killed some industrial workers [34]. Mastrangelo [35] reviewed epidemiologic signs in the laborers of textile factories. Some dyes like Remazol Red showed the tumorigenetic effect on the freshwater fish *Devario aequipinnatus* [18]. The two food additives, carmoisine and tartrazine (azo dyes), were orally injected in rats for a period of 30 days and subjected to biochemical test. Researchers found that ALP, AST, ALT, creatinine, total protein, urea, and albumin levels significantly rise in the serum of carmoisine- and tartrazine-treated rats. Some biochemical markers like SOD, catalase, and GSH activities were reduced while MDA showed higher activities in rats treated with those dyes. Both dyes showed adverse effects on the kidney, liver, and modified organ biomarkers. In some cases, metabolites produced from the parent dye are more toxic than the original dye. The toxicity depends on the chemical nature and structure of the compound [36]. The aromatic amines formed after degradation are more poisonous than model dye and will possibly have carcinogenic properties [37]. Degradation of azo dyes may cause DNA-adduct, which has a toxic effect even on dye-degrading microbes [38]. Brown 5R became toxic to mammalian HepG2 cells even at a minute quantity of 93 μg [20]. The following azo dyes are listed under the mutagenic category with the help of Ames test: Pigment Red 23; Pigment Orange 5; Solvent Yellow 14; Pigment Red 4; and Pigment Solvent Yellow 7, while Pigment Red 57 : 1 and Pigment Red 53 : 1 are not mutagenic because they form sulfate aromatic amines instead of aromatic amines [39].

8.4. TYPES OF TEXTILE WASTEWATER TREATMENT

Several industrial textile dyes fall under the xenobiotic category. These dyes are recalcitrant to the environment and remain in it for a prolonged period. Therefore, the removal of dyestuff from the environs is needed to maintain the balance of the ecosystem. A number of aspects such as the composition of wastewater, a dose of an essential chemical, dye type, cost of operation, and handling cost of generated waste decides the technical feasibility of every technique. The different physicochemical methods used for the treatment of wastewater are briefly introduced in the following subsections.

8.4.1. Physicochemical Treatment

Several physicochemical techniques such as oxidation, adsorption, ozonation, filtration, flocculation/precipitation/coagulation, AOP (advanced oxidation processes), sonolysis techniques, and UV treatment [40–44] can be used for the remediation of dye effluent. In addition to these, reverse osmosis and ultra-filtration are also employed for the same purpose [40].

Even if a number of physicochemical methods succeed in 100% color removal, there are many complications with treatment. The key problems of dye processors with physicochemical methods are as follows: secondary waste generation; the magnitude of pollution; leachate formation; cost; and other technical difficulties while managing in situ treatments. Some textile wastewater treatments with physicochemical methods are listed in Table 8.3.

Table 8.3 Physicochemical treatments for textile dye remediation.

Textile dye	Treatment	Drawbacks	Advantage	References
Acid Red	Membrane filtration	Expensive treatment cost; Intense sludge production.	Complete removal of all dye types, reuse and recovery of water and chemicals.	[45]
Reactive Black 5	Ozonation	Fails to treat aromatic dyes; Half-life of ozone is short.	Effective for azo dye removal; Applied in gaseous state no alteration of volume.	[46]
Rhodamine B	Photocatalysis	Able to treat less dyed effluent; Expensive treatment process.	Treatment can be done at ambient conditions; Inexpensive and non-toxic inputs. Complete mineralization of waste in less time.	[47]
Acid Red 398	Precipitation, flocculation/ coagulation	Multiple separation and treatment steps. Only a few operating conditions.	Inexpensive, short treatment time, good treatment efficiencies.	[48]
Methylene Blue	Sonication	Primitive experimental stage of treatment and full scale implementation not available.	Easy to use; Effective treatment with integrated systems.	[49]

(Continued)

Table 8.3 (Continued)

Textile dye	Treatment	Drawbacks	Advantage	References
Acid Red 183	Fenton process	Sludge formation, highly expensive treatment.	Able to treat insoluble and soluble colored compounds.	[50]
Acid Red 14	Irradiation	Dissolved oxygen required in higher concentration.	Successful mineralization of contaminant at lab scale.	[51]
Indigo	Electrochemical Oxidation	Lots of electric power required.	No requirement of chemicals and production of less toxic metabolites.	[52]
Reactive Blue 19	Photochemical Process	Formation of different compounds.	No sludge production.	[53]
Lanasyn Navy M-DNL	Ion exchange	Used for only selected dyes.	Reuse of chemical and adsorbent.	[54]
Red Px and Orange P	Activated carbon	Expensive treatment, costly regeneration.	Effective treatment efficiency among wide types of dyes.	[55]
Basic Blue 24	Peat	Lesser surface area than activated carbon.	No activation is necessary, cellular structure makes it a good absorbent.	[56]
Basic Blue 3; Nile Blue Sulphate and RhodamineB	Alumina and Silica	Commercial application, no side reactions.	High dye removal efficiency.	[57] [58]
Vertigo Blue 49	Coal ashes	Require maximum contact time and larger quantity. Precise adsorbing surface area is smaller than activated carbon.	Economically feasible. Good treatment efficiency.	[59]
Malachite Green	Wood chips	Huge quantity with longer retention time (RT) is required.	Cellular structure makes it a good absorbent. Economically feasible. Efficiently treats acid dyes.	[60]

8.4.2. Biological Treatment

Biological treatment played a significant role in overcoming the limitations of physico-chemical treatment. The fact that microbes can also deal with toxic pollutants is an interesting part of the research. In this type of approach, microbial cells adapt to a hazardous contaminant and naturally resistant microbial strains are produced, which helps in the transformation of a hazardous pollutant into harmless metabolites. The theory behind microbial remediation involves the action of microbial enzymatic machinery. A few examples of microbial remediation textile of dyes are listed in Table 8.4.

Table 8.4 Examples of microbial remediations.

Textile dyes	Name of orgnism	References
	Bacteria	
Malachite Green	*Kocuriarosea*	[61]
Direct Blue 6	*Pseudomonas desmolyticum*	[62]
Reactive Red 141	*Rhizobium radiobacter*	[63]
Brown 3REL	*Bacillus* sp.	[64]
Methyl Orange	*Brevibacilluslaterosporus*	[65]
Crystal Violet, Brilliant Green, Malachite Green	*Citrobacter* sp.	[66]
Crystal Violet, Red pigment V2, Red pigment 2B	*Pseudomonas luteola*	[67]
Acid Red 106, Acid Orange 7, Direct Yellow 4, Direct Orange 39, Direct Yellow 12	*Aeromonas hydrophila*	[68]
Reactive Red 22	*Escherichia coli*	[69]
Acid Orange 10, Acid Orange 8, Acid Orange 7, Acid Red 4	*Sphingomonasxenophaga*	[70]
Rubine GFL, Remazol Red	*Brevibacilluslaterosporus*	[71]
Amaranth	*Clostridium perfringens*	[72]
Red HE7B	*Pseudomonas desmolyticum*	[62]
Crystal Violet	*Stenotrophomonas maltophilia*	[73]
Navy Blue2GL Brown 3REL	*Bacillus* sp. VUS	[74, 75]
Tectilon Blue 4R-01	*Bacillus benzenovorans*	[76]
Remazol Black B	*Paenibacillusazoreducens*	[77]
Reactive Red 2 Red BL1	*Pseudomonas* sp. SUK1	[78, 79]
Direct Red 5B, Direct Blue GLL	*Sphingobacterium* sp. ATM	[80, 81]
	Fungi	
Malachite Green	*Aspergillus ochraceus*	[82]
Reactive Blue 25	*Aspergillus ochraceus*	[83]
Brown 3REL	*Galactomycesgeotrichum*	[84]
Amaranth	*Phanerochaetechrysosporium*	[85]
Cotton Blue	*Penicillium ochrochloron*	[86]
Congo Red	*Aspergillus niger*	[87]
Acid Orange 7	*Coriolus versicolor*	[88]
Reactive Blue 221, Acid Blue 74, Basic Red 9, Direct Blue 71, Acid Blue 225	*Trametesmodesta*	[89]

(*Continued*)

Table 8.4 (Continued)

Textile dyes	Name of orgnism	References
Direct Black 22	*Aspergillus ficuum*	[90]
Reactive Orange 16	*Rhizopus arrhizus*	[91]
Remazol Brilliant Blue R	*Irpexlacteus*	[92]
	Yeast	
Malachite Green	*Saccharomyces cerevisiae*	[64]
Reactive Blue 19	*Kluyveromycesmarxianus*	[93]
Methyl Red	*Galactomycesgeotrichum*	[84]
Reactive Blue 19, Reactive Black 5, Reactive Red	*Candida tropicalis*	[94]
Reactive Blue 19	*Candida lipolytica*	[93]
	Algae	
Levofloxacin	*Chlorella vulgaris*	[95]
Direct Brown NM	*Chlorella pyrenoidosa*	[96]

8.5. PLANT-BASED STRATEGIES FOR DYE REMEDIATION

8.5.1. Phytoremediation

Phytoremediation, the use of potential plants for environmental clean-up, is emerging as a true green technology [6]. The term phytoremediation comes from the Greek *phyto* = plant and *remedium* = restore or clean [97]. Scientifically, "phytoremediation can be described as, the technology uses the flora to degrade/remove organic and inorganic pollutants from sediments, soil, groundwater and surface water." [7] Phytoremediation takes advantage of the exclusive and choosy uptake ability of the root system, composed of bioaccumulation, translocation, and pollutant storage or degradation capability of the whole plant system. The words from the definition of phytoremediation, "organic and inorganic," point to the technique that can be used against an inclusive range of pollutants. Organic pollutants comprise nitroaromatics, hydrocarbons, and pesticides, etc., and inorganic pollutants comprise metalloids, salts, radionuclides, agrochemicals, and metals. The last phrase of the defining statement "in sediments, soil, groundwater and surface water" specifies that the technique can be used in an inclusive range of areas. Plants and their rhizospheric microbes can efficiently remove pollutants via rhizodegradation, biostimulation, biostabilization, bioaccumulation, phytoextraction, and phytovolatilization [7]. In situ phytoremediation is highly feasible for public authorization because it is easy to run, economical, requires low nutrient input, and is esthetically acceptable, although it is still in experimental stages and requires a lot of attention [8]. Many times, phyto-technology is less than half the price of alternative physicochemical and biological methods. Phytoremediation can be managed and maintained with very little nutrient input as it is an autotrophic system [98]. The main target of the phytoremediation technique is the entire mineralization of toxic pollutants into fewer toxic metabolites, like ammonia, nitrate, CO_2, and chlorine [99].

8.5.2. Different Phytoremediation Strategies

Phytoremediation works with different strategies depending on the type and form of the contaminant. These strategies include phytoextraction, phytodegradation, rhizofiltration, phytostimulation, phytostabilization and phytovolatilization, which are schematically presented in Figure 8.1.

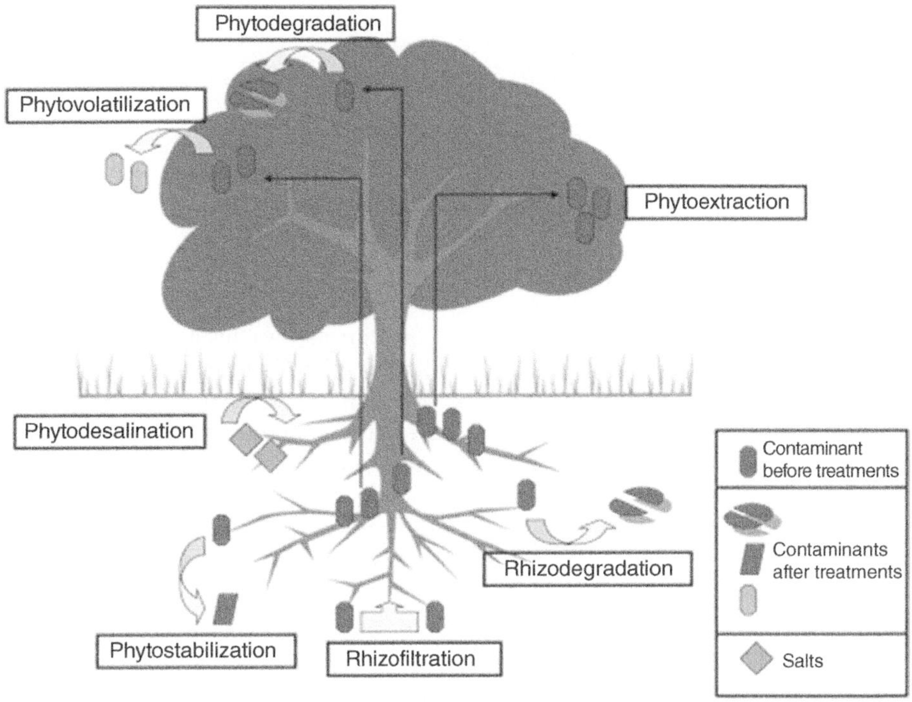

Figure 8.1 Schematic representation of different strategies of phytoremediation
Source: Adapted from Cristaldi [100]/with permission of Elsevier.

Phytoextraction is the most common strategy of phytoremediation technique, in which plants extract the pollutants from the soil and transfer them toward the upper part of the plant body (leaves and stem). These contaminants accumulate in the leaves and stem and in later stages can be harvested easily.

Phytodegradation is also known as phytotransformation, and is commonly referred to as biotransformation. In this strategy, the contaminant is brought inside and mineralized within the plant with plant metabolism. In case of contaminants nearby, the plant is degraded by the extracellular enzymatic cascade of the plant.

Rhizofiltration is also referred to as phytofiltration, which helps to filter pollutants from an aqueous solution. In this strategy, plant roots play a significant role and act as a filter for pollutants present in the liquid. This strategy acts in two ways: firstly, pollutants precipitate or adsorb on plant roots, and secondly, pollutants are absorbed inside the roots because of an abiotic or biotic procedure.

Phytostimulation is defined as stimulation of the microbial activity of the microbial community present at the rhizospheric area of the plant for faster depletion of contaminants. This also refers to rhizoremediation or rhizosphere degradation. The rhizosphere provides a site for the growth of microbes and the root exudates improve the rate of microbial remediation. The root exudate plays a vital role in phytoremediation in various ways, such as mimicking the contaminant or becoming co-metabolites essential for remediation of contaminant. Sometimes it also acts as an inducer of bacterial genes responsible for waste remediation.

Phytostabilization is commonly referred to as phytorestoration; it involves chemically or physically arresting the contaminant with help of roots adsorption, accumulation, and chemical fixation at rhizospheric soil. In this process, contaminants and their metabolites precipitate and stabilize on the root. Phytostabilization is a bit similar as phytoextraction focuses on the stabilization of contaminant and its metabolites in root rhizospheric soil, not inside the plant.

Phytovolatilization is a process of pollutant remediation in which the final product is a volatile compound. The plant uptakes pollutants from soil and biologically converts them into volatile compounds, then releases them into the atmosphere through leaves.

8.5.3. Limitations of Phytoremediation

Phytoremediation has many advantages over conventional wastewater treatment but it is also bound with some limitations. For the remediation of pollutants, plants must be grown and remediated at the geographical site of contamination. Climatic and soil conditions affect the remediation potential of the plant. To avoid these kinds of problems, it is necessary to find the native plants from the contaminated site having the high potential for phytoremediation. The length of the plant roots also limits the rate of phytoremediation. This challenge can be dealt with by pumping the contaminated water near the plant's rhizospheric area, or by implanting plants in the drilled hole which pollutants can easily access. Pollutant toxicity on plants is another limitation of this treatment. The pollutants are continuously dumped at the waste site; therefore, plants get repeated shocks of toxic chemicals and as a result the plant fails to thrive. One of the major drawbacks of phytoremediation is the slower treatment rate; as a result, it cannot sustain at large scale wastewater treatment where the discharge rate of effluent is much higher than phyto-treatment. This can be overcome by coupling the plant with another treatment like phycoremediation, mycoremediation, bioremediation, MFC, etc.

8.5.4. Selection of Effective Phytoremediator

The choice of accurate plant with some useful characteristics is one of the primitive steps in phytoremediation. Most commonly, non-edible plants should be used for large-scale application because the threat of intake of gathered toxic pollutants or its derivative from such plants by mammals and cattle is reduced. The plants that sustain a wide range of pH should be used for the treatment because industrial effluent possesses a mixture of pollutants (acidic and basic). One of the native plants, *Phragmites karka,* was found to tolerate a broad pH range as well as grow at basic, neutral, and acidic pH compared with other macrophytes. This reveals that *P. karka* is an appropriate example of the best phytoremediator. The efficient phytoremediator must have the potential to deal with structurally different chemicals. Several edible plants have already proven their great remediation efficiency with good enzymatic status. However, these plants cannot be used for on-site application. Further extraction and application of enzymes from these plants for pollutant degradation can be done. *Brassica juncea, Phaseolus mungo,* and *Sorghum vulgare* showed their degradation and detoxification potential against Reactive red 2 [65]. The enzyme cascade present in the plants defines the dye-degrading efficiency of each individual. The number of plants can remediate a single dye, but the pattern of degradation is different from plant to plant with different end products. Therefore, the plants which have non-toxic metabolites after remediation must be appropriate for treatment.

Fast-growing plants with massive root systems are ideal phytoremediators. Plants with higher biomass and circumference of the root system can easily get up to the pollutant and remove the pollutant from the solid and liquid media. Not each plant shows similar remediation potential against all types of dyes. Therefore, a large number of plants should be screened against different types of dye to find out the most efficient waste-remediating plant. During the screening, all plants biomasses must be kept equivalent with the equal shoot and root magnitudes for the reproducible results. The most appropriate way to find a perfect phytoremediator is to use the plants present at the actual site of contamination. These plants are already acclimatized to toxic pollutants and able to use the contaminant as a source of energy.

8.6. PHYTOTRANSFORMATION OF INDUSTRIAL TEXTILE DYES

Phytoremediation of heavy metals is a well-studied strategy, but the remediation of textile dye is still an unexplored research area. A few reports are present related to textile dye degradation with plants. *Ipomoea hederifolia* has earlier been reported to possess a notable potential for dye and effluent treatment [101]. A laboratory developed plant–plant consortium of *Aster amellus* and *Glandularia pulchella* was found to show efficient removal of 20 mg l^{-1} Remazol Orange 3R up to 100% within 36 hours; on the other hand, the individual plants could only remove it after 72 and 96 hours, respectively [102]. Static reactor with *P. crinitum* plantation was also found to reveal notable performance and removal of TDS (total dissolved solids), TSS (total suspended solids), BOD (biological oxygen demand), COD (chemical oxygen demand), and TOC (total organic carbon) [19]. Hybrid CW reduced the color by 70%, COD and TOC by 45%. Outcomes in terms of COD and TOC removal observed were independent of horizontal and vertical flow. Retention time, however, was found to be insignificant to alter the treatment process [103]. In another experiment, an in vitro grown consortium of *G. grandiflora* and *P. grandiflora* gave 94% color removal after 36 hours while their individual plants could achieve only 62% and 76% decolorization at 20 mg l^{-1} concentration, respectively [104]. In recent reports, *A. philoxeroides* and *S. molesta* exposed to textile effluent under static conditions were found to reduce the pH of different dye effluents toward normal [18, 21]. Combinatorial phytoreactors of *Ipomoea hederifolia* and *I. aquatica* have also shown significant treatment of textile effluent compared to the individual plant system [20]. Floating treatment of wetland using *T. domingensis* plantation reduced the COD and BOD of sewage effluent up to 87% and 87.5%, respectively [105]. An artificial neural network modeling study during degradation of acid blue 92 by *Azolla filiculoides* revealed that input variables such as decolorization time, initial dye concentration, fresh weight of the plant, initial pH, and temperature were altering the remediation efficacy [106]. The various attempts at phytoremediation are listed in Table 8.5 and the general protocol for phytoremediation of textile dyes is shown in Figure 8.2.

8.7. ENZYME CASCADE RESPONSIBLE FOR DYE REMEDIATION

The struggle for survival in any harsh condition is a key feature of all living organisms. Some species of plants, microbes, algae, yeast, and fungi have succeeded in dye-contaminated sites. For this acclimatization, they have triggered some metabolic changes in oxidoreductase enzymes. This enzyme cascade is able to mineralize textile dyes into smaller metabolites [118]. Initially, the accumulation of dye molecules takes place in the cells of the exposed organisms. Thereafter, the organisms sense the abiotic stress because of the inability to use these molecules. Further, they trigger the enzyme cascade to mineralize these complex compounds into a simpler form. As a result, the concentration of dye starts to decrease and becomes significantly lower. Accumulation and subsequent degradation are the reported mechanisms of plants while treating textile dyes. In some cases, the organism secretes extracellular enzymes and dye degradation takes place in an environment outside the living body.

Table 8.5 Phytoremediation potential of different plants for textile dyes and effluents.

Name of the plant	Concentration (ppm)	Textile Dye/Effluent	Decolorization (%)	References
Asparagus densiflorus	60	Rubine GFL	82	[107]
Ammanniabaccifera	50	Methyl Orange	89	[108]
Fimbristylisdichotoma	50	Methyl Orange	91	[108]
Paspalum scrobiculatum	100	Congo Red	73	[22]
Salvinia molesta	100	Rubin GFL	97	[21]
Ipomoea aquatica	200	Brown 5R	94	[20]
Alternanthera philoxeroides	70	Remazol Red	100	[18]
Pogonatherumcrinitum	NA	Effluent	74	[19]
Nasturtium officinale	20	Acid Blue 92	78	[109]
Ipomoea hederifolia	50	Scarlet RR	96	[101]
Typha angustifolia	75	Reactive Blue 19,	70	[110]
	100	Congo Red	80	[22]
Bouteloudadactyloides	NA	Effluent	92	[111]
Petunia grandiflora	20	Brillaint Blue G	86	[112]
Azolla filiculoide	20	Basic Red 46,	90,	[113]
		Acid Blue 92	80	[106]
Lemna minor	10	Methylene Blue,	98,	[114]
	2.5 $g L^{-1}$	Acid Blue 92	77	[106]
Portulaca grandiflora	20	Reactive Blue 172	98	[8]
Glandulariapulchella	20	Remazol Orange 3R,	100,	[102]
		Green HE4B	92	[115]
Aster amellus	20	Remazol Orange 3R,	100,	[115]
		Remazol Red	96	[8]
Typhoniumflagelliforme	20	Brilliant Blue R	65	[116]
Blumeamalcolmii	20	Direct Red 5B, Methyl orange, Red HE4B, Reactive Red 2, Malachite Green	42, 88, 76, 80, 96	[116]
Phragmites australis	750,	Acid Orange 7	68,	[117]
	100		98	[16]

Source: Modified from Khandare and Govindwar [14].

The biocatalyst remediation has some effective characteristics with good results over conventional physicochemical treatment. Enzymes have high catalytic activity, specificity, and no side product formation. In addition, they are easily degradable, reduce sludge production, can treat contaminants even at low concentrations, and are easily standardized with profitable preparation. These features of enzymes help to decrease manufacturing prices and save energy use. On the other hand, the great price of enzyme extraction, purification, and denaturing tendency beyond its optimum condition restrict the extensive application of enzyme systems. Irrespective of these drawbacks, enzyme technology is a fast-growing field of research and technical difficulties are frequently overcome. The biocatalysts responsible for dye remediation are laccase, azoreductase, tyrosinase, DCIP reductase, lignin peroxidase, aryl alcohol oxidase, and catalase.

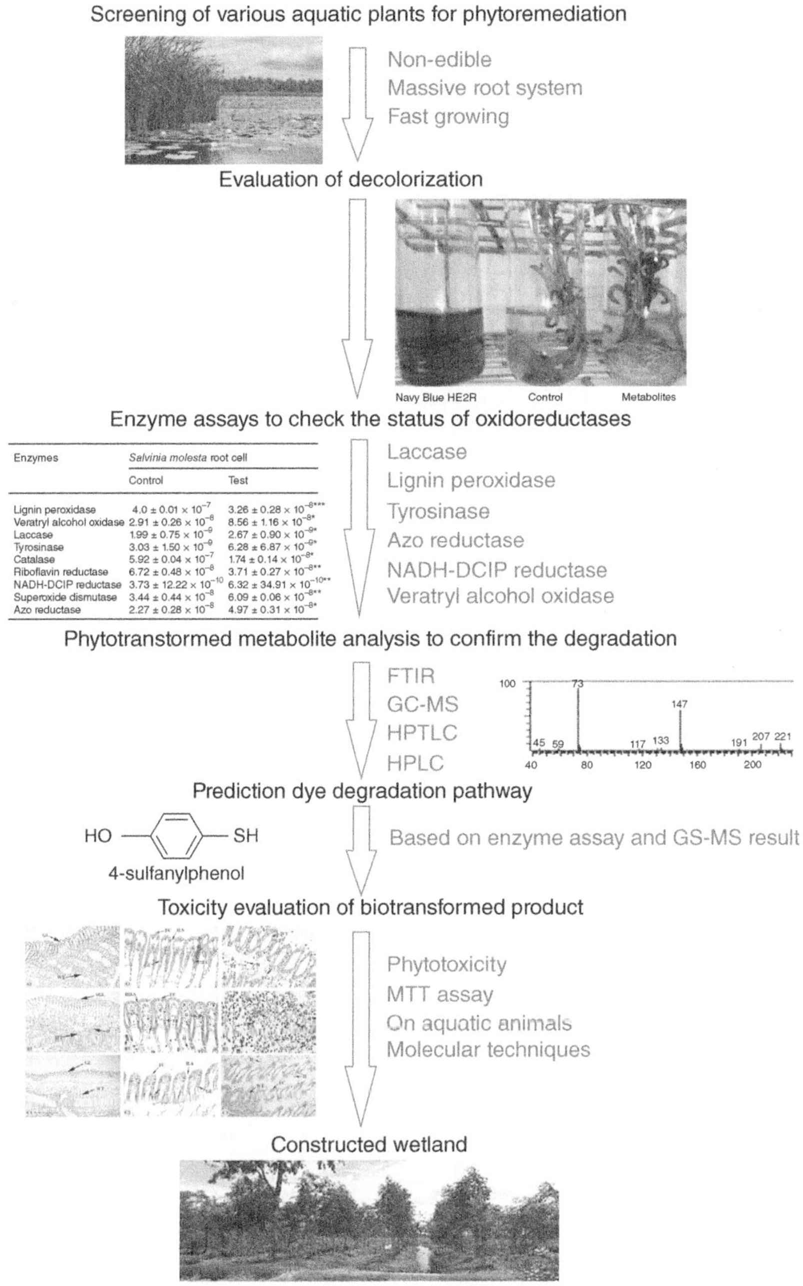

Figure 8.2 General protocol for phytoremediation. *Source:* (top photo): Artur Synenko/Adobe Stock.

The oxidoreductase-laccase belongs to the blue multicopper group, and it was described by Yoshidha [119]. Its presence was reported in fungi, bacteria, and plants [120–122]. The electron transfer in laccase is facilitated through four Cu^{2+} atoms present at copper-binding sites T1, T2, and T3. The T1 binding site is mononuclear consisting of one Cu^{2+} atom, while T2 and T3 are combined as trinuclear comprises three Cu atoms. Two tunnels permit the transfer of O_2 and release of H_2O from the T2/T3 center. The 2 β-turns are involved in the construction of ligand binding pocket near the T1 Cu atom, which states the enzyme–ligand specificity [123]. Laccase follows a one-electron transfer mechanism to oxidize substrates like polyphenols, biphenyls, diamines, aromatic amines, and ascorbic acid. The mechanism of laccase action works on the basis of one-electron oxidation for hydroxylation of the aromatic substrate beside oxygen; it is reduced in water and the substrate converts into free radicals [124] (Figure 8.3).The laccase from the genus *Basidiomycete* is an example of wood-rotting fungi that showed a significant role in lignin degradation [125]. Blue laccase purified from *Aspergillus ochraceus* NCIM-1146 showed decolorization of Methyl Orange, Reactive Yellow 84-A, and Reactive Navy Blue HER [126].

Azoreductase requires a cofactor FADH or NADPH or NADH for catalysis of azo linkage. This enzyme works according to the ping-pong mechanism and carries out the breakdown of the azo bond [127]. It is a well-recognized enzyme system to degrade dye molecules having azo linkage in their structure. For the reduction of one molecule of Methyl Red into its metabolites, the enzyme required two molecules of NADH (Figure 8.3).

Tyrosinase is also referred to as polyphenol oxidase containing copper at the active site and is preferably found in several living entities including insects, fungi, bacteria, plants, mammals, and amphibians. Tyrosinase plays a significant role in vegetables and fruit browning as well as in melanin production within vertebrates [128]. The most common

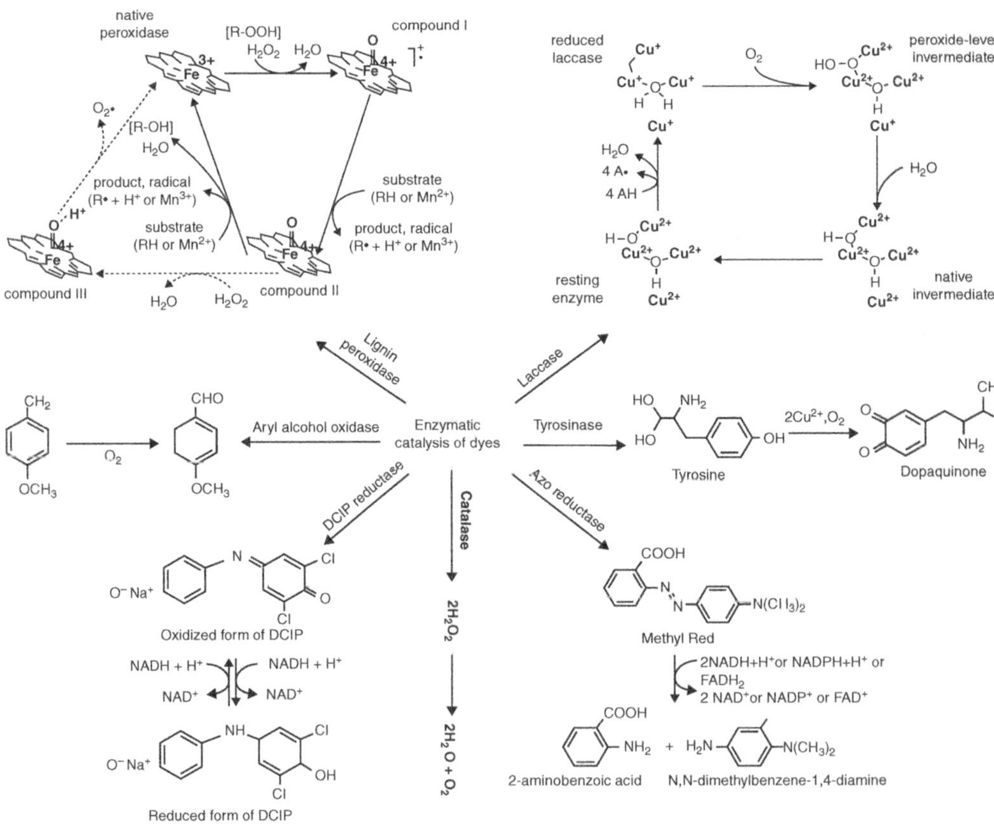

Figure 8.3 Mechanisms of catalysis of different oxidoreductase enzymes for dye degradation. *Source:* Adapted and modified from Khandare and Govindwar [14].

substrates for tyrosinase are catechol and tyrosine. The degradation of the substrate has been carried out in the presence of oxygen; a detailed mechanism is shown in Figure 8.3. Rane et al. [20] showed a substantial increase in tyrosinase from *I. aquatica* after exposure to dye Brown 5R. Rubine GFL was responsible for the induction of the tyrosinase enzyme in *S. molesta* [21].

NADH-DCIP reductase is in the family of mixed-function oxidase and is able to remediate xenobiotic chemicals [29]. This enzyme reduces DCIP by borrowing the electron from NADH (electron donor). The native oxidized state of DCIP shows a blue color and converts into a colorless form after reduction (Figure 8.3). *P. grandiflora* and *G. pulchella* showed induced DCIP reductase activity after dye exposure [113, 115].

Lignin peroxidase contains the heme group in the active site and fits into the oxidoreductase family. This enzyme is also referred to as ligninase and follows the one-electron oxidation mechanism. Degradation of phenolic compounds and production of free radicals has been carried out by utilizing one electron from hydrogen peroxide. Figure 8.3 also shows a reduced form of peroxidase from *Phanerochaete chrysosporium*. Peroxidase oxidases dye at heme-Fe and form an intermediate stage of FeIV=O (compound II). It also oxidizes methanol (3,4- di-methoxy-phenyl) (veratryl alcohol) into radical cations. The oxidative cleavage of bond C-O-C and C–C from the lignin-like compounds, arylpropane-aryl ether and diarylpropane, are proposed to carry out by veratryl alcohol radical. Hammel and Tardone [129], Hammel et al.[130], and Paszczynski and Crawford [131] demonstrated oxidation of chlorophenols, PCH, and dyes by lignin peroxidase. Numerous textile azo dyes were degraded by purified lignin peroxidase from *Acinetobacter calcoaceticus* and *Bravibacillus laterosporous* [132, 65].

Aryl-alcohol oxidase also belongs to the oxidoreductase family. Catalysis of aromatic alcohol is carried out with the help of oxygen (Figure 8.3). This enzyme oxidizes the CH-OH group of primary alcohol by reducing one molecule of oxygen. The enzyme has the chemical name aryl alcohol: oxygen oxidoreductase and several common names like aromatic alcohol oxidase, veratryl alcohol oxidase, and aryl alcohol oxidase. *I. aquatica* showed induction in veratryl alcohol oxidase enzyme after exposure to Brown 5R dye [20].

Catalase is abundantly found in aerobic organisms and protects them against oxidative damages. It mainly catalyzes hydrogen peroxide (H_2O_2-strong oxidizing agent) degradation into oxygen (O_2) and water (H_2O) [133]. The tetramer structure of catalase is composed of four polypeptides having more than 500 amino acids each [134, 135] and the heme group present at the active site [134] (Figure 8.3). The activity of catalase stimulus by the external contaminant and excess oxidative stress results in loss of activity [136]. Few studies are available on catalase dye interaction. Yang et al. [137] studied the dye Chrysoidine action on bovine liver catalase and Li et al.[138] researched the effect of dyes Sudan II and IV on catalase by molecular docking.

8.8. TOXICITY STUDIES OF TEXTILE DYES AND THEIR METABOLITES

In many cases, the degradation or mineralizations of dyes are achieved, but the toxic nature of the dyes more or less remains the same. Because of the known threats posed by dyes, it becomes inevitable to perform a toxicity assessment to check their harmfulness. Toxicity studies are not only performed to check if the particular chemical is safe or not, but also to understand the deleterious effects it can produce. Recently, the Food and Drug Administration and Environmental Protection Agency have set some regulations on toxicity testing of chemicals before they are used at the commercial level. Phytotoxicity, genotoxicity, and cytotoxicity are some methods to study the toxicity of textile dyes [139].

8.8.1. Phytotoxicity

Phytotoxicity is one of the vigorously studied methods of toxicity tests. In this method, the toxic effects of particular compounds can be tested on plant growth. Plant growth parameters like seed germination, length of shoot, and root of selected plants are observed. Several reports are found regarding phytotoxicity studies carried out using *P. mungo, S. vulgare,* and *Triticum aestivum* seeds. Phytoremediation of Congo Red dye by *T. angustifolia* and *P. scrobiculatum* has been carried out and showed confirmed detoxification by phytotoxicity in this study [22].

In this study, 10 seeds of selected plants have been placed in control and tested Petri dishes. Both dishes are nourished with 10 ml distilled water (control dish) and dye solution (test dish) every 24 hours and placed at room temperature. After five days incubation, seedlings have been removed and root and shoot length are measured. The difference in control and test seedlings roots, as well as shoots, gives a clear impression of the toxicity.

8.8.2. Aquatic Animal Toxicity

Aquatic animals are the species most affected by dye effluent, as it is frequently released into water bodies. Usually, aquatic animals like fish, bivalves, and crabs are used for toxicity studies. The histological changes occurring in the selected tissue reflects the toxic nature of the dye. The *Goodea atripinnis* fish gill histology has shown reduced toxicity of wastewater after treatment [140]. Textile dyes appear to be highly toxic to fishes like *Daphnia magna.* The toxic nature of dyes Methyl Orange and Scarlet RR has been shown by *Lamellidens marginalis* bivalve gill histology [12, 108].

In this study, toxicity assessment has been carried out in three independent plastic tubs having 5 l tap water (C), textile dye, and its biotransformed dye solution, respectively. Equal lengths of 18 (6 each) bivalves have been distributed in the three plastic tubs. The solution of each tub has changed with the same concentration of respective solution after every 12 hours. Bivalves have been dissected, and gill tissues separated after four days of acute exposure to the dye solution. The separated gill tissues have been kept in Bouin's aqueous fluid for tissue fixation for about 12 hours. After dehydration, gill tissue is embedded in wax and sectioned at five microns. The sections have been stained with hematoxelene–eosin stain and observed under a light microscope. The histological changes in the control and test gills showed the toxic nature of the dye was reduced after treatment (Figure 8.4).

8.8.3. Genotoxicity

Genotoxicity is described as a damaging effect of compounds on cell nuclear material like DNA and RNA. Compounds showing genotoxic effects are referred to as genotoxins, i.e. mutagens. We must check the genetic material of organisms to determine their integrity, which will help us to understand the toxic nature of the dye. Tests like the Comet assay, Ames test, and molecular markers like RAPD, as well as ISSR techniques, are used to find genetic alterations. The mutagenicity of 14 textile dyes has been studied by Ames test and it was revealed that 57% of dyes are directly acting as mutagens [141]. Comet assay has been performed with nuclear material of *Allium cepa* by exposing untreated and treated dye mixture and was found to reduce the mutagenic nature of dye mixture after treatment [104]. The change in the ISSR banding pattern of *L. marginalis* after exposure of Scarlet RR in increasing concentration showed that toxicity of dye magnifies with dye concentration [12].

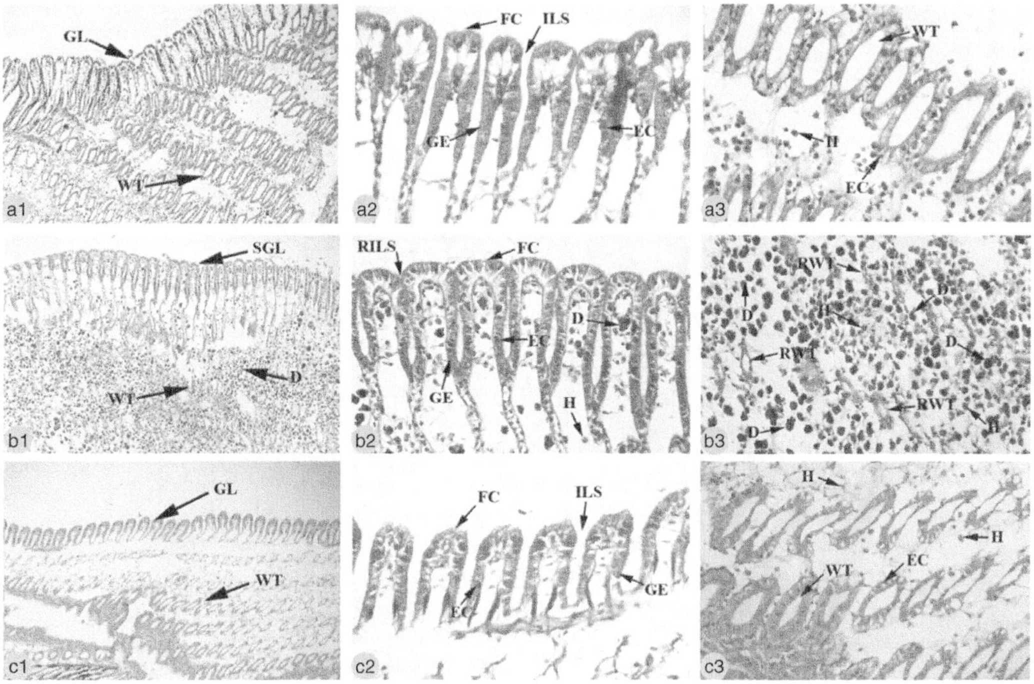

Figure 8.4 Histology of bivalve gill tissues exposed to (a) tab water; (b) Methyl Orange; (c) Phytotransformed Methyl Orange.

In the Ames test, histidine auxotrophic *Salmonella typhimurium* is used. If the subject chemical has the ability of genetic alteration, then *S. typhimurium* are able to grow on histidine-free medium [39].

In comet assay, the root tip of onion has been excised after 48 hours exposure to textile dye and homogenate with needle and prepared suspension for the comet. The prepared suspension is loaded between low melting agarose gradients present on glass slides and subjected to electrophoresis. After electrophoresis, silver staining has been carried out and observed under the microscope. The comet-shaped nuclear material represents the destruction of genetic material due to textile dye exposure.

In a molecular marker study, DNA was extracted from bivalves exposed to normal tap water and textile dye. This DNA was subjected to ISSR marker amplification using PCR and the banding pattern was observed. The change in banding pattern of DNA exposed to dye reveals the mutation in the DNA of bivalves, which results in the addition or deletion of ISSR priming sites from the genome (Figure 8.5).

8.8.4. Cytotoxicity

Compounds showing destructive effects to the cell are referred to as cytotoxic compounds and its ability is known as cytotoxicity of that compound. Most cytotoxicity studies are carried out on cell lines. MTT (3-[4,5-dimethylthiazol-2-yl]-2,5-diphenyltetrazolium bromide) assay is the most used experiment for cytotoxicity. In this assay, the MTT (Tetrazolium Day) is broken down by NADPH oxidoreductase (mitochondrial enzymes) to purple-colored formazan. If a cell is dead through the toxic effect of textile dye, then they

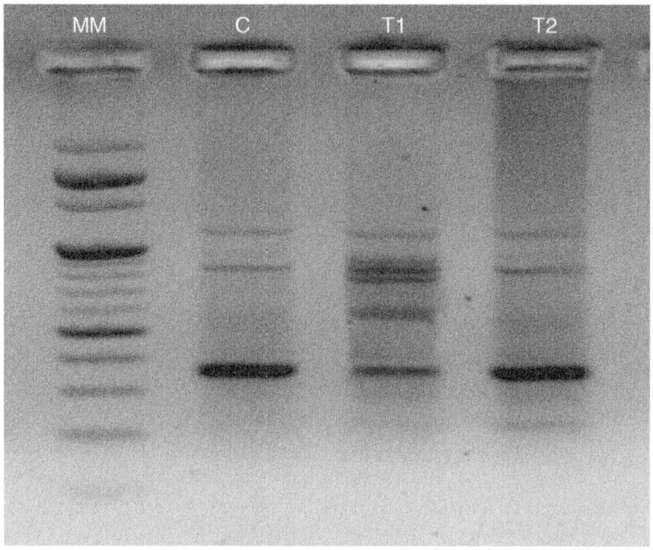

Figure 8.5 RAPD banding pattern of bivalve DNA shown as Lane (i) molecular marker, (ii) normal bivalve, (iii) bivalves exposed with untreated 50 ppm of Methyl Orange and (iv) bivalves exposed with treated dye.

are unable to produce mitochondrial enzymes. The viability of the cell is directly proportional to color intensity of formazan.

The MTT assay has been carried out on the human hepatocellular carcinoma cell line (HepG2). The 1 mg ml^{-1} and less concentration of textile dye were exposed to the HepG2 cell and incubated for 24 hours. After incubation, formed formazan crystals dissolved in DMSO and OD was taken at 570 nm. Colorless solution reflected the toxic nature of the dye. Cell viability of 68% and 93% was found for HepG2 when exposed to 1 μg ml^{-1} control and treated Rubine GFL dye, respectively [107].

8.9. PLANTS USED FOR THE TREATMENT OF TEXTILE WASTE

8.9.1. *Asparagus Densiflorus*

A. densiflorus is an evergreen of Southern African origin; this plant from genus *Asparagus* has a life span of more than two years. It represents the family Asparagaceae and physiologically resembles *Asparagus aethiopicus*. Both species are now known to be separate species, but several detailed descriptions are available under the name of *A. densiflorus*, which can also refer to *A. aethiopicus*. This is a common garden ornamental plant, unable to survive under frost conditions.

This plant has fast scrambling growth up to 1 m in height. The root system is extensive and long, composed of fibrous roots mass and grape-sized tubers. It has dark green, needle-shaped leaves that are densely arranged on stems like fern fronds. The small-sized flowers covered with foliage are light pink or white in color. These flowers are sweet-smelling and bloom once every few years. Fruits are round-shaped and resemble a red glossy berry (Figure 8.6a).

Features like a fast-growing, extensive root system and high chlorophyll content make *A. densiflorus* the best candidate for soil remediation. These plants are generally used for gardening purposes in almost every industry, therefore the same plantation can be used in waste treatment.

Figure 8.6 Photograph of (a) *Asparagus densiflorus,* (b) *Ammania baccifera,* (c) *Fimbristylis dichotoma,* (d) *Typha angustifolia,* and (e) *Chrysopogon zizanioides*. *Source:* (a) DAVIPATH/Adobe Stock; (b) Dinesh Valke/Wikimedia Commons; (c) David Eickhoff/Flickr; (d) Vanchuree/Adobe Stock; (e) Forest and KimStarr/Flickr.

8.9.2. *Ammannia Baccifera*

Ammannia baccifera, commonly referred to as blistering *Ammannia* or monarch redstem, belongs to the Lythraceae family. This is an annual herb found in the marshy and waterlogged fields of Africa, the Americas, and Asia. This plant is erect, branched, and 50 cm in height. The plant has rectangular and oblongolate-shaped leaves up to 3.5 cm long. The small flowers, 1.2 mm long, are greenish or purple in color with black seeds (Figure 8.6b). As we found this plant at a dye-contaminated place, it must have the ability to survive there. The use of fast-growing plants from dye waste places is an ideal approach to phytoremediation.

8.9.3. *Fimbristylis Dichotoma*

F. dichotoma is an annual herb commonly referred to as eight-day grass or forked fimbry. This plant belongs to the Cyperaceae family and is found in tropical regions of Asia, Africa, and Australia. Wet swamps and flooded soil are the best conditions for the growth of *F. dichotoma*. The height of the plant is up to 80 cm, of which the maximum portion is covered by numerous long stems having a diameter of 2 mm. The stems are nodeless and some portions of it are covered by inflorescence (Figure 8.6c). The *F. dichotoma* plant has the ability to grow in consortia with some other plants. This is beneficial in the co-planted treatment of textile dyes.

The characteristics of *F. dichotoma, such as* annual herbs, non-edible, massive root systems, and occurring naturally in consortia, have been hypothesized to possess noteworthy dye removal potential. This plant has previously been studied for heavy metal remediation and found to be effective in the removal of Na, K, Ca, Mg, Fe, Mn, Cr, Ni, Co, Cu, Pb, Zn, and Cd [142].

8.9.4. *Typha Angustifolia*

T. angustifolia has common names like lesser bulrush, lesser reedmace, and narrow-leaf cattail. It is from the genus *Typha* and has a life span of more than two years. *T. angustifolia* is a common herb and is found in the northern hemisphere of the Earth. This widely available common wetland plant observed feasible growth in saline localities. It is assumed that the plant was incorporated into the inland system of North America from Europe via the coastal region. This plant has an optimum height of up to 1.2 m with 12–16 narrow and flat leaves of 0.01 m width. At maturity level, different stalks have arisen from the root and become tall as leaves. The stalks are capped with finger-shaped cottony brown inflorescence (Figure 8.6d). The plants propagate very fast, with sturdy rhizomes having 0.6 m-long roots, which categorize them as weeds. The species *Typha latifolia* is very much similar to *T. angustifolia* in the case of geographic distribution. A single morphological difference is found in *T. latifolia,* with thinner leaves and distinguishable separation of flowering heads.

Features such as aquatic hydrophyte, fast-growing, and lengthy root system make *T. angustifolia* a potent phytoremediator for textile wastewater. The phytoremediation potential of *T. angustifolia* has previously been proven in vertical flow wetland, which showed a remarkable reduction in color, COD, BOD, and TDS [143]. The remediation of industrial textile wastewater has been achieved with constructed trenches (91.4 m × 1.2 m × 0.6 m; 65.8 m^3) planted with *T. angustifolia* [22].

8.9.5. *Chrysopogon Zizanioides*

C. zizanioides has the local name vetiver, and it is innate to India. They are perennial tussock grasses belonging to the Poaceae family. *C. zizanioides* closely resembles flowering plants of *Sorghum* genus; on the other hand it shares many morphological features with fragrant grasses like *Cymbopogon citratus* (lemongrass), *Cymbopogon nardus*, *Cymbopogon winterianus* (citronella), and *Cymbopogon martinii* (palmarosa).

The average height of vetiver is 1.5 m, but it can grow up to 3 m in favorable conditions. *C. zizanioides* has thin, long, and rigid stems and leaves with brownish, purple-colored flowers. The main and interesting feature of vetiver is its massive root system, which is 2–4 m in depth. Vetiver has underground shoots, which protect the plants during frost, grazing, and wildfire conditions. Leaves can grow up to 3 m, having many-branched 0.3 m inflorescence at the top. The grass flowers are present in the pair, which includes three reproductive organs named stamens (Figure 8.6e). *C. zizanioides* are able to sustain deep water levels in addition to this; they can survive up to two months under clear water. This plant also possesses characteristics like drought tolerance, sheet erosion, and new growth from buried nodes.

The remarkable features such as annual herb, fast-growing, sturdy, growing under deep water, and extensive root system makes *C. zizanioides* an effective waste soil and water remediator. In the case of dye remediation, *C. zizanioides* is a well-studied plant. Many people have implemented this grass in different wetlands to achieve better treatment. Vetiver implemented on floating platforms has shown a remarkable reduction in color, COD, BOD, and TDS [144]. The combined image of all the above-mentioned plants used in textile waste treatment has been shown in Figure 8.6.

8.10. CONSTRUCTED WETLANDS (CWS) AND THEIR TYPES

A CW (constructed wetland) is a simulated engineered wetland aimed at using naturally occurring processes to restore natural environs or to treat wastewater but in a well-organized manner [145]. Modern constructed wetlands are designed with the specific aim of enhancing wastewater treatment and universally applied for the treatment counting, landfill leachate [146, 147], agriculture waste [148], industrial effluent [149], municipal wastewater [150, 151]. This technique is now publicly recognized, as there are many advantages, such as minimum maintenance and operation cost, low sludge production, maximum nutrient absorption, required low energy, simplicity, process stability, contain high biodiversity, and low construction. Several CWs are used for textile wastewater treatment, and some of them are listed in Table 8.6.

CWs are divided into different types depending on surface water level (subsurface and free water surface), the motion of water (static and continuous flow), and a mixture of the system (hybrid) and the subsurface CW is again divided with respect to the flow direction (vertical and horizontal). Presently, the following CWs are extensively used for wastewater

Table 8.6 CWs for textile wastewater treatment.

Type of CW	Size/ Capacity of CW	Plant(s) and Bacteria Utilized	References
Vertical subsurface flow	52 × 20 × 30 cm (31.2 l volume)	*Asparagus densiflorus*	[107]
Soil beds	1 × 1 × 23 m	*Asparagus densiflorus*	[107]
Floating phytobed	1.2 × 2.7 × 1.5 m (4860 l volume)	*Fimbristylisdichotoma* and *Ammaniabaccifera*	[108]
Horizontal subsurface flow constructed drenches	91.4 × 1.2 × 0.6 m	*Typha angustifolia, Paspalum scrobiculatum*	[22]
Free water surface CW	7 × 2 × 5 m (52,500-l volume)	*Salvinia molesta*	[21]
Free water surface CW	10 × 1 × 6 m (60,000-l volume)	*Ipomoea aquatica*	[20]
VSF augmented bioreactor	40 × 20 × 30 cm	*Glandulariapulchella* augmented with *Pseudomonas monteilii* ANK	[3]
VSF augmented bioreactor	46 × 18 × 29 cm	*Portulaca grandiflora, Pseudomonas putida*	[15]
Hydroponic continuous flow-phytotunnel	1.6 × 1 cm	*Portulaca grandiflora*	[15]
CW for textile effluent	1000 ml	*Eichhornia crassipes*	[152]
CW with sand and bagasse media	20 l	*Phragmites karka, Bacillus* sp.	[153]
VSF (pilot scale)	0.8 × 0.6 × 0.6 m	Narrow-leaved cattails	[143]
Up flow CW	18 × 70 cm	*Phragmites australis*	[154]
Hybrid CW	VSF: 5 × 4 × 0.6 m HSF: 8 × 5 × 0.5 m	*P. australis*	[103]
Engineered wetland systems	200 l	Cattail, coco yam	[23]
HSF	10 × 5 m	*Phragmites australis*	[117]

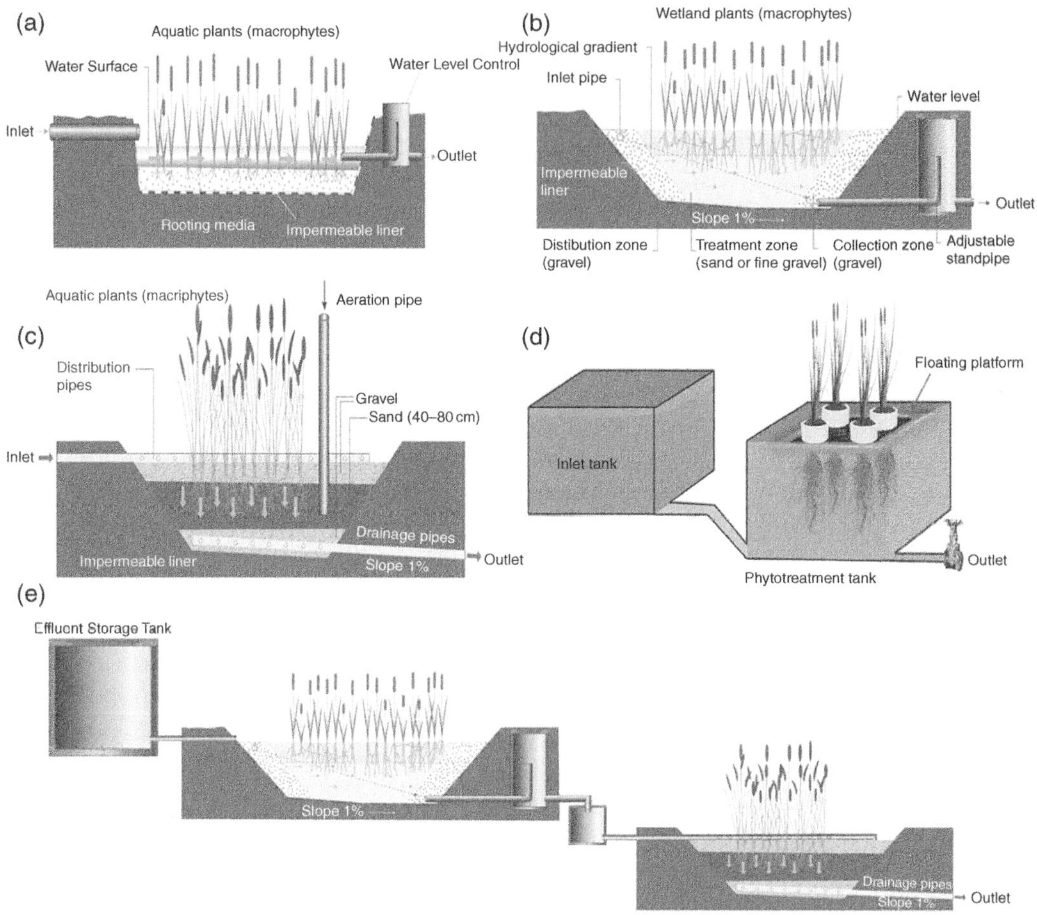

Figure 8.7 Schematic representation of constructed wetland types: (a) Free water surface CWs; (b) Horizontal subsurface flow CWs; (c) Vertical subsurface flow CWs; (d) hydroponic system; and (e) Hybrid CWs.

treatment: FWS (free water surface) [155], HSF (horizontal subsurface flow) [156], VSF (vertical subsurface flow) [157, 158], FCW (floating constructed wetlands) [105], and HCW (hybrid constructed wetlands) [159, 160] as shown in Figure 8.7.

8.10.1. Free Water Surface CWs (FWS)

The free water surface CW was the world's first reported CW built in the Netherlands in 1967 [161]. It includes an open water area, submerged or floating flora, and a rooting medium. It is very much analogous to a natural wetland. Rooting media assisted the growth of a wide variety of potential flora. The detailed schematic representation is shown in Figure 8.7a.

8.10.2. Subsurface Flow CWs (SSF)

SSF consists of porous gravel and sand media to support the emergent vegetation. The wastewater passes through sand and gravel, on which potential plants are deep-rooted. The structure of the system has been constructed so that the effluent always stays below the surface of the gravel. SSF shows higher removal efficacies for suspended solids (SS) and organic matter compared to FWS. Because of sedimentation and filtration processes carried out within the gravel and sand, the soluble organic matrix is degraded by biological processes. A SSF wetland has an advantage over FSW wetland as it does not support the mosquito population and odor formation. These types of problems arise in FSW wetlands as the water is directly exposed to the atmosphere. This SSF is again classified into two types: HSF and VSF wetlands.

8.10.3. Horizontal Subsurface Flow CWs (HSF)

In this type of SSF wetland, the flow of water is in a horizontal direction or parallel to the gravel surface. Stefanakis et al. [156] and Schierano et al. [148] recently used HSF for the treatment of wastewater. The detailed picture of HSF is shown in Figure 8.7b.

8.10.4. Vertical Subsurface Flow CWs (VSF)

In this SSF wetland, the flow direction of water is vertical, either downward from top to the bottom or upward from bottom to top of the wetland (Figure 8.7c). Seidel [162] introduced VSF in 1953 but it was not widely approved due to high maintenance, construction, and operating costs. VSF can run under more aerobic conditions than HSF when the flow direction is from top to bottom, which helps for effective nitrification.

8.10.5. Hydroponic System

This system has also been referred to as aquaculture because the plants present in this culture were grown in mineral nutrient solution in absence of soil. The concept of a hydroponic system used in the field of bioremediation is an innovative idea. It is very easy to maintain plants without soil in the case of hydroponics if the proper nutrient balance has been achieved. In fact, plants grown in hydroponics propagate much faster and become vigorous compared to the plants rooted in the soil as they get all the required nutrients. Therefore, phytoremediation with a hydroponics system will achieve better effective treatment than a soil bed wetland. Another important consideration is that soil bed CWs become infertile after a few years as they are regularly exposed to high TDS toxic effluent. These kinds of problems do not arise in the case of a hydroponics system as there is no direct contact with soil. A hydroponic system is further classified into static and continuous flow hydroponics.

8.10.6. Static Hydroponic System (SHS)

This type of hydroponic system allows the growth of plant roots in contact with the stagnant wastewater and simultaneous treatment of pollutants is achieved (Figure 8.7d). Beyond the normal wastewater treatment, SHS also deal with nitrogen and phosphorus, as they have been utilized by algal biofilm produced in the system. The biomass from vegetative flora can be used as fertilizer to enhance the economic value of the SHS.

8.10.7. Continuous Flow Hydroponic System (CFH)

Continuous flow hydroponic is a similar system as the static hydroponic system except for the continuous flow of wastewater. Floating CWs are built to develop emergent vegetation on the buoyant mat and excess growth is allowed under the floating mat [163]. Pollutants can be eliminated by emergent flora as wastewater passes beneath floating mats.

8.10.8. Hybrid System

A hybrid CW is a wetland system built up with several types of CWs together to attain precise treatment purposes or greater treatment efficacies, mainly for nitrogenous waste. The common hybrid systems consist of VSF and HSF systems in series, using the benefits of both systems in a sequential manner [164] (Figure 8.7e). The coupled VSF-HSF CWs have often been used to attain fully nitrified wastewater, as VSF systems can offer optimum situations for nitrification, while HSF systems can offer appropriate situations for denitrification.

8.11. USE OF COMPUTATIONAL TECHNIQUES IN THE FIELD OF DYE REMEDIATION

The growing field of bioinformatics has had a big impact on a number of research areas. These techniques help to predict results before performing the actual reactions. The rate of bioremediation depends upon the reaction kinetics, physicochemical characters, and specificity. This chemical characterization of enzymes depends on the chemical/physical parameters. A number of databases such as chemDplus and chemogenesis are available to provide the concerned information. In the field of enzyme remediation, it is necessary to know the 3D structure of an enzyme and its interaction with the substrate. To find the three-dimensional structure of any protein with NMR and X-ray crystallography techniques is very tedious as well as time-consuming work. As an alternative to these techniques, one more method of homology modeling is available. This method is faster than NMR and X-ray crystallography and gives truthful results. In addition, the PDB database is an online tool that provides the three-dimensional structural information of enzymes involved in bioremediation [165]. Different molecular docking helps in finding the interaction between dye and enzyme. Autodock, Patchdock, Haddock, and Gold are some tools used for docking study. The docking technique helps to find efficient enzymes to treat a particular type of dyes and it is very convenient. In silico degradation of dyes is the best method to find out the dye-degrading efficiency of an organism. The predicted fate of the metabolism of a particular dye is very important in dye remediation. Bioinformatics assist in protein, dye, and its metabolites identification, as well as characterization that further helps in the construction of degrading metabolic and regulatory pathways. BioCarta and WIT, Biocyc, BRENDA, EcoCycsystem, ExPASy, KEGG, MetaCyc, MetaRouter, PANTHER, Roche Biochemical Pathways, and UM-BBD are some databases that contain all reported metabolic pathways involved in bioremediation [165]. ChemSkech and chemspider are some other tools to predict the degradation patterns of a particular textile dye. Toxicity is another important parameter in biotransformation. ACD/TOx suite, CAESAR Comparative Toxicogenomic Database, ECOSAR, ECOTOX GENE-TOX, Hazard Expert, PBT profiler, and Toxicity Estimation Tool (TEST) are reported tools for the prediction of toxicity of a particular dye [165]. Identification of particular plants and associated bacteria or bacterial community is a must when it is employed or involved in remediation. NCBI nucleotide and MiCA PAT$^+$ database are two catalogues that help in the identification of a particular ora group of organisms, respectively. Azo dye decolorization by various bacterial enzymes was studied intensely with a computational mode like (DFT) Density Functional Theory calculation, ab initio, and CFD- Computational Fluid Dynamics modeling [166, 167]. The scientists Abbott et al. [168] and Sridhar et al. [169] carried out comparative studies between computational and in vitro enzyme-mediated dye degradation. They found that in vitro results had great support by in silico outcomes. The dye degradation potential of *Bacillus subtilis* against structurally different dyes was stated using molecular docking of laccase and catalase [170]. In the future, both techniques together can be used to find out dye enzyme interaction and binding affinity [168, 169].

8.12. COUPLING OF THE PLANT-MICROBIAL FUEL CELL (P-MFC) TO INCREASE THE DYE DEGRADATION RATE WITH SIMULTANEOUS ELECTRICITY PRODUCTION

As previously discussed, the treatment rate of phytoremediation is a little bit slower, and as a result, this technique cannot be sustained at an industrial scale even if it has many advantages over other techniques. To overcome this problem, phytoremediation coupled with MFC is the best solution. In this treatment policy, microorganisms used for electricity production are going to utilize dye as an energy source. During this treatment, microbes produce enzymes in addition to plant enzymes, and together these enzymes speed up the oxidation and reduction of dyes with electron transfer. This electron transfer is responsible for more electricity production. At the end, these coupled treatments help to enhance dye degradation efficiency as well as electricity production capability of the system. *I. aquatica* with CW-MFC for treatment of azo dye decolorization showed the plant helped in electricity production by enhancing cathode potential when placed at the cathode, and electrodes encourage dye decolorization efficiency at anode [24].

8.13. FIELD APPLICATION OF PHYTOREMEDIATION TO DEAL WITH REAL TEXTILE EFFLUENT

As the results have been obtained from lab scale phytoremediation, we carried out the on-field application. At the initial state, *A. densiflorus* was implemented on vertical subsurface flow phytoreactor (VSbF) and in-situ soil remediation. After six days treatment of textile effluent with *A. densiflorus* planted on VSbF, we saw a reduction of COD, ADMI, BOD, TSS, and TDS by 66%, 65%, 61%, 66%, and 48%, respectively. In addition to these, heavy metals like Hg, Cr, Cd, and As were also eliminated below the standard effluent discharge limit of ISW. MTT assay showed threefold reduced toxicity the effluent after treatment with *A. densiflorus*. Field cultivation of *A. densiflorus* offers an economic option for environmental management. Soil sample analysis from control (soil + effluent) and test (soil+ effluent + *A. densiflorus*) showed a remarkable reduction in water safety parameters. Significant reductions in ADMI up to 67% were observed in soil planted with *A. densiflorus*. Heavy metals such as arsenic, lead, cadmium, and chromium from test soil were reduced by 93%, 96%, 25%, and 36%, respectively. *A. densiflorus* has the additional benefit of heavy metal reduction, which makes it an appropriate candidate for phytoremediation [107].

Even though we got better phytoremediation results, some problems regarding soil bed phytoremediation remain. The continuous exposure of high TDS effluent to soil results in soil salinity and infertility of soil. To overcome this problem, we carried out textile waste treatment using co-planted floating phytobeds of *A. baccifera* and *F. dichotoma* in a cemented tank (48601). Co-plantation is beneficial in waste remediation due to the synergistic effect of enzymes from individual plants. Textile wastewater treated using this system resulted in a superior reduction in parameters such as BOD (77%), COD (72%), ADMI (79%), TSS (56%), and TDS (66%) after nine days, which is more than individual plants. Furthermore, heavy metals such as chromium, lead, cadmium, and arsenic were reduced up to 72%, 68%, 55%, and 72%, respectively. The pH of the effluent was reduced after treatment from 9.8 to 7.5. Treatment with co-planted floating phytobeds deals with limitations of conventional treatment like high TDS, odor, and color. All these results confirm that FPb is one of the best currently available solutions for industrial textile wastewater treatment [108].

At an industrial scale, the amount of influent per day is very high; if treatment requires more retention time, then the whole system fails due to insufficient storage. To overcome this issue, we coupled the co-planted floating phytobeds (FPb) with microbial fuel cells (MFC). Both the MFC and phytoremediation consist of biological systems to remediate the waste. Furthermore, microbial fuel cells make use of the degrading activity of microbes to produce electricity. Co-planted FPb-MFC with *C. zizanioides* and *T. angustifolia* treated textile wastewater exhibited a reduction in parameters such as TDS, TSS, ADMI, COD, and BOD up to 67%, 70%, 82%, 75%, and 75%, respectively. The effluent pH was decreased from 10.7 to 7.3 after treatment with FPb-MFC. Heavy metals like arsenic, lead, cadmium, and chromium were also found to be reduced up to 100%, 63%, 56%, and 46%, respectively, after treatment. In addition, the power density of 0.0769 W m^{-2} was produced during the wastewater treatment. This treatment efficiency of FPb-MFC is significantly higher than only FPb planted with *C. zizanioides* and *T. angustifolia*. Above all, experimental outcomes showing phytoremediation coupled with MFC has greater waste treatment efficiency [12].

8.14. CONCLUDING NOTE

Phytoremediation is a potent tool for textile dye effluent management. It is solar power-based technology that requires a minimal site for treatment. The plant has the ability of pollutant accumulation and subsequent degradation with the help of oxidoreductases like azo reductase, tyrosinase, LiP, veratryl alcohol oxidase, riboflavin reductase, laccase, and DCIP reductase. The annual herbs from an actual polluted site having a massive root system and that are fast-growing are assumed to be ideal for phytoremediation.

Textile dye and effluent degradation by plants has been proven and promoted in many research studies. Plants for textile effluent treatment have been carried out in different ways. Many laboratory experiments were performed but only a few field-level attempts have been reported. On-field phytoremediation is still in the experimental stages. Therefore, remediation of dye disposal sites and pilot plant studies have great research scope.

The garden ornamental plant plays a vital role in the field of soil remediation by adding aesthetic value to the treatment. Floating phytobeds have a major advantage in wastewater treatment as they prohibit soil salinity by avoiding direct contact with soil and show faster treatment efficacy. Augmentation of potent bacterial consortia can be used to enhance the bioremediation potential of plants. Partial treatment of effluent using physicochemical methods can reduce the pollutant load and enhance the longevity of plants. Industrial effluent treatment with co-planted FPb-MFC is the best solution for wastewater management. FPb-MFC is founding a new era in the field of textile waste management. This kind of combinatorial approach will reveal real tools for textile waste management.

REFERENCES

1. Chequer, F., Olivera, G.A.R., Ferraz, E. et al. (2013). Textile dyes: dyeing process and environmental impact. *Eco-Friendly Textile Dyeing and Finishing 16* (6): 151–176.
2. Khataee, A.R., Dehghan, G., Ebadi, A. et al. (2010). Biological treatment of a dye solution by macroalgae *Chara* sp.: effect of operational parameters, intermediates identification and artificial neural network modeling. *Bioresource Technology 101*: 2252–2258.

3. Kabra, A.N., Khandare, R.V., and Govindwar, S.P. (2012). Development of a bioreactor for remediation of textile effluent and dye mixture: a plant and bacterial synergistic strategy. *Water Research 23* (47): 1035–1048.
4. Central Pollution Control Board (2009). Status of water supply, wastewater generation and treatment in class-I cities & class-II towns of India.
5. Central Pollution Control Board ministry of environment and forest (2007). Advance method for treatment of textile industry wastewater.
6. Dietz, A.C. and Schnoor, J.L. (2001). Advances in phytoremediation. *Environmental Health Perspectives 109*: 163–168.
7. Pilon-Smits, E. (2005). Phytoremediation. *Annual Review of Plant Biology 11* (56): 15–39.
8. Khandare, R.V., Kabra, A.N., Kurade, M.B. et al. (2011). Phytoremediation potential of *Portulaca grandiflora* hook. (Moss-rose) in degrading a sulfonated diazo reactive dye navy blue HE2R (reactive blue 172). *Bioresource Technology 2* (102): 6774–6777.
9. Brandt, R., Merkl, N., Schultze-Kraft, R. et al. (2006). Potential of *vetiver* (*Vetiveriazizanioides* L. Nash) for phytoremediation of petroleum hydrocarbon-contaminated soils in Venezuela. *International Journal of Phytoremediation 18* (8): 273–284.
10. Carias, C.C., Novais, J.M., and Martins-Dias, S. (2007). Phragmites australis peroxidases role in the degradation of an azo dye. *Water Science and Technology 1* (56): 263–269.
11. Rao, M., Scelza, R., Scotti, R. et al. (2010). Role of enzymes in the remediation of polluted environments. *Journal of Soil Science and Plant Nutrition 10*: 333–353.
12. Kadam, S.K., Watharkar, A.D., Chandanshive, V.V. et al. (2018). Co-planted floating phyto-bed along with microbial fuel cell for enhanced textile effluent treatment. *Journal of Cleaner Production 31* (203): 788–798.
13. Doran, P.M. (2009). Application of plant tissue cultures in phytoremediation research: incentives and limitations. *Biotechnology and Bioengineering 1* (103): 60–76.
14. Khandare, R.V. and Govindwar, S.P. (2015). Phytoremediation of textile dyes and effluents: current scenario and future prospects. *Biotechnology Advances 33*: 1697–1714.
15. Khandare, R.V., Kabra, A.N., Awate, A.V. et al. (2013). Synergistic degradation of diazo dye direct red 5B by *Portulaca grandiflora* and *pseudomonas putida*. *International journal of Environmental Science and Technology 20* (10): 1039–1050.
16. Ong, S.A., Ho, L.N., Wong, Y.S. et al. (2011). Semi-batch operated constructed wetlands planted with *Phragmites australis* for treatment of dyeing wastewater. *Journal of Engineering Science and Technology 6*: 623–631.
17. Shehzadi, M., Afzal, M., Khan, M.U. et al. (2014). Enhanced degradation of textile effluent in constructed wetland system using *Typha domingensis* and textile effluent-degrading endophytic bacteria. *Water Research 24* (58): 152–159.
18. Rane, N.R., Chandanshive, V.V., Watharkar, A.D. et al. (2015). Phytoremediation of sulfonated Remazol red dye and textile effluents by *Alternanthera philoxeroides*: An anatomical, enzymatic and pilot scale study. *Water Research 4* (83): 271–281.
19. Watharkar, A.D., Khandare, R.V., Waghmare, P.R. et al. (2015). Treatment of textile effluent in a developed phytoreactor with immobilized bacterial augmentation and subsequent toxicity studies on *Etheostomaolmstedi* fish. *Journal of Hazardous Materials 11* (283): 698–704.
20. Rane, N.R., Patil, S.M., Chandanshive, V.V. et al. (2016). *Ipomoea hederifolia* rooted soil bed and *ipomoea aquatica* rhizofiltration coupled phytoreactors for efficient treatment of textile wastewater. *Water Research 11* (96): 1–11.
21. Chandanshive, V.V., Rane, N.R., Gholave, A.R. et al. (2016). Efficient decolorization and detoxification of textile industry effluent by *Salvinia molesta* in lagoon treatment. *Environmental Research 26* (150): 88–96.
22. Chandanshive, V.V., Rane, N.R., Tamboli, A.S. et al. (2017). Co-plantation of aquatic macrophytes *Typha angustifolia* and *Paspalum scrobiculatum* for effective treatment of textile industry effluent. *Journal of Hazardous Materials 12* (338): 47–56.
23. Mbuligwe, S.E. (2005). Comparative treatment of dye-rich wastewater in engineered wetland systems (EWSs) vegetated with different plants. *Water Research 39*: 271–280.

24. Fang, Z., Song, H.L., Cang, N. et al. (2013). Performance of microbial fuel cell coupled constructed wetland system for decolorization of azo dye and bioelectricity generation. *Bioresource Technology 28* (144): 165–171.
25. Fang, Z., Cheng, S., Wang, H. et al. (2017). Feasibility study of simultaneous azo dye decolorization and bioelectricity generation by microbial fuel cell-coupled constructed wetland: substrate effects. *RSC Advances 17* (7): 16542–16552.
26. Fang, Z., Song, H., Cang, N. et al. (2015). Electricity production from azo dye wastewater using a microbial fuel cell coupled constructed wetland operating under different operating conditions. *Biosensors and Bioelectronics 68*: 135–141.
27. Samuel, R.B., Jebakumar-Solomon, R.D., Prathipa, R. et al. (2013). Production of electricity from agricultural soil and dye industrial effluent soil using microbial fuel cell. *International Journal of Research in Engineering and Technology 2*: 140–148.
28. Khan, M.D., Abdulateif, H., Ismail, I.M. et al. (2015). Bioelectricity generation and bioremediation of an azo dye in a microbial fuel cell coupled activated sludge process. *PLoS One 23* (10): 1–18.
29. Saratale, R.G., Saratale, G.D., Chang, J.S. et al. (2011). Bacterial decolorization and degradation of azo dyes: a review. *Journal of the Taiwan Institute of Chemical Engineers 42*: 138–157.
30. Combes, R.D. and Haveland-Smith, R.B. (1982). A review of the genotoxicity of food, drug and cosmetic colours and other azo, triphenylmethane and xanthene dyes. *Mutation Research Genetic Toxicology 98*: 101–243.
31. Puvaneswari, N., Muthukrishnan, J., and Gunasekaran, P. (2006). Toxicity assessment and microbial degradation of azo dyes. *Indian Journal of Experimental Biology 1* (44): 618–626.
32. Garg, A., Bhat, K.L., and Bock, C.W. (2002). Mutagenicity of aminoazobenzene dyes and related structures: a QSAR/QPAR investigation. *Dyes and Pigments 27* (55): 35–52.
33. IARC (1982). *Monographs on the evaluation of the carcinogenic risk of chemicals to humans, suppl. 4. Chemicals, industrial processes and industries associated with cancer in humans.*, 1–292. Lyon.
34. Hoet, P.H., Gilissen, L.P., Leyva, M. et al. (1999). *in vitro* cytotoxicity of textile paint components linked to the "Ardystil syndrome". *Toxicological Sciences 27* (52): 209–216.
35. Mastrangelo, G., Fedeli, U., Fadda, E. et al. (2002). Epidemiologic evidence of cancer risk in textile industry workers: a review and update. *Toxicology and Industrial Health 18*: 171–181.
36. CCICED (1999). *Impacts of Environmental Standards and Requirements in EU Countries on China's Textile Industry*. Policy Research Center for Environment and Economy.
37. Bell, J., Plumb, J.J., Buckley, C.A. et al. (2000). Treatment and decolorization of dyesin as anaerobic baffled reactor. *Journal of Environmental Engineering 126*: 1026–1032.
38. Banat, I.M., Nigam, P., Singh, D. et al. (1996). Microbial decolorization of textile-dye-containing effluents: a review. *Bioresource Technology 13* (58): 217–227.
39. Mortelmans, K. and Zeiger, E. (2000). The Ames salmonella/microsome mutagenicity assay. *Mutation Research 20* (455): 29–60.
40. Avlonitis, S.A., Poulios, I., Sotiriou, D. et al. (2008). Simulated cotton dye effluents treatment and reuse by nanofiltration. *Desalination 1* (221): 259–267.
41. Bandala, E.R., Peláez, M.A., García-López, A.J. et al. (2008). Photocatalytic decolourisation of synthetic and real textile wastewater containing benzidine-based azo dyes. *Chemical Engineering and Processing Process Intensification 47*: 169–176.
42. Bestani, B., Benderdouche, N., Benstaali, B. et al. (2008). Methylene blue and iodine adsorption onto an activated desert plant. *Bioresource Technology 14* (99): 8441–8444.
43. Zhou, M., Yu, Q., and Lei, L. (2008). The preparation and characterization of a graphite-PTFE cathode system for the decolorization of C.I. acid red 2. *Dyes and Pigments 77*: 129–136.
44. Gupta, V.K. and Suhas (2009). Application of low-cost adsorbents for dye removal - a review. *Journal of Environmental Management 4* (90): 2313–2342.
45. AL-Bastaki, N., Bu-Ali, Q., AL-Aseeri, M. et al. (2007). Removal of acid red and sodium chloride mixtures from aqueous solutions using nanofiltration. *Desalination* AL-Aseeri M *206*: 407–413.
46. Colindres, P., Yee-Madeira, H., and Reguera, E. (2010). Removal of reactive black 5 from aqueous solution by ozone for water reuse in textile dyeing processes. *Desalination 10* (258): 154–158.

47. Byrappa, K., Subramani, A., Ananda, S. et al. (2006). Photocatalytic degradation of rhodamine B dye using hydrothermally synthesized ZnO. *Bulletin of Materials 6* (29): 433–438.
48. Zonoozi, M.H., Reza, M., Moghaddam, A. et al. (2008). Removal of acid red 398 dye from aqueous solutions by coagulation/flocculation process. *Environmental Engineering and Management Journal 7*: 695–699.
49. Sonawane, S., Chaudhari, P., Ghodke, S. et al. (2009). Ultrasound assisted adsorption of basic dye onto organically modified bentonitenanoclay. *Journal of Scientific and Industrial Research 68*: 162–167.
50. Rajendran, A. and Karthikeyana, C. (2011). Effective degradation of an azo dye acid red −183 by Fenton and photofenton treatment. *Acta Chimica and Pharmaceutica Indica 23* (1): 57–64.
51. Ziaefar, N., Mehrabad, J., and Arjomandirad, F. (2012). Laser-induced removal of azo dye acid Red14 using tungsten oxide. *Journal of Basic and Applied Scientific Research 2*: 1725–1727.
52. Doğan, D. and Türkdemir, H. (2005). Electrochemical oxidation of textile dye indigo. *Journal of Chemical Technology and Biotechnology 14* (80): 916–923.
53. Rezaee, A., Ghaneian, M.T., Khavanin, A. et al. (2008). Photochemical oxidation of reactive blue 19 dye (RB19) in textile wastewater by UV/K2S2O8 process. *Iranian Journal of Environmental Health Science & Engineering 11* (5): 95–100.
54. Kaušpėdienė, D., Kazlauskienė, E., and Selskienė, A. (2010). Removal of chromium complex dye from aqueous solutions using strongly basic and weakly basic anion exchangers. *Ion Exchange Letters 24* (3): 19–24.
55. Danis, U., Gurses, A., and Canpolat, N. (1999). Removal of some azo dyes from wastewater using PAC as adsorbent. *Fresenius Environmental Bulletin 8*: 358–365.
56. Sepúlveda, L., Fernández, K., Contreras, E. et al. (2004). Adsorption of dyes using peat: equilibrium and kinetic studies. *Environmental Technology 19* (25): 987–996.
57. Ahmed, M.N. and Ram, R.N. (1992). Removal of basic dye from wastewater using silica as adsorbent. *Environmental Pollution 77*: 79–86.
58. Namasivayam, C. and Arasi, D.J.S.E. (1997). Removal of Congo red from wastewater by adsorption onto waste red mud. *Chemosphere 34*: 401–417.
59. Dinçer, A.R., Güneş, Y., and Karakaya, N. (2007). Coal-based bottom ash (CBBA) waste material as adsorbent for removal of textile dyestuffs from aqueous solution. *Journal of Hazardous Materials 22* (141): 529–535.
60. Seey, T.L., Jain, M., and Mohd, N. (2012). Characterization of mangrove bark as a potentially low- cost adsorbent for reactive dye removal from aqueous solutions: equilibrium, mechanisms and kinetics. *International Journal of Pure and Applied Sciences and Technology 7* (9): 9–19.
61. Parshetti, G., Kalme, S., Saratale, G. et al. (2006). Biodegradation of malachite green by *Kocuria rosea* MTCC 1532. *Acta Chimica Slovenica 53*: 492–498.
62. Kalme, S.D., Parshetti, G.K., Jadhav, S.U. et al. (2007). Biodegradation of benzidine based dye direct Blue-6 by *pseudomonas desmolyticum* NCIM 2112. *Bioresource Technology 98*: 1405–1410.
63. Telke, A., Kalyani, D., Jadhav, J. et al. (2008). Kinetics and mechanism of reactive red 141 degradation by a bacterial isolate *rhizobium radiobacter* MTCC 8161. *Acta Chimica Slovenica 55*: 320–329.
64. Jadhav, J.P. and Govindwar, S.P. (2006). Biotransformation of malachite green by *Saccharomyces cerevisiae* MTCC 463. *Yeast 23*: 315–323.
65. Ghodake, G.S., Talke, A.A., Jadhav, J.P. et al. (2009). Potential of *Brassica juncea* in order to treat textile-effluent- contaminated sites. *International Journal of Phytoremediation 11*: 297–312.
66. An, S.Y., Min, S.K., Cha, I.H. et al. (2002). Decolorization of triphenylmethane and azo dyes by *Citrobacter* sp. *Biotechnology Letters 16* (24): 1037–1040.
67. Hu, T.L. (2001). Kinetics of azoreductase and assessment of toxicity of metabolic products from azo dyes by *Pseudomonas luteola*. *Water Science and Technology 43*: 261–269.
68. ChenKC, W.J.Y., Huang, C.C., Liang, Y.M. et al. (2003). Decolorization of azo dye using PVA-immobilized microorganisms. *Journal of Biotechnology 101*: 241–252.
69. Chang, J.S. and Lin, C.Y. (2001). Decolorization kinetics of a recombinant Escherichia coli strain harboring azo-dye-decolorizing determinants from *Rhodococcus* sp. *Biotechnology Letters 16* (23): 631–636.

70. Coughlin, M.F., Kinkle, B.K., and Bishop, P.L. (1999). Degradation of azo dyes containing aminonaphthol by *Sphingomonas* sp. strain 1CX. *Journal of Industrial Microbiology & Biotechnology 3*: 341–346.
71. Kurade, M.B., Waghmode, T.R., and Govindwar, S.P. (2011). Preferential biodegradation of structurally dissimilar dyes from a mixture by *Brevibacilluslaterosporus*. *Journal of Hazardous Materials 8* (192): 1746–1755.
72. Semdé, R., Pierre, D., Geuskens, G. et al. (1998). Study of some important factors involved in azo derivative reduction by *Clostridium perfringens*. *International Journal of Pharmaceutics 161*: 45–54.
73. Kim, J.-D., Yoon, J.-H., Park, Y.-H. et al. (2002). Identification of *Stenotrophomonas maltophilia* LK-24 and its degradability of crystal violet. *Journal of Microbiology and Biotechnology 1* (12): 437–443.
74. Dawkar, V.V., Jadhav, U.U., Jadhav, S.U. et al. (2008). Biodegradation of disperse textile dye Brown 3REL by newly isolated bacillus sp. VUS. *Journal of Applied Microbiology 105*: 14–24.
75. Dawkar, V.V., Jadhav, U.U., Telke, A.A. et al. (2009). Peroxidase from *bacillus* sp. VUS and its role in the decolorization of textile dyes. *Biotechnology and Bioprocess Engineering 14*: 361–368.
76. Walker, G.M. and Weatherley, L.R. (2000). Biodegradation and biosorption of acid anthraquinone dye. *Environmental Pollution 108*: 219–223.
77. Meehan, C., Bjourson, A.J., and McMullan, G. (2001). *Paenibacillusazoreducens* sp. nov., a synthetic azo dye decolorizing bacterium from industrial wastewater. *International Journal of Systematic and Evolutionary Microbiology 51*: 1681–1685.
78. Kalyani, D.C., Patil, P.S., Jadhav, J.P. et al. (2008). Biodegradation of reactive textile dye red BLI by an isolated bacterium *pseudomonas* sp. SUK1. *Bioresource Technology 99*: 4635–4641.
79. Kalyani, D.C., Telke, A.A., Dhanve, R.S. et al. (2009). Ecofriendly biodegradation and detoxification of reactive red 2 textile dye by newly isolated *pseudomonas* sp. SUK1. *Journal of Hazardous Materials 163*: 735–742.
80. Tamboli, D.P., Kagalkar, A.N., Jadhav, M.U. et al. (2010). Production of polyhydroxyhexadecanoic acid by using waste biomass of *Sphingobacterium* sp. ATM generated after degradation of textile dye direct red 5B. *Bioresource Technology 101*: 2421–2427.
81. Tamboli, D.P., Kurade, M.B., Waghmode, T.R. et al. (2010). Exploring the ability of Sphingobacterium sp. ATM to degrade textile dye direct blue GLL, mixture of dyes and textile effluent and production of polyhydroxyhexadecanoic acid using waste biomass generated after dye degradation. *Journal of Hazardous Materials 10* (182): 169–176.
82. Saratale, G., Kalme, S., and Govindwar, S.P. (2006). Decolorisation of textile dyes by *Aspergillusochraceus* (NCIM-1146). *Indian Journal of Biotechnology 5*: 407–410.
83. Parshetti, G.K., Kalme, S.D., Gomare, S.S. et al. (2007). Biodegradation of reactive blue-25 by *Aspergillusochraceus* NCIM-1146. *Bioresource Technology 98*: 3638–3642.
84. Jadhav, S.U., Jadhav, U.U., Dawkar, V.V. et al. (2008). Biodegradation of disperse dye brown 3REL by microbial consortium of *Galactomycesgeotrichum* MTCC 1360 and *Bacillus* sp. VUS. *Biotechnology and Bioprocess Engineering 31* (13): 232–239.
85. Swamy, J. and Ramsay, J.A. (1999). Effects of $Mn^{2+}$and NH^{4+}concentrations on laccase and manganese peroxidase production and amaranth decoloration by *Trametesversicolor*. *Applied Microbiology and Biotechnology 51*: 391–396.
86. Shedbalkar, U., Dhanve, R., and Jadhav, J. (2008 Sept 15;). Biodegradation of triphenylmethane dye cotton blue by *Penicilliumochrochloron* MTCC 517. *Journal of Hazardous Materials 157*: 472–479.
87. Fu, Y. and Viraraghavan, T. (2002). Removal of Congo red from an aqueous solution by fungus *Aspergillus niger*. *Advances in Environmental Research 7*: 239–247.
88. Sam, M. and Yeilada, O. (2001). Decolorization of Orange II dye by white-rot fungi. *Folia Microbiologica 46*: 143–145.
89. Nyanhongo, G.S., Gomes, J., Gübitz, G.M. et al. (2002). Decolorization of textile dyes by laccases from a newly isolated strain of *Trametesmodesta*. *Water Research 36*: 1449–1456.
90. Dong, X.J., Du, Z.Y., and Chen, Z. (2001). Decolorizaion of direct black 22 by *Aspergillus ficuum*. *Journal of Environmental Sciences 13*: 472–475.

91. O'Mahony T, Guibal, E., and Tobin, J. (2002). Reactive dye biosorption by *Rhizopusarrhizus* biomass. *Enzyme and Microbial Technology 19* (31): 456–463.
92. Bhatt, M., Patel, M., Rawal, B. et al. (2000). Biological decolorization of the synthetic dye RBBR in contaminated soil. *World Journal of Microbiology and Biotechnology 16*: 195–198.
93. Aksu, Z. and Dönmez, G. (2003). A comparative study on the biosorption characteristics of some yeasts for Remazol blue reactive dye. *Chemosphere 50*: 1075–1083.
94. Dönmez, G. (2002). Bioaccumulation of the reactive textile dyes by *Candidatropicalis* growing in molasses medium. *Enzyme and Microbial Technology 30*: 363–366.
95. Xiong, J.Q., Kurade, M.B., and Jeon, B.H. (2017). Biodegradation of levofloxacin by an acclimated freshwater microalga, *Chlorellavulgaris*. *Chem. Eng. J. 1* (313): 1251–1257.
96. Guolan, H., Hongwen, S., and Li, C.L. (2000). Study on the physiology and degradation of dye with immobilized algae. *Artificial Cells, Blood Substitutes, and Immobilization Biotechnology 28*: 347–363.
97. Cunningham, S.D., Shann, J.R., Crowley, D.E. et al. (1997). Phytoremediation of contaminated water and soil. *ACS Symposium Series, American Chemical Society 8* (664): 2–17.
98. Claus, H. (2004). Laccases: structure, reactions, distribution. *Micron 35*: 93–96.
99. Cunningham, S.D., Berti, W.R., and Huang, J.W. (1995). Phytoremediation of contaminated soils. *Trends in Biotechnology 13*: 393–397.
100. Cristaldi, A., Conti, G.O., Jho, E.H. et al. (2017). Phytoremediation of contaminated soils by heavy metals and PAHs. A brief review. *Environ. Technol. Innov. 8*: 309–326.
101. Rane, N.R., Chandanshive, V.V., Khandare, R.V. et al. (2014). Green remediation of textile dyes containing wastewaters by *Ipomoeahederifolia* L. *RSC Advances 24* (4): 36623–36632.
102. Kabra, A.N., Khandare, R.V., Kurade, M.B. et al. (2011). Phytoremediation of a sulphonated azo dye green HE4B by *Glandulariapulchella* (sweet) tronc. (Moss verbena). *Environmental Science and Pollution Research 18*: 1360–1373.
103. Bulc, T.G. and Ojstršek, A. (2008). The use of constructed wetland for dye-rich textile wastewater treatment. *Journal of Hazardous Materials 30* (155): 76–82.
104. Watharkar, A.D. and Jadhav, J.P. (2014). Detoxification and decolorization of a simulated textile dye mixture by phytoremediation using *Petunia grandiflora* and *Gailardia grandiflora*: a plant – plant consortialstrategy. *Ecotoxicology and Environmental Safety 19* (103): 1–8.
105. Ijaz, A., Iqbal, Z., and Afzal, M. (2016). Remediation of sewage and industrial effluent using bacterially assisted floating treatment wetlands vegetated with *Typha domingensis*. *Water Science and Technology 74*: 2192–2201.
106. Khataee, A.R., Movafeghi, A., Vafaei, F. et al. (2013). Potential ofthe aquatic fern *Azolla filiculoides* in biodegradation of an azo dye: modeling of experimentalresults by artificial neural networks. *International Journal of Phytoremediation 15*: 729–742.
107. Watharkar, A.D., Kadam, S.K., Khandare, R.V. et al. (2018). *Asparagusdensiflorus* in a vertical subsurface flow phytoreactor for treatment of real textile effluent: a lab to land approach for in situ soil remediation. *Ecotoxicology and Environmental Safety 18* (161): 70–77.
108. Kadam, S.K., Chandanshive, V.V., Rane, N.R. et al. (2018). Phytobeds with *Fimbristylisdichotoma* and *Ammanniabaccifera* for treatment of real textile effluent: An in situ treatment. *Environmental Research 160*: 1–11.
109. Torbati, S., Khataee, A.R., and Movafeghi, A. (2014). Application of watercress (Nasturtium officinale R. Br.) for biotreatment of a textile dye: investigation of some physiological responses and effects of operational parameters. *Chemical Engineering Research and Design 92*: 1934–1941.
110. Mahmood, Q., Masood, F., and Bhatti, Z.A. (2014). Biological treatment of the dye reactive blue 19 by cattails and anaerobic bacterial consortia. *Toxicological and Environmental Chemistry 27* (96): 530–541.
111. Vijayalakshmidevi, S.R. and Muthukumar, K. (2014). Phytoremediation of textile effluent pretreated with ultrasound and bacteria. *Journal of Environmental Chemical Engineering 2*: 1813–1820.

112. Watharkar, A.D., Rane, N.R., Patil, S.M. et al. (2013). Enhanced phytotransformation of navy-blue RX dye by *Petuniagrandiflora*Juss. With augmentation of rhizospheric*Bacilluspumilus* strain PgJ and subsequent toxicity analysis. *Bioresource Technology 142*: 246–254.
113. Vafaei, F., Movafeghi, A., Khataee, A.R. et al. (2013). Potential of *Hydrocotyle vulgaris* for phytoremediation of a textile dye: inducing antioxidant response in roots and leaves. *Ecotoxicology and Environmental Safety 93*: 128–134.
114. Reema, R.M., Saravanan, P., Kumar, M.D. et al. (2011). Accumulation of methylene blue dye by growing *Lemna minor*. *Separation Science and Technology 15* (46): 1052–1058.
115. Kabra, A.N., Khandare, R.V., Waghmode, T.R. et al. (2011). Differential fate of metabolism of a sulfonated azo dye Remazol Orange 3R by plants *Aster amellus* Linn., *Glandulariapulchella* (sweet) tronc. And their consortium. *Journal of Hazardous Materials 15* (190): 424–431.
116. Kagalkar, A.N., Jagtap, U.B., Jadhav, J.P. et al. (2009). Biotechnological strategies for phytoremediation of the sulfonated azo dye direct red 5B using *Blumeamalcolmii* hook. *Bioresource Technology 100*: 4104–4110.
117. Davies, T.H., Cottingham, P.D., Centre, W.S. et al. (1994). The use of constructed wetlands for treating industrial effluent (textile dyes). *Water Science & Technology 29*: 227–232.
118. Pandey, A., Singh, P., and Iyengar, L. (2007). Bacterial decolorization and degradation of azo dyes. *International Biodeterioration and Biodegradation 59*: 73–84.
119. Yoshida, B.H. (1883). LXIII.-chemistry of lacquer (Urushi). *Journal of the Chemical Society 43*: 472–486.
120. Mayer, A.M. and Staples, R.C. (2002). Laccase: new functions for an old enzyme. *Phytochemistry 2* (60): 551–565.
121. Claus, H. (2003). Laccases and their occurrence in prokaryotes. *Archives of Microbiology 7* (179): 145–150.
122. Caparrós-Ruiz, D., Fornalé, S., Civardi, L. et al. (2006). Isolation and characterisation of a family of laccases in maize. *Plant Science 18* (171): 217–225.
123. Piontek, K., Antorini, M., and Choinowski, T. (2002). Crystal structure of a laccase from the fungus *Trametes versicolor* at 1.90 - a resolution containing a full complement of coppers. *The Journal of Biological Chemistry 5* (277): 37663–37669.
124. Xu, F., Shin, W., Brown, S.H. et al. (1996). A study of a series of recombinant fungal laccases and bilirubin oxidase that exhibit significant differences in redox potential, substrate specificity, and stability. *Biochimica et Biophysica Acta, Protein Structure and Molecular Enzymology 8* (1292): 303–311.
125. Coll, P.M., Fernandez-Abalos, J.M., Villanueva, J.R. et al. (1993). Purification and characterization of a phenoloxidase (laccase) from the lignin-degrading basidiomycete PM1 (CECT 2971). *Applied and Environmental Microbiology 11* (59): 2607–2613.
126. Telke, A.A., Kadam, A.A., Jagtap, S.S. et al. (2010). Biochemical characterization and potential for textile dye degradation of blue laccase from *aspergillus ochraceus* NCIM-1146. *Biotechnology and Bioprocess Engineering 9* (15): 696–703.
127. Russ, R., Rau, J., and Stolz, A. (2000). The function of cytoplasmic flavin reductases in the reduction of azo dyes by bacteria. *Applied and Environmental Microbiology 66*: 1429–1434.
128. Fenoll, L.G., Peñalver, M.J., Rodríguez-López, J.N. et al. (2004). Deuterium isotope effect on the oxidation of monophenols and o-diphenols by tyrosinase. *Biochemical Journal 15* (380): 643–650.
129. Hammel, K.E. and Tardone, P.J. (1988). The oxidative 4-dechlorination of polychlorinated phenols is catalyzed by extracellular fungal lignin peroxidases. *Biochemistry 20* (27): 6563–6568.
130. Hammel, K.E., Kalyanaraman, B., and Kirk, T.K. (1986). Oxidation of polycyclic aromatic hydrocarbons and dibenzo[p]dioxins by Phanerochaetechrysosporiumligninase. *The Journal of Biological Chemistry 25* (261): 16948–16952.
131. Paszczynski, A. and Crawford, R.L. (1991). Degradation of azo compounds by ligninasePhanerochaetechrysosporium: involvement of veratryl alcohol. *Biochemical and Biophysical Research Communications 15* (178): 1056–1063.

132. Gomare, S.S., Jadhav, J.P., and Govindwar, S.P. (2008). Degradation of sulfonated azo dyes by the purified lignin peroxidase from *Brevibacillus laterosporus* MTCC 2298. *Biotechnology and Bioprocess Engineering 13*: 136–143.
133. Jha, V., Louis, S., Chelikani, P. et al. (2011). Modulation of heme orientation and binding by a single residue in catalase HPII of *Escherichiacoli*. *Biochemistry 18* (50): 2101–2110.
134. Murthy, M.R.N., Reid, T.J. III, Sicignano, A. et al. (1981). The structure and heme environment of beef liver catalase at 2.5 Å resolution. *Acta Crystallographica, Section A: Foundations of Crystallography 78*: 4767–4771.
135. Montavon, P. and Bortlik, K. (2004). Evolution of Robusta green coffee redox enzymatic activities with maturation. *Journal of Agricultural and Food Chemistry 4* (52): 3590–3594.
136. Wang, G., Conover, R.C., Olczak, A.A. et al. (2005). Oxidative stress defense mechanisms to counter iron-promoted DNA damage in helicobacter pylori. *Free Radical Research 39*: 1183–1191.
137. Yang, B., Hao, F., Li, J. et al. (2013). Biology binding of chrysoidine to catalase: spectroscopy, isothermal titration calorimetry and molecular docking studies. *Journal of Photochemistry and Photobiology B: Biology 5* (128): 35–42.
138. Li, T., Hao, M., Pan, J. et al. (2017). Comparison of the toxicity of the dyes Sudan II and Sudan IV to catalase. *Journal of Biochemical and Molecular Toxicology 31*: 21943–21951.
139. Haq, I. and Raj, A. (2018). Biodegradation of azure-B dye by *Serratialiquefaciens* and its validation by phytotoxicity, genotoxicity and cytotoxicity studies. *Chemosphere 196*: 58–68.
140. Ortiz-Ordoñez, E., Uría-Galicia, E., Ruiz-Picos, R.A. et al. (2011). Effect of yerbimat herbicide on lipid peroxidation, catalase activity, and histological damage in gills and liver of the freshwater fish Goodeaatripinnis. *Archives of Environmental Contamination and Toxicology 61*: 443–452.
141. Bakshi, D.K. and Sharma, P. (2003). Genotoxicity of textile dyes evaluated with Ames test and rec-assay. *Journal of Environmental Pathology, Toxicology and Oncology 22*: 101–109.
142. Muhammad, S., Shah, M.T., Khan, S. et al. (2013). Wild plant assessment for heavy metal phytoremediation potential along the mafic and ultramafic terrain in northern Pakistan. *BioMed Research International 3* (2013): 1–9.
143. Nilratnisakorn, S., Thiravetyan, P., and Nakbanpote, W. (2009). A constructed wetland model for synthetic reactive dye wastewater treatment by narrow-leaved cattails (*Typhaangustifolia* Linn.). *Water Science and Technology 60*: 1565–1574.
144. Charoenlarp, K., Surakul, K., Winitkhetkamnoun, P. et al. (2016). Textile wastewater treatment using vetiver grass cultivated with floating platform technique. *InproceedingTextile 10* (10): 51–57.
145. Vymazal, J. (2010). Constructed wetlands for wastewater treatment. *Water 27* (2): 530–549.
146. Sim, C.H., Quek, B.S., Shutes, R.B.E. et al. (2013). Management and treatment of landfill leachate by a system of constructed wetlands and ponds in Singapore. *Water Science and Technology 68*: 1114–1122.
147. Wallace, J., Champagne, P., and Monnier, A.C. (2015). Performance evaluation of a hybrid-passive landfill leachate treatment system using multivariate statistical techniques. *Waste Management 35*: 159–169.
148. Schierano, M.C., Maine, M.A., and Panigatti, M.C. (2017). Dairy farm wastewater treatment using horizontal subsurface flow wetlands with *Typha domingensis* and different substrates. *Environmental Technology 38*: 192–198.
149. Chen, T.Y., Kao, C.M., Yeh, T.Y. et al. (2006). Application of a constructed wetland for industrial wastewater treatment: a pilot-scale study. *Chemosphere 18* (64): 497–502.
150. Wang, X., Bai, X., Qiu, J. et al. (2005). Municipal wastewater treatment with pond- constructed wetland system: a case study. *Water Science and Technology 51*: 325–329.
151. Kõiv, M., Vohla, C., Mõtlep, R. et al. (2009). The performance of peat-filled subsurface flow filters treating landfill leachate and municipal wastewater. *Ecological Engineering 9* (35): 204–212.
152. Sivakumar, D., Shankar, D., Vijaya Prathima, A.J.R. et al. (2013). Constructed wetland treatment of textile industry wastewater using aquatic macrophytes. *International Journal of Environmental Sciences 6* (3): 1223–1232.

153. Goyal, S., Sharma, G., and Bhardwaj, K.K. (2009). Decolorizatton of synthetic dye (methyl red) wastewater using constructed wetlands having up flow and downflow loading formate. *Rasayan Journal of Chemistry 2*: 329–331.
154. Ong, S.A., Uchiyama, K., Inadama, D. et al. (2009). Simultaneous removal of color, organic compounds and nutrients in azo dye-containing wastewater using up-flow constructed wetland. *Journal of Hazardous Materials 15* (165): 696–703.
155. Maine, M.A., Hadad, H.R., Sánchez, G.C. et al. (2017). Long-term performance of two free-water surface wetlands for metallurgical effluent treatment. *Ecological Engineering 98*: 372–377.
156. Stefanakis, A.I., Seeger, E., Dorer, C. et al. (2016). Performance of pilot-scale horizontal sub-surface flow constructed wetlands treating groundwater contaminated with phenols and petroleum derivatives. *Ecological Engineering 9* (95): 514–526.
157. Yun, Y., Zhou, X., Li, Z. et al. (2015). Comparative research on phosphorus removal by pilotscale vertical flow constructed wetlands using steel slag and modified steel slag as substrates. *Water Science and Technology 71*: 996–1003.
158. Paing, J., Serdobbel, V., Welschbillig, M. et al. (2015). Treatment of high organic content wastewater from food-processing industry with the French vertical flow constructed wetland system. *Water Science and Technology 27* (72): 70–76.
159. Sehar, S., Sumera Naeem, S., Perveen, I. et al. (2015). A comparative study of macrophytes influence on wastewater treatment through subsurface flow hybrid constructed wetland. *Ecological Engineering 11* (81): 62–69.
160. Zhao, J., Zhao, Y., Xu, Z. et al. (2016). Highway runoff treatment by hybrid adsorptive media-baffled subsurface flow constructed wetland. *Ecological Engineering 12* (91): 231–239.
161. Jong, J. (1976). Purification of wastewater with the aid of rush or reed ponds. In: *Biological Control of Water Pollution* (ed. J. Tourbier and R.W. Pierson Jr.), 133–139. Philadelphia, PA: University of Pennsylvania.
162. Seidel, K. (1953). Pflanzungenzwischengewassern und land. In: *Mitteilungen Max-Planck Gesselschaft*, 17–20.
163. Pavlineri, N., Skoulikidis, N.T., and Tsihrintzis, V.A. (2017). Constructed floating wetlands: a review of research, design, operation and management aspects, and data meta-analysis. *Chemical Engineering Journal 15* (308): 1120–1132.
164. Cooper, P. (1999). A review of design and performance of verticle-flow and hybrid reed bed treatment system. *Water Science and Technology 40*: 1–9.
165. Khan, N.T. (2018). Integration of bioinformatics in bioremediation. *International Journal of Biomedical Data Mining 13* (7): 130–133.
166. Laurence, D., Uribe, J.F., and Utyuzhnikov, S. (2005). A robust formulation of the v2-f model. *Flow, Turbulence and Combustion 73*: 169–185.
167. Mahmoodi, N.M., Arami, M., and Gharanjig, K. (2009). Laboratory studies and CFD modeling of photocatalytic degradation of colored textile wastewater by titania nanoparticles. *Desalination and Water Treatment 1*: 312–317.
168. Abbott, L.C., Batchelor, S.N., Oakes, J. et al. (2005). Experimental and computational studies of structure and bonding in parent and reduced forms of the azo dye orange II. *The Journal of Physical Chemistry. A 24* (109): 2894–2905.
169. Sridhar, S., Chinnathambi, V., Arumugam, P. et al. (2013). In silico and in vitro physicochemical screening of Rigidoporus sp. crude laccase-assisted decolorization of synthetic dyes: approaches for a cost-effective enzyme-based remediation methodology. *Applied Biochemistry and Biotechnology 6* (169): 911–922.
170. Kadam, S.K., Tamboli, A.S., Sambhare, S.B. et al. (2018). Enzymatic analysis, structural study and molecular docking of laccase and catalase from *B. subtilis* SK1 after textile dye exposure. *Ecological Informatics 19* (48): 269–280.

9

Exploitation of Soil Amendments to Remediate Heavy Metal Toxicity for Safe Cultivation of Crops

Anandkumar Naorem[1], Abhishek Patel[1], and Shiva Kumar Udayana[2]

9.1. INTRODUCTION

Soil is a natural resource that not only provides anchorage to plants but also delivers several nutrients to the plant for their growth and development. With increasing industrialization, the soil system is now being misused and treated as a sink for the disposal of several types of wastes from various sources [1]. The increase in unhealthy anthropogenic activities such as unrestricted open disposal of industrial waste, petrochemicals leakage, over mining, metal treatment waste, lead-based paints, land application of biosolids (sewage sludge), and settlement of emanation from automobiles leads to potent soil contamination [2–4]. All these activities release large quantities of non-essential metals that have a toxic effect on various lifeforms on Earth, hence referred to as toxic elements [5]. In broader terms, the elements that are nonessential and have a specific density higher than 6 g cm^{-3} such as cadmium, lead, and mercury are termed heavy metals [5, 6]. However, some heavy metals such as cobalt, copper, chromium, manganese, and zinc are biologically essential for living organisms provided that they are present in ideal concentrations. Untreated and unmonitored disposal processes result in the accumulation of these heavy metals in the soil. This build-up in turn begins to act as an entry point for toxic heavy metals in the food supply chain (Figure 9.1). Heavy metal contamination in soils is currently a major environmental problem with adverse effects on the environment as well as human health. On a global scale, approximately 20 million hectares of land is contaminated with heavy metals [7]. Out of 10 million contaminated soils in the world, more than 50% are heavy metal contaminated soils [8]. The accumulation of heavy metals in the soil system above the threshold safe limit, such as given by CODEX standard, poses serious threats to soil, plant, animal, and human health. Soil contamination with heavy metals can negatively affect soil functions in providing ecosystem services.

[1]ICAR– Central Arid Zone Research Institute, Regional Research Station, Bhuj, Gujarat, India
[2]Department of Soil Science and Agricultural Chemistry, MS Swaminathan School of Agriculture, Centurion University of Technology and Management, Gajapati, Odhisha, India

Microbial Biotechnology: Role in Ecological Sustainability and Research, First Edition.
Edited by Pankaj Chowdhary, Sujata Mani, and Preeti Chaturvedi.

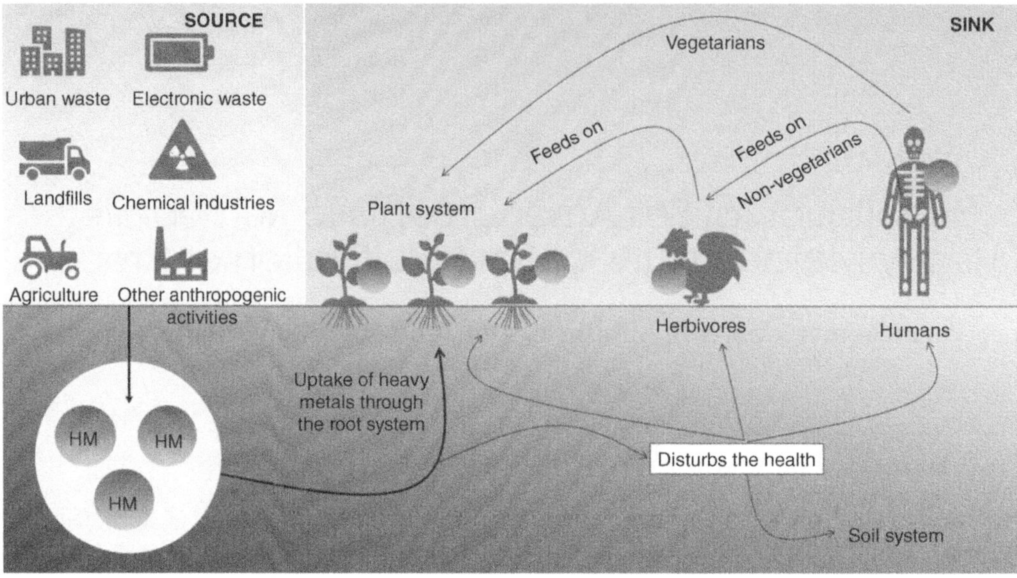

Figure 9.1 The entry of heavy metals into the food chain through the soil system.

9.2. SOURCES OF HEAVY METAL LOAD IN THE SOIL SYSTEM

Natural, as well as anthropogenic, processes load heavy metals into the soil system. Naturally governed processes, such as pedogenic and lithogenic, do not generally affect the soil heavy metal balance and remain safe [9, 10]. However, anthropogenic activities are unrestricted and rapid, adding toxic heavy metals to the soil mass and creating a risk to humans, plants, and animals in particular, and the environment as a whole [11]. Anthropogenic processes are more problematic in delivering highly mobile heavy metals than natural processes [12, 13]. There are several pathways through which the concentration of heavy metals in the soil system increases. Some of the prominent agriculture-related pathways are long-term sewage irrigation, use of fertilizers and pesticides, and use of some livestock manures. These heavy metals are then accumulated in the edible parts of plants. Consumption of these heavy metal-contaminated plants leads to the transfer of the heavy metals to animals and humans, causing health issues. Therefore, one of the main objectives of heavy metal research is the remediation of heavy metal toxicity in the soil system through in-situ or ex-situ techniques.

9.3. EFFECT OF METAL CONTAMINATION ON SOIL HEALTH

Long-term, persistent heavy metal imparts toxic effects on the soil's environment and its basic functions including biota and micro-flora [14, 15] based on soil's physical (i.e. temperature, structure, and clay minerals) and chemical (i.e. pH, ions, and organic matter) strength [16, 17]. Heavy metals target the microbial and biochemical activities in the soil leading to an alteration in soil physical and chemical properties [14] and physiological status of soil [18]. Affected soils are prone to alteration of physiological behavior in the soil environment. For example, the presence of toxic heavy metal generates reactive oxygen species (ROS) and affects chloroplast pigments, and disturbs the level of plant acids and normal functioning of the cell membrane [19]. ROS are responsible for the generation of stress in plants. The heavy metal build-up in soil forces the plant to release antioxidant enzymes such as glutathione reductase (GR), superoxide dismutase (SOD), carotenoids, and ascorbate. [20, 21]. Heavy metal in soil hinders microbial functionalities of soil enzymes in affected soils [22, 23].

9.4. REMEDIATION TECHNIQUES FOR HEAVY METAL-CONTAMINATED SOIL

Heavy metal-contaminated soil can be reclaimed through several reclamation processes but can't be freed from heavy metal completely [24, 25]. The reclamation of heavy metal-contaminated soil may be achieved through physical methods like soil washing [26] and replacement [27], as well as thermal treatment [28] (Table 9.1). Several authors and researchers had advocated for the use of soil amendments to remediate heavy metal toxicity [29–31]. These soil amendments can be classified into immobilizing agents and mobilizing agents, based on their mechanism of action (Figure 9.2). Some of the common examples are compost, vermicompost, basic slag, lime, biochar, humic substances, EDTA (ethylenediaminetetraacetic acid), and citric acid. Mobilizing agents enhance the mobility of heavy metals and then the heavy metals are removed through phytoremediation. On the other hand, immobilizing agents form insoluble complexes with heavy metals and restrict their mobility, thus limiting the plant uptake of these heavy metals. Some of the common mechanisms of immobilizing agents are precipitation, complexation, adsorption, and ion exchange.

Chemical amendments like sulfuric, hydrochloric acids ($pH < 2$) or chelators (EDTA) alone or in combination are used in the process of making heavy metals highly mobile (water-soluble complex) so that they can be leached out (in-situ process) or washed out (ex-situ process). The use of EDTA chelator is less hazardous to soil than acids [32] and perform more effectively to achieve the goal of heavy metal remediation. However, it is rather difficult to separate the leached heavy metal from the extract [24] (Figure 9.3). The metal stabilization process, i.e., making the metal highly immobile, can be carried out with the application of additives such as lime, cement, and organic matter [33–36]. These amendments bind the heavy metal and limit their entry into the food chain or deep percolation into groundwater (Figure 9.3).

The in situ remediation technique involves the addition of soil amendments to mitigate the adverse effects of heavy metals on soil health. The soil amendments are highly effective in blocking the movement of the heavy metals from soil into the food chain through various chemical processes as sorption and precipitation [29] along with dissolution, adsorption-desorption, and complexation [37].

The ex situ techniques are usually applied in soils with a very high degree of heavy metal contamination, where the soil is removed from the place of origin. Ex situ techniques have an edge over the in situ techniques since it is fast. However, it might generate contaminated byproducts during the process, which might call for another treatment. On the other hand, in situ techniques are advantageous since they may be more affordable and have minimum disturbance to the soil system. They are simpler than ex situ methods and therefore more practiced and can be applied to a broad spectrum of pollutants. Unlike ex situ techniques, they carry less risk of spreading the contaminants in nearby areas. Another reason for high public acceptance of the in situ technique is the lower energy and labour requirement compared to the ex situ technique. Some ill effects of heavy metal toxicity in the soil system are low nitrogen availability, micronutrient imbalance, low soil organic matter, and extreme soil

Table 9.1 Common types of remediation techniques in relation to heavy metal contamination in the soil system.

S. No.	Reclamation techniques	Amendment material	Process Involved	Limitation/problems
1.	i. Phyto-Extraction (No Chemical) ii. Phyto-Extraction (Chemical Assisted)	No amendment (only hyperaccumulator plant) CDTA (*trans*-1,2-cyclohexylenedi-nitrilotetraacetic acid), DTPA (diethylenetrinitrilopentaaceticacid), EDTA (ethylenedinitrilotetraaceticacid), EGTA (ethylenebis[oxyethylenetrinitrilo] tetraacetic acid), and EDDS (ethylenediamine dissuccinate)	Uptake by plant split-uptake mechanism	If not done efficiently, then it would pose a potentially high risk for contamination of groundwater
2.	Soil washing/ flushing	sulfuric, hydrochloric acids ($pH < 2$), chelators (EDTA) (alone or in combination)	Desorption and solubilization	Difficult separation of metals from the waste-chelator extract
3.	Metal stabilization	lime, cement, and organic matter	adsorption and inclusion mechanisms followed by precipitation (Immobilization)	Difficult in highly alkaline environments (mobile under these conditions)

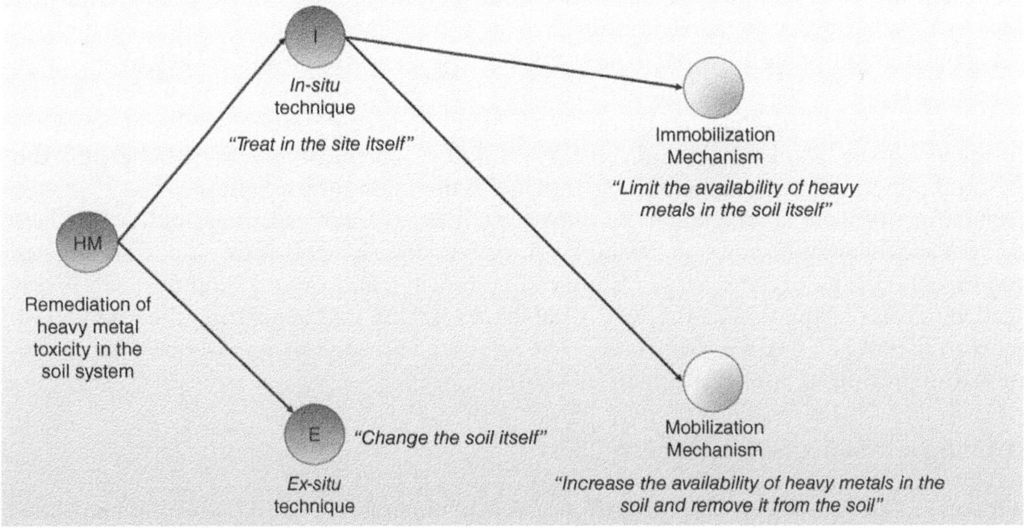

Figure 9.2 Classification of heavy metal remediation techniques.

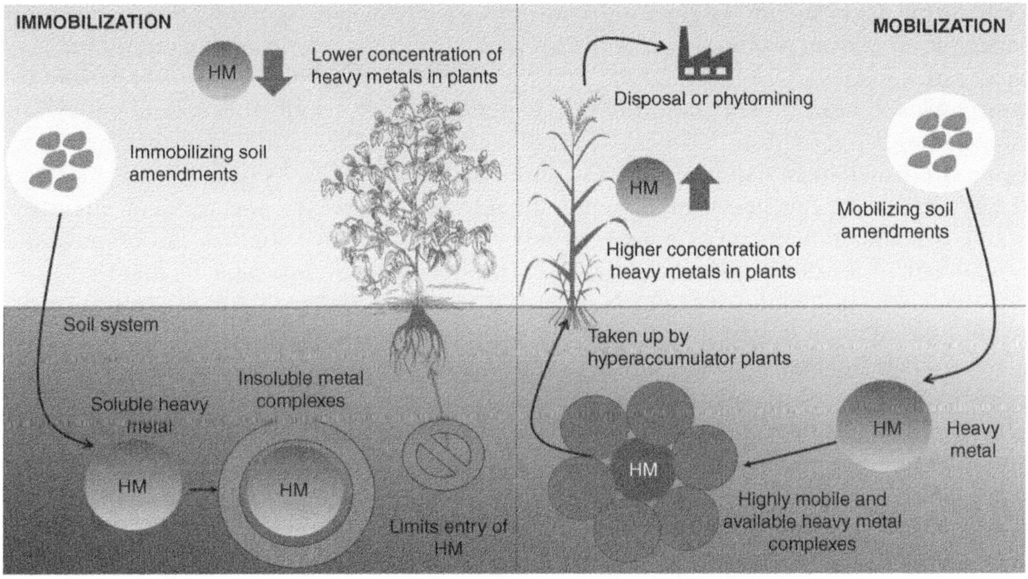

Figure 9.3 Overall mechanisms of heavy metal remediation in immobilization (left) and mobilization mechanisms (right).

pH. Heavy metal contamination in the soil system could indirectly lead to poor soil physical condition [38]. Therefore, to maintain a balance between soil physical, chemical, and biological health, remediation of heavy metal toxicity in the soil system must be addressed.

The in situ process of reclamation is better than ex situ in terms of cost effectiveness. Reclamation of contaminated soil requires the addition of amendments to the soil system to increase the mobility of the heavy metals so they are further leached out from the soil system or it decreases the mobility of the heavy metals so that there will be a smaller amount of heavy metal available for microbial activities as well as plant uptake. Chelating agents are also added to make the heavy metal more available to be extracted from contaminated soil through the hypoextraction process via plants and this process is termed chemical chelate-enhanced phytoextraction [39]. However, some hyperaccumulator plants may not require metal-chelates as they are capable of hyperaccumulating heavy metals from the contaminated soil. Since the hyperaccumulator plants do not involve any additive chemical, i.e., metal-chelates, this method is considered environmentally friendly with minimum investment [40].

Amendments such as lime materials (limestone hydroxides of calcium, poultry eggshell, etc.) release hydroxide ions to react with heavy metals and form the immobilized complexes [41]. The organic carbon sources like biochar and organic compost are highly effective in soil reclamation as they are capable of sorption and generating the ionic bondages with metals [42, 43]. Clay substance (like sepiolite and zeolite) are also used for stabilization and precipitation of heavy metals [44, 45].

9.4.1. Organic Soil Amendments

Organic manure is one of the most affordable soil amendments and it is often mentioned in heavy metal remediation. It not only provides plants available nutrients and enhances the soil physicochemical properties, but it can also be used for reducing the toxicity of heavy metals through complexation and adsorption [46], owing to its several functional groups that bind the heavy metals. Organic soil amendments are usually immobilizing agents that change the speciation of the heavy metals from highly available to insoluble fractions associated with carbonates, organic matter, and metal oxides [47]. The humic acids present in the organic soil amendments bind the heavy metals [47]. Other organic soil amendments include sawdust, wood ash, biosolids, sewage sludge, etc. The main reasons for applying organic soil amendments in heavy metal-contaminated soil are low cost and enhancement of revegetation in the heavy metal-contaminated soil. However, the residual effect of organic soil amendments on heavy metal bioavailability must be studied. Decomposition of organic matter can decrease the surface area and cation exchange capacity and thus promote the release of metals over time. Narwal and Singh [48] showed that the application of cow and pig manure in nickel-contaminated soil could reduce the available nickel in soil owing to the formation of strong complexes of nickel and organic matter. In a study conducted by Alamgir et al. [49], the application of farmyard manure could decrease lead and cadmium concentration in the plant tissues of Amaranth. Liu et al. [50] showed improved plant growth and low cadmium concentration in plant tissue with the application of compost. Therefore, the main mechanisms of mitigating heavy metal toxicity in the soil system through organic soil amendments are an increase in soil pH, complexation of heavy metals with organic matter, and co-precipitation. However, these positive effects of organic soil amendments on reducing heavy metal toxicity in a soil largely depend on the soil type, the type of heavy metal to be remediated, and the properties of the soil amendments [51]. One example of such a case is the increase in arsenic mobility in soil with the application of compost [52]. This is attributed to increased competition between dissolved organic matter and soluble arsenic for adsorption sites [53]. However, based on the types and chemistry of organic manures, soil pH can either increase or decrease, affecting the solubility of heavy metals.

One of the currently studied soil amendments in heavy metal remediation is biochar. Due to its high adsorption capacity, biochar is highly recommended by several authors in the remediation of heavy metal toxicity. Biochar is prepared from organic materials in high-temperature and anaerobic conditions [54]. The high adsorption capacity of biochar is attributed to the presence of several oxygen-containing functional groups and a large specific surface area [55]. These properties reduce the mobility and bioavailability of the heavy metals in the soil system while improving the soil's physical, chemical, and biological properties [56, 57].

Biochar is a carbon-rich product of biomass pyrolysis [58]. The efficacy of biochar treatment to the contaminated soil depends on the biochar properties, which are the functions of the biochar preparation process (method, temperature, feedstock, and time) along with soil conditions. Biochar can be derived from crop residues like millets [59], barley [60], rice [61, 62], soybean [63], wheat [64, 65], and sugar cane [66] and are shown to stabilize the heavy metals (Cd, Pb, Cu, and Zn) by reducing their bioavailability to the plants in contaminated soils. Biochar derived from broiler litter and dairy manure [67, 68], and chicken manure [69] are reportedly suitable for decreasing the mobility of heavy metals in contaminated soil. However, the effectiveness of amelioration of heavy metal-contaminated soil through biochar depends on the soil type and condition (physicochemical), application rates and quality of biochar, interaction between biochar and heavy metals, and heavy metal mobility rate. Therefore it is crucial to manage the quality of the biochar during preparation and optimal application of biochar for soil reclamation [70]. In contrast to all the benefits of biochar amendment for soil reclamation, several studies have shown the negative impacts of biochar amendments [59, 65, 71–74], which indicates that the efficiency of biochar in stabilizing the heavy metals highly depends on soil types and conditions, feedstock material, and the process of biochar preparation and application rates.

9.4.2. Inorganic Soil Amendments

Chemical amendments such as limestone (LS), and steel slag (SS) increase the pH of the soil, which produces the ions to bind or stabilize the soluble or mobile heavy metals [75–77]. The reduction efficiency of bioavailable heavy metals through the stabilization effect of chemical amendments depends on the amendment type, the release of ions to stabilize the metal, and suitability of heavy metal toward stabilization. It has been found that different amendments such as limestone (LS), and steel slag (SS), and acid mine drainage sludge (AMDS) had the reduction efficiency in the range of 4.2–92.5% based on the amendment type, their treatment and ability to release the stabilization ions, and the type of heavy metals [23]. However, studies show that the AMDS is highly capable of stabilizing or forming complexes with heavy metals through the release of FeO. Stabilization with SS involves the sorption, precipitation, and ion exchange mechanisms when added to the contaminated soils and is found to be effective in stabilizing cadmium and lead with the reduction rate of 5.2–34.9% and 4.2–11.2%, respectively [23]. LS is a potent amendment for remediation of heavy metal toxicity [76, 78–80]. However, the size of grain powder of amendment may reduce efficiency [23, 81] unless the sorption capacity of the amendment is high.

Liming materials are frequently used as remediation materials for reducing heavy metal availability in the soil system [82]. Some of the common liming materials used for mitigating heavy metal toxicity in soils are gypsum, dolomite, basic slag, and calcium carbonate. Each of these liming materials has a different acid-neutralizing capacity and therefore affects heavy metal remediation with various degrees of efficiency. Khan and Jones [3] reported a considerable reduction in DTPA-extractable copper, iron, and zinc in an abandoned copper mine tailing site. However, it is not necessarily true that soil pH changes (>7) with the application of lime will always decrease the bioavailability of the heavy metals in the soil system. It is because the bioavailability of molybdenum and metalloid arsenic increases with high soil pH through the formation of more soluble arsenite (in the case of arsenic). Liming materials not only increase the soil pH but also increase the mobilization of essential plant nutrients [83].

Wang et al. [84] tested the efficacy of biochar, hydroxyapatite, and organic manure on heavy metal accumulation in maize as a test crop grown in acidic soils. The application of hydroxyapatite and organic manure (both at the rate of 0.1% and 1%) showed a decrease in cadmium, lead, and zinc concentration in maize shoots and roots, thereby increasing plant growth. In the case of biochar treatment, it decreased the cadmium and lead concentration in the shoot while it decreased the concentration of zinc and lead in roots only at 1%. Out of these three amendments, Wang et al. [84] reported that hydroxyapatite showed the most pronounced effect in restricting the mobility of heavy metals, leading to an increase in plant growth. The application of hydroxyapatite even at a lower dosage (0.1%) showed better mitigating effects than biochar and organic manure. This result indicates that every soil amendment has different mechanisms for remediating heavy metal toxicity in the soil system. In the case of manure and organic manure, complexation and adsorption are the predominant mechanisms of action [85, 86]. On the other hand, hydroxyapatite forms insoluble phosphate precipitates with heavy metals, which are low in solubility and bioavailability [87]. Those soil amendments containing phosphorus are generally used for reducing lead mobility through precipitation [88].

The liming effect largely contributes to the effectiveness of the soil amendments in reducing the mobility and bioavailability of heavy metals. This is attributed to the strong dependence of heavy metal availability on soil pH [89]. The mobility of cadmium and lead are lowered in high pH soils, leading to lower uptake of these elements in the plant system. Therefore, in the study conducted by Wang et al. [84], hydroxyapatite, organic manure, and biochar showed promising results in immobilizing some heavy metals due to the increase in soil pH with the application of these soil amendments. The liming effect from these soil amendments also facilitates the formation of insoluble heavy metal precipitates [88]. The increase in soil pH with the application of hydroxyapatite can be associated with the release of OH^- ions [90]. Organic manure is found to increase soil pH in acidic soils due to the addition of basic cations. Another probable reason is the production of ammonia during manure decomposition that further increases soil pH [91]. Biochar is generally alkaline owing to its high content of alkaline minerals and is therefore used in increasing pH in acidic soils [92].

Some soils are highly contaminated with multiple types of heavy metals, especially industrialized areas or urban soils. In this case, competition between these heavy metals for adsorption sites also affects the immobilization efficiency of the soil amendments [88]. Therefore, Wang et al. [84] reported that since zinc is a mobile element in the soil, it is easily out-competed by other heavy metals such as lead and cadmium and therefore, the treatment of hydroxyapatite could not stabilize zinc as much as it could immobilize lead and cadmium. The presence of phosphates can considerably reduce lead and cadmium bioavailability but is slightly lower in the case of zinc [93]. On the other hand, since all soil amendments have different mechanisms of action, the application of biochar in zinc-contaminated soils could stabilize zinc, indicating that co-application of different types of soil amendments seems more effective than a single treatment.

9.5. WAY FORWARD AND CONCLUSIONS

Heavy metal (HM) pollution has become a serious environmental issue of public concern. The soil system is a major sink of HMs through which the plants take up these HMs and transfer through the food chain. The high efflux and loading of HMs in the soil alters the soil's physicochemical and biological properties, which ultimately affects soil health. This creates nutrient imbalances in the plants and leads to decreasing crop productivity. Therefore, keeping these points in view, this chapter aimed at exploiting several soil amendments that could be applied to reduce the HM-related risks and problems. These soil amendments work either through stabilizing the mobility of the HMs or increasing their mobility so that they can be removed through phytoremediation or leaching. The in situ immobilizing amendments such as lime and nanomaterials have been recently employed to reduce the bioavailability of HMs and restrict further movement and uptake of HMs through plant roots. The mobilizing amendments such as compost increase the mobility of the HMs and they are removed through other remediation processes. These two forms of amendments must be low-cost and environmentally safe so that they could not only improve the soil health but also prevent any disposal issues. This chapter critically reviews the types of soil amendments and their efficiency in remediating heavy metal toxicity in cultivated soils. This chapter focused on the types of soil amendments exploited to reclaim the heavy metal contamination in the soil system. However, there is no universal soil amendment that has an overall efficiency in remediating all types of heavy metals, owing to the different speciation of heavy metals and the interactions between heavy metals and the complex soil system. There are several factors that determine the suitable selection of a soil amendment based on the suitability of contact between immobilizing agents and heavy metal present in the soil. This is one of the crucial points that needs to be addressed in order to identify the best soil amendment. The economic feasibility of the amendment must be considered as the naturally available materials are more cost effective than preparation of new amendments. Moreover, the in situ vs. ex situ amelioration must consider the economic investment and its returns (both direct monetary gains and indirect gains through increase in crop yield or improvement in soil quality). One of the major drawbacks of the immobilization techniques is that although the entry of the heavy metals is restricted or limited by forming insoluble complexes of heavy metal, the total amount of heavy metals in the soil system remains unaltered. Therefore, there is a high chance of releasing these heavy metals from insoluble complexes with the change in the properties of soil system, environmental changes, and soil or crop management practices. Therefore, in this regard, continuous monitoring of the available heavy metal content in the soil needs to be taken into account, along with further studies on its dynamics and the kinetics of heavy metals in the soil system over the course of time.

REFERENCES

1. Power, J.F., Dick, W.A., Kashmanian, R.M. et al. (2000). *Land Application of Agricultural, Industrial, and Municipal by-Products*. Soil Science Society of America Inc.
2. Basta, N.T., Ryan, J.A., and Chaney, R.L. (2005). Trace element chemistry in residual-treated soil: key concepts and metal bioavailability. *Journal of Environmental Quality 34*: 49–63.
3. Jamal Khan, M. and Jones, D.L. (2008). Chemical and organic immobilization treatments for reducing phytoavailability of heavy metals in copper-mine tailings. *Journal of Plant Nutrition and Soil Science 171* (6): 908–916.
4. Zhang, M.K., Liu, Z.Y., and Wang, H. (2010). Use of single extraction methods to predict bioavailability of heavy metals in polluted soils to rice. *Communications in Soil Science and Plant Analysis 41* (7): 820–831.
5. Adriano, D.C. (2001). *Trace Elements in Terrestrial Environments: Biogeochemistry, Bioavailability and Risks of Metals*, 2e. New York: Springer.
6. Alloway, B.J. (1990). Cadmium. In: *Heavy Metals in Soils* (ed. B.J. Alloway), 100–124. New York Wiley.
7. Liu, L., Li, W., Song, W. et al. (2018). Remediation techniques for heavy metal-contaminated soils: principles and applicability. *The Science of the Total Environment 15* (633): 206–219.

8. He, Z., Shentu, J., Yang, X. et al. (2015). Heavy metal contamination of soils: sources, indicators and assessment. *Journal of Environmental Indicators 9*: 17–18.
9. Pierzynski, G.M., Vance, G.F., and Sims, J.T. (2005). *Soils and Environmental Quality*. CRC press.
10. Kabata-Pendias, A. and Pendias, H. (2001). *Trace Metals in Soils and Plants*. Boca Raton: CRC Press.
11. D'amore, J.J., Al-Abed, S.R., Scheckel, K.G. et al. (2005). Methods for speciation of metals in soils: a review. *Journal of Environmental Quality 34* (5): 1707–1745.
12. Kuo, S., Heilman, P.E., and Baker, A.S. (1983). Distribution and forms of copper, zinc, cadmium, iron, and manganese in soils near a copper smelter. *Soil Science 135*: 101–109.
13. Kaasalainen, M. and Yli-Halla, M. (2003). Use of sequential extraction to assess metal partitioning in soils. *Environmental Pollution 126* (2): 225–233.
14. Nannipieri, P., Badalucco, L., Landi, L. et al. (1997). Measurement in assessing the risk of chemicals to the soil ecosystem. In: *Proceedings of the OECD Workshop on Ecotoxicology: Responses and Risk Assessment*, 507–534. Fair Haven: SOS Publications.
15. Obbard, J.P. (2001). Ecotoxicological assessment of heavy metals in sewage sludge amended soils. *Applied Geochemistry 16* (11–12): 1405–1411.
16. Bååth, E. (1989). Effects of heavy metals in soil on microbial processes and populations (a review). *Water, Air, and Soil Pollution 47* (3): 335–379.
17. Giller, K.E., Witter, E., and Mcgrath, S.P. (1998). Toxicity of heavy metals to microorganisms and microbial processes in agricultural soils: a review. *Soil Biology and Biochemistry 30* (10–11): 1389–1414.
18. Ellis, R.J., Morgan, P., Weightman, A.J. et al. (2003). Cultivation-dependent and-independent approaches for determining bacterial diversity in heavy-metal-contaminated soil. *Applied and Environmental Microbiology 69* (6): 3223–3230.
19. Nehnevajova, E., Lyubenova, L., Herzig, R. et al. (2012). Metal accumulation and response of antioxidant enzymes in seedlings and adult sunflower mutants with improved metal removal traits on a metal-contaminated soil. *Environmental and Experimental Botany 76*: 39–48.
20. Dixit, V., Pandey, V., and Shyam, R. (2001). Differential antioxidative responses to cadmium in roots and leaves of pea (Pisum sativum L. cv. Azad). *Journal of Experimental Botany 52* (358): 1101–1109.
21. Gong, H., Zhu, X., Chen, K. et al. (2005). Silicon alleviates oxidative damage of wheat plants in pots under drought. *Plant Science 169* (2): 313–321.
22. Song, J., Shen, Q., Wang, L. et al. (2018). Effects of Cd, Cu, Zn and their combined action on microbial biomass and bacterial community structure. *Environmental Pollution 243*: 510–518.
23. Hong, Y.K., Kim, J.W., Lee, S.P. et al. (2020). Heavy metal remediation in soil with chemical amendments and its impact on activity of antioxidant enzymes in Lettuce (Lactuca sativa) and soil enzymes. *Applied Biological Chemistry 63* (1): 1. –0.
24. Mulligan, C.N., Yong, R.N., and Gibbs, B.F. (2001). Remediation technologies for metal-contaminated soils and groundwater: an evaluation. *Engineering Geology 60* (1–4): 193–207.
25. Yong, S.K., Shrivastava, M., Srivastava, P. et al. (2015). Environmental applications of chitosan and its derivatives. *Reviews of Environmental Contamination and Toxicology 233*: 1–43.
26. Hu, P., Yang, B., Dong, C. et al. (2014). Assessment of EDTA heap leaching of an agricultural soil highly contaminated with heavy metals. *Chemosphere 117*: 532–537.
27. Douay, F., Roussel, H., Pruvot, C. et al. (2008). Assessment of a remediation technique using the replacement of contaminated soils in kitchen gardens nearby a former lead smelter in northern France. *The Science of the Total Environment 401* (1–3): 29–38.
28. Busto, Y., Cabrera, X., Tack, F.M. et al. (2011). Potential of thermal treatment for decontamination of mercury containing wastes from chlor-alkali industry. *Journal of Hazardous Materials 186* (1): 114–118.
29. Bolan, N.S. and Duraisamy, V.P. (2003). Role of inorganic and organic soil amendments on immobilisation and phytoavailability of heavy metals: a review involving specific case studies. *Soil Research 41* (3): 533–555.
30. Mahar, A., Ping, W.A., Ronghua, L.I. et al. (2015). Immobilization of lead and cadmium in contaminated soil using amendments: a review. *Pedosphere 25* (4): 555–568.

31. Lwin, C.S., Seo, B.H., Kim, H.U. et al. (2018). Application of soil amendments to contaminated soils for heavy metal immobilization and improved soil quality—a critical review. *Soil Science and Plant Nutrition 64* (2): 156–167.
32. Xu, Y. and Zhao, D. (2005). Removal of copper from contaminated soil by use of poly (amidoamine) dendrimers. *Environmental Science and Technology 39* (7): 2369–2375.
33. Ruby, M.V., Davis, A., and Nicholson, A. (1994). In situ formation of lead phosphates in soils as a method to immobilize lead. *Environmental Science and Technology 28* (4): 646–654.
34. Walker, D.J., Clemente, R., Roig, A. et al. (2003). The effects of soil amendments on heavy metal bioavailability in two contaminated Mediterranean soils. *Environmental Pollution 122* (2): 303–312.
35. Tang, X.Y., Zhu, Y.G., Chen, S.B. et al. (2004). Assessment of the effectiveness of different phosphorus fertilizers to remediate Pb-contaminated soil using in vitro test. *Environment International 30* (4): 531–537.
36. Gadepalle, V.P., Ouki, S.K., Herwijnen, R.V. et al. (2007). Immobilization of heavy metals in soil using natural and waste materials for vegetation establishment on contaminated sites. *Soil and Sediment Contamination 16* (2): 233–251.
37. Gong, Y., Zhao, D., and Wang, Q. (2018). An overview of field-scale studies on remediation of soil contaminated with heavy metals and metalloids: technical progress over the last decade. *Water Research 147*: 440–460.
38. Ramamoorthy, D. (2015). Assessment of heavy metal pollution and its impacts on soil physical, chemical properties and β-glucosidase activities in agricultural lands, Puducherry region. *Polish Journal of Environmental Studies 25*: 1045–1051.
39. Salt, D.E., Smith, R.D., and Raskin, I. (1998). Phytoremediation. *Annual Review of Plant Biology 49* (1): 643–668.
40. McGrath, S.P., Zhao, J., and Lombi, E. (2002). Phytoremediation of metals, metalloids, and radionuclides. *Advances in Agronomy 75*: 1–56.
41. Cui, H., Fan, Y., Fang, G. et al. (2016). Leachability, availability and bioaccessibility of Cu and Cd in a contaminated soil treated with apatite, lime and charcoal: a five-year field experiment. *Ecotoxicology and Environmental Safety 134*: 148–155.
42. Bian, R., Chen, D., Liu, X. et al. (2013). Biochar soil amendment as a solution to prevent Cd-tainted rice from China: results from a cross-site field experiment. *Ecological Engineering 58*: 378–383.
43. Cui, L., Pan, G., Li, L. et al. (2016). Continuous immobilization of cadmium and lead in biochar amended contaminated paddy soil: a five-year field experiment. *Ecological Engineering 93*: 1–8.
44. Liang, X., Han, J., Xu, Y. et al. (2014). In situ field-scale remediation of Cd polluted paddy soil using sepiolite and palygorskite. *Geoderma 235*: 9–18.
45. Abad-Valle, P., Álvarez-Ayuso, E., Murciego, A. et al. (2016). Assessment of the use of sepiolite amendment to restore heavy metal polluted mine soil. *Geoderma 280*: 57–66.
46. Liu, X., Song, Q., Tang, Y. et al. (2013). Human health risk assessment of heavy metals in soil–vegetable system: a multi-medium analysis. *The Science of the Total Environment 463*: 530–540.
47. Walker, D.J., Clemente, R., and Bernal, M.P. (2004). Contrasting effects of manure and compost on soil pH, heavy metal availability and growth of Chenopodium album L. in a soil contaminated by pyritic mine waste. *Chemosphere 57* (3): 215–224.
48. Narwal, R.P. and Singh, B.R. (1998). Effect of organic materials on partitioning, extractability and plant uptake of metals in an alum shale soil. *Water, Air, and Soil Pollution 103* (1): 405–421.
49. Alamgir, M., Kibria, M.G., and Islam, M. (2011). Effects of farm yard manure on cadmium and lead accumulation in Amaranth (Amaranthus oleracea L.). *Journal of Soil Science and Environmental Management 2* (8): 237–240.
50. Liu, L., Chen, H., Cai, P. et al. (2009). Immobilization and phytotoxicity of Cd in contaminated soil amended with chicken manure compost. *Journal of Hazardous Materials 163* (2–3): 563–567.
51. Bernal, M.P., Clemente, R., and Walker, D.J. (2007). The role of organic amendments in the bioremediation of heavy metal-polluted soils. In: *Environmental research at the leading edge*, 1–57.

52. Hartley, W., Dickinson, N.M., Riby, P. et al. (2010). Arsenic mobility and speciation in a contaminated urban soil are affected by different methods of green waste compost application. *Environmental Pollution 158* (E): 3560–3570.
53. Wang, S. and Mulligan, C.N. (2006). Effect of natural organic matter on arsenic release from soils and sediments into groundwater. *Environmental Geochemistry and Health 28* (3): 197–214.
54. Lehmann, J., Rillig, M.C., Thies, J. et al. (2011). Biochar effects on soil biota–a review. *Soil Biology and Biochemistry 43* (9): 1812–1836.
55. Chang, R., Sohi, S.P., Jing, F. et al. (2019). A comparative study on biochar properties and Cd adsorption behavior under effects of ageing processes of leaching, acidification and oxidation. *Environmental Pollution 254*: 113123.
56. Rehman, M.Z., Khalid, H., Akmal, F. et al. (2017). Effect of limestone, lignite and biochar applied alone and combined on cadmium uptake in wheat and rice under rotation in an effluent irrigated field. *Environmental Pollution 227*: 560–568.
57. Igalavithana, A.D., Ok, Y.S., Niazi, N.K. et al. (2017). Effect of corn residue biochar on the hydraulic properties of sandy loam soil. *Sustainability 9* (2): 266.
58. Klinar, D. (2016). Universal model of slow pyrolysis technology producing biochar and heat from standard biomass needed for the techno-economic assessment. *Bioresource Technology 206*: 112–120.
59. Zhang, Z., Solaiman, Z.M., Meney, K. et al. (2013). Biochars immobilize soil cadmium, but do not improve growth of emergent wetland species Juncus subsecundus in cadmium-contaminated soil. *Journal of Soils and Sediments 13* (1): 140–151.
60. Khan, K.Y., Ali, B., Cui, X. et al. (2017). Impact of different feedstocks derived biochar amendment with cadmium low uptake affinity cultivar of pak choi (Brassica rapassb. chinensis L.) on phytoavoidation of Cd to reduce potential dietary toxicity. *Ecotoxicology and Environmental Safety 141*: 129–138.
61. Kim, H.S., Kim, K.R., Kim, H.J. et al. (2015). Effect of biochar on heavy metal immobilization and uptake by lettuce (Lactuca sativa L.) in agricultural soil. *Environment and Earth Science 74* (2): 1249–1259.
62. Yang, X., Liu, J., McGrouther, K. et al. (2016). Effect of biochar on the extractability of heavy metals (Cd, Cu, Pb, and Zn) and enzyme activity in soil. *Environmental Science and Pollution Research 23* (2): 974–984.
63. Moon, D.H., Park, J.W., Chang, Y.Y. et al. (2013). Immobilization of lead in contaminated firing range soil using biochar. *Environmental Science and Pollution Research 20* (12): 8464–8471.
64. Bian, R., Joseph, S., Cui, L. et al. (2014). A three-year experiment confirms continuous immobilization of cadmium and lead in contaminated paddy field with biochar amendment. *Journal of Hazardous Materials 272*: 121–128.
65. Chen, D., Guo, H., Li, R. et al. (2016). Low uptake affinity cultivars with biochar to tackle Cd-tainted rice—a field study over four rice seasons in Hunan, China. *Science of the Total Environment. 541*: 1489–1498.
66. Puga, A.P., Abreu, C., Melo, L.C. et al. (2015). Biochar application to a contaminated soil reduces the availability and plant uptake of zinc, lead and cadmium. *Journal of Environmental Management 159*: 86–93.
67. Cao, X., Ma, L., Gao, B. et al. (2009). Dairy-manure derived biochar effectively sorbs lead and atrazine. *Environmental Science and Technology 43* (9): 3285–3291.
68. Uchimiya, M., Lima, I.M., Thomas Klasson, K. et al. (2010). Immobilization of heavy metal ions (CuII, CdII, NiII, and PbII) by broiler litter-derived biochars in water and soil. *Journal of Agricultural and Food Chemistry 58* (9): 5538–5544.
69. Meier, S., Curaqueo, G., Khan, N. et al. (2017). Chicken-manure-derived biochar reduced bioavailability of copper in a contaminated soil. *Journal of Soils and Sediments 17* (3): 741–750.
70. Lu, H., Li, Z., Fu, S. et al. (2014). Can biochar and phytoextractors be jointly used for cadmium remediation? *PLoS One 9* (4): e95218.
71. Abdelhafez, A.A., Li, J., and Abbas, M.H. (2014). Feasibility of biochar manufactured from organic wastes on the stabilization of heavy metals in a metal smelter contaminated soil. *Chemosphere 117*: 66–71.

72. Ahmad, M., Ok, Y.S., Rajapaksha, A.U. et al. (2016). Lead and copper immobilization in a shooting range soil using soybean stover-and pine needle-derived biochars: chemical, microbial and spectroscopic assessments. *Journal of Hazardous Materials 301*: 179–186.
73. Shen, Z., McMillan, O., Jin, F. et al. (2016). Salisbury biochar did not affect the mobility or speciation of lead in kaolin in a short-term laboratory study. *Journal of Hazardous Materials 316*: 214–220.
74. Jin, J., Li, Y., Zhang, J. et al. (2016). Influence of pyrolysis temperature on properties and environmental safety of heavy metals in biochars derived from municipal sewage sludge. *Journal of Hazardous Materials 320*: 417–426.
75. He, H., Tam, N.F., Yao, A. et al. (2017). Growth and Cd uptake by rice (Oryza sativa) in acidic and Cd-contaminated paddy soils amended with steel slag. *Chemosphere 189*: 247–254.
76. Lahori, A.H., Zhang, Z., Guo, Z. et al. (2017). Potential use of lime combined with additives on (im) mobilization and phytoavailability of heavy metals from Pb/Zn smelter contaminated soils. *Ecotoxicology and Environmental Safety 145*: 313–323.
77. Francisca, F.M. and Glatstein, D.A. (2020). Environmental application of basic oxygen furnace slag for the removal of heavy metals from leachates. *Journal of Hazardous Materials 384*: 121294.
78. Gray, C.W., Dunham, S.J., Dennis, P.G. et al. (2006). Field evaluation of in situ remediation of a heavy metal contaminated soil using lime and red-mud. *Environmental Pollution 142* (3): 530–539.
79. Garau, G., Castaldi, P., Santona, L. et al. (2007). Influence of red mud, zeolite and lime on heavy metal immobilization, culturable heterotrophic microbial populations and enzyme activities in a contaminated soil. *Geoderma 142* (1–2): 47–57.
80. Novak, J.M., Ippolito, J.A., Ducey, T.F. et al. (2018). Remediation of an acidic mine spoil: miscanthus biochar and lime amendment affects metal availability, plant growth, and soil enzyme activity. *Chemosphere 1* (205): 709–718.
81. Cui, H., Shi, Y., Zhou, J. et al. (2018). Effect of different grain sizes of hydroxyapatite on soil heavy metal bioavailability and microbial community composition. *Agriculture, Ecosystems and Environment 15* (267): 165–173.
82. Kaitibie, S., Epplin, F.M., Krenzer, E.G. et al. (2002). Economics of lime and phosphorus application for dual-purpose winter wheat production in low-pH soils. *Agronomy Journal 94* (5): 1139–1145.
83. Haynes, R.J. and Naidu, R. (1998). Influence of lime, fertilizer and manure applications on soil organic matter content and soil physical conditions: a review. *Nutrient Cycling in Agroecosystems 51* (2): 123–137.
84. Wang, F., Zhang, S., Cheng, P. et al. (2020). Effects of soil amendments on heavy metal immobilization and accumulation by maize grown in a multiple-metal-contaminated soil and their potential for safe crop production. *Toxics 8* (4): 102.
85. Khalid, S., Shahid, M., Niazi, N.K. et al. (2017). A comparison of technologies for remediation of heavy metal contaminated soils. *Journal of Geochemical Exploration 182*: 247–268.
86. He, L., Zhong, H., Liu, G. et al. (2019). Remediation of heavy metal contaminated soils by biochar: mechanisms, potential risks and applications in China. *Environmental Pollution 252*: 846–855.
87. Sun, R.J., Chen, J.H., Fan, T.T. et al. (2018). Effect of nanoparticle hydroxyapatite on the immobilization of Cu and Zn in polluted soil. *Environmental Science and Pollution Research 25* (1): 73–80.
88. Kumpiene, J., Lagerkvist, A., and Maurice, C. (2008). Stabilization of As, Cr, Cu, Pb and Zn in soil using amendments–a review. *Waste Management 28* (1): 215–225.
89. Collins, C. (2013). Uptake of organic pollutants and potentially toxic elements (PTEs) by crops. In: *Persistent Organic Pollutants and Toxic Metals in Foods.*, 1e (ed. M. Rose and A. Fernandes), 129–144. Woodhead Publishing.
90. Zeng, G., Wan, J., Huang, D. et al. (2017). Precipitation, adsorption and rhizosphere effect: the mechanisms for phosphate-induced Pb immobilization in soils—a review. *Journal of Hazardous Materials 5* (339): 354–367.

91. Pocknee, S. and Sumner, M.E. (1997). Cation and nitrogen contents of organic matter determine its soil liming potential. *Soil Science Society of America Journal 61* (1): 86–92.
92. O'Connor, D., Peng, T., Zhang, J. et al. (2018). Biochar application for the remediation of heavy metal polluted land: a review of in situ field trials. *The Science of the Total Environment 619*: 815–826.
93. Impellitteri, C.A. (2005). Effects of pH and phosphate on metal distribution with emphasis on As speciation and mobilization in soils from a lead smelting site. *The Science of the Total Environment 345* (1–3): 175–190.

10

Microbial Proteomics: Understanding Metabolic Pathways in Microorganisms for Bioremediation of Aromatic Hydrocarbon Pollutants

Bhargab Kalita

10.1. INTRODUCTION

Since the dawn of the nineteenth century, industrialization has played a major role in persistent accumulation of hydrocarbons in our surrounding environment. The aromatic hydrocarbons, which contain one or more aromatic rings, are found to be widely distributed, recalcitrant, and are very toxic to the environment and human health [1–4]. These ubiquitous compounds are generated via natural as well as industry-driven anthropogenic activities, viz. incomplete combustion of fossil fuels, emissions from vehicles, and incineration of waste. They easily contaminate the soil and water bodies and tend to accumulate in organisms via the food chain, thereby posing a serious threat to the ecosystem [4, 5].

Several chemical methods, including ozonation, precipitation, coagulation, adsorption, and oxidation, have been employed for the remediation of these toxic aromatic hydrocarbons. However, they are often limited by secondary toxic compound production as well as high operational cost [6–10]. On the contrary, microbial bioremediation is cost-effective and considered environmentally friendly. Despite the persistent nature of the aromatic hydrocarbons, several bacterial species have remarkably evolved and developed metabolic versatilities to counteract and utilize these toxic compounds for their growth [11]. Several of these microbial metabolic pathways have already been established to render unmitigated mineralization or fragmentary transformation of aromatic hydrocarbons to dead-end intermediates [12]. Therefore, microbial bioremediation continues to appeal to microbiologists due to the metabolic diversity of bacteria. Since the use of microbial transformation and degradation has exceedingly become the principal process of hydrocarbon remediation and has tremendous commercialization potential, it is of paramount relevance to understand the pathways and their constituents involved in the metabolism of aromatic hydrocarbons.

Microbial Biotechnology and Protein Research Laboratory, Department of Molecular Biology and Biotechnology, Tezpur University, Tezpur, Assam, India; Proteomics Lab, National Centre for Cell Science, Pune, Maharashtra, India

Microbial Biotechnology: Role in Ecological Sustainability and Research, First Edition.
Edited by Pankaj Chowdhary, Sujata Mani, and Preeti Chaturvedi.

10.2. THE STUDY OF AROMATIC HYDROCARBON METABOLISM PATHWAYS – TRADITIONAL APPROACHES

There are several reports that have extensively investigated the remediation of aromatic hydrocarbons by several bacterial species, and in the process microbiologists have identified diverse microbial strains that harbor the ability to degrade aromatic hydrocarbon [13, 14]. Further, it has also been observed that microbes grown in the presence of aromatic hydrocarbons often express enzymes that are active constituents of the metabolic pathways. Among them, dioxygenase or Rieske non-heme iron ring-hydroxylating oxygenase (RHO) plays a crucial role. They catalyze the conversion of different aromatic hydrocarbons to intermediate compounds such as protocatechuate and catechol by introducing two oxygen atoms into the aromatic hydrocarbons to form cis-dihydrodiols. Thereafter, these intermediate compounds are further sequentially metabolized by several other enzyme systems to form the TCA cycle intermediates [14–16]. Aromatic hydrocarbon degradation pathways have previously been explored through identification of metabolic intermediates, enzyme characterization, and gene cloning, sequencing, and gene knockout studies [15]. Although these investigations have yielded crucial information on individual metabolic pathways and enzymes involved in microbial aromatic hydrocarbon degradation, a global picture depicting the cellular changes, as well as cross-talk among diverse metabolic pathways of aromatic hydrocarbon degradation, could not be fully deciphered. These limitations prompted microbiologists to shift toward a more holistic approach to study these microbial metabolic pathways that would result in a more comprehensive understanding of the degradation processes.

10.3. A SHIFT TOWARD PROTEOMIC APPROACHES TO INVESTIGATE MICROBIAL AROMATIC HYDROCARBON METABOLISM

In the late 1980s, the inception of matrix-assisted laser desorption/ionization (MALDI) [17] and electrospray ionization (ESI) [18] techniques has established mass spectrometry-based proteomics as an integral component in biological research. With the growth of robust database search algorithms, accumulating genomic data, and the advent of powerful analytical de-complexing tools, proteomics has emerged as a powerful tool to unravel the structural and functional aspects of proteins and enzymes related to microbial aromatic hydrocarbon metabolism.

The majority of proteomic investigations on microbial aromatic hydrocarbon metabolism employed two-dimensional sodium dodecyl sulfate-polyacrylamide gel electrophoresis (2D SDS-PAGE) for protein de-complexation followed by in-gel trypsin digestion and mass spectrometry analysis for protein identification (Table 10.1). This approach provides valuable information on the distribution of proteins present in the sample, their molecular weight, and isoelectric point (pI), as well as comparative expression profiles of protein mixtures isolated from microbes cultured under different growth conditions. Although this approach has very efficient resolution power, it is nevertheless limited by the suboptimal identification of low abundant proteins, glycoproteins, membrane proteins, and those with extreme pI and molecular weight. Similar inherent hurdles were also faced by researchers who relied on other gel-based approaches for proteomic characterization. Therefore, there has gradually been a shift from the gel-based proteomic approach toward the liquid chromatography (LC) based methods that can overcome these limitations. Subsequently, in-solution trypsin digestion of crude or de-complexed protein samples

Table 10.1 A summary of proteomic investigations on the metabolism of aromatic hydrocarbons by diverse bacterial species.

Microorganism	Aromatic hydrocarbon(s)	Proteomic strategy	Protein identification	Number of proteins identified	References
Achromobacter xylosoxidans	Pyrene	1D-SDS-PAGE and in-gel trypsin digestion	ESI-Q-TOF-MS	999	[9]
Acinetobacter baumannii	Benzoate	2D-SDS-PAGE and in-gel trypsin digestion	MALDI-TOF-MS, ESI-Q-TOF-MS	~14	[19]
Acinetobacter lwoffii	Aniline	2D-SDS-PAGE and in-gel trypsin digestion	MALDI-TOF-MS, ESI-Q-TOF-MS	>40	[20]
	Benzoate	2D-SDS-PAGE and in-gel trypsin digestion	ESI-Q-TOF-MS	2	[21]
	p-hydroxybenzoate	2D-SDS-PAGE and in-gel trypsin digestion	ESI-Q-TOF-MS, N-terminal sequencing	2	[22]
Acinetobacter radioresistens	Phenol, Benzoate	2D-SDS-PAGE and in-gel trypsin digestion	N-terminal sequencing	> 10	[23]
	Phenol	2D-SDS-PAGE and in-gel trypsin digestion	Microsequencing analysis	6	[24]
Acinetobacter sp.	Benzoate	2D-SDS-PAGE and in-gel trypsin digestion	MALDI-TOF-MS, N-terminal sequencing	~10	[25]
Arthrobacter phenanthrenivorans	Phenanthrene, Phthalate	In-solution digestion of cell extract	nano-ESI-LC-MS/MS	1197	[11]
Azoarcus evansii	Benzoate	2D-SDS-PAGE and in-gel trypsin digestion	N-terminal sequencing	> 5	[26]
Brevibacillus brevis	Pyrene	2D-SDS-PAGE and in-gel trypsin digestion	MALDI-TOF-MS	6	[4]
	Benzo[a]pyrene	In-solution digestion of cell extract	ESI-triple-TOF-MS	56	[27]
Burkholderia xenovorans	*p*-chlorobenzoate	2D-SDS-PAGE and in-gel trypsin digestion	MALDI-TOF-MS, ESI-Q-TOF-MS	> 10	[28]
	Biphenyl, Benzoate	2D-SDS-PAGE and in-gel trypsin digestion	MALDI-TOF-MS	~65	[29]
	Biphenyl, 4-chlorobenzoate	2D-SDS-PAGE and in-gel trypsin digestion	MALDI-TOF-MS	~20	[30]
Comamonas sp.	4-chloronitrobenzene	2D-SDS-PAGE and in-gel trypsin digestion	MALDI-TOF-MS	> 70	[31]

(Continued)

Table 10.1 (Continued)

Microorganism	Aromatic hydrocarbon(s)	Proteomic strategy	Protein identification	Number of proteins identified	References
Corynebacterium glutamicum	Benzoate, Gentisate, *p*-cresol, Phenol, Resorcinol	2D-SDS-PAGE and in-gel trypsin digestion	MALDI-TOF-MS, MALDI-Q-TOF-MS	> 90	[32]
Mycobacterium sp.	Pyrene	2D-SDS-PAGE and in-gel trypsin digestion	nano-ESI-LC-MS/MS	~20	[33]
	Fluoranthene	1D/2D-SDS-PAGE and in-gel trypsin digestion	nano-ESI-LC-MS/MS	> 25	[34]
Mycobacterium vanbaalenii	Pyrene, Phenanthrene, Dibenzothiophene	2D-SDS-PAGE and in-gel trypsin digestion	N-terminal sequencing	2	[35]
	Pyrene, Fluoranthene, Phenanthrene, Anthracene, Pyrene-4,5-quinone	2D-SDS-PAGE and in-gel trypsin digestion	nano-ESI-LC-MS/MS	> 10	[36]
	Pyrene	1D-SDS-PAGE and in-gel trypsin digestion	nano-ESI-LC-MS/MS	1028	[37]
	Fluoranthene	1D-SDS-PAGE and in-gel trypsin digestion	nano-ESI-LC-MS/MS	1122	[38]
Novosphingobium pentaromativorans	Phenanthrene, Pyrene, and Benzo[a]pyrene	1D-SDS-PAGE and in-gel trypsin digestion	nano-ESI-LC-MS/MS	13	[12]
Pseudomonas aeruginosa	Pyrene	In-solution digestion of cell extract	nano-ESI-LC-MS/MS	115	[1]
Pseudomonas alkaligenes	Gentisate	2D-SDS-PAGE and in-gel trypsin digestion	MALDI-TOF-MS, ESI-Q-TOF-MS, N-terminal sequencing	~15	[39]
Pseudomonas putida	Toluene	2D-SDS-PAGE and in-gel trypsin digestion	MALDI-TOF-MS, nanoLC-MALDI-Q-TOF-MS	~30	[40]
	Benzoate, *p*-hydroxybenzoate, Phenylacetate	2D-SDS-PAGE and in-gel trypsin digestion	MALDI-TOF-MS	~10	[41]

	Benzoate, *p*-hydroxybenzoate, Phenylacetate, Vanillin	2D-SDS-PAGE and in-gel trypsin digestion	MALDI-TOF-MS	> 100	[42]
	Benzoate	2D-SDS-PAGE and in-gel trypsin digestion	MALDI-TOF-MS	> 50	[43, 44]
Pseudomonas sp.	Phenol	1D-SDS-PAGE and in-gel trypsin digestion	MALDI-TOF-MS	~20	[45]
	Phenol	1D-SDS-PAGE and in-gel trypsin digestion	MALDI-TOF-MS	~50	[46]
	Phenol	1D-SDS-PAGE and in-gel trypsin digestion	MALDI-TOF-MS	~30	[47]
	Benzoate	2D-SDS-PAGE and in-gel trypsin digestion	ESI-Q-TOF-MS	40	[48]
Rhodococcus sp.	Phenylacetate	2D-SDS-PAGE and in-gel trypsin digestion	MALDI-TOF-MS	~30	[49]
	Phenol	2D-SDS-PAGE and in-gel trypsin digestion	nano-ESI-LC-MS/MS	~90	[50]
	Benzoate, Phthalate	2D-SDS-PAGE and in-gel trypsin digestion	MALDI-TOF-MS, ESI-LC-MS/MS	~60	[51]
	Naphthalene, Tetrain, Phthalate	2D-SDS-PAGE and in-gel trypsin digestion	MALDI-TOF-MS, ESI-LC-MS/MS	> 10	[52]
	Biphenyl, Benzene, Styrene, Ethylbenzene	2D-SDS-PAGE and in-gel trypsin digestion	MALDI-TOF-MS	> 150	[53]
	Phenanthrene	In-solution digestion of cell extract	nano-ESI-LC-MS/MS	2588	[54]
Sphingomonas sp.	Chrysene, Phenanthrene, Naphthalene	1D/2D-SDS-PAGE and in-gel trypsin digestion	ESI-Q-TOF-MS	8	[55]

followed by nanoLC–MS/MS for protein identification has dramatically improved the proteome coverage, thereby yielding comprehensive information on thousands more identified proteins [1, 11].

Apart from protein identification (qualitative aspect), the other important aspect of proteomic analysis is the differential protein quantitation that determines the change in expression levels of different proteins with respect to aromatic hydrocarbon metabolism. A few recent proteomic studies have successfully used isobaric tags for relative and absolute quantization (iTRAQ) technology for investigating the change in expression of the identified proteins [27, 54, 56, 57]. However, isotopic labelling strategies are limited by factors such as the need for higher amounts of starting biological material, high costs of reagents, and increased complexity of the method [58]. In the recent years, label-free quantitation (LFQ) methods have gained significant attention as feasible alternatives to these isotope labeling methods and several studies have incorporated LFQ strategy to monitor the change in protein expression of microbes grown in the presence of different aromatic hydrocarbons [1, 9, 11, 37]. Nevertheless, these label-free methods are database dependent and they yield reliable results for model organisms whose genome information is already present in the database.

10.4. METABOLISM OF LOW MOLECULAR WEIGHT (LMW) AROMATIC HYDROCARBONS

Aromatic hydrocarbons containing three or less aromatic rings are classified as low molecular weight (LMW) aromatic hydrocarbons. Several metabolic processes for degradation of these LMW aromatic hydrocarbons have been deciphered and they are available at the Kyoto Encyclopedia of Genes and Genomes (KEGG) web-based server. The proteomic investigations on degradation of LMW aromatic hydrocarbons, viz. benzoate, *p*-hydroxybenzoate, phenylacetate, vanillin, biphenyl, 4-chlorobenzoate, phenylacetate, phthalate, benzene, styrene, 4-cresol, ethylbenzene, and 4-chloronitrobenzene by several microbial species such as *Pseudomonas putida, Burkholderia xenovorans, Rhodococcus jostii, Corynebacterium glutamicum*, and *Comamonas* sp. have identified variable numbers of proteins including enzymes concerning the LMW aromatic hydrocarbon degradation pathways (Table 10.1) [26, 28–32, 42, 49, 51, 53, 59]. Previously reported data on metabolic pathways (from web-based servers) and the annotated genome sequences were then cross-referenced against these identified enzymes to shed light on the LMW aromatic hydrocarbon catabolic pathways. In addition, the LFQ data from a few proteomic studies has also yielded valuable information on the up- and downregulated protein subsets, which signify differential gene expression and their regulation in response to the aromatic hydrocarbons that are used as sole sources of carbon and energy [30, 36, 39, 41]. For instance, the microbe *B. xenovorans* switches three different benzoate metabolism pathways with variable growth kinetics when maintained on media containing succinate, benzoate, and biphenyl as sole carbon sources [29].

Apart from cross-referencing established metabolic pathways, proteome profiling can also facilitate novel pathway identification. Gescher and his coworkers postulated a novel metabolic pathway via benzoyl-CoA for aerobic benzoate oxidation by *Azoarcus evansii* and a *Bacillus stearothermophilus*-like strain [26]. Further, another proteomic investigation ended with identification of two novel proteins, a putative reductase (accession no. Q8NSW7) and a short-chain dehydrogenase (accession no. Q8NSW5), in *C. glutamicum* that are involved in the 4-cresol metabolism [32].

10.5. METABOLISM OF HIGH MOLECULAR WEIGHT (HMW) POLYCYCLIC AROMATIC HYDROCARBONS (PAH)

Polycyclic aromatic hydrocarbons (PAHs) are usually high molecular weight compounds that consist of four or more fused benzene rings [60]. Notably, high molecular weight (HMW) PAHs are complex and recalcitrant molecules and microbiologists have been able to characterize only a few bacterial species that can metabolize PAH. Therefore, compared to LMW aromatic hydrocarbons, investigation on HMW PAHs is very arduous. Further, the characteristic structural arrangement of atoms and functional groups in PAHs demand involvement of multiple enzyme machinery with heterogeneous metabolic pathways and more complex enzymatic reactions [15]. At the genome level, more genes coding for diverse enzymes are involved in the HMW PAH degradation pathways, many of which are previously uncharacterized. Further, these genes are arranged and disseminated in random clusters in the microbial genome, which suggests the existence of intricate regulation of these catabolic genes [15]. In addition, due to low sequence homology between the catabolic genes of LMW aromatic hydrocarbon degrading bacteria and HMW PAH degraders, the classical approach based on nag, nah, and phn genes is not very suitable [61]. Therefore, the study of HMW PAHs requires a thorough and systematic investigation to yield comprehensive results that can address the different metabolic pathways involved in their degradation.

Despite the challenges posed by HMW PAHs, microbiologists have been able to tap several of these recalcitrant compounds and isolate microbial species that can successfully metabolize and utilize PAHs. Notably, using the proteomic approach, several studies have demonstrated the expression of specific proteins that characterize the physiological state of the microbe in response to HMW PAH degradation [11, 12]. Further, these findings have also shed light on the diverse metabolic pathways activated in the bacterial species for PAH bioremediation [1, 4, 9, 11, 12, 27, 37]. A summary of the PAHs along with the microbial species that have been demonstrated to utilize them is included in Table 10.1. Among the HMW PAHs, pyrene that contains four benzene rings is the most widely studied PAH. Pyrene is frequently associated with water, air, and soil pollution, it enhances the carcinogenicity of benzo[a]pyrene in mouse models, and is toxic to aquatic life forms [62, 63]. Further, pyrene structurally resembles various HMW PAHs that exhibit carcinogenic effects. Therefore, pyrene has served as an important model compound in the study of PAH metabolism by microbes. Several microbial species including *Pseudomonas aeruginosa*, *Novosphingobium pentaromativorans*, *Brevibacillus brevis*, *Mycobacterium vanbaalenii*, and *Achromobacter xylosoxidans* have been shown to metabolize and use pyrene as their carbon source [1, 4, 9, 12, 37].

Shotgun proteomic analysis identified 196 and 115 intracellular proteins in *P. aeruginosa* strain ASP-53 when the microbe was cultured in medium containing glucose and pyrene as sole carbon sources, respectively [1]. Among them, 150 and 69 proteins were uniquely identified in the strain grown in glucose- and pyrene-containing media, respectively. Further, the study also identified nine enzymatic proteins that are involved in pyrene metabolism; these proteins being either upregulated (determined by LFQ proteomics) or uniquely expressed in isolates growing in pyrene-containing medium. In addition, the mass spectrometry analysis also indentified certain key enzymes essential for pyrene metabolism via the salicylate pathway. The pyrene metabolism in *P. aeruginosa* strain ASP-53 was postulated to commence with the action of a dioxygenase (identified as gi|761895334) followed by sequential transformations mediated by dihydrodiol dehydrogenase (gi|1256708), quinone oxidoreductase (gi|727775772), aldehyde dehydrogenase (gi|565900676), salicylaldehyde dehydrogenase (gi|761540016), and salicylate-5-hydroxylase (gi|226934607). Eventually, the conversion of salicylate to gentisate is catalyzed by salicylate-5-hydroxylase, and the product enters the tyrosine metabolism. Notably, the findings of the proteomic analysis were also confirmed by PCR amplification of the alkB monooxygenase gene, and C12O, PAH-RHDα and C23O dioxygenase enzyme system genes in *P. aeruginosa* strain ASP-53 [1].

10.6. CONCLUSION

With the development of high-resolution mass spectrometers and robust database search algorithms, proteomics has emerged as a useful tool for investigation of microbial metabolic pathways concerning aromatic hydrocarbon degradation. Although 2D SDS-PAGE analysis coupled to MS/MS analysis was the pioneer strategy for these studies, application of shotgun proteomic approaches have gradually gained momentum due to their ease of application, minimal sample requirement, dynamic range of protein identification, and rapid analysis of an entire proteome. On the contrary, gel-based approaches inevitably suffer the major hurdle of narrow dynamic range for comprehensive proteome analysis. Furthermore, advancement in algorithms intended for label-free quantification strategies has made shotgun proteomics an attractive and suitable approach in microbial proteome mining. For these reasons, the application of proteomics for investigating microbial aromatic hydrocarbon degradation has flourished in the last decade and subsequently an increasing number of studies have been reported that serve as valuable references for understanding microbial aromatic hydrocarbon metabolism. Nevertheless, keeping in mind the enormous diversity of

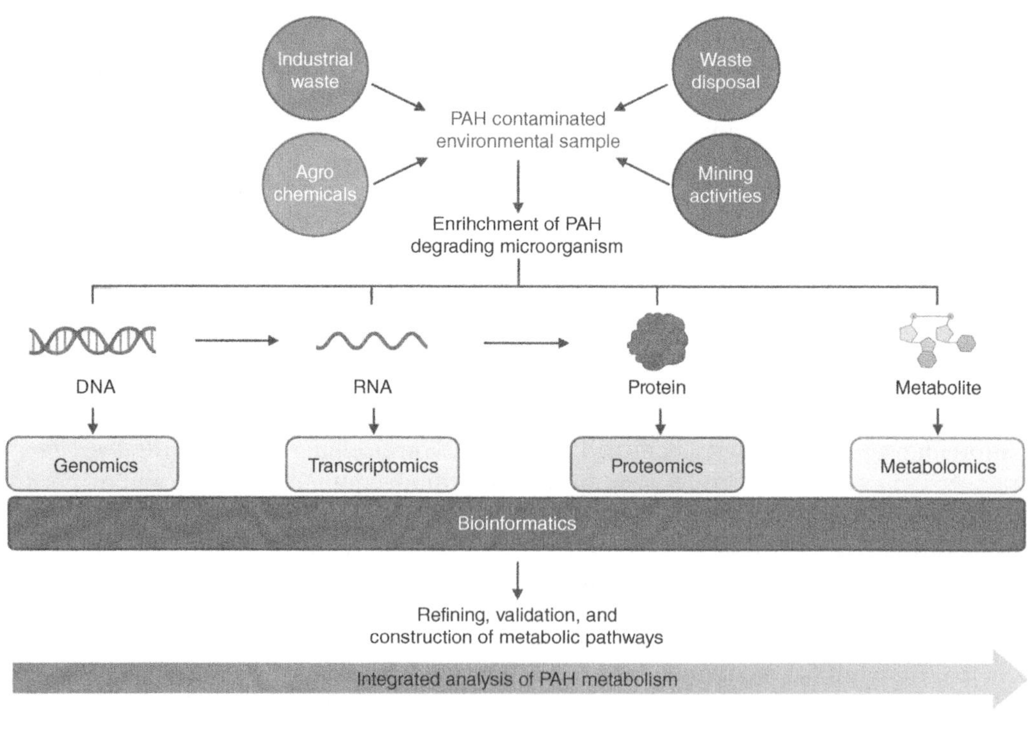

Figure 10.1 Integration of genomics, transcriptomics, proteomics and metabolomics approaches for analysis of the metabolic pathways involved in polycyclic aromatic hydrocarbon metabolism. *Source:* Bhargab Kalita.

microbes in our environment, the field of microbial proteomics is still at the infant stage. Therefore, efforts must be directed toward isolating microbes that can efficiently utilize recalcitrant HMW PAHs. In addition, development of a new data-independent acquisition strategy with even more comprehensive proteome coverage may also be extended to microbial proteomics to get a more vivid picture of the minor yet significant changes in the microbial proteome in context of aromatic hydrocarbon degradation. These advances in microbial proteomics in conjunction with genomic, transcriptomic, and metabolomic investigations (Figure 10.1) can map new avenues for deeper understanding of microbial aromatic hydrocarbon degradation in the near future.

ACKNOWLEDGEMENT

I acknowledge the financial support from the DBT-RA Program in Biotechnology and Life Sciences.

REFERENCES

1. Mukherjee, A.K., Bhagowati, P., Biswa, B.B. et al. (2017). A comparative intracellular proteomic profiling of *Pseudomonas aeruginosa* strain ASP-53 grown on pyrene or glucose as sole source of carbon and identification of some key enzymes of pyrene biodegradation pathway. *Journal of Proteomics 167*: 25–35.
2. Martorell, I., Perello, G., Marti-Cid, R. et al. (2010). Polycyclic aromatic hydrocarbons (PAH) in foods and estimated PAH intake by the population of Catalonia, Spain: temporal trend. *Environment International 36* (5): 424–432.
3. Somtrakoon, K., Suanjit, S., Pokethitiyook, P. et al. (2008). Enhanced biodegradation of anthracene in acidic soil by inoculated *Burkholderia* sp. VUN10013. *Current Microbiology 57* (2): 102–106.
4. Wei, R., Ni, J., Chen, W. et al. (2017). Variation in soil aggregate-size distribution affects the dissipation of polycyclic aromatic hydrocarbons in long-term field-contaminated soils. *Environmental Science and Pollution Research International 24* (28): 22332–22339.
5. Klankeo, P., Nopcharoenkul, W., and Pinyakong, O. (2009). Two novel pyrene-degrading *Diaphorobacter* sp. and *Pseudoxanthomonas* sp. isolated from soil. *Journal of Bioscience and Bioengineering 108* (6): 488–495.
6. Gill, R.T., Harbottle, M.J., Smith, J.W.N. et al. (2014). Electrokinetic-enhanced bioremediation of organic contaminants: a review of processes and environmental applications. *Chemosphere 107*: 31–42.
7. Luo, W., Zhu, X., Chen, W. et al. (2014). Mechanisms and strategies of microbial cometabolism in the degradation of organic compounds – chlorinated ethylenes as the model. *Water Science and Technology: a Journal of the International Association on Water Pollution Research 69* (10): 1971–1983.
8. Medellin-Castillo, N.A., Ocampo-Perez, R., Leyva-Ramos, R. et al. (2013). Removal of diethyl phthalate from water solution by adsorption, photo-oxidation, ozonation and advanced oxidation process (UV/H(2)O(2), O(3)/H(2)O(2) and O(3)/activated carbon). *The Science of the Total Environment 442*: 26–35.
9. Nzila, A., Ramirez, C.O., Musa, M.M. et al. (2018). Pyrene biodegradation and proteomic analysis in *Achromobacter xylosoxidans*, PY4 strain. *International Biodeterioration and Biodegradation 130*: 40–47.
10. Pazos, M., Rosales, E., Alcantara, T. et al. (2010). Decontamination of soils containing PAHs by electroremediation: a review. *Journal of Hazardous Materials 177* (1-3): 1–11.
11. Vandera, E., Samiotaki, M., Parapouli, M. et al. (2015). Comparative proteomic analysis of *Arthrobacter phenanthrenivorans* Sphe3 on phenanthrene, phthalate and glucose. *Journal of Proteomics 113*: 73–89.

12. Lyu, Y., Zheng, W., Zheng, T. et al. (2014). Biodegradation of polycyclic aromatic hydrocarbons by *Novosphingobium pentaromativorans* US6-1. *PLoS One 9* (7): e101438.
13. Peng, R.H., Xiong, A.S., Xue, Y. et al. (2008). Microbial biodegradation of polyaromatic hydrocarbons. *FEMS Microbiology Reviews 32* (6): 927–955.
14. van der Meer, J.R., de Vos, W.M., Harayama, S. et al. (1992). Molecular mechanisms of genetic adaptation to xenobiotic compounds. *Microbiological Reviews 56* (4): 677–694.
15. Kim, S.J., Kweon, O., and Cerniglia, C.E. (2009). Proteomic applications to elucidate bacterial aromatic hydrocarbon metabolic pathways. *Current Opinion in Microbiology 12* (3): 301–309.
16. Kweon, O., Kim, S.J., Baek, S. et al. (2008). A new classification system for bacterial Rieske non-heme iron aromatic ring-hydroxylating oxygenases. *BMC Biochemistry 9*: 11.
17. Karas, M. and Hillenkamp, F. (1988). Laser desorption ionization of proteins with molecular masses exceeding 10,000 daltons. *Analytical Chemistry 60* (20): 2299–2301.
18. Fenn, J.B., Mann, M., Meng, C.K. et al. (1989). Electrospray ionization for mass spectrometry of large biomolecules. *Science 246* (4926): 64–71.
19. Park, S.H., Kim, J.W., Yun, S.H. et al. (2006). Characterization of beta-ketoadipate pathway from multi-drug resistance bacterium, *Acinetobacter baumannii* DU202 by proteomic approach. *Journal of Microbiology 44* (6): 632–640.
20. Kim, E.A., Kim, J.Y., Kim, S.J. et al. (2004). Proteomic analysis of *Acinetobacter lwoffii* K24 by 2-D gel electrophoresis and electrospray ionization quadrupole-time of flight mass spectrometry. *Journal of Microbiological Methods 57* (3): 337–349.
21. Yoon, Y.H., Yun, S.H., Park, S.H. et al. (2007). Characterization of a new catechol branch of the beta-ketoadipate pathway induced for benzoate degradation in *Acinetobacter lwoffii* K24. *Biochemical and Biophysical Research Communications 360* (3): 513–519.
22. Kahng, H.Y., Cho, K., Song, S.Y. et al. (2002). Enhanced detection and characterization of protocatechuate 3,4-dioxygenase in *Acinetobacter lwoffii* K24 by proteomics using a column separation. *Biochemical and Biophysical Research Communications 295* (4): 903–909.
23. Giuffrida, M.G., Pessione, E., Mazzoli, R. et al. (2001). Media containing aromatic compounds induce peculiar proteins in *Acinetobacter radioresistens*, as revealed by proteome analysis. *Electrophoresis 22* (9): 1705–1711.
24. Mazzoli, R., Pessione, E., Giuffrida, M.G. et al. (2007). Degradation of aromatic compounds by *Acinetobacter radioresistens* S13: growth characteristics on single substrates and mixtures. *Archives of Microbiology 188* (1): 55–68.
25. Kim, S.I., Song, S.Y., Kim, K.W. et al. (2003). Proteomic analysis of the benzoate degradation pathway in *Acinetobacter* sp. KS-1. *Research in Microbiology 154* (10): 697–703.
26. Gescher, J., Zaar, A., Mohamed, M. et al. (2002). Genes coding for a new pathway of aerobic benzoate metabolism in *Azoarcus evansii*. *Journal of Bacteriology 184* (22): 6301–6315.
27. Zhu, Y., Chen, K., Ding, Y. et al. (2019). Metabolic and proteomic mechanism of benzo[a]pyrene degradation by *Brevibacillus brevis*. *Ecotoxicology and Environmental Safety 172*: 1–10.
28. Martinez, P., Agullo, L., Hernandez, M. et al. (2007). Chlorobenzoate inhibits growth and induces stress proteins in the PCB-degrading bacterium *Burkholderia xenovorans* LB400. *Archives of Microbiology 188* (3): 289–297.
29. Denef, V.J., Patrauchan, M.A., Florizone, C. et al. (2005). Growth substrate- and phase-specific expression of biphenyl, benzoate, and C1 metabolic pathways in *Burkholderia xenovorans* LB400. *Journal of Bacteriology 187* (23): 7996–8005.
30. Agullo, L., Camara, B., Martinez, P. et al. (2007). Response to (chloro)biphenyls of the polychlorobiphenyl-degrader *Burkholderia xenovorans* LB400 involves stress proteins also induced by heat shock and oxidative stress. *FEMS Microbiology Letters 267* (2): 167–175.
31. Zhang, Y., Wu, J.F., Zeyer, J. et al. (2009). Proteomic and molecular investigation on the physiological adaptation of *Comamonas* sp. strain CNB-1 growing on 4-chloronitrobenzene. *Biodegradation 20* (1): 55–66.
32. Qi, S.W., Chaudhry, M.T., Zhang, Y. et al. (2007). Comparative proteomes of *Corynebacterium glutamicum* grown on aromatic compounds revealed novel proteins involved in aromatic

degradation and a clear link between aromatic catabolism and gluconeogenesis via fructose-1,6-bisphosphatase. *Proteomics 7* (20): 3775–3787.

33. Liang, Y., Gardner, D.R., Miller, C.D. et al. (2006). Study of biochemical pathways and enzymes involved in pyrene degradation by *Mycobacterium* sp. strain KMS. *Applied and Environmental Microbiology 72* (12): 7821–7788.
34. Lee, S.E., Seo, J.S., Keum, Y.S. et al. (2007). Fluoranthene metabolism and associated proteins in *Mycobacterium* sp. JS14. *Proteomics 7* (12): 2059–2069.
35. Wang, R.F., Wennerstrom, D., Cao, W.W. et al. (2000). Cloning, expression, and characterization of the katG gene, encoding catalase-peroxidase, from the polycyclic aromatic hydrocarbon-degrading bacterium *Mycobacterium* sp. strain PYR-1. *Applied and Environmental Microbiology 66* (10): 4300–4304.
36. Kim, S.J., Jones, R.C., Cha, C.J. et al. (2004). Identification of proteins induced by polycyclic aromatic hydrocarbon in *Mycobacterium vanbaalenii* PYR-1 using two-dimensional polyacrylamide gel electrophoresis and de novo sequencing methods. *Proteomics 4* (12): 3899–3908.
37. Kim, S.J., Kweon, O., Jones, R.C. et al. (2007). Complete and integrated pyrene degradation pathway in *mycobacterium vanbaalenii* PYR-1 based on systems biology. *Journal of Bacteriology 189* (2): 464–472.
38. Kweon, O., Kim, S.J., Jones, R.C. et al. (2007). A polyomic approach to elucidate the fluoranthene-degradative pathway in *Mycobacterium vanbaalenii* PYR-1. *Journal of Bacteriology 189* (13): 4635–4647.
39. Zhao, B., Yeo, C.C., and Poh, C.L. (2005). Proteome investigation of the global regulatory role of sigma 54 in response to gentisate induction in *Pseudomonas alcaligenes* NCIMB 9867. *Proteomics 5* (7): 1868–1876.
40. Segura, A., Godoy, P., van Dillewijn, P. et al. (2005). Proteomic analysis reveals the participation of energy- and stress-related proteins in the response of *Pseudomonas putida* DOT-T1E to toluene. *Journal of Bacteriology 187* (17): 5937–5945.
41. Morales, G., Linares, J.F., Beloso, A. et al. (2004). The Pseudomonas putida Crc global regulator controls the expression of genes from several chromosomal catabolic pathways for aromatic compounds. *Journal of Bacteriology 186* (5): 1337–1344.
42. Kim, Y.H., Cho, K., Yun, S.H. et al. (2006). Analysis of aromatic catabolic pathways in *pseudomonas putida* KT 2440 using a combined proteomic approach: 2-DE/MS and cleavable isotope-coded affinity tag analysis. *Proteomics 6* (4): 1301–1318.
43. Cao, B., Geng, A., and Loh, K.C. (2008). Induction of ortho- and meta-cleavage pathways in pseudomonas in biodegradation of high benzoate concentration: MS identification of catabolic enzymes. *Applied Microbiology and Biotechnology 81* (1): 99–107.
44. Cao, B. and Loh, K.C. (2008). Catabolic pathways and cellular responses of *Pseudomonas putida* P8 during growth on benzoate with a proteomics approach. *Biotechnology and Bioengineering 101* (6): 1297–1312.
45. Tsirogianni, I., Aivaliotis, M., Karas, M. et al. (2004). Mass spectrometric mapping of the enzymes involved in the phenol degradation of an indigenous soil pseudomonad. *Biochimica et Biophysica Acta 1700* (1): 117–123.
46. Tsirogianni, E., Aivaliotis, M., Papasotiriou, D.G. et al. (2006). Identification of inducible protein complexes in the phenol degrader *Pseudomonas* sp. strain phDV1 by blue native gel electrophoresis and mass spectrometry. *Amino Acids 30* (1): 63–72.
47. Papasotiriou, D.G., Markoutsa, S., Meyer, B. et al. (2008). Comparison of the membrane subproteomes during growth of a new pseudomonas strain on lysogeny broth medium, glucose, and phenol. *Journal of Proteome Research 7* (10): 4278–4288.
48. Kim, S.I., Kim, J.Y., Yun, S.H. et al. (2004). Proteome analysis of *Pseudomonas* sp. K82 biodegradation pathways. *Proteomics 4* (11): 3610–3621.
49. Navarro-Llorens, J.M., Patrauchan, M.A., Stewart, G.R. et al. (2005). Phenylacetate catabolism in *Rhodococcus* sp. strain RHA1: a central pathway for degradation of aromatic compounds. *Journal of Bacteriology 187* (13): 4497–4504.

50. Santos, S.C., Alviano, D.S., Alviano, C.S. et al. (2007). Comparative studies of phenotypic and genetic characteristics between two desulfurizing isolates of *Rhodococcus erythropolis* and the well-characterized *R. erythropolis* strain IGTS8. *Journal of Industrial Microbiology & Biotechnology 34* (6): 423–431.
51. Patrauchan, M.A., Florizone, C., Dosanjh, M. et al. (2005). Catabolism of benzoate and phthalate in *Rhodococcus* sp. strain RHA1: redundancies and convergence. *Journal of Bacteriology 187* (12): 4050–4063.
52. Tomas-Gallardo, L., Canosa, I., Santero, E. et al. (2006). Proteomic and transcriptional characterization of aromatic degradation pathways in *Rhodoccocus* sp. strain TFB. *Proteomics 6* (Suppl 1): S119–S132.
53. Patrauchan, M.A., Florizone, C., Eapen, S. et al. (2008). Roles of ring-hydroxylating dioxygenases in styrene and benzene catabolism in *Rhodococcus jostii* RHA1. *Journal of Bacteriology 190* (1): 37–47.
54. Wang, H., Yang, Y., Xu, J. et al. (2019). iTRAQ-based comparative proteomic analysis of differentially expressed proteins in *Rhodococcus* sp. BAP-1 induced by fluoranthene. *Ecotoxicology and Environmental Safety 169*: 282–291.
55. Demaneche, S., Meyer, C., Micoud, J. et al. (2004). Identification and functional analysis of two aromatic-ring-hydroxylating dioxygenases from a sphingomonas strain that degrades various polycyclic aromatic hydrocarbons. *Applied and Environmental Microbiology 70* (11): 6714–6725.
56. Li, F., Song, W., Wei, J. et al. (2016). Comparative proteomic analysis of phenol degradation process by *Arthrobacter*. *International Biodeterioration and Biodegradation 110*: 189–198.
57. Liu, H., Sun, W.B., Liang, R.B. et al. (2015). iTRAQ-based quantitative proteomic analysis of *Pseudomonas aeruginosa* SJTD-1: a global response to n-octadecane induced stress. *Journal of Proteomics 123*: 14–28.
58. Choi, H., Fermin, D., and Nesvizhskii, A.I. (2008). Significance analysis of spectral count data in label-free shotgun proteomics. *Molecular & Cellular Proteomics : MCP 7* (12): 2373–2385.
59. Denef, V., Klappenbach, J., Patrauchan, M. et al. (2006). Genetic and genomic insights into the role of benzoate-catabolic pathway redundancy in *Burkholderia xenovorans* LB400. *Applied and Environmental Microbiology 72* (1): 585–595.
60. Cerniglia, C.E. (1993). Biodegradation of polycyclic aromatic hydrocarbons. *Current Opinion in Biotechnology 4* (3): 331–338.
61. Widada, J., Nojiri, H., Kasuga, K. et al. (2002). Molecular detection and diversity of polycyclic aromatic hydrocarbon-degrading bacteria isolated from geographically diverse sites. *Applied Microbiology and Biotechnology 58* (2): 202–209.
62. Belkin, S., Stieber, M., Tiehm, A. et al. (1994). Toxicity and genotoxicity enhancement during polycyclic aromatic hydrocarbons' biodegradation. *Environmental Toxicology and Water Quality 9* (4): 303–309.
63. Cancer IAfRo. (1986). IARC monographs programme on the evaluation of the carcinogenic risk of chemicals to humans. *Preamble. IARC Monograph Evaluation Carcinogenic Risk Chemical Humans 39*: 13–32.

11

Bioremediation of Problematic Soil for Sustainability

Abhishek Raj[1], Manoj Kumar Jhariya[2], Nahid Khan[2], and Arnab Banerjee[3]

11.1. INTRODUCTION

Land degradation is a key concern for plant growers and the main topic of discussion among many researchers, policy makers, and stakeholders around the world [1–3]. The productivity of agricultural land is declining enormously due to unsustainable land use practices. These practices deprive soil of health and quality by affecting the soil's physicochemical properties [4–10]. Soil salinization, sodicity, acidifications, etc., are major chemical changes that affect both the productivity and quality of soil. Soil contamination due to heavy metals and oils causes major changes in soil properties that affect various ecosystem services. As per the WMO [5], land degradation affected approx. 2.5 billion people, whereas close to one billion people in developing countries are at high risk due to poor land quality. Similarly, approximately 25% of global soil and water resources are under chunked that drag food and nutritional security at alarming level [11]. Several anthropogenic activities release various harmful gases, toxicity, and carcinogenic substances that accumulate in living organisms and the entire soil ecosystem [12, 13].

The problem of soil salinity has spread globally due to a scarcity of good-quality water and effective management practices in many countries, especially in dry regions. Uncertain rainfall and lower precipitation due to climate change leads to accumulation of salt in soils. This will affect overall soil-related ecosystem services including food security and environmental sustainability [14, 15]. Globally, petroleum oil and some hydrocarbons pollute the soil due to their toxic nature. These harmful materials change soil properties and destroy soil-inhabiting biomes by affecting their population and diversity [16–18]. Petroleum oil

[1]Pt. Deendayal Upadhyay College of Horticulture and Forestry, Dr. Rajendra Prasad Central Agriculture University, Samastipur, Bihar, India

[2]Department of Farm Forestry, Sant Gahira Guru Vishwavidyalaya, Ambikapur, Chhattisgarh, India

[3]Department of Environmental Science, Sant Gahira Guru Vishwavidyalaya, Ambikapur, Chhattisgarh, India

Microbial Biotechnology: Role in Ecological Sustainability and Research, First Edition.
Edited by Pankaj Chowdhary, Sujata Mani, and Preeti Chaturvedi.

corrodes different soil layers due to easy infiltration properties that lead to soil salinization. This activity can pollute soil layers and increase compactions [19]. Petroleum medicated soil contaminations decrease pH value and also affect the C:N ratio along with soil enzyme activity [20]. However, degradation of soil structure and its permeability are affected by long-term accumulation of oil substances. Oils also affect plant root systems by forming mucous membranes that damage plants and various microbial populations in soils. Therefore, plant yield and soil productivity are also affected by oil-related soil contaminants [21]. These oils not only affect soil health and plant productivity, but also have potential to change the nutritional composition and quality along with emanating unpleasant smells [22]. The chemical constituents of petroleum oils are key factors that destroy both soil quality and human health. For example, a polycyclic aromatic hydrocarbon (PAH) found in petroleum oil is not easily degradable and enters the food chain through accumulation into the soils. Therefore, this substance can harm humans with its carcinogenic and mutagenic properties.

Mitigating impacts of these soil contaminants are important concerns that require various biotic approaches. In this context, using novel plants and microbes is a good strategy that not only minimizes this contaminant but also improves soil health and quality [23]. Plants including leguminous trees help in rejuvenating soil fertility along with removing toxicity through the process of phytoremediation [24, 25]. Microbes also play an important role in minimizing toxicity and contaminant from the soils. Bacteria and fungi including VAM fungi are effective in alleviating salinity, sodicity, and heavy metal toxicity from the soils [26]. These fungi also help to make essential nutrients available to the plants. Management of soil contaminants through bioremediation involves various processes based on economic viability, social acceptability, and environmental sustainability. However, applying chemical amendments for amelioration of salinity and sodicity are non-ecological methods and are economically not feasible. Similarly, adopting physical and chemical measures for saline or sodic soil reclamation is not ecologically viable and economically feasible [27]. Thus, adopting biological tools and related bioremediation technology can reclaim salty and sodic soils to a greater extent. This will improve soil health, productivity, and sustainability [28]. Salt-tolerant plants are efficiently used for reclaiming soils under this biological process. Thus, bioremediation is considered a low-cost, ecofriendly practice that improves soil health and maintains fertility for better production and profitability. Nowadays, using microbes for salt reclamation has been popularized among researchers, scientists, and policy makers. For example, halophilic microorganisms play a key role due to higher tolerance capacity in salty soils and higher potential of bioremediation. These bacteria promote other vegetation to withstand salty environments by triggering recovery of salt-affected soils.

Problematic soils affect agroecosystem health and production. Many authors have reported production and economic losses in agroecosystem practices globally. For example, approximately 12 billion USD is lost due to salinity problems worldwide [29]. Similarly, 3.7 billion USD is lost yearly due to soil salinity in the United States [30]. Around 25% and 31% of crop yield reduction was observed in the regions of Canada and Pakistan due to soil salinity [31]. Higher market price and demands on selected plants will enhance the economic value of phytoremediation strategies. However, bioremediation is considered for long-term soil improvement from a social, economic, and environmental point of view. Thus, bioremediation technology could be viable for soil remediation along with improved yield and socioeconomic status of farmers.

There are very limited studies on bioremediation for soil contamination due to less comprehensive works. This means there are several opportunities and challenges for future prospects of bioremediation techniques in the context of soil reclamation and management.

This chapter provides a comprehensive discussion about plant- and microbe-related soil contaminant remediation. Further, it also compiles recent technologies for phytoremediation and micro-remediation for improving soil health for better ecosystem stability.

11.2. PROBLEMATIC SOIL: A GLOBAL COVERAGE

Soil salinity and sodicity are major environmental problems globally. This problematic soil is distributed throughout the world among major countries and has become a global threat to farmers and biodiversity. A global distribution of saline and sodic soil (Mha) is depicted in Figure 11.1 [32]. Soil salinity threatens global food, income, and environmental security. Degradation of soil quality due to salinity is also prevalent in arid and coastal regions of the world. As per the global prediction, approximately 50% of the world's arable

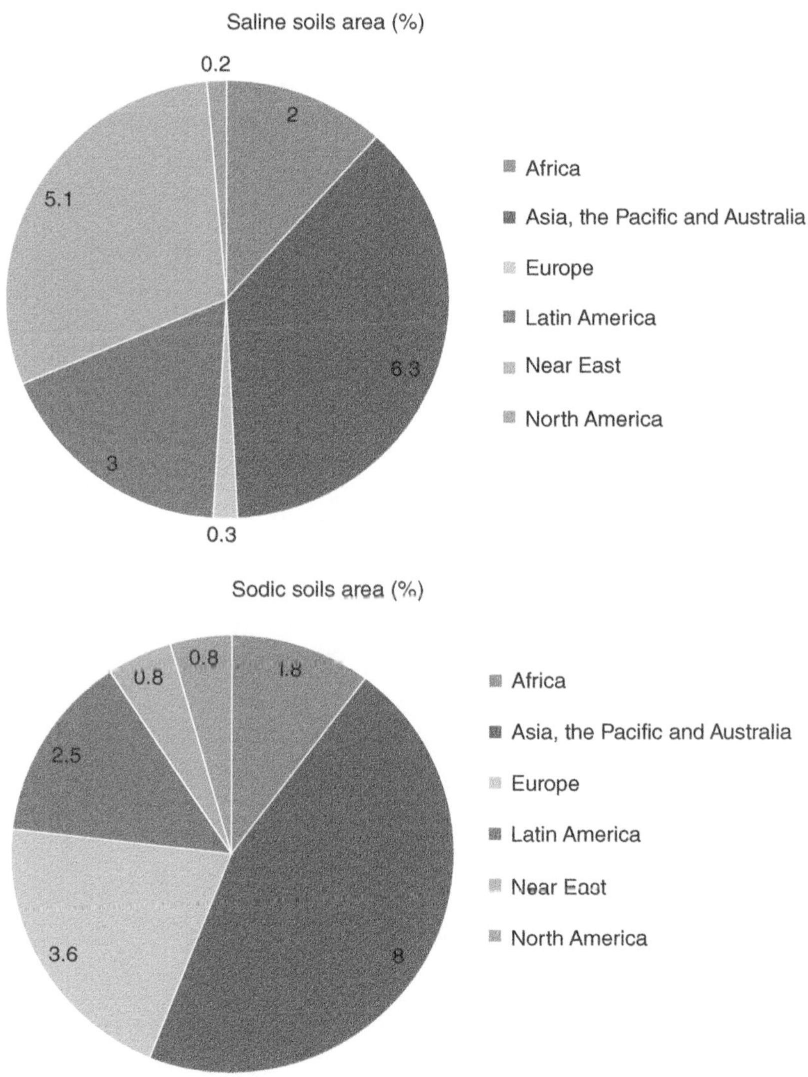

Figure 11.1 Global distribution of saline and sodic soil (Mha).

land may be affected by soil salinity by 2050. However, mitigating soil salinity and other contaminants could enhance global food production up to 57% by 2050 [33, 34].

Currently, salinity affects approximately 62 Mha (20%) of the world's irrigated land areas [35]. This soil salinity induces poor physical properties, toxicity, osmotic stress in plants, and related nutritional disorders that reduce plant productivity and profitability [36]. Around 19.0 Mha areas are prone to soil salinity in the sub-Saharan African region [37]. Approximately 11.73 M km^2 of soils are reported to be facing salinity problems across the globe, of which cropland salinity areas covered 0.16 M km^2 between 1980 and 2018 based on a predictive model of gridded dataset [38]. Likewise, heavy metals reported in contaminated soils in the world are depicted in Figure 11.2 [39–46]. However, changing areas of problematic soils vary according to varying geographical areas, topography, and climatic situations. Soil contamination due to oil and hydrocarbon is also observed in Russia, which is the leading oil producer in the world. Nearly one-eighth of soil areas are degraded due to 60 000 oil spills that occur every year. Also, soil contaminants and pollution due to oils were recorded 10 times higher near soil bank areas in Mexico. Similarly, the United Kingdom is also affected by continuous oil pollution near all established refineries [47]. Currently, 4.8 Mkm^2 of soils were polluted due to higher number of (>400) oil and gas fields, which covered 320 000 km^2 in China [48]. Therefore, these worldwide recorded data are enough to represent dramatic figures of problematic soil due to salinity, sodicity, oil, and related

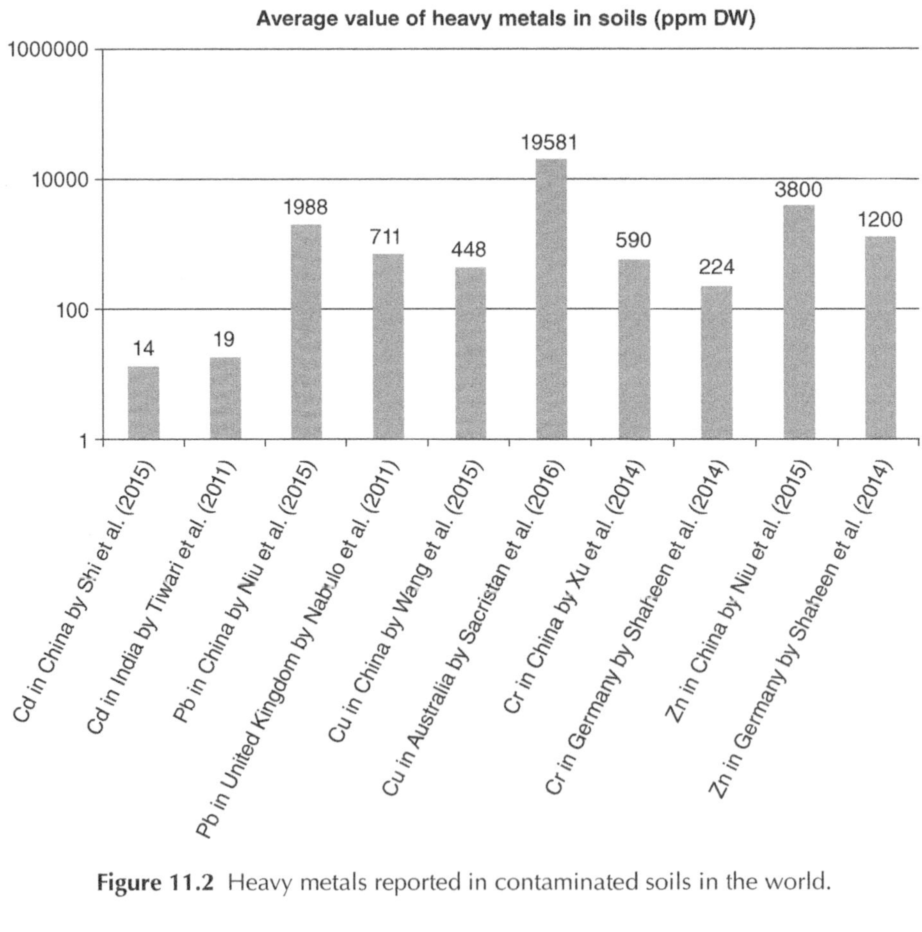

Figure 11.2 Heavy metals reported in contaminated soils in the world.

hydrocarbon substances. Several countries such as China, USA, Greece, Jordan, Spain, etc., have worked efficiently to tackle soil salinity problems in their farms for better health and productivity [49]. Thus, remediation of problematic soils ensures soil-food-climate security and promises environmental sustainability and ecological stability at a global scale.

11.3. BIOREMEDIATION: A CONCEPTUAL FRAMEWORK

Bioremediation is a new concept based on the degradation and removal of various contaminants from soil and water resources. It normalizes the life of various natural resources by involving plants (as phytoremediation) and other microbes (as microbial remediation). Fungi and bacteria are important components that alter soil contaminants naturally. Plants, microbes, and various enzymatic remediations are three key approaches that are used in bioremediation process. The concept of bioremediation is highly popularized among researchers, and it is an ecofriendly and economically viable approach for problematic soil [50]. Technically, bioremediation processes are classified into in-situ or ex-situ techniques [51]. In-situ techniques are applied on soil and water contaminant sites in natural places with minimal disturbances. The ex-situ technique involves removal of contaminants from soil and water resources and to a contained area in which the process of bioremediation is practiced.

The conceptual framework of bioremediation involves biostimulation, natural attenuation, and bioaugmentation based mitigation processes. Natural attenuation techniques do not involve any humans but are based on natural processes of contaminant remediation. Indigenous microbes are utilized for remediation of problematic soils through the process of biostimulation. Similarly, the process of bioaugmentation uses some genetically altered microorganisms for remediation of soil contaminants. Moreover, the presence of microbes, contaminants, and electron acceptors are three key ingredients for successful bioremediation. The process of bioremediation is recommended for contaminants due to quick degradation capacity rather than the chemical nature of the contaminant. Microorganisms easily degrade naturally occurring compounds in contaminants compared to chemical contaminants [26].

Incorporating designer plants including crops and tree species into contaminated soils would restore soil quality through bioremediation. Phytoremediation and microbial remediation have greater pollutant removal capacity from contaminated soils. Besides, these plant species can produce biomass and energy through C sequestration in vegetation and soils [52, 53]. Thus, incorporating plants and microbes can ensure bioremediation along with biomass and energy production. Therefore, ecosystem restoration can be possible through bioremediation mediated soil improvement environmental services. In this context, a conceptual model for ecosystem restoration through bioremediation is depicted in Figure 11.3 [54, 55].

11.4. BIOREMEDIATION OF PROBLEMATIC SOILS: FACTS AND FAITH

Plant and microbes are key components of bioremediation that justify key facts and faith of problematic soil remediation. However, soil types and nature are also important considerations for a successful bioremediation process. Soil type determines the potential of in-situ bioremediation techniques that involve the same place irrespective of ex-situ technique. Bioventing is an important process that is involved in in-situ biostimulation. Nutrients and oxygen are pumped through injection wells into the soil under the process of bioventing. However, an even distribution of nutrients and oxygen are observed throughout the soil contaminants. Soil texture and its permeability are key factors that determine the process of bioventing. For example, the process of bioventing has been checked in fine textured soil such as clay due to less permeability with improper distribution of oxygen and nutrients into

Figure 11.3 Ecosystem restoration through bioremediation.

the soils. Therefore, a coarse textured, well drained, and medium grained soil is suitable for bioventing. These are important facts about the mechanisms and process of bioremediation for problematic soils [26].

11.5. MICROBIAL VS PLANT-ASSISTED BIOREMEDIATION

Plant and microbes are key components for bioremediation of any problematic soils. These natural components are ecologically safe and economically feasible for the process of bioremediation, which involves the removal of contaminants from problematic soils. Phytoremediation involves important plant species, including leguminous trees, that potentially remove hazardous material and heavy metals from the soil. Plants involve some enzymatic reactions and detoxify the xenobiotics in soil through direct uptake of the contaminants from the soils [56]. Similarly, involving rhizospheric microorganisms with the phytoremediation process and their associations is an important tool that promotes the bioremediation process. Likewise, microbes such as bacteria and fungi are important components that ensure degradation and removal of contaminants from the soils. These microbes perform a

Table 11.1 Soil heavy metal remediation through microbes.

Heavy metal	Adsorption potential of heavy metals by microbes (mg/g)	
	Bacteria	Fungi
Zn	36 mg by Acinetobacter species [58]	43.87 mg by Trametes versicolor [59]
	30 mg by *Streptomyces rimosus* [60]	14 mg by *Rhizopus arrhizus* [61]
	133.0 mg by *Aphanothecehalophytica* [62]	0.2 mg by *Penicillium spinulosum* [63]
	172.4 mg by *Thiobacillusferrooxidans* [64]	–
Pb	92.3 mg by Bacillus species [65]	1120 mg by *Aspergillus lentulus* [66]
	467 mg by *Bacillus firmus* [67]	54.05 mg by *Aspergillus niger* [68]
	567.7 mg by *Corynebacteriumglutamicum* [69]	116 mg by *Penicillium chrysogenum* [70]
	46.1 mg by *Pseudomonas aeruginosa* [71]	165 mg by *Pleurotusostreatus* [72]
	270.4 mg by *Pseudomonas putida* [73]	166 mg by *Rhizopus nigricans* [61]
	171.8 mg by *Enterobacter cloacae* [74]	56 mg by *Rhizopus arrhizus* [61]
Cu	20.8 mg by *Bacillus subtilis* [75]	20.91 mg by *Aspergillus niger* [68]
	275 mg by Enterobacter sp. [76]	6 mg by *Aureobasidium pullulans* [77]
	65.3 mg by *Pseudomonas cepacian* [78]	24 mg by *Ganodermalucidum* [79]
	22.9 mg by *Pseudomonas stutzeri* [75]	9 mg by *Penicillium chrysogenum* [70]
	60 mg by *Sphaerotilusnatans* [80]	9.5 mg by *Rhizopus arrhizus* [61]
Cr	2 mg by *Zoogloearamigera* [81]	24.84 mg by *Termitomycesclypeatus* [82]
	–	4.33 mg by *Rhizopus sp.* [83]
	–	331 mg by *Aspergillus lentulus* [66]
Cd	320 mg by *Stenotrophomonas* sp. [76]	2.72 mg by *Aspergillus sp.* [83]
	–	7.3 mg by *Aspergilus versicolor* [84]
	–	11 mg by *Penicillium chrysogenum* [70]
	–	127 mg by Pleurotussapidus [85]
	–	84.5 mg by *Phanerochaetechrysosporium* [86]
	–	2.72 mg by *Rhizopus* sp. [83]
	–	19 mg by *Rhizopus nigricans* [61]
	–	27 mg by *Rhizopus arrhizus* [61]
U	–	636.9 mg by *Pleurotusmutilus* [87]
	–	10 mg by *Aspergillus terreus* [88]
Hg	–	287 mg by *Pleurotussapidus* [85]
Th	–	60 mg by *Aspergillus terreus* [88]
Ni	–	5 mg by *Rhizopus nigricans* [61]
	–	18 mg by Rhizopus arrhizus [61]

catalytic role in degradation and mineralization of various contaminants in the soils. They also convert toxic contaminants to non-toxic byproducts and detoxify the soils during bioremediation [57]. Soil heavy metal remediation through microbes is depicted in Table 11.1.

11.6. PLANTS FOR SOIL REMEDIATION

Using plants in phytoremediation is a good strategy that involves degradation and removal of contaminants from the soils for better environmental health and sustainability [24]. Phytoremediation is a key tool for minimizing anthropogenic toxins and restoring the soil health and quality to a large extent. That's why the term "green remediation" has evolved due to plant involvement in bioremediations. The phytoremediation process involves remediation of some organic and inorganic contaminants in the soil, water, and other natural resources. Soil contamination due to salinity, sodicity, hydrocarbons, and heavy metals

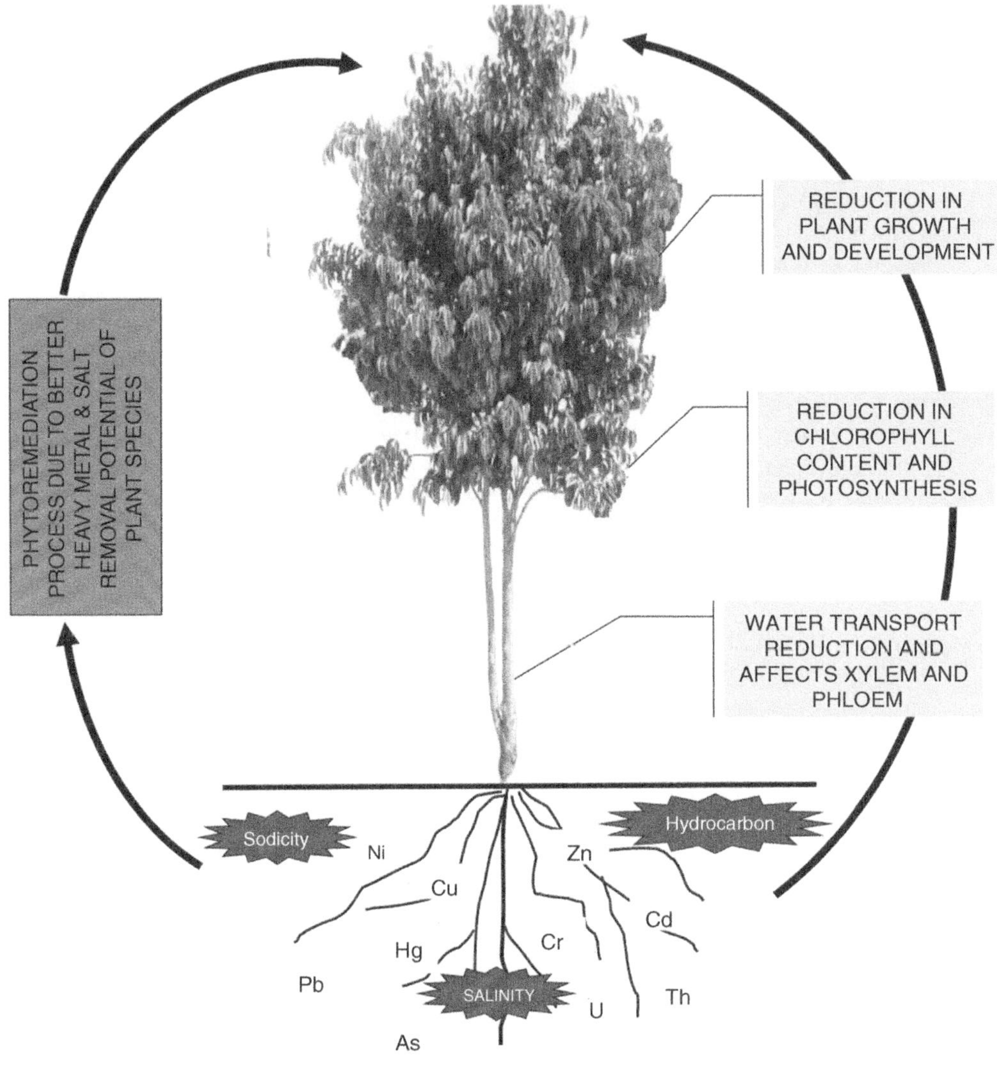

Figure 11.4 Model of soil contaminant consequences and phytoremediation.

affects plant growth and development in several ways. For example, reduction in chlorophyll content and photosynthesis activity in leaves, reduction in water transfer, and food translocation due to poor xylem and phloem quality are key consequences observed due to soil contaminants. In this context, model plants can work in the phytoremediation process that reclaims poor soils due to its greater potential for heavy metal and salt removal. Thus, a process of phytoremediation suppresses negative consequences of contaminants and improves soil quality and its sustainability. A model of soil contamination consequences and phytoremediation is depicted in Figure 11.4 [89, 90]. Phytoremediation processes and mechanisms for soil contaminants in plants are depicted in Table 11.2 [91].

Many authors have reported the potential of phytoremediation toward remediation of problematic soils. For examples, various plants such as *Viola calaminaria, Avena sativa, Thlaspicaerulescens,* and *Acacia* based leguminous tree species have the potential to accumulate heavy metals (copper, nickel, selenium, cadmium, and zinc) from the soils [92].

Likewise, several important leguminous tree species are involved in contaminant removal and accumulations into them. Grasses such as *Panicum virginatum* can potentially accumulate dangerous radionuclides such as Sr-90 and Cs-137, respectively [93]. Similarly, trees like *Populus deltoids* have the potential to reduce nitrate concentration in water and degrade the atrazine herbicide into the contaminant soils [94]. Many authors have studied the role of plants in soil remediation globally. For example, Ghazaryan et al. [95] reported phytoextraction capacity of Mo and Cu from soil by two plant species, *Amaranthus retroflexus* and *Melilotus officinalis* in Armenia. Similarly, tree species such as *Tetraclinis articulata* and *Pinus halepensis* have greater potential to reclaim heavy metal enrich soils in Spain [96]. As per Garrido et al. [97] the plant *Dittrichia viscosa* has extracted Sb from soil and accumulated into leaves in the region of Spain. Some wetlands grasses/plants such as *Iris pseudacorus*, *Phragmites australis*, and *Juncus effusus* were used to reclaim soil salinity and extract potentially toxic elements from contaminated soils [98]. Most of the plants are used for removal of harmful oils and petroleum hydrocarbons from problematic soils [99]. Therefore, plant-based bioremediation represents an innovative tool for accumulation, immobilization, and transformation of persistent contaminants in the soils [100].

11.7. LEGUMINOUS TREES FOR REMEDIATION OF SALT-AFFECTED SOIL

Leguminous plants improve soil health through fertility enhancement and removal of contaminants. These species increase nitrogen (N) status in the soil due to the inherent potential of N_2 fixing in nature. However, legumes strengthen soil physicochemical properties and microbial populations [35]. Legumes reclaim problematic soils by minimizing salt contamination and maintaining environmental sustainability. Hydrocarbons and heavy metal removal is another key potential mediated through leguminous plants. Also, legumes reduce soil alkalinity through their potential ability to increase acidity compared to other non-leguminous plant species [101]. For example, species like *Medicago sativa* have the potential to tolerate salt-affected soils and reclaim soil salinity to a greater extent [102]. Legumes have the capability to reclaim soil salinity and not only remove toxic ions but also improve N stock in the soils. For example, species like *Hedysarum carnosum* improve soil fertility under saline conditions by increasing the Na + accumulation in the roots and maintaining N status in the soil [103]. Also, leguminous tree species such as *Leucaena leucophela* and *Acacia nilotica* have the potential to restore problematic soils with greater N fixation. These species have the potential to grow on saline soils compared to non-leguminous tree species [104]. Other tree species such as *Dalbergia sissoo* and *Prosopis juliflora* have greater potential to reclaim soil salinity [105]. Similarly, *Acacia albida, Acacia tortilis*, and *Acacia luederitzil* showed greater potential to reclaim Ni and Cu contaminated soil in Botswana [106].

11.8. MICROBES FOR SOIL REMEDIATION

Microorganisms are most abundant, ubiquitous, and diversified organisms that hold versatile and effective metabolic systems that help in the degradation of toxic substances and utilize the energy in the metabolic process [107]. The use of microbes is an innovative tool for problematic soil remediation. Bacteria and fungi are key microbes that are potentially involved in reclaiming soil contaminants. Microbes can survive in high temperature, high salinity, and alkalinity conditions due to their greater adaptability and survival efficiency. Microbes are also able to develop some biological resistance against various harmful and toxic substances due to their jumping genes. Application of genetically modified or engineered microbes is a modern tool and is gaining wider popularity among scientists and

Table 11.2 Phytoremediation processes and mechanisms for soil contaminants in plants.

Process and mechanism	Contaminants	Plant species
Hyperaccumulation mechanism in phytoextraction	Metals comprising Zn, Cu, Pb, Ni and Cd	*Brassica juncea* (Indian Mustard), *Helianthus annuus* (Sunflower), and rapeseed plant
Rhizosphere accumulation in rhizofiltration	Heavy metals such as Zn, Cu, Pb, Ni and Cd, Radionuclides such as 137Cs, 90Sr and 238U, and Hydrophobic organic materials	Aquatic plants such as duck weed and pond Weed along with Hydrilla
Complexation mechanism in phytostabilization	Heavy metals such as Zn, Cu, Pb, As, Se, U and Cd, Hydrophobic organic materials such as PAHs, PCBs, dioxins, furans, pentachlorophenol and DDT.	Some phreatophytic trees
Volatilization by leaves in phytovolatilization	Tritium (3H1), Mercury (Hg), and selenium (Se)	*B. juncea* (Indian mustard), *Populusdeltoides* (Poplar), and *Brassica napus* L. (canola plant)
Degradation in plants under phytodegradation	Chlorinated aliphatics, herbicides (atrazine, alachlor), and aromatic substances such as BTEX	Phreatophyte trees, *Populusdeltoides* (poplar), Salix species, *Sorghum bicolor* (sorghum), *Trifolium* repens (clover), *Medicago sativa* L. (alfalfa), and *Vigna unguiculata* (cowpeas)

researchers. These modern microbial tools increase the degradation rate of soil contaminant and improve the soil health and quality.

The process of mycoremediation involves important fungi that remove organic and inorganic pollutants through bioremediation techniques. Fungal mycelia accumulate various soil heavy metals and toxic substances similar to plant root systems. Researchers and scientists have recommended microbial remediation due to its ecofriendly nature for treating contaminated soils. The process like bioleaching, bioventing, and bioaugmentation are used under microbe-mediated soil remediation techniques [108, 109]. For example, *Aspergillus*, *Fusarium*, *Amorphoteca*, *Penicillium*, *Neosartoria*, and *Talaromyces* are important fungi that are involved in petroleum oil treatment in soils. Similarly, yeast species like *Candida*, *Pichia*, *Pseudozyma*, and *Rhodotorula* are reported for restoration of oil-contaminated soils [110]. Likewise, several bacterial species such as *Azoarcus*, *Acinetobacter*, *Arthrobacter*, *Corynebacterium*, *Ochrobactrum*, *Marinobacter*, and *Flavobacterium* have reported for hydrocarbon degradation for soil restoration [111, 112].

11.9. BIOREMEDIATION OF HEAVY METAL IN SOIL

Toxic metal mediated soil contamination is becoming a major environmental hazard. These toxic substances pollute soils and affect organisms and related ecosystem services. Heavy metals in soils alter soil physicochemical and biological properties that influence plant growth and development. Heavy metals are major environmental pollutants that disturb the soil ecosystem and environment. Industrial effluents, mining extraction, and their combustion process release a variety of heavy metals such as arsenic (As), lead (Pb), mercury (Hg), cadmium (Cd), nickel (Ni), chromium (Cr), and aluminum (Al) in the environment. These heavy metals affect soil health and quality and enter into food chains, which disturbs the whole ecosystem's health and productivity. These heavy metals disturb plant metabolic activities and human health due to various harmful diseases and disorders [113, 114]. However, some heavy metals such as iron (Fe), copper (Cu), selenium (Se), and zinc (Zn) are present in small amounts but their higher accumulation in the soil damages the health and productivity of environment [115].

The degradation and removal of these heavy metals are an urgent priority for the betterment of soil–plant health and productivity. Various chemical methods such as heat treatment, precipitation, electroremediation, chemical leaching, and soil replacement are used for degradation of heavy metals in problematic soils. But these methods are costly and so are not preferred in any agricultural system. Thus, the chemical methods are not ecologically sound or economically feasible. In this context, using bioremediation techniques would work more efficiently in profitable ways. Phytoremediation technology is a promising tool that not only tolerates heavy amounts of toxic metals but also accumulates it into their plant parts. Therefore, green plants are used for detoxifying the heavy metal contaminants and improving soil health and fertility. The phytoremediation mechanisms of plant species for heavy metal are depicted in Table 11.3. Moreover, latest technologies of genetic engineering such as using phytohormones, adding nanoparticles, and the process of transgenic transformation are used for better results. Using plant growth-promoting bacteria and inoculation of AMF (arbuscular mycorrhiza fungi) can suppress the effects of heavy metal contamination in the soils [89].

Bioremediation includes some plant growth-promoting bacteria (PGPB), which is used for degradation of toxic substances in the contaminated soils [141]. However, these PGPB were used to enhance the phytoremediation potential to capture the maximum amount of heavy metal from the contaminated soils. Therefore, these novel bacteria were used for heavy metal decontamination due to their effective mechanisms. Bacteria release some chelating substances such as siderophores and other organic acids that decrease soil pH and enhance bioavailability of heavy metals into the soils [142]. Most of the bacteria release glomalin- and polysaccharide- based key polymeric compounds that reduce the mobility of heavy metals through the phytostabilization process [143]. Improving plant detoxification rates, modifying soil pH, secreting root enzymes, etc., are the key roles played by PGPR (plant growth-promoting rhizobacteria) for a better phytoremediation process. These activities maximize accumulation and degradation of heavy metals in contaminated soils [144].

Plant inoculation with some microbes including bacteria and fungi can help to detoxify heavy metals in the soils. For example, using the *Bacillus* XZM strain as inoculants of *Vallisneria denses errulata* plants is a good strategy for detoxification of a heavy metal such as arsenic (As) [145]. Also, inoculation of *Pteris vittata* (Chinese brake) plant with *Cupriavidus basilensis* strain r507 increased arsenic (As) accumulation as much as 170% in their plant parts [119]. Similarly, AMF play a symbiotic role with plant root systems and improve phosphorus phytoavailability [72]. Inoculating the leguminous plant species *Cassia italica* with AMF improve the cadmium (Cd) tolerancy and check its mobility to aerial plant parts [146]. Similarly, inoculating leguminous plant such as *Trigonella foenum-graecum* with AMF improved phytostabilization capacity of cadmium (Cd) under various concentration of Cd viz., 0, 2.25, and 6.25 mM $CdCl_2$. Moreover, inoculation of *Artemisia annua* (sweet wormwood) plant by the fungi *Piriformospora indica* improves tolerant capacity for arsenic (As) stress [147].

Table 11.3 Phytoremediation mechanisms of plant species for heavy metals.

Plant species	Phytoremediation methods for heavy metals	References
Alternanthera bettzickian, Boehmerianivea, and *Sedum alfredii*	Phytoextraction of Cu	[116–118]
Pennisetum purpureum, Atriplexnummularia, Atriplexlentiformis, Sesuviumportulacastrum, and *Sedum alfredii*	Phytoextraction of Cd	[119–123]
Sesuviumportulacastrum and *Cannabis sativa*	Phytoextraction of Ni	[124, 125]
Halimioneportulacoides, P. purpureum, Armeriaarenaria, and *C. sativa*	Phytoextraction of Zn	[119, 125–127]
Tamarixsmyrnensis, Brassica juncea, and *Pelargonium hortorum*	Phytoextraction of Pb	[128–130]
Acanthus ilicifolius and *Atriplexhalimus*	Phytostabilization of Cd	[131, 132]
Pistacialentiscus	Phytostabilization of Zn	[133]
B. juncea	Phytostabilization of Pb	[129]
Rhizophora mangle	Phytodegradation of PAHs	[134]
Salix viminalis	Phytodegradation of TCE	[135]
Salvinia molesta	Rhizofiltration of Cr	[136]
S. molesta	Rhizofiltration of As	[137]
Phragmites australis	Rhizofiltration of U	[138]
Arundo donax	Phytovolatilization of As	[139]
Stanleyapinnata	Phytovolatilization of Se	[140]

11.10. BIOREMEDIATION OF OIL-CONTAMINATED SOIL

Oil pollution is a major problem and needs great attention worldwide. Anthropogenic activities release oil into the soils, and this affects population and plant ecosystems. Oils, hydrocarbons, and petroleum have negative impacts on soil health and quality. Oil-related substances pollute the soil ecosystem and destroy plant health and productivity. Oil-mediated soil contamination can also affect human health and environmental quality [148]. However, the process of oil production, its processing, transportation, and utilization release large quantities of hydrocarbons into the soils. This has caused serious soil pollution problems that require proper restoration of oil-contaminated soils [149].

Several management practices have been adopted for reclaiming oil-contaminated soils. Land tillage practices are used to remediate approximately 500 kg of oil-contaminated soils in agricultural land [150]. However, some agricultural waste products such as wheat bran and swine wastewater have been used for bioremediation of oil-affected soils. Microbes also play a key role in degradation of oil contaminants into the soil. For examples, strains of *Bacillus subtilis* CICC 21312 and *Candida bombicola* ATCC 22214 have excellent capacity to decompose oil contaminants in the soil [151]. Indigenous microorganisms are also used for bioremediation of oil-contaminated soils [152]. The content of petroleum hydrocarbon decreased from 82 533 to 47 600 mg kg^{-1} after adding some oil-degrading bacteria and repairing the soil health and quality. Petroleum hydrocarbon was removed at the rate of 42.3% [153].

11.11. SOIL REMEDIATION AND SUSTAINABLE DEVELOPMENT

Soil is important natural resource that ensures many ecosystem services including food-climate security for sustainability. Remediation of problematic soil is urgently needed due to its negative consequences on biodiversity [154, 155]. However, salinity, sodicity, heavy metals, and oil contamination in soils hamper plant health and productivity [156]. These are the real menaces for sustainable development, and they affect the overall soil-food-climate security worldwide. Soil salinity and sodicity created food insecurity by affecting soil health and quality [157]. For example, problematic soils, including salinity, affect sustainable rice production and related food security in the delta region of India, Bangladesh, and Myanmar [158]. Therefore, soil remediation through plants and microbes is essential for diversified agroecosystem production, which paves the way for achieving the goal of sustainability. Soil remediation is strongly linked with sustainable development. Achieving sustainability can be possible through better soil quality and related ecosystem services. Moreover, good health and higher production from the agroecosystem are key services that are pillars for sustainable development. Thus, soil bioremediation intensifies agroecosystem production and ensure food and nutritional security along with environmental sustainability.

11.12. EMERGING TECHNOLOGIES IN SOIL REMEDIATION

Using physical and chemical technology along with incineration and bioremediation are important techniques for remediation of soil contaminants. However, the practice of bioremediation with recent advancement ensures higher efficiency of contaminant removal from the problematic soils. This technology is eco-friendly and economically feasible without any harmful effects on the environment. Novel approaches in microbial remediation work more efficiently for soil restoration. Recently, a number of technologies have been used which improve speed, reliability, and cost efficiency of bioremediations. Transferring the process of intrinsic bioremediation to recent application of genetically engineered bioremediation has been popularized. However, the microbial-based bioremediation process is often slow under extreme environmental conditions, and it needs to be changed in new momentum by recent technological development. Introducing new engineering tools in the bioremediation process can improve the efficiency of research in various dimensions. Similarly, microbe-assisted phytoremediation is a new emerging technology that promises soil health and plant productivity. Various developments and technologies are applied in the bioremediation process to eradicate the problem of contaminated soils [159]. For example, using phytoremediation with some emerging tools can enhance the degradation potential of hydrocarbons especially PAHs [160]. Electro-kinetic remediation technology is also used which is recent advancement in bioremediation process [161]. Moreover, using ultrasonic treatment in bioremediation can promote PAH removal and its degradation into the soil [162]. The latest research focuses on bioremediation of pesticide-contaminated soils for better environmental sustainability [163]. Phytoremediation, microbial-based mercury reduction, and algae-mediated mercury removal were also studied as recent developments in the bioremediation process [164]. Thus, recent advancement in bioremediation technology is needed to minimize the problematic soils (saline, alkaline, heavy metals, oil, hydrocarbon, etc.) and their deleterious impacts on soil–plant relationships and the environment.

11.13. CONSTRAINTS AND OPPORTUNITIES

Many scientific and technological constraints are observed in proper development and implementation of bioremediation techniques for problematic soil remediation. The industrial application of phytoremediation is slow and not up to the standard. Using slow-growing plants in phytoremediation is another constraint for eradication of soil contaminants. Moreover, lower efficiency of plant root systems for soil penetration and its poor contaminant removal capacity were also observed as a major constraint and limitation [165]. The addition of organic contaminants into the plant cells and tissues affects plant growth and biomass production due to higher accumulation of toxicity. This situation is more problematic due to multiple types of contaminants such as heavy metals, surfactants, emulsifiers, and organic substances from different sources, including oil refineries and mining industries.

Thus, there is a great opportunity for researchers, scientists, and policy makers to observe these constraints and accordingly adopt some minimizing strategies. Adopting plant-microbial based remediation is good practice for overcoming the issues of problematic soils. Both work simultaneously for remediation and restoration of problematic soils. Microbes generally use sugar and other metabolites, which are secreted by plant root systems. Further, the availability of some nutrients such as nitrogen, phosphorus, and others are also made possible through rhizosphere microbes. However, these beneficial microbes are able to degrade organic contaminants prior to entering plant cell and tissues. Also, plant inoculation with some microbes (bacteria and fungi) and its utility in any contaminated soils are other great opportunities for soil restoration and management [165]. In recent years, using the concept of designer plants and its utility for remediation of problematic soils has been effectively adopted worldwide. These plants along with some endophytic microbes work simultaneously to check soil problems and their remediation [166]. These approaches could be more economically viable if the plant had some commercial importance such as for biofuel and other biomass production. Thus, tree species such as *Salix* (willow), *Populus deltoids* (poplar), and *Jatropha,* etc., are used for both energy production and phytoremediation.

11.14. POLICY AND FUTURE ROADMAP

Industrialization, improper disposal of organic and inorganic waste, intensive manufacturing, and oil leakages from refineries release harmful contaminants into soils. These activities not only decrease soil health and plant productivity but also affect environmental health and sustainability. Thus, adoption of effective policies for disposal and other management practices is needed. Policy for better research and development for bioremediation adoption in any institutions and organizations are emphasized. Recent technologies for phytoremediation and microbial-assisted remediation are needed to strengthen soil restoration and health. Heavy metals, oils, and hydrocarbons are key pollutants that not only destroy soil health but also accumulate into plant parts over long periods. These hazardous materials affect overall soil organisms, plant growth, and productivity. Thus, a policy must be framed in accordance with the understanding that some metabolic and physiological changes appear after accumulation of some harmful heavy metals into plant parts and their environmental consequences. Therefore, the future roadmap should be planned to enhance bioremediation potential and positive impacts on soil–plant systems and sustainability.

11.15. CONCLUSIONS

Soil contamination due to many anthropogenic activities is a serious problem today. This affects soil health and quality and induces various physiological and metabolic abnormalities into the plant. Soil salinity and sodicity are other issues that affect plants' health and productivity. Industrialization, improper management of organic and inorganic waste, oil refineries, and related effluents release harmful heavy metals and hydrocarbons that impair soil health and productivity. In this context, plant- and microbe-based bioremediation techniques can degrade soil contaminants and improve soil–plant health and productivity. Ensuring phytoremediation practices in any contaminated sites restore soils and enhance energy and biomass productions. Microbial-assisted soil remediations are gaining wider popularity due to recent advancements and technological inputs. Thus, phytoremediation and microbial-assisted bioremediation are best suited for problematic soils and their rejuvenation for ensuring a sustainable world.

REFERENCES

1. Lal, R. (2004). Soil carbon sequestration impacts on global climate change and food security. *Science 304*: 1623–1627.
2. Khan, N., Jhariya, M.K., Raj, A. et al. (2021a). Soil carbon stock and sequestration: implications for climate change adaptation and mitigation. In: *Ecological Intensification of Natural Resources for Sustainable Agriculture* (ed. M.K. Jhariya, R.S. Meena and A. Banerjee), 461–489. Singapore: Springer https://doi.org/10.1007/978-981-33-4203-3_13.
3. Khan, N., Jhariya, M.K., Raj, A. et al. (2021b). Eco-designing for sustainability. In: *Ecological Intensification of Natural Resources for Sustainable Agriculture* (ed. M.K. Jhariya, R.S. Meena and A. Banerjee), 565–595. Singapore: Springer https://doi.org/10.1007/978-981-33-4203-3_16.
4. Bai ZG, Dent DL, Olsson L, et al. (2008). *Global assessment of land degradation and improvement. 1. Identification by remote sensing* (Report 2008/01). Wageningen: ISRIC-World Soil Information.
5. WMO (2005). *Climate and land degradation.* (WMO-No. 989, p. 32). World Meteorological Organization.
6. Jhariya, M.K., Banerjee, A., Meena, R.S. et al. (2019a). *Sustainable Agriculture, Forest and Environmental Management*, 606. Singapore: Springer Nature https://doi.org/10.1007/978-981-13-6830-1.
7. Jhariya, M.K., Yadav, D.K., and Banerjee, A. (2019b). *Agroforestry and Climate Change: Issues and Challenges*, 335. US & Canada: Apple Academic Press Inc., CRC Press - a Tayler and Francis Group https://doi.org/10.1201/9780429057274.
8. Jhariya, M.K., Meena, R.S., and Banerjee, A. (2021a). *Ecological Intensification of Natural Resources for Sustainable Agriculture*, 655. Singapore: Springer https://doi.org/10.1007/978-981-33-4203-3.
9. Jhariya, M.K., Meena, R.S., Banerjee, A. et al. (2021b). *Natural Resources Conservation and Advances for Sustainability*, 623. Elsevier, Academic Press. ISBN: 9780128229767.
10. Jhariya, M.K., Banerjee, A., Meena, R.S. et al. (2021c). *Sustainable Intensification for Agroecosystem Services and Management*, 870. Singapore: Springer https://doi.org/10.1007/978-981-16-3207-5.
11. FAO (2011). *State of the World's Land and Water Resources for Food and Agriculture (SOLAW)*. Rome, Italy: Food and Agriculture Organization of the United Nations.
12. Raj, A., Jhariya, M.K., Yadav, D.K. et al. (2020). *Climate Change and Agroforestry Systems: Adaptation and Mitigation Strategies*, 383. US & Canada: Apple Academic Press Inc., CRC Press- a Tayler and Francis Group https://doi.org/10.1201/9780429286759.
13. Banerjee, A., Jhariya, M.K., Yadav, D.K. et al. (2020). *Environmental and Sustainable Development through Forestry and Other Resources*, 400. US & Canada: Apple Academic Press Inc., CRC Press- a Tayler and Francis Group https://doi.org/10.1201/9780429276026.
14. Khan, N., Jhariya, M.K., Yadav, D.K. et al. (2020a). Herbaceous dynamics and CO_2 mitigation in an urban setup- a case study from Chhattisgarh. *India. Environ Sci Poll Res. 27* (3): 2881–2897. https://doi.org/10.1007/s11356-019-07182-8.
15. Khan, N., Jhariya, M.K., Yadav, D.K. et al. (2020b). Structure, diversity and ecological function of shrub species in an urban setup of Sarguja, Chhattisgarh, India. *Environmental Science and Pollution Research 27* (5): 5418–5432. https://doi.org/10.1007/s11356-019-07172-w.
16. Jhariya, M.K. and Singh, L. (2021a). Herbaceous diversity and biomass under different fire regimes in a seasonally dry forest ecosystem. *Environment, Development and Sustainability 23* (5): 6800–6818. https://doi.org/10.1007/s10668-020-00892-x.
17. Jhariya, M.K. and Singh, L. (2021b). Effect of fire severity on soil properties in a seasonally dry forest ecosystem of Central India. *International journal of Environmental Science and Technology 18*: 3967–3978. https://doi.org/10.1007/s13762-020-03062-8.
18. Jhariya, M.K. and Singh, L. (2021c). Productivity and carbon accumulation in tropical dry deciduous Forest ecosystem. In: *Tropical Dry Forests: Emerging Features and Ecological Perspectives* (ed. R.K. Chaturvedi, R. Singh and R. Bhadouria), 269–287. NY USA: Nova Science.

19. Tejada, M., Hernandez, M.T., and Garcia, C. (2009). Soil restoration using composted plant residues: effects on soil properties. *Soil and Tillage Research 102*: 109–117.
20. Haimi, J. (2000). Decomposer animals and bioremediation of soils. *Environmental Pollution 107*: 233–238.
21. Dhal, B., Thatoi, H.N., Das, N.N. et al. (2013). Chemical and microbial remediation of hexavalent chromium from contaminated soil and mining/metallurgical solid waste: a review. *Journal of Hazardous Materials 250*: 272–291.
22. Li, H., Zhang, Y., Zhang, C.G. et al. (2005). Effect of petroleum-containing wastewater irrigation on bacterial diversities and enzymatic activities in a paddy soil irrigation area. *Journal of Environmental Quality 34*: 1073–1080.
23. Gupta, K., Biswas, R., and Sarkar, A. (2020). Advancement of omics: prospects for bioremediation of contaminated soils. In: *Microbial Bioremediation & Biodegradation* (ed. M. Shah), 113–142. Singapore: Springer https://doi.org/10.1007/978-981-15-1812-6_5.
24. Jhariya, M.K., Yadav, D.K., and Banerjee, A. (2018a). Plant mediated transformation and habitat restoration: phytoremediation an eco-friendly approach. In: *Metallic Contamination and its Toxicity* (ed. A. Gautam and C. Pathak), 231–247. New Delhi: Daya Publishing House/ A Division of Astral International Pvt. Ltd. ISBN: 9789351248880.
25. Jhariya, M.K., Banerjee, A., Yadav, D.K. et al. (2018b). Leguminous trees an innovative tool for soil sustainability. In: *Legumes for Soil Health and Sustainable Management* (ed. R.S. Meena, A. Das, G.S. Yadav and R. Lal), 315–345. Singapore: Springer http://doi.org/10.1007/978-981-13-0253-4_10.
26. Banerjee, A., Jhariya, M.K., Yadav, D.K. et al. (2018). Micro-remediation of metals: a new frontier in bioremediation. In: *Handbook of Environmental Materials Management* (ed. C. Hussain), 1–36. Cham: Springer https://doi.org/10.1007/978-3-319-58538-3_10-1.
27. Meena, R.S., Kumar, S., Datta, R. et al. (2020). Impact of agrochemicals on soil microbiota and management: A review. *Land (MDPI) 9* (2): 34. http://doi.org/10.3390/land9020034.
28. Raj, A., Jhariya, M.K., Khan, N. et al. (2021). Ecological intensification for sustainable development. In: *Ecological Intensification of Natural Resources for Sustainable Agriculture*, 137–170. Singapore: Springer https://doi.org/10.1007/978-981-33-4203-3_5.
29. Ghassemi, F., Jakeman, A.J., and Nix, H.A. (1995). *Salinisation of Land and Water Resources: Human Causes, Extent, Management and Case Studies*. Wallingford: CAB international.
30. Dove, A. (2017). *Central California is Losing $3.7 Billion in Crop Yield Every Year. Report from Department of Civil and Environmental Engineering*. Pittsburgh, PA: Carnegie Mellon University.
31. Ilyas, F. (2017). *Sindh Suffers 31pc Crop Loss Annually Due to Waterlogging, Salinity. Dawn*. https://www.dawn.com/news/1357033.
32. FAO (2008). Land and plant nutrition management service. http://www.fao.org/ag/agl/agll/spush.
33. Wang, W., Vinocur, B., and Altman, A. (2003). Plant responses to drought, salinity and extreme temperatures: towards genetic engineering for stress tolerance. *Planta 218*: 1–14.
34. Gopalakrishnan, T. and Kumar, L. (2021). Linking long-term changes in soil salinity to paddy land abandonment in Jaffna Peninsula, Sri Lanka. *Agriculture 11* (211): 1–12. https://doi.org/10.3390/agriculture11030211.
35. Egamberdieva, D., Wirth, S., Bellingrath-Kimura, S.D. et al. (2019). Salt-tolerant plant growth promoting rhizobacteria for enhancing crop productivity of saline soils. *Frontiers in Microbiology 10*: 2791. https://doi.org/10.3389/fmicb.2019.02791.
36. Kayode, O.T., Aizebeokhai, A.P., and Odukoya, A.M. (2021). Soil salinity and its implications on sustainable agriculture in southern and northcentral states of Nigeria. IOP conf. Series: earth and environmental. *Science* 655:012077. https://doi.org/10.1088/1755-1315/655/1/012077.
37. Tully, K., Sullivan, C., Weil, R. et al. (2015). The state of soil degradation in SubSaharan Africa: baselines, trajectories, and solutions. *Sustainability. 7*: 6523–6550.
38. Hassani, A., Azapagic, A., and Shokri, N. (2020). Predicting long-term dynamics of soil salinity and sodicity on a global scale. *Proceedings of the National Academy of Sciences of the United States of America 117*: 33017–33027.
39. Shi, G.L., Zhu, S., Bai, S.N. et al. (2015). The transportation and accumulation of arsenic, cadmium, and phosphorus in 12 wheat cultivars and their relationships with each other. *Journal of Hazardous Materials 299*: 94–102.

40. Tiwari, K.K., Singh, N.K., Patel, M.P. et al. (2011). Metal contamination of soil and translocation in vegetables growing under industrial wastewater irrigated agricultural field of Vadodara, Gujarat, India. *Ecotoxicology and Environmental Safety 74*: 1670–1677.
41. Niu, L.Q., Jia, P., Li, S.P. et al. (2015). Slash-and-char, an ancient agricultural technique holds new promise for management of soils contaminated by Cd, Pb, Zn. *Environmental Pollution 205*: 333–339.
42. Nabulo, G., Black, C.R., and Young, S.D. (2011). Trace metal uptake by tropical vegetables grown on soil amended with urban sewage sludge. *Environmental Pollution 159*: 368–376.
43. Wang, Q., Liu, J., and Cheng, S. (2015). Heavy metals in apple orchard soils and fruits and their health risks in Liaodong Peninsula, Northeast China. *Environmental Monitoring and Assessment 187*: 1–8.
44. Sacristán, D., Rossel, R.A.V., and Recatalá, L. (2016). Proximal sensing of Cu in soil and lettuce using portable X-ray fluorescence spectrometry. *Geoderma 265*: 6–11.
45. Xu, X., Zhao, Y., Zhao, X. et al. (2014). Sources of heavy metal pollution in agricultural soils of a rapidly industrializing area in the Yangtze Delta of China. *Ecotoxicology and Environmental Safety 108*: 161–167.
46. Shaheen, S.M., Rinklebe, J., Rupp, H. et al. (2014). Temporal dynamics of pore water concentrations of Cd Co, Cu, Ni, and Zn and their controlling factors in a contaminated floodplain soil assessed by undisturbed groundwater lysimeters. *Environmental Pollution 191*: 223–231.
47. Iturbe, R., Castro, A., Perez, G. et al. (2008). TPH and PAH concentrations in the subsoil of polyduct segments, oil pipeline pumping stations, and right-of-way pipelines from Central Mexico. *Environmental Geology 55*: 1785–1795.
48. Gogoi, B.K., Dutta, N.N., Goswami, P. et al. (2003). A case study of bioremediation of petroleum-hydrocarbon contaminated soil at a crude oil spill site. *Advances in Environmental Research 7*: 767–782.
49. Sastre-Conde, I., Lobo, M.C., Beltran, H. et al. (2015). Remediation of saline soils by a two-step process: washing and amendment with sludge. *Geoderma 247-248*: 140–150.
50. Singh, B.K. and Walker, A. (2006). Microbial degradation of organophosphorus compounds. *FEMS Microbiology Reviews 30*: 428–471.
51. Hatzinger, P.B., Whittier, M.C., Arkins, M.D. et al. (2002). In situ and ex-situ bioremediation options for treating perchlorate in groundwater. *Remediation Journal 12*: 69–86.
52. Raj, A. and Jhariya, M.K. (2021a). Site quality and vegetation biomass in the tropical Sal mixed deciduous forest of Central India. *Landscape and Ecological Engineering* 17. https://doi.org/10.1007/s11355-021-00450-1.
53. Raj, A. and Jhariya, M.K. (2021b). Carbon storage, flux and mitigation potential of tropical Sal mixed deciduous forest ecosystem in Chhattisgarh, India. *Journal of Environmental Management 293* (1): 112829. https://doi.org/10.1016/j.jenvman.2021.112829.
54. Rayu, S., Karpouzas, D.G., and Singh, B.K. (2012). Emerging technologies in bioremediation: constraints and opportunities. *Biodegradation 23*: 917–926.
55. Arora, N.K. (2018). Bioremediation: a green approach for restoration of polluted ecosystems. *Environmental Sustainability 1*: 305–307.
56. Macek, T., Kotrba, P., Svatos, A. et al. (2008). Novel roles for genetically modified plants in environmental protection. *Trends in Biotechnology 26*: 146–152.
57. Gomez, M.J., Pazos, F., Guijarro, F.J. et al. (2007). The environmental fate of organic pollutants through the global microbial metabolism. *Molecular Systems Biology 3*: 1–11.
58. Tabaraki, R., Ahmady-Asbchin, S., and Abdi, O. (2013). Biosorption of Zn (II) from aqueous solutions by Acinetobacter sp. isolated from petroleum spilled soil. *Journal of Environmental Chemical Engineering 1* (3): 604–608.
59. Şahan, T., Ceylan, H., and Aktaş, N. (2015). Optimization of biosorption of Zn (II) ions from aqueous solutions with low-cost biomass *Trametes versicolor* and the evaluation of kinetic and thermodynamic parameters. *Desalination and Water Treatment 57*: 1–12.
60. Mameri, N., Boudries, N., Addour, L. et al. (1999). Batch zinc biosorption by a bacterial nonliving Streptomyces rimosus biomass. *Water Research 33*: 1347–1354.
61. Fourest, E. and Roux, J.C. (1992). Heavy metal biosorption by fungal mycelial by-products: mechanisms and influence of pH.Applied microbiology. *Biotechnology 37* (3): 399–403.

62. Incharoensakdi, A. and Kitjaharn, P. (2002). Zinc biosorption from aqueous solution by a halotolerant cyanobacterium *Aphanothecehalophytica*. *Current Microbiology 45* (4): 261–264.
63. Townsley, C.C. and Ross, I.S. (1985). Copper uptake by *Penicillium spinulosum*. *Microbios 44*: 125–134.
64. Liu, H.L., Chen, B.Y., Lan, Y.W. et al. (2004). Biosorption of Zn(II) and Cu(II) by the indigenous *Thiobacillusthiooxidans*. *Chemical Engineering Journal 97*: 195–201.
65. Tunali, S., Cabuk, A., and Akar, T. (2006). Removal of lead and copper ions from aqueous solutions by bacterial strain isolated from soil. *Chemical Engineering Journal 115* (3): 203–211.
66. Mishra, A. and Malik, A. (2012). Simultaneous bioaccumulation of multiple metals from electroplating effluent using *Aspergillus lentulus*. *Water Research 46* (16): 4991–4998.
67. Salehizadeh, H. and Shojaosadati, S.A. (2003). Removal of metal ions from aqueous solution by polysaccharide produced from *Bacillus firmus*. *Water Research 37* (17): 4231–4235.
68. Iskandar, N.L., Zainudin, N.A.I.M., and Tan, S.G. (2011). Tolerance and biosorption of copper (Cu) and lead (Pb) by filamentous fungi isolated from a freshwater ecosystem. *Journal of Environmental Sciences 23* (5): 824–830.
69. Choi, S.B. and Yun, Y.S. (2004). Lead biosorption by waste biomass of *Corynebacterium glutamicum* generated from lysine fermentation process. *Biotechnology Letters 26* (4): 331–336.
70. Niu, H., Xu, X.S., Wang, J.H. et al. (1993). Removal of lead from aqueous solutions by penicillium biomass. *Biotechnology and Bioengineering 42* (6): 785–787.
71. Ahmady-Asbchin, S., Safari, M., and Tabaraki, R. (2015). Biosorption of Zn (II) by *Pseudomonas aeruginosa* isolated from a site contaminated with petroleum. *Desalination and Water Treatment 54* (12): 3372–3379.
72. Zhang, X., Chen, B., and Ohtomo, R. (2015). Mycorrhizal effects on growth, P uptake and Cd tolerance of the host plant vary among different AM fungal species. *Soil Science & Plant Nutrition 61*: 359–368.
73. Uslu, G. and Tanyol, M. (2006). Equilibrium and thermodynamic parameters of single and binary mixture biosorption of lead (II) and copper (II) ions onto pseudomonas putida: effect of temperature. *Journal of Hazardous Materials 135* (1): 87–93.
74. Banerjee, G., Pandey, S., Ray, A.K. et al. (2015). Bioremediation of heavy metals by a novel bacterial strain *Enterobacter cloacae*and its antioxidant enzyme activity, flocculant production, and protein expression in presence of lead, cadmium, and nickel. *Water, Air, and Soil Pollution 226* (4): 1–9.
75. Nakajima, K., Sena, G., Nawy, T. et al. (2001). Intercellular movement of the putative transcription factor SHR in root patterning. *Nature 413*: 307–311.
76. Bestawy, E.E., Helmy, S., Hussien, H. et al. (2013). Bioremediation of heavy metal-contaminated effluent using optimized activated sludge bacteria. *Applied Water Science 3* (1): 181–192.
77. Gadd, G.M. and Mowll, J.L. (1985). Copper uptake by yeast-like cells, hyphae, and chlamydospores of *Aureobasidium pullulans*. *Experimental Mycology 9* (3).
78. Savvaidis, I., Hughes, M.N., and Poole, R.K. (2003). Copper biosorption by *Pseudomonas cepacia*and other strains. *World Journal of Microbiology and Biotechnology 19* (2): 117–121.
79. Muraleedharan, T.R. and Venkobachar, C. (1990). Mechanism of biosorption of copper (Ii) by *Ganoderma lucidum*. *Biotechnology and Bioengineering 35* (3): 320–325.
80. Beolchini, F., Pagnanelli, F., Toro, L. et al. (2006). Ionic strength effect on copper biosorption by *Sphaerotilusnatans*: equilibrium study and dynamic modelling in membrane reactor. *Water Research 40* (1): 144–152.
81. Nourbakhsh, M., Sag, Y., Özer, D. et al. (1994). A comparative study of various biosorbents for removal of chromium (VI) ions from industrial waste waters. *Process Biochemistry 29* (1): 1–5.
82. Fathima, A., Aravindhan, R., Rao, J.R. et al. (2015). Biomass of *Termitomycesclypeatus* for chromium (III) removal from chrome tanning wastewater. *Clean Technologies and Environmental Policy 17* (2): 541–547.
83. Zafar, S., Aqil, F., and Ahmad, I. (2007). Metal tolerance and biosorption potential of filamentous fungi isolated from metal contaminated agricultural soil. *Bioresource Technology 98* (13): 2557–2561.

84. Fazli, M.M., Soleimani, N., Mehrasbi, M. et al. (2015). Highly cadmium tolerant fungi: their tolerance and removal potential. *Journal of Environmental Health Science and Engineering 13* (1): 1.
85. Yalçinkaya, Y., Arica, M.Y., Soysal, L. et al. (2002). Cadmium and mercury uptake by immobilized *Pleurotussapidus. Turkish Journal of Chemistry 26* (3): 441–452.
86. Gabriel, J., Vosahlo, J., and Baldrian, P. (1996). Biosorption of cadmium to mycelial pellets of woodrotting fungi. *Biotechnology Techniques 10* (5): 345–348.
87. Mezaguer, M., El Hayet, K.N., Lounici, H. et al. (2013). Characterization and properties of Pleurotusmutilus fungal biomass as adsorbent of the removal of uranium (VI) from uranium leachate. *Journal of Radioanalytical and Nuclear Chemistry 295* (1): 393–403.
88. Tsezos, M. and Volesky, B. (1981). Biosorption of uranium and thorium. *Biotechnology and Bioengineering 23* (3): 583–604.
89. Nedjimi, B. (2021). Phytoremediation: a sustainable environmental technology for heavy metals decontamination. *SN Applied Sciences 3*: 286. https://doi.org/10.1007/s42452-021-04301-4.
90. Wei, Z., Van Le, Q., Peng, W. et al. (2021). A review on phytoremediation of contaminants in air, water and soil. *Journal of Hazardous Materials 403*: https://doi.org/10.1016/j.jhazmat.2020.
91. Mukhopadhyay, S. and Maiti, S.K. (2009). Phytoremediation of metal mine waste. *Applied Ecology and Environmental Research 8*: 207–222.
92. Rascio, N. and Navari-Izzo, F. (2011). Heavy metal hyperaccumulating plants: how and why do they do it? And what makes them so interesting? *Plant Science 180*: 169–181.
93. Eutry, J.A., Watrud, L.C., and Relves, M. (1991). Accumulation of 137 Cs and 90Sr from contaminated soil by three grass species inoculated with mycorrizal fungi. *Environmental Pollution 104*: 449–457.
94. Burken, J.G. and Schnoor, J.L. (1997). Uptake and metabolism of atrazine by poplar trees. *Environmental Science & Technology 31* (5): 1399–1406. https://doi.org/10.1021/es960629v.
95. Ghazaryan, K.A., Movsesyan, H.S., Minkina, T.M. et al. (2019). The identification of phytoextraction potential of *Melilotus officinalis* and *Amaranthus retroflexus* growing on copper- and molybdenum-polluted soils. *Environmental Geochemistry and Health 43*: 1327–1335. https://doi.org/10.1007/s10653-019-00338-y.
96. Conesa, H.M. and Párraga-Aguado, I. (2019). Effects of a soil organic amendment on metal allocation of trees for the phytomanagement of mining impacted soils. *Environmental Geochemistry and Health* 1–12. https://doi.org/10.1007/s10653-019-00479-0.
97. Garrido, I., Ortega, A., Hernández, M. et al. (2020). Effect of antimony in soils of an Sb mine on the photosynthetic pigments and antioxidant system of *Dittrichiaviscosa* leaves. *Environmental Geochemistry and Health 17*: https://doi.org/10.1007/s10653-020-00616-0.
98. Hernández-Pérez, C., Martínez-Sánchez, M.J., García-Lorenzo, M.L. et al. (2021). Phytoremediation of potentially toxic elements using constructed wetlands in coastal areas with a mining influence. *Environmental Geochemistry and Health* 1–16. https://doi.org/10.1007/s10653-021-00843-z.
99. Reynolds, C.M., Wolf, D.C., Gentry, T.J. et al. (1999). Plant enhancement of indigenous soil micro-organisms: a low-cost treatment of contaminated soils. *Polar Record 35* (192): 33–40. https://doi.org/10.1017/S0032247400026310.
100. Pulford, I. and Watson, C. (2003). Phytoremediation of heavy metal contaminated land by trees—a review. *Environment International 29*: 529–540.
101. Ferguson, B.T., Lin, M.H., and Gresshoff, P.M. (2013). Regulation of legume nodulation by acidic growth conditions. *Plant Signalling Behaviour 8*: e23426. https://doi.org/10.4161/psb.23426.
102. Munns, R. and Tester, M. (2008). Mechanisms of salinity tolerance. *The Annual Review of Plant Biology 59*: 651–681. https://doi.org/10.1146/annurev.arplant.59.032607.092911.
103. Kouas, S., Slatni, T., Ben Salah, I. et al. (2010). Eco-physiological responses and symbiotic nitrogen fixation capacity of salt-exposed *Hedysarumcarnosum* plants. *African Journal of Biotechnology 9*: 7462–7469. https://doi.org/10.5897/AJB10.211.

104. Singh, Y., Singh, G., Singh, B. et al. (2017). Restoration of degraded salt affected lands to productive forest ecosystem. *Geophysical Research Abstracts 19*: EGU 2017–1473. http://meetingorganizer.copernicus.org/EGU2017/EGU2017-1473.pdf.
105. Basavaraja, P.K., Sharma, S.D., Badrinath, M.S. et al. (2007). *Prosopis juliflora*: an efficient tree species for reclamation of salt affected soils. *Karnataka Journal Of Agricultural Sciences 20* (4): 727–731.
106. Ultra, V.U. and Manyiwa, T. (2020). Influence of mycorrhiza and fly ash on the survival, growth and heavy metal accumulation in three *Acacia* species grown in Cu–Ni mine soil. *Environmental Geochemistry and Health* 1–17. https://doi.org/10.1007/s10653-020-00627-x.
107. Singh, B.K., Campbell, C., Sorensen, S.J. et al. (2009). Soil genomics is the way forward. *Nature Reviews. Microbiology 7*: 756.
108. Chen, Y., Yang, W., Chao, Y. et al. (2017). Metal-tolerant *Enterobacter* sp. strain EG16 enhanced phytoremediation using *Hibiscus cannabinus* via siderophore-mediated plant growth promotion under metal contamination. *Plant and Soil 413*: 203–216.
109. Koul, B. and Taak, P. (2018). Soil remediation through microbes. In: *Biotechnological Strategies for Effective Remediation of Polluted Soils*, 101–128. Singapore: Springer https://doi.org/10.1007/978-981-13-2420-8_6.
110. Wilkes, H., Buckel, W., Golding, B.T. et al. (2016). Metabolism of hydrocarbons in n-alkane utilizing anaerobic bacteria. *Journal of Molecular Microbiology and Biotechnology 26*: 138–151.
111. Meckenstock, R.U., Boll, M., Mouttaki, H. et al. (2016). Anaerobic degradation of benzene and polycyclic aromatic hydrocarbons. *Journal of Molecu alar Microbiology and Biotechnology 26*: 92–118.
112. Varjani, S.J. and Upasani, V.N. (2016). Biodegradation of petroleum hydrocarbons by oleophilic strain of Pseudomonas aeruginosa NCIM 5514. *Bioresource Technology 222*: 195–201.
113. Awa, S.H. and Hadibarata, T. (2020). Removal of heavy metals in contaminated soil by phytoremediation mechanism: a review. *Water, Air, and Soil Pollution 231*: 47.
114. Nedjimi, B. (2009). Calcium can protect *Atriplexhalimus* subsp. schweinfurthii from cadmium toxicity. *Acta Botanica Gallica 156* (3): 391–397.
115. Ashraf, S., Ali, Q., Zahir, Z.A. et al. (2019). Phytoremediation: environmentally sustainable way for reclamation of heavy metal polluted soils. *Ecotoxicology and Environmental Safety 174*: 714–727.
116. Khalid, A., Farid, M., Zubair, M. et al. (2020). Efficacy of *Alternanthera bettzickiana* to remediate copper and cobalt contaminated soil physiological and biochemical alterations. *International Journal of Environmental Research 14*: 243–255.
117. Rehman, M., Maqbool, Z., Peng, D. et al. (2019). Morpho-physiological traits, antioxidant capacity and phytoextraction of copper by ramie (*Boehmerianivea* L.) grown as fodder in copper-contaminated soil. *Environmental Science and Pollution Research 26*: 5851–5861.
118. Liu, D., Islam, E., Li, T. et al. (2008). Comparison of synthetic chelators and low molecular weight organic acids in enhancing phytoextraction of heavy metals by two ecotypes of Sedum alfredii Hance. *Journal of Hazardous Materials 153*: 114–122.
119. Yang, W., Gu, J., Zhou, H. et al. (2020). Effect of three Napier grass varieties on phytoextraction of Cd- and Zn-contaminated cultivated soil under mowing and their safe utilization. *Environmental Science and Pollution Research 27*: 16134–16144.
120. Nedjimi, B. (2018). Heavy metal tolerance in two Algerian saltbushes: a review on plant responses to cadmium and role of calcium in its mitigation. In: *Plant Nutrients and Abiotic Stress Tolerance* (ed. M. Hasanuzzaman, M. Fujita, H. Oku, et al.), 205–220. Berlin: Springer.
121. Eissa, M.A. (2017). Phytoextraction mechanism of Cd by *Atriplexlentiformis* using some mobilizing agents. *Ecological Engineering 108*: 220–226.
122. Ghnaya, T., Nouairi, I., Slama, I. et al. (2005). Cadmium effects on growth and mineral nutrition of two halophytes: *Sesuviumportulacastrum* and *Mesembryanthemumcrystallinum*. *Journal of Plant Physiology 162*: 1133–1140.

123. Zhou, W. and Qiu, B. (2005). Effects of cadmium hyperaccumulation on physiological characteristics of *Sedum alfredii* hance (Crassulaceae). *Plant Science 169*: 737–745.
124. Fourati, E., Wali, M., Vogel-Mikuš, K. et al. (2016). Nickel tolerance, accumulation and subcellular distribution in the halophytes *Sesuviumportulacastrum* and *Cakile maritime*. *Plant Physiology and Biochemistry 108*: 295–303.
125. Meers, E., Ruttens, A., Hopgood, M. et al. (2005). Potential of *Brassica rapa, Cannabis sativa, Helianthus annuus* and *Zea mays* for phytoextraction of heavy metals from calcareous dredged sediment derived soils. *Chemosphere 61*: 561–572.
126. Cambrollé, J., Mancilla-Leytón, J.M., Muñoz-Vallés, S. et al. (2012). Zinc tolerance and accumulation in the salt-marsh shrub Halimioneportulacoides. *Chemosphere 86*: 867–874.
127. Frérot, H.,, Lefèbvre, C., Gruber, W. et al. (2006). Specific interactions between local metallicolous plants improve the phytostabilization of mine soils. *Plant and Soil 282*: 53–65.
128. Kadukova, J., Manousaki, E., and Kalogerakis, N. (2008). Pb and Cd accumulation and phytoexcretion by salt cedar (*Tamarixsmyrnensis* Bunge). *International Journal of Phytoremediation 10*: 31–46.
129. Meyers, D.E.R., Auchterlonie, G.J., Webb, R.I. et al. (2008). Uptake and localization of lead in the root system of *Brassica juncea*. *Environmental Pollution 153*: 323–332.
130. Manzoor, M., Gul, I., Manzoor, A. et al. (2020). Lead availability and phytoextraction in the rhizosphere of *Pelargonium* species. *Environmental Science And Pollution Research 27*: 39753–39762. http://doi.org/10.1007/s11356-020-08226-0.
131. Shackira, A.M. and Puthur, J.T. (2019). Cd2+ influences metabolism and elemental distribution in roots of Acanthus ilicifolius L. *International Journal of Phytoremediation 21*: 866–877.
132. Nedjimi, B. and Daoud, Y. (2009). Cadmium accumulation in Atriplexhalimus subsp. schweinfurthii and its influence on growth, proline, root hydraulic conductivity and nutrient uptake. *Flora 204*: 316–324.
133. Concas, S., Lattanzi, P., Bacchetta, A. et al. (2015). Zn, Pb and Hg contents of *Pistacialentiscus* L. grown on heavy metal-rich soils: implications for phytostabilization. *Water, Air, and Soil Pollution 226* (340).
134. Sampaio, C.J.S., de Souza, J.R.B., Damião, A.O. et al. (2019). Biodegradation of polycyclic aromatic hydrocarbons (PAHs) in a diesel oil-contaminated mangrove by plant growth-promoting rhizobacteria. *3 Biotech 9* (155): https://doi.org/10.1007/s13205-019-1686-8.
135. Clausen, L.P.W., Broholm, M.M., Gosewinkel, U. et al. (2017). Test of aerobic TCE degradation by willows (Salix viminalis) and willows inoculated with TCE-cometabolizing strains of Burkholderiacepacia. *Environmental Science and Pollution Research 24*: 18320–18331.
136. Kodituwakku, K.A.R.K. and Yatawara, M. (2020). Phytoremediation of industrial sewage sludge with *Eichhorniacrassipes*, *Salvinia molesta* and *Pistia stratiotes* in batch fed free water flow constructed wetlands. *Bulletin of Environmental Contamination and Toxicology 104*: 627–633.
137. da Silva, A.A., de Oliveira, J.A., de Campos, F.V. et al. (2018). Phytoremediation potential of Salvinia molesta for arsenite contaminated water: role of antioxidant enzymes. *Theoretical and Experimental Plant Physiology 30*: 275–286.
138. Wang, W. and Dudel, E. (2017). Fe plaque-related aquatic uranium retention via rhizofiltration along a redox-state gradient in a natural Phragmites australisTrin ex Steud. wetland. *Environmental Science and Pollution Research 24* (13): 24.10.1007/s11356-017-8889-5.
139. Guarino, F., Miranda, A., Castiglione, S. et al. (2020). Arsenic phytovolatilization and epigenetic modifications in Arundo donax L. assisted by a PGPR consortium. *Chemosphere 251* (126310): https://doi.org/10.1016/j.chemosphere.2020.126310.
140. Parker, D.R., Feist, L.J., Varvel, T.W. et al. (2003). Selenium phytoremediation potential of *Stanleyapinnata*. *Plant and Soil 249*: 157–165.
141. Ullah, A., Heng, S., Munis, M.F.H. et al. (2015). Phytoremediation of heavy metals assisted by plant growth promoting (PGP) bacteria: a review. *Environmental and Experimental Botany 117*: 28–40.

142. Chen, X., Liu, X., Zhang, X. et al. (2017). Phytoremediation effect of Scirpustriqueter inoculated plant-growth-promoting bacteria (PGPB) on different fractions of pyrene and Ni in co-contaminated soils. *Journal of Hazardous Materials 325*: 319–326.
143. Rajkumar, M., Sandhya, S., Prasad, M. et al. (2012). Perspectives of plant-associated microbes in heavy metal phytoremediation. *Biotechnology Advances 30*: 1562–1574.
144. Liu, S., Yang, B., Liang, Y. et al. (2020). Prospect of phytoremediation combined with other approaches for remediation of heavy metal-polluted soils. *Environmental Science and Pollution Research 27*: 16069–16085.
145. Irshad, S., Xie, Z., Wang, J. et al. (2020). Indigenous strain bacillus XZM assisted phytoremediation and detoxification of arsenic in *Vallisneriadenseserrulata*. *Journal of Hazardous Materials 381*: 120903.
146. Hashem, A., AbduAllah, E.F., Alqarawi, A.A. et al. (2016). Bioremediation of adverse impact of cadmium toxicity on *Cassia italica* mill by arbuscular mycorrhizal fungi. *Saudi Journal of Biological Sciences 23*: 39–47.
147. Rahman, S.U., Khalid, M., Kayani, S.I. et al. (2020). The ameliorative effects of exogenous inoculation of *Piriformospora indica* on molecular, biochemical and physiological parameters of *Artemisia annua* L. under arsenic stress condition. *Ecotoxicology and Environmental Safety 206* (111202).
148. Wang, S., Wang, D., Yu, Z. et al. (2021). Advances in research on petroleum biodegradability in soil. *Environmental Science: Processes & Impacts 23*: 9–27.
149. Das, N. and Chandran, P. (2011). *Microbial Degradation of Petroleum Hydrocarbon Contaminants: An Overview*, 941810. Biotechnology Res International.
150. Besalatpour, A., Hajabbasi, M.A., Khoshgoftarmanesh, A.H. et al. (2011). Landfarming process effects on biochemical properties of petroleum-contaminated soils. *Journal of Soil Contamination 20*: 15.
151. Zhang, C., Wu, D., and Ren, H. (2020). Bioremediation of oil contaminated soil using agricultural wastes via microbial consortium. *Scientific Reports 10*: 9188. https://doi.org/10.1038/s41598-020-66169-5.
152. Banks, M.K., Schwab, P., Liu, B. et al. (2003). The effect of plants on the degradation and toxicity of petroleum contaminants in soil: a field assessment. *Advances in Biochemical Engineering/Biotechnology 78*: 75–96.
153. Wu, M.L., Yuan, J., Li, W. et al. (2016). Bioremediation of petroleum contaminated soil and activity of hydrocarbon degrading bacteria. *Chinese Journal of Applied & Environmental Biology 22* (5): 878–883.
154. Raj, A., Jhariya, M.K., and Harne, S.S. (2018a). Threats to Biodiversity and Conservation Strategies. In: *Forests, Climate Change and Biodiversity* (ed. K.K. Sood and V. Mahajan), 304–320. India: Kalyani Publisher.
155. Raj, A., Jhariya, M.K., and Bargali, S.S. (2018b). Climate smart agriculture and carbon sequestration. In: *Climate Change and Agroforestry: Adaptation Mitigation and Livelihood Security* (ed. C.B. Pandey, M.K. Gaur and R.K. Goyal), 1–19. New Delhi/India: New India Publishing Agency (NIPA).
156. Ahmad, P. (2014). *Oxidative Damage to Plants: Antioxidant Networks and Signaling*. Cambridge, MA: Academic Press.
157. FAO and ITPS (2015). *Status of the World's Soil Resources (SWSR) – Main Report*. Rome: Food and Agriculture Organization of the United Nations and Intergovernmental Technical Panel on Soils.
158. Szabo, S., Hossain, M.S., Adger, W.N. et al. (2016). Soil salinity, household wealth and food insecurity in tropical deltas: evidence from south-west coast of Bangladesh. *Sustainability Science 11* (3): 411–421.
159. Maceiras, R. (2016). Emerging Technologies for Soil Remediation of hydrocarbons. *Pharmaceutical Analytical Chemistry: Open Access 2*: e102. https://doi.org/10.4172/2471-2698.1000e102.

160. Balasubramaniyam, A. (2015). The influence of plants in the remediation of petroleum hydrocarbon-contaminated sites. *Pharmaceutical Analytical Chemistry: Open Access 1*: 1–11.
161. Duarte, M.I., Bustos, E., and Teutli, L.M. (2014). Compost aided electrokinetic remediation of an hydrocarbon polluted soil. *Journal of the Mexican Chemical Society 58*: 343–347.
162. Zhang, W., Li, J., Han, Y. et al. (2010). Experimental study on the remediation of oil contaminated soil using ultrasound. IEEE. *Explorer* 1–4.
163. Morillo, E. and Villaverde, J. (2017). Advanced technologies for the remediation of pesticide-contaminated soils. *Science of the Total Environment 586*: 576–597.
164. Wang, L., Hou, D., Cao, Y. et al. (2020). Remediation of mercury contaminated soil, water, and air: a review of emerging materials and innovative technologies. *Environmental International 134*: 105281. https://doi.org/10.1016/j.envint.2019.105281.
165. Juwarkar, A., Singh, S., and Mudhoo, A. (2010). A comprehensive overview of elements in bioremediation. *Reviews in Environmental Science and Biotechnology 9*: 215–288.
166. Abhilash, P.C., Singh, H.B., Powell, J.R. et al. (2012). Plant–microbe interactions: novel applications for exploitation in multi-purpose remediation technologies. *Trends in Biotechnology 30*: 416–420.

12

Recent Advances in Biosensors for Rapid Identification of Antibiotics in Dairy Products

Shraddha Chauhan[1], Vikas Sharma[1], Digvijay Dahiya[2], and Preeti Chaturvedi Bharagava[1]

12.1. INTRODUCTION

Food obtained from animals is considered to be healthy and natural but large amounts of antibiotics that are being used these days to increase its yield have led to an increase in the concentration of such antibiotic residues in food, which is resulting in a rise of potential allergies among humans. On consumption, these antibiotic residues build resistance against drugs that are consumed worldwide and this developed resistance is said to have toxic effects. Microorganism resistance to important antibiotics has resulted in high mortality rates. The presence of these antibiotic residues in food products is contrary to the misleading advertising messages that portray these products to be good and healthy for the consumers. The use of veterinary drugs such as antibiotics given to diseased milk cows are essential for animal safety, but even with minimal use and with the observance of all protective measures, this treatment is still intimately linked with the risk of harmful residues in milk. These injections deteriorate the quality of milk and dairy products. Banned drugs like oxytocin, which is administered to cattle in many areas across the country to increase milk production, have harmful effects not only for animals but also for humans who consume the milk. Over time, this drug affects the reproductivity of the cattle and reduces their lifespan. Studies have linked the consumption of milk contaminated with oxytocin to the early onset of puberty among children, which is an alarming phenomenon these days. With the increasing incidents of antibiotic contamination in food, dairy products, and agriculture, their regular monitoring should be of primary importance. Figure 12.1 describes the cyclic procedure of the transfer of antibiotics and the formation of resistant microorganisms. Figure 12.1 explains the route of antibiotic transmission from animals and to patients subjected to various medications that act as transmitters.

[1]*Aquatic Toxicology Laboratory, Environmental Toxicology Group, CSIR-Indian Institute of Toxicology Research, Lucknow, Uttar Pradesh, India*

[2]*Department of Biotechnology, National Institute of Technology, Tadepalligudem, Andhra Pradesh, India*

Microbial Biotechnology: Role in Ecological Sustainability and Research, First Edition.
Edited by Pankaj Chowdhary, Sujata Mani, and Preeti Chaturvedi.

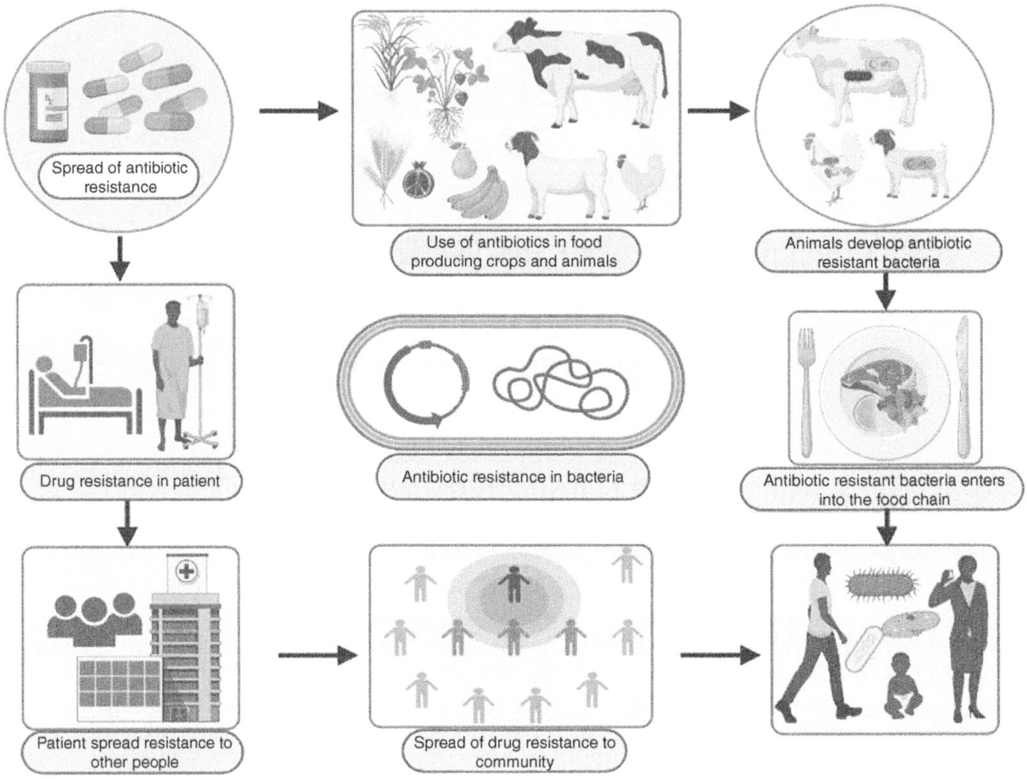

Figure 12.1 How antibiotic resistance spreads.

12.1.1. Use of Antibiotics and Associated Risks

Antibiotics are among the most widely prescribed and studied of the antimicrobial agents. The biggest strength of antibiotics is fighting against microbial infection. Antibiotics have made a significant contribution to curbing old infections that used to be among the top reasons for disease and mortality in humans [1]. The use of veterinary antibiotics has contributed to the improvement of animal health and wellbeing, which led to a significant improvement in the productivity of livestock for human use [2]. In 2005, the antibiotics market accounted for approximately US$25 billion of global sales, which increased to US$42 billion by 2009, which is a 5% share of the global market for pharmaceuticals and 46% share of anti-infective agent sales [3].

Antibiotics are effective against pathogenic bacteria as well as against several commensal bacteria, leading to a build-up of resistant strains [4, 5]. In the 1940s the first instances of antibiotic resistance were reported shortly after the large-scale use of antibiotics [6]. It was unavoidable because most bacteria have some degree of inherent resistance to antibiotics, and most bacteria have specific genes to develop resistance against a particular antibiotic reagent with time [7]. Bacteria also acquire antibiotic resistance through a horizontal gene transfer mechanism [8]. Due to increasing numbers, range, and the diversity of resistant strains, it is now a massive clinical issue with the existence of superbugs such as vancomycin-resistant *enterococci* and methicillin-resistant *Staphylococcus aureus* [9].

Antibiotics are often misused for the treatment of viral infections in the form of self-medication due to over-the-counter availability, and doctor's unnecessary prescriptions, making them less safe [10]. Misuse of veterinary medicine occurs in form of non-antibiotic usage in metaphylaxis and growth promotion [4]. According to estimates, around 50% of human use, and up to 80% of veterinary use of antibiotics can be eradicated without any severe consequences [11].

Upwards of 80% of the veterinary antibiotic administration in animals is through oral flock treatment, which entails long-term delivery of broad-spectrum antibiotics at low levels to animal herds, whatever the prevalent infection might be [12]. Moreover, the scenario becomes complicated when waste from hospitals, farms, and homes containing antibiotic-resistant microbiota are disposed of openly without proper monitoring and processing, which leads to the spread of the antibiotic-resistant gene in the environment [7]. In the past few decades, there has been widespread agreement that antibiotic treatment for promotion of growth and yield enhancement increases the chances of the spread of antibiotic resistance in microbiota associated with animals [13]. Additionally, there is a sufficient amount of evidence available that proves that consumption of animal food contaminated by resistant bacteria may cause transfer of resistant elements to the human gut microbiota, causing disease [7].

12.1.2. Current State of Food Inspection

For consumer protection, legislation has imposed strict regulations for the treatment of animals with drugs/antibiotics. At the same time, the law has determined maximum residue levels (MRL) for possible residues in foods of animal origin. Even though there has been an EU Regulation, 2377/901, which has been in place for the last 20 years, it now clear that the effective regulation of MRL values in food items can only be applied if techniques for single substance qualification and quantification are available online for routine monitoring [14].

12.1.3. Conventional Testing Methods

Generally, a cost-efficient microbiological residue test by the name of BRT, i.e. Brilliant Black Residue Test, is performed in dairy farms. This test allows the detection of a broad range of residues but fails to identify the specific residue in case of a positive reaction. Furthermore, this detection method is not suitable for time-bound areas of application due to the long duration of testing hours, often spanning a minimum of two hours.

Dairies mostly use receptor tests, also known as screening tests, for the inspection and analysis of the milk before delivering it from trucks. The tests, which are obtained within minutes, use an antibody to detect a single or a group of antibiotics with reasonable detection limits. However, some antibiotics remain undetected. Most screening tests are effective on commonly used antibiotics such as the beta-lactam group, whereas the non-beta-lactam group remains undetected/unidentified. However, few antibiotic agents are not covered by any of the analytical tests performed in the inspection of the dairy. This could be aided by performing analytical methods like liquid chromatography and mass spectrometry. In recent times, these methods could be relied upon to identify and quantify antibiotics but it requires both expensive equipment and a high level of technical expertise. They are also very time-consuming and cost-intensive due to complex sample preparation, hence making their usage mostly restricted to application areas in special laboratories.

12.2. BIOSENSORS FOR DETECTION OF ANTIBIOTICS IN FOOD

In the past few years, biosensors have been regarded as an alternative to conventional analytical methods. Biosensors are evolving into a suitable method for detection and screening purposes. A biosensor capitalizes on a biological recognition element such as enzyme, DNA,

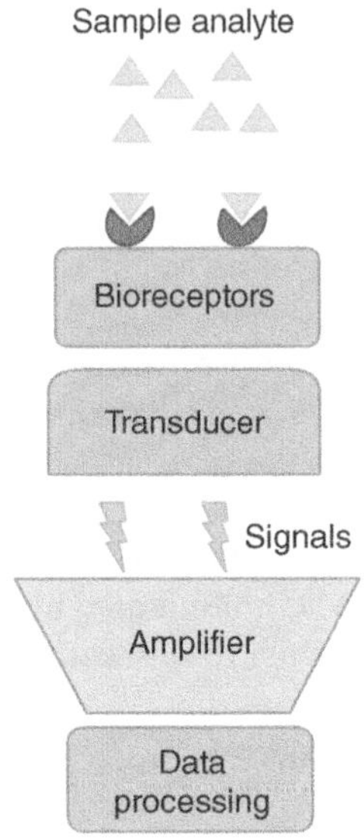

Figure 12.2 Schematic representation of the configuration of a biosensor.

protein, etc. It works on the principle of recognition of a target molecule based on the different signals produced such as thermal, piezoelectric, optical, electrochemical, etc., as shown in Figure 12.2 [15]. In comparison to other conventional analytical methods of detection, biosensors are portable, feasible, quick, practical, and do not require special skills. The performance of the biosensors is crucially dependent upon the binding affinity and specificity of the molecular recognition elements. Every biosensor pairs a transducer with a biological recognition element to generate a detectable signal in proportion to the analyte concentration [16].

Mostly biosensors are used for the detection of toxic chemicals present in the environment, but they are also increasingly being used in food screening for antibiotic traces [17]. Biosensor assays have a high automation and throughput capacity, quick result generation ability, and require negligible or very simple samples pre-treatment [18]. Unlike other instruments, biosensors have a limitation: there are stability issues in the biorecognition component due to parameters like ionic strength, pH, and temperature [17]. Biosensors are sufficiently robust to permit regeneration, allowing consecutive analysis cycles to be completed using the same element [17]. They can be divided into several groups depending on the differences in their biological elements and transducer.

In a microbial biosensor, the whole microorganism is used as the biological recognition element. Only a few biosensors are available for antibiotic detection. These biosensors function on the principle of sensing the change in the enzymatic function of a microorganism in the presence of antibiotic residues. These systems work in microbial inhibition tests, the detection of inhibition in growth of microbes in the presence of an antibiotic. A hybrid biosensor, reported by Ferrini et al., detects the presence of antibacterial residues as well as generates electrochemical detection [19]. This system was constructed using *Bacillus stearothermophilusvar* Calidolactis strains and detects the inhibition in growth of microorganisms, electrochemically estimating the amount of CO_2 produced. The antibiotic residues present in the sample act as inhibitors for microorganisms, thus suppressing the growth of microorganisms and decreasing the production rate of CO_2. The rate of CO_2 production was recorded for over 120 minutes in comparison with a residue-free milk sample. Another bacterial sensor, developed by Das et al. using *Bacillus cereus* 66, was based on the β-lactamatic activity and used iodine as the indicator of reaction [20]. Its sensitivity and selectivity were examined by using different β-Ls and antibiotics. No color change was observed in the culture in the absence of antibiotics, while in the presence of antibiotics, color change was observed and the growth of *B. cereus* 66 was inhibited.

An electrochemical microbial biosensor was developed for the identification of tetracyclines (TCs) and quinolones (Qs) by Pellegrini et al. [21]. It also works on the inhibition mechanism of microbial growth and CO_2 production. *Escherichia coli* (ATCC 11303) is used in this study because it has good sensitivity for Qs and TCs. After 120 minutes the inhibition rate is recorded. The detection limit of these residues is 25 μg l^{-1}. This type of biosensor is used sensitively and specifically for these antibiotics only.

Most of the biosensors are based on the use of immunochemical biorecognition reactions. Generally, immunosensors are either electrochemical or optical. Afterward, surface plasmon resonance (SPR) biosensors came into use. They permit the easy handling, on-time result, and label-free reviewing of biomolecular interactions. Among approximately 49% of the developed detection methods reported, SPR was shown to be the best transducer [22]. Biosensor detection methods have been authenticated according to 2002/657/EC and are evolving rapidly as a transmission device [23]. Methods using biosensors have been regarded as the best suitable method for detection due to their cross-reactivity within antibiotic groups. Another SPR biosensor was developed for the detection of β-L residues present in milk, which was very sensitive to open β-L rings [24]. To increase the sensitivity of the method, two pretreatment methods were tested; one was enzymatic and another was chemical. Pre-treatment detection limit of AMP was 33, and 12.5 μg l^{-1} was achieved [24].

One study by Zhang et al. described one SPR immunosensor for the detection of AMP. In this biosensor, AMP was immobilized covalently on the surface of the sensor [25]. The sample that contains AMP was added to monoclonal anti-AMP antibodies. This sensor then detected the free anti-AMP antibody present on the surface of the sensor. The detection limit of this sensor was 2.5 μg l^{-1}. One SPR biosensor for the identification of three antibiotics was proposed by Fernandez et al. This immunosensor was designed to detect fluoroquinolone (FQ) antibiotics present in milk, like ciprofloxacin, enrofloxacin, and norfloxacin [26]. This sensor also works in inhibition format but in this case performed indirect inhibition. The polyclonal anti-FQ-haptenized protein (FQ-BSA) antibody was bound on the surface of the SPR and it was activated using FQ-BSA. The binding was inhibited by FQS presence. The detection limit for this sensor was 2.0 μg l^{-1}. Before this, another SPR biosensor was constructed, and that biosensor was also able to detect three different antibiotics [26]. The detection limits of this sensor were 1.7, 2.1, and 1.1 μg l^{-1} for ENRO, SPY, and CAP, respectively.

For the detection of CAP residues in milk, an inhibition-based immunosensor was developed by Ferguson et al. [27]. However, it was not as sensitive and the detection limit of this sensor was also very low: 0.05 μg l^{-1}. Haasnoot et al. identified SPR immunoassays for streptomycin residue detection in milk [28]. The immunoassays, based upon monoclonal anti-dihydro streptomycin antibodies, have their detection limit at 20 μg l^{-1}. Another streptomycin immunosensor, reported by Ferguson et al., uses Qflex TM antibodies, which gives the detection limit of 30 μg l^{-1} in whole bovine milk [27]. A parallel affinity immunosensor array (PASA), reported by Knecht et al., is used for detecting 10 different antibiotic residues in samples of milk [29]. It uses multi-analyte immunoassays based on the format of an indirect competitive ELISA. Its detection limit was from 0.12 to 32 μg l^{-1}. The improved form of the PASA system, proposed by Kloth et al., allows the study of 13 antibiotics [14]. Conzuelo et al. reported an amperometric magneto-immunosensor for the detection of TC, which immobilizes a polyclonal anti-tetracycline antibody from sheep, attached to magnetic bead surface, which is functionalized by protein G and screen-printed carbon electrodes. The detection limit of this sensor was $8.9\,\mu g\,l^{-1}$ for TC [30]. A similar immunosensor was proposed by Conzuelo et al. for sulfonamide (SAS). This immunosensor consists of polyclonal antibodies of rabbit, immobilized on the surface of the electrode, its detection limit being $0.15\,\mu g\,l^{-1}$ [30]. An immunoassay for the identification of sulfapyridine was designed in which the antibody was immobilized on glassy carbon plates modified with protein G, with a detection limit of $0.13\,\mu g\,l^{-1}$ [31]. An SPR biosensor detection method along with LC–MS/MS confirmation were reported for the detection of chloramphenicol residues in four different food mediums. The authentication of this method was achieved as mentioned in 2002/657/EC [14]. Marchesini et al. established a double SPR biosensor assay, which inspects the sample, after the initial run of SPR analysis, by exposing them to an HPLC fractionation followed by the second run of SPR and lastly by an LC–electrospray ionization-time of flight–MS confirmatory investigation to recognize and measure the remains in positive portions of protecting the fluoroquinolone receptor binding activity [32]. Biosensors containing assay multiplexing are the future of antibiotic estimation. They are also transportable devices that can be used for field purposes [18]. Recently developed SPR biosensor microarrays by Bremer et al. concurrently identify two aminoglycosides or mixtures of four widely used antibiotic families – aminoglycosides, amphenicols, sulfonamides, and fluoroquinolones – on a single sensor chip [33]. Similarly, a wavelength-interrogated optical system transducer-based biosensor can concurrently sense β-lactam, sulfonamide, fluoroquinolone, and tetracycline-based antibiotics on a multi-analyte biosensor [34]. Transportable, multiplex SPR biosensors have also been designed for field study of milk samples for the presence of fluoroquinolone family composites, sulfonamide, fluoroquinolone residues, and chloramphenicol [35].

Another class of recognition element known as DNA-based aptamer was recently reported to be used in tetracycline detection [36]. Aptamers are small sequences of single-stranded (ss) DNA or RNA oligonucleotides acquired in vitro procedure. Aptamers are identified as systemic evolution of ligands by exponential enrichment (SELEX), which chooses and amplifies aptamers from very huge and random DNA or RNA libraries [37]. Aptamers are specific; they have strong affinity toward the targeted molecules. As recognition elements, aptamers also provide the benefits of greater selectivity, facile labeling thermal stability, and cost-effective alteration, which makes them perfect candidates for the advancement of new biosensing approaches for the detection and identification of specific target molecules [38]. Despite the great potential of aptamer-based sensing approaches in the regions of therapeutics and clinical diagnostics, only limited aptamer-based products are found in the market [39].

A few corporations have released aptamers as an element of aptamer assay kits such as the NeoVentures Ochratoxin A and aflatoxin ELISA-like microwell plate assays with few affinity columns [40]. Based on the above characteristics and with sensibly designed sequences, the aptamers function as effective sensing elements in the design and development of constant, cost-effective, strong, and practical biosensing platforms for the recognition of target analytes. In the presence of its target molecule, the aptamer bends into a particular 3D conformation. A well-known example of such a 3D conformation is the antiparallel-G-quadruplex aptamer-target complex structure. The complex construction of this target aptamer results in numerous signal generations, contingent upon the transduction principle applied. Electrochemical recognition methods have various features such as automation, easy modification, disposable assays, sensors, high detection speed, and low volume consumption of reagents, which is cost-effective. Thus, based on these benefits, electrochemical aptamer sensors are well-suited for practical work and field applications [41].

Aptamer-based biosensors have gained greater importance in the detection of antibiotic traces in the milk sample. Aptamers are very stable and remain unaffected by temperature or pH. Compared to antibodies, aptamers are smaller in size, hence they reach the target site faster than antibodies [42]. There are two major techniques used for the immobilization of aptamers: one way is the direct attachment on the surface of the sensor via linkers, and the second method is noncovalent conjugation. For direct immobilization, some terminal functional groups are used to functionalize aptamers [43]. Occasionally, an additional linker is required to generate flexible movement between the terminal functional group and the aptamer [43, 44]. The linker, generally, has a series of thymidines, so that nonspecific binding could be minimized. The milk sample contains fat, proteins, and some non-transparent material that hamper the optical detection method, and therefore, pretreatment of the milk sample is necessary.

Several aptamer-based biosensors have been reported for the identification of β-L antibiotics. One polymer biosensor chip was reported by Dapra et al. that includes a microfluidic structure with aptamers immobilized [45]. Polymers can convert biological signals into measurable signals. Sometimes, they are used in place of the electrode and are cost-effective. Dapra et al. suggested an aptamer-based sensor in which ssDNA was functionalized using fluorescein amidite (FAM) [45]. This sensor can detect very low-level ampicillin (AMP). One aptasensor was proposed for sulfadimethoxine (SDMX), which entailed a ssDNA aptamer, modified by FAM and it was linked to the polymer nanobelts (CPNBs) [46]. The limit of detection of this sensor is $10\,\mu g\,l^{-1}$. Sometimes, the presence of other components in milk can also affect the specificity of the biosensor. To overcome this problem, 10-fold diluted milk sample should be used. This antibiotic is banned for animals because of chloramphenicol (CAP) detection. Ample aptasensors are designed for their detection. A fluorescence-based aptasensor, designed by Wu et al., was conjugated with magnetic nanoparticles (MNPs) to check the presence and concentration of CAP [47]. When the antibiotic is absent, the complex of magnetic nanoparticles and aptamer hybridizes to its complementary DNA (cDNA). This cDNA is then modified with up-conversion nanoparticles (UCNPs) to display fluorescence. Upon the addition of CAP, the aptamer tends to bind with CAP, and their binding causes separation of a few cDNA and production of a UCNPs-cDNA complex that decreases the fluorescence signal. This gives the detection limit of $0.01–1\,\mu g\,l^{-1}$. For the detection of CAP, an aptasensor was designed by Alibolandi et al., in which aptamers were associated with Cd-Te quantum dots (QDs). This system shows high quantum yield, stable fluorescence, and wide adsorption spectra. The limit of detection of this sensor was $0.2\,\mu g\,l^{-1}$ [48]. For tetracyclin (TCs) detection in milk, numerous aptasensors have been designed. Jeong and Paeng designed two enzyme-linked aptamer-based sensors [49]. They can be based either on an RNA-aptamer or on an ssDNA aptamer. Both the systems are based on the competitive assay. Kim et al. also developed a similar system, in which ssDNA aptamer was used for ox tetracycline (OTC) detection. They proposed an aptasensor for TC detection [50]. In this system, aptamers were attached by electrostatic interaction on the surface of gold nanoparticles (AUNPs) [51]. The presence of TCs was detected based on the color change phenomenon. The solution changes its color from blue to wine red in the presence of TC. The CTAB method, though it has a higher detection limit (2.7 times) than PDDA, has some disadvantages and the pre-treatment of milk is obligatory. Zhang et al. also proposed a biosensor for TC that works on a electrochemical method without the use of label, consisting of ssDNA-aptamer and modified glassy carbon (GC) electrode [52]. This system does not need any sample pretreatment and provides the detection limit of $0.1–100\,\mu g\,l^{-1}$. Luo et al. defined an ultrasensitive resonance scattering process for the recognition of TC in milk [53]. In this biosensor, gold nanoparticles were coated with

Table 12.1 Biosensors for identification of antibiotics.

Type of biosensor	Antibiotic	Detection Limit	Reference
Hybrid whole cell biosensor	–	–	[19]
Whole cell biosensor	Quinolone, Tetracycline (TCs)	25 μg l^{-1}	[21]
SPR-based biosensor	B-lactam	33 μg l^{-1}	[24]
SPR immunosensor	Antimicrobial peptides	2.5 μg l^{-1}	[25]
SPR immunosensor	fluoroquinolone	2.0 μg l^{-1}	[26]
SPR	aminoglycosides	–	[33]
Immunosensor	Chloramphenicol (CAP)	0.05 μg l^{-1}	[27]
SPR immunosensor	Streptomycin	2.0 μg l^{-1}	[28]
Affinity immunosensor	–	0.12–32 μg l^{-1}	[29]
Amperometric immunosensor	TC, Sulfonamide	8.9 μg l^{-1}, 0.15 μg l^{-1}	[30]
Aptamer biosensor	CAP	0.01–1 μg l^{-1}	[47]
Aptamer biosensor	CAP	0.2 μg l^{-1}	[48]
ssDNA-aptamer biosensor	TC	0.1–100 μg l^{-1}	[52]
Aptamer biosensor	TC	22 μg l^{-1}	[53]
Photoelectrochemical biosensor	–	0.22pM	[56]
Paper-based biosensor	–	–	[57]
Antibody immunosensor	TC	288.9 ng ml^{-1}	[58]
Antibody immunosensor	TC	12.44 ng ml^{-1}	[59]
Aptamer	TC	0.0073 ng ml^{-1}	[60]
Aptamer	Kanamycin	0.014 ng ml^{-1}	[61]

ssDNA-aptamer through Van Der Waals force that prevents gold nanoparticles from aggregation. The detection of TC is based on the aggregation of gold nanoparticles. In the presence of TC, aptamer binds with TC as they have a high affinity toward each other. The detection limit of this system was 22 μgl^{-1}. Protein-based sensor performance can also be enhanced with modifications: a fluorescein-labeled b-galactosidase mutant was used as a biorecognition element having compact catalytic activity in a fluorescence-based biosensor for b-lactams detection [54].

Further commercial success of biosensors needs wireless technology, miniaturization, and atomization [55]. In 2018, Sui, Zhou, et al. developed a photoelectrochemical biosensor, fabricated with graphite-like carbon nitride nanosheet, with the detection limit of 0.22 pM [56]. Nowadays, microfluidic sensors and paper-based sensors are gaining much attention for their rapid and point-of-care detection properties. Recently, a paper-based biosensor was reported to provide colorimetric detection of antibiotics that inhibit protein synthesis in bacteria. The biosensor generates a change in color induced by b-galactosidase, synthesized on paper discs that were freeze-dried and contained an in-vitro transcription/translation system. In the samples lacking any antibiotic, when applied to the paper discs, b-galactosidase will be produced, which hydrolyses a color producing substrate thus generating a change in color from yellow to purple. But when antibiotics are present, the color change will be hindered as the synthesis of b-galactosidase will be inhibited [57]. The types of biosensors that have been developed for the identification of antibiotics in the medium are listed in Table 12.1.

12.3. FUTURE PROSPECTS

The fate of biosensors looks extremely promising. Further investigations are needed to work on the exactness of recognition. Along these lines, for future improvement, the initial step is to enhance the strategy and desire to more continuously locate numerous antitoxins. Furthermore, the correct determination of optoelectronic materials confirms the accuracy and effectiveness of the test [62]. In addition, the perception of signs is likewise vital, which can frame the internet of things through the interconnection application with man-made brainpower. Simultaneously, we can use micromechanical frameworks to assemble useful stages to scale down the whole trial and testing process. The transportability and integrated nature of the chip is an important part in the biosensor used for antitoxin or other small particle recognition.

12.4. CONCLUSION

Antibiotic resistance has emerged as a threat and a major future challenge. The problem regarding antibiotic agent contamination and drug resistance is not limited to any one country but is a global challenge. To avoid any antibiotic traces in food items, EU legislation has enforced the establishment of monitoring plants at the national level, which will be responsible for detecting the amounts of contaminants and antibiotic traces present in animal-based food items. Many conventional techniques for the analysis of antibiotics residues in agricultural, food, and clinical therapy have been reported. These analytical techniques for antimicrobials are divided into three categories: physicochemical, microbiological, or immunochemical. Various techniques from each of these categories have been authenticated by following the guidelines. Screening for antibiotic residue present in food is mainly done using microbial growth-inhibition assays. Biosensors are emerging as screening devices, which provide quick and sensitive analysis. Currently, biosensor assays are used only for antibiotic detection. It is a sensitive, cost-effective, robust, and specific screening method. Algorithm-based methods can improve the specificity to the level where individual members of any antibiotic families can also be directly identified. In light of these findings, biosensors could become prevalent among the screening methods available for antibiotic residue analysis, possibly replacing the less specific and sensitive, more laborious, and voluminous microbial growth inhibition assay.

ACKNOWLEDGMENTS

The authors are highly thankful to the Director, CSIR-Indian Institute of Toxicology Research (IITR), Lucknow (India), for his valuable encouragement and intra institutional financial support.

REFERENCES

1. Centers for Disease Control, & Prevention (US) (2005). *Surveillance for Dental Caries, Dental Sealants, Tooth Retention, Edentulism, and Enamel Fluorosis: United States, 1988–1994 and 1999–2002*, vol. *54*. Department of Health and Human Services, Centers for Disease Control and Prevention.
2. Healy, E.M. and Kaitin, K.I. (1999). The European Agency for the Evaluation of medicinal Products' centralized procedure for product approval: current status. *Drug Information Journal 33* (4): 969–978.
3. Hamad, B. (2010). The antibiotics market. *Nature Reviews Drug Discovery 9* (9): 675.
4. Carlet, J., Collignon, P., Goldmann, D. et al. (2011). Society's failure to protect a precious resource: antibiotics. *The Lancet 378* (9788): 369–371.
5. Wright, G.D. (2007). The antibiotic resistome: the nexus of chemical and genetic diversity. *Nature Reviews Microbiology 5* (3): 175–186.
6. Waksman, S.A., Reilly, H.C., and Schatz, A. (1945). Strain specificity and production of antibiotic substances: V. Strain resistance of bacteria to antibiotic substances, especially to streptomycin. *Proceedings of the National Academy of Sciences of the United States of America 31* (6): 157.

7. Allen, H.K., Donato, J., Wang, H.H. et al. (2010). Call of the wild: antibiotic resistance genes in natural environments. *Nature Reviews Microbiology 8* (4): 251–259.
8. Partridge, S.R. (2011). Analysis of antibiotic resistance regions in gram-negative bacteria. *FEMS Microbiology Reviews 35* (5): 820–855.
9. Gøtzsche, P.C. and Jørgensen, A.W. (2011). Opening up data at the European medicines agency. *BMJ 342*: 1–4.
10. Dryden, M.S., Cooke, J., and Davey, P. (2009). Antibiotic stewardship—more education and regulation not more availability? *Journal of Antimicrobial Chemotherapy 64* (5): 885–888.
11. Wise, R., Hart, T., Cars, O. et al. (1998). Resistance is a major threat to public health. *British Medical Journal 317*: 609–610.
12. Ungemach, F.R., Müller-Bahrdt, D., and Abraham, G. (2006). Guidelines for prudent use of antimicrobials and their implications on antibiotic usage in veterinary medicine. *International Journal of Medical Microbiology 296*: 33–38.
13. Aarestrup, F.M., Kruse, H., Tast, E. et al. (2000). Associations between the use of antimicrobial agents for growth promotion and the occurrence of resistance among Enterococcus faecium from broilers and pigs in Denmark, Finland, and Norway. *Microbial Drug Resistance 6* (1): 63–70.
14. Kloth, K., Rye-Johnsen, M., Didier, A. et al. (2009). A regenerable immunochip for the rapid determination of 13 different antibiotics in raw milk. *Analyst 134* (7): 1433–1439.
15. Metkar, S.K. and Girigoswami, K. (2019). Diagnostic biosensors in medicine–a review. *Biocatalysis and Agricultural Biotechnology 17*: 271–283.
16. McGrath, T.F., Elliott, C.T., and Fodey, T.L. (2012). Biosensors for the analysis of microbiological and chemical contaminants in food. *Analytical and Bioanalytical Chemistry 403* (1): 75–92.
17. Cháfer-Pericás, C., Maquieira, A., and Puchades, R. (2010). Fast screening methods to detect antibiotic residues in food samples. *TrAC Trends in Analytical Chemistry 29* (9): 1038–1049.
18. Huet, A.C., Delahaut, P., Fodey, T. et al. (2010). Advances in biosensor-based analysis for antimicrobial residues in foods. *TrAC Trends in Analytical Chemistry 29* (11): 1281–1294.
19. Ferrini, A.M., Mannoni, V., Carpico, G. et al. (2008). Detection and identification of β-lactam residues in milk using a hybrid biosensor. *Journal of Agricultural and Food Chemistry 56* (3): 784–788.
20. Das, S., Kumar, N., Vishweswaraiah, R.H. et al. (2014). Microbial based assay for specific detection of β-lactam group of antibiotics in milk. *Journal of Food Science and Technology 51* (6): 1161–1166.
21. Pellegrini, G.E., Carpico, G., and Coni, E. (2004). Electrochemical sensor for the detection and presumptive identification of quinolone and tetracycline residues in milk. *Analytica Chimica Acta 520* (1–2): 13–18.
22. Reder-Christ, K. and Bendas, G. (2011). Biosensor applications in the field of antibiotic research—a review of recent developments. *Sensors 11* (10): 9450–9466.
23. Ashwin, H.M., Stead, S.L., Taylor, J.C. et al. (2005). Development and validation of screening and confirmatory methods for the detection of chloramphenicol and chloramphenicol glucuronide using SPR biosensor and liquid chromatography–tandem mass spectrometry. *Analytica Chimica Acta 529* (1–2): 103–108.
24. Gaudin, V., Fontaine, J., and Maris, P. (2001). Screening of penicillin residues in milk by a surface plasmon resonance-based biosensor assay: comparison of chemical and enzymatic sample pre-treatment. *Analytica Chimica Acta 436* (2): 191–198.
25. Zhang, Z.F., Liu, J., Xu, K.X. et al. (2009). Development of a miniature optical biosensor-based inhibition immunoassay for detection of ampicillin residues in milk. *Nanotechnology and Precision Engineering 7*: 483–489.
26. Fernandez, F., Pinacho, D.G., Sanchez-Baeza, F. et al. (2011). Portable surface plasmon resonance immunosensor for the detection of fluoroquinolone antibiotic residues in milk. *Journal of Agricultural and Food Chemistry 59* (9): 5036–5043.
27. Ferguson, J., Baxter, A., Young, P. et al. (2005). Detection of chloramphenicol and chloramphenicol glucuronide residues in poultry muscle, honey, prawn and milk using a surface plasmon resonance biosensor and Qflex® kit chloramphenicol. *Analytica Chimica Acta 529* (1–2): 109–113.

28. Haasnoot, W., Loomans, E.E., Cazemier, G. et al. (2002). Direct versus competitive biosensor immunoassays for the detection of (dihydro) streptomycin residues in milk. *Food and Agricultural Immunology 14* (1): 15–27.
29. Knecht, B.G., Strasser, A., Dietrich, R. et al. (2004). Automated microarray system for the simultaneous detection of antibiotics in milk. *Analytical Chemistry 76* (3): 646–654.
30. Conzuelo, F., Gamella, M., Campuzano, S. et al. (2012). Disposable and integrated amperometric immunosensor for direct determination of sulfonamide antibiotics in milk. *Biosensors and Bioelectronics 36* (1): 81–88.
31. Conzuelo, F., Stratmann, L., Grützke, S. et al. (2014). Detection and quantification of sulfonamide antibiotic residues in milk using scanning electrochemical microscopy. *Electroanalysis 26* (3): 481–487.
32. Marchesini, G.R., Haasnoot, W., Delahaut, P. et al. (2007). Dual biosensor immunoassay-directed identification of fluoroquinolones in chicken muscle by liquid chromatography electrospray time-of-flight mass spectrometry. *Analytica Chimica Acta 586* (1–2): 259–268.
33. Raz, S.R., Bremer, M.G., Giesbers, M. et al. (2008). Development of a biosensor microarray towards food screening, using imaging surface plasmon resonance. *Biosensors and Bioelectronics 24* (4): 552–557.
34. Adrian, J., Pasche, S., Voirin, G. et al. (2009). Wavelength-interrogated optical biosensor for multi-analyte screening of sulfonamide, fluoroquinolone, β-lactam and tetracycline antibiotics in milk. *TrAC Trends in Analytical Chemistry 28* (6): 769–777.
35. Fernández, F., Hegnerová, K., Piliarik, M. et al. (2010). A label-free and portable multichannel surface plasmon resonance immunosensor for on site analysis of antibiotics in milk samples. *Biosensors and Bioelectronics 26* (4): 1231–1238.
36. Kim, Y.J., Kim, Y.S., Niazi, J.H. et al. (2010). Electrochemical aptasensor for tetracycline detection. *Bioprocess and Biosystems Engineering 33* (1): 31. https://doi.org/10.1007/s00449-009-0371-4.
37. Song, S.P., Seifert, J.L., Connor, R.E. et al. (2008). TrAC. *Trends in Analytical Chemistry 27*: 108–117.
38. Swensen, J.S., Xiao, Y., Ferguson, B.S. et al. (2009). Continuous, real-time monitoring of cocaine in undiluted blood serum via a microfluidic, electrochemical aptamer-based sensor. *Journal of the American Chemical Society 131* (12): 4262–4266.
39. Bruno, J.G. (2015). Predicting the uncertain future of aptamer-based diagnostics and therapeutics. *Molecules 20* (4): 6866–6887. https://doi.org/10.3390/molecules20046866.
40. Penner, G. (2012). Commercialization of an aptamer-based diagnostic test. *IVD Technologies 18* (4): 31–37.
41. Sharma, A., Chandra Singh, A., Bacher, G. et al. (2016). Recent advances in aptamer-based biosensors for detection of antibiotic residues. *Aptamers Synth. Antibodies 2*: 43–54.
42. Toh, S.Y., Citartan, M., Gopinath, S.C. et al. (2015). Aptamers as a replacement for antibodies in enzyme-linked immunosorbent assay. *Biosensors and Bioelectronics 64*: 392–403.
43. Balamurugan, S., Obubuafo, A., Soper, S.A. et al. (2008). Surface immobilization methods for aptamer diagnostic applications. *Analytical and Bioanalytical Chemistry 390* (4): 1009–1021.
44. Centi, S., Tombelli, S., Minunni, M. et al. (2007). Aptamer-based detection of plasma proteins by an electrochemical assay coupled to magnetic beads. *Analytical Chemistry 79* (4): 1466–1473.
45. Daprà, J., Lauridsen, L.H., Nielsen, A.T. et al. (2013). Comparative study on aptamers as recognition elements for antibiotics in a label-free all-polymer biosensor. *Biosensors and Bioelectronics 43*: 315–320.
46. Song, K.M., Jeong, E., Jeon, W. et al. (2012). A coordination polymer nanobelt (CPNB)-based aptasensor for sulfadimethoxine. *Biosensors and Bioelectronics 33* (1): 113–119.
47. Wu, S., Zhang, H., Shi, Z. et al. (2015). Aptamer-based fluorescence biosensor for chloramphenicol determination using upconversion nanoparticles. *Food Control 50*: 597–604.
48. Alibolandi, M., Hadizadeh, F., Vajhedin, F. et al. (2015). Design and fabrication of an aptasensor for chloramphenicol based on energy transfer of CdTe quantum dots to graphene oxide sheet. *Materials Science and Engineering: C 48*: 611–619.
49. Jeong, S. and Rhee Paeng, I. (2012). Sensitivity and selectivity on aptamer-based assay: the determination of tetracycline residue in bovine milk. *The Scientific World Journal 2010*: 10.

50. Kim, C.H., Lee, L.P., Min, J.R. et al. (2014). An indirect competitive assay-based aptasensor for detection of oxytetracycline in milk. *Biosensors and Bioelectronics 51*: 426–430.
51. He, L., Luo, Y., Zhi, W. et al. (2013). A colorimetric aptamer biosensor based on gold nanoparticles for the ultrasensitive and specific detection of tetracycline in milk. *Australian Journal of Chemistry 66* (4): 485–490.
52. Zhang, J., Zhang, B., Wu, Y. et al. (2010). Fast determination of the tetracyclines in milk samples by the aptamer biosensor. *Analyst 135* (10): 2706–2710.
53. Luo, Y., He, L., Zhan, S. et al. (2014). Ultrasensitive resonance scattering (RS) spectral detection for trace tetracycline in milk using aptamer-coated nanogold (ACNG) as a catalyst. *Journal of Agricultural and Food Chemistry 62* (5): 1032–1037.
54. Chan, P.H., Liu, H.B., Chen, Y.W. et al. (2004). Rational design of a novel fluorescent biosensor for β-lactam antibiotics from a class A β-lactamase. *Journal of the American Chemical Society 126* (13): 4074–4075.
55. Luong, J.H., Male, K.B., and Glennon, J.D. (2008). Biosensor technology: technology push versus market pull. *Biotechnology Advances 26* (5): 492–500.
56. Sui, C., Zhou, Y., Wang, M. et al. (2018). Aptamer-based photoelectrochemical biosensor for antibiotic detection using ferrocene modified DNA as both aptamer and electron donor. *Sensors and Actuators B: Chemical 266*: 514–521.
57. Duyen, T.T.M., Matsuura, H., Ujiie, K. et al. (2017). Based colorimetric biosensor for antibiotics inhibiting bacterial protein synthesis. *Journal of Bioscience and Bioengineering 123* (1): 96–100.
58. Devkota, L., Nguyen, L.T., Vu, T.T. et al. (2018). Electrochemical determination of tetracycline using AuNP-coated molecularly imprinted overoxidized polypyrrole sensing interface. *Electrochimica Acta 270*: 535–542.
59. Starzec, K., Cristea, C., Tertis, M. et al. (2020). Employment of electrostriction phenomenon for label-free electrochemical immunosensing of tetracycline. *Bioelectrochemistry 132*: 107405.
60. Huang, Y., Yan, X., Zhao, L. et al. (2019). An aptamer cocktail-based electrochemical aptasensor for direct capture and rapid detection of tetracycline in honey. *Microchemical Journal 150*: 104179.
61. Zhou, Y., Li, F., Wu, H. et al. (2019). Electrochemical aptasensing strategy for kanamycin detection based on target-triggered single-strand DNA adsorption on MoS2 nanosheets and enzymatic signal amplification. *Sensors and Actuators B: Chemical 296*: 126664.
62. Sun, Y., Zhao, J., and Liang, L. (2021). Recent development of antibiotic detection in food and environment: the combination of sensors and nanomaterials. *Microchimica Acta 188* (1): 1–22.

13

Application of Microbes in Dye Decolorization

Kuruvalli Gouthami[1], Lavanya Lakshminarayana[1], Vadamalai Veeraraghavan[1], Muhammad Bilal[2], Ram N. Bharagava[3], Luiz F. R. Ferreira[4], Abbas Rahdar[5], Paul O. Bankole[6], Juliana H-P Américo-Pinheiro[7], and Sikandar I. Mulla[1]

13.1. INTRODUCTION

Microbes are ubiquitous in the biosphere. In the ecosystem, the key strategy for colonization and interactions between the host–microorganism and microorganism–microorganism occurs under a variety of different environmental conditions [1, 2]. Microbes' existence still affects the ecosystem in which they thrive. Microbes' impacts on their surroundings may be positive, negative, or undetectable by observation and measurements in the ecosystems [3]. On the other hands, human beings build industries and urban cities to live at a sophisticated level, and in doing so they are disturbing the natural green environment, which in turn impacts the environment with various contaminants (chlorinated compounds, dyes, nitroaromatics, pesticides, fungicides, etc.) released in the air, soil, and water [4–16]. Literature revealed that several of these have carcinogenic and mutagenic activities [17–20]. Azole dyes are a group of dyes that produce azo compounds [21]. Azo compounds are organic chemical compounds that have the general formula (R-N=N-R′), where R′ and R are either alkyl (aliphatic) or aryl (aromatic), and (N=N) is a nitrogen atom [22]. Azo dyes are perhaps the most prevalent and flexible type of dye, responsible for more than half of all dyes produced annually. Currently, more than 2000 different azo dyes are used to color a variety of materials. To make azo dyes, an aromatic primary amine is

[1]*Department of Biochemistry, School of Allied Health Sciences, REVA University, Bangalore, India*
[2]*School of Life Science and Food Engineering, Huaiyin Institute of Technology, Huai'an, China*
[3]*Department of Environmental Microbiology, School for Environmental Sciences, Babasaheb Bhimrao Ambedkar University (A Central University), Lucknow, Uttar Pradesh, India*
[4]*Waste and Effluent Treatment Laboratory, Institute of Technology and Research (ITP), Tiradentes University, Aracaju, Sergipe, Brazil*
[5]*Department of Physics, University of Zabol, Zabol, Iran*
[6]*Department of Pure and Applied Botany, College of Biosciences, Federal University of Agriculture, Abeokuta, Ogun, Nigeria*
[7]*Post-graduate Program in Environmental Sciences, Brazil University, São Paulo, Brazil*

Microbial Biotechnology: Role in Ecological Sustainability and Research, First Edition.
Edited by Pankaj Chowdhary, Sujata Mani, and Preeti Chaturvedi.

diazotized and then coupled, which contains one or more nucleophiles rich in electrons, such as hydroxyl and amino [23].

There are different methods of azo dye synthesis. In an alkaline medium, nitroaromatic derivatives are reduced; quinone and hydrazine condensation, and condensation of primary amine with nitroso compounds, primary amine oxidation by lead tetra acetate or permanganate potassium, nitroso compounds are reduced by $LiAlH_4$, etc. [23]. Color-display groups and polar groups are found on the ring of the dyes, which are usually aromatic and heterocyclic compounds. Their structure is more complicated and stable, making degrading printing and dyeing wastewater more difficult. Benzene rings, naphthalene, aromatic ring heterocyclic, and enolizable aliphatic groups can all be bonded to the azo. These are essential for the dye's color, as they come in a variety of shades and intensities. Microbes have been used widely in dyes that are used in many manufacturing applications all over the world. Dyes have been an important part of the socioeconomic structure of our cultures; dyes have been used since antiquity. Although natural dyes have been used, synthetic dyes have been an essential component of contemporary economy since 2000 BCE [24]. The azo dye causes the color to appear in polluted water that is dyed, and this hue filters sunlight, which is detrimental to numerous photo-initiated chemical processes that are necessary for aquatic life [25]. Azo dyes are commonly used in a variety of applications such as coloring compounds in the textile, pharmaceutical, and culinary sectors, but they have a high level of toxicity, mutagenicity, and carcinogenicity. There are many technical solutions used to treat dyes such as advanced oxidation and coagulation-flocculation [26], but they are not practicable due to the high price and intricate processes required. In addition, some developing nations with large textile industries but unreliable energy sources, such as China and Bangladesh, can benefit from less energy-intensive wastewater treatment methods [27]. Dyes remain in discharged wastewaters due to their recalcitrance, posing health risks and inducing color pollution. A wide number of dyes can be recolored by a variety of fungi and bacteria [28]. Azo dyes are now the most common type of dye synthesis, and their importance is expected to increase in the coming years. They have an important role in the regulation of the dye and printmaking sectors. A basic coupling and diazotization process is used to make these dyes [29]. To achieve the desired yield, particle size, and color properties of the dye for increased dispensability, various routes and modifications are used [30]. Bioremediation is the transformation or breakdown of pollutants by microorganisms into non-hazardous or less hazardous compounds. Bioremediation is mostly based on microorganisms that attack contaminants enzymatically and transform them into harmless compounds. The chemical is transformed by microbial organisms via metabolic or enzymatic activities. It works on the basis of two methods: co-metabolism and growth. An organic contaminant is used during periods of rapid expansion as the sole source of energy and carbon.

During the process, organic contaminants are completely degraded (mineralized). Phytoremediation refers to the use of plants in the bioremediation of contaminants. Phytoremediation is a new green technique that allows hazardous compounds to be removed or degraded. Phytoremediation occurs in a different environmentally friendly method for removing or degrading harmful substances from soils, sediments, groundwater, surface water, and air are all examples of natural resources (RTDF) [31]. Plants that have been genetically modified are also used. As a result, nature's method of useful materials, or transforming biodegradable materials into resources that other species may ingest and utilize, is described as biodegradation.

13.2. POLLUTION OF AZO DYES IN THE ENVIRONMENT

Azo dyes are frequently used in clothing, nutrition, cosmetics, and pharmaceutical sectors as coloring agents [32]. Some of these and other metabolic byproducts are mutagenic, carcinogenic, and poisonous, reducing their presence in industrial wastewaters has become a pressing issue. According to the World Health Organization, polluted water is responsible for 75% of diseases and over 8.29 lakhs death per year. Azo dyes account for 65–75% of all textile dyes. In 10–25% of textiles during the dyeing process, colors are lost, and 2–20% are released as fluid effluents in various environmental components. Textile dyeing procedures are the most ecologically harmful production processes because the chemical reagents employed include a high concentration of inorganic and organic chemical components. Furthermore, one of the most serious issues is the appearance of color in effluent textile wastewater. It has been seen that the colored waters used for dyeing processes are highly contaminated with chemicals, clothing auxiliaries, and dyes. Textile wastewater has different properties depending on the manufacturing process, technology, and chemicals used. Dye wastes are the most well-known substances in fabric wastewater, and those substances are dangerous to the organic environment, as well as blocking sunlight, which causes serious problems in biological populations (Table 13.1) [35].

Azo dyes contribute more than 70% of all industrial output worldwide (approximately 9 million tons). The yearly disposal of 4 500 000 piles of dyes and/or their byproducts was both an environmentally and economically damaging issue due to their carcinogenic or genotoxic potential. Microbe(s)-mediated dye degradation/decolorization is considered cost-effective and environmentally friendly with lower input than physicochemical approaches [35]. However, associations with chemically numerous components under a variety of environmental situations, dyes with metabolically different bacteria create metabolites with varying toxicity. Furthermore, most microbial dye-degradation research focuses on decolorization, with minimal consideration given to detoxification.

Textile wastewater pollutes the atmosphere significantly. The main difficulties are high levels of organic matter and specialized enduring colorants (dyes) that must be resistant to perspiration, detergent, light, water, and other oxidizing agents. As a result, the azo dyes such as BR46 and AB113 are commonly used in the weaving trade. Dye concentrations in textile industry effluents usually range from 10 to 200 mgl. At concentrations as low as 1 $mg l^{-1}$, most textile dyes are visible to the naked eye [33].

Table 13.1 Effects of azo dyes on human health and treatment methods.

Name of the Azo dyes	Effects on human health	Physical treatment	Chemical treatment	Biological treatment	References
Reactive brilliant red	In human beings, it inhibits the function of serum albumin	Sedimentation	Neutralization	Stabilization	[25]
Acid violet	Lipid peroxidation	Filtration	Reduction	Aerated lagoons	[34]
Disperse red-1	Effects of micronuclei in human lymphocytes	Flotation	Oxidation	Trickling filters	[30]
Reactive black 5	Decrease urease activity	Reverse Osmosis	Ion exchange	Anaerobic digestion	[34, 82]
Dispersive orange 291	Mutagenic, Cytotoxic, genotypic effects	Solvent extraction	Electrolysis	Fungal treatment	[30]

13.2.1. Toxic Effects of Azo Dyes

Today's synthetic dye delivery chemicals are usually exceedingly harmful, cancer-causing, or even irritating [36]. The compound aniline in the sources of azo dyes are a common category of dyes. They are extremely toxic substances and hazardous to work with, as well as being extremely flammable. Dioxin is one of the other harmful compounds used in the dyeing process; it is carcinogenic and may alter hormones. Formaldehyde, which is a potential carcinogen, and heavy metals that are proven carcinogens, such as chromium, zinc, copper, etc., are also found [37]. The release of brightly colored sewage from colored synthetic dyes in inland and coastal waterways is an increasing environmental issue. Synthetic dyes are used in a variety of sectors, including photography, textiles, paper printing, pharmaceuticals, cosmetics, food, etc. These industries employ approximately 10000 different dyes and pigments, and over 0.7 million tons of synthetic dyes are produced worldwide. Azo dyes are only hazardous when the azo link is broken and aromatic amines are produced, which is mostly by anaerobic gut microbes. Aromatic amines are converted to reactive electrophilic compounds that bind DNA covalently by metabolic reduction. Azo dyes may be rapidly taken by oxidizing the azo bond to highly reactive electrophilic diazonium ions. Following azo (decolorization), both compounds produce sulphanilic acid, but with differing amino naphthols. However, after reducing the binding by azonaphthol, food Yellow's toxicity reduced marginally, whereas Acid Orange 7's toxicity surged 100-fold. When compared to its sulphonated analog, 1-amino-2-naphthol (1-amino-2-naphthol-6-sulphonate) was very toxic. Sulphanilic acid's toxicity was equivalent to that of the dyes that had not been lowered. As a result, the release was most likely responsible for the increased toxicity of Acid Orange 7 when it was minimized to 1-amino-2-naphthol [38].

13.2.2. Azo Dye Effects on Human Systems

Azo dyes are difficult to remove from wastewater using traditional wastewater treatment procedures because they are light-stable and resistant to microbial deterioration or fading via washing. Around 10% of the reactive dyes used during dyeing of fabrics is thought never to attach to the fibers and is consequently released into the environment. In humans, several azo dye components, such as benzidine, have been associated to bladder cancer. In addition, dyers exposed to azo dyes were at increased risk of developing bladder cancer, and the dyes' byproducts were carcinogenic to living organisms including humans [39]. Humans can be exposed to azo dyes by food, inhalation, or skin contact. Within the body, azo dyes are biotransformed into aromatic amines. Human skin microbiota, including *Staphylococcus*, *Corynebacterium*, *Kocuria*, *Kytococcus*, *Micrococcus*, and *Dermacoccus* cleaved azo dyes including Orange II and Methyl Red. Hundreds of organisms live on human skin, and many of them produce azo reductase [39]. This information should be important to anybody who uses tattoo inks, fabrics, or cosmetics. Without being broken into aromatic amines, some azo dyes, such as o-Aminoazotoluene (also known as C.I., Fast Garnet GBC base, or Solvent Yellow 3), might be carcinogenic. Methyl Yellow has been discovered to be a powerful cancer agent. Other azo dyes, such as 3, 3'-Dichlorobenzidine, have been linked to human bladder cancer. 4-Aminobiphenyl can be found in contaminated food dyes, cigarette smoke, frying oil vapors, etc. It is a highly potent human carcinogen that induces bladder cancer.

13.2.3. Azo Dye Effects on Animals

Various industries release toxic effluents including azo dyes, which have an implication for water resources, soil fertility, aquatic species, and ecosystem stability. They are hazardous to both animals and mammals. Tumors were found in rats and mice exposed to 2,4-diaminotoluene. The most prevalent forms of tumors include kidney carcinomas, hepatocellular carcinoma, mammary gland tumors, fibroma, etc. [40]. In female B6C3F1 mice, dietary 2-nitro-p-phenylenediamine was found to be carcinogenic, causing an increase in the prevalence of primary adenomas and hepatocellular neoplasms. Several azo dyes have been shown in studies with microbes and mammalian cells to be genotoxic, mutagenic, and carcinogenic. These effects can be attributed to the direct impact of dyes on cells or, more specifically, to the creation of metabolic products resulting from the breakdown of the azo link, which are capable of interacting with and disrupting the DNA molecule [41]. When mice and rats were fed modest dosages of the azo dye p-dimethylaminobenzene (p-DAB) over lengthy periods of time, the hepatocarcinogenic effect of this chemical was demonstrated. Chronic p-DAB administration to these animals resulted in a considerable increase in mitotic activity of liver parenchyma cells compared to the negative control. Prenatal exposure to Congo Red, for example, permanently decreased the amount of germ cells in male and female rats and mice.

13.2.4. Azo Dye Effects on the Aquatic System

Under natural environmental circumstances, azo dyes basically do not deteriorate. When wastewater from the industry is discharged into the environment, its bioaccumulates. This causes problems not just in the water into which it has been discharged, but also across the environment. Azo dyes, for example, have been linked to growth inhibition, neurosensory impairment, metabolic stress, and fish mortality as well as plant growth and productivity. Therefore, the water pollution caused by these companies has an impact on the use of downstream water, not only in terms of drinking water for humans and animals, but also in all agricultural, fishing, tourism, and leisure industries [42]. The textile industry generates a huge volume of liquid waste. Organic and inorganic chemicals are present in these textile effluents. Not all colors applied to garments are fixed on them during the dyeing process, and some portion of these colors always remain unfixed to the textiles and are rinsed off. Unfixed dyes have been detected in significant amounts in textile effluents. The amount of water used and discharged varies according to the type of cloth manufactured [42]. To make 1 kg of fabric, 0.15 m^3 of water is required. One of the supplies of agent activity discovered within the Cristais watercourse in Brazil, a source of drinking water for 60 000 residents, was the Associate in Nursing group coloring facility. Despite the fact that the drinking water was treated at a plant 6 km downstream of the discharge location, tests proved the presence of carcinogenic aromatic amines after it was discharged from the treatment plant, and it was deemed safe. This was unknown before the tests, and it might have had a negative influence on the health of the 60 000 people who had previously relied on this water [43].

13.3. IMPORTANCE OF MICROBES IN THE ENVIRONMENT

The dissolution of all organic materials by a wide variety of microbes is referred to as "biodegradation." In the microbiological sense, a diverse group of organisms, mostly bacteria, yeast, and fungus, as well as perhaps other creatures, are involved. Bioremediation is a long and drawn-out procedure. Only a few fungal and bacteria species are proven to be effective pollution degraders. Many strains are successful as bioremediation agents in the laboratory, but only in that setting. Temperature, humidity, oxygen, pH value, soil structure, and the corresponding amount of nutrients, as well as the low bioavailability of pollutants and the presence of other pollutants, influence the development of bacteria [44]. One of the primary processes that determines the environmental fate and transit of organic pollutants is transformation or degradation, which also includes abiotic degradation and biodegradation. Organic contaminants are converted into degradation products or totally mineralized into a carbon field throughout these processes. Microorganisms may degrade organic pollutants in both anaerobic and aerobic environments. However, aerobic conditions allow rapidly and thorough biodegradation of the majority of organic contaminants [45]. The initial intracellular charge of an organic pollutant, such as a hydrocarbon, is an oxidative process, with oxygen activation and incorporation mediated by oxygenases and peroxidases. Hydrocarbons such as oil, radionuclides, polyaromatic hydrocarbons (PAHs), polychlorinated biphenyls (PCBs), and other metals are among the compounds that bioremediation and biotransformation methods aim to degrade, transform, or accumulate using the amazing, naturally occurring microbial catabolic diversity [46].

In the last several decades, very hazardous organic substances have been produced and released into the environment for long-term direct or indirect usage. Some of these substances include fuels, polychlorinated biphenyls (PCBs), polycyclic aromatic hydrocarbons (PAHs), dyes, pesticides, etc. Addtionally, chemicals such as metals and radionuclides that are enormously present in natural organic compounds are more resistant to degradation in indigenous plants, which disintegrate quickly once introduced into the environment.

13.3.1. Microbes in the Carbon Cycle

Microbes play a crucial role in the carbon balance, which is a necessary component of all living things. In marine habitats, microbes found in oxygen-free places like marshes and deep muck in lakes anaerobically process carbon. The most prevalent source of carbon that reaches a carbon cycle is carbon dioxide (CO_2). CO_2 is a chemical that is soluble in water and is found in the atmosphere. CO_2 is used by plants and photosynthetic algae to synthesize carbohydrates throughout photosynthesis. Animals and plants utilize carbon to form carbs, proteins, and lipids, which are subsequently used to build internal structures or receive electricity. CO_2 is used by chemoautotrophs, such as bacteria and archaea, to produce carbohydrates. The tricarboxylic acid cycle converts this carbon, which is present in the form of sugar, into energy via respiration. Carbon may also be used by microbes to create energy in anaerobic environments by a mechanism known as fermentation [47]. A terrestrial ecosystem's principal contributions are plants; but, in certain habitats, free-living planktons, symbionts, and cyanobacteria including lichens often help fix carbon in the living ecosystem. Heterotrophic bacteria and fungi recycle non-living organic matter, while saprobes use organic matter and while breathing, they emit CO_2 and add it to the carbon cycle. Higher species, such as herbivores and carnivores, use gut bacteria in their digestive tracts to absorb organic materials for energy; this mechanism is known as decomposition, and it results in inorganic compounds such as ammonia, CO_2, water, and so on [Figure 13.1] [48].

Similarly, bacteroidetes degrade more recalcitrant carbon molecules, such as chitin, lignin, and cellulose, which are involved in a higher amount of usable energy. Nitrogen aids in the development of transport and extracellular enzymes [49].

13.3.2. Microbes in Methane Production

Methane is a simple substance that occurs as a gas in the atmosphere and is one of the most significant fossil fuels. Heat is produced when a methane molecule decomposes [49]. By anaerobic or fermentative breakdown, certain microorganisms convert organic molecules into organic acids and gases like CO2 and hydrogen. Under anaerobic conditions, methanogens can use hydrogen to convert CO_2 to methane. Under aerobic circumstances, methane-oxidizing bacteria like methanotrophs convert methane to CO_2, water, and energy, completing the cycle [50]. Purple sulfur and green bacteria, for example, when producing

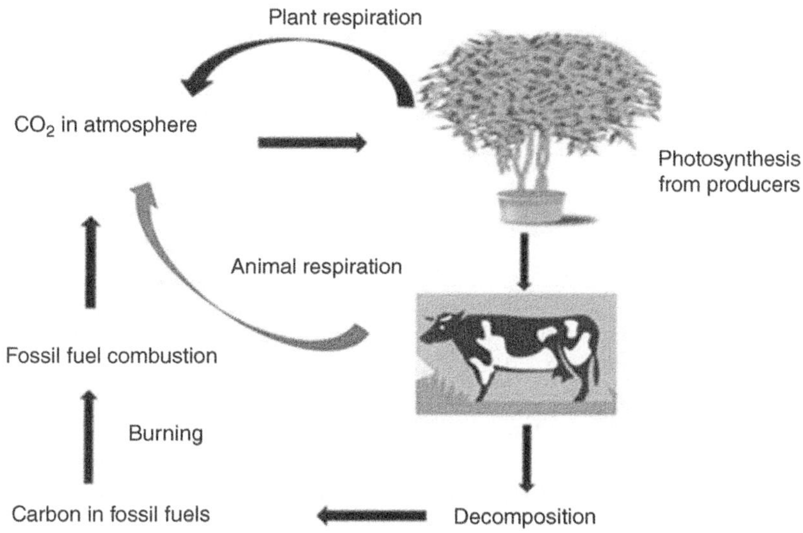

Figure 13.1 Representation of the importance of the carbon cycle.

energy, break down hydrogen sulfide (H_2S) into carbon-containing molecules, contributing to the carbon cycle. *Thiobacillus ferrooxidans*, for example, as a result of the oxidation of ferrous iron to ferric iron, obtains energy and contributes to the carbon cycle [51].

13.3.3. Microbes in the Nitrogen Cycle

The element nitrogen is essential and is found in the structure of nucleic acids, proteins, chlorophyll, and amino acids. Microorganisms play a significant part in the nitrogen cycle via numerous processes such as nitrification, denitrification, nitrogen fixation, etc. [52]. Hence, the method of converting nitrogen from the atmosphere into nitrite, nitrate, and ammonium is referred to as the nitrogen cycle. While the availability of plant biomass output is limited by nitrogen, ecosystem productivity is constrained by microbial activities [53]. Ammonification is the process of converting organic nitrogen into ammonia. Organic nitrogen is produced during decomposition and is then converted to ammonia by fungi and bacteria (aerobic/anaerobic). Three enzymes are involved in this process: glutamate dehydrogenase, glutamate oxoglutarate aminotransferase, and glutamine synthetase. Bacteria and archaea can also fix nitrogen from the atmosphere by converting it to ammonium. Ammonia can also be formed directly through plant and animal decomposition, as well as by animal waste collection. Nitrogenase is an oxygen-sensitive enzyme that catalyzes nitrification when there isn't much oxygen available. N-fixation necessitates each mole of nitrogen fixed; energy is used in the form of ATP (16 mol) [54].

Free-living microbes including *Clostridium*, *Burkholderia*, *Azotobacter*, etc., can fix nitrogen and have symbiotic associations with plants such as frankia, rhizobium, and mesorhizobium in the rhizosphere [55]. These bacteria can be found in the nodules of legume roots, such as beans and peas. They have a nitrogenase enzyme that produces ammonia by combining gaseous atmospheric nitrogen with hydrogen. Native legumes Clianthus and Sophora develop a symbiotic relationship with *Rhizobium* or *Mesorhizobium leguminosarum* [55]. As a result, the relationship is mutualistic and symbiotic. The ability to fix nitrogen through symbiotic rhizobia is the concentration of free-living soil microorganisms two to three orders of magnitude greater. On an industrial scale, nitrogen may even be artificially fixed. The Haber process, as it is known today, is primarily used to produce fertilizer [56]. Nitrification is the conversion of ammonium to nitrite, which is then converted to nitrate. The conversion of ammonium to nitrate is the first step. Some soil bacteria, such as *Crenarchaeum*, *Nitrosomonas*, *Nitrososphaera*, *Nitrosospira*, convert ammonia to nitrite, and other bacteria subsequently oxidize it to nitrate, such as *Nitrospira* and *Nitrobacter*. Various bacteria species, including Nitrosomonas, are responsible for this. The release of energy from the oxidation of ammonia to nitrite, which is utilized by nitrifying bacteria to ingest CO_2, also transforms the ionic phase of soil from negative to positive [56].

The nitrogen cycle is completed by denitrification. Nitrate is converted back to nitrogen gas in this process. Denitrification is the process of converting nitrite (NO_2^-), nitrate ($NO^{3-)}$, and nitric oxide (NO) to nitrous oxide ($_{N2O}$), mild nitrogen gas (N_2), or a greenhouse gas. This part of the nitrogen cycle is anaerobic, which means it takes place without the presence of oxygen. The amount of oxygen present in the soil reduces as it occurs mostly in water-logged areas. This causes facultative anaerobic bacteria to catalyze denitrification. To complete the nitrogen cycle, fixed nitrogen from soil and water through the process of denitrification returns nitrogen to the atmosphere [57].

A variety of soil microorganisms carry out denitrification; *Firmicutes*, *Actinobacteria*, and *Proteobacteria*, as well as other soil eukaryotes, decompose nitrogen. Most bacteria lack one or more denitrification enzymes, making them imperfect denitrifiers. For example, most bacteria and fungi lack nitrous oxide reductase, resulting in the production of N_2O as a final product. As a result of incomplete denitrification, greenhouse gases are released [Figure 13.2] [58].

13.3.4. Microbes in the Sulfur Cycle

Sulfur is found in all living things in the form of amino acids, coenzymes, vitamins, and other compounds. Cysteine and methionine are the two amino acids that contain sulfur [59]. In certain cases, the amount of sulfur found in the dry weight of living matter is less than 2%. Despite its paucity of organisms, it is a critical component in all living systems. Microbes,

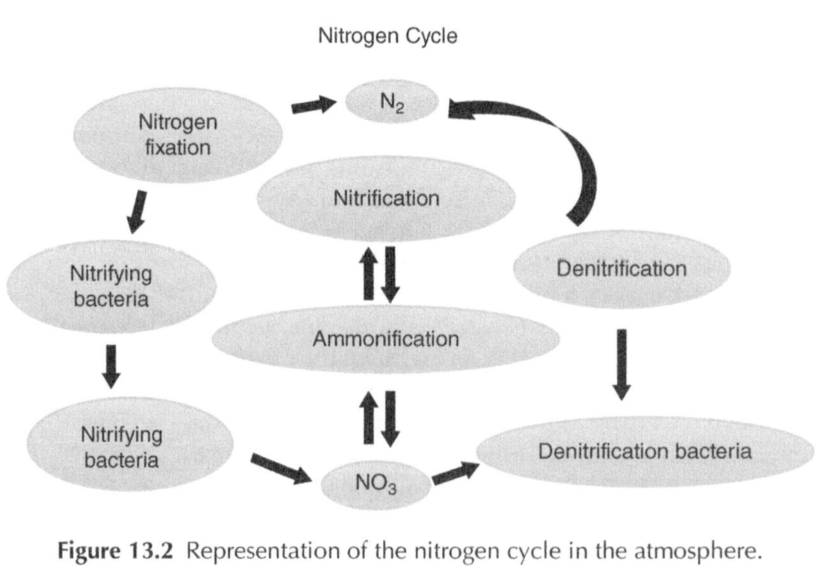

Figure 13.2 Representation of the nitrogen cycle in the atmosphere.

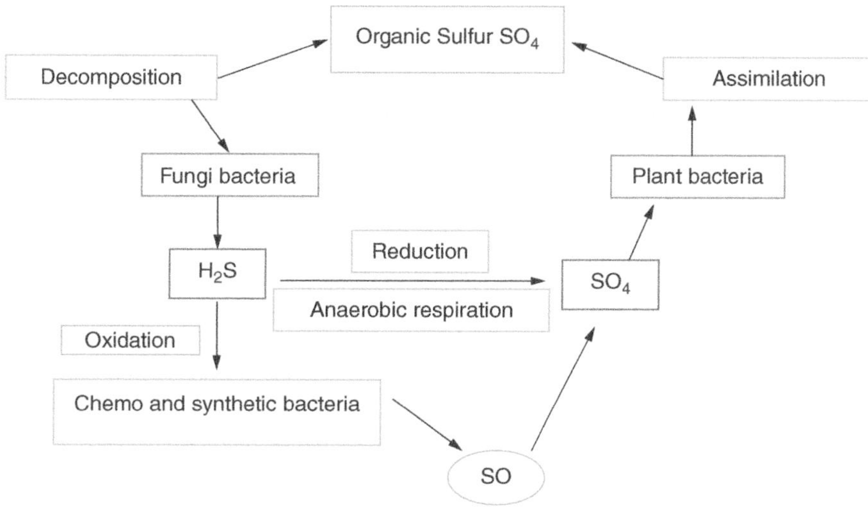

Figure 13.3 Schematic representation of microbes' role in the sulfur cycle.

including carbon and nitrogen, are able to change sulfur from its highly oxidized state (sulfate or SO_4) to its most reduced state (sulfide or H_2S) [Figure 13.3] [60].

Some prokaryotes in particular are involved in the sulfur cycle. H_2S is oxidized to S and further from S to SO_4 by two unrelated classes of prokaryotes. The first is oxygenic photosynthetic green and purple sulfur bacteria that use H_2S as an electron supply during cyclic photophosphorylation.

The second is the "colorless sulfur bacteria" (which is now an inaccuracy since the group includes several archaea) that utilizes Sand H_2S as energy resources. The species will normally intervene the full oxidation of H_2S to SO_4 in either case [61]:

$$H_2S \rightarrow S \rightarrow SO_4 \rightarrow \text{phototrophic sulfur oxidation or litho}$$

As thermophiles, sulfur-oxidizing prokaryotes can be found in hot (volcanic) springs and near H_2S-rich deep-sea magma chambers. They may also be acidophiles, since they produce sulfuric acid which acidifies their atmosphere. During anaerobic respiration, sulfate-reducing bacteria produce H_2S, which is similar to denitrification since S and SO_4 can be used as electron acceptors for respiration. Only anaerobic conditions require the use of SO_4 for respiration, so they can be employed as electron acceptors. In anaerobic sediments, soils, bogs, etc., or anywhere it arises, the process produces a distinct odor of H_2S. Plants and bacteria absorb sulfur as SO_4 for application and removal of sulfide. During decomposition, bacteria and animals may extract the proteins with a sulfide group to provide sulfur. The sulfur cycle is completed through these steps [62].

13.4. REMOVAL AND DEGRADATION OF AZO DYES

Degradation and decolorization of azo dyes can be accomplished in two ways: the chemicals are adsorbing through biodegradation of the dyes or microbial biomass (biosorption) through living cells. Biochemical azo dye treatment is the most effective approach because it can degrade almost all dye materials while still overcoming many of the physicochemical processes. The color can be removed from dye-containing wastewaters using a variety of biological, physical, and mechanical treatment methods. Each strategy has its own set of benefits and drawbacks. As a result, the majority of dye reduction methods include a variety of processes [63]. Microbial interactions are dependent on the maintenance and establishment of microbial populations in the environment. These associations are triggered by environmental recognition, which is accompanied by the transmission of genetic and molecular material, which includes a variety of pathways and molecule groups. These processes enable microorganisms to develop themselves in a population, which may result in a lot of diversity based on the multitrophic interaction [64]. It was previously assumed that relationships between microbes are only inhibitory. The role of various chemical, physical, genetic, and biological factors in controlling various forms of microbial interactions is defined [65]. Antimicrobial host defenses and environmental conditions play significant roles as well. Microorganism interaction allows a population of bacteria to control gene expression collectively in response to host and environmental signals provided by the same or even different organisms. As a result, this causes a coordinated reaction in the microbial community, resulting in good pathogenic outcomes that individual cells will not be able to achieve [65]. Owing to the difficulty of degradation, as well as the high cost and technical complexity of the different processes involved, many methods developed to treat these dyes are not feasible. Other treatment options must be used in countries with an unstable energy supply. Aside from being recalcitrant and resistant to various degradation processes, not only is the release of dye clothing wastewater into receiving streams unacceptably ugly and harmful to marine life but many of these are the dyes also mutagenic, allergenic, poisonous, carcinogenic, etc. [66]. These are commonly contemplated xenobiotic composites that are resistant to rate of degradation. Nonetheless, some microorganisms have been shown in recent years to be capable of converting under specific climatic circumstances, and azo dyes can be added to non-color items or even entirely mineralized. Using manganese peroxidases, laccases, and ligninases, various lignolytic fungi were observed to decolorize azo dyes.

Engineered habitats and vertical-flow built wetlands are planned to eliminate contaminants from wastewater. To simulate the treatment that happens in natural wetlands, these systems employ heterotrophic bacteria, aquatic plants, and a combination of naturally occurring processes. The study mainly focused on efficiency of ponds and in particular, built wetlands, in the treatment of textile wastewater [67]. Performance, on the other hand, rarely covers all seasons [68]. Textile production is the primary source of water contamination due to the release of azo dye by textile effluent, causing severe consequences to environmental and human health [69]. Numerous experiments focusing on the use of microorganisms to dissolve dyes indicate that biodegradation is an environmentally safe and cost-effective process for treating dye-containing wastewater. Textile wastewater is biologically treated in a variety of ways, ranging from fungal, algae, yeast, and bacterial culture to consortia. Since enzymes have catalytic activity, they can be used in very small amounts to speed up reactions. Through aerobic, anaerobic, and concurrent anaerobic-aerobic treatment cycles, plants and their enzymes can successfully eliminate the color of a variety of azo dyes. As a result, microorganisms that generate enzymes may be an effective solution for reducing water emissions [69].

13.4.1. Decolorization of Azo Dyes by Bacteria

Bacteria may degrade azo dye to a substantial quantity in both anaerobic and aerobic situations. The reductive cleavage of the –N=N– bond is the initial step in the bacterial degradation of azo dyes. Anaerobes and facultative anaerobes also provide strong degradation results [70]. The function of various bacteria classes in azo dyes that are decolorized needs to be studied broadly. This bacterium can be found in a variety of ecological roles, including water, soil, human and animal excreta, and even contaminated food. Since bacteria are simple to cultivate and grow quickly, they are ideal for decolorized and azo dye mineralization [71]. A bacterial system's decolorization mechanism may be anaerobic, aerobic or a mixture of both. Enzymes produced by bacteria, including laccase, peroxidase, azoreductase, etc., can degrade azo dye in significant amounts.

13.4.1.1. Azo Dye Decolorization under Anaerobic Conditions

Methanogens, acetogenic bacteria, and acidogenic bacteria are among the many trophic groups of bacteria that are needed to participate in methanogenesis from complex organic compounds [71]. Bacterial decolorization occurs under anaerobic conditions when several soluble cytoplasmic azoreductase with low substrate specificity reduce the dye molecule's strongly electrophilic azo bond. While many of the bacterial cultures might grow in an aerobic environment, only anaerobic conditions allowed them to decolorize. Under these conditions, decolorized dye needs a natural carbon energy supply. Under methanogenic conditions, simple substrates such as acetate, fructose ethanol, and glucose as well as further intricate substrates such as tapioca and whey have been used for decolorized dye [61–71]. Extensive study has been done to determine the function of distinct bacteria groups in the decolorization of azo dyes. Other researchers discovered that both acidogenic and methanogenic bacteria lead to dye decolorization. Members of the gamma proteobacteria, as well as sulfate-reducing bacteria (SRB), in anaerobic-baffled reactors processing industrial dye waste, are important members of the mixed bacteria population, according to molecular processes used to describe the microbial populations. A methanogenic population governed by *Methanosaeta* and *Methanomethylovoranshollandica* genera have helped with industrial wastewater treatment [71]. 2-Bromoethanesulfonic acid (BES), using a methanogen-specific inhibitor had no effect on the amount of methane produced by the decolorization of Orange96.

This contradicts the findings of several other researchers, implying that methanogens were not involved in the decolorization. However, in the manifestation of sulfate, molybdate, acetate, a unique SRB agent, significantly reduced the rate of discoloration. The majority of a diverse group of azo compounds discolor under anaerobic conditions, although the rate of discoloration depends on the cause of the added organic carbon and the formation of the dye. Likewise, there remains certainly no correlation between molecular weight and decolorization rate, implying that decolorization is not a distinct process and that cell penetrability has no bearing on decolorization. As a result, anaerobic azo dye decolorization is a random coincidence in which dye serves as an acceptor for electrons provided by electron transport system transporters. Decolorization may also occur because of non-specific extracellular reactions between anaerobic biomass's reduced compounds. For a few dyes, such as acid orange 7(AO7), autocatalysis by quinone-like compounds formed during azo dye reduction adds considerably to the total reduction process AO7.

13.4.1.2. Azo Dye Decolorization under Aerobic Conditions

Several bacterial strains capable of aerobically decolorizing azo dyes have been discovered in recent years. Organic carbon sources are required since dye cannot be employed as a growth substrate for many of these bacteria. Because oxygen inhibits azo bond breakdown activity, azo dyes are not easily digested and remain durable under aerobic environments. Under aerobic circumstances, *Pseudomonas aeruginosa* decolored a commercial textile dye, tannery, Navitan fast blue S5R, in the presence of glucose. Some bacterial strains, however, may decrease the azo linkage by reductive processes when grown in aerobic circumstances. This organism may also decolorize a variety of other azo dyes [72]. Only a few bacteria can thrive completely on azo compounds as a source of carbon. These bacteria reductively break –N=N– bonds and grow by devouring amines as an energy and carbon source. [73]. Under aerobic situations, the adapted bacteria produce an oxygen-insensitive or aerobic azoreductase specific for azo compounds that can reductively cleave the azo community.

These species are substrate-specific for their environment. *Xenophilus azovorans* KF 46 (previously *Pseudomonas* sp. KF46) and *Pigmentiphaga kullae* K24 (previously *Pseudomonas* sp. K24) are two bacteria strains that can acquire aerobically on carboxy-orange I and carboxy-orange II, respectively [74]. The sulfonated dyes acid AO7 and orange 20 (Orange I), which are structurally similar, did not allow these organisms to grow. Long-term variation of 4-aminobenzenesulfonate (4-ABS), which degrades the S1 intermediate strain of Hydrogenophaga, to progress on 4-carboxy-40-sulfoazobenzene (COD) as the only source of organic carbon resulted in the separation of the S5 strain, reduced COD, and used both amine metabolites. The isolation of a *Sphingomonas* sp. strain 1CX, an obligate aerobe that can develop solely on the azo dye AO7 for energy, nitrogen, and carbon [75]. Only one of the component amines produced during AO7 decolorization was degraded by this strain (1-amino 2-naphthol). However, the existence of an unidentified strain, SAD4i, was needed for 4-aminobenzesulfonate (4-ABS) degradation. *Sphingomonas* ICX could also decolorize azo dyes made up of 1-amino-2-naphthol or 2-amino-1-naphthol linked to a phenyl or naphthyl moiety through an azo bond. *Pseudomonas*, *Bacillus*, *Listeria*, and *Sphingomonas* species have been found to decolorize azo dyes in aerobic conditions. *Bacillus* sp. OY1-2, *Pseudomonas* sp. PR41-1 and *Xanthomona*s sp. NR25-2 were recognized as bacteria that could utilize azo dye after being identified from soil and sewage samples (acid red 88 or AO 7) as their only source of carbon [76]. Using solely methyl red as a carbon source, four bacterial species were recently identified. *Pseudomonas nitroreducens* and *Vibrio logei* have been described as two of the strains. The absence of amine products in the culture medium indicates that they have degraded.

13.4.2. Decolorization of Azo Dyes by Fungus

Fungi have been shown to be an effective organism for treating textile wastewater and removing color. These investigations may be classified into three types based on the mechanism involved: bioaccumulation, biodegradation, and biosorption [77]. Fungi can produce lignin-modifying enzymes to mineralize and/or decolorize azo dyes such as manganese peroxidase (MnP), laccase, lignin peroxidase (LiP), etc. [78]. In static and/or agitated decolorization techniques, fungal cultures are used as free or immobilized cultures. Some azo dyes, such as Congo red, Sudan 1, Amaranth, Acid Black 1, Reactive black 5, Basic blue 41, Acid Orange 6, Acid Orange 7, Acid Violet 7 were decolored using cell cultures [79] (Table 13.2).

13.5. CURRENT AND FUTURE PERSPECTIVE

In the environment, microorganisms live in clusters, where they interact with a variety of hosts and other microorganisms. Furthermore, they are subjected to changes in environmental factors, which influence the relationship between host and microorganisms [80].

Microbes are usually found in diverse ecosystems of several other organisms. These microbial species are crucial for humans because they exist in and on our bodies, influencing our health and well-being [81]. Many different factors, such as resource competition or toxin development, can drive microbial interactions. Even though these aspects are all unique, relations are mostly mediated by the climate; microbes alter the environment, and they and other microbes must adapt to this new environment [82].

Table 13.2 Decolorization of azo dyes by fungal cultures.

Name of the Azo dye with concentration	Media condition [pH, temp]	Fungal culture	% of removal/ degradation	References
Direct red 180 (100 mg L^{-1})	4.5, 30 °C	*Phanerochaete chryososporium* (Culture filtrates)	100	[57, 78]
Blue 49 (200 mg L^{-1})	3.5, 26 °C	*Trametes versicolor* (Purified laccase)	94	[45]
Reactive Orange II (0.5 g L^{-1})	5.6, 28 °C	*Cunninghamella elegans* (Whole cultures)	93	[55, 78]
Solar brilliant red8 (50 mg L^{-1})	4.5, 30 °C	*Schizophyllum commune* IBL-06 (Purified laccase)	100	[70, 76]
Congo Red (10 mg L^{-1})	3.0, 60 °C	*Aspergillus niger* (Whole cultures)	99	[76]
Reactive Black 5; (1 g L^{-1})	4.0, 40 °C	*Armillaria* (Whole cultures)	80	[70]
Amaranth (33 mg L^{-1})	3.5, 45 °C	*Trametes versicolor* (Whole cultures)	97	[70, 76]

There are still some research ideas that need to be pursued to completely comprehend the field of textile dye biodegradation by isolates. For future research, a series of experiments are suggested.

- To increase the mineralization of toxic aromatic amines, anaerobic-aerobic treatment is used.
- In future projects, the hunt for less expensive supplementary sources will be critical. The purpose of this study was to see how a mediator affected dye decolorization.
- Furthermore, to Methyl red degradation, the biodegradation mechanisms of various dyes used by bacterial isolates need to be identified. Biodegradation pathways of textile dyes are predicted at gene level.
- Purification and characterization of dye-degrading enzymes, as well as a demonstration of their function in the treatment process, using an immobilized cell and enzyme reactor for dye decolorization. Prospects for using immobilized cells to better understand the biodegradation process of a mixture of textile dyes are being investigated.
- To fix human health concerns, researchers are extending the toxicity of biodegraded samples on cell lines, performing Daphnia assays, and testing fish toxicity, among other things.

REFERENCES

1. Al-Sakkaf, B.M., Nasreen, S., and Ejaz, N. (2020). Degradation pattern of textile effluent by using bio and sono chemical reactor. *Journal of Chemistry 2020*: Article ID 8965627. https://doi.org/10.1155/2020/8965627.
2. Rajashekar, K.K., Mahadevan, G.D., Neelagund, S.E. et al. (2017). Decolorization of amaranth R.I and fast red E azo dyes by thermophilic *Geobacillus thermoleovorans* KNG 112. *Journal of Chemical Technology and Biotechnology* online first. https://doi.org/10.1002/jctb.6834.
3. Hoskeri, R.S., Mulla, S.I., and Ninnekar, H.Z. (2014). Biodegradation of chloroaromatic pollutants by bacterial consortium immobilized in polyurethene foam and other matrices. *Biocatalysis and Agricultural Biotechnology 3*: 390–396.

4. Mulla, S.I., Hoskeri, R.S., Shouche, Y.S. et al. (2011). Biodegradation of 2-nitrotoluene by *Micrococcus* sp. strain SMN-1. *Biodegradation 22*: 95–102.
5. Mulla, S.I., Manjunatha, T.P., Hoskeri, R.S. et al. (2011). Biodegradation of 3-nitrobenzoate by *Bacillus flexus* strain XJU-4. *World Journal of Microbiology and Biotechnology 27*: 1587–1592.
6. Mulla, S.I., Bangeppagari, M.D., Mahadevan, G.D. et al. (2016). Biodegradation of 3-chlorobenzoate and 3-hydroxybenzoate by polyurethane foam immobilized cells of *Bacillus* sp. OS13. *Journal of Environmental Chemical Engineering 4* (2): 1423–1431.
7. Mulla, S.I., Sun, Q., Hu, A. et al. (2016). Evaluation of sulfadiazine degradation in three newly isolated pure bacterial cultures. *PLoS One 11*: e0165013.
8. Mulla, S.I., Wang, H., Sun, Q. et al. (2016). Characterization of triclosan metabolism in *Sphingomonas* sp. strain YL-JM2C. *Scientific Reports 6*: 21965.
9. Mulla, S.I., Hu, A., Wang, Y. et al. (2016). Degradation of triclocarban by a triclosan-degrading *Sphingomonas*sp. strain YL-JM2C. *Chemosphere 144*: 292–296.
10. Mulla, S.I., Ameen, F., Tallur, P.N. et al. (2017). Aerobic degradation of fenvalerate by a gram-positive bacterium, *Bacillus flexus* strain XJU-4. *3 Biotech 7*: 320.
11. Mulla, S.I., Hu, A., Sun, Q. et al. (2018). Biodegradation of sulfamethoxazole in bacteria from three different origins. *Journal of Environmental Management 206*: 93–102.
12. Megadi, V.B., Tallur, P.N., Mulla, S.I. et al. (2010). Bacterial degradation of Fungicide captan. *Journal of Agricultural and Food Chemistry 58*: 12863–12868.
13. Mulla, S.I., Bharagava, R.N., Belhaj, D. et al. (2019). A review on micropollutants removal by microalgae. In: *Application of Microalgae in Wastewater Treatment: Volume I: Domestic and Industrial Wastewater Treatment* (ed. S.K. Gupta and F. Bux), 41–55. Switzerland: Springer International. ISBN: eBook ISBN: 978-3-030-13913-1.
14. Mulla, S.I., Ameen, F., Talwar, M.P. et al. (2020). Organophosphate pesticides: impact on environment, toxicity, and their degradation bioremediation of industrial waste for environmental safety. In: *Bioremediation of Industrial Waste for Environmental Safety.* (ed. G. Saxena and R. Bharagava), 265–290. Singapore: Springer.
15. Mulla, S.I., Asefi, B., Bharagava, R.N. et al. (2020). Processes for the removal of triclosan in the environment and engineered systems—a review. *Environmental Reviews 28*: 55–66.
16. Mulla, S.I., Bagewadi, Z.K., Faniband, B. et al. (2021). Various strategies applied for the removal of Emerging Micropollutant Sulfamethazine: a systematic review. *Environmental Science and Pollution Research*, online first https://doi.org/10.1007/s11356-021-14259-w.
17. Mulla, S.I., Bharagava, R.N., Belhaj, D. et al. (2019). An overview of nitro group-containing compounds and herbicides degradation in microorganisms. In: *Microbial Metabolism of Xenobiotic Compounds* (ed. P.K. Arora), 319–335. Springer International https://doi.org/10.1007/978-981-13-7462-3_16. ISBN: 978-981-13-7461-6.
18. Mulla, S.I., Bharagava, R.N., Belhaj, D. et al. (2019). Microbial degradation of phenolic compounds. In: *Microbes and Enzymes in Soil Health and Bioremediation, Microorganisms for Sustainability* (ed. A. Kumar and S. Sharma), 305–320. Singapore: Springer Nature https://doi.org/10.1007/978-981-13-9117-0_13. ISBN: 978-981-13-9116-3.
19. Edalli, V.A., Patil, K.S., Le, V.V. et al. (2018). An overview of aniline and chloroaniline compounds as environmental pollutants. *Significances of Bioengineering & Biosciences 1* (4): SBB.000519.2018.
20. Mulla, S.I., Talwar, M.P., Hoskeri, R.S. et al. (2012). Enhanced degradation of 3- nitrobenzoate by immobilized cells of *Bacillus flexus* strain XJU-4. *Biotechnology and Bioprocess Engineering 17*: 1294–1299.
21. Tallur, P.N., Mulla, S.I., Megadi, V.B. et al. (2015). Biodegradation of cypermethrin by immobilized cells of *Micrococcus* sp. strain CPN 1. *Brazilian Journal of Microbiology 46*: 667–672.
22. Benkhaya, S., M'rabet, S., and El Harfi, A. (2020). Classifications, properties, recent synthesis and applications of azo dyes. *Heliyon 6* (1): e03271. https://doi.org/10.1016/j.heliyon.2020.e03271.
23. Deng, D., Lamssali, M., Aryal, N. et al. (2020). Textiles wastewater treatment technology: a review. *Water Environment Research 92* (10): 1805–1810. https://doi.org/10.1002/wer.1437.

24. Haque, M.M., Haque, M.A., Mosharaf, M.K. et al. (2021). Decolorization, degradation and detoxification of carcinogenic sulfonated azo dye methyl orange by newly developed biofilm consortia. *Saudi Journal of Biological Sciences 28* (1): 793–804. https://doi.org/10.1016/j.sjbs.2020.11.012.
25. Khan, M.N., Parmar, D.K., and Das, D. (2020). Recent applications of Azo dyes: a paradigm shift from medicinal chemistry to biomedical sciences. *Mini Reviews in Medicinal Chemistry* https://doi.org/10.2174/1389557520999201123210025.
26. Bezerra, F.M., Lis, M.J., Firmino, H.B. et al. (2020). The role of β-cyclodextrin in the textile industry-review. *Molecules 25* (16): 3624. https://doi.org/10.3390/molecules25163624.
27. Li, H.H., Wang, Y.T., Wang, Y. et al. (2019). Bacterial degradation of anthraquinone dyes. *Journal of Zhejiang University. Science. B 20* (6): 528–540. https://doi.org/10.1631/jzus.B1900165.
28. Bagewadi, Z.K., Mulla, S.I., and Ninnekar, H.Z. (2017). Purification and immobilization of laccase from *Trichodermaharzianum* strain HZN10 and its application in dye decolourization. *Journal, Genetic Engineering & Biotechnology 15*: 139–150.
29. Ledakowicz, S. and Paździor, K. (2021). Recent achievements in dyes removal focused on advanced oxidation processes integrated with biological methods. *Molecules 26* (4): 870. https://doi.org/10.3390/molecules26040870.
30. Ali, Y., Hamid, S.A., and Rashid, U. (2018). Biomedical applications of aromatic azo compounds. *Mini Reviews in Medicinal Chemistry 18* (18): 1548–1558. https://doi.org/10.2174/1389557518666180524113111.
31. Saxena, G., Purchase, D., Mulla, S.I. et al. (2019). Phytoremediation of heavy metal-contaminated sites: eco-environmental concerns, field studies, sustainability issues, and future prospects. *Reviews of Environmental Contamination and Toxicology 249*: 71–131.
32. Javaid, R. and Qazi, U.Y. (2019). Catalytic oxidation process for the degradation of synthetic dyes: an overview. *International Journal of Environmental Research and Public Health 16* (11): 2066. https://doi.org/10.3390/ijerph16112066.
33. Hussein, A. and Scholz, M. (2018). Treatment of artificial wastewater containing two azo textile dyes by vertical-flow constructed wetlands. *Environmental Science and Pollution Research 25* (7): 6870–6889. https://doi.org/10.1007/s11356-017-0992-0.
34. Smita, V., Kumar, V., and Abdur, R.Q. (2017). Dye decomposition by combined ozonation and anaerobic treatment: cost effective technology. *Journal of Applied Research and Technology 15* (4): 340–345.
35. Bharagava, R.N., Mani, S., Mullab, S.I. et al. (2018). Degradation and decolourization potential of an ligninolytic enzyme producing *Aeromonashydrophila* for crystal violet dye and its phytotoxicity evaluation. *Ecotoxicology and Environmental Safety 156*: 166–175.
36. Abe, F.R., Machado, A.L., Soares, A.M.V.M. et al. (2019). Life history and behavior effects of synthetic and natural dyes on *Daphnia magna*. *Chemosphere 236*: 124390. https://doi.org/10.1016/j.chemosphere.2019.124390.
37. Ismail, M., Akhtar, K., Khan, M.I. et al. (2019). Pollution, toxicity and carcinogenicity of organic dyes and their catalytic bio-remediation. *Current Pharmaceutical Design 25* (34): 3645–3663. https://doi.org/10.2174/1381612825666191021142026.
38. Bastaki, M., Farrell, T., Bhusari, S. et al. (2017). Lack of genotoxicity in vivo for food color additive Allura Red AC. *Food and Chemical Toxicology 105*: 308–314. https://doi.org/10.1016/j.fct.2017.04.037.
39. Ajaz, M., Shakeel, S., and Rehman, A. (2020). Microbial use for azo dye degradation-a strategy for dye bioremediation. *International Microbiology 23* (2): 149–159. https://doi.org/10.1007/s10123-019-00103-2.
40. Adhikari, J.R. and Lohani, S.P. (2019). Design, installation, operation and experimentation of septic tank – UASB wastewater treatment system. *Renewable Energy 143*: 1406–1415. https://doi.org/10.1016/j.renene.2019.04.059.
41. Hernández-Chover, V., Bellver-Domingo, Á., and Hernández-Sancho, F. (2018). Efficiency of wastewater treatment facilities: the influence of scale economies. *Journal of Environmental Management 228*: 77–84. https://doi.org/10.1016/j.jenvman.2018.09.014.

42. Cristaldi, A., Fiore, M., Zuccarello, P. et al. (2020). Efficiency of wastewater treatment plants (WWTPs) for microplastic removal: a systematic review. *International Journal of Environmental Research and Public Health 17* (21): 8014. https://doi.org/10.3390/ijerph17218014.
43. Gao, Y., Yang, B., and Wang, Q. (2018). Biodegradation and decolorization of dye wastewater: a review. *IOP Conference Series: Earth and Environmental Science 178*: 012013. https://doi.org/10.1088/1755-1315/178/1/012013.
44. Gustave, W., Yuan, Z., Liu, F. et al. (2021). Mechanisms and challenges of microbial fuel cells for soil heavy metal(loid)s remediation. *Science of the Total Environment 2021* (756): 143865. https://doi.org/10.1016/j.scitotenv.2020.143865.
45. Katiyar, P., Pandey, N., and Sahu, K.K. (2020). Biological approaches of fluoride remediation: potential for environmental clean-up. *Environmental Science and Pollution Research 27* (12): 13044–13055. https://doi.org/10.1007/s11356-020-08224-2.
46. Abd El-Rahim, W.M., Moawad, H., Abdel Azeiz, A.Z. et al. (2017). Optimization of conditions for decolorization of azo-based textile dyes by multiple fungal species. *Journal of Biotechnology 260*: 11–17. https://doi.org/10.1016/j.jbiotec.2017.08.022.
47. Castañon, D., Alvarez, L.H., Peña, K. et al. (2019). Azo dye biotransformation mediated by AQS immobilized on activated carbon cloth in the presence of microbial inhibitors. *Environmental Pollution 252* (Pt B): 1163 1169. https://doi.org/10.1016/j.envpol.2019.06.050.
48. Hussain, I., Aleti, G., Naidu, R. et al. (2018). Microbe and plant assisted-remediation of organic xenobiotics and its enhancement by genetically modified organisms and recombinant technology: a review. *Science of the Total Environment 628-629*: 1582–1599. https://doi.org/10.1016/j.scitotenv.2018.02.037.
49. Defrenne, C.E., Abs, E., Longhi Cordeiro, A. et al. (2021). The Ecology Underground coalition: building a collaborative future of belowground ecology and ecologists. *The New Phytologist 229* (6): 3058–3064. https://doi.org/10.1111/nph.17163.
50. Abs, E., Leman, H., and Ferrière, R. (2020). A multi-scale eco-evolutionary model of cooperation reveals how microbial adaptation influences soil decomposition. *Communications Biology 3* (1): 520. https://doi.org/10.1038/s42003-020-01198-4.
51. Chatragadda, R. and Dufossé, L. (2021). Ecological and biotechnological aspects of pigmented microbes: a way forward in development of food and pharmaceutical grade pigments. *Microorganisms 9* (3): 637. https://doi.org/10.3390/microorganisms9030637.
52. Franca, R., Vieira, A., Carvalho, G. et al. (2020). Oerskoviapaurometabola can efficiently decolorize azo dye Acid Red 14 and remove its recalcitrant metabolite. *Ecotoxicology and Environmental Safety 191*: 110007. https://doi.org/10.1016/j.ecoenv.2019.110007.
53. Gupta, R., Anand, G., Gaur, R. et al. (2021). Plant-microbiome interactions for sustainable agriculture: a review. *Physiology and Molecular Biology of Plants 27* (1): 165–179. https://doi.org/10.1007/s12298-021-00927-1.
54. McGee, C.F. (2020). The effects of silver nanoparticles on the microbial nitrogen cycle: a review of the known risks. *Environmental Science and Pollution Research 27* (25): 31061–31073. https://doi.org/10.1007/s11356-020-09548-9.
55. Mahmood, S., Khalid, A., Arshad, M. et al. (2016). Detoxification of azo dyes by bacterial oxidoreductase enzymes. *Critical Reviews in Biotechnology 36* (4): 639–651. https://doi.org/10.3109/07388551.2015.1004518.
56. Legerská, B., Chmelová, D., and Ondrejovič, M. (2018). Decolourization and detoxification of monoazo dyes by laccase from the white-rot fungus *Trametes versicolor*. *Journal of Biotechnology 285*: 84–90. https://doi.org/10.1016/j.jbiotec.2018.08.011.
57. Koistinen, J., Sjöblom, M., and Spilling, K. (2020). Determining inorganic and organic nitrogen. *Methods in Molecular Biology 2020* (1980): 71–80. https://doi.org/10.1007/7651_2018_128.
58. Kiayi, Z., Lotfabad, T.B., Heidarinasab, A. et al. (2019). Microbial degradation of azo dye carmoisine in aqueous medium using *Saccharomyces cerevisiae* ATCC 9763. *Journal of Hazardous Materials 373*: 608–619. https://doi.org/10.1016/j.jhazmat.2019.03.111.
59. Jørgensen, B.B. (2019). Unravelling the sulphur cycle of marine sediments. *Environmental Microbiology 21* (10): 3533–3538. https://doi.org/10.1111/1462-2920.14721.

60. Fuentes-Lara, L.O., Medrano-Macías, J., Pérez-Labrada, F. et al. (2019). From elemental sulfur to hydrogen sulfide in agricultural soils and plants. *Molecules 24* (12): 2282. https://doi.org/10.3390/molecules24122282.
61. Maniyam, M.N., Ibrahim, A.L., and Cass, A. (2020). Decolourization and biodegradation of azo dye methyl red by *Rhodococcus* strain UCC 0016. *Environmental Technology 41* (1): 71–85. https://doi.org/10.1080/09593330.2018.1491634.
62. Koprivova, A., Schmalenberger, A., and Kopriva, S. (2020). Sulfatase assay to determine influence of plants on microbial activity in soil. *Bio-protocol 10* (2): e3490. https://doi.org/10.21769/BioProtoc.3490.
63. Kaur, P., Kushwaha, J.P., and Sangal, V.K. (2018). Transformation products and degradation pathway of textile industry wastewater pollutants in Electro-Fenton process. *Chemosphere 207*: 690–698. https://doi.org/10.1016/j.chemosphere.2018.05.114.
64. Azzam, S.Z., Cayme, G.J., and Martinez, L.R. (2020). Polymicrobial interactions involving fungi and their importance for the environment and in human disease. *Microbial Pathogenesis 140*: 103942. https://doi.org/10.1016/j.micpath.2019.103942.
65. Kostylev, M., Kim, D.Y., Smalley, N.E. et al. (2019). Evolution of the *Pseudomonas aeruginosa* quorum-sensing hierarchy. *Proceedings of the National Academy of Sciences 116* (14): 7027–7032. https://doi.org/10.1073/pnas.1819796116.
66. Brüschweiler, B.J. and Merlot, C. (2017). Azo dyes in clothing textiles can be cleaved into a series of mutagenic aromatic amines which are not regulated yet. *Regulatory Toxicology and Pharmacology 88*: 214–226. https://doi.org/10.1016/j.yrtph.2017.06.012.
67. Chen, X., Deng, Q., Lin, S. et al. (2017). A new approach for risk assessment of aggregate dermal exposure to banned azo dyes in textiles. *Regulatory Toxicology and Pharmacology 91*: 173–178. https://doi.org/10.1016/j.yrtph.2017.10.022.
68. Chen, Y., Zhang, L., Feng, L. et al. (2019). Exploration of the key functional strains from an azo dye degradation microbial community by DGGE and high-throughput sequencing technology. *Environmental Science and Pollution Research 26* (24): 24658–24671. https://doi.org/10.1007/s11356-019-05781-z.
69. Guo, G., Hao, J., Tian, F. et al. (2020). Decolorization and detoxification of azo dye by halo-alkaliphilic bacterial consortium: systematic investigations of performance, pathway and metagenome. *Ecotoxicology and Environmental Safety 204*: 111073. https://doi.org/10.1016/j.ecoenv.2020.111073.
70. Abbas, A., Mushtaq, A., Cheema, A.I. et al. (2020). Heterologous expression of azoreductase-encoding gene azrS of *Bacillus* sp. MR-1/2 for enhanced azo dye decolorization and wastewater treatment. *Archives of Microbiology 202* (8): 2135–2145. https://doi.org/10.1007/s00203-020-01940-w.
71. He, X.L., Song, C., Li, Y.Y. et al. (2018). Efficient degradation of Azo dyes by a newly isolated fungus Trichodermatomentosum under non-sterile conditions. *Ecotoxicology and Environmental Safety 150*: 232–239. https://doi.org/10.1016/j.ecoenv.2017.12.043.
72. Cui, D., Cui, M.H., Liang, B. et al. (2020). Mutual effect between electrochemically active bacteria (EAB) and azo dye in bio-electrochemical system (BES). *Chemosphere 239*: 124787. https://doi.org/10.1016/j.chemosphere.2019.124787.
73. Ding, P., Chu, L., and Wang, J. (2018). Advanced treatment of petrochemical wastewater by combined ozonation and biological aerated filter. *Environmental Science and Pollution Research 25* (10): 9673–9682. https://doi.org/10.1007/s11356-018-1272-3.
74. Haddad, M., Abid, S., Hamdi, M. et al. (2018). Reduction of adsorbed dyes content in the discharged sludge coming from an industrial textile wastewater treatment plant using aerobic activated sludge process. *Journal of Environmental Management 223*: 936–946. https://doi.org/10.1016/j.jenvman.2018.07.009.
75. Ahmad, M., Pataczek, L., Hilger, T.H. et al. (2018). Perspectives of microbial inoculation for sustainable development and environmental management. *Frontiers in Microbiology 9*: 2992. https://doi.org/10.3389/fmicb.2018.02992.
76. Al-Tohamy, R., Sun, J., Fareed, M.F. et al. (2020). Ecofriendly biodegradation of Reactive Black 5 by newly isolated *Sterigmatomyceshalophilus* SSA1575, valued for textile azo dye wastewater

processing and detoxification. *Scientific Reports 10* (1): 12370. https://doi.org/10.1038/s41598-020-69304-4.

77. Ajaz, M., Rehman, A., Khan, Z. et al. (2019). Degradation of azo dyes by *Alcaligenesaquatilis* 3c and its potential use in the wastewater treatment. *AMB Express 9*: 64. https://doi.org/10.1186/s13568-019-0788-3.
78. Hassani, M.A., Durán, P., and Hacquard, S. (2018). Microbial interactions within the plant holobiont. *Microbiome 6* (1): 58. https://doi.org/10.1186/s40168-018-0445-0.
79. Meerbergen, K., Willems, K.A., Dewil, R. et al. (2018). Isolation and screening of bacterial isolates from wastewater treatment plants to decolorize azo dyes. *Journal of Bioscience and Bioengineering 125* (4): 448–456. https://doi.org/10.1016/j.jbiosc.2017.11.008.
80. Liu, G., Zhu, J., Jin, R. et al. (2019). Accelerating effects of humin on sulfide-mediated azo dye reduction. *Ecotoxicology and Environmental Safety 2019* (175): 102–109. https://doi.org/10.1016/j.ecoenv.2019.03.047.
81. Huang, T., Zhang, K., Qian, Y. et al. (2018). Ultrasound enhanced activation of peroxydisulfate by activated carbon fiber for decolorization of azo dye. *Environmental Science and Pollution Research 25* (15): 14407–14414. https://doi.org/10.1007/s11356-018-1442-3.
82. Chen, G., An, X., Feng, L. et al. (2020). Genome and transcriptome analysis of a newly isolated azo dye degrading thermophilic strain *Anoxybacillus* sp. *Ecotoxicology and Environmental Safety 203*: 111047. https://doi.org/10.1016/j.ecoenv.2020.111047.

14

Removal Potential of Microplastics in Organic Solid Wastes via Biological Treatment Approaches

Sartaj A. Bhat[1], Guangyu Cui[2], Naik Yaseera[3], Xuyang Lei[4], Fuad Ameen[5], and Fusheng Li[1]

14.1. INTRODUCTION

In recent years, the fate, occurrence, and behavior of microplastics in wastewater treatment plants (WWTPs) have been investigated by many researchers [1–3]. Usually microplastics are plastic particles < 5 mm, which can be dangerous to aquatic ecosystems as well as to humans and animals [4]. Plastic industries may release microplastics into the surrounding areas through leakage and transportation mishaps. Other sources include washing of synthetic textiles, polymer manufacturing and processing factories, and personal care and cosmetic products [5–8]. Recently, many researchers have revealed that wastewater treatment sludge is a significant reservoir for microplastics, with a large quantity of up to 200 particles/g (dry sludge) [9, 10]. According to Li et al. [11] microplastics can easily absorb a variety of pollutants, including toxic metals and persistent organic contaminants (antibiotics), demonstrating that microplastics are a breeding ground of antibiotic resistance to gene transmission. WWTPs are recognized as receptors of microplastics contamination and are efficient in capturing the greater part of microplastics in the sludge throughout treatment procedures [9]. Many researchers have shown that biological treatment approaches like composting and anaerobic digestion have the ability to decrease the load of microplastics in organic solid wastes through plastic degrading bacteria [12, 13]. Wastewater treatment sludge and farm animal manure are usually hotbeds of various pathogens carrying antibiotic resistant

[1]*River Basin Research Center, Gifu University, Gifu, Japan*

[2]*State Key Laboratory of Pollution Control and Resource Reuse, Tongji University, Shanghai, China*

[3]*Department of Environmental Sciences, Government Degree College Anantnag, Jammu and Kashmir, India*

[4]*Department of Resource and Environmental Engineering, Hebei Vocational University of Technology and Engineering, Xingtai, China*

[5]*Department of Botany and Microbiology, College of Science, King Saud University, Riyadh, Saudi Arabia*

Microbial Biotechnology: Role in Ecological Sustainability and Research, First Edition.
Edited by Pankaj Chowdhary, Sujata Mani, and Preeti Chaturvedi.

genes (ARGs), and can pose severe threats to community and environmental health. Many researchers have revealed that microplastics in WWTPs decrease the quantity of many microorganisms such as nitrifying bacteria, denitrifying bacteria, and bacteria capable of degrading phenolic compounds [14, 15]. Solid organic wastes signify a major source of microplastics to the surroundings, in spite of being understudied compared to wastewater sources.

To better identify and manage microplastic contamination in organic wastes, it is important that all microplastics sources as well as the fate of microplastics after biotreatment technologies are accurately comprehended. Even though solid waste is a key basis of microplastics pollution, microplastics in various organic solid waste sources – their fate, degradation, and management – remain inadequately understood. In this chapter, we document microplastics in typical organic solid wastes: their sources, types, abundances, and analytical methods in solid organic wastes. The fate of microplastics through biotreatment technologies including aerobic composting, anaerobic digestion, thermal drying, and vermicomposting are discussed in depth. Additionally, potential research views are also discussed.

14.2. MICROPLASTICS IN TYPICAL ORGANIC WASTES

14.2.1. Sources, Types, and Abundances

WWTPs are principal contributors to microplastics pollution since wastewater holds a different combination of polymers derived from textiles and cosmetics, and are weakly degraded in WWTP systems [16]. Concentrations of microplastics in wastewater sludge differ by geographic site, probably owing to diverse population densities and waste managing strategies [17]. Larger population density results in high use of personal care goods, a greater quantity of laundry wastewater, and a larger surge of wastewater to treatment plants, in contrast to lightly populated nations, consequently raising the concentration of microplastics in wastewater sludge in highly populated areas [18]. Global studies of microplastics in sludge and landfill leachate are shown in Table 14.1. Various researchers have observed that the concentration of microplastics is generally higher in primary sludge compared to secondary sludge, demonstrating a larger removal of microplastics throughout primary sedimentation [19, 27, 28]. Landfills are furthermore a major sink for microplastics. Various researchers have detected microplastics in sediments, water bodies, and water animals near landfills [9], which indicates that landfills are performing as a base for microplastics into the water ecosystem. Microplastics can be classified into two major types: primary and secondary. Primary microplastics in the surrounding areas are generally emitted from toothpaste, cosmetics, and cleansers, whereas secondary microplastics are decomposed from primary ones and are broken down from bigger products such as clothes, plastic packaging, and toys [29]. Microplastics particles are also classified based on the microplastic forms such as film, fiber, fragment, foam, flake, line, and pellets [30]. Human activities play a significant role in microplastics pollution.

14.3. ANALYTICAL METHODS OF MICROPLASTICS IN THE ENVIRONMENT

Substantial hard work has been done to recognize and measure microplastics in the environment. Small microplastics particles can be recognized visually, but authentication of the type of microplastics requires thermal as well as spectroscopic instruments such as thermogravimetric analysis (TGA), differential scanning calorimetry (DSC), scanning electron microscopy combined with energy-dispersive X-ray spectroscopy (SEM–EDX), optical

Table 14.1 Global studies of microplastics in sludge and landfill leachate.

Country	Microplastics concentration	Microplastics size range	Microplastics shape	References
Sludge				
Canada	8600–21 200 items/kg	–	Fiber, fragment	[19]
China	1565–271 700 items/kg	0.025–5 mm	Fiber, fragment	[20]
Finland	8.2–301.4 items/kg	<1 mm	–	[1]
France	14 930–17 330 items/kg	0.02–0.5 mm	Fiber, fragment	[21]
Germany	1000–24 000 items/kg	<5 mm	–	[22]
Ireland	4196–15 385 items/kg	0.25–4 mm	Fiber, fragment, film	[13]
Italy	56 000–170 000 items/kg	0.5–0.1 mm	Fiber, fragment	[23]
United Kingdom	~ 2000 items/kg	1.34–1.62 mm	–	[24]
USA	800–4000 items/kg	<5 mm	Fiber	[9]
Landfill leachate				
China	0.4–24.6 items/l	0.07–1 mm	Fragment, fiber, film, granule	[25]
France	50–1110 items/kg (sediment)	0.02–0.08 mm	Fragment, fiber, film, sphere	[21]
Thailand	13.5–27.5 items/kg	–	Fiber, film, granule, irregular	[26]

Table 14.2 Visual and analytical methods of microplastics identification.

S. No	Method	Characteristics
1	Optical microscopes	Observation of size, shape, and surface texture
2	SEM–EDX	Identification of elemental composition
3	TGA/DSC	Monitoring characteristics of thermolytic behavior of microplastics
4	FTIR spectroscopy	Identification of functional group on microplastics
5	Raman spectroscopy	Identification of functional group and structure of microplastics

microscopy, Fourier-transform infrared spectroscopy (FTIR), and Raman spectroscopy. Visual and analytical methods of microplastics identification are presented in Table 14.2. Optical microscopes (OMs) are commonly used detection tools for minor plastic particles, since OMs enlarge images of sub-millimeter-sized microplastics in ecological samples by means of an objective lens [31–33]. SEM can identify hundreds of nanometer-sized microplastic particles. SEM/EDX can give additional information about the elemental composition of various organic and inorganic species. TGA examines the thermal stability and small portion of volatile mixtures in a material [34]. TGA measures the mass change of materials or samples as a result of thermolysis temperature to measure the volatile substances, and the derived mass-change curve determines the mass-loss rate. Various plastic materials show definite thermolytic behaviors in a wide temperature range, allowing for the identification of exact kinds of plastics when waste substances are completely divided and purified from the environmental medium: organic waste substances in an environmental medium of plants, or soils that are volatilized in the temperature range between 300 and 500 °C. DSC determines the quantity of the energy necessary to raise the temperature of a material and a blank reference as a temperature function [35, 36]. Plastic substances have exclusive thermolytic outlines for glass transition regions, and the melting points are much lower than those of inorganic substances. The recognition of microplastics in the samples can hence be attained by means of recognized libraries of the thermal degradation patterns of microplastics. The thermolytic performances of microplastics are allegedly exaggerated by thermolytic circumstances and the position of plastic particles such as particle size, additives, and level of polymerization [37–39]. FTIR and Raman spectroscopy are generally recognized analytical instruments that can give information on the functional groups and molecular structure of organic compounds [40, 41]. Since microplastics have carbon-based functional groups associated with covalent bonds, the kinds of microplastics can be recognized by means of these analytical instruments. Chemical bonds and structures of different plastics display distinctive spectra all the way through FTIR and Raman spectroscopy.

14.4. FATE OF MICROPLASTICS IN ORGANIC WASTES THROUGH BIOTREATMENT TECHNOLOGIES

Microplastics present in the effluents are removed by WWTPs and are retained in biosolids. It is a common practice to use biosolids as fertilizer in agricultural soils. The fate of microplastics in organic wastes through biological treatment methods are presented in Figure 14.1. Microplastics have been observed to hinder various physiological and gut microbiota attributes; as well, they are bioaccumulated in soil fauna [42]. Microplastics also transfer from the environment to humans via ingestion, inhalation, and dermal exposure of microplastics in packed foods [43]. Microplastics are persistent and can stay in the soil for a very extended time, which can be potentially dangerous to environmental functions, soil biodiversity, global food production, and human health [44]. WWTPs play a considerable

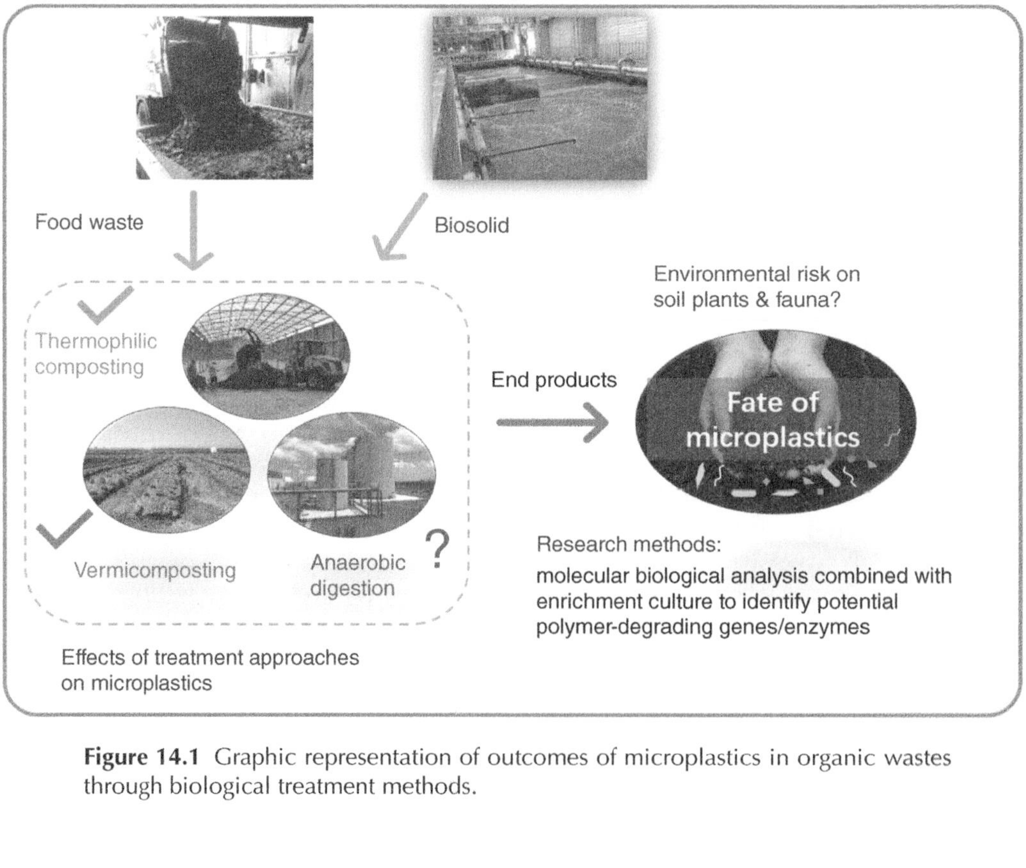

Figure 14.1 Graphic representation of outcomes of microplastics in organic wastes through biological treatment methods.

role in discharging microplastics to the surrounding areas. The bulky amounts of waste/sludge generated from biological WWTPs need to be managed. Activated sludge is a significant reservoir for microplastics and the research on removal of microplastics in sludge via composting, vermicomposting, biodrying, and anaerobic digestion is scarce.

14.4.1. Composting

The composting method has been acknowledged as an environmentally friendly technique of managing and recycling organic wastes such as activated sludge, animal manure, food waste, and municipal waste [45, 46]. In addition to dealing with organic waste, the composting treatment method is able to deal with microplastics [47, 48]. Various researchers have revealed that biological treatment approaches like composting and anaerobic digestion can decrease the load of microplastics in organic solid wastes through plastic-degrading bacteria [12, 13].

Chen et al. [48] revealed that composting treatment of sewage sludge reduced the large quantity of microplastics by 43.7%. Hosni et al. [49] also indicated that the polycaprolactone (PCL) microplastic can be decomposed without difficulty using the composting treatment method instead of in a natural ecosystem. The ample temperature and humidity, and sufficient oxygen settings in the composting method can speed up the oxidation of plastics and enhance the surface roughness and oxygen-holding functional group content of plastics [48, 50]. In the meantime, the composting process might also support the splitting of the C—C bond of plastics and as a result provide a favorable environment for additional biodegradation of plastics by microorganisms. Sun et al. [51] investigated the degradation characteristics of various microplastics like polyethylene, polyvinyl chloride, and polyhydroxyalkanoates and their effects on the bacterial population during the composting process of a combination of cattle dung and sawdust. The authors revealed that the large quantity and minor size of microplastics in all treatments reduced after composting process, except in polyvinyl chloride treatment. In addition, all microplastic contact decreased the richness and diversity of the microbe population at the thermophilic phase. In a recent study, hyperthermophilic composting was shown to be a feasible treatment method that could efficiently eliminate the abundances of microplastics in municipal sludge by 43.7% in 45 days, which is 10 times more degradation efficiency than conventional thermophilic composting [48]. Furthermore, the authors confirmed that hyperthermophilic bacteria such as *Thermus*, *Bacillus*, and *Geobacillus* played a critical role in microplastic biodegradation. However, high power consumption should be considered when the treatment approach is applied in a real composting environment.

14.4.2. Anaerobic Digestion

Municipal WWTPs are receptors of microplastics from numerous sources of human behavior [52]. Wastewater and its derivatives such as waste-activated sludge have been significantly generated, with an annual average increase rate of 13% [53]. Taking into account that the key composition of waste-activated sludge is the organic matter [54], anaerobic digestion is the favored and most common method to attain sludge resource recovery as well as sludge reduction. Mahon et al. [13] revealed that the anaerobic digestion treatment procedure may decrease microplastics, based on the categorization of biosolids samples from seven different wastewater plant systems. Many researchers have investigated various types of microplastics, such as polyethylene, polyvinyl chloride, polystyrene, etc., and found that they can unfavorably affect the anaerobic digestion process in both short- and long-term exposures [55, 56]. The existence of microplastics can considerably influence the performance of the anaerobic digestion process, and the parallel impact is very much connected to the content of microplastics.

14.4.3. Vermicomposting

Vermicomposting is a treatment process that promotes agricultural production through the synergistic effect of earthworms and microorganisms [57]. In the current scenario, with the nonstop growth of the plastic industry and its broad functions in various fields, the inhibitory consequence of microplastics on earthworm and microbial activity has been paid more and more consideration [55]. However, earthworms such *Lumbricus terrestris* (Oligochaeta), have shown that the gut microbes of the earthworms could transform low-density polyethylene (LDPE) into long chain carbon-contained compounds like octadecane, eicosane, docosane, and tricosane [58], which may indicate that composting with the help of earthworms, especially epigeic species such as *Eisenia fetida,* is a promising way to reduce the abundance of microplastics in solid organic waste like sewage sludge. Nevertheless, conventional enrichment culture for the isolation of microplastic-degrading bacteria from the earthworm gut has several drawbacks, such as ignored uncultured degrading microbes and enzymes. Therefore, metagenomic analysis for the identification of potential functional genes related to biodegradation of microplastics (LDPE, PP and PS, etc.) in the environment should be used. In addition, Sanchez-Hernandez et al. [59] reviewed the potential of earthworms and vermicomposting in biodegradable plastics waste (mulching films) from several angles, including earthworm ecological behaviors (bioturbation), their gut microbes, and polymer-degrading enzymes. All in all, more research, especially for microbial mechanisms like biofilm formation on the surface of microplastics, is required to better understand how vermicomposting and earthworms enhance microplastics degradation and mineralization [60].

14.5. CONCLUSIONS AND FUTURE PERSPECTIVES

The potential removal of microplastics in organic solid wastes through biological treatment approaches is well documented. Outcomes of treating microplastics through biological treatment of organic wastes have been provided. The WWTP sludge is a vital reservoir for microplastics and ARGs, through larger abundance compared to that of soil and other organic wastes. Biological treatment methods such as thermophilic composting, anaerobic digestion, and vermicomposting have the ability to lessen microplastics, but the degree of diminution is dissimilar. Additional research is needed to elucidate the underlying mechanism of biological treatments on microplastics, so as to diminish the abundance and diversity of microplastics in the treated sludge.

REFERENCES

1. Lares, M., Ncibi, M.C., Sillanpää, M. et al. (2018). Occurrence, identification and removal of microplastic particles and fibers in conventional activated sludge process and advanced MBR technology. *Water Research 133*: 236–246.
2. Hou, L., Kumar, D., Yoo, C.G. et al. (2021). Conversion and removal strategies for microplastics in wastewater treatment plants and landfills. *Chemical Engineering Journal 406*: 126715.
3. Zhang, Z., Su, Y., Zhu, J. et al. (2021). Distribution and removal characteristics of microplastics in different processes of the leachate treatment system. *Waste Management 120*: 240–247.

4. Thompson, R.C. (2015). Microplastics in the marine environment: sources, consequences and solutions. In: *Marine Anthropogenic Litter* (ed. M. Bergmann, L. Gutow and M. Klages), 185–200. Cham: Springer.
5. Browne, M.A., Crump, P., Niven, S.J. et al. (2011). Accumulation of microplastic on shorelines woldwide: sources and sinks. *Environmental Science & Technology 45* (21): 9175–9179.
6. Lechner, A. and Ramler, D. (2015). The discharge of certain amounts of industrial microplastic from a production plant into the River Danube is permitted by the Austrian legislation. *Environmental Pollution 200*: 159–160.
7. Napper, I.E., Bakir, A., Rowland, S.J. et al. (2015). Characterisation, quantity and sorptive properties of microplastics extracted from cosmetics. *Marine Pollution Bulletin 99* (1–2): 178–185.
8. van Wezel, A., Caris, I., and Kools, S.A. (2016). Release of primary microplastics from consumer products to wastewater in the Netherlands. *Environmental Toxicology and Chemistry 35* (7): 1627–1631.
9. Carr, S.A., Liu, J., and Tesoro, A.G. (2016). Transport and fate of microplastic particles in wastewater treatment plants. *Water Research 91*: 174–182.
10. Wu, X., Pan, J., Li, M. et al. (2019). Selective enrichment of bacterial pathogens by microplastic biofilm. *Water Research 165*: 114979.
11. Li, X., Mei, Q., Chen, L. et al. (2019). Enhancement in adsorption potential of microplastics in sewage sludge for metal pollutants after the wastewater treatment process. *Water Research 157*: 228–237.
12. Zhang, L., Xie, Y., Liu, J. et al. (2020). An overlooked entry pathway of microplastics into agricultural soils from application of sludge-based fertilizers. *Environmental Science & Technology 54* (7): 4248–4255.
13. Mahon, A.M., O'Connell, B., Healy, M.G. et al. (2017). Microplastics in sewage sludge: effects of treatment. *Environmental Science & Technology 51* (2): 810–818.
14. McCormick, A., Hoellein, T.J., Mason, S.A. et al. (2014). Microplastic is an abundant and distinct microbial habitat in an urban river. *Environmental Science & Technology 48* (20): 11863–11871.
15. Sun, X., Chen, B., Li, Q. et al. (2018). Toxicities of polystyrene nano-and microplastics toward marine bacterium Halomonas alkaliphila. *Science of the Total Environment 642*: 1378–1385.
16. De Falco, F., Gullo, M.P., Gentile, G. et al. (2018). Evaluation of microplastic release caused by textile washing processes of synthetic fabrics. *Environmental Pollution 236*: 916–925.
17. Lebreton, L.C., Van Der Zwet, J., Damsteeg, J.W. et al. (2017). River plastic emissions to the world's oceans. *Nature Communications 8* (1): 15611.
18. Talvitie, J., Mikola, A., Koistinen, A. et al. (2017). Solutions to microplastic pollution–Removal of microplastics from wastewater effluent with advanced wastewater treatment technologies. *Water Research 123*: 401–407.
19. Gies, E.A., LeNoble, J.L., Noël, M. et al. (2018). Retention of microplastics in a major secondary wastewater treatment plant in Vancouver, Canada. *Marine Pollution Bulletin 133*: 553–561.
20. Yang, L., Li, K., Cui, S. et al. (2019). Removal of microplastics in municipal sewage from China's largest water reclamation plant. *Water Research 155*: 175–181.
21. Kazour, M., Terki, S., Rabhi, K. et al. (2019). Sources of microplastics pollution in the marine environment: importance of wastewater treatment plant and coastal landfill. *Marine Pollution Bulletin 146*: 608–618.
22. Mintenig, S.M., Int-Veen, I., Löder, M.G. et al. (2017). Identification of microplastic in effluents of waste water treatment plants using focal plane array-based micro-Fourier-transform infrared imaging. *Water Research 108*: 365–372.
23. Magni, S., Binelli, A., Pittura, L. et al. (2019). The fate of microplastics in an Italian wastewater treatment plant. *Science of the Total Environment 652*: 602–610.
24. Murphy, F., Ewins, C., Carbonnier, F. et al. (2016). Wastewater treatment works (WwTW) as a source of microplastics in the aquatic environment. *Environmental Science & Technology 50* (11): 5800–5808.

25. He, P., Chen, L., Shao, L. et al. (2019). Municipal solid waste (MSW) landfill: a source of microplastics?-Evidence of microplastics in landfill leachate. *Water Research 159*: 38–45.
26. Puthcharoen, A. and Leungprasert, S. (2019). Determination of microplastics in soil and leachate from the Landfills. *Thai Environmental Engineering Journal 33* (3): 39–46.
27. Zhang, J., Wang, L., Halden, R.U. et al. (2019). Polyethylene terephthalate and polycarbonate microplastics in sewage sludge collected from the United States. *Environmental Science & Technology Letters 6* (11): 650–655.
28. Corradini, F., Meza, P., Eguiluz, R. et al. (2019). Evidence of microplastic accumulation in agricultural soils from sewage sludge disposal. *Science of the Total Environment 671*: 411–420.
29. Rochman, C.M., Browne, M.A., Underwood, A.J. et al. What do we know about the ecological impacts of microplastic debris? *MICRO 2016. Fate and Impact of Microplastics in Marine Ecosystems 2016 Nov 29* (56).
30. Su, Y., Zhang, Z., Wu, D. et al. (2019). Occurrence of microplastics in landfill systems and their fate with landfill age. *Water Research 164*: 114968.
31. Desforges, J.P., Galbraith, M., Dangerfield, N. et al. (2014). Widespread distribution of microplastics in subsurface seawater in the NE Pacific Ocean. *Marine Pollution Bulletin 79* (1–2): 94–99.
32. Eriksen, M., Lebreton, L.C., Carson, H.S. et al. (2014). Plastic pollution in the world's oceans: more than 5 trillion plastic pieces weighing over 250,000 tons afloat at sea. *PLoS One 9* (12): e111913.
33. Kang, J.H., Kwon, O.Y., and Shim, W.J. (2015). Potential threat of microplastics to zooplanktivores in the surface waters of the Southern Sea of Korea. *Archives of Environmental Contamination and Toxicology 69* (3): 340–351.
34. Lee, T., Nam, I.H., Jung, S. et al. (2020). Synthesis of nickel/biochar composite from pyrolysis of Microcystis aeruginosa and its practical use for syngas production. *Bioresource Technology 300*: 122712.
35. Freire, E. (1995). Differential scanning calorimetry. *Protein Stability and Folding* 191–218.
36. Jung, S., Lee, J., and Kim, I.W. (2013). Structures and physical properties of the cocrystals of adefovir dipivoxil with dicarboxylic acids. *Journal of Crystal Growth 373*: 59–63.
37. Choi, D., Jung, S., Lee, S.S. et al. (2020). Leveraging carbon dioxide to control the H2/CO ratio in catalytic pyrolysis of fishing net waste. *Renewable and Sustainable Energy Reviews 138*: 110559.
38. Chialanza, M.R., Sierra, I., Parada, A.P. et al. (2018). Identification and quantitation of semi-crystalline microplastics using image analysis and differential scanning calorimetry. *Environmental Science and Pollution Research 25* (17): 16767–16775.
39. Guaita, M., Chiantore, O., and Costa, L. (1985). Changes in degree of polymerization in the thermal degradation of polystyrene. *Polymer Degradation and Stability 12* (4): 315–332.
40. Koenig, J.L. (1984). Fourier transform infrared spectroscopy of polymers. *Spectroscopy: NMR, Fluorescence, FT-IR* 87–154.
41. Everall, N. and King, B. (1999). Raman spectroscopy for polymer characterization in an industrial environment. In: *Macromolecular Symposia*, vol. *141*(1), 103–116. Weinheim, Germany: WILEY-VCH Verlag GmbH & Co. KGaA.
42. Ju, H., Zhu, D., and Qiao, M. (2019). Effects of polyethylene microplastics on the gut microbial community, reproduction and avoidance behaviors of the soil springtail, *Folsomia candida*. *Environmental Pollution 247*: 890–897.
43. Walker, T.R. (2021). (Micro) plastics and the UN sustainable development goals. *Current Opinion in Green and Sustainable Chemistry 30*: 100497.
44. Scheurer, M. and Bigalke, M. (2018). Microplastics in Swiss floodplain soils. *Environmental Science & Technology 52* (6): 3591–3598.
45. Wu, J., Zhao, Y., Qi, H. et al. (2017). Identifying the key factors that affect the formation of humic substance during different materials composting. *Bioresource Technology 244*: 1193–1196.
46. Ren, X., Wang, Q., Awasthi, M.K. et al. (2019). Improvement of cleaner composting production by adding Diatomite: from the nitrogen conservation and greenhouse gas emission. *Bioresource Technology 286*: 121377.

47. Mercier, A., Gravouil, K., Aucher, W. et al. (2017). Fate of eight different polymers under uncontrolled composting conditions: relationships between deterioration, biofilm formation, and the material surface properties. *Environmental Science & Technology 51* (4): 1988–1997.
48. Chen, Z., Zhao, W., Xing, R. et al. (2020). Enhanced in situ biodegradation of microplastics in sewage sludge using hyperthermophilic composting technology. *Journal of Hazardous Materials 384*: 121271.
49. Al Hosni, A.S., Pittman, J.K., and Robson, G.D. (2019). Microbial degradation of four biodegradable polymers in soil and compost demonstrating polycaprolactone as an ideal compostable plastic. *Waste Management 97*: 105–114.
50. Zhang, M., Zhao, Y., Qin, X. et al. (2019). Microplastics from mulching film is a distinct habitat for bacteria in farmland soil. *Science of the Total Environment 688*: 470–478.
51. Sun, Y., Ren, X., Rene, E.R. et al. (2021). The degradation performance of different microplastics and their effect on microbial community during composting process. *Bioresource Technology 332*: 125133.
52. Kang, H.J., Park, H.J., Kwon, O.K. et al. (2018). Occurrence of microplastics in municipal sewage treatment plants: a review. *Environmental Health and Toxicology 33* (3): e2018013.
53. Yang, G., Zhang, G., and Wang, H. (2015). Current state of sludge production, management, treatment and disposal in China. *Water Research 78*: 60–73.
54. He, Z.W., Liu, W.Z., Tang, C.C. et al. (2019). Performance and microbial community responses of anaerobic digestion of waste activated sludge to residual benzalkonium chlorides. *Energy Conversion and Management 202*: 112211.
55. Fu, S.F., Ding, J.N., Zhang, Y. et al. (2018). Exposure to polystyrene nanoplastic leads to inhibition of anaerobic digestion system. *Science of the Total Environment 625*: 64–70.
56. Wei, W., Hao, Q., Chen, Z. et al. (2020). Polystyrene nanoplastics reshape the anaerobic granular sludge for recovering methane from wastewater. *Water Research 182*: 116041.
57. Bhat, S.A., Singh, S., Singh, J. et al. (2018). Bioremediation and detoxification of industrial wastes by earthworms: vermicompost as powerful crop nutrient in sustainable agriculture. *Bioresource Technology 252*: 172–179.
58. Lwanga, E.H., Thapa, B., Yang, X. et al. (2018). Decay of low-density polyethylene by bacteria extracted from earthworm's guts: a potential for soil restoration. *Science of the Total Environment 624*: 753–757.
59. Sanchez-Hernandez, J.C., Capowiez, Y., and Ro, K.S. (2020). Potential use of earthworms to enhance decaying of biodegradable plastics. *ACS Sustainable Chemistry & Engineering 8* (11): 4292–4316.
60. Sun, M., Chao, H., Zheng, X. et al. (2020). Ecological role of earthworm intestinal bacteria in terrestrial environments: a review. *Science of the Total Environment 740*: 140008.

15

Role of Microbes in Wastewater Treatment and Energy Generation Potentials: A Sustainable Approach

Ashish Kumar[1], Tania Chalotra[1], Atin K. Pathak[2], and Neelu Raina[1]

15.1. INTRODUCTION

The need for water and energy resources has increased, due to rise in livelihood standards and huge consumer demand in recent years. An increased need for these has led to increased wastage of resources, which may cause scarcity and various environmental problems with water, air, and soil pollution. Freshwater demand has increased drastically since the industrial revolution, which has resulted in decreased sources of natural freshwater. The use of water in different industries, agriculture, and domestic sectors has increased the wastewater generation rate. Wastewater created due to human activities is mostly disastrous to the environment. The main sources of wastewater come from pharmaceutical, paper, textile, polymer, food industries, municipalities, etc. [1–5]. Wastewater comprises a high amount of organic and inorganic salts that may intoxicate the water streams and environment if released without proper treatment [6].

Numerous advancements have been established for the treatment of wastewater such as lime-soda process, magnetic nanoparticles (NPs), ozonation, UV radiation, and activated carbon-based technology. Chemical methods are quick and require much more energy, but they may also be harmful to the environment, thus, a biological method is preferred even though it's a slow process. Currently available technologies include solar, hydro, and wind energy, but these are site and climate-dependent technologies that come with increased costs [7, 8]. Therefore, there's an immense need for on-site technology that is less expensive and environmentally friendly. Biological wastewater treatment (WWT) is considered the best way to treat wastewater in an efficient and eco-friendly way to degrade toxic compounds and simultaneously produce energy.

[1]*School of Biotechnology, Shri Mata Vaishno Devi University, Jammu and Kashmir, India*
[2]*School of Energy Management, Shri Mata Vaishno Devi University, Jammu and Kashmir, India*

Microbial Biotechnology: Role in Ecological Sustainability and Research, First Edition.
Edited by Pankaj Chowdhary, Sujata Mani, and Preeti Chaturvedi.

Biological WWT uses microbial activity (bacteria, fungi, algae, etc.) to eliminate toxic elements from various wastewater sources and treat them. Application of certain microbial species can help in the elimination of fats, oils, grease, suspended particles, biological oxygen demand (BOD), and chemical oxygen demand (COD), which in turn improves the state of wastewater and could be reused for various activities. Bacteria play an essential role in WWT by degrading the contaminants and purifying water. *Lactobacillus delbrueckii* [9] and *Bacillus megaterium* [10] are two bacteria that are used in the treatment of sewage sludge and domestic wastewater, respectively. The most widely present bacterial types include *Hyphomicrobium, Tetrasphaera, Candidatus microthrix, Rhodoferax, Trichococcus,* and *Rhodobacter*. Another bacterium, *Candidatus accumulibacter phosphatis* (CAP), also shows the property to eliminate phosphorus from wastewater [11]. Similarly, algae can break down contaminants present in wastewater and accumulate nutrients and essential elements from it. Algal turf scrubbers (ATS) are part of a technical approach that is helpful in the elimination of contaminated water from various wastewater sources with improved nutrient removal and high biomass production [12]. Some algal species such as *Chlorella, Cladophora, Spirogyra, Oedogonium*, etc., have been found to help in the treatment of wastewater [13]. *Aspergillus carbonarius* M33, a fungal strain, efficiently treats textile industry wastewater [14]. *Trametes versicolor* and *Aspergillus luchuensis* efficiently remove almost all the phosphorus, ammonia nitrogen, and organic carbon content from municipal wastewater. Certain species of yeast can degrade organic pollutants. *Candida* and *Hansenula polymorpha* are examples of oxidized yeast applied in the petroleum industry WWT [15]. Protozoa tend to reduce the turbidity of effluents. Protozoa and metazoa enhance the water quality of effluent and help in dispersing bacterial species within the wastewater for efficient treatment.

Another positive aspect of biological WWT methods is the production of energy. This makes wastewater a great renewable and economical source for energy production. Various energy products include bioethanol, biogas, bioelectricity, biohydrogen, and biodiesel. Biogas is generated by the biological breakdown of organic matter in the absence of oxygen and is produced from biomass, sewage, plant, and other waste material. Biogas comprises CO_2, hydrogen, CH_4, nitrogen, H_2S, oxygen, and sulfur [16]. *Mucor hiemalis* and *Mucor indicus* generated around 3.16 m^3 of methane from each m^3 of fermented wastewater [17]. In bioelectricity, the oxidation of organic matter can be catalyzed by microbes in microbial fuel cells (MFCs) to generate electricity. MFCs not only treat wastewater but also generate electricity [18]. *Chlorococcum* sp. and *Synechococcus* sp. were seen to generate a power density of 30.2 and 41.5 mW m^{-2}, respectively.

Biodiesel is another energy product formed of vegetable oils and fats derived from animals. *Nostoc* sp. consists of high lipid content that is beneficial in biodiesel production [19]. Microalgae contain 30% oil content by weight and yield about 58 7001 ha^{-1}. Other microalgal species such as *Schizochytrium* and *Botryococcus* yield about 70% of total dry biomass [20]. Biohydrogen produces water as a byproduct of combustion and is the cleanest renewable energy source with zero carbon footprints. Hydrogen gas may yield up to 2.75 times more energy than hydrocarbon fuels. *Bacillus coagulans* MO11 and *Bacillus coagulans* MO21 were found to produce 0.33 ml of H_2 and 0.66 ml of hydrogen gas, respectively [21]. Another alternative to transportation fuel is bioethanol. *Euchema spinosum* consists of carbohydrates that could be converted into bioethanol with a 75% productivity rate [22]. *Gelidium amansii* was found to enhance bioethanol yield to 76.9% [23]. The present chapter is designed with these views in mind.

The present chapter elaborates on up-to-date knowledge on various wastewater generation sources and their treatment technology options. The microbes involved in WWT have been discussed. Simultaneously, we also focus on biological treatment options and their byproducts in the form of bioenergy (biogas, bioethanol, biodiesel etc.).

15.2. WASTEWATER AND THEIR GENERATION SOURCE

Wastewater refers to the contaminated water retrieved from various distinct sources such as industries, municipalities, storm runoff, agricultural areas, and households (Figure 15.1). Wastewater may include distinct pollutants or contaminants including deleterious substances such as oils, human excreta, and soaps.

15.2.1. Black Water (BW)

Black water refers to the used water from toilets and domestic sewage systems [24]. It may also be called domestic or sanitary sewage. It comprises around 99.9% water by weight. BW accounts for about 12–33% of domestic sewage by volume [25, 26]. BW holds a high amount of COD, nitrogen, and phosphate. Its composition varies with the variation in the amount of urine, feces, and flushed water present in it. BW consists of higher COD, total nitrogen (TN), and total phosphorous (TP) than that present in gray water and a mixture of domestic wastewater [27].

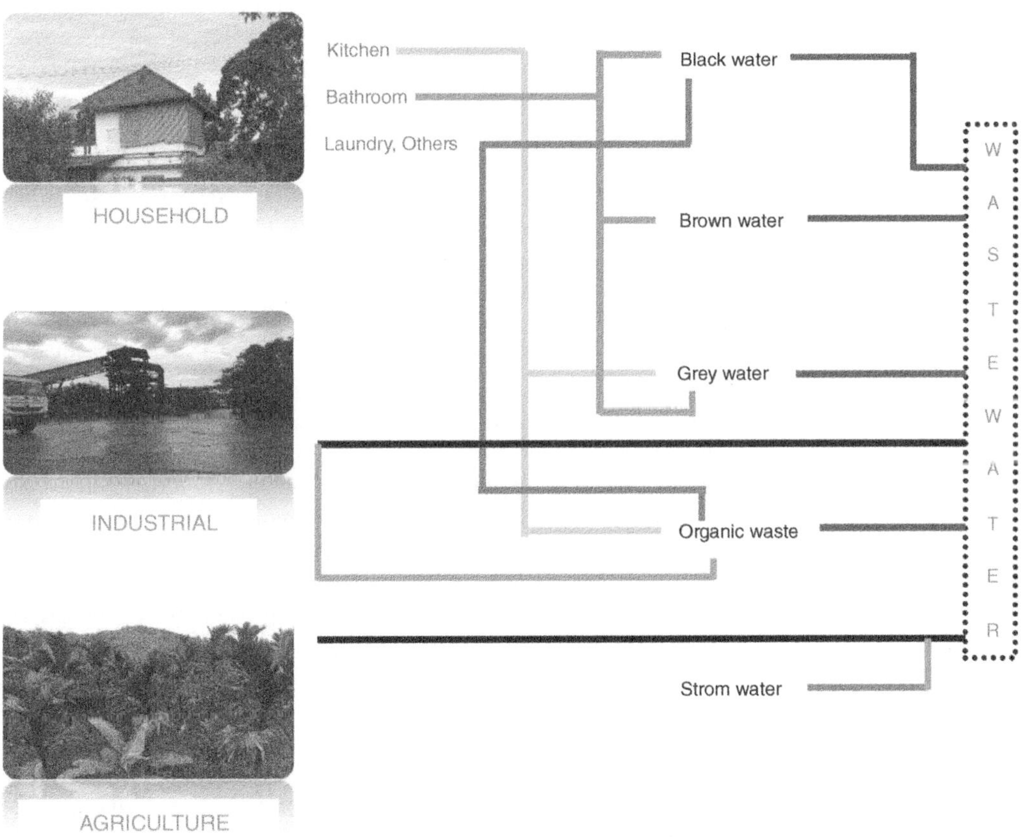

Figure 15.1 An overview of diagrammatical presentation of sources and routes of different types of wastewater production. *Source:* Photos by Wajan/Adobe Stock; Danicek/Adobe Stock; Olly/Adobe Stock.

15.2.2. Gray Water (GW)

Wastewater produced from bathing and kitchen activities is referred to as gray water. It includes water from dishwashers, sinks, and showers. It comprises lesser contaminants and can be easily treated for reuse in different activities. Its quality can degrade quickly as it is warm and comprises pathogens as well as organic matter. It makes up to 65% of the total wastewater from households. It may also be referred to as foul water [28].

Other sources of wastewater include industrial wastewater, agricultural wastewater, and storm sewage. The water that is left behind after all the chemical and manufacturing processes in industrial settings is known as industrial water. It may contain heavy metals such as arsenic (As), cadmium (Cd), and mercury (Hg). The runoff water collected in pipelines and other channels after precipitation is referred to as storm runoff [29].

15.3. ADVANCEMENTS IN WWT TECHNOLOGIES

This section deals with the new and advanced technologies aside from microbes in recent years for WWT.

Traditional treatment of wastewater from municipal plants is generally based on an activated sludge process that is neither cost-effective nor energy efficient. Numerous technologies that can increase the removal of pollutants from wastewater have been brought to light. These technologies include ozonation, magnetic nanoparticles (MNPs), UV radiation, activated carbon, and lime soda process.

UV radiation damages the genetic material of the microorganisms by penetrating through their cell wall and creating thymine dimers in between adjacent thymine molecules. Around 99.9% of microorganisms can be destroyed by using UV radiation present in wastewater. Rosińska [30] carried out studies showing how UV radiation and UV/chlorination affect polycyclic aromatic hydrocarbons (PAHs) present in wastewater. The study revealed that after 30 minutes treatment of UV radiation resulted in a 66% decrease, whereas with using the UV/chlorine process a 68% decrease in PAH concentration was obtained.

Industrial as well as local water can be treated with the help of activated carbon-based technology. Activated carbon has a high surface area which consists of porous material and thus can adsorb contaminants over it [31–36]. Herein, the pollutants get adsorbed over the margins of the activated carbon upon reacting with the carbon material and then move into the carbon pores. This way they are adsorbed onto the inner carbon walls [37]. It was found in a study that around 96% of indium could be removed using mesoporous activated carbon (MC01) and it could be recovered by adsorbing it and applying nitric acid (HNO_3) and hydrofluoric acid (HF) solutions to it. It was seen that lower pH (0.5) of HNO_3 solutions enhanced the recovery to around 82% [38].

Shukla et al. [39] reported that heavy metals can be removed from wastewater using rice husk. They detected the recovery of 67.91, 87.17, 96.95, 98.18, and 99.25% of cadmium (Cd), lead (Pb), nickel (Ni), copper (Cu), and iron (Fe), respectively, from their respective solutions of varying concentrations.

Another technology which is in demand nowadays is ozonation which is another technology that involves an advanced oxidation method. It involves the production of highly reactive oxygen species (ROS). Free radicals are highly reactive and have a short life span under standard conditions (pH 11, 32–35 °C). ROS can degrade organic and inorganic contaminants and attack numerous pathogens [40–43].

In one study, sulfur-doped copper-yttrium bimetallic oxides (S-CuYO) were utilized in the catalytic ozonation of aniline. S-CuYO showed the highest activity in the degradation and mineralization of aniline among several other catalysts. The rate of degradation in aniline was found to be 96% and around 57% total organic matter was removed from the system containing 2.0 g/l S-CuYO and ozone (O_3) flow rate of 9.65 mg in 15 minutes [44].

Magnetic nanoparticles (MNPs) are a simple and economical method for WWT. Fe_3O_4 MNPs manifested a promising pathway for the elimination of pollutants including metals. Adsorbents made out of MNPs show increased absorbance to contaminants due to the magnetic behavior and high surface-to-volume ratio. Numerous organic compounds and biomolecules that alter the surface of MNPs increase adsorption to varying toxic contaminants in wastewater [45, 46]. Maia et al. [46] found in their research that Fe_3O_4/Au adsorbent could be an economical method to eliminate mercury (II) in wastewater. In another work by Saadat et al. [47], they created amino-functionalized silica NPs that could eliminate lead and cadmium. Chen et al. used chitosan/magnetic NPs (ChM) in the elimination of coke powders present in coking diesel [48].

Lime soda (LS) process is also known as lime softening or Clark's process. This technique helps in the reduction of the hardness of water with the help of precipitation. The appropriate amount of lime and soda are mixed together in a reaction tank that consists of hard water as well as superheated steam. The chemicals react with the salts (that result in the hardness of water) under the steam of high temperatures. After the reaction, the salts are precipitated as hydroxide and carbonate into a sedimentation tank in the form of sludge, which can be collected in a sludge outlet. The wastewater after the elimination of hardness moves into the sand filter, where it interacts with sand particles of distinct sizes to eliminate the suspended matter [49–51]. It was observed that the application of hydrated lime along with high magnesium content enhances the natural organic matter (NOM) during the process of softening. It was also found that calcium hydroxide with a magnesium content of 0.32 and 0.45% in the form of MgO resulted in 20 and 21% removal of dissolved organic carbon (DOC) present in raw water, respectively. Increased magnesium content improved the removal rate because of enhanced co-precipitation [52].

Another treatment technology that came into existence is zeolites. These are crystalline particles comprising silicon, aluminum, and oxygen. They help in softening water [53]. Hard water including calcium and magnesium ions is passed through a bed of sodium zeolite wherein the Ca^{2+} and Mg^{2+} ions are replaced by sodium ions [54, 55]. The reaction that takes place during hardwater softening is as follows:

$$Na_2Ze + Ca(HCO_3)_2 \rightarrow 2NaHCO_3 + CaZe \tag{i}$$

$$Na_2Ze + Mg(HCO_3)_2 \rightarrow 2NaHCO_3 + MgZe \tag{ii}$$

$$Na_2Ze + CaSO_4 \rightarrow Na_2SO_4 + CaZe \tag{iii}$$

$$Na_2Ze + MgSO_4 \rightarrow Na_2SO_4 + MgZe \tag{iv}$$

The dewaterability of sludge can be upgraded with the help of persulfate oxidation, polyelectrolyte, and zeolites [56]. Loiola et al. [57] found that the dewaterability of wastewater-activated sludge (WAS) improved by the use of zeolites along with Fe(II)/2-S_2O_8. This happened because zeolites led to the formation of channels that helped water escape. Zeolites resulted in the formation of a porous medium between the flocs of WAS and filtration cloth that enhanced the dewaterability of WAS [58].

The nitrification process has been studied by using natural zeolite as a carrier of biofilm to sequence moving bed biofilm bioreactor. This enhanced nitrifying system resulted in the efficient removal of total ammonia, nitrogen, and COD to more than 90% [59]. These technologies show extraordinary achievement decreases the use of the technology, and suggests using some new and eco-friendly technology.

15.4. MICROBIAL TECHNOLOGY/MICROBES INVOLVED IN WWT

Wastewater is contaminated with microbiological populations all the time, which may include bacteria, fungi, algae, protozoa, and metazoa. These microorganisms can be a boon as well as a curse for wastewater. Contaminants may result in detrimental effects on living organisms as well as the environment. Various microorganisms of this population can cause some disease. In contrast, a few others can be of great help in eliminating these contaminants/pollutants and treating wastewater, thereby generating byproducts of great value. Application of certain microbial species can help in the elimination of fats, oils, grease, suspended particles, BOD, and COD, which in turn enhances the state of wastewater that could be reused for various activities. This section discusses the use of bacteria, algae, fungi, protozoa, and other microorganisms that help in the process of WWT.

15.4.1. Bacteria

About 95% of the total wastewater comprises bacterial species, which may be facultative, aerobic, or anaerobic. Anaerobic bacteria do not require oxygen; instead, inorganic as well as organic molecules act as electron acceptors. These bacteria produce methane gas that could be used as a biofuel by breaking down sludge. A study reported that the anammox process efficiently removed inorganic nitrogen present in WWTPs (wastewater treatment plants) of the swine industry. The first stage resulted in around 99.99% ammonium-nitrogen and 94% nitrogen elimination. The most widely present anammox bacteria were found to be *Candidatus* and *Brocadia,* which accounted for 16.5% of the other bacteria that removed nitrogen [60]. On the other hand, aerobic bacteria require oxygen to break down the pollutants or contaminants to generate energy for them for reproduction. Facultative bacteria are those that can survive under both the presence and absence of oxygen in the environment. In another study, it was observed that Cr (III) adsorption protein was present on the surface of *Escherichia coli* that enhanced chromium adsorption using a magnetic stirrer. A genetically engineered strain of *E. coli* MBL21 resulted in 2.38 mmol g^{-1} of adsorption ability. This was found to be beneficial in leather industries as it comprises toxic chemicals such as trivalent chromium [9].

In classical WWT, oxygen is supplied to enhance the action of aerobes over anaerobes. Various factors such as pH, agitation, settling, and other manageable factors influence the reduction of organic matter by the action of bacteria.

In urban WWT plants, *proteobacter*, Gram-negative bacteria, are dominant, around 21–65%. Various Gram-negative bacteria are said to have heavy metal tolerance genes that are used in WWT. These include *arsB, czcA, merA, terF, pcoD,* and *silA*. Among these heavy metal tolerance genes, *silA* and *pcoD* were found to be dominant in wastewater [10]. The most widely present bacterial types include *Hyphomicrobium, Tetrasphaera, Candidatus microthrix, Rhodoferax, Trichococcus,* and *Rhodobacter*.

Anaerobic culture of *E. coli* and *Pseudomonas fluorescence* [24] was used for the treatment of effluent from coffee industries resulting in the decrease of BOD, COD, and total solids (TS) present in wastewater effluent. There was a decrease in BOD, TS, and COD from 320.26 mg l^{-1}, 3545 mg l^{-1}, and 1261 mg l^{-1} to 58.37 mg l^{-1}, 1198 mg l^{-1}, and 152 mg l^{-1}, respectively [61].

Toxic elements can be immobilized with the help of ureolytic bacteria in microbially induced carbonate precipitation. A strain of *Lysinibacillus* sp. was able to tolerate 100 ppm of copper and 1000 ppm of lead. In the study, it was found that metal affects well-adapted urease activity [62].

Acidithiobacillus ferrooxidans and *Acidithiobacillus thiooxidans* were used individually and together for bioleaching. It was found that the mixed culture could help extract 100% Zn, 98.5% Ni, 41.9% Ba, and 97.8% Li under optimal conditions after 14 days with 0.5% n (w/v) pulp density [63].

Higher lead tolerance was seen in *Kleibsiella pneumoniae* Kpn555 that was isolated from the waste pulp of coffee along with 900 mg l^{-1} of inhibitory concentration.

About 84.2% of supplemented PAHs were degraded via carrier systems driven by bacterial species. It was found that about 100, 89.5, and 86.3% of crude oil was removed via coconut fiber-bacteria, polyurethane foam-bacteria, and cinder beats-bacteria system [64]. Table 15.1 describes the use of different bacterial strain for WWT and pollutant removal efficiency.

Candidatus accumulibacter phosphatis (CAP) belongs to the class betaproteobacter and has been of prime importance in WWT. It is a rod-shaped Gram-negative bacterium that is considered ecologically crucial in the elimination of phosphorus from sewage or wastewater. It settles down sludge and other sedimentary particles in the treatment plants. Wastewater effluent can lead to a plethora of phosphorus compounds in the environment that could lead to the formation of algal blooms. Enhanced biological phosphorus removal (EPBR) and chemical removal processes are classical methods for the elimination of phosphorus [80]. EPBR has a low operating cost, decreases the production of sludge, allows the sludge to be reused easily, and reduces the use of chemical byproducts applied in chemical treatment methods. In domestic sewage, CAP is found to be about 4–18% [81].

15.4.1.1. Metabolism

CAP is a chemoheterotroph and because of its unique metabolism, it can adapt itself to survive under aerobic as well as anaerobic conditions. Depending upon the various environmental factors it runs molecules for storage and energy production. CAP acts as an acetate oxidizer. Figure 15.2 represents the enhanced biological removal process of a mixture of wastewater with aerobic sludge.

15.4.2. Aerobic Conditions

15.4.2.1. Anabolic Pathways

An energy-rich phosphate chain, polyphosphate is produced from orthophosphate upon its uptake from the environment by CAP under aerobic conditions. If aerobic conditions persist polyphosphates are stored in CAP. Likewise, glycogen is also stored in it.

15.4.2.2. Catabolic Pathway

In this step, the reduction of poly-beta-hydroxy alkanoates (PHAs) takes place to generate energy. These are stored under anaerobic conditions [82].

15.4.3. Anaerobic Conditions

15.4.3.1. Anabolic Pathways

CAP intakes volatile fatty acids (VFA) and stores the carbon in them as PHAs in the cell.

Table 15.1 Pollutants removed from different wastewater by various bacterial species.

S. No.	Species of bacterial strain used	Pollutant removed	Removal percentage	Type of wastewater	References
1.	*Lactobacillus delbrueckii* and *Streptococcus thermophillus*	Fe, Zn	48.2, 51%	Sewage sludge	[65]
2.	*Acinetobacter* sp., *Bacillus megaterium,* and *Sphingobacterium* sp.	Fe, Mn	62–89%	Domestic wastewater	[66]
3.	*Anoxybacillus flavithermus*	Fe, Cu	75.9%	Floating wetlands	[67]
4.	*Leptothrix, Pseudomonas, Hyphomicrobium,* and *Planctomyces*	Mn	89%	Domestic wastewater	[68]
5.	*Pseudodonghicola xiamenensis*	Ca, K, Na	45–49%	Datesyrup wastewater	[69]
6.	Bacterial consortium composed of *Pseudomonas, Lysinibacillus, Lactococcus,* and *Dysgonomonas*	Metanil yellow dye	84–90.3%	Textile wastewater	[70]
7.	*Candidatus Brocadia*	NH4 + –N & inorganic N	98.9, 94.7%	Swine wastewater	[60]
8.	*Bacteroides ovatus*	indicate the occurrence of fecal contamination	76.1–83.6%	Sewage sludge, dairy wastewater	[71]
9.	*Clostridium XVIII*	COD	92.7–95.23%	sulfate-laden wastewater	[72]
10.	*Aulosira fertilissima*	Ammonia, nitrite, and phosphate reduction	78.5%	Aquaculture wastewater	[73]
11.	*Chryseomonas luteola*	Co, Cu, Ni	79.7%	Sewage sludge	[74]
12.	*Pseudomonas aeruginosa* SVM16	Reactive orange 16 dye	98.9%	Textile wastewater	[75]
13.	*Anabaena spiroides*	Mn	67%		[68]
14.	*Ralstonia solanacearum*	Pb	96%	Industrial wastewater	[76]
15.	*Pseudomonas aeruginosa* SVM16	Reactive blue 19 dye	94.8%	Textile wastewater	[75]
16.	*Candidatus Scalindua, Candidatus Kuenenia*	N	9.55–27.75%	Saline wastewater	[11]
17.	*Proteobacteria* and *Bacteroidetes*	Fe, Al, S, NH_3–N	18.1%	Domestic waste	[68]
18.	*Bacillus licheniformis*	Pb, PO4	83–97%	Industrial wastewater	[77]
19.	Purple nonsulfur bacteria	COD and N reduction		Winery wastewater	[78]
20.	*Enterobacter ludwigii, Bacilllus cereus, Enterobacter aerogenes* and *Enterobacter cloacae*	TDS and COD	37.6–40.9% and 40.1–48.9%	coffee cherry pulping wastewater	[79]

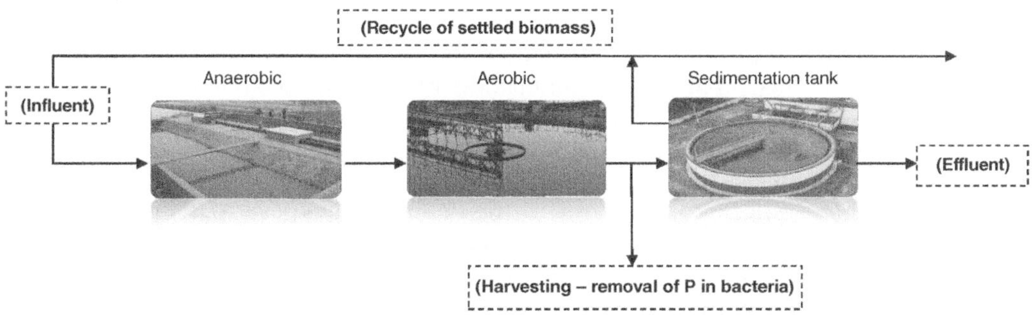

Figure 15.2 Enhanced biological removal process representing a mixture of wastewater with aerobic sludge. *Source:* Fernando butcher/Flickr and DedMityay/ Adobe Stock.

15.4.3.2. Catabolic Pathway

Under aerobic conditions polyphosphates are stored and then later reduced into orthophosphate for energy production. There is another pathway that reduces glycogen by glycolysis, this results in the production of ATP and NAD reduction equivalents [82].

EBPR comprises bacteria that gather the phosphorus present in the wastewater inside its cells and then help in their removal. Bacteria accumulate in the sludge layer. The sludge layer comprises a layer of solid particles which are present at the bottom of WWT bioreactors. Initially, a culture of CAP is grown in the sludge layers of aerobic wastewater. A small amount of sludge is added either into the beginning of the system or mixed with incoming anaerobic wastewater to initiate a new culture of microorganisms that eliminates phosphate. These microorganisms will reproduce, eat up, and condense phosphorus in the biomass and assemble in the sludge layer of the aerobic system [83].

Glycogen accumulating organisms and CAP both compete with each other as they rely on the same carbon source, i.e. acetate [84]. Along with CAP, the sludge layers also contain a population of *Gammaproteobacteria, Actinobacteria,* and *Betaproteobacteria* [85].

As the anaerobic phase goes on, CAP uptakes VFA and stores them intracellularly as PHAs. Simultaneously, intracellular phosphates degrade ATP resulting in the release of phosphate into the system. Following this in the aerobic phase, PHAs are utilized to produce energy, and phosphate present in the system helps in the formation of polyphosphate. Genomic remodeling from an EBPR reactor that comprised CAP IIA showed that the reactor consisted of two distinct phosphate transporter types. These were high and low-affinity pit transporters [13].

15.4.4. Algae

Algae generate oxygen in the same way as other photosynthetic organisms and incorporate carbon dioxide [86]. In this way, a beneficial cycle can run between bacteria and algae as the oxygen produced by algae can be utilized by the bacteria to oxidize organic carbon and then the algae can incorporate the carbon dioxide produced through bacterial respiration. The requirements for aeration and the carbon dioxide emissions associated with it can be decreased by assimilating algae into WWT plants. During the growth period, algae incorporate nitrogen, fixed carbon, and phosphorus into the system. This reduces the demand for the removal of bacterial nitrogen and phosphorus as well as the demand for aeration and nitrous oxide emissions [87]. Biological treatment with microalgae has several advantages including its photosynthetic capabilities that help in the conversion of solar energy into useful biomass and assimilate nutrients causing eutrophication [88].

The self-adhesive property of filamentous algae to develop clumps provides ease in retention and harvesting and protects from predation. Because of the adhesive properties, they can adhere to pond walls, piping outlets, and paddle wheels that need regular cleaning. The formation of huge mats on the water surface causes self-shading that may result in the decrease of light utilized by the deepest areas of the pond.

ATS is a technical approach to WWT based on periphyton, which is an attached complex population of microorganisms. Periphytons include a self-seeding microbial population that comprises filamentous algae that adhere to the screen. The screen is submerged in shallow flowing wastewater so that the organisms in the periphyton can uptake nutrients. Later the biomass is mechanically removed using a scrapper or vacuum and used for biofuel applications. ATS is helpful in the elimination of contaminated water including dairy, aquaculture, agricultural runoff, manure effluent, and secondary effluent with improved nutrient removal and high biomass production [12].

Various freshwater filamentous algae are present in nature, but only a few have been observed to be of importance in WWT. *Microspora, Stigeoclonium, Cladophora, Rhizoclonium, Klebsormium, Spirogyra,* and *Oedogonium* are a few filamentous algae that are predominant in algal WWT. Some other algal species such as *Scenedesmus obliquus* [89–91] and *Chlorella vulgaris* [92–94] have been utilized in the treatment of brewery and textile effluents, respectively. Various other species of *Chlorella* have been reported in the treatment of animal waste to remove nutrients [95–97]. The removal rate of microalgal species varies with the species involved in the treatment [98].

Chlorella variabilis TH03-bacterial consortia were used to study its ability in the removal of pollutants from domestic wastewater. The consortia were found to remove 99.7–100% COD, 85.1–96.8% total nitrogen, and 64.7–90.7% total phosphorous from the wastewater [99]. Phyco-mitigation of numerous pollutants and heavy metals in wastewater from urban areas was carried out using *Pseudochlorella pringsheimii – Ind-Jiht-1*, which efficiently reduced 66.7% alkalinity, 83.2% COD, and 69.6% hardness [100]. The list of algal species utilized to degraded wastewater is described in Table 15.2.

Different nutrient uptake mechanisms of algae are discussed here (Figure 15.3):

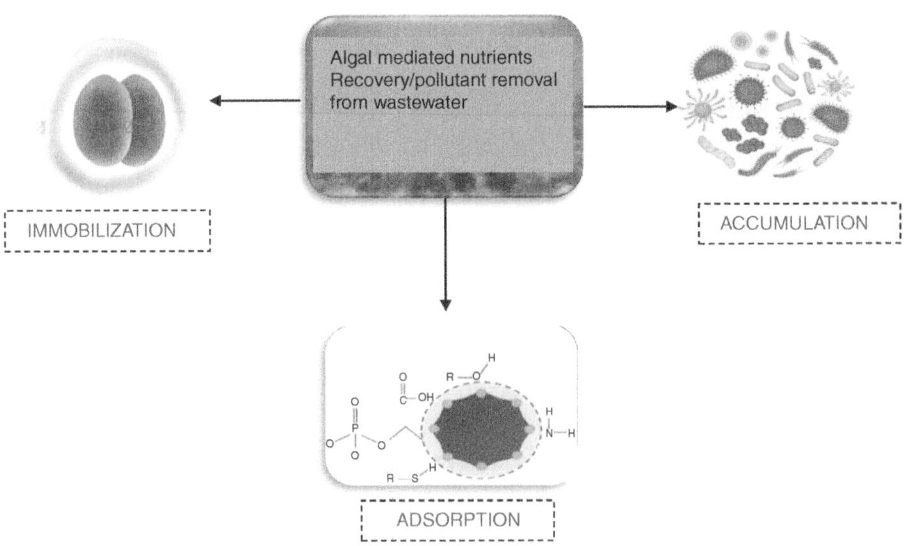

Figure 15.3 Different mechanisms of nutrient uptake by algal species.

Table 15.2 Pollutants removed from different wastewater by algal species.

S. No.	Species of algal strain used	Pollutant removed	Removal percentage	Type of wastewater	References
1.	*Lemna minor*	N/NH3-N, COD, P	84, 74.9, 84.1%	Swine wastewater	[101]
2.	*Lemna* sp.	P, N/NH3-N	60.1, 36%	Industrial wastewater	[102]
3.	*Chlorella* sp.	Cu, Zn	56.5, 86.7%	Domestic wastewater	[103]
4.	*Chlorella pyrenoidosa*	N and P	94.2, 92.58%	Cheese whey wastewater	[104]
5.	*Chlorella sorokiniana* and macrophyte *Lemna minor*	COD, TKN, and PO4-P	99, 90, 91%	Municipal wastewater	[105]
6.	*Desmodesmus* sp.	Ni, Cu	7.5, 94%	Industrial wastes	[106]
7.	*Synecdatehocystis* sp.	Cr (VI), Cu (II), Pb (II) and Cd (II)	69.9, 62, 42.6, 38.06%	Industrial water	[107]
8.	*Scenedesmus* sp. L-1	COD, NH3-N, NO3, TP	90.7, 90.4, 90.2, 79.8%	Cattle manure wastewater	[108]
9.	*Scenedesmus acutus*	COD, NO_3, NH4, PO4	77.3, 71.1, 93.6, 66.2%	Municipal wastewater	[109]
1o.	*Synechocystis salina*	NO_3, NO_2	82.5, 92.8%	Chemical wastewater	[109]
11.	*C. vulgaris* and *N. oleoabundans*	COD, Ammonia, P	51–80%, 63–72%, 70–84%, 84%	Industrial wastewater	[110]
12.	*Galdieria sulphuraria*	NH3-N, P	30–40%	Industrial wastes	[111]
13.	*Scenedsmus sp.*	N/NH3-N	85.2%	Swine wastewater	[112]
14.	*Scenedesmus obliquus*	Ammonium, phosphate, organic carbon	99.2, 91.8, 83.2%	Municipal wastewater	[113]
15.	*Chlamydomonas polypyrenoideum*	Phosphate, nitrate, ammonia	70, 74, 90%	Dairy wastewater	[114]

15.4.4.1. Adsorption

Adsorption is the key physical method in which the contaminants or pollutants are bound to removal agents that could either be microalgae or anything else by passive binding. Microalgae have distinct functional groups including -OH, $-PO_3O_2$, thiol, $-NH_2$, -COOH, -SH, and collection of polymers including cellulose, hemicellulose, and proteins over the cell surface [115]. Microalgal functional groups have specific properties of charge and affinity that make them an important aspect of tertiary and quaternary WWT. Contaminants or pollutants adhere to the algal blooms and are separated due to the existence of receptors that are capable of attracting and binding cations [116].

15.4.4.2. Accumulation

Elimination of pollutants can also take place when microalgae accumulate into their intracellular spaces. The utilization of personal care and pharmaceutical substances including antibiotics is reviewed as a key source of harmful compounds that end up in the wastewater streams. Around 74% of antibiotic triclosan (TCS) has been reported to be accumulated straight away by the micro algal species of *Nannochlorpsis* [117]. After seven days the same species accumulate the entire 100% of antibiotic TCS from the media. This is followed by biodegradation that includes the breakdown of pollutants inside or outside the cells of algae [118] (Figure 15.4).

15.4.4.3. Immobilization

In this technique, microalgae are immobilized within a gel of required porosity that allows the movement of the wastewater through this pose to the microalgae that absorb the pollutants in the wastewater. Pellets of microalgae can be prepared using sodium alginate. Encapsulated microcapsules were applied to the membrane bioreactor and were found to be efficient in the removal of COD and ammonia nitrogen [119]. In comparison to algae/sodium alginate beads and free algae, algae-encapsulating microcapsules removed 62.23% COD and 97.38% ammonia nitrogen (NH_3–N) [14]. Enhanced removal of raw sewage has been observed with the species *Chlorella kessleri, C. vulgaris,* and *Scenedesmus* sp. researched and found that *Chlorella coloniales* acts as an efficient biosorbent in wastewater [120]. It was identified that *C. coloniales* removed 95% Cd, Cr, As, Fe, and Co from the wastewater.

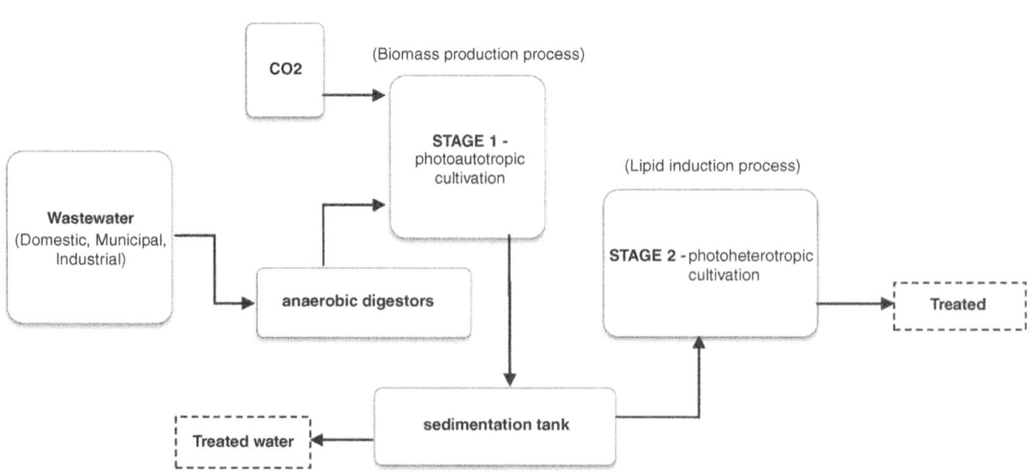

Figure 15.4 Flow chart representing different routes for biomass production.

Calcium alginate can also be used to trap microalgae in the form of beads for the elimination of pollutants from wastewater. This technique reported a 72% decrease in nitrate and 99% in orthophosphate content. The immobilization technique has been proved to be efficient in the prevention of cell contamination, energy overuse, and cell wash out [121].

15.4.5. Fungi

Fungal WWT provides various advantages including hydrates for degradation of complex organic compounds because of the presence of specific fungal species that degrade them, enhanced separation of solids of fungal biomass from mixed liquor, and the ability to retrieve fungal byproducts. Deleterious or xenobiotic organic contaminants can be treated through fungal WWT. Enzymes are developed throughout all the phases of the life cycle and are also present at low pollutant concentrations. Researchers have observed that high molecular mass organic pollutants can be degraded by specific and non-specific extracellular enzymes of fungal biomass [122].

For example, extracellular enzymes that catalyze the degradation of PAHs with the help of non-specific oxidation reactions result in the formation of different quinones and hydroxylated aromatic compounds. Highly oxidative enzymes including ligninase, phenol oxidase, and manganese peroxidase that can degrade lignin, phenol, dyes, and various other xenobiotic contaminants or pollutants are produced by white-rot fungi. *Aspergillus carbonarius* M333, a filamentous fungus, was used to form a bioactive ultra-filtration membrane that could be used to treat wastewater from textile industries. This bioactive membrane was found to reduce 73.2% COD and 91% decolorization [123, 124]. They found in their studies that fungal consortia enhanced the degradation of various xenobiotics present in WWPs. Some other fungi that play roles in the degradation of pollutants include *Aspergillus niger,* which degrades apple distillery waste; *Pleurotus ostreatus* that degrades lignocellulosic biomass; *Myrothecium verrucaria* and *Trametes hirsuta* that degrade cellulose-rich waste; *Phanerochate chrysosporium* that degrades lignin; and *Humicola grisea* that degrades raffinose. [125]. In another study, it was found that fungal consortia of *Mucor circinelloides, Trametes polyzona, A. niger, Trichoderma longibrachiatum,* and *Rhizopus* microspores were able to degrade PhCs and their byproducts such as ibuprofen, diclofenac, etc. [126]. In a study involving the use of *Tinea versicolor* monoculture for the treatment of municipal wastewater, it was found that it eliminated >99.9% diclofenas for an incubation period of three hours. This mechanism was based on the biosorption and laccase enzyme production ability of *Tinea versicolor.* In research, it was found that fungi derived from mangrove-growing regions of Sunderban were able to eliminate heavy metals using biosorption present in wastewater. It was identified that *Alternaria alternata* was able to remove 84.64% of cadmium and 98.27% of lead from a solution containing metal ions [127]. The pollutant removal efficiency of different algal sp. is tabulated in Table 15.3.

15.4.6. Yeast

Yeast is a single-celled fungus that may be circular, ovoid, or cylindrical. They may be divided into two categories, namely fermented yeast and oxidized yeast. Fermented yeasts are those that ferment only six-carbon sugars into carbon dioxide and alcohol [140]. Fermented yeast is used in the preparation of bread, wine, steamed bread, etc. On the other hand, oxidized yeast comprises strong oxidizing properties and weak or no fermentation capabilities. *Candida* and *Hansenula polymorpha* are examples of oxidized yeast applied in the petroleum industry and WWT. *Candida tropicalis* strain SDP-1 was found to be tolerant

Table 15.3 Pollutants removed from different wastewater by various fungal species.

S. No.	Species of fungal strain used	Pollutant removed	Removal percentage	Type of water	References
1.	*T. versicolor*	Diclofenac	99.7%	Municipal wastewater	[128]
2.	*Trametes versicolor* and *Aspergillus luchuensis*	Phosphorus, ammonia, nitrogen, and organic Carbon	98, 99.2, 98.9%	Municipal wastewater	[128]
3.	*Sordaria macrospora* k-hell and *M. thermophila*	Lignin	15–20%	Paper industry wastewater	[129]
4.	*Pleurotus ostreatus*	Lignin and color dye	37.7–46.5%	Paper industry wastewater	[130]
5.	*Phanerochaete chrysosporium* MTCC No. 787	Lignin and color dye	86, 71%	Paper industry wastewater	[131, 132]
6.	*Talaromyces islandicus*	Pb	80.72%	Dye industry and electroplating industry wastewater	[124]
7.	*B. adusta* and *P. ostreatus*	2-Naphthalenesulfonic acid	30–60%	Chemical industry wastewater	[133]
8.	*Achaetomium strumarium*	Acid Red 88	99%	Dye industry wastewater	[134]
9.	*Cylindrocephalum aurelium* RY06	Mordant Orange-1	85%	Dye industry wastewater	[135]
10.	*Penicillium glabrum*	COD	66.6–90.3%	Pistachio processing wastewater	[136]
11.	*Aspergillus flavus* CR500	As, Pb, Cr, and Ni	97.5, 93.3, 82.2, 46.6%	Multimetal contaminated wastewater	[137]
12.	*Fusarium oxysporum*	Terbuthylazine, difenoconazole, diflufenican, pendimethalin	70, 72.1. 67.7, 100%	Agrochemical wastewater	[15]
13.	*Trichoderma viride*	Terbuthylazine, difenoconazole, diflufenican, pendimethalin	57.7, 51.3, 45.2, 100%	Agrochemical wastewater	[15]
14.	*Ganoderma applanatum* and *Laetiporus sulphurous*	Celecoxib, diclofenac and ibuprofen	92,87, 79, and 89, 80, 66%	Pharmaceutical wastewater	[138]
15.	*Aspergillus terreus* GS28	Direct Blue-1 (Azo dye)	98.4%	Carpet industry wastewater	[139]

against Zn^{2+}, Mn^{2+}, and Cr^{3+} in the aquatic phase and could enhance phenol removal under various ranges of temperature (29–40 °C), pH (3.0–9.0), and NaCl. Immobilized cells present in coking (wastewater from coke industries) wastewater were found to have the ability to remove 383 mg l^{-1} of phenol and reduce COD by 50.38% [141].

Organic poisons, as well as refractory compounds, can be easily degraded by yeast. *Trichothecium roseum* is perfect for the treatment of wastewater due to its ability to eliminate nitrogen and phosphorus from the wastewater and its protein content [142].

In the late 1970s, Yoshizawa [143] came up with a yeast WWT method. Non-toxic and nutritious single-cell proteins can yield from organic matter by the action of yeast; it has high efficiency of WWT. Low pH (5.6) is ideal for the optimum growth of yeast in wastewater. Yeast cultures in wastewater were observed to eliminate 100% nitrate and total ammonium nitrogen (TAN) along with 92.6% orthophosphate. Yeast cultures obtained higher biomass concentrations of around 3.7 ± 0.1 g l^{-1} and 4.2 ± 0.1 g l^{-1} and were found to be beneficial for aerobic fermentation leading to bioethanol production and WWT [144]. High BOD to nitrogen and high BOD to phosphorus ratio was observed to enhance the growth of yeast, usually pointing out nutrient limiting conditions. Under aerated conditions with high BOD (comprising of simple sugars like sucrose, maltose, glucose, fructose, etc.) yeast multiplies quickly converting these substrates into biomass and increasing the volume of sludge. Alcohol and low molecular weight acids are produced from biomass during anaerobic conditions. Filamentous yeast has a higher surface area which helps them to scavenge nutritional elements much easier than bacteria [145].

15.4.7. Protozoa

Secretion of minerals by protozoa results in faster consumption of carbon sources by the bacteria. Protozoa cannot increase carbon mineralization under limited carbon conditions with the help of these indirect factors. Around 4% of wastewater comprises the population of protozoa. The most widely found protozoa include amoeba, ciliates, and flagellates. Protozoa also increase the clarity of effluents by consuming unsettled floc and free bacteria [146, 147].

Amoebas are primarily present during young sludge age (SA) as they require high nutritional levels and low competition to grow. Amoeba can also play a dominant role under toxic conditions, low dissolved oxygen (DO), shock loads of BOD, and high concentration of particulate matter. Amoeba also forms a gelatinous shell in the wastewater.

The growth of flagellates is influenced during younger sludge ages before bacteria have a chance to populate. Wastewater comprising of high soluble nutrient levels or high food to mass ratio consists of a high population of flagellates. *Peranema* sp. are the only flagellate applicable as a bioindicator in MBR (membrane bioreactor) systems [148].

Ciliates are influenced by healthy sludge composition. They are an indicator of healthy floc formation as they do not consume organic matter but do feed on bacteria; as a result, they are used as clarifying agents. In the absence of ciliates, algal and bacterial populations may grow out of control in the wastewater system [149]. Aerobic biological treatment system comprises of various ciliated protists. The majority of ciliates feed on dispersed bacterial populations in microbial WWT [150]. These may be crawling, attached, or free-swimming ciliates. Ciliates feed on bacteria that are present in the mixed liquor, which helps in the regulation of bacterial biomass in the effluent and improves its quality. The most important group of ciliates found in WWTPs are the subclass *Peritrichia* [150–154].

Curds et al. [151] observed in their experiment that effluents turned turbid in the absence of protozoa, as there was a high content of suspended bacterial species. However, with the addition of protozoa into the effluent, the turbidity disappeared.

Table 15.4 Pollutant removal by protozoa species in different wastewater.

S. No.	Species of protozoa strain used	Pollutant Removed	Removal Percentage	Type of wastewater	References
1.	*Paramecium* sp., *Opercularia* sp.	COD	25–98%	Petroleum industry wastes	[160]
2.	*Cinetochilum margaritaceum*	bacterivorous grazers, carbon wastes	4×10^{-7} to 1×10^{-6} ml hours	Textile wastewater	[155]
3.	*Vorticella* sp. and *Epistylis* sp.	COD	15–95%	Petroleum industry wastewater	[160]
4.	*Aspidisca* sp., *Paramecium* sp., *Peranema* sp., *Trachelophyllum* sp.	Phosphate, nitrate and Ni^{2+}	66.4–99.36%, 56.19–99.88%, and 45.98–85.69%	Industrial wastes	[161]
5.	*Euplotes affinis*	N, bacterivorous grazers	73.6%, 4×10^{-7} to 1×10^{-6} ml hours	Domestic wastes	[162]

The findings of Sridhar, Pillai, and Macek [155] concluded that few species of protozoa including *Chilodonella uncinata, Epistylis articulate, Stylonychia putrina, Aspidisca cicada, Colpidium camylum,* and *Vorticella* sp. [156] tend to reduce COD and the content of suspended matter [157–159]. Table 15.4 lists the different protozoa species utilized in different wastewater removal.

15.4.8. Others

Metazoans are multicellular organisms that have evolved from unicellular ancestors. They feed on protozoa, algal, and bacterial populations. Metazoans account for about 3.9% of the wastewater. Rotifers, nematodes, tardigrades (water bear), annelids, ostracods (daphnia), and copepods (water flies and mites) are among the most widely found metazoan populations in WWT.

Rotifers play a role in the consumption and elimination of remaining matter in wastewater. They secrete a certain sticky substance that aids floc to stay firm and clump together. In the presence of toxic substances in the wastewater, these are the first to be impacted [163].

Nematodes are found to be present in older sludge matter. This may feed on fungus, protozoa, and bacterial populations, and in some cases, they may also feed upon other nematodes. Certain nematodes have teeth that penetrate their prey and then consume the food out of it. The fastest-growing species (*Paroigolaimella bernensis* and *Diplogasteritus nudicapitatus*) were observed in the pilot reactor in contrast to the more slowly growing species (*Diploscapter coronatus* and *Acrostichus* sp.), which dominated in the laboratory reactors [164].

Tardigrade feed upon smaller protozoa and algae. If oxygen is scarce, *Tardigrade* may swell up like a balloon and float around for a few days. As soon as the environment dries up, they shrink up like a raisin. Nevertheless, they are quite sensitive to a toxic environment. Thus, their presence may indicate little to no ammonia content.

Annelids may refer to ringed worms. They also include leeches, earthworms, and ragworms. *Tubifex,* which is also known as sludge worm, is present in the sediments of lakes, rivers, etc. It is commonly found around sewer outlets and acts as an indicator of water pollution [165].

Copecods are true aerobes and are sensitive to a toxic environment. They are mostly found in stable activated sludge systems.

Ostracods indicate the quality of surface water. They are found to be scavengers of organic matter in wastewater. *Cypress teresa* was found to dominate ostracod populations in areas polluted by untreated industrial and domestic waste in Melah lagoon, Egypt [166]. Low ostracod diversities were found in silty sediments with heavy metal concentrations that are likely toxic [167].

15.5. DIFFERENT REMEDIATION TECHNIQUES

This section deals with the different remediation techniques utilized to remove various pollutants from wastewater. This section is subdivided in three sections: bioremediation; mycoremediation; and phycoremediation.

15.5.1. Bioremediation

Bioremediation is a biologically mediated treatment method wherein, microorganisms are applied to eliminate or neutralize contaminants or pollutants through the metabolism of microorganisms [168]. *Bacillus thuringensis* has been used to biologically degrade diesel oil [169]. *Haloalkane dehalogenase, alanine* has been used to transfer heavy groups of amino acids and an alternative was found to carry dichlorination of dichlorohexane. Another microbe named *Deinococcus radiodurans*, which is a radioactivity resistant microbe has been found to endure radiations beyond natural standards and hence, is used in cleaning up radioactive waste [168].

Microorganisms help in converting various organic contaminants into end products such as carbon dioxide and water or into intermediates of metabolism that may be used as chief substrates in the growth of cells in bioremediation. They help in maintaining a bidirectional defense system. Firstly, they produce certain enzymes that help in the degradation of specific contaminants or pollutants. Secondly, they provide resistance toward related heavy metal substances. Few ways to eliminate or degrade pollutants include immobilization, transformation, binding, oxidation, and converting heavy metals into volatile substances. Table 15.5 list the different heavy metals accumulated through various microorganisms. *Trichoderma lixii* CR700 was found to be tolerant to Pb, Cd, Zn, Cr, Ni, As, and Cu. *T. lixii* CR700 had high tolerance against Cr, Zn, Ni, and As, whereas it was less tolerant to Pb and Cd. *T. lixii* CR700 had a tolerance index of 0.15 ± 0.02 and tolerated around $1200\,mg\,l^{-1}$ of Cu^{2+} [171]. In another study, Chen et al. [172] found that *Penicillium simplicissimum* was able to

Table 15.5 Heavy metals accumulated by different microorganisms [170].

S. No.	Microorganisms	Heavy metals
1.	*E. coli strain*	Cd^{2+}
2.	*Deinococcus radiodurans*	Hg
3.	*Caulobacter crescentus*	Cd^{2+}
4.	*Methylococcus capsulatus*	Cr^{4+}
5.	*P. ostreatus*	Pb
6.	*Pseudomonas fluorescens OS8*	Cd, Hg, Zn, Pb
7.	*Staphylococcus aureus*	Cd, Hg, Zn, Pb
8.	*Sphingomonas desiccabilis*	As
9.	*Chlorella vulgaris*	Cu, Pb

eliminate 63% of Cu^{2+}. *Penicillium janthinellum, Fusarium solani, P. simplicissimum,* and *Aspergillus flavus* CR500 was also reported to eliminate various contaminants present in the wastewater [171–174]. Resistance toward As, Cd, and Cr has been observed in various bacterial populations due to the presence of chemiosmotic ion/proton pumps and ATPases [170]. Kumar and Dwivedi [137] reported that *Aspergillus flavus* CR500 removed around 99% As, 97.7% Pb, 74.9% Cr, and 73% Ni present in single metal amended liquid medium. The growth of microbes that utilize discharged chemical contaminants or pollutants as their source for energy as well as food is restored by bioremediation. In nontoxic environmental conditions, numerous organisms can break down organic matter. The microorganisms that feed on pollutants are provided with fertilizer, oxygen, and other factors through bioremediation to enhance their growth rate. This helps them to break down organic matter at a better rate.

Biosorption is referred as the greater affinity of a biosorbent toward a sorbate that helps in the maintenance of equilibrium between the two. A biosorbent is a microorganism and sorbate may be any metal ion. *Saccharomyces cervisiae* and *Pseudomonas aeruginosa* B237 are applied in the ion-exchange method which aids in the degradation of cadmium (Cd^{2+}) and zinc (Zn^{2+}) ions [175, 176]. *Ulva fasciata* and *Colpomenia sinuosa* showed enhanced biosorption at 50 and 80 mg l^{-1} of Co (II) concentration, respectively. Similarly, *C. vulgaris* [177], *Ulva lactuca*, and *U. fasciata* showed a higher ability to biosorb cadmium at 75 mg l^{-1} and the latter at 20 mg l^{-1} [178]. *Synechocystis pevalekii* and *Scenedesmus bernardii* are other microorganisms used to degrade heavy metals (e.g. cobalt) found in wastewater from industry effluent [179]. In a research study, an integrated shrimp-macroalgae system was used to remove nutrients from wastewater. *Gracilaria changii* and *Gracilaria edulis* were used and their efficiency in nutrient removal was studied. It was found that *G. changii* removed 58.8% ammonium and 56.8% nitrate. On the other hand, *G. edulis* removed 72.5% ammonium and 71% nitrate [180]. Microorganisms also obtain energy by oxidizing organic components with manganese (Mn^{4+}) and iron (Fe^{3+}) ions, which act as electron acceptors. Microbes are also helpful in altering the state of metals and the solubility of metals. *Acinetobacter* is used in biodegradation as it reduces the soluble state of uranium (U^{6+}) into its insoluble state (U^{4+}) and helps in the removal of uranium from waste water [181]. In 1981, a strain of *Pseudomonas putida* was declared the first patent as a bioremediation agent. *Nitrosomonas, Penicillium, Pseudomonas, Xanthofacter, Bacillus, Flavobacterium,* and *Mycobacterium* are a few microorganisms that belong to the group that has been applied in the biodegradation of distinct compounds or substances [182]. Table 15.6 describes the widely used bioremediation methods worldwide.

15.5.2. Mycoremediation

The application of fungi to eliminate toxic or deleterious contaminants from the environment is referred to as mycoremediation [184]. Mycoremediation is applied to heavy metals, agriculture waste, dyes, detergents, etc., to treat them (Figure 15.5). The most appropriate mycoremediators include mushrooms and macro fungi. Oyster mushrooms or *Pleurotus* sp. is among the most widely grown varieties in the world and are said to be accredited with higher yields and low cost of production [185].

Mushrooms are an efficient biosorption tool as they can accumulate heavy metals in their body even above maximum legitimate concentrations in small lifetime. Leo'n Santiesteban et al. [186] reported that *Rhizopus oryzae* CDBB-H-1877 was able to biosorb pentachlorophenol. This was possible through methylation and dechlorination. Heavy metals are widely present in areas with higher traffic, battery waste polluted areas, areas with high pollutant

Table 15.6 Widely applied bioremediation methods.

Bioaugmentation	Inclusion of bacterial cultures to a contaminated medium to decompose different pollutants; often used in ex situ systems and bioreactors [183].
Biofilters	Application of certain microorganisms to an organic or inorganic medium to degrade volatile pollutants in the air.
Biosparging	Air is injected into the saturated region under pressure; this may result in the upward movement of volatile substances into the unsaturated region and thus enhance biodegradation.
Biostimulation	Alteration of the environment to restore the pre-existing microorganisms to carry bioremediation. It may occur either in ex situ or in situ.
Bioreactors	Tanks or systems that degrade contaminants present in the soil or groundwater or slurries using microorganisms.
Bioventing	Biodegradation of contaminants is restored by the incorporation of air or oxygen into existing microbes in the soil.
Composting	An anaerobic process involving the decaying of organic waste. Here, the organic matter is decomposed into compost, which is a mixture of substances that enhances the quality of the soil.
Land farming	It is an ex-situ above-ground restoration technology. It is of great significance in case of lands with liquid hydrocarbon or gasoline contamination.

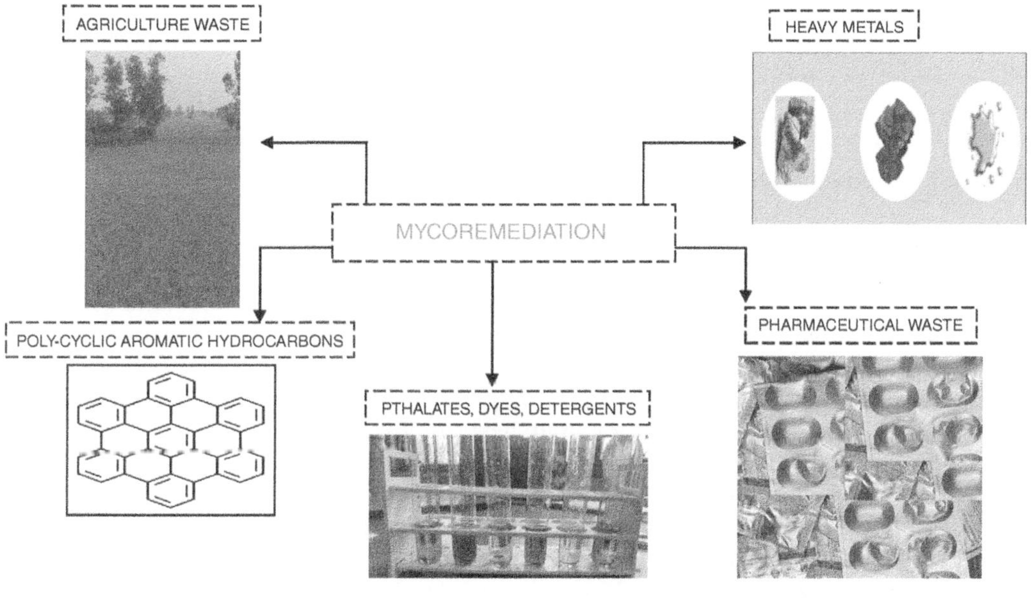

Figure 15.5 Diagrammatical representation of different applications of mycoremediation.

emissions, etc. *Pleurotus* sp. growing around these areas can accumulate heavy metals in their body in higher concentrations. *Pleurotus sajor-caju* has also been found to accumulate cadmium content above legitimate concentrations [187]. Chatterjee et al. [188] identified that super-paramagnetic Fe_3O_4 NPs formed from *Aspergillus niger* BSC-1 could be used to eliminate hexavalent chromium from wastewater.

Mushrooms use distinct methods to cleanse polluted areas and restore environmental conditions. These methods include biodegradation, biosorption, and bioconversion. Initially, organic matter is degraded by the application of microorganisms, which is termed biodegradation. White-rot fungal strains obtained from *Bjerkandera adusta*, *Ganoderma resinaceum*, *Phlebia rufa*, *Irpex lacteus*, and *T. versicolor* were reported to show increased dephenolization and decreased total organic carbon (TOC). The aim was to study the impact of these strains on vinasse biodegradation. It was also seen that among all the strains, *P. rufa* was able to decolorize better and had a better ability to tolerate vinasse toxicity [189].

Uptake of contaminants or pollutants may occur either through biosorption or bioaccumulation. Bioaccumulation refers to the accumulation of contaminants or pollutants inside the body of the microorganism. It is a metabolism-dependent system. Bioconversion refers to the conversion of organic matter into valuable products or energy with the help of microorganisms. It further involves pre-treatment, which allows cellulose to be available for cellulolytic enzymes. Hydrolysis of cellulose and hemicellulose occurs and in the end, hydrolytic enzymes help in fermentation to produce valuable products such as ethanol [190].

Flammulina velutipes, *P. ostreatus*, and *Agaricus blazei* have been utilized in wine production. Mushroom species such as *P. ostreatus* degraded oxo-biodegradable plastic and grew on it, and *Pleurotus pulmonarius* degraded radioactive cellulosic-based waste and crude oil. *Schizophyllum commune* and *Polyporous* sp. [190] were found to degrade malachite green. Ortiz-Monsalve et al. [191] observed that *Trametes villosa SCS-10* resulted in an 80% reduction in COD and TOC. It also showed an enhanced ability to decolorize dyes such as Acid Blue 161, Acid Black 210, and Acid red 357 in aqueous solution. Mushrooms are capable of eliminating poly-R-478, which tells us that it is appropriate for degrading PAHs [184]. *Trametes hirsute* and *Pleurotus dryinus* were used in the mycoremediation of biorefinery wastewater (BRW) and it was found that they were able to eliminate phenol up to 94 and 100%, respectively. These species also resulted in the simultaneous production of laccase in BRW [192].

15.5.3. Phycoremediation

Phycoremediation involves the application of algal species to treat wastewater. Wastewater comprises of nutrients and hazardous chemicals that may harm the environment if left untreated. It is an emerging technology applied in the treatment of wastewater. Various micro- and macroalgae have the natural ability to degrade chemicals present in wastewater. Various factors influence the process of phycoremediation including pH, temperature, ion concentration, chemical composition of microalgal biomass, etc. Microalgae have also been found to help in carbon sequestration that helps to fight global warming and climate change [193]. The principle behind phycoremediation is the biosorption ability of algae. Certain algal species are able to biosorp toxic contaminants into their cell wall and then convert them into forms that are non-toxic to the environment [194]. This ability of algal species makes them an ideal contestant for remediation of contaminants present in wastewater. In contrast to treating wastewater, algal biomass produced during phycoremediation can be used for energy generation purposes.

Algae remove organic as well as inorganic pollutants through physical, biochemical and biological methods. The biochemical method involves cation/anion exchange, precipitation, oxidation/reduction and absorption. On the other hand, biological process includes open pond system/waste stabilization pond.

- **Cation/anion exchange:** Algal cell wall comprises of –COOH, –OH, PO_3^{2-}, $-NH_2$, –SH, $-P_2O_3$ etc. that act as strong metal-cation binding site and help in metal exchange.
- **Absorption:** Wastewater comprises inorganic ions such as NO_3^-, PO_4^{3-}, and heavy metals. Microalgae convert inorganic nitrogen into organic nitrogen through the process of

assimilation. During assimilation, inorganic nitrogen moves into the cytoplasm. A number of oxidation and reduction reactions occur in the cytoplasm resulting in the conversion of nitrogen to ammonia in the presence of NO_2^- and NO_3^-. Later, ammonia is absorbed in the cytoplasm. In the case of phosphorus, it is consumed in the form of $H_2PO_4^-$ and HPO_4^{2-}. Further, PO_4^{3-} is converted into organic state through phosphorylation. Metal ions with higher electronegativity and smaller ionic radii are immediately absorbed by algal biomass.

- **Precipitation:** Wastewater secretes numerous chemicals and secondary metabolites that lead to decrease in pH. This influences precipitation of various pollutants. Precipitation helps in the enhancement of phosphorous reduction. Metal cations are left with no sites to bind due to the binding of protons to the active sites at low pH. With the increase in pH, negatively charged sites increase and thus, metal cations adsorb on the cell surface, resulting in the decline of bioavailability.

Microalgae have the natural ability to convert cadmium, mercury and other compounds to form sulphides that are less toxic to the environment. *Chlamydomonas reinhardtii, Cyanidioschyzon merolae,* and *Synechoccocus leopoliensi,* when exposed to cadmium were found to convert it into cadmium sulfide [195]. In another report, *Hydrodictyon* sp., *Oedogonium* sp., and *Rhizoclonium* sp. were found to accumulate vanadium and arsenic from the wastewater derived from coal industries. *Nitella pseudoflabellata,* green algae, was also found in a study to remediate Cr (IV) [196]. Another algal species, *Chlorella pyrenoidosa,* reduced nitrate, COD, and phosphate present in synthetic wastewater to 99.2, 61.0, and 70.1%, respectively [197]. Various algal species including *Haematococcus* sp., *C. vulgaris, U. lactuca,* and *Oedogonium subplagiostomum* AP1were found to reduce Malachite green (67%), Aniline blue (58%), Methylene blue (4.012%), and Methyl orange (97%), respectively [198].

15.6. ADVANTAGE AND DISADVANTAGE OF USING BIOLOGICAL WWT METHODS

This section summarizes the use of microbes as a biological method as boon or ban for WWT.

15.6.1. Benefits of Biological WWT

(i) Employment of microbes for biological degradation of organic pollutants is a simple, cost-effective approach.
(ii) It is a self-supporting system.
(iii) It provides us with a beneficial approach to cleanse water bodies and reduce energy intake.
(iv) Implementation of microbes in WWT greatly reduces aeration time and enhances the efficiency of the treatment.
(v) It helps in the major elimination of deleterious pathogens, odor and improves the quality of air.
(vi) Significantly eliminates biodegradable organic compounds and other harmful elements.
(vii) Various microorganisms present in the biofilm break down organic matter, etc. after the removal, the treated water is passed through biofilters and then reused for various activities.
(viii) Assists in controlling the rise in BOD, COD, DO, and turbidity.

15.6.2. Detriments of Biological WWT

(i) Management and conservation of microorganisms, as well as a physiochemical pre-treatment, is essential.
(ii) It is a somewhat cumbersome process.
(iii) Certain substances, as in the case of dyes, have a low degradability rate.
(iv) There is a possibility that the sludge may bulk up and the formation of foam may take place during the treatment.
(v) Biological WWT requires sufficient knowledge about the way enzymatic processes control the degradation of substances.
(vi) Expensive as it requires high operating costs.

15.7. POTENTIAL OF WASTEWATER AS AN ENERGY GENERATION OPTION

Wastewater retrieved from municipal areas can be a great source of chemical energy. We can retrieve organic carbon from the sludge and recover chemical energy by concentrating organic carbon, etc. On the other hand, local or domestic wastewater can be a source of thermal energy. Wastewater can yield up to 10 times the energy required in its treatment. Wastewater is a good source of nutrients and plays a significant role in energy generation. There are several ways in which energy can be generated and applied in various aspects of life. Here in this section, we will be reviewing different energy products such as biogas, biohydrogen, bioethanol, biodiesel, and bioelectricity obtained from wastewater. Table 15.7 summarizes the energy generation potential in form of biogas, biohydrogen, bioethanol, bioelectricity, and biodiesel from different wastewater which is used by various microorganisms.

15.7.1. Biogas

Biogas primarily refers to a gas that is generated by the biological breakdown of organic matter without oxygen. Biogas is produced from sewage, biomass, plant material, and other waste materials. Biogas can be utilized as fuel for cooking purposes. It can also be added into an anaerobic digester where the energy in the gas is transformed into heat and electricity. Compressed biogas can also be used as natural gas to generate power for motor vehicles. Biogas mainly comprises CO_2 (25–50%), hydrogen (0–1%), CH_4 (50–75%), nitrogen (0–10%), H_2S (0–3%), oxygen (O), and sulfur (S) [16].

It is generated as digester gas or landfill gas. Landfill gas mainly comprises about 50% methane. Modern WWT can generate biogas with concentrations of CH_4 up to 55–75%. In certain cases, biogas generates siloxanes, which are formed as a result of the decomposition of substances under anaerobic conditions. Numerous factors are responsible to affect methane production which includes retention time, toxics, completion between sulfur-reducing bacteria and methane-producing bacteria, temperature, and pH. It may also depend upon the type of WWTP used [216].

Various bacterial populations are said to be involved in the anaerobic production of methane:

- Hydrogen-producing acetogens (HPA) cause the catalysis of some fatty acids, etc.
- Hydrolytic bacteria are said to be responsible for the catalysis of lipids, carbohydrates, proteins, and other biomass components.
- Methanogens consume carbon dioxide, hydrogen, and acetate to form methane.
- Homoacetogens that use acetic acid to produce acetate [217].

Table 15.7 Summarizes the use of different microorganisms for energy generation by utilizing different wastewater.

S. No.	Type of microorganism	Wastewater	Output (Energy generation)	References
		Biogas		
1.	*Acutodesmus* sp. *AARL G023*	Kitchen wastewater	1.44 ± 0.07 ml g^{-1} d^{-1}	[199]
2.	*Arthrospira platensis*	Municipal wastewater	481 ml g^2-VS	[200]
3.	*Chlorella vulgaris* or *Scenedesmus abundans*	Industrial wastewater	248 ± 10 ml CH_4/g VS	[201]
4.	*Chlorella vulgaris*	Sulfate-laden wastewater	06.36 m l-CH_4 g^{-1}	[202]
		Bioelectricity		
5.	*Shewanella putrefaciens* and *Acinetobacter calcoaceticus*	VBR wastewater	194.8 mW m^{-3}	[203]
6.	*Clostridium, Tetrathiobacter,* and *Desulfovibrio* sp.	Sulfate-laden wastewater	1188 mW m^{-3}	[204]
7.	*Psathyrella candolleana*	Septic tank wastewater	110 ± 3 mW m^{-2}	[205]
8.	*Synechococcus* sp. and *Chlorococcum* sp.	Kitchen wastewater	41.5 ± 1.2 mW m–^{2}	[206]
		Biodiesel		
9.	*Graesiella emersonii*	Garden-based wastewater	13.8 mg l^{-1} day^{-1}	[202]
10.	*Monoraphidium* species	Industrial wastewater	78.9 mg l^{-1} day^{-1}	[207]
11.	*Oleaginous bacteria DS-7*	Dairy wastewater	1.2 gl^{-1} day^{-1}	[208]
12.	*Chlorella minutissima*	Dairy wastewater	5.76 ± 0.06 mgl^{-1} day^{-1}	[209]
		Biohydrogen		
13.	*B. coagulans* MO11	Molasses and agricultural wastewater	79 ml l^{-1}	[21]
14.	*Clostridium beijerinckii* CN	Agricultural wastewater	125 ml l^{-1}	[21]
15.	*Clostridium beijerinckii* DSM 791	Rice mill wastewater	1.5 mg l^{-1}	[210]
16.	*Enterobacter aerogenes*	Sago wastewater	630.67 μmol l^{-1}	[211]
		Bioethanol		
17.	*H. tetrachotoma ME03*	Municipal wastewater	11.2 ± 0.3 gl^{-1}	[212]
18.	*Chlorella* sp.	Industrial wastewater	10.4 g l^{-1}	[213]
19.	*Wickerhamomyces* sp.	Marine wastewater	27.31 ± 1.40 g l^{-1}	[214]
20.	*S. cerevisiae*	Industrial wastewater	24.53 ± 0.68 g l^{-1}	[215]

Hydrolytic, HPA, and homoacetogens include *Cellulomonas, Clostridium, bacillus, bacteroides, Ruminococcus, Eubacterium,* etc., and on the other hand, methanogens include *Methanosarcina, Methanothrix, Methanobacterium,* and *Methanospirillum.*

Purple phototropic bacteria enhanced biogas in piggery WWT under a photo-anaerobic environment. The use of the algal-bacterial group assisted methane concentrations to increase biogas by 73.6%, while purple phototropic bacteria increased it to 93.3% that confirming the ability of purple phototropic bacteria for biogas increment coupled to piggery WWT [218].

On account of using a low amount of black water, there is a reduction in carbon footprint. When black water is combined with reduced water, it may gain 38 kg CO_2 per household every year [219]. When we combine black water with kitchen water in an upflow-anaerobic sludge blanket (UASB) reactor, it converts 40% of COD into biogas and yields average energy of 65 700 KWh (thermal) every year, which is similar to 8200 Nm^3 of natural gas, 80% methane, and this yields an additional 60 450 KWh (electrical) every year. It was reported that on pretreatment of thermal sludge hydrolysis, biogas production raised to 150% [17]. It was reported that when yeast wastewater was put through digestion under anaerobic conditions and fungal fermentation it led to the reduction in COD and produced valuable products. *Aspergillus oryzae* grown over 1 m^3 brewer's yeast wastewater (BYW) reduced COD by 36.3% and yielded 5.13 kg of fungal biomass. After cultivation of *Mucor hiemalis* and *Mucor indicus* with wastewater, around 3.16 m^3 of methane was generated from each m^3 of fermented wastewater [220].

Another microorganism, *C. pyrenoidosa,* grew in synthetic wastewater as well as real sewage treatment plant and was found to reduce nitrate by 99.25%, phosphate by 94.2%, and COD by 87.1%. It has been utilized in the production of biogas on digestion with anaerobic microbes. Pretreatment of algal biomass either through thermal, hydrothermal, ultrasonic, or other methods may enhance biogas production. Untreated algal biomass yields about 479 ml of biogas and 147 ml per volatile solid (g) of methane. Biogas generation by hypothermal, fenton, thermal and ultrasonic pre-treatment was found to be about 14.19, 11.06, 10.43, and 7.93%, respectively [197]. Likewise, Arias et al. [221] found that micro algal biomass diluted with secondary wastewater enhanced micro algae growth. Complete uptake of N-NH^{4+} and P-PO_4^{-3} was observed and 1.1 g TSS l^{-1} of algal biomass was constantly generated. It is predominantly comprised of *Scenedesmus sp.* with low CH_4 yield (82 ml CH_4/g VS), improved by 130% on auto hydrolysis treatment and co-digestion with waste-activated sludge [221].

15.7.2. Floating Gas Holder Type Biogas Plant

In this, a steel dome-shaped holder is present to collect biogas. The holder is movable and keeps on floating over the slurry in the digester tank. On the other hand, anaerobes break down complex carbon compounds present in cattle dung into simpler ones in the presence of water in the digester tank. The anaerobic decomposition to yield biogas takes around 60 days [222]. The schematic representation of floating gas holder type biogas plant is shown in Figure 15.6.

15.7.3. Fixed Dome Type Biogas Plant

The slurry formed by the combination of equal proportions of biomass and water is maintained in the digester tank at 35 °C. Microorganisms decompose the slurry to yield biogas, which are collected in the dome-shaped holder. As the amount of biogas in the dome starts increasing, biogas builds up pressure and forces the depleted slurry to move out of the outlet chamber into the overflow tank [222]. The schematic representation of a fixed dome type biogas plant is shown in Figure 15.7.

Biogas is an energy-rich renewable source that is being used to meet the growing energy demands of the world. Biogas acts as thermal as well as electrical energy source for various distinct activities. Along with being an energy source, biogas also helps in the uptake of certain ions such as -NH^{4+} and P-PO_4^{-3}.

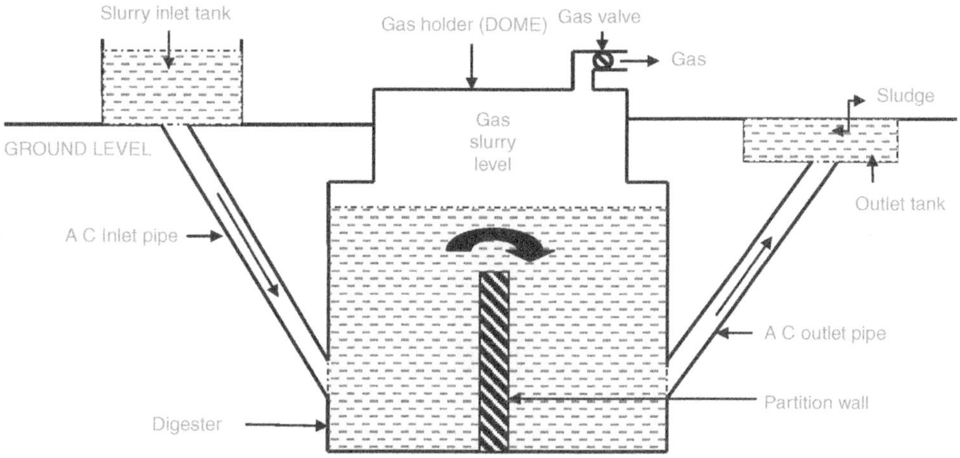

Figure 15.6 Schematic representation of floating gas holder type biogas plant.

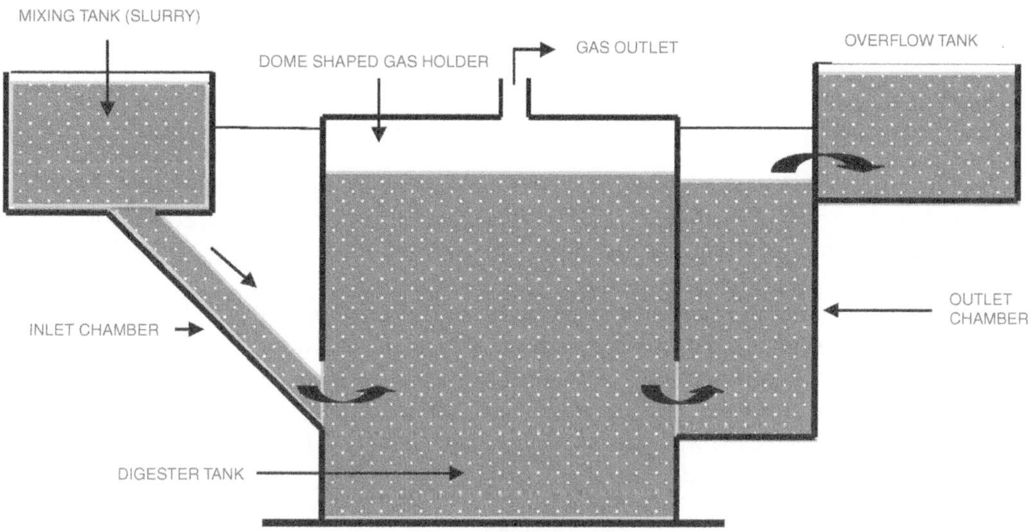

Figure 15.7 Schematic representation of fixed dome type biogas plant.

15.7.4. Bioelectricity

Certain microorganisms catalyze the oxidation of organic matter to generate electrical energy in microbial fuel cells or MFCs. MFC is a bioelectrochemical system (BES) that can not only treat wastewater but also generate bioelectricity [18] (Figure 15.8). Potter [223] in 1911 became the first person to observe that bacterial energy can be used to generate electricity. Mokhtarian et al. [224] discovered that membranes and mediators are not an essential part of MFCs. Other researchers also reported that the constriction of microbial reactors influenced the performance of MFCs [225].

MFCs consist of a proton exchange chamber that separates the anode from the cathode chamber. The rapid movement of electrons from anode to cathode results in the generation

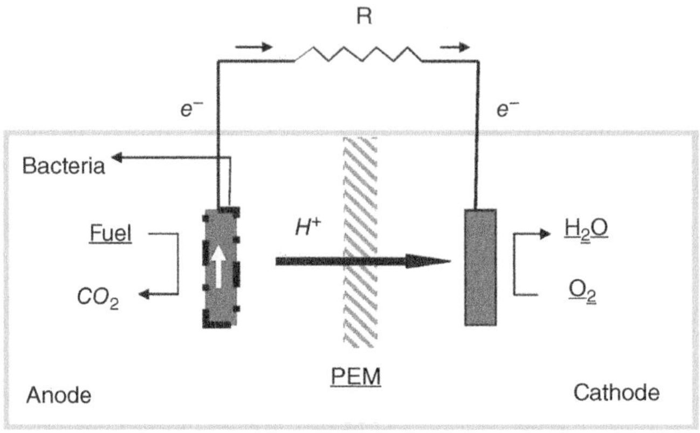

Figure 15.8 Representation of simple microbial fuel cell.

of electric current in the system. In an MFC of acetate as their substrate, the following reactions occur:

$$\text{Reaction at anode}: CH_3COOH + H_2O \rightarrow 2CO_2 + 8H + +8e^- \quad \text{(v)}$$

$$\text{Reaction at cathode}: 8H + +8e^- + 2O_2 \rightarrow 4H_2O \quad \text{(vi)}$$

$$\text{Total reaction}: CH_3COOH + 2O_2 \rightarrow 2H_2O + 2CO_2 \quad \text{(vii)}$$

The generation of electricity based on the electrons released from biochemical reactions facilitated by microbes is evaluated. MFC represents an eco-friendly approach for generating electricity while purifying wastewater concurrently, thus achieving up to 50% COD removal and power densities in the range of 420–460 mW m^{-2} [226].

Microbial species referred to as exoelectrogens aid in the transfer of electrons, whereas microbes in inactive states aid in the degradation of complex substances [227]. *Clostridium* is responsible for the conversion of complex substrates, *Deltaproteobacteria* is found to transfer electrons directly to the anode. Another microbe, *Geobacter sulfurreducens* KN400, produces up to 3.9 W m^{-2} of electricity, whereas *Shewanella putrefaciencs* yield 4.4 W m^{-2} of electricity. NASA applies *Shewanella oneidensis* to gather energy for its spaceships. Other microbes with exoelectrogenic capabilities include *Saccharomyces cerevisiae, Rhodopseudomonas palustris* DX1, *E. coli* DH5α, *and Candida melibiosica* [228].

MFCs have been applied with microalgae – *Synechococcus* sp. and *Chlorococcum* sp. – to process kitchen wastewater [206]. A power density of 30.2 mW m^{-2} was generated by the use of *Chlorococcum* sp., and 41.5 mWm^{-2} in case of *Synechococcus* sp. The electricity generation capacity of *Chlorella sorokiniana* was also studied using a double-chambered flat panel. The power density generated was 2.32 Wm^{-3} under 12:12 light: dark photoperiod, 12 hours inoculum, and light intensity of 140 μm [229]. *Choricystis* sp., green algae, along with boron has the potential to convert light into electricity. Growth mediums with distinct boron concentrations were treated with photosynthetic algae and then put into BEFCs (bio-enzymatic fuel cells) to develop a biofilm. Green algae with 60 mM boron resulted in 33 mV of voltage. Higher potential of 15 mV was observed in the boron system, while the boron-deficient system lost 20% of starting an activity. Boron systems retained about 95% of total activity. Maximum power density in the case of *Choricystis* sp., in a BEFC system, was 45.2 mWm^{-2} at 154 mAm^{-2} current density [230].

Bhagchandanii et al. stated that some physical and biological factors govern the performance of MFCs. These physical factors include configuration of electrode material and separator, whereas biological factors refer to inoculum choice, concentration, and type of substrate, etc. [231].

MFCs can either be mediator MFCs or mediator-less MFCs. Unlike mediator MFCs, in a mediator, some MFCs electron mediators are replaced by microorganisms that help in the transfer of electrons to electrodes resulting in the generation of electricity using nanowires.

Widely applied substrates in electricity generation include glucose, acetate, pyruvate, lactase, etc. The existence of glucose in the wastewater sludge improves the conducting ability of MFCs.

The study of Umar et al. [232] showed that petrochemical and chemical industries release a compound named xylene that may cause chronic problems. Xylene is firstly oxidized into 3-methylbenzoic acid, which further gets converted into carbon dioxide. At 1 kΩ of cell potential, BMFC could generate 410 mV of voltage within 23–90 days. BMFC (Benthic microbial fuel cells) reactor generated a maximum density of power of about 63 mW m^{-2} and 0.4 mA of current under the resistance of 20–1000 Ω. *Staphylococcus edaphicus* and *Staphylococcus saprophyticus* were seen to be the predominant species in the control and BMFC electrode in association with xylene biodegradation.

Thulasinathan et al. [205] employed *Psathyrella candolleana* for the degradation of PAHs due to their high ligninolytic ability. Different conditions were implemented with six distinct MFC systems for the experiment. In MFC 1 system, $K_3[Fe(CN)_6]$ was applied as cathode buffer and septic tank wastewater at the anode, this resulted in 110 ± 10 mW m^{-2} of power density and 90 ± 10 mA m^{-2} current density. In rest of the 5 MFC systems, basidiomycetes fungi were applied at anode and septic tank wastewater at anode. A current density of 497 ± 17 mA m^{-2}, 519 ± 10 mA m^{-2}, 522 ± 2 mA m^{-2}, and 525 ± 20 mA m^{-2} was observed in MFC2, MFC3, MFC4, MFC5, and MFC6, respectively. On day 14, *P. candolleana* achieved a power density of 525 ± 20 mA m^{-2}, 77% COD elimination, and 62 ± 1.13% of anthracene biodegradation.

Bioelectricity helps in WWT and energy generation with the help of numerous microbial species including *Clostridium, Deltaproteobacteria, S. oneidensis,* etc. MFCs have been found to remove up to 50% COD and generate 420–460 mW m^{-2} of power with the exchange of electrons along the chambers.

15.7.5. Biodiesel

Biodiesel is formed from the oils of vegetables and fats derived from animals. It acts as a substitute for diesel. Biodiesel is considered the most important biofuel in Europe [233]. It makes up about 80% of the market for transport biofuels. Oils and fats can be used to generate biodiesel in four distinct ways: thermal cracking or pyrolysis, direct blending, transesterification, and emulsification. Direct application of fats and oils in diesel engines is not possible as it may cause damage due to its high viscosity, fatty acid content, acidity, polymerization, and oxidation during its storage and combustion [234].

Silambarasan et al. [19] isolated seven distinct cyanobacterial species (LS01-LS07), among which *Nostoc* sp. (LS04) was capable of growing over wastewater from municipal treatment plants. It propagated in 75% wastewater and had the highest nutrient elimination ability of about 81.02–95.17%. Its lipid content was 14.85 ± 0.86% (dry cell weight) and productivity of 19.46 ± 8.08 mg l^{-1} d^{-1}. The higher amount of lipid in *Nostoc* sp. was the reason for its application in biodiesel production [19]. Furthermore, Behera et al. observed in their findings that oleaginous bacterium DS-7 could gather a higher amount of lipids (>80%) on

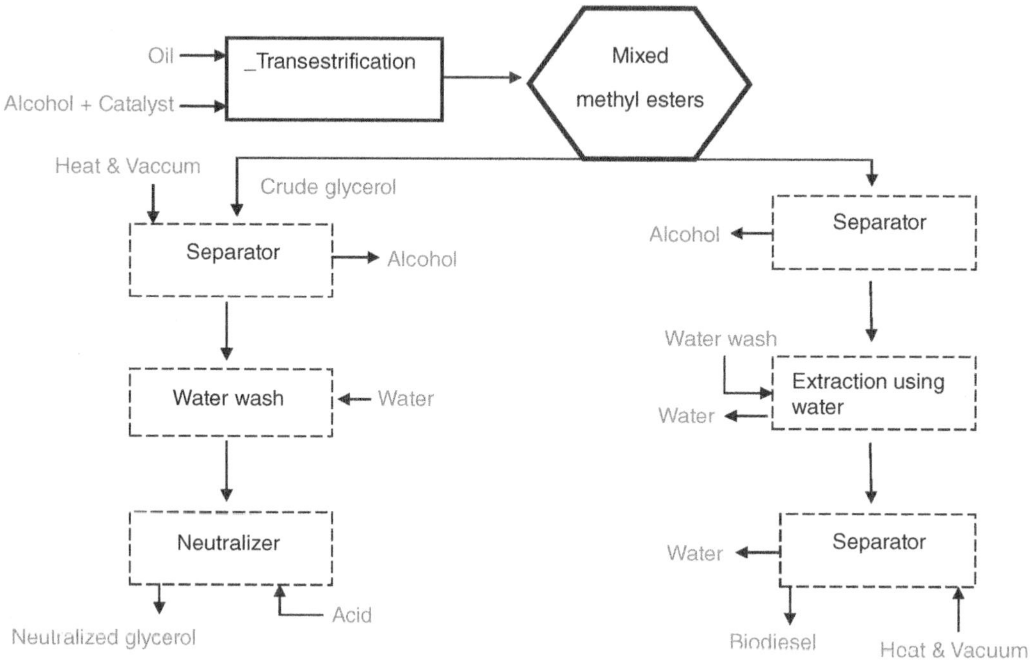

Figure 15.9 Flow representing the process of transesterification.

being grown over distinct carbon sources. The carbon sources may be glucose, lactose, sucrose, starch, etc. This observation indicated that the bacterium was able to use dairy wastewater as an efficient feedstock for biodiesel production [208].

In transesterification (Figure 15.9), alcohol reacts with triglycerides in the presence of catalysts (generally alkali catalysts) to yield fatty acid alkyl esters. Waste cooking oil (WCO), waste from slaughterhouses, edible oils like microalgae, algae, and jatropha oils are utilized in the generation of biodiesel. The wastewater for biodiesel comprises certain characteristics such as high COD, suspended matter, grease, and oil. Untreated biodiesel wastewater has a high concentration of grease and oil that could clog pipelines if released into public sewage systems. Biodiesel wastewater also affects the microbial activity of activated sludge [235].

The ester bond between the fatty acids and hydroxyl group of glycerol are broken down and free fatty acids form new ester bonds with alcohol molecules, yielding biodiesel and glycerol in raw form. Fatty acid methyl esters (FAME) are produced when methanol is used in the transesterification mixture and fatty acid ethyl esters (FAEE) are produced in the case of ethanol. FAME as well as FAEE can be used as biodiesel. The glycerol produced during processed refrigeration can be purified and then be used in the food and cosmetic industries as well as in the oleochemical industry [236].

In order to separate glycerol from alcohol, water is added to both of them, which helps in the removal of unwanted side products. Solvent extraction is used to separate wash water (any water used to clean or wash materials) and the trace water (leftover water) is evaporated out of the biodiesel. To get neutralized glycerol, acid is added to glycerol. Arif et al. [237] stated that micro algal lipids are comprised of polar as well as non-polar lipids. Non-polar lipids like triacylglyceride can be transformed into biodiesel, whereas polar lipids cannot do so. Polar lipids thus increase the unsaturation that results in insufficient and expensive production. Microalgae are grown under numerous stress conditions to enhance biodiesel production. *S. obliquus, Ourococcus multisporus,* and *Chlamydomonas pitschmannii* are efficient biodiesel producers because of their high lipid content.

Transesterification can also be done with the help of enzymes. Lipases can be used to catalyze transesterification reactions. Triacylglycerol lipases are responsible for the hydrolysis of triglycerides comprising of long-chain fatty acids to glycerol and fatty acids. In contrast, esteraces prefer to hydrolyze only those triglycerides that are composed of short-chain fatty acids. Enzymes derived from bacteria, yeast, and filamentous fungi also help in the production of biodiesel. *Rhizopus, Pseudomonas,* and *Candida* are among the most widely used enzymes for biodiesel production. *Candida antarctica, Rhizopusoryzae, Burkholderia cepacia,* and *Mucormiehei* are the main lipase-producing organisms. The disadvantage of using soluble enzymes is that they are stabilized in water solutions. Water hydrolyzes triglycerides so to overcome this, enzymes can be freeze-dried, but this may also reduce their activity [238].

Algae help in biofuel production by using a co-generative method. Along with treating wastewater, it also forms algal biomass that could be used as a form of liquid or gaseous biofuel [239]. Biodiesel derived from algal biomass commonly comprises FAME.

Microalgae have also been used to produce biofuel. They contain about 30% of oil content by weight of dry biomass that yields around 58 7001 ha^{-1}. Other species of microalgae such as *Botryococcus* and *Schizochytrium* yield around 70% of total dry biomass [20]. Zhou et al. [240] cultivated microalgal bacterial culture of *Chlorella pyroidosa* and heterotrophic strains of ammonia-oxidizing microbes in municipal wastewater. They observed enhancement in biomass and lipid concentration to 14.8 and 13.6%, respectively. The concentration of PUFAs, MUFAs, and SFAs was 43.9, 37.1, and 19.0%, respectively.

Biodiesel production from wastewater would be an immense chance to deliver power to the world in a renewable, eco-friendly way. Bioethanol helps in decreasing the greenhouse gas emissions as the road transport accounts for about 22% of it with no carbon footprint.

15.7.6. Biohydrogen

Microbiological generation of biohydrogen relies on the application of renewable energy sources by using microbes. Biohydrogen produces only water as a byproduct on its combustion compared to classical fossil fuels, and therefore is the cleanest renewable energy source with no carbon footprint. Hydrogen is commonly utilized as an energy source in fuel cells to produce electricity. Hydrogen gas may yield up to 2.75 times more energy than hydrocarbon fuels.

15.7.6.1. Primary Treatment

This may be done using microwave radiation or heat shock treatment (HST). Microwave pretreatment involves irradiation of microwave waves of 300 MHz to 300 GHz frequency that generates heat in fluids, disrupts the cell wall of biomass, and increases the solubility of the medium. This heat generated by microwaves extracts the contaminants present in the inoculum and inactivates non-hydrogen producers. HST refers to the heat treatment in which the growth of hydrogen producing bacteria is increased and activity of methanogens is suppressed. Hydrogen-producing bacteria form spores that help in differentiating them from methanogens. In this way, non-sporing bacterium are killed by HST [241].

15.7.6.2. Mechanical Treatment

This involves ultrasonication of microbes. Ultrasonication involves passing a wave through the medium that leads to bubble formation. When these bubbles collapse they produce highly active radicals, high temperature, pressure, and immense force. Ultrasonication of *Rhodopseudomonas palustris* QK01 enhanced biohydrogen yield to 66.7% [242].

15.7.6.3. Chemical Treatment

It involves the treatment with inhibitors of methanogens such as acetylene, iodopropane, and 2-bromo ethanesulfonic acid (BESA). BESA enhances the destruction of methanogens without interrupting hydrogen producing bacteria. Acid treatment has also been used to kill methanogens which survive at a small pH range, whereas acidogenic bacteria (AB) can survive at an extreme pH range [241]. Yin and Wang [243] found that Fe^{2+} supplementation with macroalgae enhanced hydrogen production. With 400 mg l^{-1} of iron supplementation, hydrogen yield of 19.47 ml g^{-1} VS_{added} was achieved. Depth analysis of microbial distribution, substrate degradation, and metabolite formation showed that the iron-supplemented group consisted predominantly of *Ruminococcus gnavus* (24.2%) and *Clostridium butyricum* (67.2%), while the iron-deficient group consisted of *Clostridium stricto* 13 (23.4%), *Exiguobacterium* sp. (29.0%), and *Acinetobacter lwoffii* (24.5%) that lead to enhanced mineralization and hydrolysis of biomass.

15.7.6.4. Biological Treatment

This treatment employs bacteria or enzymes to enhance the hydrolysis rate. *Clostridium* species have been observed to produce hydrogen during their exponential growth phase. It has been reported that during the sludge digestion there was around 64.6% clostridia species. *Aeromonos* sp., *pseudomonas* sp., and *vibrio* sp. can also help in the production of biohydrogen. Twelve species from the genera *Bacillus, Sporolactobacillus, Paenibacillus, Lactobacillus, Pediococcus,* and *Chronobacter* were found on screening the bacteria that were capable of producing hydrogen. These species were produced from seven different substrates (molasses, coconut husk, coconut milk waste, refinery, palm oil factory, and ethanol distilled wastewater) obtained from siz different locations. 0.33 ml of H_2 was produced by *Bacillus coagulans* MO11 restored from defined medium. On the other hand, *Bacillus coagulans* MO21 produced 0.66 ml of hydrogen gas [21].

Distinct biological methods have been a part of generating biohydrogen; these may be categorized as light-dependent and light independent processes.

Light-dependent processes [244] include biophotolysis and photofermentation.

15.7.6.5. Biophotolysis or Water-Splitting Photosynthesis

This process requires sunlight and water from photosynthetic microbes such as cyanobacteria and green algae to help in the production of hydrogen. Green alga generally uses Fe-Fe hydrogenase, on the other hand, cyanobacteria use nitrogenase to generate hydrogen. It may either occur directly or indirectly. Direct biophotolysis involves photosystems I and II. In it, the electrons derived from certain microbes are transported through both the photosystems and then to ferredoxin that reduces hydrogenase, which is responsible for hydrogen production. Indirect biophotolysis involves the formation of carbohydrates by the conversion of light energy to chemical energy. These carbohydrates then help in the production of hydrogen.

15.7.6.6. Photofermentation

Purple non-sulfur bacteria are used here to generate hydrogen. Acetic acid, butyric acid, lactic acid may be used as a substrate in photofermentation.

The light-independent process involves dark fermentation. Dark fermentation [245] refers to the anaerobic fermentation in the absence of light that breaks down carbohydrates to create hydrogen and other intermediates such as VFA and alcohol. *Clostridia* sp., and *Enterobacter* sp., have been used in dark fermentation of carbohydrates. Various factors influence hydrogen production methods that should be maintained for optimum production. These include pH, temperature, substrate, etc. *C. butyricum* has helped in the production of biohydrogen using wheat starch [246]. *Synechocystis sp. PCC 6803* generated 0.037 mmol H_2 mg^{-1} Chl hr^{-1} amount of hydrogen within 120 hours in the dark. On the other hand, *Desertifilum* sp. *IPPAS B1220* generated 0.229 mmol H_2 mg^{-1} Chl hr^{-1}within 166 hours of incubation in the light. DCMU concentration of about 10 mM enhanced hydrogen production to 0.348 mmol H_2 mg^{-1} Chl hr^{-1} with the use of *Desertifilum sp. IPPAS B1220* by 1.5 fold [247].

In a control system, it was seen that *Romboutsia* (5.9%) was predominantly present, followed by *Fusibacter* (1.7%) and *Clostridium* (0.5%). Less hydrogen production for control group may be a result of slow progression of hydrogen producing bacteria. On treatment with 0.3 g SS^{-1} of sodium citrate, hydrogen producing bacterial population was enriched to 32.6%, i.e. four times of that of the control system [248].

Biohydrogen production depends on various factors including the kind of substrate, operating conditions, and the supplements used. It has been a part of green sustainable energy source. Biohydrogen would be an important fuel in the coming future because of its property to be inexhaustible, renewable and economical.

15.7.7. Bioethanol

Bioethanol is a biologically synthesized substitute for transportation fuel. As per the statistics of 2016, yearly ethanol production around the world was about 32 billion liters. USA and Brazil are the leading producers of ethanol.

Several microbes help in the production of bioethanol such as *Zymomonas mobilis, S. cerevisiae,* and *E. coli* [249]. These are homo ethanol producers that only consume glucose as their substrate. *Z. mobilis* follows Entner-doudoroff (ED) pathway. On the other hand, *E. coli* and *S. cerevisiae* follow the Embden-meyerhoff-parnas (EMP) pathway. *Z. mobilis* produces a high amount of ethanol than *S. cerevisiae* [250]. Alcohol dehydrogenase and pyruvate decarboxylase play an important role in ethanol production. *Z. mobilis* has a disadvantage in that it cannot utilize pentose sugar such as lignocellulose to produce ethanol [251]. The other two microorganisms can transform pentose sugar into alcohol *E. coli* has been genetically engineered to possess the genes encoding for pyruvate decarboxylase and alcohol dehydrogenase taken from *Z. mobilis*.

It has been known that gasoline can be replaced by bioethanol or can be blended together. The blends can be E10 (10% bioethanol, 90% gasoline), E85 (85% bioethanol, 15% gasoline), E5 (5% bioethanol, 95% gasoline mixture), etc. Melikoglu and Turkmen [249] observed that 230–250 l of ethanol would be required to supply around 10% of Turkey's petrol consumption. It has been calculated that growing wheat, potato, and sugar beet on unused land that is accessible for agriculture can produce around 13.7, 5.8, and 8.7 billion liters of ethanol, respectively.

Hindakia tetrachotoma ME03 was grown at 0, 25, 50, 75, and 100% concentrations of wastewater to yield bioethanol. Apart from this *S. cerevisiae* was used to study distinct microalgal biomass for degrading complex carbohydrates into monosaccharides and generate bioethanol. *Hindakia tetrachotoma* ME03 produced $11.2 \pm 0.3\,gl^{-1}$ of bioethanol content and percentage yield of $94 \pm 2.2\%$, on fermentation of *S. cerevisiae* in 50% wastewater. Thus, the studies suggested *Hindakia tetrachotoma ME03* be efficient at bioethanol production once fungal and bacterial contamination problems are eliminated [212].

In research, industrial wastewater was used for the production of bioethanol. The products of hydrolysis with a high amount of reducing sugar were utilized in the fermentation of ethanol with the help of *S. cerevisiae*. A yield of 11.6 g_{EtOH}/g_{algae} was obtained with an acid/dried algae ratio of 7 and a sulfuric acid concentration of 9% by weight. Present studies show that the wastewater of *Eucheuma spinosum* comprises carbohydrates that could be converted into bioethanol with a productivity rate up to 75% [22].

15.7.7.1. Macroalgae and Bioethanol Production

Certain writings reveal that macroalgae can produce a high amount of bioethanol around 0.43 g per substrate in grams in comparison to earthly feedstock, and it also takes less time to produce bioethanol. Conversion of microalgae into bioethanol involves three steps. In the first step, macroalgae are cultivated by collecting the biomass from artificial environments or onshore areas. In the case of laboratory cultivation, biomass can be collected from the culture. Then this biomass is washed, dried, and converted to powder. Powdered biomass is then subjected to compositional analysis. Biomass is treated before its main treatment to obtain sugars that could be fermented. In the final step, monomeric sugars are transformed into bioethanol using certain microbial species. A study reported that *G. amansii,* an algal strain increases the yield of bioethanol to 76.9% [23].

Bioethanol can also be produced with the use of green seaweeds as feedstock. Carbohydrates can be extracted from algal biomass by treatment with dilute sulphuric acid. This results in higher sugar yield on enzyme hydrolysis. In *U. lactuca* and *Encephalitozoon intestinalis*, neutralization using Na_2CO_3 led to low sugar elimination around 39.8 and 14.7%, respectively. *U. lactuca* has a carbohydrate content of 62.15% and *E. intestinalis* of 40.1%, this makes them an efficient feedstock for bioethanol production [252].

15.7.7.2. Banana and Bioethanol Production

Banana waste has also been extensively used for the production of bioethanol. Banana pseudostem has been exploited as a feedstock for the production of bioethanol by *Saccharomyces cerevisiae NCIM 3570* in the solid-state fermentation method which requires a longer fermentation time of about 72 hours and fungal pre-treatment [253]. Rotten banana has also been used with *S. cerevisiae*. Two enzymes, pectinase and cellulose, are used for the hydrolysis of rotten bananas to produce bioethanol [254].

15.7.7.3. Actinomycetes and Bioethanol Production

Streptomyces thermodiastaticus [255], *Streptomyces setonii, Thermomonospora curvata, Streptomyces viridosporus,* and *Thermomonospora fusca* are some actinomycetes that have been reported to have the cellulolytic ability [256]. *Streptomyces griseus* B1 was isolated from leaf debris to study its role in ethanol production [257]. The studies involved the use of softwood besides hardwood fermented at 37 °C and *S. griseus* B1 lead to a much lower drop in the level of klason lignin compared to control experiments. *S. griseus* B1 favored hardwood substrates over softwood substrates. Hardwood substrate led to 23.4% lignin loss, whereas softwood led to 10.5% loss in lignin.

Reports claim that the bacterial species of the Enterobacteriaceae family comprises cellulase enzyme that degrades cellulosic waste [258, 259]. A member of the Enterobacteriaceae family (EtK3) was isolated and was found to have the cellulolytic ability to degrade cellulosic wastes [259]. EtK3 was said to break down dried banana fruit peel waste without any pretreatment. The fermentation of dried banana peel resulted in the production of ethanol. At optimum fermentation, the maximum yield was said to be 23.6%. The 1l culture was taken into account to generate a mass balance. On fermentation of 155 g of dried banana peel waste, 7.62 g of waste residue was obtained and a total of 2.07 g of ethanol with 4.672 g of biomass were recovered [260].

Bioethanol production by using *Z. mobilis, S. cerevisiae,* and *E. coli* has been found to be beneficial in energy generation. Enzymes such as alcohol dehydrogenase and pyruvate decarboxylase play crucial role in the whole production mechanism. Microalgae and other microbial species can also be utilized to produce bioethanol in order to meet world energy demand.

15.8. ADVANTAGE AND DISADVANTAGE OF USING MICROBES AS AN ENERGY GENERATION METHOD FROM WASTEWATER

There are both advantages and disadvantage to using microbes to generate energy from wastewater. The following subsections list the benefits and drawbacks.

15.8.1. Benefits of Microbial Energy Production from Wastewater

- Bioenergy products help generate electricity that may be used for various activities.
- Biofuels act as a substitute for diesel resulting in lesser damage to the environment.
- Biofuels can be beneficial in the transportation sector.
- The energy produced from biomass is carbon neutral.
- Biomass products lead to less garbage production.
- Biogas production has aided in reducing deforestation, pollution, and in improving the lifestyle of people living in rural areas.
- Biohydrogen can prove to reduce our dependence on fossil fuels and reduction in the emission of greenhouse gases.
- Bioenergy products are renewable in nature.

15.8.2. Detriments of Microbial Energy Production from Wastewater

- These are less efficient than classical approaches.
- Deforestation may be caused.
- Biomass-generated products are not completely clean.
- Constructing biological plants is quite expensive.

15.9. CONCLUSION

In this chapter, an overview of the role of microbes in WWT and energy generation has been highlighted. Certain microbes in nature are capable of decreasing contaminants from domestic sewages, factories, and pesticide fluxes from agricultural lands without sacrificing environmental factors. Wastewater is contaminated with microbiological populations all the time, which may include bacteria, fungi, algae, protozoa, and metazoa. These microorganisms can be a boon as well as a curse for wastewater. Microbial species can help in the elimination of fats, oils, grease, suspended particles, BOD, and COD, which in turn enhances the state of wastewater and this could be reused for various activities. In this respect, wastewater is also observed as an energy generation material. Thus, microbes help in combined energy redemption from wastewater and appear to be the only current sustainable renewable source that can potentially and completely proxy of the energy balance and problems arising of WWT and water crisis. There are various ways in which energy can be generated and applied in direction of life. Biogas, biohydrogen, bioethanol, biodiesel, and bioelectricity are all energy products produced from wastewater with the help of microbes. Biological WWT and energy generation holds a significant place in the world and offers a great promise for future.

ACKNOWLEDGMENTS

Ashish Kumar is grateful to Department of Biotechnology, Ministry of Science and Technology, Govt. of India for providing the Master scholarship in the programme on DBT-Fellowship scheme at SoBT, Shri Mata Vaishno Devi University, Katra (J&K). Atin Kumar Pathak (IF150422) is grateful to DST (Govt. of India) for proving the Senior Research Fellowship in the programme on DST-Inspire Fellowship.

REFERENCES

1. de Andrade, J.R., Oliveira, M.F., da Silva, M.G. et al. (2018). Adsorption of pharmaceuticals from water and wastewater using nonconventional low-cost materials: a review. *Industrial & Engineering Chemistry Research 57* (9): 3103–3127.
2. Freyria, F.S., Geobaldo, F., and Bonelli, B. (2018). Nanomaterials for the abatement of pharmaceuticals and personal care products from wastewater. *Applied Sciences 8* (2): 170.
3. Carlos, F.S., Schaffer, N., Andreazza, R. et al. (2018). Treated industrial wastewater effects on chemical constitution maize biomass, physicochemical soil properties, and economic balance. *Communications in Soil Science and Plant Analysis 49* (3): 319–333.
4. Rincón-Llorente, B., la Lama-Calvente, D., Fernández-Rodríguez, M.J. et al. (2018). Table olive wastewater: problem, treatments and future strategy. A review. *Frontiers in Microbiology 9*: 1641.
5. Abbas, A., Valek, L., Schneider, I. et al. (2018). Ecotoxicological impacts of surface water and wastewater from conventional and advanced treatment technologies on brood size, larval length, and cytochrome P450 (35A3) expression in Caenorhabditis elegans. *Environmental Science and Pollution Research 25* (14): 13868–13880.
6. Pathak, A.K., Kothari, R., Tyagi, V. et al. (2020). Integrated approach for textile industry wastewater for efficient hydrogen production and treatment through solar PV electrolysis. *International Journal of Hydrogen Energy 45* (48): 25768–25782.
7. Kaewkannetra, P., Chiwes, W., and Chiu, T. (2011). Treatment of cassava mill wastewater and production of electricity through microbial fuel cell technology. *Fuel 90* (8): 2746–2750.
8. Singh, H.M., Pathak, A.K., Chopra, K. et al. (2019). Microbial fuel cells: a sustainable solution for bioelectricity generation and wastewater treatment. *Biofuels 10* (1): 11–31.
9. Wang, J., Zhao, S., Ling, Z. et al. (2021). Enhanced removal of trivalent chromium from leather wastewater using engineered bacteria immobilized on magnetic pellets. *Science of the Total Environment 775*: 145647.
10. Zagui, G.S., Moreira, N.C., Santos, D.V. et al. (2021). High occurrence of heavy metal tolerance genes in bacteria isolated from wastewater: a new concern? *Environmental Research 196*: 110352.
11. Li, J., Chen, X., Liu, W. et al. (2020). Biostimulation of a marine anammox bacteria-dominated bioprocess by Co (II) to treat nitrogen-rich, saline wastewater. *Science of the Total Environment 749*: 141489.
12. Liu, J., Pemberton, B., Lewis, J. et al. (2020). Wastewater treatment using filamentous algae–a review. *Bioresource Technology 298*: 122556.
13. Martin, H.G., Ivanova, N., Kunin, V. et al. (2006). Metagenomic analysis of two enhanced biological phosphorus removal (EBPR) sludge communities. *Nature Biotechnology 24* (10): 1263–1269.
14. Qin, L., Gao, M., Zhang, M. et al. (2020). Application of encapsulated algae into MBR for high-ammonia nitrogen wastewater treatment and biofouling control. *Water Research 187*: 116430.
15. Pinto, A., Teixeira, D., Caldeira, A. et al. (2020). Biodegradation of pesticides by adapted fungi. Potential use on biopurification systems? In: *Agrochemicals Detection, Treatment and Remediation* (ed. P. MNV), 1–23. Elsevier.

16. Nzihou, A. (2020). *Handbook on characterization of biomass, biowaste and related by-products*. Springer Nature.
17. Kamyab, B. and Zilouei, H. (2021). Investigating the efficiency of biogas production using modelling anaerobic digestion of baker's yeast wastewater on two-stage mixed-UASB reactor. *Fuel 285*: 119198.
18. Das, I., Das, S., and Ghangrekar, M. (2020). Application of bimetallic low-cost CuZn as oxygen reduction cathode catalyst in lab-scale and field-scale microbial fuel cell. *Chemical Physics Letters 751*: 137536.
19. Silambarasan, S., Logeswari, P., Sivaramakrishnan, R. et al. (2021). Cultivation of Nostoc sp. LS04 in municipal wastewater for biodiesel production and their deoiled biomass cellular extracts as biostimulants for Lactuca sativa growth improvement. *Chemosphere 280*: 130644.
20. ASCE, AWWA, EPA U (1996). *Technology transfer handbook: management of water treatment plant residuals*. New York & Denver, CO: ASCE AWWA.
21. Lertsriwong, S. and Glinwong, C. (2020). Newly-isolated hydrogen-producing bacteria and biohydrogen production by Bacillus coagulans MO11 and Clostridium beijerinckii CN on molasses and agricultural wastewater. *International Journal of Hydrogen Energy 45* (51): 26812–26821.
22. Alfonsín, V., Maceiras, R., and Gutiérrez, C. (2019). Bioethanol production from industrial algae waste. *Waste Management 87*: 791–797.
23. Dave, N., Selvaraj, R., Varadavenkatesan, T. et al. (2019). A critical review on production of bioethanol from macroalgal biomass. *Algal Research 42*: 101606.
24. Gao, M., Guo, B., Zhang, L. et al. (2019). Microbial community dynamics in anaerobic digesters treating conventional and vacuum toilet flushed Blackwater. *Water Research 160*: 249–258.
25. Thibodeau, C., Monette, F., and Glaus, M. (2014). Comparison of development scenarios of a black water source-separation sanitation system using life cycle assessment and environmental life cycle costing. *Resources, Conservation and Recycling 92*: 38–54.
26. Jin, Z., Lv, C., Zhao, M. et al. (2018). Black water collected from the septic tank treated with a living machine system: HRT effect and microbial community structure. *Chemosphere 210*: 745–752.
27. Xu, R., Xu, S., Florentino, A.P. et al. (2019). Enhancing Blackwater methane production by enriching hydrogenotrophic methanogens through hydrogen supplementation. *Bioresource Technology 278*: 481–485.
28. Al Jahmani, A.K., Rao, L.N., Varghese, M. et al. (2021). Treatment of kitchen grey water by ceramic filter and optimization using response surface methodology. *Materials Today: Proceedings 57*: Part 6.
29. Lakho, F.H., Le, H.Q., Mattheeuws, F. et al. (2021). Decentralized grey and black water reuse by combining a vertical flow constructed wetland and membrane based potable water system: full scale demonstration. *Journal of Environmental Chemical Engineering 9* (1): 104688.
30. Rosińska, A. (2021). The influence of UV irradiation on PAHs in wastewater. *Journal of Environmental Management 293*: 112760.
31. Toke, L.K., Bhadane, S.G., Metkar, T.P. et al. (2019). Design and development of activated carbon-based purification system for automobile emissions. *Recent Trends in Automation and Automobile Engineering 2* (2).
32. Ligotski, R., Sager, U., Schneiderwind, U. et al. (2019). Prediction of VOC adsorption performance for estimation of service life of activated carbon based filter media for indoor air purification. *Building and Environment 149*: 146–156.
33. Cao, B., Wang, R., Zhang, W. et al. (2019). Carbon-based materials reinforced waste activated sludge electro-dewatering for synchronous fuel treatment. *Water Research 149*: 533–542.
34. Propolsky, D., Romanovskaia, E., Kwapinski, W. et al. (2020). Modified activated carbon for deironing of underground water. *Environmental Research 182*: 108996.
35. Periyasamy, S., Kumar, I.A., and Viswanathan, N. (2020). Activated carbon from different waste materials for the removal of toxic metals. In: *Green Materials for Wastewater Treatment* (ed. M. Naushad and E. Lichtfouse), 47–68. Springer.
36. Talebi, A., Razali, Y.S., Ismail, N. et al. (2020). Selective adsorption and recovery of volatile fatty acids from fermented landfill leachate by activated carbon process. *Science of the Total Environment 707*: 134533.

37. Heidarinejad, Z., Dehghani, M.H., Heidari, M. et al. (2020). Methods for preparation and activation of activated carbon: a review. *Environmental Chemistry Letters 18* (2): 393–415.
38. Díez, E., Gómez, J., Rodríguez, A. et al. (2020). A new mesoporous activated carbon as potential adsorbent for effective indium removal from aqueous solutions. *Microporous and Mesoporous Materials 295*: 109984.
39. Shukla, S. (2020). Rice husk derived adsorbents for water purification. In: *Green Materials for Wastewater Treatment* (ed. M. Naushad and E. Lichtfouse), 131–148. Springer.
40. Hu, E. (2016). Catalytic ozonation for textile dyeing wastewater treatment and reuse. PhD Thesis.
41. Wu, Q.-Y., Zhou, Y.-T., Li, W. et al. (2019). Underestimated risk from ozonation of wastewater containing bromide: both organic byproducts and bromate contributed to the toxicity increase. *Water Research 162*: 43–52.
42. Song, Z., Wang, M., Wang, Z. et al. (2019). Insights into heteroatom-doped graphene for catalytic ozonation: active centers, reactive oxygen species evolution, and catalytic mechanism. *Environmental Science & Technology 53* (9): 5337–5348.
43. Borrelli, E. (2020). The use of ozone as redox modulator in the treatment of the chronic obstructive pulmonary disease (COPD). *Oxidative Stress in Lung Diseases: Springer* 413–426.
44. Chen, C., Jia, N., Song, K. et al. (2020). Sulfur-doped copper-yttrium bimetallic oxides: a novel and efficient ozonation catalyst for the degradation of aniline. *Separation and Purification Technology 236*: 116248.
45. Guo, W., Fu, Z., Zhang, Z. et al. (2020). Synthesis of Fe3O4 magnetic nanoparticles coated with cationic surfactants and their applications in Sb (V) removal from water. *Science of the Total Environment 710*: 136302.
46. Maia, L.F.O., Santos, M.S., Andrade, T.G. et al. (2018). Removal of mercury (II) from contaminated water by gold-functionalised Fe3O4 magnetic nanoparticles. *Environmental Technology 41*: 959–970.
47. Saadat, A., Hajiaghababaei, L., Badiei, A. et al. (2019). Amino functionalized silica coated Fe3O4 magnetic nanoparticles as a novel adsorbent for removal of Pb2+ and Cd2+. *Pollution 5* (4): 847–857.
48. Chen, Y., Zhang, Z., Chen, D. et al. (2019). Removal of coke powders in coking diesel distillate using recyclable chitosan-grafted Fe3O4 magnetic nanoparticles. *Fuel 238*: 345–353.
49. Alexander, H.J. and McClanahan, M.A. (1975). Kinetics of calcium carbonate precipitation in lime–soda ash softening. *Journal - American Water Works Association 67* (11): 618–621.
50. Gedda, G., Kolli Balakrishna, R.U.D., and Shah, K.J. (2021). Introduction to conventional wastewater treatment technologies: limitations and recent advances. *Advances in Wastewater Treatment I* (91): 1–36.
51. Samarasiri, K.T., Rathnadheera, C.K., Ishan, S. et al. Process Optimization for the Lime Soda Water Softening Method for Reducing the Hardness in Well Water.
52. Bob, M. and Walker, H.W. (2006). Lime-soda softening process modifications for enhanced NOM removal. *Journal of Environmental Engineering 132* (2): 158–165.
53. Applebaum, S. (1925). Characteristic properties of zeolites for water softening. *Journal (American Water Works Association) 13* (2): 213–220.
54. Applebaum, S. and Behrman, A. (1938). Applications of Carbonaceous Zeolites to Water Softening [with Discussion]. *Journal (American Water Works Association) 30* (6): 947–978.
55. Cinar, S. and Beler-Baykal, B. (2005). Ion exchange with natural zeolites: an alternative for water softening? *Water Science and Technology 51* (11): 71–77.
56. Arrigo, I., Catalfamo, P., Cavallari, L. et al. (2007). Use of zeolitized pumice waste as a water softening agent. *Journal of Hazardous Materials 147* (1-2): 513–517.
57. Loiola, A., Andrade, J., Sasaki, J. et al. (2012). Structural analysis of zeolite NaA synthesized by a cost-effective hydrothermal method using kaolin and its use as water softener. *Journal of Colloid and Interface Science 367* (1): 34–39.
58. de Andrade, B.R., de Sousa, C.L., Oliveira, C.P. et al. (2017). Kaolin-based magnetic zeolites A and P as water softeners. *Microporous and Mesoporous Materials 245*: 64–72.

59. Montalvo, S., Huiliñir, C., Borja, R. et al. (2020). Application of zeolites for biological treatment processes of solid wastes and wastewaters–a review. *Bioresource Technology 301*: 122808.
60. Pan, Z., Dai, R., Liao, J. et al. (2020). Spontaneous formation and mechanism of anaerobic ammonium oxidation (anammox) bacteria in swine wastewater treatment system. *International Biodeterioration & Biodegradation 154*: 105058.
61. Hailemariam, F.A., Velmurugan, P., and Selvaraj, S.K. (2021). Treatment of wastewater from coffee (coffea arebica) industries using mixed culture Pseudomonas florescence and Escherichia coli bacteria. *Materials Today: Proceedings 46* (17): 7396–7401.
62. Li, W., Fishman, A., and Achal, V. (2021). Ureolytic bacteria from electronic waste area, their biological robustness against potentially toxic elements and underlying mechanisms. *Journal of Environmental Management 289*: 112517.
63. Pirsaheb, M., Zadsar, S., Rastegar, S.O. et al. (2021). Bioleaching and ecological toxicity assessment of carbide slag waste using Acidithiobacillus bacteria. *Environmental Technology and Innovation 22*: 101480.
64. Mai, C.T.N., Linh, N.V., Lich, N.Q. et al. (2021). Advanced materials for immobilization of purple phototrophic bacteria in bioremediation of oil-polluted wastewater. *Chemosphere 278*: 130464.
65. Sofu, A., Sayilgan, E., and Guney, G. (2015). Experimental design for removal of Fe (II) and Zn (II) ions by different lactic acid bacteria biomasses. *International Journal of Environmental Research 9* (1): 93–100.
66. Li, C., Wang, S., Du, X. et al. (2016). Immobilization of iron-and manganese-oxidizing bacteria with a biofilm-forming bacterium for the effective removal of iron and manganese from groundwater. *Bioresource Technology 220*: 76–84.
67. Franzblau, R.E., Daughney, C.J., Swedlund, P.J. et al. (2016). Cu (II) removal by Anoxybacillus flavithermus–iron oxide composites during the addition of Fe (II) aq. *Geochimica et Cosmochimica Acta 172*: 139–158.
68. Guo, Y., Huang, T., Wen, G. et al. (2017). The simultaneous removal of ammonium and manganese from groundwater by iron-manganese co-oxide filter film: the role of chemical catalytic oxidation for ammonium removal. *Chemical Engineering Journal 308*: 322–329.
69. Mostafa, Y.S., Alrumman, S.A., Alamri, S.A. et al. (2020). Bioplastic (poly-3-hydroxybutyrate) production by the marine bacterium Pseudodonghicola xiamenensis through date syrup valorization and structural assessment of the biopolymer. *Scientific Reports 10* (1): 1–13.
70. Guo, G., Tian, F., Zhang, C. et al. (2019). Performance of a newly enriched bacterial consortium for degrading and detoxifying azo dyes. *Water Science and Technology 79* (11): 2036–2045.
71. Ajonina, C., Buzie, C., Rubiandini, R.H. et al. (2015). Microbial pathogens in wastewater treatment plants (WWTP) in Hamburg. *Journal of Toxicology and Environmental Health, Part A 78* (6): 381–387.
72. Biswal, B.K., Wang, B., Chen, L. et al. (2020). Rational design of sulfidogenic granular sludge reactor with clostridia as dominant bacteria for energy-efficient sulfate-laden wastewater treatment. *Bioresource Technology 317*: 124017.
73. Samantaray, S. and Mallick, N. (2012). Production and characterization of poly-β-hydroxybutyrate (PHB) polymer from Aulosira fertilissima. *Journal of Applied Phycology 24* (4): 803–814.
74. Liu, Y., Dong, Q., and Shi, H. (2015). Distribution and population structure characteristics of microorganisms in urban sewage system. *Applied Microbiology and Biotechnology 99* (18): 7723–7734.
75. Mishra, S., Mohanty, P., and Maiti, A. (2019). Bacterial mediated bio-decolourization of wastewater containing mixed reactive dyes using jack-fruit seed as co-substrate: process optimization. *Journal of Cleaner Production 235*: 21–33.
76. Pramanik, K., Mitra, S., Sarkar, A. et al. (2018). Alleviation of phytotoxic effects of cadmium on rice seedlings by cadmium resistant PGPR strain Enterobacter aerogenes MCC 3092. *Journal of Hazardous Materials 351*: 317–329.

77. Wen, X., Du, C., Zeng, G. et al. (2018). A novel biosorbent prepared by immobilized Bacillus licheniformis for lead removal from wastewater. *Chemosphere 200*: 173–179.
78. Policastro, G., Luongo, V., and Fabbricino, M. (2020). Biohydrogen and poly-β-hydroxybutyrate production by winery wastewater photofermentation: effect of substrate concentration and nitrogen source. *Journal of Environmental Management 271*: 111006.
79. Chandran, T., Muthunarayanan, V., Ravindran, B. et al. (2020). Evaluation of efficacy of indigenous acidophile-bacterial consortia for removal of pollutants from coffee cherry pulping wastewater. *Bioresource Technology Reports 11*: 100533.
80. Zeng, W., Wang, A., Li, C. et al. (2017). Population dynamics of "Candidatus Accumulibacter phosphatis" under the modes of complete nitrification and partial nitrification (nitritation) in domestic wastewater treatment system. *Biochemical Engineering Journal 124*: 69–77.
81. Kim, J.M., Lee, H.J., Lee, D.S. et al. (2013). Characterization of the denitrification-associated phosphorus uptake properties of "Candidatus Accumulibacter phosphatis" clades in sludge subjected to enhanced biological phosphorus removal. *Applied and Environmental Microbiology 79* (6): 1969–1979.
82. Nurmiyanto, A., Kodera, H., Kindaichi, T. et al. (2017). Dominant Candidatus Accumulibacter phosphatis enriched in response to phosphate concentrations in EBPR process. *Microbes and Environments 47*: ME17020.
83. Wisniewski, K., Kowalski, M., and Makinia, J. (2018). Modeling nitrous oxide production by a denitrifying-enhanced biologically phosphorus removing (EBPR) activated sludge in the presence of different carbon sources and electron acceptors. *Water Research 142*: 55–64.
84. Schroeder, S., Ahn, J., and Seviour, R.J. (2008). Ecophysiology of polyphosphate-accumulating organisms and glycogen-accumulating organisms in a continuously aerated enhanced biological phosphorus removal process. *Journal of Applied Microbiology 105* (5): 1412–1420.
85. Blackall, L.L., Crocetti, G.R., Saunders, A.M. et al. (2002). A review and update of the microbiology of enhanced biological phosphorus removal in wastewater treatment plants. *Antonie Van Leeuwenhoek 81* (1): 681–691.
86. Mohsenpour, S.F., Hennige, S., Willoughby, N. et al. (2021). Integrating micro-algae into wastewater treatment: a review. *Science of the Total Environment 752*: 142168.
87. Casagli, F., Zuccaro, G., Bernard, O. et al. (2021). ALBA: a comprehensive growth model to optimize algae-bacteria wastewater treatment in raceway ponds. *Water Research 190*: 116734.
88. Zhang, Y., Wei, D., Morrison, L. et al. (2019). Nutrient removal through pyrrhotite autotrophic denitrification: implications for eutrophication control. *Science of the Total Environment 662*: 287–296.
89. Ferreira, A., Ribeiro, B., Marques, P.A. et al. (2017). Scenedesmus obliquus mediated brewery wastewater remediation and CO2 biofixation for green energy purposes. *Journal of Cleaner Production 165*: 1316–1327.
90. Mata, T.M., Melo, A.C., Simões, M. et al. (2012). Parametric study of a brewery effluent treatment by microalgae Scenedesmus obliquus. *Bioresource Technology 107*: 151–158.
91. Subramaniyam, V., Subashchandrabose, S.R., Ganeshkumar, V. et al. (2016). Cultivation of chlorella on brewery wastewater and nano-particle biosynthesis by its biomass. *Bioresource Technology 211*: 698–703.
92. Lim, S. L., Chu, W.-L., and Phang, S.-M. (2010). Use of Chlorella vulgaris for bioremediation of textile wastewater. *Bioresource Technology 101* (19): 7314–7322.
93. Lin, C.-Y., Nguyen, M.-L.T., and Lay, C.-H. (2017). Starch-containing textile wastewater treatment for biogas and microalgae biomass production. *Journal of Cleaner Production 168*: 331–337.
94. Robledo-Padilla, F., Aquines, O., Silva-Núñez, A. et al. (2020). Evaluation and predictive modeling of removal condition for bioadsorption of indigo blue dye by Spirulina platensis. *Microorganisms 8* (1): 82.
95. Cheng, P., Chen, D., Liu, W. et al. (2020). Auto-flocculation microalgae species Tribonema sp. and Synechocystis sp. with T-IPL pretreatment to improve swine wastewater nutrient removal. *Science of the Total Environment 725*: 138263.

96. Wang, Y., Wang, S., Sun, L. et al. (2020). Screening of a Chlorella-bacteria consortium and research on piggery wastewater purification. *Algal Research 47*: 101840.
97. Zheng, M., Ji, X., He, Y. et al. (2020). Simultaneous fixation of carbon dioxide and purification of undiluted swine slurry by culturing Chlorella vulgaris MBFJNU-1. *Algal Research 47*: 101866.
98. López-Pacheco, I.Y., Silva-Núñez, A., García-Perez, J.S. et al. (2021). Phyco-remediation of swine wastewater as a sustainable model based on circular economy. *Journal of Environmental Management 278*: 111534.
99. Van Do, T.C., Nguyen, T.N.T., Tran, D.T. et al. (2020). Semi-continuous removal of nutrients and biomass production from domestic wastewater in raceway reactors using Chlorella variabilis TH03-bacteria consortia. *Environmental Technology and Innovation 20*: 101172.
100. Kumar, V., Jaiswal, K.K., Verma, M. et al. (2021). Algae-based sustainable approach for simultaneous removal of micropollutants, and bacteria from urban wastewater and its real-time reuse for aquaculture. *Science of the Total Environment 774*: 145556.
101. Dinh, T.T.U., Soda, S., Nguyen, T.A.H. et al. (2020). Nutrient removal by duckweed from anaerobically treated swine wastewater in lab-scale stabilization ponds in Vietnam. *Science of the Total Environment 722*: 137854.
102. Sudiarto, S.I.A., Renggaman, A., and Choi, H.L. (2019). Floating aquatic plants for total nitrogen and phosphorus removal from treated swine wastewater and their biomass characteristics. *Journal of Environmental Management 231*: 763–769.
103. Mishra, A., Medhi, K., Maheshwari, N. et al. (2018). Biofuel production and phycoremediation by Chlorella sp. ISTLA1 isolated from landfill site. *Bioresource Technology 253*: 121–129.
104. Sathya, A.B., Sivashankar, R., Velmurugan, S. et al. (2018). Green algae as source for polyhydroxyalkanoate synthesis from food processing wastewater. *Research Journal of Chemistry and Environment 22*: 66–74.
105. Kotoula, D., Iliopoulou, A., Irakleous-Palaiologou, E. et al. (2020). Municipal wastewater treatment by combining in series microalgae Chlorella sorokiniana and macrophyte Lemna minor: preliminary results. *Journal of Cleaner Production 271*: 122704.
106. Rugnini, L., Costa, G., Congestri, R. et al. (2018). Phosphorus and metal removal combined with lipid production by the green microalga Desmodesmus sp.: an integrated approach. *Plant Physiology and Biochemistry 125*: 45–51.
107. Shen, L., Wang, J., Li, Z. et al. (2020). A high-efficiency Fe2O3@ Microalgae composite for heavy metal removal from aqueous solution. *Journal of Water Process Engineering 33*: 101026.
108. Luo, L., Ren, H., Pei, X. et al. (2019). Simultaneous nutrition removal and high-efficiency biomass and lipid accumulation by microalgae using anaerobic digested effluent from cattle manure combined with municipal wastewater. *Biotechnology for Biofuels 12* (1): 1–15.
109. Emparan, Q., Harun, R., and Danquah, M. (2019). Role of phycoremediation for nutrient removal from wastewaters: a review. *Applied Ecology and Environmental Research 17* (1): 889–915.
110. AlMomani, F. and Örmeci, B. (2020). Assessment of algae-based wastewater treatment in hot climate region: treatment performance and kinetics. *Process Safety and Environmental Protection 141*: 140–149.
111. Tchinda, D., Henkanatte-Gedera, S., Abeysiriwardana-Arachchige, I. et al. (2019). Single-step treatment of primary effluent by Galdieria sulphuraria: removal of biochemical oxygen demand, nutrients, and pathogens. *Algal Research 42*: 101578.
112. Cheng, P., Osei-Wusu, D., Zhou, C. et al. (2020). The effects of refractory pollutants in swine wastewater on the growth of Scenedesmus sp. with biofilm attached culture. *International Journal of Phytoremediation 22* (3): 241–250.
113. Qu, Z., Duan, P., Cao, X. et al. (2019). Comparison of monoculture and mixed culture (Scenedesmus obliquus and wild algae) for C, N, and P removal and lipid production. *Environmental Science and Pollution Research 26* (20): 20961–20968.
114. Prajapati, S.K., Choudhary, P., Malik, A. et al. (2014). Algae mediated treatment and bioenergy generation process for handling liquid and solid waste from dairy cattle farm. *Bioresource Technology 167*: 260–268.

115. Rai, L., Gaur, J., and Kumar, H. (1981). Phycology and heavy-metal pollution. *Biological Reviews 56* (2): 99–151.
116. Arita, C.E.Q., Peebles, C., and Bradley, T.H. (2015). Scalability of combining microalgae-based biofuels with wastewater facilities: a review. *Algal Research 9*: 160–169.
117. Bai, X. and Acharya, K. (2016). Removal of trimethoprim, sulfamethoxazole, and triclosan by the green alga Nannochloris sp. *Journal of Hazardous Materials 315*: 70–75.
118. Yu, Y., Zhou, Y., Wang, Z. et al. (2017). Investigation of the removal mechanism of antibiotic ceftazidime by green algae and subsequent microbic impact assessment. *Scientific Reports 7* (1): 1–11.
119. Travieso, L., Benitez, F., Weiland, P. et al. (1996). Experiments on immobilization of microalgae for nutrient removal in wastewater treatments. *Bioresource Technology 55* (3): 181–186.
120. Jaafari, J. and Yaghmaeian, K. (2019). Optimization of heavy metal biosorption onto freshwater algae (Chlorella coloniales) using response surface methodology (RSM). *Chemosphere 217*: 447–455.
121. Shaker, S., Nemati, A., Montazeri-Najafabady, N. et al. (2015). Treating urban wastewater: nutrient removal by using immobilized green algae in batch cultures. *International Journal of Phytoremediation 17* (12): 1177–1182.
122. Isik, Z., Arikan, E.B., Bouras, H.D. et al. (2019). Bioactive ultrafiltration membrane manufactured from Aspergillus carbonarius M333 filamentous fungi for treatment of real textile wastewater. *Bioresource Technology Reports 5*: 212–219.
123. Gullotto, A., Lubello, C., Mannucci, A. et al. (2015). Biodegradation of naphthalenesulphonate polymers: the potential of a combined application of fungi and bacteria. *Environmental Technology 36* (4): 538–545.
124. Sharma, R., Talukdar, D., Bhardwaj, S. et al. (2020). Bioremediation potential of novel fungal species isolated from wastewater for the removal of lead from liquid medium. *Environmental Technology and Innovation 18*: 100757.
125. Sankaran, S., Khanal, S.K., Jasti, N. et al. (2010). Use of filamentous fungi for wastewater treatment and production of high value fungal byproducts: a review. *Critical Reviews in Environmental Science and Technology 40* (5): 400–449.
126. Kasonga, T.K., Coetzee, M.A., Kamika, I. et al. (2020). Data on the degradation of pharmaceuticals and their metabolites by a fungal consortium in a non-sterile stirred fluidized bioreactor. *Data in Brief 28*: 105057.
127. Das, P., Mahanty, S., Ganguli, A. et al. (2019). Role of Manglicolous fungi isolated from Indian Sunderban mangrove forest for the treatment of metal containing solution: batch and optimization using response surface methodology. *Environmental Technology and Innovation 13*: 166–178.
128. Dalecka, B., Juhna, T., and Rajarao, G.K. (2020). Constructive use of filamentous fungi to remove pharmaceutical substances from wastewater. *Journal of Water Process Engineering 33*: 100992.
129. Yang, X., Gu, C., and Lin, Y. (2020). A novel fungal laccase from Sordaria macrospora k-hell: expression, characterization, and application for lignin degradation. *Bioprocess and Biosystems Engineering 43* (7): 1133–1139.
130. Lee, S., Kang, M., Bae, J.-H. et al. (2019). Bacterial valorization of lignin: strains, enzymes, conversion pathways, biosensors, and perspectives. *Frontiers in Bioengineering and Biotechnology 7*: 209.
131. Chopra, A. and Singh, P.P. (2012). Removal of color, COD and lignin from pulp and paper mill effluent by Phanerochaete chrysosporium and Aspergillus fumigatus. *Journal of Chemical and Pharmaceutical Research 4* (10): 4522–4532.
132. Khatami, S., Deng, Y., Tien, M. et al. (2019). Formation of water-soluble organic matter through fungal degradation of lignin. *Organic Geochemistry 135*: 64–70.
133. Bulkan, G., Ferreira, J.A., and Taherzadeh, M.J. (2020). Removal of organic micro-pollutants using filamentous fungi. *Current Developments in Biotechnology and Bioengineering* 363–395.

134. Bankole, P.O., Adekunle, A.A., and Govindwar, S.P. (2018). Enhanced decolorization and biodegradation of acid red 88 dye by newly isolated fungus, Achaetomium strumarium. *Journal of Environmental Chemical Engineering 6* (2): 1589–1600.
135. Mostafa, A.A.-F., Elshikh, M.S., Al-Askar, A.A. et al. (2019). Decolorization and biotransformation pathway of textile dye by Cylindrocephalum aurelium. *Bioprocess and Biosystems Engineering 42* (9): 1483–1494.
136. Isik, Z., Arikan, E.B., Ozay, Y. et al. (2020). Electrocoagulation and electrooxidation pretreatment effect on fungal treatment of pistachio processing wastewater. *Chemosphere 244*: 125383.
137. Kumar, V. and Dwivedi, S. (2020). Multimetal tolerant fungus Aspergillus flavus CR500 with remarkable stress response, simultaneous multiple metal/loid removal ability and bioremediation potential of wastewater. *Environmental Technology and Innovation 20*: 101075.
138. Bankole, P.O., Adekunle, A.A., Jeon, B.-H. et al. (2020). Novel cobiomass degradation of NSAIDs by two wood rot fungi, Ganoderma applanatum and Laetiporus sulphureus: ligninolytic enzymes induction, isotherm and kinetic studies. *Ecotoxicology and Environmental Safety 203*: 110997.
139. Singh, G. and Dwivedi, S. (2020). Decolorization and degradation of Direct Blue-1 (Azo dye) by newly isolated fungus Aspergillus terreus GS28, from sludge of carpet industry. *Environmental Technology and Innovation 18*: 100751.
140. Azhar, S.H.M., Abdulla, R., Jambo, S.A. et al. (2017). Yeasts in sustainable bioethanol production: a review. *Biochemistry and Biophysics Reports 10*: 52–61.
141. Gong, Y., Ding, P., Xu, M.-J. et al. (2021). Biodegradation of phenol by a halotolerant versatile yeast Candida tropicalis SDP-1 in wastewater and soil under high salinity conditions. *Journal of Environmental Management 289*: 112525.
142. Wang, Y., Qiu, L., and Hu, M. (ed.) (2018). Application of yeast in the wastewater treatment. In: *E3S Web of Conferences*. EDP Sciences.
143. Yoshizawa, K. (1978). Treatment of waste water discharged from a sake brewery using yeast. *Hakko Kogaku Zasshi;(Japan) 56* (4).
144. Walls, L.E., Velasquez-Orta, S.B., Romero-Frasca, E. et al. (2019). Non-sterile heterotrophic cultivation of native wastewater yeast and microalgae for integrated municipal wastewater treatment and bioethanol production. *Biochemical Engineering Journal 151*: 107319.
145. Yoshizawa, K. (1981). Development of the new treating method of waste water from food industry using yeast. *Journal of the Agricultural Chemical Society of Japan (Japan) 55*: 705–711.
146. Curds, C. and Cockburn, A. (1970). Protozoa in biological sewage-treatment processes—II. Protozoa as indicators in the activated-sludge process. *Water Research 4* (3): 237–249.
147. Curds, C.R. and Cockburn, A. (1970). Protozoa in biological sewage-treatment processes—I. A survey of the protozoan fauna of British percolating filters and activated-sludge plants. *Water Research 4* (3): 225–236.
148. Parada-Albarracín, J., Pérez, J., and Gómez, M. (2017). Bioindicator value of flagellates in urban wastewater treatment using membrane bioreactors. *Water Research 122*: 526–535.
149. Sydenham, D. (1971). A re-assessment of the relative importance of ciliates, rhizopods and rotatorians in the ecology of activated sludge. *Hydrobiologia 38* (3): 553–563.
150. Hughes, D., Stafford, D., and Gaudy, A.F. Jr. (1976). The microbiology of the activated-sludge process. *Critical Reviews in Environmental Science and Technology 6* (3): 233–257.
151. Curds, C.R. (1982). The ecology and role of protozoa in aerobic sewage treatment processes. *Annual Reviews in Microbiology 36* (1): 27–28.
152. Madoni, P. (1983). Growth and succession of ciliate populations during the establishment of a mature activated sludge. *Acta Hydrobiologica 24* (3): 223–232.
153. Aescht, E. and Foissner, W. (1992). Biology of a high-rate activated sludge plant of a pharmaceutical company. *Archiv für Hydrobiologie Supplementband Untersuchungen des Elbe-AEstuars 90* (2): 207–251.

154. Salvado, H. (1994). Effect of mean cellular retention time on ciliated protozoan populations in urban wastewater treatment plants based on a proposed model. *Water Research 28* (6): 1315–1321.
155. Madoni, P. (2011). Protozoa in wastewater treatment processes: a minireview. *The Italian Journal of Zoology 78* (1): 3–11.
156. Weisse, T., Jezberova, J., and Moser, M. (2021). Picoplankton feeding by the ciliate Vorticella similis in comparison to other peritrichs emphasizes their significance in the water purification process. *Ecological Indicators 121*: 106992.
157. Kinner, N. and Curds, C. (1987). Development of protozoan and metazoan communities in rotating biological contactor biofilms. *Water Research 21* (4): 481–490.
158. Madoni, P. (1994). Microfauna biomass in activated sludge and biofilm. *Water Science and Technology 29* (7): 63–66.
159. Madoni, P. (1994). Estimates of ciliated protozoa biomass in activated sludge and biofilm. *Bioresource Technology 48* (3): 245–249.
160. Kachieng'a L, Momba M. (2018). The synergistic effect of a consortium of protozoan isolates (Paramecium sp., Vorticella sp., Epistylis sp. and Opercularia sp.) on the biodegradation of petroleum hydrocarbons in wastewater. *Journal of Environmental Chemical Engineering 6* (4): 4820–4827.
161. Kamika, I. and Momba, M.N. (2015). Effect of nickel on nutrient removal by selected indigenous protozoan species in wastewater systems. *Saudi Journal of Biological Sciences 22* (2): 147–156.
162. Pauli, W., Jax, K., and Berger, S. (2001). Protozoa in wastewater treatment: function and importance. In: *Biodegradation and Persistence* (ed. B. Beek), 203–252. Springer.
163. Lapinski, J. and Tunnacliffe, A. (2003). Reduction of suspended biomass in municipal wastewater using bdelloid rotifers. *Water Research 37* (9): 2027–2034.
164. Bergtold, M., Mayr, G., and Traunspurger, W. (2007). Nematodes in wastewater biofilms—appearance and density of species in three biofilter reactors. *Water Research 41* (1): 145–151.
165. Ito, M., Ito, K., Ohta, K. et al. (2016). Evaluation of bioremediation potential of three benthic annelids in organically polluted marine sediment. *Chemosphere 163*: 392–399.
166. Ruiz, F., Abad, M., Galán, E. et al. (2006). The present environmental scenario of El Melah Lagoon (NE Tunisia) and its evolution to a future sabkha. *Journal of African Earth Sciences 44* (3): 289–302.
167. Tan, C., Gouramanis, C., Pham, T. et al. (2021). Ostracods as pollution indicators in Lap An Lagoon, central Vietnam. *Environmental Pollution 278*: 116762.
168. Giovanella, P., Vieira, G.A., Otero, I.V.R. et al. (2020). Metal and organic pollutants bioremediation by extremophile microorganisms. *Journal of Hazardous Materials 382*: 121024.
169. Abtahi, H., Parhamfar, M., Saeedi, R. et al. (2020). Effect of competition between petroleum-degrading bacteria and indigenous compost microorganisms on the efficiency of petroleum sludge bioremediation: field application of mineral-based culture in the composting process. *Journal of Environmental Management 258*: 110013.
170. Pal, A.K., Singh, J., Soni, R. et al. (2020). The role of microorganism in bioremediation for sustainable environment management. In: *Bioremediation of Pollutants* (ed. V.C. Pandey and V. Singh), 227–249. Elsevier.
171. Kumar, V. and Dwivedi, S. (2019). Hexavalent chromium stress response, reduction capability and bioremediation potential of Trichoderma sp. isolated from electroplating wastewater. *Ecotoxicology and Environmental Safety 185*: 109734.
172. Chen, S.H., Cheow, Y.L., Ng, S.L. et al. (2019). Mechanisms for metal removal established via electron microscopy and spectroscopy: a case study on metal tolerant fungi Penicillium simplicissimum. *Journal of Hazardous Materials 362*: 394–402.
173. Feng, X., Xu, J., Liang, Y. et al. (2017). A proteomic-based investigation of potential copper-responsive biomarkers: proteins, conceptual networks, and metabolic pathways featuring Penicillium janthinellum from a heavy metal-polluted ecological niche. *MicrobiologyOpen 6* (4): e00485.

174. Kumar, V., Singh, S., Singh, G. et al. (2019). Exploring the cadmium tolerance and removal capability of a filamentous fungus Fusarium solani. *Geomicrobiology Journal 36* (9): 782–791.
175. Luo, W., Lu, Y., Wang, G. et al. (2008). Distribution and availability of arsenic in soils from the industrialized urban area of Beijing, China. *Chemosphere 72* (5): 797–802.
176. Limcharoensuk, T., Sooksawat, N., Sumarnrote, A. et al. (2015). Bioaccumulation and biosorption of Cd2+ and Zn2+ by bacteria isolated from a zinc mine in Thailand. *Ecotoxicology and Environmental Safety 122*: 322–330.
177. El-Sheekh, M., El Sabagh, S., Abou El-Souod, G. et al. (2019). Biosorption of cadmium from aqueous solution by free and immobilized dry biomass of Chlorella vulgaris. *International Journal of Environmental Research 13* (3): 511–521.
178. El-Sheekh, M., El-Sabagh, S., Abou Elsoud, G. et al. (2020). Efficacy of immobilized biomass of the seaweeds Ulva lactuca and Ulva fasciata for cadmium biosorption. *Iranian Journal of Science and Technology, Transactions A: Science 44* (1): 37–49.
179. Fawzy, M.A., Hifney, A.F., Adam, M.S. et al. (2020). Biosorption of cobalt and its effect on growth and metabolites of Synechocystis pevalekii and Scenedesmus bernardii: isothermal analysis. *Environmental Technology and Innovation 19*: 100953.
180. Mawi, S., Krishnan, S., Din, M.F.M. et al. (2020). Bioremediation potential of macroalgae Gracilaria edulis and Gracilaria changii co-cultured with shrimp wastewater in an outdoor water recirculation system. *Environmental Technology and Innovation 17*: 100571.
181. Zeng, T., Mo, G., Hu, Q. et al. (2020). Microbial characteristic and bacterial community assessment of sediment sludge upon uranium exposure. *Environmental Pollution 261*: 114176.
182. Psaltou, S. and Zouboulis, A. (2020). Catalytic ozonation and membrane contactors—a review concerning fouling occurrence and pollutant removal. *Water 12* (11): 2964.
183. Shivalkar, S., Singh, V., Sahoo, A.K. et al. (2021). Bioremediation: a potential ecological tool for waste management. In: *Bioremediation for Environmental Sustainability*, 1–21. Elsevier.
184. Akhtar, N. and Mannan MA-u. (2020). Mycoremediation: expunging environmental pollutants. *Biotechnology Reports 26*: e00452.
185. Roshandel, F., Saadatmand, S., Iranbakhsh, A. et al. (2021). Mycoremediation of oil contaminant by Pleurotus florida (P. Kumm) in liquid culture. *Fungal Biology 125*: 667–678.
186. Koul, B., Ahmad, W., and Singh, J. (2021). Mycoremediation: a novel approach for sustainable development. In: *Microbe Mediated Remediation of Environmental Contaminants* (ed. A. Kumar, V.K. Singh and V.K. Mishrap), 409–420. Elsevier.
187. Sadañoski, M.A., Benítez, S.F., Fonseca, M.I. et al. (2019). Mycoremediation of high concentrations of polychlorinated biphenyls with Pleurotus sajor-caju LBM 105 as an effective and cheap treatment. *Journal of Environmental Chemical Engineering 7* (6): 103453.
188. Chatterjee, S., Mahanty, S., Das, P. et al. (2020). Biofabrication of iron oxide nanoparticles using manglicolous fungus Aspergillus niger BSC-1 and removal of Cr (VI) from aqueous solution. *Chemical Engineering Journal 385*: 123790.
189. Fernandes, J.M., Sousa, R.M.O., Fraga, I. et al. (2020). Fungal biodegradation and multi-level toxicity assessment of vinasse from distillation of winemaking by-products. *Chemosphere 238*: 124572.
190. Barh, A., Kumari, B., Sharma, S. et al. (2019). Mushroom mycoremediation: kinetics and mechanism. In: *Smart Bioremediation Technologies* (ed. P. Bhatt), 1–22. Elsevier.
191. Ortiz-Monsalve, S., Valente, P., Poll, E. et al. (2019). Biodecolourization and biodetoxification of dye-containing wastewaters from leather dyeing by the native fungal strain Trametes villosa SCS-10. *Biochemical Engineering Journal 141*: 19–28.
192. Ariste, A.F., Batista-García, R.A., Vaidyanathan, V.K. et al. (2020). Mycoremediation of phenols and polycyclic aromatic hydrocarbons from a biorefinery wastewater and concomitant production of lignin modifying enzymes. *Journal of Cleaner Production 253*: 119810.
193. Biswas, J.K., Kumar, A., Biswas, S. et al. (2022). Win-win wastewater phycoremediation: coupled carbon sequestration and heavy metal removal. In: *An Integration of Phycoremediation Processes in Wastewater Treatment* (ed. M. Shah, S. Rodriguez-Couto and J. Biswas), 529–548. Elsevier.

194. Shackira, A., Jazeel, K., and Puthur, J.T. (2021). Phycoremediation and phytoremediation: promising tools of green remediation. In: *Sustainable Environmental Clean-up* (ed. V.K. Mishra and A. Kumar), 273–293. Elsevier.
195. Phang, S.-M., Chu, W.-L., and Rabiei, R. (2015). Phycoremediation. In: *The Algae World* (ed. Sahoo D and J. Seckbach), 357–389. Springer.
196. Gomes, P.I. and Asaeda, T. (2009). Phycoremediation of Chromium (VI) by Nitella and impact of calcium encrustation. *Journal of Hazardous Materials 166* (2-3): 1332–1338.
197. Kumari, P., Varma, A.K., Shankar, R. et al. (2021). Phycoremediation of wastewater by Chlorella pyrenoidosa and utilization of its biomass for biogas production. *Journal of Environmental Chemical Engineering 9* (1): 104974.
198. Ayele, A., Getachew, D., Kamaraj, M. et al. (2021). Phycoremediation of synthetic dyes: an effective and eco-friendly algal technology for the dye abatement. *Journal of Chemistry 2021*: 14.
199. Moungmoon, T., Chaichana, C., Pumas, C. et al. (2020). Quantitative analysis of methane and glycolate production from microalgae using undiluted wastewater obtained from chicken-manure biogas digester. *Science of the Total Environment 714*: 136577.
200. Zabed, H.M., Akter, S., Yun, J. et al. (2020). Biogas from microalgae: technologies, challenges and opportunities. *Renewable and Sustainable Energy Reviews 117*: 109503.
201. Kube, M., Mohseni, A., Fan, L. et al. (2020). Energy and nutrient recovery by treating wastewater with fluidised-beds of immobilised algae. *Journal of Water Process Engineering 38*: 101585.
202. Kumar, V.S., Sarkar, S.D., Das, B.K. et al. (2021). Sustainable biodiesel production from microalgae Graesiella emersonii through valorization of garden wastes-based vermicompost. *Science of the Total Environment 807*: 150995.
203. Wu, L.-C., Chen, C.-Y., Lin, T.-K. et al. (2020). Highly efficient removal of Victoria blue R and bioelectricity generation from textile wastewater using a novel combined dual microbial fuel cell system. *Chemosphere 258*: 127326.
204. Kumar, S.S., Kumar, V., Gude, V.G. et al. (2020). Alkalinity and salinity favor bioelectricity generation potential of Clostridium, Tetrathiobacter and Desulfovibrio consortium in microbial fuel cells (MFC) treating sulfate-laden wastewater. *Bioresource Technology 306*: 123110.
205. Thulasinathan, B., Jayabalan, T., Sethupathi, M. et al. (2021). Bioelectricity generation by natural microflora of septic tank wastewater (STWW) and biodegradation of persistent petrogenic pollutants by basidiomycetes fungi: an integrated microbial fuel cell system. *Journal of Hazardous Materials 412*: 125228.
206. Mohamed, S.N., Hiraman, P.A., Muthukumar, K. et al. (2020). Bioelectricity production from kitchen wastewater using microbial fuel cell with photosynthetic algal cathode. *Bioresource Technology 295*: 122226.
207. Lin, Y., Song, G., Ling, H. et al. (2021). Isolation of a high-ammonium-tolerant Monoraphidium sp. and evaluation of its potential for biodiesel production. *Process Biochemistry 111*: 297–304.
208. Behera, A.R., Dutta, K., Verma, P. et al. (2019). High lipid accumulating bacteria isolated from dairy effluent scum grown on dairy wastewater as potential biodiesel feedstock. *Journal of Environmental Management 252*: 109686.
209. Chandra, R., Pradhan, S., Patel, A. et al. (2021). An approach for dairy wastewater remediation using mixture of microalgae and biodiesel production for sustainable transportation. *Journal of Environmental Management 297*: 113210.
210. Rambabu, K., Bharath, G., Thanigaivelan, A. et al. (2021). Augmented biohydrogen production from rice mill wastewater through nano-metal oxides assisted dark fermentation. *Bioresource Technology 319*: 124243.
211. Ulhiza, T.A., Puad, N.I.M., and Azmi, A.S. (2018). Optimization of culture conditions for biohydrogen production from sago wastewater by Enterobacter aerogenes using response surface methodology. *International Journal of Hydrogen Energy 43* (49): 22148–22158.
212. Onay, M. (2019). Bioethanol production via different saccharification strategies from H. tetrachotoma ME03 grown at various concentrations of municipal wastewater in a flat-photobioreactor. *Fuel 239*: 1315–1323.

213. Lee, O.K., Oh, Y.-K., and Lee, E.Y. (2015). Bioethanol production from carbohydrate-enriched residual biomass obtained after lipid extraction of Chlorella sp. KR-1. *Bioresource Technology 196*: 22–27.
214. Greetham, D., Zaky, A., Makanjuola, O. et al. (2018). A brief review on bioethanol production using marine biomass, marine microorganism and seawater. *Current Opinion in Green and Sustainable Chemistry 14*: 53–59.
215. Favaro, L., Jansen, T., and van Zyl, W.H. (2019). Exploring industrial and natural Saccharomyces cerevisiae strains for the bio-based economy from biomass: the case of bioethanol. *Critical Reviews in Biotechnology 39* (6): 800–816.
216. Tetteh, E.K., Rathilal, S., Chetty, M. et al. (2019). Treatment of water and wastewater for reuse and energy generation-emerging technologies. In: *Water and Wastewater Treatment* (ed. M. Eyvaz), 53–80. IntechOpen.
217. Atelge, M., Atabani, A., Banu, J.R. et al. (2020). A critical review of pretreatment technologies to enhance anaerobic digestion and energy recovery. *Fuel 270*: 117494.
218. Marín, D., Posadas, E., García, D. et al. (2019). Assessing the potential of purple phototrophic bacteria for the simultaneous treatment of piggery wastewater and upgrading of biogas. *Bioresource Technology 281*: 10–17.
219. Frijns, J., Hofman, J., and Nederlof, M. (2013). The potential of (waste) water as energy carrier. *Energy Conversion and Management 65*: 357–363.
220. Hashemi, S.S., Karimi, K., and Taherzadeh, M.J. (2021). Integrated process for protein, pigments, and biogas production from baker's yeast wastewater using filamentous fungi. *Bioresource Technology 337*: 125356.
221. Arias, D.M., Solé-Bundó, M., Garfí, M. et al. (2018). Integrating microalgae tertiary treatment into activated sludge systems for energy and nutrients recovery from wastewater. *Bioresource Technology 247*: 513–519.
222. Nguyen, L.N., Kumar, J., Vu, M.T. et al. (2020). Biomethane production from anaerobic co-digestion at wastewater treatment plants: a critical review on development and innovations in biogas upgrading techniques. *Science of the Total Environment 765*: 142753.
223. Potter, M.C. (1911). Electrical effects accompanying the decomposition of organic compounds. *Proceedings of the Royal Society of London Series B, Containing Papers of a Biological Character 84* (571): 260–276.
224. Kim, H.J., Moon, S.H., and Byung, H.K. (1999). A microbial fuel cell type lactate biosensor using a metal-reducing bacterium, Shewanella putrefaciens. *Journal of Microbiology and Biotechnology 9* (3): 365–367.
225. Nader, M., Wan, R., Rahimnejad, M. et al. (2012). Bioelectricity generation in biological fuel cell with and without mediators. *World Applied Sciences Journal 18* (4): 559–567.
226. Obileke, K., Onyeaka, H., Meyer, E.L., and Nwokolo, N. (2021). Microbial fuel cells, a renewable energy technology for bio-electricity generation: a mini-review. *Electrochemistry Communications 125*: 107003.
227. Yan, X., Lee, H.-S., Li, N. et al. (2020). The micro-niche of exoelectrogens influences bioelectricity generation in bioelectrochemical systems. *Renewable and Sustainable Energy Reviews 134*: 110184.
228. Liu, D.-F. and Li, W.-W. (2020). Potential-dependent extracellular electron transfer pathways of exoelectrogens. *Current Opinion in Chemical Biology 59*: 140–146.
229. Varanasi, J.L., Prasad, S., Singh, H. et al. (2020). Improvement of bioelectricity generation and microalgal productivity with concomitant wastewater treatment in flat-plate microbial carbon capture cell. *Fuel 263*: 116696.
230. Cevik, E., Tombuloglu, H., Anıl, I. et al. (2020). Direct electricity production from Microalgae Choricystis sp. and investigation of the boron to enhance the electrogenic activity. *International Journal of Hydrogen Energy 45* (19): 11330–11340.
231. Sarma, H., Bhattacharyya, P., Jadhav, D.A. et al. (2021). Fungal-mediated microbial electrochemical system: exciting development, possible application and certain challenges. *Current Research in Microbial Sciences 2*: 100041.

232. Umar, M.F., Rafatullah, M., Abbas, S.Z. et al. (2021). Bioelectricity production and xylene biodegradation through double chamber benthic microbial fuel cells fed with sugarcane waste as a substrate. *Journal of Hazardous Materials 419*: 126469.
233. García-Olivares, A., Solé, J., and Osychenko, O. (2018). Transportation in a 100% renewable energy system. *Energy Conversion and Management 158*: 266–285.
234. Ajay, C., Mohan, S., Dinesha, P. et al. (2020). Review of impact of nanoparticle additives on anaerobic digestion and methane generation. *Fuel 277*: 118234.
235. Li, Y., Tian, G., and Xu, H. (2012). Application of biodiesel in automotive diesel engines. In: *Biodiesel: Feedstocks, Production and Applications* (ed. Z. Fang). London: IntechOpen https://doi.org/10.5772/53222.
236. Wu, M., Wu, G., Han, L. et al. (2005). Low-temperature fluidity of bio-diesel fuel prepared from edible vegetable oil. *Petroleum Processing and Petrochemicals 36*: 57–60.
237. Arif, M., Bai, Y., Usman, M. et al. (2020). Highest accumulated microalgal lipids (polar and non-polar) for biodiesel production with advanced wastewater treatment: role of lipidomics. *Bioresource Technology 298*: 122299.
238. Martínez, A., Mijangos, G.E., Romero-Ibarra, I.C. et al. (2019). In-situ transesterification of Jatropha curcas L. seeds using homogeneous and heterogeneous basic catalysts. *Fuel 235*: 277–287.
239. Sialve, B., Bernet, N., and Bernard, O. (2009). Anaerobic digestion of microalgae as a necessary step to make microalgal biodiesel sustainable. *Biotechnology Advances 27* (4): 409–416.
240. Zhou, X., Jin, W., Wang, Q. et al. (2020). Enhancement of productivity of Chlorella pyrenoidosa lipids for biodiesel using co-culture with ammonia-oxidizing bacteria in municipal wastewater. *Renewable Energy 151*: 598–603.
241. Usman, T.M., Banu, J.R., Gunasekaran, M. et al. (2019). Biohydrogen production from industrial wastewater: an overview. *Bioresource Technology Reports 7*: 100287.
242. Hay, J.X.W., Wu, T.Y., Ng, B.J. et al. (2016). Reusing pulp and paper mill effluent as a bioresource to produce biohydrogen through ultrasonicated Rhodobacter sphaeroides. *Energy Conversion and Management 113*: 273–280.
243. Yin, Y. and Wang, J. (2019). Mechanisms of enhanced biohydrogen production from macroalgae by ferrous ion: insights into correlations of microbes and metabolites. *Bioresource Technology 291*: 121808.
244. Lamb, J.J., Hillestad, M., Rytter, E. et al. (2020). Traditional routes for hydrogen production and carbon conversion. In: *Hydrogen, Biomass and Bioenergy* (ed. J.J. Lamb and B.G. Pollet), 21–53. Elsevier.
245. Sivagurunathan, P., Kumar, G., Mudhoo, A. et al. (2017). Fermentative hydrogen production using lignocellulose biomass: an overview of pre-treatment methods, inhibitor effects and detoxification experiences. *Renewable and Sustainable Energy Reviews 77*: 28–42.
246. Banu, J.R., Kavitha, S., Kannah, R.Y. et al. (2020). Industrial wastewater to biohydrogen: possibilities towards successful biorefinery route. *Bioresource Technology 298*: 122378.
247. Kossalbayev, B.D., Tomo, T., Zayadan, B.K. et al. (2020). Determination of the potential of cyanobacterial strains for hydrogen production. *International Journal of Hydrogen Energy 45* (4): 2627–2639.
248. Yang, G. and Wang, J. (2019). Enhancing biohydrogen production from waste activated sludge disintegrated by sodium citrate. *Fuel 258*: 116177.
249. Melikoglu, M. and Turkmen, B. (2019). Food waste to energy: forecasting Turkey's bioethanol generation potential from wasted crops and cereals till 2030. *Sustainable Energy Technologies and Assessments 36*: 100553.
250. Tan, J.S., Phapugrangkul, P., Lee, C.K. et al. (2019). Banana frond juice as novel fermentation substrate for bioethanol production by Saccharomyces cerevisiae. *Biocatalysis and Agricultural Biotechnology 21*: 101293.
251. Di Donato, P., Finore, I., Poli, A. et al. (2019). The production of second generation bioethanol: the biotechnology potential of thermophilic bacteria. *Journal of Cleaner Production 233*: 1410–1417.

252. Ramachandra, T. and Hebbale, D. (2020). Bioethanol from macroalgae: prospects and challenges. *Renewable and Sustainable Energy Reviews 117*: 109479.
253. Ingale, S., Joshi, S.J., and Gupte, A. (2014). Production of bioethanol using agricultural waste: banana pseudo stem. *Brazilian Journal of Microbiology 45* (3): 885–892.
254. Alshammari, A.M., Adnan, F.M., Mustafa, H. et al. (2011). Bioethanol fuel production from rotten banana as an environmental waste management and sustainable energy. *African Journal of Microbiology Research 5* (6): 586–598.
255. Crawford, D.L. and McCoy, E. (1972). Cellulases of Thermomonospora fusca and Streptomyces thermodiastaticus. *Applied Microbiology 24* (1): 150–152.
256. Deng, Y. and Fong, S.S. (2010). Influence of culture aeration on the cellulase activity of Thermobifida fusca. *Applied Microbiology and Biotechnology 85* (4): 965–974.
257. Arora, A., Nain, L., and Gupta, J. (2005). Solid-state fermentation of wood residues by Streptomyces griseus B1, a soil isolate, and solubilization of lignins. *World Journal of Microbiology and Biotechnology 21* (3): 303–308.
258. Alikkunju, A.P., Sainjan, N., Silvester, R. et al. (2016). Screening and characterization of cold-active β-galactosidase producing psychrotrophic Enterobacter ludwigii from the sediments of Arctic Fjord. *Applied Biochemistry and Biotechnology 180* (3): 477–490.
259. Akintola, A.I., Oyedeji, O., Adewale, I.O. et al. (2021). Production and physicochemical properties of thermostable, crude cellulase from Enterobacter cloacae IP8 isolated from plant leaf litters of Lagerstroemia indica Linn. *Journal of Microbiology, Biotechnology and Food Sciences 2021*: 989–994.
260. Sarkar, D., Prajapati, S., Poddar, K. et al. (2019). Production of ethanol by Enterobacter sp. EtK3 during fruit waste biotransformation. *International Biodeterioration & Biodegradation 145*: 104795.

16

Actinobacteria from Soils and their Applications in Environmental Bioremediation

Nurul H. Adenan[1,2] and Adeline S-Y Ting[1]

16.1. INTRODUCTION

Actinobacteria, also known as actinomycetes, are one of the largest groups of Gram-positive bacteria with high guanine-cytosine content in their DNA [1]. The predominant genus is *Streptomyces* (i.e. 50–70%) while other genera such as *Nocardia*, *Micromonospora*, *Actinoplanes*, *Micromonospora*, *Streptosporangium* [2–4], *Actinomadura*, *Microbispora*, *Nonomurea*, *Mycobacterium*, *Frankia*, *Saccharopolyspora,* and *Verrucosispora* are lesser-known genera of actinobacteria [5]. Actinobacteria are filamentous bacteria, mostly aerobic, and are classified based on the size and type of spores, as well as the colony color [6, 7]. Another unique feature of actinobacteria is the production of a secondary metabolite known as geosmin, which exudes an "earthy" odor. Geosmins are most notably produced by the genus *Streptomyces* [7]. Actinobacteria inhabit terrestrial and aquatic environments, though they are more commonly found in soil. The presence of actinobacteria in soil mainly contributes to the decomposition of organic matter and the nutrient cycling processes [8].

Historically, actinobacteria are known to produce important bioactive compounds such as vitamins [9] and compounds with the following properties: antibacterial, antitumor, antifungal [10], anticancer, antiviral, antiprotozoal, and immunosuppressant [11]. The genus *Streptomyces* is well-known for producing differing types of secondary metabolites. These include polyene macrolides, actinomycins, aminoglycosides, streptothricins, anthracyclines, cyclopolylactones, and quinoxaline peptides. Other genera of actinobacteria (i.e. non-*Streptomyces*) are also involved in the production of beneficial compounds, e.g. glycopeptides and orthosomycins [5]. According to Sathya et al. [5], secondary metabolites of actinobacteria have a major advantage as they have lower toxicity compared to fungal metabolites (i.e. only 2% of actinobacterial products are toxic). These bioactive metabolites contribute to the production of two-thirds of the existing antibiotics in the market [12, 13]. This is followed by the production of enzymes, which are reported as the second-most essential compound produced by actinobacteria after antibiotics [14].

[1]*School of Science, Monash University Malaysia, Selangor, Malaysia*

[2]*School of Biology, Universiti Teknologi MARA (UiTM), Negeri Sembilan, Malaysia*

Microbial Biotechnology: Role in Ecological Sustainability and Research, First Edition.
Edited by Pankaj Chowdhary, Sujata Mani, and Preeti Chaturvedi.

In recent years, actinobacteria have also become known for their ability to remove both organic and inorganic pollutants [15]. In addition, actinobacteria have also been found to tolerate extreme conditions such as high temperature, pH, and salinity. Hence, actinobacteria are considered desirable for bioremediation as alternative bioagents [16]. This chapter will focus on the origin of actinobacteria, their roles in the soil ecosystem, their production of enzymes/beneficial compounds, and application in bioremediation, which includes the removal of heavy metals, dyes, hydrocarbons, and pesticides.

16.2. ORIGIN OF ACTINOBACTERIA

Actinobacteria or actinomycetes are unicellular organisms with branched filaments [17]. They form one of the phyla from the 18 major lineages found in the Kingdom Eubacteria. There are 5 subclasses, 6 orders, and 14 suborders in the phyla [1, 2, 18]. Over the years, many species have been identified, with most species predominantly belonging to the genus *Streptomyces*, *Frankia*, and *Corynebacterium* [2]. In the early years, the characterization of actinobacteria was divided between their classification as bacteria (*Eubacteriales,* higher bacteria) or fungi (*Hyphomycetes,* lower fungi) as similarities to both groups were found in actinobacteria [19]. The term actinomycetes itself originated from two Greek words: *atkis* and *mykes*, which mean "ray" and "fungus," respectively [20]. Actinomycetes are now commonly known as actinobacteria to reflect their bacterial origin [21]. The class Actinobacteria, proposed by Stackebrandt et al. [22], was suggested to be changed to Actinomycetia [23]. However, the former term is still widely used in recent literature [24, 25].

Waksman [26] reported that the first observation of actinobacteria was from the tear duct of the human eye by Ferdinand Cohn in 1875, which he later named as *Streptothrix foersteri*. This was followed by the discovery of *Actinomyces bovis* from an infectious disease known as actinomycosis (i.e. lump jaw in cattle) by Harz in 1877 [27]. According to Sharma et al. [8], research on actinobacteria was primarily the work of Waksman, between 1919 until 1939. In the early 1940s, Waksman successfully discovered the new antibiotic streptomycin from *Streptomyces* sp. [8]. This discovery propelled the value of actinobacteria as producers of beneficial compounds. To date, more than 16 000 species have been identified from the genus *Streptomyces* alone [28], with 16 000 bioactive compounds known to be produced from actinobacteria [29].

16.3. DISTRIBUTION AND ROLE OF ACTINOBACTERIA IN SOIL

Actinobacteria are ubiquitous in the environment, found more often in soil than in any other environment [30, 31]. In soil, actinobacteria have roles as decomposers of organic matter: they secrete extracellular enzymes for the breakdown of organic substrates such as polysaccharides, cellulose, hemicelluloses, and polyphenolic lignin [32, 33]. This indirectly contributes to soil improvement as the decomposition of organic matter enhances the nutrient levels in the soil through recycling of essential nutrients and humus formation [34]. Another major role of actinobacteria in the soil is their ability to inhibit pathogens [35]. Many plants suffer from diseases caused by fungal pathogens such as rusts, smuts, rots, wilt, and anthracnose [36]. According to Zhao et al. [37], actinobacteria, especially *Streptomyces* spp., produce vital defense-related enzymes such as chitinase and an array of bioactive compounds. This leads to growth improvements in plants associated with actinobacteria, and the protection of plants against biotic and abiotic stresses [38].

Researchers have successfully isolated a number of actinobacterial species from different soil types (e.g. hill soil, clay loam soil, and desert soil), different environmental settings (e.g. pristine or polluted), and from various geographical distributions (e.g. India, Slovakia, Israel, Algeria, and China). They are found from hill soils (in India) [39], heavy metal-polluted soils (in Slovakia) [9], clay loam soils (in Israel) [14], desert soils (in Algeria) [13], and quinclorac polluted soils (in China) [40]. The distribution of actinobacteria isolated from soils in different countries is summarized in Table 16.1. It is interesting to note that actinobacteria from polluted soils have tremendous potential for bioremediation, especially for hydrocarbon and metal [50]. Nevertheless, the discovery from unexplored soils (e.g. desert soils) is also crucial as novel actinobacteria and bioactive metabolites can be discovered [47, 51], for example, the discovery of *Saccharothrix* sp. from desert soil in Southern Algeria. Saccharothrix *sp.* produces two novel bioactive compounds known as cyanogriside I and cyanogriside J with significant activities against several pathogens [52].

Table 16.1 Examples of actinobacteria isolated from various soil types and geographical regions.

Genera	Soil type	Location	Country	References
Actinomadura *Actinopolyspora* *Elylrosporangium* *Nocardia* *Nocardiopsis* *Streptomyces*	Clay loam and pristine	Al-Aqsa mosque	Israel	Mansour et al. [14]
Streptomyces	Hill and pristine	Nahargarh hills	India	Sharma et al. [39]
Streptomyces	Desert and pristine	Sahara	Algeria	Mohamed et al. [13]
Micromonospora *Streptomyces*	Saline and pristine	Lake Elton	Russia	Zenova et al. [41]
Actinopolyspora *Catellospora* *Glycomyces* *Kitasatosporia* *Nocordia* *Streptomyces*	Dye-polluted soil	Tirupur	India	Vijayakumar and Malathi [6]
Nocardia *Streptomyces* *Streptosporangium*	Forest and pristine	Eastern Himalaya	India	Das et al. [42]
Actinokineospora *Amycolatopsis* *Kribbella* *Micromonospora* *Nocardia* *Nocardiopsis* *Nonomuraea*	Mountain, forest and cave, and pristine	Kermanshah	Iran	Azarakhsh et al. [43]
Saccharothrix *Streptosporangium* *Streptomyces*	Mountain, forest and cave, and pristine	Kermanshah	Iran	Azarakhsh et al. [43]
Kocuria *Streptomyces*	Agricultural and pristine	Egyptian governorates	Egypt	Elbendary et al. [44]
Streptomyces	Anthropogenic and polluted; Conserved and pristine	Sundarbans	India	Sengupta et al. [45]

(Continued)

Table 16.1 (Continued)

Genera	Soil type	Location	Country	References
Gordonia *Leifsonia* *Mycobacterium* *Microbacterium* *Micromonospora* *Nocardia* *Nocardioides* *Sinomonas* *Streptacidiphilus* *Streptomyces* *Terrabacter*	Mangrove and pristine	Tanjung Lumpur	Malaysia	Lee et al. [4]
Streptomyces	Quinclorac polluted soil	Xuancheng	China	Lang et al. [40]
Actinomadura *Kribbella* *Microbacterium* *Micromonospora* *Nocardia*	Plateau soil and pristine	Eastern Black Sea	Turkey	Isik et al. [3]
Streptomyces	High altitude and pristine	Kashmir	India	Shah et al. [46]
Actinomadura *Amycolatopsis* *Kribbella* *Lentzea* *Microbispora* *Micromonospora* *Nocardia* *Nonomuraea* *Streptomyces*	Red and pristine	Jiangxi	China	Guo et al. [47]
Actinomadura *Kibdelosporangium* *Kitasatosporia* *Nocardiopsis* *Pseudonocardia* *Streptomyces* *Streptoverticillium*	Paddy and pristine	Tiruchirappalli	India	Priyadharsini and Dhanasekaran [34]
Actinomycetales (Order) *Arthrobacter* *Catenuloplanes* *Glycomyces* *Micrococcus* *Pseudonocardia* *Rubrobacter* *Streptomyces*	Heavy metal polluted soil	Sereď	Slovakia	Remenár et al. [9]
Micromonospora	Desert and pristine	Atacama Desert	Chile	Carro et al. [48]
Phytoactinopolyspora	Saline and pristine	Xinjiang	China	Ji et al. [49]

16.4. ACTINOBACTERIAL ENZYMES FOR BIOREMEDIATION

Living organisms are known as producers of a wide array of enzymes [53]. However, enzymes obtained from bacteria are preferred as they are easy to grow, cost-effective [53], reliable, and in some cases, capable of tolerating different types of extreme environments [16]. In fact, more than 50% of enzymes are produced by bacteria, including actinobacteria [29]. To date, actinobacterial strains have been reported to produce an array of valuable enzymes. This is mostly studied in species of *Streptomyces* such as *Streptomyces scabrisporus*, *Streptomyces sparsogenes*, *Streptomyces misakiensis*, *Streptomyces cirratus*, *Streptomyces lincolnensis*, *Streptomyces endophyticus*, *Streptomyces chartreusis*, and *Streptomyces alboniger* [11].

In general, there are two types of enzymes responsible for the degradation of various pollutants: oxidative, also known as oxidoreductases (e.g. oxygenases, laccases, peroxidases), and hydrolases (e.g. cellulases, lipases, protease, amylase) [54, 55]. Actinobacteria have been reported to produce almost all the important enzymes required for various applications. These include peptidase, ligninase, sugar isomerase, pectinase, hemicellulose, keratinase [56], amylase, cellulase, protease, tyrosinase, lipase, catalase, phosphatase [14], chitinase, xylanase [53], gelatinase, pectinase, and urease [11]. For many years, these enzymes have been commercially used in industries including bioremediation [16].

Peroxidases (i.e. lignin peroxidase, manganese peroxidase, and versatile peroxidase) belong to the oxidoreductase group. Peroxidases are used to catalyze the oxidation of molecules. The focus on actinobacterial peroxidases has gained momentum in the last 20 years due to their capacity to degrade lignin and other pollutants such as polycyclic aromatic hydrocarbons (PAHs), pesticides, and dyes [57]. Known actinobacteria species that produce peroxidases include *Streptomyces albus, Streptomyces albidoflavus,* and *Thermobifida fusca* [57]. Some actinobacteria species produce specific enzymes (strain-specific) that are not typically produced by other actinobacteria. One such example is the novel dye-decolorizing peroxidase (DyP) known as SaDyP2 by *Streptomyces avermitilis* [58]. DyP was able to degrade anthraquinone dyes, unlike other types of peroxidases.

Cellulases are a group of hydrolytic enzymes that are responsible for breaking down β-1,4-glycosidic bonds of cellulose [16]. Many actinobacterial strains are able to produce cellulases including *Streptomyces plicatus* [56], *Cellulomonas fimi*, *Microbispora bispora*, *Thermobifida fusca* [59], *Streptomyces ruber*, *Streptomyces lividans*, and *Streptomyces rutgersensis* [16]. The cellulase (CMCase, FPase, and β-glucosidase) activities were somewhat higher when cultured in submerged fermentations. Thomas et al. [60] observed this when *Promicromonospora* sp. was cultivated on untreated lignocellulosic waste (i.e. wheat straw and sugarcane bagasse). In their findings, *Promicromonospora* sp. was also tolerant against various pollutants (i.e. antibiotics, heavy metals, and organic solvents), further suggesting that this microbial strain has good potential for bioremediation.

Lipase enzymes are also valuable in bioremediation as these enzymes are able to hydrolyze oils and fats [16]. There are several strains of lipase-producing actinobacteria that have been reported, such as *Rhodococcus erythropolis* [53], *Streptomyces exfoliates*, *Nocardiopsis alba* [16], and *Actinomadura sediminis* [61]. Lipase-producing strains have been isolated from oil-contaminated soils and they exhibit strong lipase activities, even in alkaline pH conditions and with resistance to organic solvents and detergents, as well as oxidizing and reducing agents [61]. In addition, *Micrococcus* sp. isolated from nickel-polluted soil has also shown high lipase activities during the removal (i.e. 98% decolorization) of azo dye (i.e. Reactive red 5B) [9].

Amylases represent 25% of the industrial enzymes in the market [53] and are responsible for extracellular digestion occurring outside of cells [8]. The following species of actinobacteria are reported to produce amylases; *Streptomyces hygroscopicus*, *Streptomyces limosus*, *Streptomyces praecox*, *Thermomonospora vulgaris* [53], *Streptomyces erumpens,* and *Thermobifidafusca* [16]. In a study by Lamilla et al. [62], species from the genus *Brevibacterium*, *Arthrobacter,* and *Streptomyces* were deemed to produce the most amylase activities. These amylase enzymes are useful in the degradation of organic pollutants such as pesticides [63].

16.5. ACTINOBACTERIA AS BIOREMEDIATION AGENTS

In recent years, actinobacteria have been discovered to have great potential in degrading different types of pollutants [15]. In addition, they have been reported to be able to endure extreme conditions (e.g. high temperature, low moisture, and limited nutrient) as well as demonstrate tolerance toward various pollutants [64]. The adaptive capacity of actinobacteria is attributed to its resistance mechanism [64] and diverse metabolisms [65]. Actinobacteria are also able to colonize substrates rapidly [66] and produce different types of enzymes (e.g. superoxide dismutase) [64], which are essential in bioremediation.

The degradation of pollutants occurs via enzymatic reaction with the involvement of actinobacteria as a biocatalyst. The degradation rate generally relies on the expression of enzymes produced [67]. The efficacy of bioremediation, however, also involves other factors such as chemical nature and concentration of pollutants, physicochemical characteristics of the environment, and the compatibility of actinobacteria with the contaminants [67]. To date, actinobacteria have demonstrated potential to remove toxic heavy metals, carcinogenic dyes, hydrocarbons, and pesticides [14, 68]. The various actinobacterial strains used for the removal of common pollutants are listed in Table 16.2.

16.5.1. Actinobacteria for Metal Removal

Metals are known as important micronutrients and in minute amounts, they are essential co-enzyme factors responsible for catalytic processes [86]. However, when present in high concentrations, they become toxic [87]. There are numerous metals found in the environment as residues from industrial, agricultural, and anthropogenic activities. These include nickel (Ni), copper (Cu), antimony (Sb) [64], cadmium (Cd), manganese (Mn), arsenic (As), lead (Pb), zinc [41], aluminum (Al), gold (Au), mercury [44], and silver (Ag) [88]. The pollution from toxic metals in the environment is caused by urbanization and human activities such as excessive use of fertilizer and agrochemicals; polluted wastewaters from industries such as mining, smelting, electroplating [64]; pollution from spillage of petroleum chemicals; and coal combustion residues [65]. Metals, when at high concentrations, lead to many health and environmental hazards. These lead to adverse effects such as growth inhibition, cancer, organ damage, nervous system impairment, and death [87]. Timková et al. [89] have highlighted that metal toxicity is dependent on the concentration of metal, admission or contact route, and the duration of exposure.

Several conventional techniques have been used to remove heavy metals, such as reverse osmosis [90], ion exchange, evaporation, filtration [91], adsorption, precipitation, and chemical oxidation–reduction [92]. Unfortunately, these methods are inefficient, expensive, complex, and produce toxic waste (i.e. secondary pollutant) [93]. The reverse osmosis method, for example, requires high pressure and is known to have low water permeability [87]. Filtration often generates sludge, while ion-exchange techniques often result in incomplete removal of metal ions [94]. The adsorption approach may fail to absorb the water-soluble type of metals [94]. These methods are, therefore, generally effective when

Table 16.2 List of actinobacteria used for the removal of heavy metals, dyes, hydrocarbons, and pesticides.

Actinobacterial strain	Pollutant	Types of pollutant	Mechanism	References
Streptomyces roseisederoticus	Heavy metals	Chromium (Cr)	Bioaccumulation	Vinod et al. [69]
Streptomyces flavochromogenes		Lead (Pb)		
Streptomyces aureofacienes		Zinc (Zn)	Biosorption	El-Sayed et al. [70]
Nocardiopsis sp.		Zinc (Zn)	Biosorption	El-Gendy and El-Bondkly [71]
Streptomyces sp.		Lead (Pb)	Bioaccumulation	El Baz et al. [66]
Streptomyces coelicolor		Nickel (Ni)	Accumulation	Lu et al. [72]
Nocardia corallina	Dyes	Crystal violet	Biodegradation	Yatome et al. [73]
Kocuria rosea		Malachite green	Biodegradation	Parshetti et al. [74]
Rhodococcus qingshengii		Crystal violet Methyl violet	Biodegradation	Li et al. [75]
Streptomyces fulvissimus		Crystal violet	Biosorption Biodegradation	Buntić et al. [76]
Streptomyces ruber		Congo red	Biosorption	El-Sersy et al. [77]
Streptomyces globosus		Acid-fast red		
Amycolatopsis sp.	Hydrocarbons	Naphthalene	Biodegradation	Bourguignon et al. [78]
Rhodococcus sp.		Phenanthrene		
Streptomyces sp		Pyrene		
Streptomyces coelicolor	Hydrocarbons	n-alkanes	Biodegradation	Gallo et al. [79]
Streptomyces spp.		Crude oil	Biodegradation	Burghal et al. [80]
Rhodococcus erythropolis		Alkanes	Degradation	Margesin et al. [81]
Rhodococcus cercidiphyllus		Phenol		
Arthrobacter sulfureus, Pimelobacter simplex		Anthracene		
Streptomyces parvulus	Pesticides	Cypermethrin	Biodegradation	Lin et al. [82]
Streptomyces sp.		Lindane	Biodegradation	Philip et al. [83]
Rhodococcus sp.		p-Nitrophenol	Biodegradation	Ningthoujam et al. [84]
Brevibacterium casei				
Streptomyces spp.		Pentachlorophenol Chlorpyrifos	Biodegradation	Fuentes et al. [85]

high concentrations of metal ions are present, and less effective for the removal of low concentrations of metals (i.e. less than 100 ppm) [95].

In recent years, researchers introduced biological approaches that involved microorganisms (e.g. actinobacteria) as biosorbents for the removal of metals, known as biosorption and bioaccumulation. Gram-positive microorganisms such as actinobacteria typically perform greater sorption of metals as they have a thicker layer of cell wall [89]. They are cost-effective, efficient, environmentally friendly [96], and both live and dead cells can be used [87].

To date, there are 35 genera of actinobacteria that are reported to be resistant to metals [15]. *Streptomyces* spp. are the most widely studied for bioremediation purposes including for metal removal. El Baz et al. [66] tested 27 strains of actinobacteria (i.e. live cells) belonging to the genus *Streptomyces* and *Amycolatopsis* for the removal of lead (Pb), copper (Cu), zinc (Zn), and cadmium (Cd). Several strains are highly resistant to lead (Pb) (0.55 mg ml^{-1}) and chromium (Cr) (0.10 mg ml^{-1}). Further observations revealed that the removal of lead (Pb) (i.e. 600 mg) by *Streptomyces* sp. BN3 occurred via bioaccumulation. In a separate study, four strains of *Streptomyces* spp. (i.e. *Streptomyces roseisederoticus*, *Streptomyces flavochromogenes*, *Streptomyces vastus,* and *Streptomyces praguaeneses*) were tested for heavy metal removal (i.e. chromium, cadmium, zinc, and lead). Results showed that *S. roseisederoticus* removed 73% of chromium (Cr) while *S. flavochromogenes* removed 60% of lead (Pb) via bioaccumulation by metal-binding proteins. Both strains are also resistant to chromium (Cr) and lead (Pb) at 8000 mg l [69]. Other strains of *Streptomyces* were also examined for their tolerance to zinc (Zn) where *Streptomyces aureofacienes* showed the highest absorption of zinc (Zn) (i.e. 734.8 μg g^1 of Zn). This is followed by *S. viridodiastaticus*, *Streptomyces badius,* and *Streptomyces diastaticus* (i.e. 696.5, 684.0, and 663.2 μg g^{-1} of Zn, respectively) [70]. Aburas [97] demonstrated the removal of *Streptomyces* sp. on different metals. The results showed 99.5%, 97.0%, and 92.5% removal of chromium (Cr), cadmium (Cd), and copper (Cu), respectively. In another study, live cells of *Promicromonospora* sp. were examined to remove zinc (Zn) and cadmium (Cd). Results indicated 140 mm and 9.2 mm decolorized zone, respectively [98]. Another study used both live and dead cells of two strains (i.e. *Nocardiopsis* sp. MORSY1948 and *Nocardia* sp. MORSY2014) for the removal of zinc (Zn). Their findings indicated higher removal by live cells of *Nocardia* sp. MORSY2014 (i.e. 67.91%) compared to *Nocardiopsis* sp. MORSY1948 (i.e. 46.91%). A similar trend was observed using dead cells where *Nocardia* sp. MORSY2014 exhibited 90.37% removal, while the removal rate by *Nocardiopsis* sp. MORSY1948 was 84.15%. The removal of zinc (Zn) in this study was carried out at pH8 via biosorption [71].

Biosorption of heavy metals is a passive process where the biosorbate (i.e. metal ions) binds to the cell surface of actinobacteria, which is mostly composed of functional groups such as carboxyl, phosphonate, amine, and hydroxyl [89]. The efficacy of binding between the functional groups and heavy metals is based on the physicochemical condition of biosorption [99]. The biosorption process only involves dead cells [91], which can be obtained either by autoclaving or boiling [99]. The use of dead cells has several advantages, like the fact that they are not influenced by the toxicity of heavy metals, can be stored for the long term, and are reusable [92]. However, the improvement for the removal of heavy metals by dead cells is only limited to chemical modifications, as biological processes are not involved [99]. The removal of heavy metals via biosorption is easy to operate, cost-effective (e.g. no nutrients are required for bacterial growth), and generates low sludge. Most importantly, this approach can remove heavy metals in low concentrations, unlike conventional techniques [96].

On the contrary, bioaccumulation is defined as an active process where live cells of actinobacteria are used [91]. Bioaccumulation involves two phases where the heavy metal ions first bind to the actinobacterial cell surface (i.e. biosorption). This is followed by the transportation and accumulation of heavy metal ions intracellularly [100]. Since live cells of actinobacteria are used in bioaccumulation, the optimization of nutritional requirements as well as the incubation conditions (e.g. pH and temperature) is important to improve the effectiveness of heavy metal removal. Bioaccumulation has an advantage over biosorption as this process involved the conversion of heavy metals into less toxic products. Moreover, the toxicity of heavy metals can also be reduced via efflux pumping, intracellular binding, and metal-resistant strains [92].

Several improvements for the removal of heavy metals can be done through the following methods: physical modifications (e.g. boiling, freezing, drying, and lyophilization); immobilization (e.g. calcium alginate) [91]; and genetic modification (e.g. protein engineering) [101]. Actinobacteria are also able to produce metal-binding proteins (e.g. metallothioneins) [65] and/or enzymes that produce polymers such as peptides that can bind effectively to metals [101]. For example, *Streptomyces coelicolor* produced a novel nickel (Ni) binding protein known as SCO4226 (UniProt Q9FCE4), which binds multiple nickel (Ni) ions with low affinity. Further analysis of this study has shown the removal of nickel ions was demonstrated via self-accumulation [72].

16.5.2. Actinobacteria for Dye Removal

There are approximately 700 000 t of dyes produced annually throughout the world, as reported in Color Index (i.e. a reference database of color products) [102]. The classification of dyes is based on the origin (e.g. natural or synthetic), types of dyes (e.g. nitroso dyes), and their application (e.g. dispersive dyes) [103, 104]. Dyes can also be characterized as anionic (i.e. acid, direct, and reactive dyes), cationic (i.e. basic dyes), or non-ionic (i.e. disperse dyes) [105]. To date, there are approximately 20–30 classes of dyes identified based on their chemical and chromophore structure, and more than 100 000 dyes that are commercially available [103]. The most commonly used are azo, triphenylmethane (TPM), anthraquinone, nitro, nitroso, and indigoid dyes [103]. The application of dye is primarily for clothing, textiles, leather, foods, paper, cosmetics, and pharmaceuticals [106].

The extensive use of dyes is worrying, as the improper discharge of the effluents into the environment is rampant [107]. In general, dyes can be detected with as low as 1 mg ml^{-1} in water, and unfortunately, dyes from many industries are found at a higher range of 10–200 mg ml^{-1} [108]. The dispersion and presence of dyes in water reduce light penetration and increase the level of biological oxygen demand (BOD) as well as chemical oxygen demand (COD) [102]. With high levels of BOD and COD, soil fertility and the growth of living organisms are compromised [109]. Dyes are a serious threat as dye molecules or their byproducts from degradation are toxic, carcinogenic, and mutagenic [106]. This is due to the existence of NO_2, NO, N·N bonds, and aromatic amine, which are difficult to degrade [105, 110]. As a result, living organisms may suffer from organ damage (e.g. brain, liver, kidneys) [105] and allergic reactions [111] due to exposure to the dye pollutants.

As a measure to remove dye pollutants from the environment, several physicochemical techniques have been attempted such as advanced oxidation, photocatalysis [112], adsorption, membrane filtration, coagulation, and ozonation [113]. However, they are reported to have several drawbacks, i.e. expensive and producing secondary sludge [112]. Physical removal of dyes, which often employ adsorption and membrane filtration methods, decolorize dyes merely by transferring the dye molecules. Physical removal is limited by the fact that dye degradation does not occur. Moreover, physical removal is also not practical for the treatment of large volumes of dye effluents. The chemical approach of advanced oxidation also produces solid waste, incurring additional costs for waste disposal [105]. Hence, biological treatment using microorganisms such as *Aspergillus* sp. [114], *Nocardia* sp. [115], and *Bacillus* sp. [116] is preferred as these methods are simple, sustainable, cost-effective, and able to convert organic compounds to non-toxic products [116]. The efficiency of the biological approach involves several factors such as carbon and nitrogen source, pH, temperature, presence of oxygen, initial dye concentration, type of dye [117], biomass dosage, and rate of agitation [104].

Over the years, several actinobacteria species have been explored for dye removal potential. Live cells of *Nocardia corallina* have removed 80% of crystal violet (i.e. 2.3 $\mu mol\,dm^{-3}$) via biodegradation after 90 minutes of incubation [73]. Another rare actinobacterial strain, *Kocuria rosea,* completely decolorized (100% decolorization) malachite green (50 mg l) when live cells were incubated with the dye for five hours. The degradation of malachite green was attributed to dichlorophenol indophenol (DCIP) reductase and malachite green (MG) reductase [74]. In a study done by Li et al. [75], *Rhodococcus qingshengii* (live cells) was tested for the removal of crystal violet and methyl violet. Results indicated that *R. qingshengii* removed 79.6% of crystal violet and 85.7% of methyl violet, where the production of lignin peroxidase (LiP) and NADH-DCIP reductase were also observed. In a separate study, *S. badius* and *Thermomonospora mesophile* were used for the removal of polymeric dye Poly-R. The decrease in optical density (OD) was observed daily. Results showed 0.96 and 0.89 of OD, respectively, after incubation for 12 days [118]. Five actinobacterial strains (i.e. *Streptomyces globosus*, *Streptomyces alanosinicus*, *S. ruber*, *Streptomyces gancidicus*, and *Nocardiopsisaegyptia*) were tested for decolorization of acid-fast red and congo red dyes. *S. globosus* has shown high removal of acid-fast red dye with 81.6 and 70.2% decolorization under static and shaking conditions, respectively. Meanwhile, *S. ruber* exhibited 72.7 and 60.7% decolorization of congo red (i.e. static and shaking conditions) via biosorption [77]. Another *Streptomyces* sp. (i.e. *S. fulvissimus*) indicated complete decolorization of crystal violet under its optimal conditions (i.e. shaking, 30 °C, and neutral pH). The dye removal primarily involved biosorption, which was then followed with biodegradation [76]. Complete decolorization was also reported using *Amycolatopsisorientalis* on crystal violet (i.e. 30 $\mu g\,ml^{-1}$) [119]. Another removal of crystal violet was demonstrated using *Streptomyces* sp. and *Nocardia* sp. (live cells). In their findings, *Streptomyces* sp. exhibited a 603 mm^2 decolorized zone, while *Nocardia* sp. indicated a 83 mm^2 decolorized zone, in which 0.012 $U\,ml^{-1}$ laccase was also observed [57]. In recent years, soil actinobacteria, particularly species of *Streptomyces* and *Nocardiopsis*, have demonstrated capacity in the removal of triphenylmethane dyes (malachite green, methyl violet, crystal violet, and cotton blue), with higher decolorization efficiency by live cells (16.2–97.0%) compared to dead cells (11.5–96.3%) [120–122].

The removal of dyes using actinobacteria typically involves mechanisms known as biosorption, bioaccumulation, and/or biodegradation [103]. This is similar to mechanisms employed for metal removal. Biosorption occurs when live or dead cells of actinobacteria are used [103]. In biosorption, the functional groups (e.g. amino or carboxyl) present on the actinobacterial cell wall are binding sites to the dye molecules. The use of live cells, unfortunately, is not suggested for continuous treatment of dye removal as it is expensive for long-term operations. Dead cells are preferred as they are cheaper since nutrient supply is not required, and dead cells are more flexible in different environmental conditions [105]. Bioaccumulation of dyes by actinobacteria primarily involves the biosorption process (metabolism-independent), followed by the accumulation of dyes in the cytoplasm of actinobacterial cells (metabolism-dependent) [117]. However, neither biosorption nor bioaccumulation is able to reduce the toxicity of the dye parent compound. This is largely because both mechanisms only entrap and accumulate the dye molecules [103]. Biodegradation of dyes, therefore, offers a better solution as this mechanism involves the production of extracellular and intracellular enzymes of actinobacteria. The oxidative (e.g. manganese peroxidase, laccase, and tyrosinase) and reductive (e.g. NADH-dependent reductases) enzymes produced by actinobacteria play a key role in the degradation of dyes [106]. With biodegradation, complete mineralization (i.e. carbon dioxide, water, and inorganic products) is possible with fewer harmful products produced [103]. Nevertheless, it is important to note that biodegradation requires a high cost for continuous treatment [105]. It is important to note that these mechanisms can be operated either sequentially or simultaneously [117].

The efficiency of dye removal can be enhanced by increasing the biomass surface using pre-treatments such as physical methods (e.g. autoclave) and chemical methods (e.g. alkalis, chelating agent). In addition, the immobilization of actinobacterial cells in alginate beads could improve the biosorption capacity [117]. The combination of different techniques (e.g. physical and biological treatment) [102], or utilization of more than one actinobacterial strain as a biosorbent [104], is also suggested for the acceleration of dye removal. On the other hand, molecular biology techniques such as genetic engineering are reportedly beneficial for the degradation of dyes, as the expression of the desired gene or enzyme can be increased [123]. Additionally, a genetically modified *Aspergillus oryzae* has successfully decolorized malachite green (92% decolorization) in 120 minutes through over-expression activities by laccase [124].

16.5.3. Actinobacteria for Hydrocarbon Removal

Hydrocarbon is defined as a compound that is composed of carbon and hydrogen and includes PAHs [125]. PAHs, which consist of 16 types of contaminants (e.g. naphthalene, phenanthrene, and pyrene) are known as the most toxic and difficult to degrade [126]. In general, there are four classifications of hydrocarbons based on their structures, i.e. alkanes (saturated hydrocarbons), alkenes (unsaturated hydrocarbons), cycloalkanes, and aromatics (benzene, toluene, ethylbenzene, and xylenes) [126]. Hydrocarbons, with a larger number of carbon and greater chain length on the aromatic rings, have a higher level of toxicity [126]. Most hydrocarbons are harmful, easily identified at low concentrations (i.e. $\mu g l^{-1}$ or ppb), and have poor water solubility [125]. Hydrocarbons are derived from anthropogenic activities, e.g. oil spills, leakage of an oil pipeline, exploration of gas, refining, storage of hydrocarbons, steamers, and abandoned industrial sites [127]. As a result, hydrocarbons that are discharged into the environment affect the survival of living organisms and may lead to mutation and death [128].

Various physicochemical methods have been used to remove the hydrocarbons such as evaporation, oil dispersion, soil washing, thermal treatment, vapor extraction, solidification, and stabilization [129]. However, these methods are expensive and ineffective [127] as most of the pollutants are not degraded completely [125]. Therefore, bioremediation technology that employs microorganisms to remove hydrocarbons is preferred due to its cost-effectiveness and eco-friendly nature. Bacteria, including actinobacteria, are known as the largest group of microorganisms involved in the removal of hydrocarbons [130]. Among the various species of actinobacteria, the genus *Streptomyces* is known as a good bioremediation agent for hydrocarbons [7]. Other actinobacterial genera involved in hydrocarbon removal include *Amycolatopsis* [78], *Rhodococcus*, *Dietzia*, and *Gordonia* [131].

Bourguignon et al. [78] demonstrated 15 actinobacterial strains for the degradation of PAHs (i.e. naphthalene, phenanthrene, and pyrene). Results showed strains (i.e. live cells) from genus *Amycolatopsis*, *Rhodococcus*, and *Streptomyces* having the highest percentage removal of naphthalene (i.e. 73.0–76.6%), while *Amycolatopsis* sp. removed 36.2% of phenanthrene. *Streptomyces* sp. however, exhibited lower degradation of phenanthrene and pyrene (i.e. 20 and 4.3% of removal, respectively). In a separate study, *Actinomyces israelii* exhibited 98% removal of crude oil (i.e. 10 g l). This study proposed that peroxidases were involved in the degradation when 1.54 U mg^{-1} of peroxidase activity was observed [132]. Two actinobacterial strains, i.e. *R. erythropolis* and *Rhodococcus cercidiphyllus*, were tested for degradation of alkanes and anthracene. Findings indicated that *R. cercidiphyllus* removed 93% of alkanes after 14 days, whereas *R. erythropolis* showed complete degradation of anthracene after 28 days at a concentration of 20 mg^{-1} [81].

The degradation of hydrocarbons by actinobacteria involves the use of live cells, therefore the efficacy relies on microbial population, type of enzyme, temperature, pH, nutrient, and the presence of oxygen [133]. The degradation of hydrocarbon by actinobacteria may lead to mineralization where the hydrocarbons are transformed into less harmful and non-hazardous compounds (i.e. carbon dioxide, water, and inorganic products) [125]. This mechanism can take place in aerobic or anaerobic conditions, or in the presence of electron acceptors such as nitrate, ferric iron, and sulfate [134]. Most of the hydrocarbons, however, are reportedly degraded under aerobic conditions, where the degradation of hydrocarbons occurs intracellularly via the oxidative process, incorporation, and activation of enzymes (e.g. oxygenases and peroxidases) [127]. In contrast, the anaerobic condition often exhibited a slower degradation rate, especially in sub-optimal conditions. This is due to nutrient limitation [135], high temperature, and intense irradiance from sunlight [129]. Although biodegradation is known as a primary mechanism for hydrocarbon removal, there are also a number of limited studies reporting on hydrocarbon biosorption. Hydrocarbons generally undergo the biosorption process when the molecules are adsorbed onto the surface of microorganisms (e.g. actinobacteria) [128].

There are several approaches that can be applied to speed up the process of bioremediation: bioattenuation, bioventing, biopiling, biostimulation, and bioaugmentation [67]. The latter two are commonly used for the removal of hydrocarbons [133, 136]. Biostimulation allows the introduction of nutrient sources (e.g. growth supplements and trace minerals) or the introduction of oxygen to stimulate the rapid growth of the actinobacterial population; thus, increasing the degradation rate [67]. Biostimulation is, therefore, an approach that creates a suitable environment for biodegradation through the manipulation of abiotic factors. On the other hand, bioaugmentation is an approach that involves the biotic element. It entails the addition of microorganisms (e.g. natural or genetically engineered strain of actinobacteria) to enhance the biodegradation process. In a study done by Burghal et al. [80], six strains from genus *Streptomyces* were tested for crude oil removal. The findings exhibited 79.6% of degradation after 90 days through bioaugmentation.

Genetically engineered actinobacteria also have potential where essential genes can be integrated, and over-expression of certain enzymes can be demonstrated. For example, the recombinant strain of *Streptomyces celicolor* (M145-AH) was developed by recombining the *alkB* gene encoding for the alkane monooxygenase enzyme, which successfully degraded n-alkanes after 48 hours [79]. Nevertheless, the application of genetic engineering modification has a few disadvantages. These include cell death and growth delay [67].

16.5.4. Actinobacteria for Pesticide Removal

Pesticides are defined as any material or mixtures that are used to prevent, destroy, or repel any pest; these include insecticides, herbicides, and fungicides [137]. Pesticides are categorized into carbamates [138], organochlorides, organophosphates, pyrethroids [139], and neonicotinoids [140]. They are also can be classified based on their application, mode of action, or chemical function [139]. The utilization of pesticides began in the 1950s when people used chemicals to protect their livestock and crops [141]. Since then, they have had a positive impact on agriculture as food production has considerably increased [137]. The prolonged utilization of pesticides, however, has caused harm to the environment as they are highly toxic and can bioaccumulate in the food chain [85]. This causes contamination in soil, crops, marine environments, and they also affect human health [142], leading to various side effects such as cancer, hormone imbalance, abnormal fetal growth, hyperexcitation, aggressiveness, skin allergy, and seizures [141].

There are various physicochemical technologies used for the removal of pesticides including adsorption, percolator filters, and advanced oxidation. However, these methods are expensive and have several drawbacks as they require specific conditions for the treatment process. They also often result in secondary pollutants where the byproducts produced are toxic [137]. As such, researchers divert their attention to the biological process where pesticides are degraded into simple inorganic compounds using microorganisms, including actinobacteria [142]. The implementation of this method has several advantages, such as fewer hazardous compounds produced, environmentally friendly, and cost-effective [143].

In recent years, many studies focused on using soil microorganisms such as actinobacteria to degrade pesticides [142]. Several genera of actinobacteria used for the removal of pesticides including *Arthrobacter*, *Brevibacterium*, *Corynebacterium*, *Micromonospora*, *Mycobacterium*, *Nocardia*, *Rhodococcus*, and *Streptomyces* have been reported [141]. Among all, the genus *Streptomyces* is the most extensively used for the removal of pesticides [55]. *Streptomyces parvulus* removed the insecticide cypermethrin (50 mg l) with 92% degradation occurring within 24 hours and complete removal (100% degradation) in 30 hours [82]. A study done by Philip et al. [83] showed the capability of *Streptomyces* sp. in utilizing organochlorine-type (i.e. lindane) as a carbon source, and its resistance toward toxicity of lindane. Results indicated significant growth of *Streptomyces* sp. in high concentration (i.e. 400 μg/lit.) of lindane after 14 days. Janaki [144] tested the potential of *Streptomyces cacaoi* as a biopesticide by inhibiting the synthesis of chitin in an insect. Findings showed that *S. cacaoi* removed 98% of *Culex quinquefasciatus* after 24 hours. Other genera of actinobacteria, for example, the *Rhodococcus* sp., completely removed p-Nitrophenol (PNP) at 350 mg l in 87 hours while *Brevibacterium casei* fully degraded PNPat 270 mg l in 113 hours [84].

The removal of pesticides by actinobacteria involves intracellular (e.g. cytochrome P450) [145] and extracellular enzymes (e.g. hydrolases, peroxidases, oxygenases) [137]. Cytochrome P450 is produced by different genera of actinobacteria [145] where the degradation process takes place via these three phases: the conversion of parent compound into water-soluble and lesser toxic compounds via oxidation, reduction, and hydrolysis (Phase 1); the conjugation of pesticides to amino acid and sugar (Phase 2); and the transformation of the previous compound into harmless secondary conjugates (Phase 3) [137].

Currently, various environmental parameters such as temperature, pH, moisture content, and nutrients play an important role in improving the bioremediation of hydrocarbons [140]. In addition, genetic modification [143] and cell immobilization [15] of actinobacteria can be applied to enhance the degradation rate of pesticides. Cell immobilization offers stability, where actinobacterial cells can be reused numerous times without affecting their significant activity [15]. For example, several strains of genus *Streptomyces* were tested for the removal of pesticides (i.e. pentachlorophenol and chlorpyrifos) as a free and immobilized consortium. Results revealed a better removal percentage of pentachlorophenol and chlorpyrifos by *Streptomyces* spp. using immobilized cells compared to free cells (i.e. 71.05 and 14.72%, respectively) [85]. According to Verma et al. [143], mixed strains and naturally occurring bacteria (i.e. actinobacteria) also performed efficient removal of pesticides compared to genetically modified (GM) strains. Application of GM often encountered environmental problems after being introduced into the polluted sites, as the transferred gene in the GM strain resulted in changes of genotypic diversity, which may lead to different phenotypic expressions [143].

16.6. CONCLUSION AND FUTURE PROSPECTS

It is evident that actinobacteria have the potential for bioremediation applications. They produce a wide range of enzymes that are responsible for the biodegradation of heavy metals, dyes, hydrocarbons, and pesticides. The genus *Streptomyces* is the most extensively studied for the removal of pollutants. The removal of these pollutants using conventional techniques is inefficient, expensive, and often produces toxic waste. Therefore, biological treatment, which involves microorganisms including actinobacteria as a bioremediation agent, is preferred as it is simple, cheaper, and pollutants can be converted to carbon dioxide, water, and inorganic products. Bioremediation using actinobacteria involves the use of live cells or dead cells, and it occurs primarily via biosorption, bioaccumulation, or biodegradation. Numerous strategies such as genetic modification, cell immobilization, and a combination of multiple approaches can be adopted to increase the efficiency of bioremediation.

ACKNOWLEDGMENTS

The authors express gratitude to Monash University Malaysia for the research funding and facilities.

REFERENCES

1. Wahab, A., Shumaila, S., Subhan, S. et al. (2015). Isolation and identification of actinomycetes isolated from Karachi soil and screening of antimicrobial compounds. *International Journal of Current Research 7* (2): 12760–12765.
2. Ventura, M., Canchaya, C., Tauch, A. et al. (2007). Genomics of Actinobacteria: tracing the evolutionary history of an ancient phylum. *Microbiology and Molecular Biology Reviews 71* (3): 495–548.
3. Isik, K., Gencbay, T., Özdemir-Kocak, F. et al. (2014). Molecular identification of different actinomycetes isolated from East Black Sea region plateau soil by 16S rDNA gene sequencing. *African Journal of Microbiology Research 8* (9): 878–887.
4. Lee, L.-H., Zainal, N., Azman, A.-S. et al. (2014). Diversity and antimicrobial activities of actinobacteria isolated from tropical mangrove sediments in Malaysia. *Scientific World Journal 2014*: 698178.
5. Sathya, A., Vijayabharathi, R., and Gopalakrishnan, S. (2017). Plant growth-promoting actinobacteria: a new strategy for enhancing sustainable production and protection of grain legumes. *3 Biotech 7* (2): 1–10.
6. Vijayakumar, R. and Malathi, R. (2014). Isolation, characterization and antibacterial activity of actinobacteria from dye polluted soils of Tirupur. *Facta Universitatis Series: Medicine and Biology 16* (1): 43–48.
7. Anandan, R., Dharumadurai, D., and Manogaran, G.P. (2016). An introduction to actinobacteria. In: *Actinobacteria-Basics and Biotechnological Applications* (ed. D. Dhanasekaran and Y. Jiang), 3–37. United Kingdom: Intechopen.
8. Sharma, M., Dangi, P., and Choudhary, M. (2014). Actinomycetes: source, identification, and their applications. *International Journal of Current Microbiology and Applied Sciences 3* (2): 801–832.

9. Remenár, M., Karelová, E., Harichová, J. et al. (2014). Actinobacteria occurrence and their metabolic characteristics in the nickel-contaminated soil sample. *Biologia 69* (11): 1453–1463.
10. Subhan, S.A., Wahab, A., Mujahid, T.Y. et al. (2015). Screening of Actinomycetes from indigenous soil for production of extracellular metabolites. *International Journal of Current Research 7* (1): 12078–12083.
11. Bhatti, A.A., Haq, S., and Bhat, R.A. (2017). Actinomycetes benefaction role in soil and plant health. *Microbial Pathogenesis 111*: 458–467.
12. Montalvo, N.F., Mohamed, N.M., Enticknap, J.J. et al. (2005). Novel actinobacteria from marine sponges. *Antonie Van Leeuwenhoek 87* (1): 29–36.
13. Mohamed, H., Miloud, B., Zohra, F. et al. (2017). Isolation and characterization of Actinobacteria from Algerian Sahara soils with antimicrobial activities. *International Journal of Molecular and Cellular Medicine 6* (2): 109.
14. Mansour, S.R., Abdel-Azeem, A.M., and Abo-Deraz, S.S.S. (2015). A new record of Actinobacteria isolated from soil in Jerusalem and their enzymatic potential. *F1000Research 4*: 11.
15. Alvarez, A., Saez, J.M., Costa, J.S.D. et al. (2017). Actinobacteria: current research and perspectives for bioremediation of pesticides and heavy metals. *Chemosphere 166*: 41–62.
16. Mukhtar, S., Zaheer, A., Aiysha, D. et al. (2017). Actinomycetes: a source of industrially important enzymes. *Journal of Proteomics and Bioinformatics 10*: 12.
17. Li, Q., Chen, X., Jiang, Y. et al. (2016). Morphological identification of actinobacteria. In: *Actinobacteria-Basics and Biotechnological Applications* (ed. D. Dhanasekaran and Y. Jiang), 59–86. UK: Intechopen.
18. Barka, E.A., Vatsa, P., Sanchez, L. et al. (2016). Taxonomy, physiology, and natural products of Actinobacteria. *Microbiology and Molecular Biology Review 80* (1): 1–43.
19. Mahajan, G.B. and Balachandran, L. (2012). Antibacterial agents from actinomycetes-a review. *Frontiers in Bioscience (Elite Edition) 4* (4): 240–253.
20. Das, S., Lyla, P., and Khan, S.A. (2008). Distribution and generic composition of culturable marine actinomycetes from the sediments of Indian continental slope of Bay of Bengal. *Chinese Journal of Oceanology and Limnology 26* (2): 166–177.
21. Azman, A.-S., Othman, I., Velu, S.S. et al. (2015). Mangrove rare actinobacteria: taxonomy, natural compound, and discovery of bioactivity. *Frontiers in Microbiology 6*: 856.
22. Stackebrandt, E., Rainey, F.A., and Ward-Rainey, N.L. (1997). Proposal for a new hierarchic classification system, Actinobacteria classis nov. *International Journal of Systematic Bacteriology 47* (2): 479–491.
23. Salam, N., Jiao, J.-Y., Zhang, X.-T. et al. (2020). Update on the classification of higher ranks in the phylum Actinobacteria. *International Journal of Systematic and Evolutionary Microbiology 70* (2): 1331–1355.
24. Presentato, A., Piacenza, E., Turner, R.J. et al. (2020). Processing of metals and metalloids by actinobacteria: cell resistance mechanisms and synthesis of metal (loid)-based nanostructures. *Microorganisms 8* (12): 2027.
25. Hazarika, S.N. and Thakur, D. (2020). Actinobacteria. In: *Beneficial Microbes in Agro-Ecology* (ed. N. Amaresan, M. Senthil Kumar, K. Annapurna, et al.), 443–476. Cambridge: Academic Press.
26. Waksman, S. (1959). *The Actinomycetes: Nature, Occurrence and Activities*, vol. *I*, 113. London: Balliere, Tindall and Cox.
27. Sowani, H., Kulkarni, M., Zinjarde, S. et al. (2017). Gordonia and related genera as opportunistic human pathogens causing infections of skin, soft tissues, and bones. In: *The Microbiology of Skin, Soft Tissue, Bone and Joint Infections* (ed. K. Kon and M. Rai), 105–121. United Kingdom: Elsevier.
28. Lewin, G.R., Carlos, C., Chevrette, M.G. et al. (2016). Evolution and ecology of Actinobacteria and their bioenergy applications. *Annual Review of Microbiology 70*: 235–254.
29. Hamedi, J., Poorinmohammad, N., and Wink, J. (2017). The role of actinobacteria in biotechnology. In: *Biology and Biotechnology of Actinobacteria* (ed. J. Wink, F. Mohammadipanah and J. Hamedi), 269–238. Cham: Springer.

30. Moraga, N.B., Poma, H.R., Amoroso, M.J. et al. (2014). Isolation and characterization of indigenous *Streptomyces* and *Lentzea* strains from soils containing boron compounds in Argentina. *Journal of Basic Microbiology 54* (6): 568–577.
31. Trujillo, M.E., Riesco, R., Benito, P. et al. (2015). Endophytic actinobacteria and the interaction of *Micromonospora* and nitrogen fixing plants. *Frontiers in Microbiology 6*: 1341.
32. Větrovský, T., Steffen, K.T., and Baldrian, P. (2014). Potential of cometabolic transformation of polysaccharides and lignin in lignocellulose by soil Actinobacteria. *PLoS One 9* (2): e89108.
33. Wang, C., Dong, D., Wang, H. et al. (2016). Metagenomic analysis of microbial consortia enriched from compost: new insights into the role of Actinobacteria in lignocellulose decomposition. *Biotechnology for Biofuels 9* (1): 1–17.
34. Priyadharsini, P. and Dhanasekaran, D. (2015). Diversity of soil allelopathic Actinobacteria in Tiruchirappalli district, Tamilnadu, India. *Journal of the Saudi Society of Agricultural Sciences 14* (1): 54–60.
35. Agadagba, S.K., Bash, E., Chitte, R. et al. (2014). Isolation of actinomycetes from soil. *Journal of Microbiology Research 4* (3): 136–140.
36. Ashokvardhan, T., Rajithasri, A., Prathyusha, P. et al. (2014). Actinomycetes from *Capsicum annuum* L. rhizosphere soil have the biocontrol potential against pathogenic fungi. *International Journal of Current Microbiology and Applied Sciences 3* (4): 894–903.
37. Zhao, K., Li, J., Zhang, X. et al. (2018). Actinobacteria associated with Glycyrrhiza inflata Bat. are diverse and have plant growth promoting and antimicrobial activity. *Scientific Reports 8* (1): 1–13.
38. Qin, S., Miao, Q., Feng, W.-W. et al. (2015). Biodiversity and plant growth promoting traits of culturable endophytic actinobacteria associated with *Jatropha curcas* L. growing in Panxi dry-hot valley soil. *Applied Soil Ecology 93*: 47–55.
39. Sharma, C., Agarwal, A., Mehtani, P. et al. (2014). Characterization of actinomycetes from soil sample of Nahargarh Hill Area, Jaipur, India for production of antimicrobial compounds and enzymes. *International Journal of Advanced Biotechnology and Research 5* (3): 304–312.
40. Lang, Z., Qi, D., Dong, J. et al. (2018). Isolation and characterization of a quinclorac-degrading Actinobacteria *Streptomyces* sp. strain AH-B and its implication on microecology in contaminated soil. *Chemosphere 199*: 210–217.
41. Zenova, G., Dubrova, M., Gracheva, T. et al. (2016). Ecological and taxonomic features of actinomycetes complex of the soils of the Lake Elton region. *Moscow University Soil Science Bulletin 71* (4): 174–177.
42. Das, R., Romi, W., Das, R. et al. (2018). Antimicrobial potentiality of actinobacteria isolated from two microbiologically unexplored forest ecosystems of Northeast India. *BMC Microbiology 18* (1): 1–16.
43. Azarakhsh, Y., Mohammadipanah, F., Nassiri, S. et al. (2017). Isolation and screening of proangiogenic and antiangiogenic metabolites producing rare actinobacteria from soil. *Journal of Applied Microbiology 122* (6): 1595–1602.
44. Elbendary, A.A., Hessain, A.M., El-Hariri, M.D. et al. (2018). Isolation of antimicrobial producing actinobacteria from soil samples. *Saudi Journal of Biological Sciences 25* (1): 44–46.
45. Sengupta, S., Pramanik, A., Ghosh, A. et al. (2015). Antimicrobial activities of actinomycetes isolated from unexplored regions of Sundarbans mangrove ecosystem. *BMC Microbiology 15* (1): 1–16.
46. Shah, A.M., Hussain, A., Mushtaq, S. et al. (2017). Antimicrobial investigation of selected soil actinomycetes isolated from unexplored regions of Kashmir Himalayas, India. *Microbial Pathogenesis 110*: 93–99.
47. Guo, X., Liu, N., Li, X. et al. (2015). Red soils harbor diverse culturable actinomycetes that are promising sources of novel secondary metabolites. *Applied and Environmental Microbiology 81* (9): 3086–3103.
48. Carro, L., Castro, J.F., Razmilic, V. et al. (2019). Uncovering the potential of novel micromonosporae isolated from an extreme hyper-arid Atacama Desert soil. *Scientific Reports 9* (1): 4678.
49. Ji, Y., Chunyu, W.-X., Li, E.-Y. et al. (2018). Phytoactinopolyspora halotolerans sp. nov., a halotolerant actinobacterium isolated from a saline soil in Xinjiang, northwest of China. *Antonie Van Leeuwenhoek 111* (1): 27–34.

50. Tiku, D., Asikong, B., and Bassey, I. (2016). Heavy metal tolerance profile among bacteria from auto-mechanic workshop and pristine soil. *Microbiology Research Journal International 12* (6): 1–10.
51. Fatima, A., Aftab, U., Shaaban, K.A. et al. (2019). Spore forming actinobacterial diversity of Cholistan Desert Pakistan: Polyphasic taxonomy, antimicrobial potential and chemical profiling. *BMC Microbiology 19* (1): 1–17.
52. Lahoum, A., Sabaou, N., Bijani, C. et al. (2019). Antimicrobial activities of novel bipyridine compounds produced by a new strain of *Saccharothrix* isolated from Saharan soil. *Saudi Pharmaceutical Journal 27* (1): 56–65.
53. Vaijayanthi, G., Vijayakumar, R., and Dhanasekaran, D. (2016). Actinobacteria—a biofactory of novel enzymes. In: *Actinobacteria-Basics and Biotechnological Applications* (ed. D. Dhanasekaran and Y. Jiang), 329–352. London: IntechOpen.
54. Gianfreda, L. and Rao, M.A. (2004). Potential of extra cellular enzymes in remediation of polluted soils: a review. *Enzyme and Microbial Technology 35* (4): 339–354.
55. Karigar, C.S. and Rao, S.S. (2011). Role of microbial enzymes in the bioremediation of pollutants: a review. *Enzyme Research 2011*: 805187.
56. Sinha, K., Hegde, R., and Kush, A. (2014). Exploration on native actinomycetes strains and their potential against fungal plant pathogens. *International Journal of Current Microbiology and Applied Sciences 3* (11): 37–45.
57. Le Roes-Hill, M., Rohland, J., and Burton, S. (2011). Actinobacteria isolated from termite guts as a source of novel oxidative enzymes. *Antonie Van Leeuwenhoek 100* (4): 589–605.
58. Sugawara, K., Nishihashi, Y., Narioka, T. et al. (2017). Characterization of a novel DyP-type peroxidase from *Streptomyces avermitilis*. *Journal of Bioscience and Bioengineering 123* (4): 425–430.
59. Saini, A., Aggarwal, N.K., Sharma, A. et al. (2015). Actinomycetes: a source of lignocellulolytic enzymes. *Enzyme Research 2015*: 279381.
60. Thomas, L., Ram, H., Kumar, A. et al. (2016). Production, optimization, and characterization of organic solvent tolerant cellulases from a lignocellulosic waste-degrading actinobacterium, *Promicromonospora* sp. VP111. *Applied Biochemistry and Biotechnology 179* (5): 863–879.
61. Imanparast, S., Hamedi, J., and Faramarzi, M.A. (2018). Enzymatic esterification of acylglycerols rich in omega-3 from flaxseed oil by an immobilized solvent-tolerant lipase from *Actinomadura sediminis* UTMC 2870 isolated from oil-contaminated soil. *Food Chemistry 245*: 934–942.
62. Lamilla, C., Pavez, M., Santos, A. et al. (2017). Bioprospecting for extracellular enzymes from culturable Actinobacteria from the South Shetland Islands, Antarctica. *Polar Biology 40* (3): 719–726.
63. Briceño, G., Fuentes, M.S., Saez, J.M. et al. (2018). *Streptomyces* genus as biotechnological tool for pesticide degradation in polluted systems. *Critical Reviews in Environmental Science and Technology 48* (10–12): 773–805.
64. Hema, T., Getha, K., Tan, G. et al. (2014). Actinobacterial isolates from tin tailings and forest soil for bioremediation of heavy metals. *Journal of Tropical Forest Science 26* (1): 153–162.
65. Bankar, A. and Nagaraja, G. (2018). Recent trends in biosorption of heavy metals by Actinobacteria. In: *New and Future Developments in Microbial Biotechnology and Bioengineering* (ed. V. Gupta), 257–275. New Delhi: Elsevier.
66. El Baz, S., Baz, M., Barakate, M. et al. (2015). Resistance to and accumulation of heavy metals by actinobacteria isolated from abandoned mining areas. *Scientific World Journal 2015*: 761834.
67. Endeshaw, A., Birhanu, G., Zerihun, T. et al. (2017). Application of microorganisms in bioremediation-review. *Journal of Environmental Microbiology 1* (1): 2–9.
68. Lakshmipathy, T.D., Prasad, A.A., and Kannabiran, K. (2010). Production of biosurfactant and heavy metal resistance activity of *Streptomyces* sp. VITDDK3-a novel halo tolerant actinomycetes isolated from saltpan soil. *Biological Research 4* (2): 108–115.
69. Vinod, K., Jaiprakash, C., Thamizhmani, R. et al. (2014). High metal resistance and metal removal properties of antibiotics producing actinobacteria isolated from rhizosphere region of Casuarina equisetifolia. *International Journal of Current Microbiology and Applied Sciences 3* (7): 803–811.

70. El-Sayed, O.H., Refaat, H.M., Swellam, M.A. et al. (2011). Bioremediation of zinc by *Streptomyces aureofacienes*. *Research Journal of Applied Sciences 11* (5): 873–877.
71. El-Gendy, M.M.A.A. and El-Bondkly, A.M.A. (2016). Evaluation and enhancement of heavy metals bioremediation in aqueous solutions by *Nocardiopsis* sp. MORSY1948, and *Nocardia* sp. MORSY2014. *Brazilian Journal of Microbiology 47*: 571–586.
72. Lu, M., Jiang, Y.-L., Wang, S. et al. (2014). *Streptomyces coelicolor* SCO4226 is a nickel binding protein. *PLoS One 9* (10): e109660.
73. Yatome, C., Yamada, S., Ogawa, T. et al. (1993). Degradation of crystal violet by *Nocardia corallina*. *Applied Microbiology and Biotechnology 38* (4): 565–569.
74. Parshetti, G., Kalme, S., Saratale, G. et al. (2006). Biodegradation of Malachite Green by *Kocuria rosea* MTCC 1532. *Acta Chimica Slovenica 53* (4): 492–498.
75. Li, G., Peng, L., Ding, Z. et al. (2014). Decolorization and biodegradation of triphenylmethane dyes by a novel *Rhodococcus qingshengii* JB301 isolated from sawdust. *Annales de Microbiologie 64* (4): 1575–1586.
76. Buntić, A., Pavlović, M., Šiler-Marinković, S. et al. (2016). Biological treatment of colored wastewater by *Streptomyces fulvissimus* CKS 7. *Water Science and Technology 73* (9): 2231–2236.
77. El-Sersy, N.A., Abou-Elela, G.M., Hassan, S.W. et al. (2011). Bioremediation of acid fast red dye by *Streptomyces globosus* under static and shake conditions. *African Journal of Biotechnology 10* (17): 3467–3474.
78. Bourguignon, N., Isaac, P., Alvarez, H. et al. (2014). Enhanced polyaromatic hydrocarbon degradation by adapted cultures of actinomycete strains. *Journal of Basic Microbiology 54* (12): 1288–1294.
79. Gallo, G., Piccolo, L.L., Renzone, G. et al. (2012). Differential proteomic analysis of an engineered *Streptomyces coelicolor* strain reveals metabolic pathways supporting growth on n-hexadecane. *Applied Microbiology and Biotechnology 94* (5): 1289–1301.
80. Burghal, A.A., Al-Mudaffarand, N.A., and Mahdi, K.H. (2015). Ex situ bioremediation of soil contaminated with crude oil by use of actinomycetes consortia for process bioaugmentation. *European Journal of Experimental Biology 5* (5): 24–30.
81. Margesin, R., Moertelmaier, C., and Mair, J. (2013). Low-temperature biodegradation of petroleum hydrocarbons (n-alkanes, phenol, anthracene, pyrene) by four actinobacterial strains. *International Biodeterioration and Biodegradation 84*: 185–191.
82. Lin, Q., Chen, S., Hu, M. et al. (2011). Biodegradation of cypermethrin by a newly isolated actinomycetes HU-S-01 from wastewater sludge. *International journal of Environmental Science and Technology 8* (1): 45–56.
83. Philip, C., Shrivastava, A., and Shrivastava, R. (2014). Effect of organochlorine pesticide on growth of Actinomycetes. *Indian Journal of Science Research 4* (1): 191–195.
84. Ningthoujam, D.S., Sanasam, S., and Mutum, A. (2012). Characterization of p-nitrophenol degrading actinomycetes from Hundung limestone deposits in Manipur, India. *African Journal of Biotechnology 11* (44): 10210–10220.
85. Fuentes, M., Benimeli, C., Cuozzo, S. et al. (2010). Isolation of pesticide-degrading actinomycetes from a contaminated site: bacterial growth, removal and dechlorination of organochlorine pesticides. *International Biodeterioration & Biodegradation 64* (6): 434–441.
86. Diaz-Ochoa, V., Jellbauer, S., Klaus, S. et al. (2014). Transition metal ions at the crossroads of mucosal immunity and microbial pathogenesis. *Frontiers in Cellular and Infection Microbiology 4*: 2.
87. Javanbakht, V., Alavi, S.A., and Zilouei, H. (2014). Mechanisms of heavy metal removal using microorganisms as biosorbent. *Water Science and Technology 69* (9): 1775–1787.
88. Igiri, B.E., Okoduwa, S.I., Idoko, G.O. et al. (2018). Toxicity and bioremediation of heavy metals contaminated ecosystem from tannery wastewater: a review. *Journal of Toxicology 2018*: 2568038.
89. Timková, I., Sedláková-Kaduková, J., and Pristaš, P. (2018). Biosorption and bioaccumulation abilities of actinomycetes/streptomycetes isolated from metal contaminated sites. *Separations 5* (4): 54.
90. Mishra, A. and Malik, A. (2013). Recent advances in microbial metal bioaccumulation. *Critical Reviews in Environmental Science and Technology 43* (11): 1162–1222.

91. Zabochnicka-Świątek, M. and Krzywonos, M. (2014). Potentials of biosorption and bioaccumulation processes for heavy metal removal. *Polish Journal of Environmental Studies 23* (2): 551–561.
92. Chen, S.H., Cheow, Y.L., Ng, S.L. et al. (2019). Mechanisms for metal removal established via electron microscopy and spectroscopy: a case study on metal tolerant fungi *Penicillium simplicissimum*. *Journal of Hazardous Materials 362*: 394–402.
93. Jin, Y., Luan, Y., Ning, Y. et al. (2018). Effects and mechanisms of microbial remediation of heavy metals in soil: a critical review. *Applied Sciences 8* (8): 1336.
94. Joshi, N.C. (2017). Heavy metals, conventional methods for heavy metal removal, biosorption and the development of low cost adsorbent. *European Journal of Pharmaceutical and Medical Research 4*: 388–393.
95. Algureiri, A.H. and Abdulmajeed, Y.R. (2016). Removal of heavy metals from industrial wastewater by using RO membrane. *Iraqi Journal of Chemical and Petroleum Engineering 17* (4): 125–136.
96. Kanamarlapudi, S., Chintalpudi, V.K., and Muddada, S. (2018). Application of biosorption for removal of heavy metals from wastewater. *Biosorption 18*: 69–116.
97. Aburas, M. (2016). Bioremediation of toxic heavy metals by waste water actinomycetes. *International Journal of Current Research 8* (1): 24870–24875.
98. Hamedi, J., Dehhaghi, M., and Mohammdipanah, F. (2015). Isolation of extremely heavy metal resistant strains of rare Actinomycetes from high metal content soils in Iran. *International Journal of Environmental Research 9* (2): 475–480.
99. Ting, A.S.Y. (2015). Microbial cells dead or alive: prospect, potential and innovations for heavy metal removal. In: *Advances in Biodegradation and Bioremediation of Industrial Waste* (ed. R. Chandra), 31–66. Boca Raton: CRC Press.
100. Rosca, M., Hlihor, R.-M., Cozma, P. et al. (ed.) (2015). Potential of biosorption and bioaccumulation processes for heavy metals removal in bioreactors. *E-Health and Bioengineering Conference (EHB) 2015*: 1–4.
101. Diep, P., Mahadevan, R., and Yakunin, A.F. (2018). Heavy metal removal by bioaccumulation using genetically engineered microorganisms. *Frontiers in Bioengineering and Biotechnology 6*: 157.
102. Ahmad, A., Mohd-Setapar, S.H., Chuong, C.S. et al. (2015). Recent advances in new generation dye removal technologies: novel search for approaches to reprocess wastewater. *RSC Advances 5* (39): 30801–30818.
103. Jasińska, A., Paraszkiewicz, K., Słaba, M. et al. (2015). Microbial decolorization of triphenylmethane dyes. In: *Microbial Degradation of Synthetic Dyes in Wastewaters* (ed. S.N. Singh), 169–186. Cham: Springer.
104. Vikrant, K., Giri, B.S., Raza, N. et al. (2018). Recent advancements in bioremediation of dye: current status and challenges. *Bioresource Technology 253*: 355–367.
105. Vijayaraghavan, J., Basha, S.S., and Jegan, J. (2013). A review on efficacious methods to decolorize reactive azo dye. *Journal of Urban and Environmental Engineering 7* (1): 30–47.
106. Singh, P.K. and Singh, R.L. (2017). Bio-removal of azo dyes: a review. *International Journal for Research in Applied Sciences and Biotechnology 5* (2): 108–126.
107. Meerbergen, K., Willems, K.A., Dewil, R. et al. (2018). Isolation and screening of bacterial isolates from wastewater treatment plants to decolorize azo dyes. *Journal of Bioscience and Bioengineering 125* (4): 448–456.
108. Jafari, N., Kasra-Kermanshahi, R., and Soudi, M.R. (2013). Screening, identification and optimization of a yeast strain, *Candida palmioleophila* JKS4, capable of azo dye decolorization. *The Iranian Journal of Microbiology 5* (4): 434.
109. Shanmuga, P.M. (2016). Decolorization of azodyes by marine actinomycetes. *Advanced Research Journal of Life Sciences 2* (2): 1–5.
110. Dawood, S. and Sen, T. (2014). Review on dye removal from its aqueous solution into alternative cost effective and non-conventional adsorbents. *Journal of Chemical and Process Engineering 1* (104): 1–11.

111. Piaskowski, K., Świderska-Dąbrowska, R., and Zarzycki, P.K. (2018). Dye removal from water and wastewater using various physical, chemical, and biological processes. *Journal of AOAC International 101* (5): 1371–1384.
112. Zuraida, M.S., Nurhaslina, C., and Ku, H.K. (2013). Removal of synthetic dyes from wastewater by using bacteria, *Lactobacillus delbruckii*. *International Journal of Engineering Science 2* (5): 1–7.
113. Chen, S.H. and Ting, A.S.Y. (2015). Biosorption and biodegradation potential of triphenylmethane dyes by newly discovered *Penicillium simplicissimum* isolated from indoor wastewater sample. *International Biodeterioration & Biodegradation 103*: 1–7.
114. Jadhav, A., Bura, D., Borage, A. et al. (2014). Microbial decolourisation of Rathiline Navy Blue dye using immobilized fungal and actinomycetal biomass. *Nature, Environment and Pollution Technology 13* (3): 535–540.
115. Chakravarthy, B., Vijayasree, J., Swathi, V. et al. (2015). Screening and exploration of azo dye decolorizing actinomycetes from marine sediments. *International Journal of Scientific and Engineering Research 6* (2): 27–30.
116. Karim, M.E., Dhar, K., and Hossain, M.T. (2018). Decolorization of textile reactive dyes by bacterial monoculture and consortium screened from textile dyeing effluent. *Journal, Genetic Engineering & Biotechnology 16* (2): 375–380.
117. Kaushik, P. and Malik, A. (2015). Mycoremediation of synthetic dyes: an insight into the mechanism, process optimization and reactor design. In: *Microbial Degradation of Synthetic Dyes in Wastewaters* (ed. S.N. Singh), 1–25. Cham: Springer.
118. Ball, A. and Colton, J. (1996). Decolorisation of the polymeric dye Poly R by *Streptomyces viridosporus* T7A. *Journal of Basic Microbiology 36* (1): 13–18.
119. Chengalroyen, M. and Dabbs, E. (2013). Identification of genetic elements from A. orientalis contributing to triphenylmethane decolorization. *International Journal of Current Microbiology and Applied Sciences 2*: 41–56.
120. Adenan, N.H., Lim, Y.Y., and Ting, A.S.Y. (2020). Discovering decolorization potential of Triphenylmethane dyes by actinobacteria from soil. *Water, Air, and Soil Pollution 231* (12): 1–13.
121. Adenan, N.H., Lim, Y.Y., and Ting, A.S.Y. (2021). Identification and optimization of triphenylmethane dyes removal by *Streptomyces* sp. from forest soil. *Sustainable Environment Research 31* (1): 1–15.
122. Adenan, N.H., Lim, Y.Y., and Ting, A.S.Y. (2021). *Nocardiopsis* sp. for the removal of triphenylmethane dyes: decolorization and optimization studies. *Water, Air, and Soil Pollution 232* (10): 1–17.
123. Mahmood, R., Sharif, F., Ali, S. et al. (2012). Isolation of indigenous bacteria and consortia development for decolorization of textile dyes. *Biologia Pakistan 58*: 53–60.
124. Huang, J., Yang, Y., Wang, Y. et al. (2018). Immobilization of a laccase/2, 2′-azino-bis-(3-ethylbenzothiazoline)-6-sulfonic acid system to layered double hydroxide/alginate biohybrid beads for biodegradation of malachite green dye. *BioMed Research International 2018*: 5471961.
125. Blanco-Enríquez, E.G., Zavala-Díaz de la Serna, F.J., Peralta-Pérez, M.D.R. et al. (2018). Characterization of a microbial consortium for the bioremoval of polycyclic aromatic hydrocarbons (PAHs) in water. *International Journal of Environmental Research and Public Health 15* (5): 975.
126. Habib, S., Johari, W.L.W., Shukor, M.Y. et al. (2017). Screening of hydrocarbon-degrading bacterial isolates using the redox application of 2, 6-DCPIP. *Bioremediation Science and Technology Research 5* (2): 13–16.
127. Yuniati, M. (ed.) (2018). Bioremediation of petroleum-contaminated soil: a review. *IOP Conference Series: Earth and Environmental Science 118* (1): 1–7.
128. Ding, J., Chen, B., and Zhu, L. (2013). Biosorption and biodegradation of polycyclic aromatic hydrocarbons by *Phanerochaete chrysosporium* in aqueous solution. *Chinese Science Bulletin 58* (6): 613–621.
129. Hamoudi-Belarbi, L., Hamoudi, S., Belkacemi, K. et al. (2018). Bioremediation of polluted soil sites with crude oil hydrocarbons using carrot peel waste. *Environments 5* (11): 124.

130. Mahjoubi, M., Cappello, S., Souissi, Y. et al. (2018). Microbial bioremediation of petroleum hydrocarbon–contaminated marine environments. In: *Recent Insights in Petroleum Science and Engineering* (ed. M. Zoveidavianpoor), 325–350. United Kingdom: Intechopen.
131. Xu, X., Liu, W., Tian, S. et al. (2018). Petroleum hydrocarbon-degrading bacteria for the remediation of oil pollution under aerobic conditions: a perspective analysis. *Frontiers in Microbiology 9*: 2885.
132. Olajuyigbe, F.M. and Ehiosun, K.I. (2016). Assessment of crude oil degradation efficiency of newly isolated actinobacteria reveals untapped bioremediation potentials. *Bioremediation Joural 20* (2): 133–143.
133. Das, N. and Chandran, P. (2011). Microbial degradation of petroleum hydrocarbon contaminants: an overview. *Biotechnology Research International 2011*: 941810.
134. Sierra-Garcia, I.N. and de Oliveira, V.M. (2013). Microbial hydrocarbon degradation: efforts to understand biodegradation in petroleum reservoirs. In: *Biodegradation-Engineering and Technology* (ed. R. Chamy and F. Rosenkranz), 47–72. United Kingdom: Intechopen.
135. Brown, D.M., Bonte, M., Gill, R. et al. (2017). Heavy hydrocarbon fate and transport in the environment. *Quarterly Journal of Engineering Geology & Hydrogeology 50* (3): 333–346.
136. Adams, G.O., Fufeyin, P.T., Okoro, S.E. et al. (2015). Bioremediation, biostimulation and bioaugmention: a review. *International Journal of Environmental Bioremediation & Biodegradation 3* (1): 28–39.
137. Ortiz-Hernández, M.L., Sánchez-Salinas, E., Dantán-González, E. et al. (2013). Pesticide biodegradation: mechanisms, genetics and strategies to enhance the process. In: *Biodegradation-Life of Science* (ed. R. Chamy and F. Rosenkranz), 251–287. United Kingdom: Intechopen.
138. Vijayabharathi, R., Kumari, B.R., Sathya, A. et al. (2014). Biological activity of entomopathogenic actinomycetes against lepidopteran insects (Noctuidae: Lepidoptera). *Canadian Journal of Plant Science 94* (4): 759–769.
139. Javaid, M.K., Ashiq, M., and Tahir, M. (2016). Potential of biological agents in decontamination of agricultural soil. *Scientifica 2016*: 1598325.
140. Parte, S.G., Mohekar, A.D., and Kharat, A.S. (2017). Microbial degradation of pesticide: a review. *African Journal of Microbiology Research 11* (24): 992–1012.
141. Kumar, T.T. and Jahangir, H.S. (2018). Isolation and characterization of efficient Lambdacyhalothrin degrading actinobacteria from Paddy field soil. *Pharmaceutical, Biological and Chemical Sciences 6* (3): 199–208.
142. Huang, Y., Xiao, L., Li, F. et al. (2018). Microbial degradation of pesticide residues and an emphasis on the degradation of cypermethrin and 3-phenoxy benzoic acid: a review. *Molecules 23* (9): 2313.
143. Verma, J.P., Jaiswal, D.K., and Sagar, R. (2014). Pesticide relevance and their microbial degradation: a-state-of-art. *Reviews in Environmental Science and Biotechnology 13* (4): 429–466.
144. Janaki, T. (2016). Larvicidal activity of *Streptomyces cacaoi* subsp. cacaoi-M20 against Culex quinquefasciatus (III Instar). *International Journal of Mosquito Research 3*: 47–51.
145. Schrijver, A.D. and Mot, R.D. (1999). Degradation of pesticides by actinomycetes. *Critical Reviews in Microbiology 25* (2): 85–119.

Part III
Current Trends and Future Possibilities

17

Current Opinion and Trends for Use of Biochar in Agriculture Sustainability

Anuradha Singh and Preeti Chaturvedi Bharagava

17.1. INTRODUCTION

The concept "sustainable agriculture" refers to the full integration of biological, physical, chemical, ecological, economic, and social disciplines to produce new agricultural systems that are safe and environmentally friendly. The world's population is growing at a rapid pace, and it is predicted to reach 9.8 billion by 2050; as a result, in order to feed the rising population food demand also increases globally and regionally, especially in developing countries. The increasing demand for food and livelihood puts a lot of pressure on the agricultural sector. Consequently, chemical fertilizer usage is projected to rise in order to meet agricultural demands. Extensive use of chemical fertilizers disturbs the natural nitrogen-related processes, thus soil acidification and N_2O emission occur [1]. These chemical fertilizers also affect the soil microbial activity and reduce the microbial biomass, as well as microbial community composition [2, 3]. Greater than 80% of soil microbial composition is sensitive to NPK fertilizers. Soil degradation such as soil erosion, acidity, low fertility, inorganic or organic pollutants, and salinization challenges mostly create a problem for food [4]. Chemical fertilizers are not environmentally sustainable and cause a number of problems; innovative and environmentally friendly methods are required to maintain soil nutrient balance and increase agricultural productivity.

Other studies focused on efficient use of different natural and synthetic fertilization methods for the agricultural sustainability. Use of biochar in agricultural soil has received increased interest over the past few decades for sustainable agriculture. Biochar improves soil quality and productivity along with various advantages such as waste management and climate change mitigation [5–8]. Biochar has a number of beneficial properties, including high stability, aromatization, high in carbon content, and it is widely used in sustainable agriculture, environmental remediation, energy generation, and other applications [9–11]. Biochar is a pyrogenic carbon-rich product that is obtained from various feedstocks. It is

Aquatic Toxicology Laboratory, Environmental Toxicology Group, CSIR-Indian Institute of Toxicology Research, Lucknow, Uttar Pradesh, India

Microbial Biotechnology: Role in Ecological Sustainability and Research, First Edition.
Edited by Pankaj Chowdhary, Sujata Mani, and Preeti Chaturvedi.

one of the best biosorption agents that has received much attention recently due to its high adsorption capacity toward various environmental contaminants [12, 13]. Pyrolysis methods (slow pyrolysis, fast pyrolysis, gasification, and torrefaction) are processes for making biochar (product), gases, and bio-oils (co-products) [14]. It is also known as dry distillation, which alludes to the thermal decomposition of a carbon amendment biomass in the environment with remarkably less oxygen than necessary for combustion [15, 16]. Biochar is a product obtained by thermal decay of organic matter under the supply of limiting oxygen and at relatively low temperatures [17]. High and low temperatures have unequivocal effect on biochar yield [90]. Figure 17.1 summarizes the different methods of pyrolysis, the reaction conditions, and the quantity of biochar obtained.

It is difficult to degrade and remains in the soil for a long time (minimum 10 years) [18]. The addition of biochar in the soil may cause various alterations in soil properties like physical, chemical, and biological (microbial) load [8, 19]. Biochar has physicochemical qualities for improvement of various field applications such as the ecological environmental quality [20]. Biochar not only enhances soil structure, nutrient content, organic matter content, moisture retention capacity, and crop yield, but it also provides a favorable habitat for microbial growth [21, 22]. An important use of biochar is the possibility to hold and store carbon in the soil. Biochar helps to remediate contaminated soil, enhance carbon sequestration and productivity of soil [23, 24], and reduce soil greenhouse gas (N_2O, CO_2, and CH_4) [25]. Biochar had a propensity to remove harmful environmental impacts like climate change and global warming in a sustainable manner [8, 26, 27].

In terms of biochar application on agricultural sustainability, this chapter focuses on available information on biochar and its application for soil amendment for nutrient source,

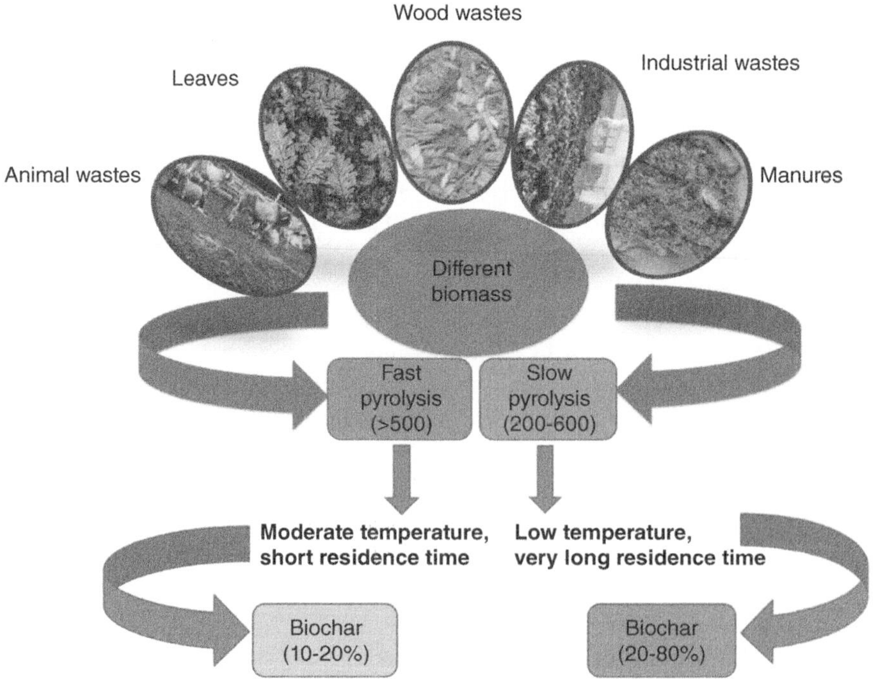

Figure 17.1 Overview of biochar preparation.

improving soil health, crop yield, soil remediation, and as a method for increasing agricultural productivity. It also discusses the various agricultural wastes of biochar production, properties, and its future prospects.

17.2. USE OF FEEDSTOCK

17.2.1. Agri Residues

Agricultural residues and agriwastes are used in the process of making biochar. These play important roles in the fertility of soil and plant growth. Various agricultural residues such as corn cob, corn stalk [28, 29], wheat residues [30], rice straw [31], and sugar beet, acacia green, Miscanthus [32], cassava rhizomes, corncobs, and cassava stems are used for making biochar. The amount and properties of different bio-molecules like cellulose, lignin and hemicelluloses, and nutrient content are varied according to biomass [33]. The high content of Si and ash present in straw biochar is more than in other biochar [34]. Wheat straw was used by Gai et al. [35] for making biochar at different temperatures such as 300, 400, 500, 600, and 700 °C. The C content of these biochars was 57.8, 70.3, 73.4, and 73.9%, respectively, and the N content was 1.5, 1.4, 1.4, and 1.2%, respectively. In another study, Mubarik et al. [36] reported sugarcane bagase as a feedstock for making biochar at temperatures of 300, 350, 400, and 500 °C. Corn stover was used for the making of biochar at 300, 400, and 500 °C. The C content of this biochar was 69.9, 74.8, and 78.1%, respectively, and N content was 0.94, 1.05, and 0.92%, respectively [37, 109]. In a study, rice husks were reported at temperatures of 400, 500, 600, 700, and 800 °C, respectively [101]. Peanut shells were also reported in a study for making biochar at temperatures of 300, 400, 500, 600, and 700 °C, respectively [35, 103]. In another study, rice straw was reported at temperatures of 300, 350, 400, 500, 600, 700, 800, and 900 °C, respectively, for making biochar [104–106]. Various feedstock and pyrolysis temperatures are summarized in Table 17.1.

17.2.2. Manures

Activated charcoal (biochar) is made by the process of pyrolysis of either animal manure or plant biomass it is using in soil applications [38]. Biochar has a high content of ash because animal manures have contained high concentrations of organic and inorganic compounds [39, 40]. Livestock manure (straw and water mix) used in biochar feedstock contains nutrient elements like nitrogen and phosphorus [41]. Animal manure or plant biomass is used for the preparation of biochar by pyrolysis and the product is primarily used for heating [17]. Biochar produced from pyrolysis of seaweeds, manure, and agricultural wastes contains higher nutrient content and pH [42]. Wu et al. [43] reported in a study about eight biochars made from waste biomass such as woody, tailing, herbaceous, manure, and sludge at 600 °C in an N_2-rich environment. The pyrolysis temperatures of various manures such as poultry manure (250, 300, 350, 400, and 450 °C), horse manure (400, 500, 600, 700C, and 800 °C), and dairy manure (350 and 700 °C) are reported in various studies [113–115].

17.2.3. Sludge

Wastewater treatment plants (WWTP) produce sludge, which is a solid waste. The high concentrations of heavy metals along with pathogens in sewage sludge are difficult to manage. The pyrolysis changeover of sewage sludge into biochar is the best method to control

Table 17.1 Feedstock type and pyrolysis temperature induce changes in biochar composition and properties.

Sources	Pyrolysis Temp (°C)	pH	Composition of chemicals										References
			C	H	N	O	S	Ash	Ca	Mg	K	P	
Rice husk	400	6.84	44.6	2.50	0.69	16.30	–	–	–	–	–	–	[101]
	500	8.99	45.1	1.27	0.47	7.12	–	–	–	–	–	–	
	600	9.41	40.3	0.85	0.37	9.23	–	–	–	–	–	–	
	700	9.52	38.8	0.46	0.26	12.70	–	–	–	–	–	–	
	800	9.62	40.4	0.28	0.22	2.69	–	–	–	–	–	–	
Wheat straw	300	7.70	51.7	–	1.38	–	–	25.0	0.63	0.45	3.0	0.26	[35]
	400	8.20	57.8	3.20	1.50	21.60	–	11.0	–	–	–	–	
	500	8.30	70.3	2.90	1.50	21.60	–	11.0	–	–	–	–	
	600	9.20	73.4	2.10	1.40	14.90	–	12.0	–	–	–	–	
	700	9.20	73.9	1.30	1.20	14.60	–	15.0	–	–	–	–	
Sugarcane bagasse	300	4.97	–	–	–	–	–	4.2	–	–	–	–	[36]
	350	4.96	57.0	–	0.34	–	0.03	–	–	–	0.48	–	
	400	5.93	–	–	–	–	–	4.2	–	–	–	–	
	500	6.12	–	–	–	–	–	4.1	–	–	–	–	
Corn straw	300	–	80.9	–	1.18	17.20	–	–	–	–	–	–	[102]
	400	10.20	56.1	4.30	2.40	22.00	–	14.0	–	–	–	–	[35]
	500	10.40	58.0	2.70	2.30	21.50	–	17.0	–	–	–	–	
	600	10.40	58.6	2.00	2.00	18.70	–	18.0	–	–	–	–	
	700	10.40	59.5	1.50	1.60	16.60	–	18.0	–	–	–	–	
Peanut shell	300	–	69.4	3.51	1.22	18.70	0.18	–	–	–	–	–	[35, 103]
	400	9.30	58.4	3.50	1.80	21.00	–	9.0	–	–	–	–	
	500	9.40	64.5	2.80	1.70	18.50	–	10.0	–	–	–	–	
	600	9.60	71.9	2.00	1.60	15.00	–	11.0	–	–	–	–	
	700	9.90	74.4	1.40	1.40	14.20	–	12.0	–	–	–	–	
Rice straw	300	8.00	45.2	–	1.15	–	–	31.0	0.91	0.81	3.6	0.11	[104]
	350	7.80	44.5	2.69	1.64	22.10	–	29.1	–	–	–	–	[105]
	400	–	51.6	2.49	0.79	26.40	–	–	–	–	–	–	[106]
	500	–	53.1	1.74	0.68	23.60	–	–	–	–	–	–	
	600	–	53.1	1.39	0.62	24.30	–	–	–	–	–	–	
	700	–	52.5	1.26	0.58	23.20	–	–	–	–	–	–	
	800	–	48.3	1.21	0.79	25.80	–	–	–	–	–	–	
	900	–	46.9	1.18	0.78	25.00	–	–	–	–	–	–	

Maize straw	350	7.90	58.9	3.36	1.00	21.30	–	15.4	–	–	–	–	[105]
	700	10.60	43.8	1.47	1.27	31.60	–	21.9	–	–	–	–	
Maize straw (air limited)	300	–	68.2	4.97	1.68	24.90	0.21	15.4	–	–	–	–	[106]
Wheat	450	8.40	52.9	–	2.34	–	–	–	–	–	–	–	[107]
	550	9.00	38.5	–	1.42	–	–	–	–	–	–	–	
Cotton straw	300	–	76.8	–	0.93	21.20	–	–	–	–	–	–	[102]
	600	–	84.6	–	0.58	12.70	–	–	–	–	–	–	
Bagasse	300	7.30	69.5	4.20	0.90	24.40		–	0.46	0.14	0.27	0.05	[108]
	450	7.50	78.6	3.52	0.90	15.50	–	–	0.83	0.18	0.25	0.07	
	600	7.50	76.4	2.93	0.79	18.30	–	–	0.91	0.21	0.15	0.08	
Canola straw	300	6.48	61.6	–	0.19	–	–	10.7	–	–	–	0.16	[34]
	500	9.39	63.4	–	0.04	–	–	18.4	–	–	–	–	
	700	10.76	54.9	–	0.04	–	–	28.6	–	–	–	–	
Corn stover	300	7.31	69.9	4.04	0.94	13.40	–	11.7	–	–	–	–	[109]
	400	7.79	74.8	3.15	1.05	6.18	–	14.8	–	–	–	–	
	500	8.10	78.1	2.11	0.92	3.77	–	15.1	–	–	–	–	
Maize straw	350	7.90	58.9	3.36	1.00	21.30	–	15.4	–	–	–	–	[105]
	700	10.60	43.8	1.47	1.27	31.60	–	21.9	–	–	–	–	
Apple tree	400	7.02	70.2	4.12	0.76	20.60	–	–	–	–	–	–	[101]
	500	9.64	79.1	2.65	0.34	12.00	–	–	–	–	–	–	
	600	10.04	81.5	1.96	0.46	13.60	–	–	–	–	–	–	
	700	10.03	82.3	1.21	0.41	16.30	–	–	–	–	–	–	
Conocarpus	200	7.37	64.2	3.96	0.69	26.60	2.28	4.5	43.40	3.43	0.38	0.84	[110]
	400	9.64	76.8	2.83	0.87	14.20	1.72	5.3	51.80	3.98	0.54	0.88	
	600	10.04	83.0	1.28	0.71	6.55	0.91	8.6	64.70	4.79	0.9	1.11	
	800	10.03	85.0	0.62	0.90	4.87	0.58	8.6	67.50	7.81	1.15	1.34	
Oak tree	400	6.43	70.6	3.70	0.69	21.50	–	–	–	–	–	–	[101]
	500	8.10	77.6	2.51	0.51	17.70	–	–	–	–	–	–	
	600	8.85	81.2	1.92	0.48	16.00	–	–	–	–	–	–	
	700	9.54	83.2	1.16	0.31	15.00	–	–	–	–	–	–	
	800	9.68	82.8	0.69	0.32	17.30	–	–	–	–	–	–	

(Continued)

Table 17.1 (Continued)

Sources	Pyrolysis Temp (°C)	pH	Composition of chemicals										References
			C	H	N	O	S	Ash	Ca	Mg	K	P	
Pinewood	250	–	58.3	5.20	0.20	36.30	–	0.9	–	–	–	–	[111]
	350	–	70.2	2.70	0.30	26.80	–	1.2	–	–	–	–	
	500	–	86.1	2.90	0.40	10.60	–	2.2	–	–	–	–	
	700	–	91.5	1.10	0.60	6.70	–	1.9	–	–	–	–	
Oak wood	400	–	70.9	3.60	0.40	12.90	–	12.1	2.70	0.20	0.9	0.10	[112]
	600	–	79.2	2.00	0.30	3.50	–	13.4	0.31	0	0.2	0.10	
Bamboo	300	7.90	66.2	4.70	0.40	27.70	–	–	0.22	0.14	0.3	0.24	[108]
	450	8.50	76.9	3.55	0.23	16.10	–	–	0.29	0.19	0.35	0.36	
	600	9.20	80.9	2.43	0.15	14.90	–	–	0.34	0.23	0.52	0.54	
Poultry manure	250	7.80	–	–	–	–	–	–	–	1.14	4.97	1.86	[113]
	300	10.10	–	–	–	–	–	–	–	1.25	5.35	2.06	
	350	11.10	–	–	–	–	–	–	–	1.50	6.45	2.35	
	400	11.50	–	–	–	–	–	–	–	1.76	7.01	2.55	
	450	8.10	43.8	2.36	1.80	12.80	0.01	37.9	–	–	–	–	[114]
Horse manure	400	–	53.1	2.72	0.65	16.10	–	–	–	–	–	–	[114]
	500	–	66.7	1.54	0.53	11.50	–	–	–	–	–	–	
	600	–	69.2	1.20	0.51	8.26	–	–	–	–	–	–	
	700	–	68.9	1.36	0.52	12.00	–	–	–	–	–	–	
	800	–	70.1	1.00	0.48	10.70	–	–	–	–	–	–	
Dairy manure	350	9.20	–	–	2.60	–	–	24.2	–	–	1.43	1.00	[115]
	700	9.90	–	–	1.51	–	–	39.5	–	–	2.31	1.69	
Date palm	300	8.32	58.0	4.08	0.54	20.80	2.14	14.4	4.85	1.53	2.18	–	[116]
	400	9.25	66.9	3.54	0.45	11.40	1.36	16.3	6.04	1.57	2.17	–	
	500	9.59	72.3	2.11	0.42	4.50	1.02	19.7	5.81	1.93	2.23	–	
	600	9.57	73.0	1.74	0.39	3.28	0.98	20.7	7.77	1.90	2.58	–	
	700	11.50	73.4	1.14	0.35	3.19	0.85	21.1	7.65	1.92	2.69	–	
	800	11.49	75.0	0.89	0.31	2.27	0.54	21.4	8.08	2.02	2.71	–	

this waste. Sludge from wastewater can positively affect soil–plant properties because it contains nutrient sources like phosphorous, nitrogen, micronutrients, and organic matter [44]. A supplementary benefit of using biochar from sewage sludge may be the immobilization of heavy metals [45]. In addition to sewage sludge, in a parallel manner as AC (activated charcoal) is used, it should reduce the bioavailability, toxicity, and mobility of organic pollutants [46]. Approximately in the range of 0.1–0.3% w/w metals like Cr, Pb, Cu, and Ni present in sludge. At the time of wastewater treatment, around 50–80% of the heavy metal content is present in wastewater. Yuan et al. [47] found that biochar (sewage sludge) made at 700 °C (high temperature) contain less total N and water-soluble N but more P and K compared to 300 °C (low temperature) biochar.

17.2.4. Algal Biochar

Algae are important feedstock for the generation of bioenergy and valuable chemicals. After the extraction of essential products, algal waste can be further converted into biogas, biofuel, and biochar. Algal feedstock, being of a thermochemical nature, takes a few days to be converted to biochar [48]. At high temperature, algal biochar that is carbonized shows good performance for the use as supercapacitors, CO_2 adsorbents, persulfate activation, and high electron transport due to its graphitic carbon structure and specific surface area [37]. Seaweeds (microalgae) are ecologically important, as well as give important ecosystem services, soil additives, animal feeds, and nutraceuticals [49]. Algae are also used in bioremediation especially in aquaculture systems [50]. Soil microbial communities are responsible for biochar amendment because it increases microbial abundance and activities [38, 51, 52]. Algal biomass is frequently participating in increment of aggregate production and decrement of environmental impacts [53].

17.3. USE OF BIOCHAR IN AGRICULTURE SUSTAINABILITY

17.3.1. Ameliorating Soil Structure and Improve Soil Fertility

Biochars are used in soil to improve soil quality for more crop yield. Biochar enhances the soil quality by increasing pH, cation exchange capacity (CEC), and water holding capacity along with soil ventilation. Biochar also affects the pore structure of soil, helps growth, and provides a good environment for microorganisms. In some studies, it has also been proven that biochar plays an important role in the formation and stabilization of soil aggregates. In biochar, the aromatic structure is present that makes it highly stable against decay [54]. The large quantity of feedstock with nutrients increases crop production and soil quality [55]. High pyrolysis temperature (HPT) helps in the yielding of biochar with large surface area and high pH value [56]. Biochar is widely used in agricultural land, environmental, and bio-ordinal activities. It plays an important role in increasing soil carbon storage, decreasing atmospheric greenhouse gas emission [57, 58]. The rectification of the feedstock with soil notably increases soil quality along with crop production [55]. Biochars have the ability to retain macronutrients directly, such as nitrogen [59–63]. In the precursor biomass, by providing soil nutrients biochar behaves as organic fertilizer [59]. Figure 17.2 highlights the role of biochar in plant nutrient cycling, including improvement in soil fertility and quality, enhancement of water retention capacity, immobilization of organic pollutants, and reduction of leaching [61]. Laghari et al. [64] investigated the use of biochar in low fertility

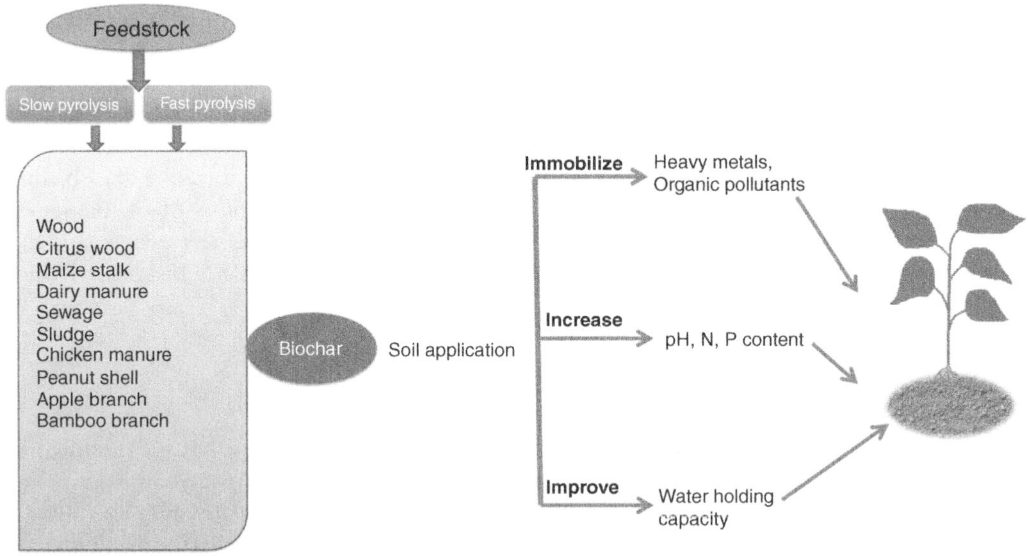

Figure 17.2 Diagrammatic representation of biochar in the plant nutrition.

soils (sandy), improving TC (7–11%), K (37–42%), P (68–70%), Ca (69–75%). It was reported in a study that the pyrolysis of biochar at low temperature strongly supports the availability of N, P, and K in a low fertility acidic soil [60].

17.3.2. Soil Alkalinity

Ash and alkali matter content (Na, Mg, K and Ca) are connected to the pH of biochar [65]. Wollastonite ($CaSiO_3$) reacts with carbon dioxide (CO_2), and after converting in low temperature it is formed carbonate [66]. At high temperature, pyrolysis of biochar was found to be capable of neutralizing soil acidity, increasing soil nutrients, and improving soil CEC [67, 68]. However, the addition of alkalinity in biochar can increase the pH value of compost [69]. In a study, Abujabhah et al. [70] used Acacia green waste biochar at 550 °C, which enhanced the organic content in the soil and promoted crop production. Verheijien et al. [71] reported that the functional group and chemical composition of biochar enhanced the pH, CEC, and nutrient content of the soil. Senbayram et al. [29] reported the various concentration of corn straw biochar such as 2, 4, and 8% w/w can enhance the soil property like WHC, pH, and electrical conductivity (EC). Biochar derived from cacao shell and rice husk at 600 and 500 °C, respectively, enhance pH and release dissolved organic matter (OM) from the soil [72].

17.3.3. Moisture Retention

Biochar makes soil highly porous and affects soil bulk density, soil moisture content (SMC), WHC, hydraulic conductivity, and aggregated stability; it also affects nutrient dynamics [67, 73–75]. Porosity is developed in physical activation through using CO_2 or treatment of steam at high temperature. Phosphoric acid, zinc chloride, sodium hydroxide, or potassium hydroxides are used for chemical activation [76]. According to the study of Omondi et al. [77], soil enhances the availability of WHC by 15.1% and aggregate stability by 8.2%

with the addition of biochar. Biochar shows a positive effect on soil properties, mainly in coarse-textured and low fertility soils [64, 77]. Biochar works as an extremely important soil conditioner, as it alters the soil's physicochemical property i.e., increases soil moisture retention and enhances air permeability, resulting in an increase in soil microbial activity [78]. Poultry manure (10%) derived biochar and cow manure derived biochar used as a mixture enhanced C content in humic and fulvic acids [79]. In sandy soil, addition of biochar (maize straw and sewage sludge) with compost enhances the available water content [80].

17.3.4. Nutrient Content

In addition to alkaline matter, macro and micronutrient content such as Ca, Mg, K, Na, P, Fe, Zn, Co, Mo, and B present in significant concentrations in biochar in the form of SiO_2, $CaCO_3$, KCl, and $CaSO_4$ as well as nitrates, oxides, and hydroxides [81, 82]. Feedstock of biochar contains different concentrations of salt and ions according to the nature of material [65]. Wheat straw biochar contains Ca, P, Fe, Mg, and K, i.e. 1, 0.3, 0.4, 0.6, and 2.6%, respectively. Micronutrients such as Fe, B, Cr, Cu, Ni, and Mo are considered important sources of biochar.

Biochars are used in the soil for the mitigation of persistent organic contaminations (POCs) for the fertility of soil. Feedstock such as rice husk, fir wood chips, mix wood, corn straw, and bamboo were reported for the mitigation of oxyfluorfen, Cd, 2,4-dichlorophenol, phenenthrene, pyrene, polychlorinated biphenyl and DDE, and PAHs [24, 92, 93, 95, 96]. In another study, sewage sludge was reported for the mitigation of Cd, Pb, and Zn in Brazilian soil [97]. Poultry, cow, and sheep manure were also reported for the removal of Cr(VI) in farmland soil [72]. In a study, bamboo, rice straw, and Chinese walnut shells were reported for the removal of Cu in industrially polluted soil [100]. Table 17.2 highlights the application of biochar in mitigation of POCs in soil.

17.3.5. Enhancement of Microbial Abundance

Biochar is a nutritive source for soil microorganisms. The microbes influence community structure and enzyme activities by modifying surface area, pore structure, pH, and mineral substances. Microorganism abundance in biochar-amended soil systems is a function of soil–biochar interaction [38].

The combination of biochar and soil (25% + 75%), (50% + 50%), and (75% + 25%) ratio was used to study the effect of biochar on soil microflora. The 16S rRNA metagenomic analysis revealed the dominance of phyla: *Protobacteria, Actinobacteria,* and *Acidobacteria* that influence the soil nutrient cycle when applied at ratio of 1:3. The soil microbial communities are responsible for enhancing seed germination, plant growth, and crop yield [38, 52]. Bacterial community structure is implicated, as is greater nutrient retention in soil and efficient nutrient transfer to crops [59].

17.3.6. Improving Crop Productivity

Biochar shows good effect on plant growth and removal of environmental contamination [83]. Biochar enhances crop growth along with soil nutrition and microbial communities. In addition, biochar provides nutrients (P, K, Ca, Mg, N) directly to the crop. Biochar pore structure provides a good environment for the soil microflora, and is helpful in crop growth and yield [60, 64]. For example, biochar from peanut hulls enhances plant growth, increases soil microbe properties and enzyme activity, and boosts agricultural production.

Table 17.2 The use of biochar to mitigate persistent POCs in soil.

Resources	Tested soil	Hazardous substance	Bioremediation effect	References
Rice hull	Loamy, sandy, loam, clay, dry land soil, Saturated soil,	Oxyfluorfen, Cd	Oxyfluorfen reduce (18–63%) Adsorption of Cd on soil (saturated) enhance (21% - 41%), dry land soil enhance (38% -54%).	[24, 92]
Fir wood chips	Rice soil	2,4-dichlorophenol, phenenthrene	Decrease the degradation and mineralization of both hazardous substances. Enhance the accumulation of their metabolites in soil.	[93]
Orchard pruning biomass	Vineyard	PAHs	Concentrations of PAH reduce with time.	[94]
Mixed wood shavings rice husk	Loamy agricultural soil	Pyrene, polychlorinated biphenyl, and DDE (dichlorodiphenyldichloroethylene)	Biochar dose (10%), bioavailability and accessibility (37 and 41%)	[95]
Corn straw, bamboo	Soil contaminated with PAHs	PAHs	The bioaccumulation of PAHs in rice roots was decreased.	[96]
SS (Sewage sludge)	Brazil soil	Cd, Pb, and Zn	The concentration and bio available levels of Cd, Pb, and Zn of in the leachates decreased	[97]
Poultry litter	Paddy soil (near Zn and Pb mines)	Cd, Cu, Zn, Pb	Cd decrease (8–10%)	[98]
Wheat straw	Acidic soil	Cd, Cu	Cu decrease (40.9%), Cd and Cu decreased (18.8 and 18.6%)	[99]
Poultry manure, cow and sheep manure	Farmland soil	Cr(VI)	Poultry manure decreased (61.54 $mg\,kg^{-1}$) Cr(VI) in acidic soil, Cr(VI) (73.93 $Mg\,kg^{-1}$) reduce in alkaline soil. Cow and sheep manure reduced alkaline soil (57.81, 68.15 $mg\,kg^{-1}$)	[72]
Bamboo, rice straw, Chinese walnut shell	Industrial polluted soil	Cu	Cu decreased (15, 35, and 26%).	[100]

Application of biochar is an emerging trend [84, 85]. In a study, the liquid digestate (LD), LD + BC, and pelleted digestate (PD) were used and compared to the application of biochar. Researchers found that increased plant number in the presence of LD + BC, BC, PD, and LD. The number of fruit increased per plant by 15%, and fruit weight increased by 24% compared to untreated control. Digestate and biochar are useful options to increase yield and quality of processing tomato production [86]. Biochar improves crop productivity for the removal of hazardous substance from the soil. In crop productivity, modified biochar was used in a wide range of instances. In a study, modified rice husk was reported for the removal of Hg in contaminated soil and corn straw for As in red soil [117, 118]. Eucalyptus wood, poultry litter, sheep manure, and bamboo hardwoods were used for the removal of Cd, Cu, and Cr in soil [24, 72, 98]. The remediation of soil pollution by various types of modified biochar is shown in Table 17.3.

17.3.7. Enhancing Composting Process

Biochar showed environmental benefits for making aerobic composting bulking agents. The characteristics of biochar such as huge specific surface area, higher porosity, different functional groups, and high CEC, as well as adsorption capacity, shows similarity between biochar and compost [87, 88, 91]. The property for making environmentally friendly biochar shows a most important role in aerobic composting and increasing physicochemical properties. The biochar amendment in soil, including in the reduction of greenhouse gas emission, showing the effect of NH_4 emission and TN loss, reduces the useful microorganism's life activities and improves composting maturity. The addition of biochar has been found to improve the physical and chemical properties of compost and also promote humification [89].

17.4. CONCLUSION AND FUTURE PROSPECTS

In recent years, biochar has emerged as an efficient medium for agricultural sustainability. It represents a potential, low-cost renewable biomass for agricultural growth. The addition of biochar to soil improves beneficial microbes and nutrient cycles that aid in crop productivity. The physical and chemical properties of biochar enhance microbial abundance in the soil. Various types of biochar are important for the growth of agricultural productivity, and it helps in the maintenance of soil nutrients as well as soil textures. Biochar plays an important role as a habitat and refuge for soil microorganisms such as bacteria (0.3–3 mm) and protozoa (7–30 mm), which protects them from predatory soil micro-arthropods. Various lab- and pilot-scale studies have investigated the use of biochar for soil amendment and crop production. There are existing challenges for the wider application of biochar at a large scale to satisfy agricultural sustainability. Biochar also contributes to sustainable goals by reducing hazardous solid waste; valorized organic waste is transformed into valuable biofertilizer. This chapter has provided insight into the numerous complexities of biochar, as well as its advantages and diverse impacts; nevertheless, further research should be undertaken to provide a better understanding of the processes of biochar interactions and their effects on soil health and agricultural productivity.

Table 17.3 The remediation of soil pollution by various types of modified biochar.

Resources	Modifications	Hazardous substance	Tested soil	Bioremediation effect	References
Rice husk	Sulfur	Hg	Hg contaminated soil	Modification increases the Hg^{2+} adsorptive capacity of biochar by 73%, 67.11 $mg\,g^{-1}$.	[116]
Corn straw	MnO	As	Red soil	Increase sorption capacity for As (III)	[105]
Corn straw	Fe-Mn	As	Paddy soil	Reduce the content of available As	[118]
Coconut shell	HCL+ ultrasonication	Cd, Ni and Zn	Topsoil of paddy fields	Cd, Ni, and Zn reduced by 30.1, 57.2, and 12.7%.	[54]
Eucalyptus wood and poultry litter	Iron	Cd, Cu, Zn, Pb	Paddy soil near Zn and mines	Poultry litter biochar enhance the Cd (8–10%), Zn (27–29%), and Cu (59–63%) in soils.	[97]
Poultry, sheep manure	Chitosan, ZVI	Cr	Uncontaminated surface soil	Sheep manure biochar modified decrease Cr(VI) by 55%, and poultry manure modification biochar reduced Cr(VI) by 48%	[72]
Bamboo hardwoods	Sulphur-iron	Cr	Plant farmland	Cd in soil was reduce BC (12.54%), S-BC (29.71%), SF-BC (18.53%).	[24]

ACKNOWLEDGMENTS

This research did not receive any specific grant from funding agencies in the public, commercial, or not-for-profit sectors. Further, authors are highly thankful to Director CSIR-Indian Institute of Toxicology Research (IITR), Lucknow (India) for his valuable encouragement and support to complete this manuscript. The CSIR-IITR manuscript no. is IITR/SEC/2022-2023/24.

REFERENCES

1. Liu, L., Zhu, B., and Wang, G.X. (2015). Azoxystrobin-induced excessive reactive oxygen species (ROS) production and inhibition of photosynthesis in the unicellular green algae Chlorella vulgaris. *Environmental Science and Pollution Research 22* (10): 7766–7775.
2. Geisseler, D. and Scow, K.M. (2014). Long-term effects of mineral fertilizers on soil microorganisms–a review. *Soil Biology and Biochemistry 75*: 54–63.
3. Van Breemen, N., Mulder, J., and Driscoll, C.T. (1983). Acidification and alkalinization of soils. *Plant and Soil 75* (3): 283–308.
4. Pradhan, A., Chan, C., Roul, P.K. et al. (2018). Potential of conservation agriculture (CA) for climate change adaptation and food security under rainfed uplands of India: a transdisciplinary approach. *Agricultural Systems 163*: 27–35.
5. Shukla, P., Giri, B.S., Mishra, R.K. et al. (2021). Lignocellulosic biomass-based engineered biochar composites: A facile strategy for abatement of emerging pollutants and utilization in industrial applications. *Renewable and Sustainable Energy Reviews 152*: 111643.
6. Peiris, C., Gunatilake, S.R., Wewalwela, J.J. et al. (2019). Biochar for sustainable agriculture: Nutrient dynamics, soil enzymes, and crop growth. *Biochar from biomass and waste Jan 1*: 211–224. Elsevier.
7. El-Naggar, A.H., Usman, A.R., Al-Omran, A. et al. (2015). Carbon mineralization and nutrient availability in calcareous sandy soils amended with woody waste biochar. *Chemosphere 138*: 67–73.
8. Vickers, N.J. (2017). Animal communication: when i'm calling you, will you answer too? *Current Biology 27* (14): R713–R715.
9. Ahmad, A., Singh, A.P., Khan, N. et al. (2021). Bio-composite of Fe-sludge biochar immobilized with Bacillus sp. in packedcolumn for bio-adsorption of Methylene blue in a hybrid treatment system:Isotherm and kinetic evaluation. *Environmental Technology & Innovation 23*: 101734.
10. Lee, D.J., Cheng, Y.L., Wong, R.J. et al. (2018). Adsorption removal of natural organic matters in waters using biochar. *Bioresource Technology 260*: 413–416.
11. Yang, F., Lee, X., Theng, B.K. et al. (2017). Effect of biochar addition on short-term N 2 O and CO 2 emissions during repeated drying and wetting of an anthropogenic alluvial soil. *Environmental Geochemistry and Health 39* (3): 635–647.
12. Ilyas, T., Chowdhary, P., Chaurasia, D. et al. (2021). Sustainable green processing of grape pomace for the production of value-added products: An overview. *Environmental Technology & Innovation 23*: 101592
13. Ayaz, M., Feizienė, D., Tilvikienė, V. et al. (2021). Biochar role in the sustainability of agriculture and environment. *Sustainability 13* (3): 1330.
14. Khan, N., Chowdhary, P., Gnansounou, E., and Chaturvedi, P. (2021). Biochar and environmental sustainability: emerging trends and techno-economic perspectives. *Bioresource Technology 332*: 125102.
15. Semida, W.M., Beheiry, H.R., Sétamou, M. et al. (2019). Biochar implications for sustainable agriculture and environment: a review. *South African Journal of Botany 127*: 333–347.
16. Ajeng, A.A., Abdullah, R., Ling, T.C. et al. (2020). Bioformulation of biochar as a potential inoculant carrier for sustainable agriculture. *Environmental Technology and Innovation 20*: 101168.

17. Major, J., Steiner, C., Downie, A. et al. (2009). Biochar effects on nutrient leaching. In: *Biochar for Environmental Management: Science and technology*, 271.
18. Verheijen, F., Jeffery, S., Bastos, A.C. et al. (2010). Biochar application to soils. A critical scientific review of effects on soil properties, processes, and functions. *JRC Scientific and Technical Report, Europe 24099*: 162.
19. Chen, Z.L., Zhang, J.Q., Huang, L. et al. (2019). Removal of Cd and Pb with biochar made from dairy manure at low temperature. *Journal of Integrative Agriculture 18* (1): 201–210.
20. Ruan, X., Sun, Y., Du, W. et al. (2019). Formation, characteristics, and applications of environmentally persistent free radicals in biochars: a review. *Bioresource Technology 281*: 457–468.
21. Li, Q., Zhang, H., Guo, S. et al. (2020). Groundwater pollution source apportionment using principal component analysis in a multiple land-use area in southwestern China. *Environmental Science and Pollution Research 27* (9): 9000–9011.
22. Chen, T., Luo, L., Deng, S. et al. (2018). Sorption of tetracycline on H3PO4 modified biochar derived from rice straw and swine manure. *Bioresource Technology 267*: 431–437.
23. Bandara, T., Franks, A., Xu, J. et al. (2020). Chemical and biological immobilization mechanisms of potentially toxic elements in biochar-amended soils. *Critical Reviews in Environmental Science and Technology 50* (9): 903–978.
24. Wu, C., Shi, L., Xue, S. et al. (2019). Effect of sulfur-iron modified biochar on the available cadmium and bacterial community structure in contaminated soils. *Science of the Total Environment 647*: 1158–1168.
25. Deng, B., Zheng, L., Ma, Y. et al. (2020). Effects of mixing biochar on soil N 2 O, CO 2, and CH 4 emissions after prescribed fire in alpine meadows of Wugong Mountain, China. *Journal of Soils and Sediments 20* (8): 3062–3072.
26. Shen, Y., Song, S., Thian, B.W. et al. (2020). Impacts of biochar concentration on the growth performance of a leafy vegetable in a tropical city and its global warming potential. *Journal of Cleaner Production 264*: 121678.
27. Qiu, Z., Zhai, Y., Li, S. et al. (2020). Catalytic co-pyrolysis of sewage sludge and rice husk over biochar catalyst: bio-oil upgrading and catalytic mechanism. *Waste Management 114*: 225–233.
28. Feng, Y., Xu, Y., Yu, Y. et al. (2012). Mechanisms of biochar decreasing methane emission from Chinese paddy soils. *Soil Biology and Biochemistry 46*: 80–88.
29. Senbayram, M., Saygan, E.P., Chen, R. et al. (2019). Effect of biochar origin and soil type on the greenhouse gas emission and the bacterial community structure in N fertilised acidic sandy and alkaline clay soil. *Science of the Total Environment 660*: 69–79.
30. Qayyum, M.F., Abid, M., Danish, S. et al. (2015). Effects of various biochars on seed germination and carbon mineralization in an alkaline soil. *Pakistan Journal of Agricultural Sciences 51*: 977–982.
31. Peng, X.Y., Ye, L.L., Wang, C.H. et al. (2011). Temperature-and duration-dependent rice straw-derived biochar: characteristics and its effects on soil properties of an Ultisol in southern China. *Soil and Tillage Research 112* (2): 159–166.
32. Głąb, T., Palmowska, J., Zaleski, T. et al. (2016). Effect of biochar application on soil hydrological properties and physical quality of sandy soil. *Geoderma 281*: 11–20.
33. Inyang, M., Gao, B., Pullammanappallil, P. et al. (2010). Biochar from anaerobically digested sugarcane bagasse. *Bioresource Technology 101* (22): 8868–8872.
34. Yuan, J.H., Xu, R.K., and Zhang, H. (2011). The forms of alkalis in the biochar produced from crop residues at different temperatures. *Bioresource Technology 102* (3): 3488–3497.
35. Gai, X., Wang, H., Liu, J. et al. (2014). Effects of feedstock and pyrolysis temperature on biochar adsorption of ammonium and nitrate. *PLoS One 9* (12): 113888.
36. Mubarik, S., Saeed, A., Athar, M.M. et al. (2016). Characterization and mechanism of the adsorptive removal of 2, 4, 6-trichlorophenol by biochar prepared from sugarcane baggase. *Journal of Industrial and Engineering Chemistry 33*: 115–121.
37. Chen, Y.D., Liu, F., Ren, N.Q. et al. (2020). Revolutions in algal biochar for different applications: state-of-the-art techniques and future scenarios. *Chinese Chemical Letters 31* (10): 2591–2602.

38. Neogi, S., Sharma, V., and Khan, N. et al. (2021). Sustainable biochar: a facile strategy for soil and environmental restoration, energy generation, mitigation of global climate change and circular bioeconomy. *Chemosphere*: 133474.
39. Cantrell, K.B., Hunt, P.G., Uchimiya, M. et al. (2012). Impact of pyrolysis temperature and manure source on physicochemical characteristics of biochar. *Bioresource Technology 107*: 419–428.
40. Wan, Q., Yuan, J.H., Xu, R.K. et al. (2014). Pyrolysis temperature influences ameliorating effects of biochars on acidic soil. *Environmental Science and Pollution Research 21* (4): 2486–2495.
41. Atienza-Martínez, M., Ábrego, J., Gea, G. et al. (2020). Pyrolysis of dairy cattle manure: evolution of char characteristics. *Journal of Analytical and Applied Pyrolysis 145*: 104724.
42. Novak, J.M., Lima, I., Xing, B. et al. (2009). Characterization of designer biochar produced at different temperatures and their effects on a loamy sand. *Annals of Environmental Science. 3* (1): 195–206.
43. Wu, H., Zeng, G., Liang, J. et al. (2016). Responses of bacterial community and functional marker genes of nitrogen cycling to biochar, compost and combined amendments in soil. *Applied Microbiology and Biotechnology 100*: 8583–8591.
44. Ahmad, A., Chowdhary, P., Khan, N. et al. (2022). Effect of sewage sludge biochar on the soil nutrient, microbial abundance, and plant biomass: A sustainable approach towards mitigation of solid waste. *Chemosphere 287*: 132112.
45. Beesley, L. and Marmiroli, M. (2011). The immobilisation and retention of soluble arsenic, cadmium and zinc by biochar. *Environmental Pollution 159* (2): 474–480.
46. Gupta, A., Singh, A., Ilyas, T., Chowdhary, P., and Chaturvedi, P. (2022). Production and environmental applications of activated sludge biochar. In *Biomass, Biofuels, Biochemicals,* Elsevier 387–406.
47. Yuan, H., Lu, T., Wang, Y. et al. (2016). Sewage sludge biochar: nutrient composition and its effect on the leaching of soil nutrients. *Geoderma 267*: 17–23.
48. Lee, X.J., Ong, H.C., Gan, Y.Y. et al. (2020). State of art review on conventional and advanced pyrolysis of macroalgae and microalgae for biochar, bio-oil and bio-syngas production. *Energy Conversion and Management 210*: 112707.
49. Chopin, T. and Sawhney, M. (2009). Seaweeds and their mariculture. In: *The Encyclopedia of Ocean Sciences* (ed. J.H. Steele, S.A. Thorpe and K.K. Turekian), 317–326.
50. Barrington, K., Chopin, T., and Robinson, S. (2009). Integrated multi-trophic aquaculture (IMTA) in marine temperate waters. Integrated mariculture: a global review. *FAO Fisheries and Aquaculture Technical Paper 529*: 7–46.
51. Chan, K.Y., Van Zwieten, L., Meszaros, I. et al. (2008). Using poultry litter biochars as soil amendments. *Soil Research 46* (5): 437–444.
52. Ameloot, N., De Neve, S., Jegajeevagan, K. et al. (2013). Short-term CO2 and N2O emissions and microbial properties of biochar amended sandy loam soils. *Soil Biology and Biochemistry 57*: 401–410.
53. Prein, M. (2002). Integration of aquaculture into crop–animal systems in Asia. *Agricultural Systems 71* (1–2): 127–146.
54. Liu, H., Xu, F., Xie, Y. et al. (2018). Effect of modified coconut shell biochar on availability of heavy metals and biochemical characteristics of soil in multiple heavy metals contaminated soil. *Science of the Total Environment 645*: 702–709.
55. Parmar, A., Nema, P.K., and Agarwal, T. (2014). Biochar production from agro-food industry residues: a sustainable approach for soil and environmental management. *Current Science 107* (10): 1673–1682.
56. Kaur, V. and Sharma, P. (2019). Physico-chemical characterization of biochar derived from different raw materials. *International Journal of Scientific Research and Reviews 8* (2): 451–468.
57. Li, H., Qu, R., Li, C. et al. (2014). Selective removal of polycyclic aromatic hydrocarbons (PAHs) from soil washing effluents using biochars produced at different pyrolytic temperatures. *Bioresource Technology 163*: 193–198.

58. Xiong, X., Iris, K.M., Cao, L. et al. (2017). A review of biochar-based catalysts for chemical synthesis, biofuel production, and pollution control. *Bioresource Technology 246*: 254–270.
59. Gul, S., Yanni, S.F., and Whalen, J.K. (2014). *Lignin Controls on Soil Ecosystem Services: Implications for Biotechnological Advances in Biofuel Crops*, 375–416. New York: Biochemistry Research Trends. Nova Science Publishers.
60. Chaurasia, D., Singh,A., Shukla, P., and Chaturvedi, P. (2022). Biochar: A sustainable solution for the management of agri-wastes and environment. *In Biochar in Agriculture for Achieving Sustainable Development Goals*, 361–379, Academic Press.
61. Randolph, P., Bansode, R.R., Hassan, O.A. et al. (2017). Effect of biochars produced from solid organic municipal waste on soil quality parameters. *Journal of Environmental Management 192*: 271–280.
62. Zhibin, L.I., Qi, L.I., Gang, L.I. et al. (2017). Effects of different biochars on Pinus elliottii growth, N use efficiency, soil N2O and CH4 emissions and C storage in a subtropical area of China. *Pedosphere 27* (2): 248–261.
63. Yue, Y., Cui, L., Lin, Q. et al. (2017). Efficiency of sewage sludge biochar in improving urban soil properties and promoting grass growth. *Chemosphere 173*: 551–556.
64. Laghari, M., Mirjat, M.S., Hu, Z. et al. (2015). Effects of biochar application rate on sandy desert soil properties and sorghum growth. *Catena 135*: 313–320.
65. Enders, A., Hanley, K., Whitman, T. et al. (2012). Characterization of biochars to evaluate recalcitrance and agronomic performance. *Bioresource Technology 114*: 644–653.
66. Bruun, E.W., Ambus, P., Egsgaard, H. et al. (2012). Effects of slow and fast pyrolysis biochar on soil C and N turnover dynamics. *Soil Biology and Biochemistry 46*: 73–79.
67. Mukherjee, A. and Lal, R. (2013). Biochar impacts on soil physical properties and greenhouse gas emissions. *Agronomy 3* (2): 313–339.
68. Dai, Z., Zhang, X., Tang, C. et al. (2017). Potential role of biochars in decreasing soil acidification-a critical review. *Science of the Total Environment 581*: 601–611.
69. Godlewska, P., Schmidt, H.P., Ok, Y.S. et al. (2017). Biochar for composting improvement and contaminants reduction. A review. *Bioresource Technology 246*: 193–202.
70. Abujabhah, I.S., Bound, S.A., Doyle, R. et al. (2016). Effects of biochar and compost amendments on soil physico-chemical properties and the total community within a temperate agricultural soil. *Applied Soil Ecology 98*: 243–253.
71. Verheijen, F.G., Zhuravel, A., Silva, F.C. et al. (2019). The influence of biochar particle size and concentration on bulk density and maximum water holding capacity of sandy vs sandy loam soil in a column experiment. *Geoderma 347*: 194–202.
72. Mandal, S., Pu, S., Adhikari, S. et al. (2021). Progress and future prospects in biochar composites: application and reflection in the soil environment. *Critical Reviews in Environmental Science and Technology 51* (3): 219–271.
73. Prendergast-Miller, M.T., Duvall, M., and Sohi, S.P. (2014). Biochar–root interactions are mediated by biochar nutrient content and impacts on soil nutrient availability. *European Journal of Soil Science 65* (1): 173–185.
74. Haider, G., Steffens, D., Moser, G. et al. (2017). Biochar reduced nitrate leaching and improved soil moisture content without yield improvements in a four-year field study. *Agriculture, Ecosystems & Environment 237*: 80–94.
75. Grunwald, D., Kaiser, M., and Ludwig, B. (2016). Effect of biochar and organic fertilizers on C mineralization and macro-aggregate dynamics under different incubation temperatures. *Soil and Tillage Research 164*: 11–17.
76. Hayashi, J., Horikawa, T., Takeda, I. et al. (2002). Preparing activated carbon from various nutshells by chemical activation with K2CO3. *Carbon NY 40*: 2381–2386.
77. Omondi, M.O., Xia, X., Nahayo, A. et al. (2016). Quantification of biochar effects on soil hydrological properties using meta-analysis of literature data. *Geoderma 274*: 28–34.
78. Mclennon, E., Solomon, J.K., Neupane, D. et al. (2020). Biochar and nitrogen application rates effect on phosphorus removal from a mixed grass sward irrigated with reclaimed wastewater. *Science of the Total Environment 715*: 137012.

79. Jain, M.S., Jambhulkar, R., and Kalamdhad, A.S. (2018). Biochar amendment for batch composting of nitrogen rich organic waste: effect on degradation kinetics, composting physics and nutritional properties. *Bioresource Technology 253*: 204–213.
80. Jindo, K., Sonoki, T., Matsumoto, K. et al. (2016). Influence of biochar addition on the humic substances of composting manures. *Waste Management 49*: 545–552.
81. Wu, W., Yang, M., Feng, Q. et al. (2012). Chemical characterization of rice straw-derived biochar for soil amendment. *Biomass and Bioenergy 47*: 268–276.
82. Xu, X., Cao, X., Zhao, L. et al. (2013). Removal of Cu, Zn, and Cd from aqueous solutions by the dairy manure-derived biochar. *Environmental Science and Pollution Research 20* (1): 358–368.
83. Mollinedo, J., Schumacher, T.E., and Chintala, R. (2016). Biochar effects on phenotypic characteristics of "wild" and "sickle" Medicago truncatula genotypes. *Plant and Soil 400* (1–2): 1–4.
84. Zhang, Y., Tan, Q., Hu, C. et al. (2015). Differences in responses of soil microbial properties and trifoliate orange seedling to biochar derived from three feedstocks. *Journal of Soils and Sediments 15* (3): 541–551.
85. Zhang, F., Wang, X., Yin, D. et al. (2015). Efficiency and mechanisms of Cd removal from aqueous solution by biochar derived from water hyacinth (Eichornia crassipes). *Journal of Environmental Management 153*: 68–73.
86. Ronga, D., Caradonia, F., Parisi, M. et al. (2020). Using digestate and biochar as fertilizers to improve processing tomato production sustainability. *Agronomy 10* (1): 138.
87. Wu, S., He, H., Inthapanya, X. et al. (2017). Role of biochar on composting of organic wastes and remediation of contaminated soils—a review. *Environmental Science and Pollution Research 24* (20): 16560–16577.
88. Wu, H., Lai, C., Zeng, G. et al. (2017). The interactions of composting and biochar and their implications for soil amendment and pollution remediation: a review. *Critical Reviews in Biotechnology 37* (6): 754–764.
89. Guo, W., Lu, S., Shi, J. et al. (2019). Effect of corn straw biochar application to sediments on the adsorption of 17α-ethinyl estradiol and perfluorooctane sulfonate at sediment-water interface. *Ecotoxicology and Environmental Safety 174*: 363–369.
90. Rawat, J., Saxena, J., and Sanwal, P. (2019). Biochar: a sustainable approach for improving plant growth and soil properties. *Biochar-An Imperative Amendment for Soil and the Environment* 1–7.
91. Khan, M.A., Khan, S., Ding, X. et al. (2018). The effects of biochar and rice husk on adsorption and desorption of cadmium on to soils with different water conditions (upland and saturated). *Chemosphere 193*: 1120–1126.
92. Gu, J., Zhou, W., Jiang, B. et al. (2016). Effects of biochar on the transformation and earthworm bioaccumulation of organic pollutants in soil. *Chemosphere 145*: 431–437.
93. Rombolà, A.G., Fabbri, D., Baronti, S. et al. (2019). Changes in the pattern of polycyclic aromatic hydrocarbons in soil treated with biochar from a multiyear field experiment. *Chemosphere 219*: 662–670.
94. Bielská, L., Škulcová, L., Neuwirthová, N. et al. (2018). Sorption, bioavailability and ecotoxic effects of hydrophobic organic compounds in biochar amended soils. *Science of the Total Environment 624*: 78–86.
95. Li, H., Dong, X., da Silva, E.B. et al. (2017). Mechanisms of metal sorption by biochars: biochar characteristics and modifications. *Chemosphere 178*: 466–478.
96. Penido, E.S., Martins, G.C., Mendes, T.B. et al. (2019). Combining biochar and sewage sludge for immobilization of heavy metals in mining soils. *Ecotoxicology and Environmental Safety 172*: 326–333.
97. Lu, H.P., Li, Z.A., Gasco, G. et al. (2018). Use of magnetic biochars for the immobilization of heavy metals in a multi-contaminated soil. *Science of the Total Environment 622*: 892–899.
98. Jia, W., Wang, B., Wang, C. et al. (2017). Tourmaline and biochar for the remediation of acid soil polluted with heavy metals. *Journal of Environmental Chemical Engineering 5* (3): 2107–2114.
99. Wang, Y., Zhong, B., Shafi, M. et al. (2019). Effects of biochar on growth, and heavy metals accumulation of moso bamboo (Phyllostachy pubescens), soil physical properties, and heavy metals solubility in soil. *Chemosphere 219*: 510–516.

100. Jindo, K., Mizumoto, H., Sawada, Y. et al. (2014). Physical and chemical characterization of biochars derived from different agricultural residues. *Biogeosciences 11* (23): 6613–6621.
101. Lian, F., Sun, B., Chen, X. et al. (2015). Effect of humic acid (HA) on sulfonamide sorption by biochars. *Environmental Pollution 204*: 306–312.
102. Ghaffar, A., Ghosh, S., Li, F. et al. (2015). Effect of biochar aging on surface characteristics and adsorption behavior of dialkyl phthalates. *Environmental Pollution 206*: 502–509.
103. Naeem, M.A., Khalid, M., Arshad, M. et al. (2014). Yield and nutrient composition of biochar produced from different feedstocks at varying pyrolytic temperatures. *Pakistan Journal of Agricultural Sciences 51* (1): 75–82.
104. Ren, X., Zhang, P., Zhao, L. et al. (2016). Sorption and degradation of carbaryl in soils amended with biochars: influence of biochar type and content. *Environmental Science and Pollution Research 23* (3): 2724–2734.
105. Yu, L., Yuan, Y., Tang, J. et al. (2015). Biochar as an electron shuttle for reductive dechlorination of pentachlorophenol by Geobacter sulfurreducens. *Scientific Reports 5* (1): 16221.
106. Dang, T., Mosley, L.M., Fitzpatrick, R. et al. (2015). Organic materials differ in ability to remove protons, iron and aluminium from acid sulfate soil drainage water. *Water, Air, & Soil Pollution 226* (11): 1–3.
107. Sun, Y., Gao, B., Yao, Y. et al. (2014). Effects of feedstock type, production method, and pyrolysis temperature on biochar and hydrochar properties. *Chemical Engineering Journal 240*: 574–578.
108. Tariq, M., Ali, S., Ahmad, F. et al. (2011). Identification, FT-IR, NMR (1H and 13C) and GC/MS studies of fatty acid methyl esters in biodiesel from rocket seed oil. *Fuel Processing Technology 92* (3): 336–341.
109. Al-Wabel, M.I., Al-Omran, A., El-Naggar, A.H. et al. (2013). Pyrolysis temperature induced changes in characteristics and chemical composition of biochar produced from conocarpus wastes. *Bioresource Technology 131*: 374–379.
110. Kupryianchyk, D., Hale, S., Zimmerman, A.R. et al. (2016). Sorption of hydrophobic organic compounds to a diverse suite of carbonaceous materials with emphasis on biochar. *Chemosphere 144*: 879–887.
111. Takaya, C.A., Fletcher, L.A., Singh, S. et al. (2016). Phosphate and ammonium sorption capacity of biochar and hydrochar from different wastes. *Chemosphere 145*: 518–527.
112. Gunes, A., Inal, A., Sahin, O. et al. (2015). Variations in mineral element concentrations of poultry manure biochar obtained at different pyrolysis temperatures, and their effects on crop growth and mineral nutrition. *Soil Use and Management 31* (4): 429–437.
113. Tsai, W.T., Huang, C.N., Chen, H.R. et al. (2015). Pyrolytic conversion of horse manure into biochar and its thermochemical and physical properties. *Waste and Biomass Valorization 6* (6): 975–981.
114. Hunt, P.G., Cantrell, K.B., Bauer, P.J. et al. (2013). Phosphorus fertilization of ryegrass with ten precisely prepared manure biochars. *Transactions of the ASABE 56* (6): 1317.
115. Usman, A.R., Abduljabbar, A., Vithanage, M. et al. (2015). Biochar production from date palm waste: charring temperature induced changes in composition and surface chemistry. *Journal of Analytical and Applied Pyrolysis 115*: 392–400.
116. O'Connor, D., Peng, T., Li, G. et al. (2018). Sulfur-modified rice husk biochar: a green method for the remediation of mercury contaminated soil. *Science of the Total Environment 621*: 819–826.
117. Yu, Z., Zhou, L., Huang, Y. et al. (2015). Effects of a manganese oxide-modified biochar composite on adsorption of arsenic in red soil. *Journal of Environmental Management 163*: 155–162.
118. Lin, L., Li, Z., Liu, X. et al. (2019). Effects of Fe-Mn modified biochar composite treatment on the properties of As-polluted paddy soil. *Environmental Pollution 244*: 600–607.

18

Environmentally Sustainable Elimination of Microbes Using Boron-Doped Diamond Electrodes: From Water Treatment to Medical Applications

Maximilian Koch[1], Stefan Rosiwal[2], and Andreas Burkovski[1]

18.1. INTRODUCTION

Traditionally, microbes such as bacteria and yeasts play an important role in ecologically sustainable production processes. For thousands of years, they were used for baking and brewing or fermentation of traditional foods such as sauerkraut, kimchi, soy sauce, tempeh, and others. Starting in the late 1950s, the production of metabolites like amino acids, vitamins, and proteins became the focus of industry and research, and today's bio-economy concepts include the replacement of oil-derived chemicals and products by equivalent or new products relying – at least partially – on biomass as part of green chemistry [1].

In contrast to biotechnological production processes, which typically work at ambient pressure and moderate temperatures, procedures for the inactivation of microorganisms are often characterized by harsh physical or chemical conditions. Based on the high resistance of many microbes against environmental stress, inactivation protocols are applied using, for example, high-energy demanding heat and pressure treatment for sterilization or toxic chemicals for disinfection. Also, the application of antibiotics for treatment of human and animal infections and elimination of pathogens may have unwanted side effects. It is well-documented that antibiotics influence microbial communities and lead to a shift of bacterial populations toward multi-resistance, e.g. sewage treatment plants are known to be hotspots of antibiotic-resistant bacteria [2]. Taken together, alternative approaches for the inactivation of microbes may be beneficial for the environment, especially when these are based on sustainable, biology-inspired mechanisms.

[1]*Department of Biology, Friedrich-Alexander-Universität Erlangen-Nürnberg, Erlangen, Germany*
[2]*Department of Material Sciences, Friedrich-Alexander-Universität Erlangen-Nürnberg, Erlangen, Germany*

Microbial Biotechnology: Role in Ecological Sustainability and Research, First Edition.
Edited by Pankaj Chowdhary, Sujata Mani, and Preeti Chaturvedi.

18.2. INACTIVATION OF MICROORGANISMS BY REACTIVE OXYGEN SPECIES

The human immune system may work as a template for an environmentally friendly procedure of microbe inactivation. To defend the body against infection, phagocytes of the human immune system ingest potentially harmful microbes and inactivate these using reactive oxygen – and nitrogen – species, which damage nucleic acids, proteins, and lipids [3, 4]. As these are active against all kind of bacteria, fungi, and viruses, reactive oxygen species (ROS) are discussed as novel antimicrobials and alternatives to antibiotics, which become more and more ineffective due to the present global antibiotic resistance crisis [5].

ROS are free radicals and other highly reactive molecules derived from molecular oxygen. Examples are superoxide, hydrogen peroxide, or hydroxyl radicals. While hydroxyl radicals are the most reactive molecules of this group due to their high standard potential ($E°'[HO•/H_2O] = 2.34\,V$), the superoxide anion ($E°'[O_2•^-/H_2O_2] = 0.93\,V$) is more stable and only reacts with a few targets, e.g. nitric oxide, resulting in the formation of peroxynitrite, a reactive nitrogen species. The perhydroxyl radical, the conjugated acid of the superoxide anion, is again more reactive ($E°'[HO_2•/H_2O_2] = 1.48\,V$) and can perform a variety of organic and inorganic oxidation reactions [6]. In human phagocytic cells, ROS are produced byenzymatic reactions, mainly by NADPH oxidase and complex 1 of the mitochondrial redox chain [7–10].

Technically, a variety of chemical and electrochemical processes can produce ROS. One of these possibilities is the ROS production by photocatalytic reactions. Photosensitizers applied in photodynamic therapy are, for example, activated by visible light, converting oxygen into highly reactive compounds, which evoke oxidative stress on cancer cells or pathogenic bacteria and fungi [11, 12]. Photosensitizers achieved greater attention as active compounds against microbial infections in recent years [13–15]. In addition, zinc oxide and nitric oxide-releasing nanoparticles [16–18] and quantum dots [19] were described, which exert an antimicrobial activity based on the production of hydrogen peroxide and other ROS. Moreover, some bactericidal antibiotics can induce oxidative stress [20].

18.3. BORON-DOPED DIAMOND ELECTRODES

A relatively new concept exploiting ROS production for disinfection is an electrochemical oxidation based on boron-doped diamond (BDD) electrodes in a so-called advanced oxidation p3rocess (AOP) [21]. In this case, the high anodic over potential (approximately 2.8 V) of BDD against O_2 generation leads to the formation of different ROS [22, 23]. These are sequentially produced by discharging water molecules under release of protons and formation of anode-adsorbed hydroxyl radicals, which then can further react to form free hydroxyl radicals, ozone, and hydrogen peroxide in solution (Figure 18.1) [24–27]. At constant potential, the inactivation rate of *Escherichia coli* and *Enterococcus faecalis,* was reasonably higher by BDD than the inactivation provided by a platinum electrode [27]. Furthermore, compared to other metal-based electrodes the high anodic over potential of up to 2.8 V in acidic solution of niobium-based BDD electrodes can reduce the oxygen formation as a side product and further enhance the disinfection effect [22, 23].

The ROS produced in the process can react with any carbon-containing molecule, which is present in the solution, resulting in the formation of carbon monoxide and carbon dioxide as endproducts of the oxidation process. This principle can be applied successfully for the inactivation of microorganisms and in the last five years, some pioneering work was published, describing the use of this new technique for disinfection purposes.

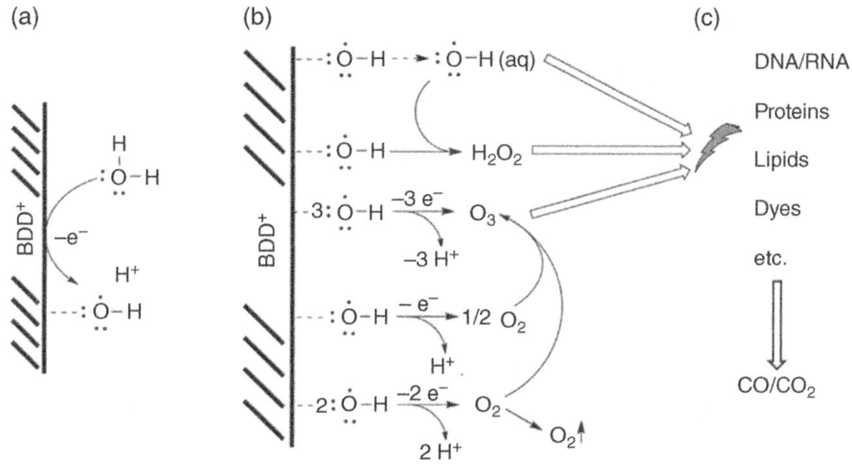

Figure 18.1 Schematic representation of BDD electrode function and effect. (a) Water discharge and adsorption of hydroxyl radical to the anode surface (BDD+). (b) Examples of possible reactions forming ROS and oxygen at the anode surface. (c) Targets of highly reactive ROS, which may be oxidized causing the disinfective effect on cells [24–27].

18.3.1. Application in Water Treatment

According to UNESCO, access to water is a human right [28]. At the same time, water consumption has been rising at a rate of approximately 1% per year since the 1980s. Under these circumstances, water is a critical resource for human life and welfare; however, according to the United Nations, more 30% of the human population lacks sufficient access to potable water and more than 60% does not have access to proper sanitation [28]. Consequently, waterborne pathogens pose a serious risk to human health.

Several water disinfection protocols have been developed based on physical methods such as filtration or ultraviolet irradiation [29–31]. Since these techniques have relatively high energy and investment costs, chemical approaches such as chlorination or the addition of ozone and hydrogen peroxide have also been described for the elimination of microbes [32, 33], which have the drawback of relying on the handling and application of hazardous chemicals.

BDD electrode treatment may be an ecologically sustainable solution to provide safe potable water and to protect aquatic ecosystems. An advantage of this technique is the low voltage necessary for BDD electrode applications, typically lower than 10 V [23]. Without the need of a standard electricity grid, a combination of solar panels and/or batteries is possible, thus providing a cost-effective, reliable, and quick solution for the disinfection of polluted water, especially in rural areas of developing countries lacking appropriate power infrastructure (an example for a small-scale BDD electrode for fast drinking water disinfection is shown in Figure 18.2). Scale-up of the method is easy and treatment times can be adapted to the quality of the water source.

For the protection of aquatic ecosystems and human health, not only the potable water, but also the effluent of sewage plants, should be as free as possible of pathogenic microorganisms. Unfortunately, a recent next-generation sequencing study revealed that bacterial pathogens may be not effectively reduced by UV treatment, and even worse, the nosocomial pathogen *Pseudomonas aeruginosa* was even enriched [34]. In addition, biological wastewater treatment plants seem to stimulate the emergence of antibiotic-resistant bacteria [2, 35].

Figure 18.2 Small-scale BDD electrode. The electrode is powered by a mobile phone via USB for the disinfection of potable water (*Source:* kindly provided by R. Borchardt, Erlangen).

When the influence of electrochemical disinfection using BDD electrodes was tested depending on the chloride concentration in the water, fast inactivation of *P. aeruginosa* was observed [36]. This was based on synergetic effects of in-situ ROS and free chlorine generation from chloride by the BDD electrode [36, 37]. However, even without chloride addition, BDD electrodes alone are highly effective in the elimination of detrimental microorganisms [23]. *E. coli*, *Pseudomonas fluorescens*, a freshwater bacterium, and *P. aeruginosa* were inactivated within minutes, while spores of the Gram-positive model organism *Bacillus subtilis* were more resistant and a complete disinfection was obtained only after increased treatment times when the same conditions were applied [23]. When sewage plant effluent after tertiary treatment was analyzed, high numbers of Gram-negative and -positive bacteria were identified, e.g. *Actinobacter towneri*, *Aeromonas caviae*, *E. coli*, and *Pseudomonas stutzeri*, while after BDD electrode treatment, both the number of bacteria and bacterial species were significantly reduced [23]. As suggested recently, BDD electrodes may even be applied to blackwater disinfection, although a significant inactivation of microbes in undiluted blackwater samples was not achieved until now [38]. This result may be simply explained by the fact that high amounts of ROS are necessary for the electrochemical disinfection in high carbon-containing medium such as blackwater. In this case, treatment times have to be adapted to make BDD electrode treatment not only suitable for the eliminations of bacteria in potable water, but also for the elimination of pathogens from blackwater.

Recently, BDD electrodes were also tested for treatment of seawater used as ship's ballast water in order to avoid introduction of putative invasive or harmful species [39, 40]. When BDD electrodes were in a continuous flow system with real seawater, a 4-Log reduction of natural marine bacteria was observed. Analysis of bacterial regrowth after treatment revealed a higher bacteria damage resulting from BDD electrode treatment compared to chlorination. Furthermore, a change in bacterial population diversity was observed in this study. Interestingly, with a residual oxidant in the seawater, complete disinfection can be achieved in three days [40].

18.3.2. Reduction of Biofilms on Medical Devices

In addition to its application for elimination of free-living microbes in water, BDD electrode treatment has been tested recently with respect to its capacity to inactivate microbial biofilms. Biofilms are consortia of microbes growing on surfaces and protected by an extracellular polymer matrix. A main advantage for microbes growing in biofilms is an enhanced rate of horizontal gene transfer, resulting in fast acquisition of antibiotic resistance markers, virulence factors, and consequently the emergence and successful colonization of surfaces by pathogens [41]. Inactivation of *P. aeruginosa* biofilms on different surfaces in an aquatic environment was reported by Simcox and coworkers [42], while biofilm inactivation on implant surfaces was tested by Koch, Göltz, and coworkers [43, 44]. These authors investigated the inactivation of biofilms growing on dental implant surfaces and evoking inflammatory processes adjacent to the implants designated as peri-implantitis. While the exact patho-mechanism is not fully understood, the central role of microbial biofilms on dental implants as causative agent of the inflammatory reaction and subsequent loss of peri-implant bone is well recognized [45, 46]. Consequently, a plethora of literature with respect to dental implant disinfection is available, describing different techniques for biofilm inactivation including treatment with antimicrobial agents and antibiotics [47, 48], cold plasma and laser application [49–51], or photodynamic therapy [52, 53]. Recently an approach for the removal of biofilms was published, which uses a platinum-based anode, the implant as cathode, and lactic acid- and iodide-containing buffers as electrolytes [54]. Through this electrochemical approach, microbial biofilms are inactivated by the production of reactive iodine (e.g. HIO) and oxygen species (H_2O_2) and detached by hydrogen formation at the implant serving as cathode. A severe drawback of the method is the possibility that titanium hydrides can be formed from the implant material in the presence of hydrogen. This may weaken implant durability against mechanical stress and even cause serious damage to the implant due to the reduced fatigue strength [55, 56].

In this respect, BDD electrodes may provide a useful alternative. Application of an easy to use and durable hand-held prototype (Figure 18.3) electrochemical disinfection of biofilm-colonized implants was compared to air abrasive and chemo-mechanical curette treatment as established standard methods [43]. The obtained results showed clear advantages to BDD electrode application. In this case, complete disinfection after a maximum treatment time of 20 minutes was achieved for yeasts (*Candida albicans*, *Candida dubliniensis*) and different bacteria such as (*En. faecalis*, *Roseomonas mucosa*, *Staphylococcus epidermidis*, and *Streptococcus sanguinis*). In case of spore-forming bacteria such as *Bacillus pumilus* and *B. subtilis* only an incomplete disinfection was achieved, most likely due to the high stress resistance of the spores. By comparison, air abrasion and curette treatment gave inferior results and independent of the microbial species tested, a complete disinfection was never achieved [43]. Furthermore, scatter electron microscopy (SEM) analyses revealed that no changes of the implant occurred during BDD electrode treatment while damages of the implant surface were observed after chemo-mechanical treatment with metal curettes and air abrasion led to particle deposition on the implant surface [44]. Subsequent analysis of chemical and physical parameters revealed only minor temperature and pH alterations upon BDD electrode application [43].

18.3.3. Dental Applications

Microbes cannot only colonize implants, but also form biofilms in the root canal of infected teeth [57]. Consequently, the first step of treatment of root canal infections focuses

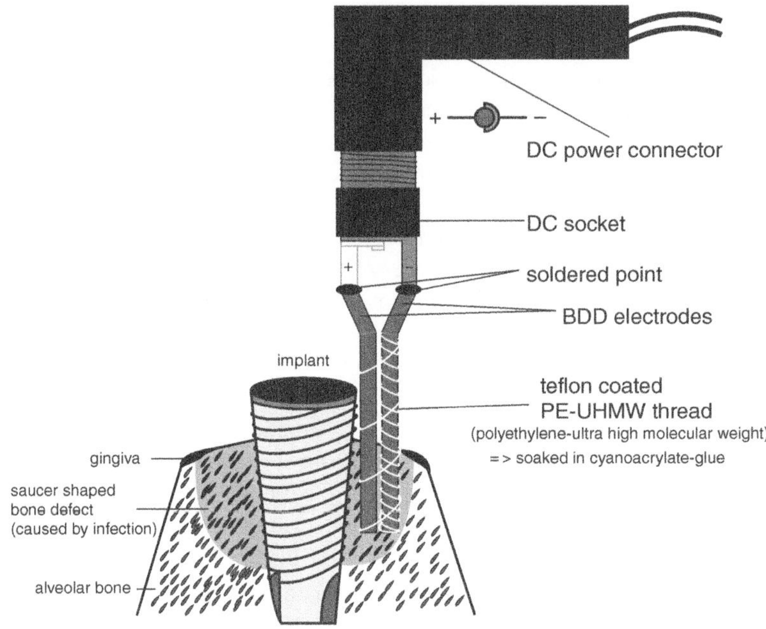

Figure 18.3 BDD electrode prototype for implant disinfection. The handheld instrument is moved around the contaminated implant. Peri-implantitis evoking microbes are inactivated by in-situ production of reactive oxygen species.

on the removal of bacteria from the canal and disinfection of the accessible space. The low accessibility of significant parts of the dental canal systems poses a major problem in this respect [58–60]. As shown recently, antimicrobial irrigants such as chlorhexidine, citric acid, or hydrogen peroxide solutions can only penetrate the root canal and the adjacent volume of dentin tubules [59]. Unfortunately, this constitutes only a minor fraction of free dental space [61]. In addition, it was shown that microorganisms forming mature, established biofilms within dentin canals cannot be inactivated easily by endodontic medicaments [62]. To address these problems, a considerable number of studies were carried out and published. For example, besides chemo-mechanical preparation of root canals [63–65], various irrigation protocols, including ultrasonic-assisted irrigation [66, 67], have been described. However, even advanced treatment approaches such as photodynamic therapy [68], laser-activated irrigation [69] or ozone application [70] alone or in combination were not sufficient to reach a complete elimination of microorganisms from root canals [71].

BDD electrodes may have the potential to solve the described problems of root canal disinfection. As shown recently [72], biofilms of *S. epidermidis* grown in the root canals of extracted human teeth were completely removed after 3.5 minutes treatment time, while *B. subtilis* biofilms were eliminated after 8.5 minutes using a simple wire-type electrode. With a more sophisticated, clinically applicable prototype electrode array, a complete disinfection was achieved after 10 minutes for *S. epidermidis* and 25 minutes for *B. subtilis*. Obviously, the tested BDD electrodes were highly suitable for disinfecting root canals and dentin tubules based on their continuous production of ROS and enhanced penetration of dentin tubules due the formation of a continuous stream of small gas bubbles. For future clinical applications, treatment times required will be even shorter compared to those *in vitro* experiments described here, since the number of microorganisms will be reduced by debridement and mechanical shaping of the canal system already before the BDD disinfection process starts.

18.4. CONCLUSIONS

Although only a limited number of studies is available, it is obvious that BDD electrodes have tremendous potential to provide tailor-made solutions for the inactivation of microbes in various settings. Described applications have in common that the in-situ generation of ROS is achieved by low voltage, securing a safe and energy-efficient process, which is sustainable and environmentally friendly.

ACKNOWLEDGEMENTS

The authors wish to thank all current and former members of their groups for excellent contributions. The financial support of our work by the ITI Foundation (Switzerland) in frame of grant 1328_2018 and the Dr. Rolf M. Schwiete Stiftung are gratefully acknowledged.

REFERENCES

1. Philp, J.C., Ritchie, R.J., and Allan, J.E.M. (2013). Biobased chemicals: the convergence of green chemistry with industrial biotechnology. *Trends in Biotechnology 31* (4): 219–221.
2. Rizzo, L., Manaia, C., Merlin, C. et al. (2013). Urban wastewater treatment plants as hotspot for antibiotic resistant bacteria and genes spread into the environment: a review. *Science of the Total Environment 447*: 345–360.
3. Kashmiri, Z.N. and Mankar, S.A. (2014). Free radicals and oxidative stress in bacteria. *International Journal of Current Microbiology and Applied Sciences 3* (9): 34–40.
4. Phaniendra, A., Jestadi, D.B., and Periyasamy, L. (2015). Free radicals: properties, sources, targets, and their implication in various diseases. *Indian Journal of Clinical Biochemistry 30* (1): 11–26.
5. Dryden, M. (2018). Reactive oxygen species: a novel antimicrobial. *International Journal of Antimicrobial Agents 51* (3): 299–303.
6. Collin, F. (2019). Chemical basis of reactive oxygen species reactivity and involvement in neurodegenerative diseases. *International Journal of Molecular Sciences 20* (10): 2407.
7. Bedard, K., Lardy, B., and Krause, K.-H. (2007). NOX family NADPH oxidases: not just in mammals. *Biochimie 89* (9): 1107–1112.
8. Kawahara, T., Quinn, M.T., and Lambeth, J.D. (2007). Molecular evolution of the reactive oxygen-generating NADPH oxidase (Nox/Duox) family of enzymes. *BMC Evolutionary Biology 7*: 109.
9. West, A.P., Brodsky, I.E., Rahner, C. et al. (2011). TLR signalling augments macrophage bactericidal activity through mitochondrial ROS. *Nature 472* (7344): 476–480.
10. Imlay, J.A. (2019). Where in the world do bacteria experience oxidative stress? *Environmental Microbiology 21* (2): 521–530.
11. Reddy, M.P., Venugopal, A., and Subrahmanyam, M. (2007). Hydroxyapatite-supported ag-TiO_2 as *Escherichia coli* disinfection catalyst. *Water Research 41* (2): 379–386.
12. Misba, L., Zaidi, S., and Khan, A.U. (2017). A comparison of antibacterial and antibiofilm efficacy of phenothiazinium dyes between gram positive and gram negative bacterial biofilm. *Photodiagnosis and Photodynamic Therapy 18*: 24–33.
13. Kang, S.-M., Jung, H.-I., and Kim, B.-I. (2019). Susceptibility of oral bacteria to antibacterial photodynamic therapy. *Journal of Oral Microbiology 11* (1): 1644111.
14. Krüger, M., Richter, P., Strauch, S.M. et al. (2019). What an *Escherichia coli* mutant can teach us about the antibacterial effect of chlorophyllin. *Microorganisms 7*: 59.
15. Richter, P., Krüger, M., Prasad, B. et al. (2019). Using colistin as a Trojan horse: inactivation of gram-negative bacteria with chlorophyllin. *Antibiotics 8* (4): 158.

16. Weller, R.B. (2009). Nitric oxide-containing nanoparticles as an antimicrobial agent and enhancer of wound healing. *Journal of Investigative Dermatology 129* (10): 2335–2337.
17. Dastjerdi, R. and Montazer, M. (2010). A review on the application of inorganic nano-structured materials in the modification of textiles: focus on anti-microbial properties. *Colloids and Surfaces. B, Biointerfaces 79* (1): 5–18.
18. Shaikh, S., Nazam, N., Rizvi, S.M.D. et al. (2019). Mechanistic insights into the antimicrobial actions of metallic nanoparticles and their implications for multidrug resistance. *International Journal of Molecular Sciences 20* (10): 2468.
19. Rajendiram, K., Zhao, Z., Pei, D.-S., and Fu, A. (2019). Antimicrobial activity and mechanism of functionalized quantum dots. *Polymers 11* (10): 1670.
20. Vatansever, F., de Melo, W.C., Avci, P. et al. (2013). Antimicrobial strategies centered around reactive oxygen species-bactericidal antibiotics, photodynamic therapy, and beyond. *FEMS Microbiology Reviews 37* (6): 955–989.
21. Kraft, A. (2007). Doped diamond: a compact review on a new, versatile electrode material. *International Journal of Electrochemical Science 2*: 355–385.
22. Tröster, I., Fryda, M., Herrmann, D. et al. (2002). Electrochemical advanced oxidation process for water treatment using DiaChem® electrodes. *Diamond and Related Materials 11* (3–6): 640–645.
23. Schorr, B., Ghanem, H., Rosiwal, S. et al. (2019). Elimination of bacterial contaminations by treatment of water with diamond electrodes. *World Journal of Microbiology and Biotechnology 35* (3): 48.
24. Marselli, B., Garcia-Gomez, J., Michaud, P.-A. et al. (2003). Electrogeneration of hydroxyl radicals on boron-doped diamond electrodes. *Journal of the Electrochemical Society 150* (3): D79–D83.
25. Michaud, P.-A., Panizza, M., Ouattara, L. et al. (2003). Electrochemical oxidation of water on synthetic boron-doped diamond thin film anodes. *Journal of Applied Electrochemistry 33* (2): 151–154.
26. Furuta, T., Rychen, P., Tanaka, H. et al. (2005). Application of diamond electrodes for water disinfection. In: *Diamond Electrochemistry* (ed. A. Fujishima, Y. Einaga, T.N. Rao and D.A. Tryk), 525–542. Burlington: Elsevier Science.
27. Ochiai, T., Ishii, Y., Tago, S. et al. (2013). Application of boron-doped diamond microelectrodes for dental treatment with pinpoint ozone-water production. *ChemPhysChem 14* (10): 2094.
28. WWAP (UNESCO World Water Assessment Programme) (2019). *The United Nationsworld Water Development Report 2019: Leaving no One Behind.* Paris: UNESCO.
29. Barkhudarov, E.M., Christof, N., Kossyi, I.A. et al. (2008). Killing bacteria present on surfaces infilms or droplets using microwave UV lamps. *World Journal of Microbiology and Biotechnology 24* (6): 761–769.
30. O'Flaherty, E., Membré, J.-M., and Cummins, E. (2018). Meta-analysis of the reduction of sensitive and antibiotic resistant *Escherichia coli* as a result of low and medium pressure UV lamps. *Water Science and Technology 2017* (2): 612–620.
31. Schwermer, C.U., Krzeminski, P., Wennberg, A.I. et al. (2018). Removal of antibiotic resistant *E. coli* in two Norwegian wastewater treatment plants and by nano- and ultrafiltration processes. *Water Science and Technology 77* (3–4): 1115–1126.
32. Cho, M., Kim, J., Kim, Y.J. et al. (2010). Mechanisms of *Escherichia coli* inactivation by several disinfectants. *Water Research 44* (11): 3410–3418.
33. Martirosyan, V., Moosavi, E., and Ayrapetyan, S. (2011). The study of the effects of carbon dioxide-induced elevation of hydrogen peroxide toxicity on microbes as a novel tool for water purification. *World Journal of Microbiology and Biotechnology 27*: 1091–1098.
34. Huang, K., Mao, Y., Zhao, F. et al. (2018). Free-living bacteria and potential pathogens in sewage treatment plants. *Applied Microbiology and Biotechnology 102* (5): 2455–2464.
35. Michael, I., Rizzo, L., McArdell, C.S. et al. (2013). Urban wastewater treatment plants as hotspots for the release of antibiotics in the environment: a review. *Water Research 47* (3): 957–995.

36. Rajab, M., Heim, C., Letzel, T. et al. (2015). Electrochemical disinfection using boron-doped diamond electrode --the synergetic effects of in situ ozone and free chlorine generation. *Chemosphere 121*: 47–53.
37. Hubler, D.K., Baygents, J.C., Chaplin, B.P. et al. (2014). Understanding chlorite and chlorate formation associated with hypochlorite generation at boron doped diamond film anodes. *Journal of the Electrochemical Society 161* (12): 182–189.
38. Thostenson, J.O., Mourouvin, R., Hawkins, B.T. et al. (2018). Improved blackwater disinfection using potentiodynamic methods with oxidized boron-doped diamond electrodes. *Water Research 140*: 191–199.
39. Tsolaki, E. and Diamadopoulos, E. (2010). Technologies for ballast water treatment: a review. *Journal of Chemical Technology and Biotechnology 85* (2): 19–32.
40. Moreno-Andrés, J., Ambauen, N., Vadstein, O. et al. (2018). Inactivation of marine heterotrophic bacteria in ballast water by an electrochemical advanced oxidation process. *Water Research 140*: 377–386.
41. Watnick, P. and Kolter, R. (2000). Biofilm, city of microbes. *Journal of Bacteriology 182*: 2675–2679.
42. Simcox, L.J., Pereira, R.P.A., Wellington, E.M.H. et al. (2019). Boron doped diamond as a low biofouling material in aquatic environments: assessment of *Pseudomonas aeruginosa* biofilm formation. *ACS Applied Materials & Interfaces 11* (28): 25024–25033.
43. Koch, M., Göltz, M., Xiangjun, M. et al. (2020). Electrochemical disinfection of dental implants experimentally contaminated with microorganisms as a model for periimplantitis. *Journal of Clinical Medicine 9* (2): 475.
44. Göltz, M., Koch, M., Detsch, R. et al. (2019). Influence of in-situ electrochemical oxidation on implant surface and colonizing microorganisms evaluated by scatter electron microscopy. *Materials 12* (23): 3977.
45. Pesce, P., Canullo, L., Grusovin, M.G. et al. (2015). Systematic review of some prosthetic risk factors for periimplantitis. *The Journal of Prosthetic Dentistry 114* (3): 346–350.
46. Al-Ahmad, A., Muzafferiy, F., Anderson, A.C. et al. (2018). Shift of microbial composition of peri-implantitis-associated oral biofilm as revealed by 16S rRNA gene cloning. *Journal of Medical Microbiology 67* (3): 332–340.
47. Gershenfeld, L., Kalos, A., Whittle, T. et al. (2018). Randomised clinical trial of the effects of azithromycin use in the treatment of peri-implantitis. *Australian Dental Journal 63* (3): 374–381.
48. Patianna, G., Valente, N.A., D'Addona, A. et al. (2018). in vitro evaluation of controlled-release 14% doxycycline gel for decontamination of machined and sandblasted acid-etched implants. *Journal of Periodontology 89* (3): 325–330.
49. Rupf, S., Idlibi, A.N., Marrawi, F.A. et al. (2011). Removing biofilms from microstructured titanium ex vivo: a novel approach using atmospheric plasma technology. *PLoS One 6* (10): e25893.
50. Idlibi, A.N., Al-Marrawi, F., Hannig, M. et al. (2013). Destruction of oral biofilms formed in situ on machined titanium (Ti) surfaces by cold atmospheric plasma. *Biofouling 29* (4): 369–379.
51. Takagi, T., Aoki, A., Ichinose, S. et al. (2018). Effective removal of calcified deposits on microstructured titanium fixture surfaces of dental implants with erbium lasers. *Journal of Periodontology 89* (6): 680–690.
52. Bombeccari, G.P., Guzzi, G., Gualini, F. et al. (2013). Photodynamic therapy to treat periimplantitis. *Implant Dentistry 22* (16): 631–638.
53. Fraga, R.S., Antunes, L.A.A., Fontes, K.B.F.D.C. et al. (2018). Is antimicrobial photodynamic therapy effective for microbial load reduction in peri-implantitis treatment? A systematic review and meta-analysis. *Photochemistry and Photobiology 94* (4): 752–759.
54. Schneider, S., Rudolph, M., Bause, V. et al. (2018). Electrochemical removal of biofilms from titanium dental implant surfaces. *Bioelectrochemicals and Bioenergy 121*: 84–94.
55. Tal-Gutelmacher, E. and Eliezer, D. (2004). Hydrogen-assisted degradation of titanium based alloys. *Materials Transactions 45* (5): 1594–1600.
56. Tal-Gutelmacher, E. and Eliezer, D. (2005). The hydrogen embrittlement of titanium-based alloys. *JOM 57* (9): 46–49.

57. Mohammadi, Z., Palazzi, F., Giardino, L. et al. (2013). Microbial biofilms in endodontic infections: an update review. *Biomedical Journal 36* (2): 59–70.
58. Hope, C.K., Garton, S.G., Wang, Q. et al. (2010). A direct comparison between extracted tooth and filter-membrane biofilm models of endodontic irrigation using *Enterococcus faecalis*. *Archives of Microbiology 192* (9): 775–781.
59. Gänsbauer, M., Burkovski, A., Karl, M. et al. (2017). Comparison of simplistic biofilm models for evaluating irrigating solutions. *Quintessence International 48* (7): 521–526.
60. Pacheco-Yanes, J., Provenzano, J.C., Marceliano-Alves, M.F. et al. (2020). Distribution of sodium hypochlorite throughout the mesial root canal system of mandibular molars after adjunctive irrigant activation procedures: a micro-computed tomographic study. *Clinical Oral Investigations 24* (2): 907–914.
61. Burkovski, A. and Karl, M. (2019). Lack of evidence for the necessity of root canal obturation. *Quintessence International 50* (1): 22–28.
62. Wang, Z., Shen, Y., and Haapasalo, M. (2012). Effectiveness of endodontic disinfecting solutions against young and old *Enterococcus faecalis* biofilms in dentin canals. *Journal of Endodontia 38* (10): 1376–1379.
63. Lin, J., Shen, Y., and Haapasalo, M. (2013). A comparative study of biofilm removal with hand, rotary nickel-titanium, and self-adjusting file instrumentation using a novel in vitro biofilm model. *Journal of Endodontia 39* (5): 658–663.
64. Martinho, F.C., Gomes, A.P., Fernandes, A.M. et al. (2014). Clinical comparison of the effectiveness of single-file reciprocating systems and rotary systems for removal of endotoxins and cultivable bacteria from primarily infected root canals. *Journal of Endodontia 40* (5): 625–629.
65. Sasanakul, P., Ampornaramveth, R.S., and Chivatxaranukul, P. (2019). Influence of adjuncts to irrigation in the disinfection of large root canals. *Journal of Endodontia 45* (3): 332–337.
66. Beus, C., Safavi, K., Stratton, J. et al. (2012). Comparison of the effect of two endodontic irrigation protocols on the elimination of bacteria from root canal system: a prospective, randomized clinical trial. *Journal of Endodontia 38* (11): 1479–1483.
67. Layton, G., Wu, W.I., Selvaganapathy, P.R. et al. (2015). Fluid dynamics and biofilm removal generated by syringe-delivered and 2 ultrasonic-assisted irrigation methods: a novel experimental approach. *Journal of Endodontia 41* (6): 884–889.
68. Xhevdet, A., Stubljar, D., Kriznar, I. et al. (2014). The disinfecting efficacy of root canals with laser photodynamic therapy. *The Journal of Lasers in Medical Sciences 5* (1): 19–26.
69. Gokturk, H., Ozkocak, I., Buyukgebiz, F. et al. (2017). Effectiveness of various irrigation protocols for the removal of calcium hydroxide from artificial standardized grooves. *Journal of Applied Oral Science 25* (3): 290–298.
70. Huth, K.C., Quirling, M., Maier, S. et al. (2009). Effectiveness of ozone against endodontopathogenic microorganisms in a root canal biofilm model. *International Endodontic Journal 42* (1): 3–13.
71. Azim, A.A., Aksel, H., Zhuang, T. et al. (2016). Efficacy of 4 irrigation protocols in killing bacteria colonized in dentinal tubules examined by a novel confocal laser scanning microscope analysis. *Journal of Endodontia 42* (6): 928–934.
72. Böhm, A.-L., Koch, M., Rosiwal, S. et al. (2019). Electrochemical disinfection of experimentally infected teeth by boron-doped diamond electrode treatment. *Journal of Clinical Medicine 8* (12): 2037.

19

Enzymatic Intervention as an Ecofriendly Approach in Industries: An Update

Supriya Gupta, Shruti Dwivedi, Kanchan Yadav, Aiman Tanveer, Sangeeta Yadav, and Dinesh Yadav

19.1. INTRODUCTION

Enzymes are biocatalysts produced by living organisms that assist in biochemical reactions. They are vital components of animals, plants, and microbes since they biocatalyze and coordinate the multifaceted reactions of cellular metabolism. Some of the enzymes play important roles in several industrial applications and are largely used in commercial processing. Since the initial usage of microbial enzymes, efforts have been made to improve the catalytic properties and scaling up of the industrially important biocatalyst. Selected microorganisms, including bacteria, fungi, and yeasts, have been explored all over the globe for the biosynthesis of cost-effective enzyme preparations.

Enzymes are advantageous over typical chemical catalysts both in terms of sustainability and process efficiency. Industrial enzymes are nontoxic, easy to handle, and reduce water and fuel consumption, and fewer byproducts are formed compared to chemical catalysts [1, 2]. Microbial enzymes are the most commonly used enzymes in industries as their catalytic efficiency, stability, and production efficiency are much higher than plant and animal-based enzymes.

Various enzymes have been developed or purposefully engineered according to the requirements of a process in recent biotechnology breakthroughs. Various well-known enzymes have been characterized for their functions in several bioprocesses. Protein engineering, biochemical reaction engineering, and metagenomics have led to the discovery of a huge number of novel enzymes. Molecular approaches have also been used to improve the catalytic properties of microbial enzymes so that they can be used in a large number of industries. Currently, almost 75% of industrial enzymes are obtained from a microorganism. They are an important component in the animal feed and food industry, the detergent industry, textile-based ventures, the pharmaceutical industry, paper and pulp processing units, leather processing factories, and agriculture-based ventures, among others. These enzymes are gradually taking over traditional processing owing to their efficiency, sustainability, and better production quality [3]. In this chapter, we have reviewed some of these

Department of Biotechnology, Deen Dayal Upadhyaya Gorakhpur University, Gorakhpur, Uttar Pradesh, India

Microbial Biotechnology: Role in Ecological Sustainability and Research, First Edition.
Edited by Pankaj Chowdhary, Sujata Mani, and Preeti Chaturvedi.

industrially important enzymes. We then discuss the methods that are used for enzyme improvement such as directed evolution, mutation, nanotechnology, as well as a metagenomics approach for discovery of the enymes with novel catalytic properties from environmental samples. We end the chapters by discussing the role of industrially important enzymes in different industrial applications.

19.2. INDUSTRIALLY IMPORTANT MICROBIAL ENZYMES

Enzymes are cornerstones for catalyzing the biofunctional activities of living organisms. The microorganisms are an important source of several enzymes that are used in diverse industries. The search for potential microbial sources of different enzymes is one major area of research that has been substantially improved with the advent of the metagenomics approach. These microbial enzymes are extensively used in several industries concerned with the manufacture of fine chemicals, including pharmaceuticals, agrochemicals, etc. [3–5]. The soil microbiome serves as a good source of enzymes: one gram of soil may contain billions of microbes comprising bacteria, fungi, protists, and some microorganisms. One microbe may be a source of thousands of types of enzymes that can be used to improve human welfare. The microorganisms serve as important sources of industrially relevant enzymes as they facilitate the large-scale production of enzymes by fermentation. Their biochemical diversity and susceptibility to gene manipulation make them important sources of industrially important enzymes [6]. In the global industrial market, a plethora of enzymes exists, classified according to the type of enzyme, its applications, and origin source.

Microbial enzymes showing stability over a wide range of temperatures and pH conditions are preferred for industrial applications. The existing enzymatic activity of the microbes can be enhanced by modulating the fermentation conditions; further, the fermentation process can be made more cost-effective by using low-cost agricultural residues as substrates. If multiple enzymes are needed for processing, they all can be produced together in a single fermentation reaction leading to cost reduction along with the retention of enzyme stability [1, 7]. Some of the advantages of microbial enzymes are listed here:

Advantages of microbial enzymes in different industries:

- Currently, microbial enzymes are more extensively used than plant and animal-based enzymes as the production is more eco-friendly, economical, and devoid of ethical issues that need to be addressed with animal-based enzymes.
- The ease of enhancement of both the quantity and quality of enzymes is quite feasible for microbial sources.
- The ease of extraction and purification of enzymes from microbial sources owing to extracellular secretion imparts several biotechnological applications.
- Enzymes from plant and animal sources vary in their yield and production at different times of the year, while none of these complications are associated with microbial enzymes.
- Enzymes isolated from microbial sources are comparatively more stable than animal and plant-based enzymes. For instance, enzymes from thermophilic microorganisms are often useful when the processing occurs at high temperatures, e.g. during starch saccharification, a highly thermostable α-amylase from *Bacillus amyloliquefaciencs* is used [7].
- Microbes can be easily subjected to genetic alteration to produce novel or transformed enzymes whereas the genetic manipulation of animals and plants is much more difficult and expensive.

There exist several enzymes of industrial significance that have been reported from diverse groups of microorganisms. Some of these microbial enzymes are listed in Table 19.1. The microbial enzymes find applications in diverse sectors like food, textiles, leather, pharmaceuticals, cosmetics, energy biomaterials, paper, fine chemicals, and detergents. Some of the important industrially important enzymes include amylases, xylanases, pectinase, and cellulases. Some of these enzymes are discussed in the following subsections with respect to their application in different industries.

19.2.1. Amylases

Amylase acts on polysaccharide (starch) and breaks it into sugars, and it generally exists in three different types, namely α, *β*, and *γ*-amylases [3, 7, 16, 17]. α-amylases hydrolyze α-1, 4 linkages between adjacent glucose subunits in polysaccharides and release small chains of oligosaccharides and dextrins. They are used in starch removing detergents especially dishwashing, ethanol production, and fruit juice clarification [17]. *β*-amylases hydrolyze second α-1, 4 glycosidic bonds of polysaccharide starch from the non-reducing end into maltoses (β-glucose). These are used in brewing and bakery-based ventures. *γ*-amylases cleave (1-6) glycosidic linkages and last (1-4) glycosidic linkages at the nonreducing end of amylose and amylopectin, and release glucose. Amylases are extensively used in food-based sectors, drug delivery, chemical and pharmaceutical industries, and in environmental and agricultural engineering [16].

Amylases have been reported from animals, plants, and microbes. Microbial sources are predominant and several bacteria like *Bacillus licheniformis*, *Bacillus stearothermophilus*, *B. amyloliquefaciens*, and *Bacillus subtilis* are the major source of thermostable amylases. Similarly, *Chromohalobacter* sp., *Halobacillus* sp., *H. hispanica, H. meridian, and Bacillus dipsosauri* are important sources of halophilicamlyases. Amylases have proved to be important in an array of applications in several sectors like paper and pulp, bread making, textiles, sweeteners, detergents, fructose and glucose syrups, alcoholic beverages, etc. [3].

19.2.2. Cellulases

Cellulases are a pool of three major classes of enzymes, namely (i) endo-(1,4)-β-D-glucanase (EC 3.2.1.4), (ii) exo-(1,4)-β-D-glucanase (EC 3.2.1.91), and (iii) β-glucosidases (EC 3.3.1.21). These depolymerize cellulose completely and liberate glucose [16]. Cellulases have a complex crystalline structure with different amino acid series that is typically connected with the hemicellulose and lignin network; hence, they find application in sectors where thermostable and halophilic cellulases are required [17]. Endoglucanases cleave β-1,4 bonds between cellobiose and glucose of cellulose in a random manner. These enzymes have been reported from plants, animals, and microbes. Some of the cellulase-producing microbes are *Ignisphaera, Thermoproteus, Metallosphaera,* and *Pyrococcus* such as *Metallosphaera cuprina*, *Ignisphaera aggregans*, *Acidilobus saccharovorans*, *Sulfolobus solfataricus*. β-glucosidases (Cellobiases; EC 3.2.1.21) break soluble cellodextrins and cellobiose and liberate β-D glucose. Important sources of these enzymes include bacteria, fungi, plants, animals, and archaea belonging to the genera *Pyrococcus, Sulfolobus,* and *Thermofilum* [17]. *Trichoderma* sp., *Aspergillus niger, Penicillium* sp., *Humicola grisea*, *Aspergillus* sp., *Chrysosporium lucknowense*, and *Acremonium* sp. are important fungal sources [3, 11, 24]. Endoglucan reduces pulp viscosity and fiber coarseness, mends quality, assists in fermentation, improves yields of wines and beers, enhances softness and water holding capacity of fiber, and provides cleaner and soft surface structure. Cellulases also help in improving the

Table 19.1 A list of some important microbial enzymes and their industrial uses.

S. No.	Enzyme	Microbial sources	Application	References
1.	Amylase	*Aspergillus niger, A. oryzae, Rhizopus* (Fungi) *Bacillus* sp., *Pseudomonas stutzeri, Thermonospora* sp. (Bacteria)	Food and beverage, paper, textiles, and detergent industries. Production of fuel ethanol from starches.	[1, 8].
2.	Proteases	*Pyrococcus, Thermococcus, Pyrobaculum* (Archaea) *Bacillus licheniformis* (Bacteria) *(Aspergillus niger, Aspergillus oryzae, A. niger, Pseudomonas* sp., *Penicillium chrysosporium, Rhizopusoligosporus* (Fungi), *Actinomycetes strains*	Laundry, detergents, textiles, peptide synthesis, in organic media, food, and feeding industries.	[9, 10].
3.	Cellulases	*Trichoderma* sp., Aspergillus sp., Penicillium sp., Humicola grisea, Chrysosporium lucknowense, Acremonium sp. (Fungi)	Paper and pulp, detergent, textiles, fruit juice, and biofuel industries.	[3, 11, 12].
4.	Laccase	*Fungus, Bacteria (Azospirillium lipoferum)*	Pulp and paper and textiles industries, food-based sectors, cosmetics ventures, and soil bioremediation.	[13–16].
5.	Xylanases	*Myeciliophthora thermophila, Bacillus* sp., *A. oryzae, Trichoderma* sp.	Paper and pulp and brewing industries, fruit juice clarification, and pentose production.	[5, 17].
6.	Pectinase	*Aspergillus* sp., *Rhizopus oryzae, Fusarium* sp., *Penicillium oxalicum*	Fruit juice clarification, coffee and tea leaf fermentation, retting of natural fibers, bioremediation, oil extraction.	[18, 19].
7.	L-Asprginases	*Bacteria*	Pharmaceutical and nutraceutical sector	[7, 20].
8.	Lipase	*Bacillus* sp., *Pseudomonas* sp., *Burkholderia* sp. (Bacteria) *Candida rugosa, Candida antarctica, Galactomyces geotricum, S. cerevisiae, Yarrowia lipolytica, Trichosporon fermantans, Cryptococcus albidus, A. flavus, Thermomyces lanuginosus* and Rhizopus oryzael (Fungi)	Detergent, pulp and paper, food, and leather industries, starches, and fuels.	[21, 22]

9.	Esterase	*Bacillus* sp., *Pseudomonas* sp., *Burkholderia* sp. (Bacteria)	Detergents, food, pulp and paper, organic synthesis, leather.	[21]
		C. rugosa, C. antarctica, Galactomyces geotricum, S. cerevisiae, Yarrowia lipolytica, Trichosporon fermantans, C. albidus, A. s flavus, Thermomyces lanuginosus and R. oryzae (Fungi)		
10.	Phytase	*Aspergillus* sp., *Aspergillus ficuum, P. funiculosum, Bacillus* sp., *Pseudomonas, Xanthomonas oryzae*	Animal feedstock	[23]
11.	Catalase	*Aspergillus* sp.	Rubber	[24]
12.	Peroxidases	*Aspergillus* sp. *Phanerochaete chrysosporium*	Biopolymer/plastic, textile industry	[7, 25]
13.	Lipases	*Aspergillus oryzae, Aspergillus terreus, Pseudomonas* sp., *Alcaligenes* sp., *Staphylococcus* sp., *Candida albicans, Rhizopus* sp., *Mucor*	Biopolymer/plastic, oil and fat, baking, pharmaceuticals, biofuel, paper and pulp, and detergent sectors.	[5, 26]
14.	Lactases	*Yeast (Kluyveromyces lactis). fungus and bacteria*	Dairy industry	[7]
15.	Chitinases	*Haloferax, Halobacterium, Pyrococcus, Sulfolobus Thermococcus*	Drug and pharmaceutical industry, anti-fungal	[17]

bleachability of softwood kraft pulp. Along with xylanases, they help release the ink from the fiber's surface. In the food industry, they improve juice yield and crop nutritional quality. Additionally, they assist in the biostoning process of jeans production and biopolishing of cotton, and they improve the nutritional quality of animal feed. β-glucosidases enhance the texture, aroma, and flavor of vegetables and fruits.

19.2.3. Xylanase

Xylanases (EC 3.2.1.8) cleave the 1,4-glycosidic bond of xylan, which is a part of plant hemicellulose. It includes a group of enzyme-like endoxylanases, exoxylanases, and β-xylosidases. Endoxylanases (EC 3.2.1.8) break down the β-1,4 bonds of the xylan backbone while exoxylanases (EC 3.2.1.37) cleaves β-1,4 bonds of xylan from the non-reducing ends into xylooligosaccharides. β-xylosidaseshydrolyse the xylobiose and xylooligosaccharides to release xylose. Xylanases have been reported from *H. utahensis, Sulfolobussol fataricus, Streptomyces* sp., *Bacillus* sp., *Pseudomonas* sp., *Aspergillus* sp., *Fusarium* sp., and *Penicillium* sp.

Xylanases find applications in diverse industrial sectors such as food-based industries, biomedical applications, animal feed production, and bioethanol manufacturing. It is an effective biobleaching enzyme that has the potential to replace the chemical-based chlorine bleaching of the cellulose pulp. It also improves the yield of essential oils, pigments in vegetables and fruits, and helps in the clarification of fruit juices. Xylanases and proteases are also used as nutritional supplements [7, 16, 17].

19.2.4. Laccases

Laccases (EC 1.10.3.2) represent multicopper oxidases that contain multiple copper at the active site of enzymes that perform redox reactions. They can degrade lignin by oxidizing one electron of the phenolic compound with an associated reduction of oxygen to water as a byproduct. Laccase oxidizes both phenolic aromatic compounds and non-phenolic aromatic compounds in lignin to form phenoxy-free radicals, which find important application in paper and pulp sectors. Fungal laccases consist of two disulfide bonds and four copper atoms distributed in three copper centers, namely T1, T2, and T3:

Type 1 – laccases contain a single copper atom that is associated with 2 his-residues and 1 cys-residues, except in some cases, where it is a trinuclear center with ligand and a methionine motif. In this type, Cu does not bind with oxygen molecules but acts only as an electron transfer site.

Type 2 – Cu center contains 2 his-residues and H_2O molecules that work as ligands.

Type 3 – possess two Cu-centers each containing 3 his-ligands and are linked to one another with hydroxide bridging ligands.

Laccases are reported mostly from fungus like rot fungus, soft rot fungus, and brown rot fungus. It is an industrially important enzyme owing to its ability to degrade both non-phenolic as well as phenolic lignin along with refractory pollutants. Laccases are mostly used in biopulping, biobleaching, and decolorization of dyes and their synthesis in the paper and pulp sector, textile sector, and baking in food sectors, as well as degradation of xenobiotics and effluent treatments of industrial wastes and pollutants [13–16].

19.2.5. Proteases

Proteases (EC 3.4.21.62) cause protein-lysis by hydrolyzing and cleaving the peptide bonds of amino acids that link together in polypeptide branches. Protease leads total enzyme sales and comprises almost 60% of the market share. It is primarily used as a detergent as it is highly stable and biologically active at alkaline pH. Serine proteases, especially Neutrase®, subtilisin A, and trypsin are commercially important enzymes. Genetically modified *Bacillus* and *Aspergillus* strains are important sources of proteases that are industrially used [16, 27].

19.2.6. Pectinases

Pectin forms the middle lamella and primary cell wall of higher plants and chemically heteropolysaccharides with a high molecular weight largely composed of (1-4α) linked d-galacturonic acid residues. The pectinase group of enzymes is characterized by its tendency to breakdown pectic substances and includes important enzymes like pectin lyases, polygalacturonases, pectatelyases, and pectin methyl esterases. Microbial pectinases occupy around 25% of the total global food and industrial enzymes market with continually increasing demand [19].

Pectinase is one of the upcoming enzymes in the current biotechnology development period, with a steadily growing market. They increased at a 2.86% annual pace from \$27.6 million in 2013 to \$30.0 million in 2016, and the market for pectinase should have reached \$35.5 million by 2021. Because pectin-degrading enzymes play an important role in phytopathogenesis, their synthesis has been extensively documented and researched in bacteria and filamentous fungus. According to several reports, pectinase enzymes are produced by microbes such as bacteria, fungus, yeast, and actinomycetes. They are also reported to be present in higher plants and in some protozoa, nematodes, and insects, but they are not found in higher animals [18].

Different species of *Aspergillus*, *Erwinia*, *Bacillus*, and *Penicillium* serve as important sources for commercial pectinase production. These enzymes play an important role in the food industry. Fruit juice extraction and wine clarity are two applications for these enzymes, along with vegetable oil extraction, jam, jellies, and pickle preparation. Pectinases are extensively used in the paper and pulp and textile industries, wastewater management, fruit juice-based ventures for clarification of fruit juices, poultry feed additives, and bioenergy generation [28]. The use of novel enzymes with acceptable biochemical and physicochemical properties as well as low-cost synthesis in industrial processes has traditionally been considered for critical research [19].

19.3. INNOVATIONS IN MICROBIAL ENZYMES

Several technological innovations for the improvement of microbial enzymes enhancing the production, specificity, catalytic activity, and diverse applicability have been reported (Figure 19.1). Some of the innovations in microbial enzyme technology are discussed in the following subsections.

19.3.1. Mutations

The chance mutation leads to evolution that occurs continuously, gradually, and uniformly in the genome regardless of environmental factors. Mechanisms of mutation deciphered in bacteria revealed a tightly controlled process that is upregulated in response to stress in the environment. Many simultaneous mutations fall in discrete clusters in genomic space, which leads to concerted evolution. Several modifications are required to adjust protein functions and protein machinery. The molecular mechanisms of stress-inducible mutations challenge evolutionary theories and provide new approaches to modeling and addressing cancer growth, infectious disease, and evolution in general [29, 30]. Single-base-pair

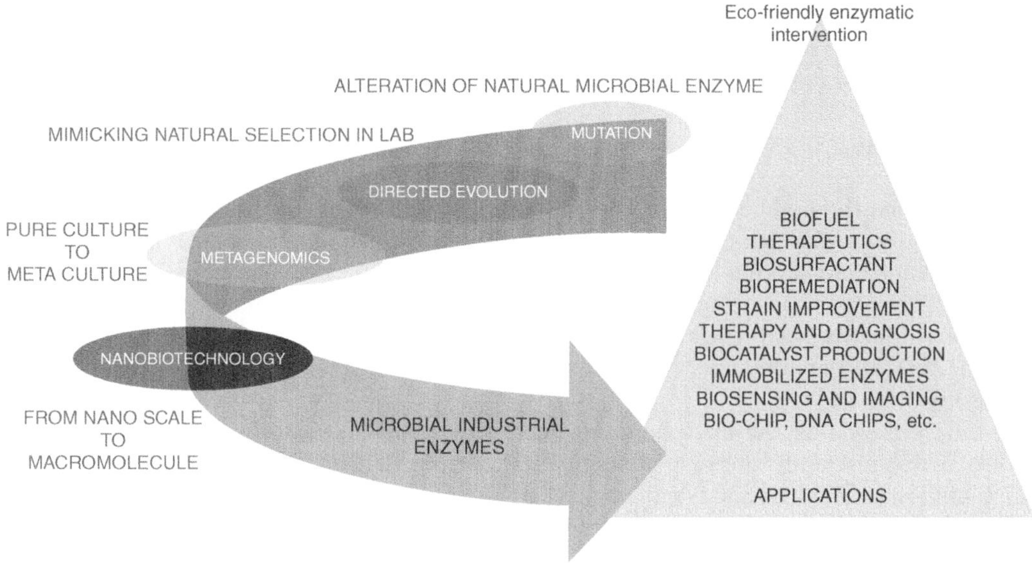

Figure 19.1 Some of the technological innovations of microbial enzymes.

changes to mega base-pair deletions, insertions, duplications, and inversions are all examples of mutations. It plays a key role in a many-fold increase in microbial metabolite production, which in turn is an important factor in the fermentation process [31]. Both physical and chemical methods of mutagenesis are being used for manipulations of microbial enzymes.

Ionizing radiation like gamma rays, X-rays, and alpha particles, as well as non-ionizing radiation like UV rays, are the most prevalent physical mutagens that have both direct and indirect methods of action. In the direct methods, the electrons are ejected or excited to a higher energy level, causing DNA strands to ionize [32]. Shifting of electrons generates free radicals (OH and H), which ionize to form OH^- and H^+ ions that cause modification in the bases and/or breakage in a single or both of the strands of DNA. Double strand breaks (DSBs) can occur as a result of these free radicals causing deletions and translocations in the DNA strand. Single-strand breaks (SSBs) can cause point mutations in specific situations, such as when pyrimidine bases are converted to 5-(hydroxymethyl) uracil, 5-formyluracil, 5-hydroxycytosine, and 5-hydroxyuracil.

Chemical mutagenesis is accomplished by several chemicals acting as base analogs, deaminating agents, alkylating agents, and intercalating agents. Transition mutations can occur when a base analog replaces a DNA base during replication. Intercalating agents, on the other hand, are compounds that can be introduced between DNA bases during replication, causing frame shift mutations. Other chemical mutagens may cause reactive oxygen species (ROS), deamination, or alkylation, among other things [33].Chemical mutagens cause random mutation. Mutagenesis has been employed in a variety of microorganisms for a long time to improve the performance and productivity of single or multiple gene characteristics. Penicillin antibiotic production from *Penicillium chrysogenum* has been enhanced more than threefold after many years of extensive research using many different mutagenic approaches [33, 34].

The role of mutation in the productivity of industrial products is well documented [35, 36]. It can be used to change the number of different metabolites made in a fermentation broth, assist in revealing secondary metabolism pathways, and synthesize novel chemicals. Some important mutagens used widely are nitrosoguanidine (NTG), 4-nitroquinolone-1-oxide, methylmethanesulfonate (MMS), ethylmethanesulfonate (EMS), hydroxylamine (HA), and ultraviolet light (UV). Individual cells or spores are preferred for mutagenesis, but non-spore producing filamentous organisms have also been successfully mutated by producing protoplasts and then regenerating on a solid medium [37]. In the case of *Streptomyces* mycelia, sonication is intermittently employed for better incorporation of the mutation. After fragmentation or protoplast production, filamentous organisms that do not sporulate well can be mutagenized [38].

Mutation followed by a screening of microbial strains has largely contributed to the enormous advances in fermentation productivity and cost savings. *P. chrysogenum* X-1612, the first superior penicillin-producing mutant, was identified through X-ray mutagenesis more than 60 years ago. This has given way to the intervention of mutational genetics in industrial microbiology [39]. Alterations in the catalytic properties of the enzyme have been discovered after subjecting the microbes to mutagenesis and random screening of the mutants. Enzyme native features could be altered per industrial requirements [31].

19.3.2. Directed Evolution Approach

Natural evolution, in which a huge number of variations are produced via mutation and then the "fittest" form is selected, is nature's way of screening. Directed evolution is a strong tool for improving biocatalysts with novel features without knowing about enzyme structure or catalytic mechanisms. Directed evolutions have emerged as a critical approach for creating enzymes with new or better features that are critical to the biotechnology sector [40, 41]. Both *in vivo* and *in vitro* evolution processes are linked through genetic variation and phenotypic selection. Random mutagenesis and gene recombination are two methods for creating protein-encoding DNA libraries [42]. Directed evolution is a simple and comprehensive approach for improving the enzyme properties as the changes introduced are independent of the complicated structure–function relationship of the enzyme and permit mutations to occur across the protein structure [41].

It is a powerful approach for obtaining enzymes with desirable characteristics by altering their catalytic and functional properties. One of the most successful strategies for the evolution or modification of any enzyme, metabolic pathway, or even an entire organism has been directed evolution. The most significant benefit of directed evolution is that information regarding protein structure, function, or amino acid substitutions is not required. Directed evolution combined with metabolic engineering has been used to devise whole new metabolic pathways for creating novel whole-cell biocatalysts useful for industrial applications [43].

The major limitation of directed evolution is the need for intensive screening of the variants several times until the desired clone is obtained. Both time and resources are needed to carry out this approach. In spite of this limitation, directed evolution has produced multiple examples in which remote mutations have proven to be beneficial, like enhancing thermostability, protein expression, and catalytic efficiency through conformational enrichment, and active site remodeling [44, 45]. Directed evolution entails a series of genetic diversification rounds that begin with the generation of a library of mutants followed by high-throughput screening, selection, and identification of mutants with desired characteristics [40, 41, 43].

Directed evolution made a swift transition from academia to industry. Biofuels, materials, bulk and fine chemicals, detergents, consumer items, laboratory reagents, and pharmaceuticals, as well as intermediates for the pharmaceutical sector, are all made with enzymes produced with directed evolution. Many companies have their scientific teams working on improving catalysts or protein-based treatments in terms of stability, activity, specificity, and other features using directed evolution methodologies. Taste enhancers, diabetes and vascular plaque-fighting medications, and lipid-lowering pharmaceuticals are only a few examples of developed enzymes and products. Directed evolution produces some enzymes in high quantities; lipases found in detergents are one example. Biocatalysts developed by directed evolution are used to create massive amounts of industrial chemicals [45, 46].

Directed evolution produces enzymes with a high level of specificity, resulting in biocatalysts that are ecologically beneficial. Enzymes created through directed evolution have replaced harsh industrial processes without any need for hazardous metals or organic solvents. In asymmetric synthesis, for example, enzymes produced by directed evolution have replaced chemical catalysts, providing a green alternative that uses less organic solvents and produces fewer waste products. Short-chain alkanes are used to make alcohols, and 2-methylpropan-1-ol is a promising biofuel contender (isobutanol). Arnold and colleagues employed directed evolution to change the enzymes' co-factor reliance to NADH, making the enzymes and hence the organism suited for biofuel generation [46–48].

DNA shuffling has become a powerful tool in protein engineering since first reported in 1994. Additional strategies for avoiding the bias in inherited sequences have been developed throughout the last decade. Truncated metagenomic gene-specific PCR (TMGS-PCR) was created in the field of metagenomics, for example. As a proof of concept, a lipase was isolated by functional metagenome screening. For the amplification of homologous genes from various ambient soil samples, shortened gene-specific primers were created and employed based on their sequence. DNaseI digestion and PCR recombination were used to recombine the recovered genes. Although a functional diversity of chimeric genes was produced, full-length gene amplification remains a stumbling block in most cases. Some of the methods commonly used for directed evolution are shown in Table 19.2.

19.3.3. Metagenomics Approach

The ability of environmental microbiology to significantly boost biotechnology's commercial potential has been tremendously enhanced by the development of molecular tools and metagenomics. Metagenomics allows for the discovery of novel enzymes whose activities match the natural enzymes. This new approach helps to discover entirely new enzymes in microbial communities without culturing them individually, which is technically challenging [51, 52]. The metagenomic-based discovery and identification of microbial biodiversity, with a focus on microorganisms from severe habitats, has made recent advancements [53–56].

Metagenomics is a relatively new technique that overcomes the shortcomings of conventional culture-based methods. In the metagenomics-based study, DNA is retrieved directly from environmental samples in metagenomics, without the need for laboratory culture [57–59]. When DNA is used to examine the diversity of bacteria, a representative and comprehensive result is obtained [60–62]. Metagenomics has been employed in a variety of domains, including human gut microbial communities, sugarcane bagasse waste, and hypersaline environments. This method can read the diversity of microorganisms in environmental samples up to 99% of total microorganisms. Metagenomics emerges as a novel concept in microbiology research, allowing scientists to expand their horizons in search of new biochemical molecules found in nature that may be used in biotechnology. In ecology and biotechnology investigations, two techniques, structural and functional metagenomics, are used to investigate microbial communities.

Table 19.2 Some important tools and advancements in directed evolution.

Tools	Methods	Applications
Site-directed mutagenesis	Uses overlapping extension PCR to change a gene sequence at a specific spot. DNA primers harboring the desired alteration in the target sequence are combined with a DNA polymerase in an amplification procedure to introduce point mutations, insertions, or deletions [43].	Improving the catalytic properties of enzymes.
Random mutagenesis	The easiest library-building technique randomly modifies the entire gene coding for an enzyme for the desired function. Library of sequences with point mutants from a single parent sequence. The library is expressed for the variant proteins in the first step in random mutagenesis. The successful candidates are then identified using a high-throughput screen for the target trait [49].	Requires no structural or mechanistic knowledge and can lead to the discovery of unexpectedly advantageous mutations.
Targeted/focused mutagenesis	Phylogenetic analyses of homologous proteins are utilized in targeted mutagenesis to discover particular amino acid alterations that are expected to increase substrate binding or catalysis. The creation of the desired function is likely affected by targeted mutagenesis to a subset of residues. Targeted mutagenesis has also been used to create new DNA and RNA polymerases [49].	*in vivo* targeted mutagenesis, followed by its selection, paves the way to a high-efficiency continuous evolution platform for protein and metabolic engineering.
Gene recombination	Gene recombination involves larger fragments of DNA. Homologous recombination, non-homologous recombination, reciprocal recombination and site-specific recombination are some of the important strategies [41].	Incorporates large volume of sequence change.
Incremental truncation for the creation of hybrid genes (ITCHY)	It is a very efficient method that works on random fusion of the domains of two parent enzymes without the need of homology in the DNA sequence. The major constraint of this technique is random fusion of shortened segments from parents at the recombination sites that are not structurally linked, which results in libraries with a large number of inactive clones [41].	A breakthrough in recombination efficiency.

(Continued)

Table 19.2 (Continued)

Tools	Methods	Applications
DNA shuffling	DNA shuffling allows the blocks of parental sequences, which should be conserved in the progeny sequences. Recombination takes place only in regions that share significant sequence identity [41]	*in vitro* recombination of homologous genes is a simple method for creating sequence libraries.
Synthetic shuffling	It allows genes to recombine autonomously, assuring unique amino acid variety. A set of oligonucleotides is created that encodes all variants found in two parental genes. Degeneracies in the oligonucleotides or alternate non-degenerate oligonucleotides are used to include differences in sequences between these genes [50].	Recombination may occur in regions exhibiting low or no sequence identity.
Error prone PCR (epPCR)	This method takes advantage of the absence of proofreading activity in the thermostable polymerase employed. Several protocols have been developed to increase the error rate of Taq polymerase, which can be infinitely varied by increasing the concentration of $MgCl_2$, uneven nucleotide concentrations, using triphosphate nucleoside analogs or a combination of all of these to achieve higher mutation rates [43].	The approach for generating molecular diversity by redrafting "mistake." When only a few mutation cycles are required, it is extremely effective, but when used for several cycles, it quickly becomes restrictive.
Cassette mutagenesis	Cassette mutagenesis is a method of mutagenesis that is used for mutagenesis of confined and defined gene regions. Short gene segments that are mutated are either hot area discovered using epPCR library screening or key domains of enzymes examined using structural data [43].	It is only beneficial when the targeted amino acids are in the same primary sequence segment.
Random chimeragenesis technique (RACHITT).	It uses complex chimeratic procedures to replace random priming of DNA pieces [41].	
Sequence homology-Independent protein recombination (SHIPREC)	Because crossovers occur at physically comparable sites and sequence length rather than sequence similarity, it can be employed to construct hybrid genes of distantly related sequences [41].	
Staggered extension protocol (STEP)	This approach avoids the need for DNaseI fragmentation and produces chimeric genes by switching templates. Denaturation and extremely short-duration annealing/polymerase-catalyzed extension cycles are repeated on the template sequences. The developing fragments anneal to different templates based on sequence complementarity in each cycle, allowing them to stretch further [41].	Offers advantage over DNA shuffling

Table 19.2 (Continued)

Tools	Methods	Applications
Fusion	By combining the catalytic and substrate domains of distinct enzymes, many chimeric enzymes have been created. The catalytic efficiency, thermostability, product selectivity, and substrate specificity of chimeric enzymes created using fusion method have all increased. For the creation of novel chimeric enzymes, oligopeptides and functional genes can be employed [43].	These extra functional features added to improvements in directed evolution.
Random or directed truncation	Unwanted domains of proteins that obstruct enzyme activity have been removed using various techniques. With changed enzyme characteristics, a truncated library or enzymes can be created [43].	

It normally takes several years from the moment a gene is discovered to the time an industrial method is built, as industrial operations must meet certain criteria:

(i) The reaction conditions are often harsh with a broad range of temperature and ph. Usually the substrate load is very high, and the solvent conditions are extreme in industrial processing.
(ii) A high degree of stereo selectivity and a rapid rate of turnover.

Metagenomics paves the way for researchers to look into enzymes with significant industrial potential. The possibilities of metagenomics for mining microbial enzymes for possible industrial uses are extensive [62–64]. Researchers that want to investigate new enzymes face a hurdle when it comes to choosing sampling locations. The role of enzyme characterization is determined by the location. Each place has its own ecological niche where new enzymes can be discovered. Functional interactions between the microbial population and its surroundings generally establish unique niches [65]. In sugarcane bagasse samples, a significant concentration of cellulolytic bacteria was reported [66]. Similarly, Nie et al. [67] found genes encoding hydrocarbon breakdown enzymes in the microbial community in the oil environment. The presence of hydrocarbon breakdown enzymes is inextricably linked to a hydrocarbon-rich oil environment. Several enzymes such as cellulases, proteases, lipases, chitinases, oxoflavin-degrading enzyme, and transaminases, to name a few, have all been discovered using a metagenomic approach. Metagenomics has aided researchers in discovering enzymes with novel properties from the environment that are useful to industry [62].

Esterase/lipase is one of the most commonly discovered new enzymes in soil metagenomes. Esterases and lipases are lipolytic enzymes that are essential biocatalysts in biotechnological applications. Lipolytic enzymes have a number of unique characteristics, including the absence of cofactors, stability in organic solvents, broad substrate specificity, stereo, and positional selectivity [68]. Cellulases are cell wall-degrading enzymes and are also biotechnologically interesting enzymes, owing to their ability to degrade biomass for bioenergy production [69]. Homology-based screening also assists in unraveling the bioactivities seen in the soil metagenome. It is a rich source for the investigation of gene clusters for the production of the bioactive molecule [70].

In addition to unique enzymes, soil metagenomes also contain many other bioactive molecules, such as antibiotics and pharmaceutically useful compounds. Turbomycins, glycopeptides, cyanobactins, type II polyketides, trans-acyltransferase polyketides, and the anticancer drug ET-743 are some of the bioactive molecules obtained through the metagenomics-based approach [70, 71]. A recent study has reported a bio surfactant-producing clone by functional screening of metagenomic library [61–63, 72].

19.3.4. Nanobiotechnology

Nanotechnology is the control of matter on the nanoscale to create and use materials, devices, and systems. Nanobiotechnology integrates the use of structure, function, and processes of biological molecules, their complexes, and nanosystems to develop innovative functional biological materials at nano levels [73, 74]. It is useful in designing materials like DNA origami, DNA nanomachines, DNA scaffolds, DNA and RNA aptamers, ribozymes, and riboswitches [74–77].Further, this technology has been used to create 3D structures, functional protein complexes, nanofilms, and other nanostructures that could be used for the large-scale production of programmable nanomaterials. Several innovative biomaterials for biosensing, bioimaging, diagnostic, and drug delivery have been produced by nanotechnology [74, 77]. The unique properties of the nanobiomaterials such as a high volume-to-surface ratio, enhanced solubility, quantum size, macroscopic quantum tunnel, and multifunctionality make them unique and useful in a wide range of applications [78–81]. Several nanomaterials used as biosensors have been used in diagnostics, pathological and food testing, environmental monitoring, drug discovery, genomics, and proteomics [82–85].

Nanomaterials are quite useful in enzyme immobilization and stabilization owing to their high electron transfer rate, low transfer resistance, protection from proteolytic digestion, and ease of separation and reuse. Different nanomaterials like polymer NPs, polymer nanofibers, GO nanosheets, porous silica NPs, sol–gel NPs, and viral NPs are used for immobilization [86, 87]. Artificial multi-enzyme systems may also be immobilized on nanomaterials to replicate natural multienzyme complexes [88–90].

19.4. ENZYMATIC INTERVENTION IN INDUSTRIAL PROCESSES

Enzymes are useful in a wide range of industrial applications due to their efficiency, particular action, gentle working conditions, and excellent biodegradability. Industrial enzymes must be capable of operating under atmospheric pressure and in mild temperature and acidic conditions. Most enzymes work best at temperatures ranging between 30 and 70 °C and close to neutral pH. Enzyme processes have the potential to save energy and money by avoiding the need for specialized equipment that is resistant to heat, pressure, or corrosion. Biocatalysts have been scaled up for commercial operations in the pharmaceutical, food and beverage, fruit juice, paper and pulp, and textile industries (Figure 19.2). However, for effective biocatalytic processes in the energy sector, more improvements in stability and biocatalyst functionality are necessary.

19.4.1. Enzymes Used in the Textile Industry

Enzymes speed up reactions, act on specified substrates, and work in mild conditions that are safe, biodegradable, and easy to manage. Hence they are a better replacement for harsh chemicals used in several industries. The textile industry is one of the major polluters of the environment, generating massive amounts of trash from fabric desizing, bleaching chemicals,

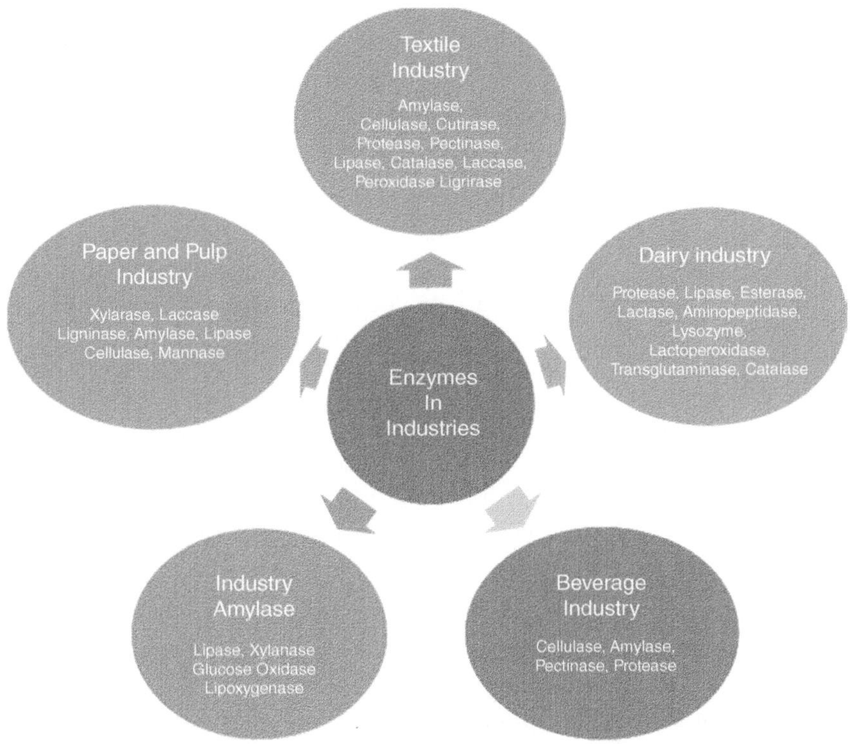

Figure 19.2 Microbial enzymes used in different industrial sectors.

and dye [91]. Enzymes assist in the development of eco-friendly fiber processing technology to influence the quality of the final product [92]. Hydrolase and oxidoreductases are mainly utilized in the pre-treatment and finishing of cotton fibers. Similarly, amylase, cellulase, cutinase, protease, pectinase, and lipase/esterase are used in several processes associated with textile industries, namely biopolishing and scouring, wool antifelting, cotton softening, denim finishing, desizing, wool finishing, and modification of synthetic fiber modification [93, 94]. Similarly enzymes, namely catalase, laccase, peroxidase, and ligninase, assist in biobleaching [95].

There are different steps in textile preparation where separate enzymes are involved. The processing involves the following steps:

a. **Enzymatic desizing** – The textile industry uses amylases that break down starch into short-chain sugars such as dextrin and maltose in the sizing process. This causes better and more uniform wet processing without harming the support fabric. At low temperatures (30–60 °C) and a pH range of 5.5–6.5, amylase can be employed for the desizing process [96].
b. **Enzymatic Scouring or (Bioscouring)** – It is used to remove non-cellulosic particles from cotton's surface. Cellulase and pectinase are the main enzymes used in this process. When compared to alkaline scouring, the biological oxygen demand (BOD) and chemical oxygen demand (COD) of the enzymatic scouring process are 20–45% lower than that of alkaline scouring which is 100%. Total Dissolved Solid (TDS) by the enzymatic route is 20–50% with respect to the alkaline scouring (100%). The enzyme-based scouring is safe for the fabric and the environment by reducing the health hazards [91, 94].

c. **Enzymatic Bleaching** – Cotton bleaching is used to remove natural colors from the fibers and give them a clean white appearance. Traditionally hydrogen peroxide is the most preferred industrial bleaching agent and the processing uses a lot of alkaline chemicals. This results in the use of a lot of water and reduction in the degree of polymerization, and causes serious damage to the fibers. Alternately, replacing harsh chemicals with enzyme-based bleaching would improve the quality of the product quality, and prevent the huge water wastage required to remove hydrogen peroxide. Several enzyme combinations have been used like amyloglucosidases, pectinases, and glucose oxidases, all having compatible pH and temperature ranges. Tzanov et al. [97] reported the use of laccases at low concentrations to improve the bleaching impact on cotton garments. Furthermore, less time is required for the enzymatic pre-treatment to improve fabric whiteness. Pereira et al. [98] have discovered that laccase from *Trametes hirsute* can improve cotton whiteness as a result of flavonoid oxidation. Basto et al. [99] presented a cotton bleaching technique that combines ultrasound along with laccase treatment. They discovered that providing low ultrasonic energy increased laccase bleaching efficiency on cotton fabrics. Catalase is an enzyme that converts hydrogen peroxide bleaching fluid into water molecules and less reactive gaseous oxygen, which results in cleaner wastewater or lower water use, as well as a reduction in energy and time.
d. **Biopolishing** – improves the quality of fabrics by eliminating woolliness and pilling in cellulosic fibers. The cellulase enzyme helps in the elimination of cotton microfibrils. Biopolishing treatment imparts a cleaner surface, luster, softer, and fabric's proclivity toward pilling [100].
e. **Enzymatic treatment to denim** – Denim is made of high-quality cotton. The dye is mostly absorbed on the fiber's surface, thus fading can be obtained without a significant loss of strength. Pumice stones (sodium hypochlorite or potassium permanganate) are traditionally employed for the treatment of denim, but it causes back-staining and a lot of wear and strain on the equipment. These drawbacks resulted in the employment of enzymes for processing. The old process is replaced by the technique known as "Bio-Stonewashing" by cellulase which works by loosening the indigo color on the denim. Several kilograms of pumice stones can be replaced with a tiny amount of enzyme. Less pumice stone use means less damage to the clothing and machine, as well as less pumice dust in the laundry environment. Laccase enzyme is also used as a stone-washing agent for denim fabric [101].

19.4.2. Enzymes Used in Paper and Pulp Industries

An increase in paper consumption has led to the emergence of more pulp and paper industries [84, 102]. Effluents from these industries generally pollute the environment, necessitating the employment of green technologies in paper production [103, 104]. This has forced us to consider various cost-effective and environmentally friendly alternatives to chemical-based methods. Biotechnology serves both cost-effective and environmentally beneficial alternatives. Along with industrial processing, enzymes also assist in the waste treatment of the effluents and manage BOD and COD [105].

Pulping, bleaching, and papermaking are the three primary phases in industry. Pulping is the process of turning wood or agricultural products into a flexible fiber that may be used to make paper. Among the various pulping methods, the most prevalent method is Kraft pulping, which is a hot alkaline sulfide digestion of wood as raw material that eliminates the majority of the lignin and leaves a flexible yet strong cellulosic fiber. Printing and writing paper grades are also made with acid sulfite pulping. These chemical pulping methods provide the foundation for the printing industry. Mechanical, thermo-mechanical, and chemo-thermomechanical pulping give higher yields but inferior strength and optical qualities. Newsprint or magazine stock can be made with such fibers. If the end application requires white paper, the fiber is cleaned and then bleached after pulping. After that, other manipulations such as size, filler addition, and color addition can be done to make the final paper product. Enzymes are useful in paper recycling which involves the use of enzymes at different steps like, pulping, washing, screening, flotation, bleaching, and wastewater treatment.

Enzymes provide sustainable alternatives in industrial processing. The enzymes can be used for dissolving pulp, delignification, bleaching, and improving the brightness of pulp alone or in combination with physical and chemical methods [102]. Xylanases and ligninases eliminate lignin and hemicelluloses, enhancing the pulp quality [106], whereas amylases are used for starch coating, deinking, enhancing paper cleanliness, and improving drainage properties [11]. Lipases perform deinking and pitch control, whereas cellulases assist in deinking, softening, and improving drainage, and repurposing old printed papers. Laccase is used in place of chlorine for the chemical pulping process. Mannases are also employed in the paper industry to degrade glucomannan to improve paper brightness [107–109].

19.4.3. Enzymes in the Food Industry

Since ancient times, microbes have been used for fermentation in numerous foods [110]. Enzymes are used in the beverage and food industries to control the brewing process. This improves the functional and nutritional properties of animal and vegetable proteins. Several microbial enzymes are used in baking, brewing, dairy, and juice production units.

19.4.3.1. Baking Industry

Enzymes are used in baking to improve flour quality, dough stability, texture, volume, and color. They also extend crumb softness and consistent crumb structure, and they enhance bread freshness. Bread baking is one of the most widely used food processing methods and enzymatic intervention is needed to enhance quality control and production efficiency. The amylase enzyme is added to bread flour, either alone or in combination with other enzymes to retain moisture, softness, freshness, and enhance the shelf life. Furthermore, lipase and xylanase are added to improve dough strength and whiteness, while glucose oxidase and lipoxygenase improve dough stability and conditioning. Similarly, transglutaminase is used to improve the quality of flour and bread, and the texture of cooked pasta. Lipases extend the shelf life of bakery products using esterifying short-chain fatty acids [111, 112].

19.4.3.2. Dairy Industry

Microbial enzyme intervention in the dairy industry improves yield and organoleptic features such as aroma, flavor, and color of milk products. Important enzymes used in the dairy industries include proteases, lipases, esterases, lactase, amino peptidase, lysozyme, lacto peroxidase, transglutaminase, and catalases. Cheese, yogurt, and other milk products are made with the help of enzymes [113, 114]. Rennet comprising of a mixture of chitosan and pepsin is used for cheese making. Other proteases are also involved in cheese processing that speeds up the process and eliminates allergenic components in the milk products [114]. Lipases are useful in flavor enhancement, faster cheese preparation, customized milk product preparation, and milk fat lipolysis. Transglutaminase catalyzes the polymerization of milk proteins and enhances functional activity. Lactose intolerance is a condition in which a person cannot digest lactose. Lactases catalyze the hydrolysis of lactose into glucose and galactose and are thus used as a digestive aid to improve the solubility and sweetness of milk products [115–117].

19.4.3.3. Beverage Industry

The beverage industry includes non-alcoholic and alcoholic beverages. Soft drinks, syrups, packaged water, fruit juices, tea, and coffee industries are non-alcoholic beverages. Distilled spirits, wine, and beer are included in the alcoholic group. Microbial enzymes in the brewing industry digest cell walls during the extraction of plant material and assist in improving yield, color, flavor, and clearer products [118]. Enzymes enhance the efficiency of processing operations and improve product quality in the fruit and vegetable juice industries [119]. Cellulases, amylases, and pectinases are used in fruit juice processing to improve yield and cost-effectiveness by macerating, liquefying, and clarifying the juice [120, 121]. The use of enzymes improves the quality and stability of juices generated. Enzymes break down pectin, carbohydrates, proteins, and cellulose in fruits and vegetables, allowing for higher yields and faster processing. Amylases are helped to clarify liquids and increase the amount of clear or murky juice produced. Cellulases and pectinases improve extraction yield, cloud stability, and texture in the extracted fruit juices [121–123]. The addition of proteases to the brewing process can help control the formation of chill hazes in beer [124, 125].

19.5. CONCLUSION

Microbial enzymes have been involved in diverse industries and are preferred in many cases based on their ease of isolation, purification, and manipulation over enzymes from plants and animal sources. Interventions of microbial enzymes in industries have drastically influenced the production level and they are also ecofriendly, reducing the level of contamination resulting from conventional physical and chemical methods. The replacement of conventional methods by microbial enzymes have been documented in a large number of industries like pharmaceuticals, food and fodder, detergents, textile, leather processing, and paper and pulp. Cost-effective production of microbial enzymes, alteration of specificity, and enhanced catalytic activity have been attempted as per the need of industries. Several state-of-the-art technologies in microbial enzyme technology like metagenomics, directed evolution, omics, mutagenesis, nanobiotechnology, immobilization, and bioinformatics have revolutionized the diverse industrial applications. This chapter reflects the relevance of microbial enzymes in different industries and highlights the diverse microbial sources, innovations, and applications.

REFERENCES

1. Niyonzima, F.N. (2019). Production of microbial industrial enzymes. *Acta Scientific Microbiology 2*: 75–89.
2. Aarti, C., Khusro, A., and Agastian, P. (2021). Lignocellulosic biomass as potent feedstock resource for bioethanol production: recent updates. *World News of Natural Sciences 37*: 164–181.
3. Gurung, N., Ray, S., Bose, S. et al. (2013). A broader view: microbial enzymes and their relevance in industries, medicine, and beyond. *BioMed Research International 1*: 2013.
4. Binod, P., Palkhiwala, P., Gaikaiwari, R. et al. (2013). Industrial enzymes-present status and future perspectives for India. *Journal of Scientific and Industrial Research 72*: 271–286.
5. Chapman, J., Ismail, A.E., and Dinu, C.Z. (2018). Industrial applications of enzymes: recent advances, techniques, and outlooks. *Catalysts 8* (6): 238.
6. Anbu, P., Gopinath, S.C., Cihan, A.C. et al. (2013). Microbial enzymes and their applications in industries and medicine. *Hindawi* 1–2.
7. Raveendran, S., Parameswaran, B., BeeviUmmalyma, S. et al. (2018). Applications of microbial enzymes in food industry. *Food Technology and Biotechnology 56* (1): 16–30.

8. Lim, S.J. and Oslan, S.N. (2021). Native to designed: microbial α-amylases for industrial applications. *PeerJ 9*: e11315.
9. Razzaq, A., Shamsi, S., Ali, A. et al. (2019). Microbial proteases applications. *Frontiers in Bioengineering and Biotechnology 7*: 110.
10. Naveed, M., Nadeem, F., Mehmood, T. et al. (2021). Protease – a versatile and ecofriendly biocatalyst with multi-industrial applications: an updated review. *Catalysis Letters 151* (2): 307–323.
11. Kuhad, R.C., Gupta, R., and Singh, A. (2011). Microbial cellulases and their industrial applications. *Enzyme Research 2011*: https://doi.org/10.4061/2011/280696.
12. Sarwan, J. and Bose, J.C. (2021). Importance of microbial cellulases and their industrial applications. *Annals of the Romanian Society for Cell Biology 1*: 3568–3575.
13. Singh, D. and Gupta, N. (2020). Microbial laccase: a robust enzyme and its industrial applications. *Biologia 75* (8): 1183–1193.
14. Unuofin, J.O., Falade, A.O., and Aladekoyi, O.J. (2021). Applications of microbial laccases in bioremediation of environmental pollutants: potential issues, challenges, and prospects. In: *Bioremediation for Environmental Sustainability* (ed. G. Saxena, V. Kumar and M.P. Shah), 519–540. Elsevier.
15. Bhandari, S., Poudel, D.K., Marahatha, R. et al. (2021). Microbial enzymes used in bioremediation. *Journal of Chemistry 9*: 2021.
16. Ravindran, R., Hassan, S.S., Williams, G.A. et al. (2018). A review on bioconversion of agro-industrial wastes to industrially important enzymes. *Bioengineering 5* (4): 93.
17. Cabrera, M.Á. and Blamey, J.M. (2018). Biotechnological applications of archaeal enzymes from extreme environments. *Biological Research 51*: 1–15.
18. Kavuthodi, B. and Sebastian, D. (2018). Review on bacterial production of alkaline pectinase with special emphasis on Bacillus species. *Bioscience Biotechnology Research Communications 11*: 18–30.
19. Sudeep, K.C., Upadhyaya, J., Joshi, D.R. et al. (2020). Production, characterization, and industrial application of pectinase enzyme isolated from fungal strains. *Fermentation 6* (2): 59.
20. Robinson, P.K. (2015). Enzymes: principles and biotechnological applications. *Essays in Biochemistry 59*: 1.
21. Ramnath, L., Sithole, B., and Govinden, R. (2017). Identification of lipolytic enzymes isolated from bacteria indigenous to Eucalyptus wood species for application in the pulping industry. *Biotechnology Reports 15*: 114–124.
22. Choudhury, P. and Bhunia, B. (2015). Industrial application of lipase: a review. *Biopharmaceuticals Journal 1* (2): 41–47.
23. Song, H.Y., El Sheikha, A.F., and Hu, D.M. (2019). The positive impacts of microbial phytase on its nutritional applications. *Trends in Food Science & Technology 86*: 553–562.
24. Liu, X. and Kokare, C. (2017). Microbial enzymes of use in industry. In: *Biotechnology of Microbial Enzymes* (ed. G. Brahmachari), 267–298. Academic Press.
25. Falade, A.O., Nwodo, U.U., Iweriebor, B.C. et al. (2017). Lignin peroxidase functionalities and prospective applications. *Microbiology Open 6* (1): e00394.
26. Chandra, P., Singh, R., and Arora, P.K. (2020). Microbial lipases and their industrial applications: a comprehensive review. *Microbial Cell Factories 19* (1): 1–42.
27. Hassan, S.S., Williams, G.A., and Jaiswal, A.K. (2019). Lignocellulosic biorefineries in Europe: current state and prospects. *Trends in Biotechnology 37* (3): 231–234.
28. Kubra, K.T., Ali, S., Walait, M. et al. (2018). Potential applications of pectinases in food, agricultural and environmental sectors. *Journal of Pharmaceutical, Chemical and Biological Sciences 6*: 23–34.
29. Dillon, M.M., Sung, W., Lynch, M. et al. (2018). Periodic variation of mutation rates in bacterial genomes associated with replication timing. *MBio 9* (4): e01371–e01318.
30. Fitzgerald, D.M. and Rosenberg, S.M. (2019). What is mutation? A chapter in the series: how microbes "jeopardize" the modern synthesis. *PLoS Genetics 15* (4): e1007995.
31. Adrio, J.L. and Demain, A.L. (2006). Genetic improvement of processes yielding microbial products. *FEMS Microbiology Reviews 30* (2): 187–214.

32. Ravanat, J.L. and Douki, T. (2016). UV and ionizing radiations induced DNA damage, differences and similarities. *Radiation Physics and Chemistry 128*: 92–102.
33. Sanghavi, G., Gupta, P., Rajput, M. et al. (2020). Microbial strain engineering. In: *Engineering of Microbial Biosynthetic Pathways* (ed. V. Singh, A. Singh, P. Bhargava, et al.), 11–32. Singapore: Springer.
34. Demain, A.L. (2010). History of industrial biotechnology. *Industrial Biotechnology: Sustainable Growth and Economic Success 21*: 17–77.
35. Vinci, V.A. and Byng, G. (1999). Strain improvement by non-recombinant methods. In: *Manual of Industrial Microbiology and Biotechnology*, 2e (ed. A.L. Demain and J.E. Davies), 103–113. Washington, DC: ASM Press.
36. Parekh, S., Vinci, V.A., and Strobel, R.J. (2000). Improvement of microbial strains and fermentation processes. *Applied Microbiology and Biotechnology 54* (3): 287–301.
37. Keller, U. (1983). Highly efficient mutagenesis of Clavicepspurpurea by using protoplasts. *Applied and Environmental Microbiology 46* (3): 580–584.
38. Kim, K.S., Cho, N.Y., Pai, H.S. et al. (1983). Mutagenesis of Micromonospora rosaria by using protoplasts and mycelial fragments. *Applied and Environmental Microbiology 46* (3): 689–693.
39. Hersbach, G.J. (1984). *The Penicillins: Properties, Biosynthesis and Fermentation*, Biotechnology of Industrial Antibiotics, 45–140. Marcel Dekker AG.
40. Kaur, J. and Sharma, R. (2006). Directed evolution: an approach to engineer enzymes. *Critical Reviews in Biotechnology 26* (3): 165–199.
41. Lutz, S. and Iamurri, S.M. (2018). Protein engineering: past, present, and future. *Methods Journal of Molecular Biology 1685*: 1–12.
42. Gabor, E., Heinze, B., and Eck, J. (2016). Hints from nature: metagenomics in enzyme engineering. In: *Understanding Enzymes* (ed. A. Svendsen), 667–680. Jenny Stanford Publishing.
43. Sharma, A., Gupta, G., Ahmad, T. et al. (2021). Enzyme engineering: current trends and future perspectives. *Food Reviews International 37* (2): 121–154.
44. Zeymer, C. and Hilvert, D. (2018). Directed evolution of protein catalysts. *Annual Review of Biochemistry 87*: 131–157.
45. Wilding, M., Hong, N., Spence, M. et al. (2019). Protein engineering: the potential of remote mutations. *Biochemical Society Transactions 47* (2): 701–711.
46. Hernandez, K.E., Renata, H., Lewis, R.D. et al. (2016). Highly stereoselective biocatalytic synthesis of key cyclopropane intermediate to ticagrelor. *ACS Catalysis 6* (11): 7810–7813.
47. Glieder, A., Farinas, E.T., and Arnold, F.H. (2002). Laboratory evolution of a soluble, self-sufficient, highly active alkane hydroxylase. *Nature Biotechnology 20* (11): 1135–1139.
48. Bastian, S., Liu, X., Meyerowitz, J.T. et al. (2011). Engineered ketol-acid reductoisomerase and alcohol dehydrogenase enable anaerobic 2-methylpropan-1-ol production at theoretical yield in *Escherichia coli*. *Metabolic Engineering 13* (3): 345–352.
49. Chowdhury, R. and Maranas, C.D. (2020). From directed evolution to computational enzyme engineering—a review. *AIChE Journal 66* (3): e16847.
50. Ness, J.E., Kim, S., Gottman, A. et al. (2002). Synthetic shuffling expands functional protein diversity by allowing amino acids to recombine independently. *Nature Biotechnology 20* (12): 1251–1255.
51. Kyrpides, N.C., Hugenholtz, P., Eisen, J.A. et al. (2014). Genomic encyclopedia of bacteria and archaea: sequencing a myriad of type strains. *PLoS Biology 12* (8): e1001920.
52. Yarza, P., Yilmaz, P., Pruesse, E. et al. (2014). Uniting the classification of cultured and uncultured bacteria and archaea using 16S rRNA gene sequences. *Nature Reviews Microbiology 12* (9): 635–645.
53. Feller, G. (2013). Psychrophilic enzymes: from folding to function and biotechnology. *Scientifica 17*: 2013.
54. Vester, J.K., Glaring, M.A., and Stougaard, P. (2014). Discovery of novel enzymes with industrial potential from a cold and alkaline environment by a combination of functional metagenomics and culturing. *Microbial Cell Factories 13* (1): 1–4.
55. Alcaide, M., Stogios, P.J., Lafraya, Á. et al. (2015). Pressure adaptation is linked to thermal adaptation in salt-saturated marine habitats. *Environmental Microbiology 17* (2): 332–345.

56. Ferrer, M., Martínez-Martínez, M., Bargiela, R. et al. (2016). Estimating the success of enzyme bioprospecting through metagenomics: current status and future trends. *Microbial Biotechnology 9* (1): 22–34.
57. Tanveer, A., Yadav, S., and Yadav, D. (2016). Comparative assessment of methods for metagenomic DNA isolation from soils of different crop growing field. *3 Biotech 6* (2): 1–5.
58. Aggarwala, V., Liang, G., and Bushman, F.D. (2017). Viral communities of the human gut: metagenomic analysis of composition and dynamics. *Mobile DNA 8* (1): 1–10.
59. Pucker, B., Schilbert, H.M., and Schumacher, S.F. (2019). Integrating molecular biology and bioinformatics education. *Journal of Integrative Bioinformatics 16*: 1–7.
60. Neelakanta, G. and Sultana, H. (2013). The use of metagenomic approaches to analyze changes in microbial communities. *Microbiology Insights 6*: 37–48.
61. Alves, L.D., Westmann, C.A., Lovate, G.L. et al. (2018). Metagenomic approaches for understanding new concepts in microbial science. *International Journal of Genomics 2018*: 1–15.
62. Prayogo, F.A., Budiharjo, A., Kusumaningrum, H.P. et al. (2020). Metagenomic applications in exploration and development of novel enzymes from nature: a review. *Journal, Genetic Engineering & Biotechnology 18* (1): 1–10.
63. Mallick, H., Franzosa, E.A., Mclver, L.J. et al. (2019). Predictive metabolomic profiling of microbial communities using amplicon or metagenomic sequences. *Nature Communications 10* (1): 1.
64. Berini, F., Casartelli, M., Montali, A. et al. (2019). Metagenome-sourced microbial chitinases as potential insecticide proteins. *Frontiers in Microbiology 10*: 1358.
65. Berini, F., Casciello, C., Marcone, G.L. et al. (2017). Metagenomics: novel enzymes from non-culturable microbes. *FEMS Microbiology Letters 364* (21): fnx211.
66. Mhuantong, W., Charoensawan, V., Kanokratana, P. et al. (2015). Comparative analysis of sugarcane bagasse metagenome reveals unique and conserved biomass-degrading enzymes among lignocellulolytic microbial communities. *Biotechnology for Biofuels 8* (1): 1–7.
67. Nie, Y., Zhao, J.Y., Tang, Y.Q. et al. (2016). Species divergence vs. functional convergence characterizes crude oil microbial community assembly. *Frontiers in Microbiology 7*: 1254.
68. Iqbal, H.A., Feng, Z., and Brady, S.F. (2012). Biocatalysts and small molecule products from metagenomic studies. *Current Opinion in Chemical Biology 16* (1–2): 109–116.
69. Escuder-Rodríguez, J.J., DeCastro, M.E., Cerdán, M.E. et al. (2018). Cellulases from thermophiles found by metagenomics. *Microorganisms 6* (3): 66.
70. Lee, M.H. and Lee, S.W. (2013). Bioprospecting potential of the soil metagenome: novel enzymes and bioactivities. *Genomics & informatics 11* (3): 114.
71. Lee, S.W., Won, K., Lim, H.K. et al. (2004). Screening for novel lipolytic enzymes from uncultured soil microorganisms. *Applied Microbiology and Biotechnology 65* (6): 720–726.
72. Jackson, S.A., Borchert, E., O'Gara, F. et al. (2015). Metagenomics for the discovery of novel biosurfactants of environmental interest from marine ecosystems. *Current Opinion in Biotechnology 33*: 176–182.
73. Gencoglu, A. and Minerick, A.R. (2014). Electrochemical detection techniques in micro-and nanofluidic devices. *Microfluidics and Nanofluidics 17* (5): 781–807.
74. Nagamune, T. (2017). Biomolecular engineering for nanobio/bionanotechnology. *Nano Convergence 4* (1): 1–56.
75. Guo, P. (2010). The emerging field of RNA nanotechnology. *Nature Nanotechnology 5*: 833–842.
76. Pinheiro, A.V., Han, D., Shih, W.M. et al. (2011). Challenges and opportunities for structural DNA nanotechnology. *Nature Nanotechnology 6*: 763–772.
77. Kumar, D., Kant, R., Kumar Sahu, A. et al. (2020). Role of nanobiotechnology in drug discovery, development and molecular diagnostic. Chapter in handbook of nanomaterials in analytical chemistry: modern trends in analysis. *Appl. NanoBiotechnology* https://doi.org/10.5772/intechopen.92796.
78. Jain, K.K. (2005). The role of nanobiotechnology in drug discovery. *Drug Discovery Today 10* (21): 1435–1442.

79. Colombo, M., Carregal-Romero, S., Casula, M.F. et al. (2012). Biological applications of magnetic nanoparticles. *Chemical Society Reviews 41* (11): 4306–4334.
80. Prasad, S. (2014). Nanobiosensors: the future for diagnosis of disease. *Nanobiosensors in Disease Diagnosis 3*: 1–10.
81. Alejandro, C. and Arben, M. (2016). Nanobiosensors in diagnostics. *Nanobiomedicine 3* (10): 1177. https://doi.org/10.1177/1849543516663574.
82. Theron, J., Cloete, T.E., and Kwaadsteniet, M.D. (2010). Molecular techniques for determining microbial diversity and community structure in natural environments. *Critical Reviews in Microbiology 36*: 318–339.
83. Stanley, M. (2016). Nanodiagnostics: A revolution in biomedical nanotechnology. *MOJ Proteomics & Bioinformatics 3* (2): 34–36.
84. Bajpai, V.K., Kamle, M., Shukla, S. et al. (2018). Prospects of using nanotechnology for food preservation, safety, and security. *Journal of Food and Drug Analysis 26* (4): 1201–1214.
85. Anwar, A., Masri, A., Rao, K. et al. (2019). Antimicrobial activities of green synthesized gums-stabilized nanoparticles loaded with flavonoids. *Scientific Reports 9* (1): 1–2.
86. Jampilek, J., Kos, J., and Kralova, K. (2019). Potential of nanomaterial applications in dietary supplements and foods for special medical purposes. *Nanomaterials 9* (2): 296.
87. Dhananjayan, N., Jeyaraj, W., and Karuppasamy, G. (2019). Interactive studies on synthetic nanopolymer decorated with edible biopolymer and its selective electrochemical determination of L-tyrosine. *Scientific Reports 9* (1): 1–2.
88. Obermeyer, A.C. and Olsen, B.D. (2015). Synthesis and application of protein containing block copolymers. *ACS Macro Letters 4*: 101–110.
89. Yang, H., Liu, L., and Xu, F. (2016 Oct). The promises and challenges of fusion constructs in protein biochemistry and enzymology. *Applied Microbiology and Biotechnology 100* (19): 8273–8281.
90. Lee, S., Kaku, Y., Inoue, S. et al. (2016). Growth signal body selects functional intrabodies in the mammalian cytoplasm. *Journal of Biotechnology 11*: 565–573.
91. Ahuja, S.K., Ferreira, G.M., and Moreira, A.R. (2004). Utilization of enzymes for environmental applications. *Critical Reviews in Biotechnology 24* (2–3): 125–154.
92. Choi, J.M., Han, S.S., and Kim, H.S. (2015). Industrial applications of enzyme biocatalysis: current status and future aspects. *Biotechnology Advances 33* (7): 1443–1454.
93. Araujo, R., Casal, M., and Cavaco-Paulo, A. (2008). Application of enzymes for textile fibres processing. *Biocatalysis and Biotransformation 26* (5): 332–349.
94. Chen, S., Su, L., Chen, J. et al. (2013). Cutinase: characteristics, preparation, and application. *Biotechnology Advances 31* (8): 1754–1767.
95. Mojsov, K. (2011). Application of enzymes in the textile industry: a review. In: *2nd international congress: Engineering, Ecology and Materials in the Processing Industry: Jahorina, Bosnia and Herzegovina*, 230–239. TehnoloskiFakultet Zvornik.
96. Cavaco-Paulo, A. and Gubitz, G. (ed.) (2003). *Textile Processing with Enzymes*. Elsevier.
97. Tzanov T, Basto C, Gübitz G, Cavaco-Paulo A. Laccases to improve the whiteness of a conventional cotton bleaching. In 227th ACS Meeting 2004 Mar 28.
98. Pereira, L., Bastos, C., Tzanov, T. et al. (2005). Environmentally friendly bleaching of cotton using laccases. *Environmental Chemistry Letters 3* (2): 66–69.
99. Basto, C., Tzanov, T., and Cavaco-Paulo, A. (2007). Combined ultrasound-laccase assisted bleaching of cotton. *Ultrasonics Sonochemistry 14* (3): 350–354.
100. Mojsov, K. (2014). Biopolishing enzymes and their applications in textiles: a review. *Tekstilnaindustrija 61* (2): 20–24.
101. Maryan, A.S., Montazer, M., and Damerchely, R. (2015). Discoloration of denim garment with color free effluent using montmorillonite based nano clay and enzymes: nano bio-treatment on denim garment. *Journal of Cleaner Production 91*: 208–215.
102. Singh, G., Kaur, S., Khatri, M. et al. (2019). Biobleaching for pulp and paper industry in India: emerging enzyme technology. *Biocatalysis and Agricultural Biotechnology 17*: 558–565.

103. Lin, X., Wu, Z., Zhang, C. et al. (2018). Enzymatic pulping of lignocellulosic biomass. *Industrial Crops and Products 120*: 16–24.
104. Gangwar, A.K., Prakash, N.T., and Prakash, R. (2014). Applicability of microbial xylanases in paper pulp bleaching: a review. *BioResources 9* (2): 3733–3754.
105. Hossain, K. and Ismail, N. (2015). Bioremediation and detoxification of pulp and paper mill effluent: a review. *Research Journal of Environmental Toxicology 9* (3): 113.
106. Maijala, P., Kleen, M., Westin, C. et al. (2008). Biomechanical pulping of softwood with enzymes and white-rot fungus Physisporinusrivulosus. *Enzyme and Microbial Technology 43* (2): 169–177.
107. Ram, C., Rani, P., Gebru, K.A. et al. (2020). Pulp and paper industry wastewater treatment: use of microbes and their enzymes. *Physical Sciences Reviews 5* (10): 20190050.
108. Joshi, N. and Mahawar, P. (2020). Review on microbial cellulase-a potential group of enzymes with the influential capabilities in various industrial applications. *Journal of Advanced Scientific Research 2*: 11.
109. Moreno, A.D., Ibarra, D., Eugenio, M.E. et al. (2020). Laccases as versatile enzymes: from industrial uses to novel applications. *Journal of Chemical Technology & Biotechnology 95* (3): 481–494.
110. Zhang, Y., He, S., and Simpson, B.K. (2018). Enzymes in food bioprocessing – novel food enzymes, applications, and related techniques. *Current Opinion in Food Science 1* (19): 30–35.
111. Dura, A. and Rosell, C.M. (2017). Enzymes in baking. In: *Microbial Enzyme Technology in Food Applications* (ed. R.C. Ray and C.M. Rosell), 295–314. CRC Press.
112. Park, S.H., Na, Y., Kim, J. et al. (2018). Properties and applications of starch modifying enzymes for use in the baking industry. *Food Science and Biotechnology 27* (2): 299–312.
113. Pai, J.S. (2003). Application of microorganisms in food biotechnology. *Indian Journal of Biotechnology 2*: 382–386.
114. Abada, E.A. (2019). Application of microbial enzymes in the dairy industry. In: *Enzymes in Food Biotechnology* (ed. M. Kuddus), 61–72. Academic Press.
115. Neelakantan, S., Mohanty, A.K., and Kaushik, J.K. (1999). Production and use of microbial enzymes for dairy processing. *Current Science 10*: 143–148.
116. Deeth, H.C. (2021). Heat-induced inactivation of enzymes in milk and dairy products. A review. *International Dairy Journal 15*: 105104.
117. Mir Khan, U. and Selamoglu, Z. (2020). Use of enzymes in dairy industry: a review of current progress. *Archives of Razi Institute 75* (1): 131–136.
118. Kårlund, A., Moor, U., Sandell, M. et al. (2014). The impact of harvesting, storage and processing factors on health-promoting phytochemicals in berries and fruits. *Processes 2* (3): 596–624.
119. Law, B.A. (2002). The nature of enzymes and their action in foods. *Enzymes in Food Technology 1*: 1–8.
120. Shiv, K. (2015). Role of enzymes in fruit juice processing and its quality enhancement. *Advances in Applied Science Research 6* (6): 114–124.
121. Garg, G., Singh, A., Kaur, A. et al. (2016). Microbial pectinases: an ecofriendly tool of nature for industries. *3 Biotech 6* (1): 1–3.
122. Toushik, S.H., Lee, K.T., Lee, J.S. et al. (2017). Functional applications of lignocellulolytic enzymes in the fruit and vegetable processing industries. *Journal of Food Science 82* (3): 585–593.
123. Sharma, H.P., Patel, H., and Sugandha (2017). Enzymatic added extraction and clarification of fruit juices – a review. *Critical Reviews in Food Science and Nutrition 57* (6): 1215–1227.
124. Mamo, J. and Assefa, F. (2018). The role of microbial aspartic protease enzyme in food and beverage industries. *Journal of Food Quality 3*: 2018.
125. Kumar, V.A., Kurup, R.S., Snishamol, C. et al. (2019). Role of cellulases in food, feed, and beverage industries. *Green Bio-processes.* 323–343.

20

The Potential of Sulfate-Reducing Microorganisms for the Bioconversion of Dissolved Sulfates to Sulfides Precipitating Metals of a Mine Liquid Effluent

Manuel J. Leal-Gutiérrez, Marisela Bernal-González, Ángel E. Chávez-Castellanos, Julio A. Solís-Fuentes, Enrique R. Bazúa-Rueda, and María-del-Carmen Durán-Domínguez-de-Bazúa

20.1. INTRODUCTION

Sulfate-reducing microorganisms (SRM) are a group of anaerobic communities widely found in anoxic environments with significant concentrations of organic matter and sulfates [1]. These microorganisms obtain the necessary energy for their biological processes by coupling the oxidation of organic compounds with the reduction of sulfate ions into sulfide ones. This bioconversion can be used to produce hydrogen sulfide (H_2S), to be used for precipitating divalent metals [2, 3]. These features of SRM have made possible the development of several systems to treat aqueous effluents of the mining industry, eliminating dissolved sulfates and metals [1, 4–8].

SRM exist in a wide range of environmental conditions, from psychrophilic up to thermophilic environments [3], as well as in acidic or basic environments with pH intervals between 2.5 and 11 [8, 9]. In addition, these microorganisms have the characteristic that their main product, hydrogen sulfide, in its undissociated form, is an inhibitory agent that can diffuse through the cell membrane, generating protein denaturation and interfering with the correct functioning of the metabolism [10]. Other important inhibitory agents of SRM are heavy metals, which have the ability to replace metal cofactors in their enzymes, altering their functions and structures. Given that each metal has a different level of affinity for binding to enzymes, the concentrations that have an inhibitory effect vary according to the metal. In the cases of copper, zinc, and lead, these effects have been demonstrated at concentrations higher than 12, 20, and 75 mg/L, respectively [10, 11].

The precipitation of metals produced by the dissolved metal ions with hydrogen sulfide has a positive impact on the performance of UASB (upflow anaerobic sludge blanket) reactor by avoiding the possible inhibitory effects of both hydrogen sulfide and heavy metals on SRM. Therefore, suitable conditions favoring this reaction have been analyzed. Choi et al. [12]

Laboratorios de Ingeniería Química Ambiental y de Química Ambiental, Departamento de Ingeniería Química, Facultad de Química, Universidad Nacional Autónoma de México, UNAM, Mexico City, Mexico

Microbial Biotechnology: Role in Ecological Sustainability and Research, First Edition.
Edited by Pankaj Chowdhary, Sujata Mani, and Preeti Chaturvedi.

found that the change in the Gibbs free energy (ΔG°) of the reaction becomes more negative as the temperature increases; that is, at a higher temperature, the precipitation reaction has a higher spontaneity, resulting in an increase in the reaction rate. However, this increase in the reaction rate does not generate a significant change in the precipitation efficiency, which allows the process to be carried out at a lower temperature with a respective increase in the residence time to reach an equivalent efficiency. Regarding pH, it affects metal precipitation in two ways. First, it determines the level of dissociation of hydrogen sulfide, so the precipitation reaction is favored with increases in pH since it is there where there is a greater dissociation of this compound [12]. Second, pH influences the solubility of the different metal species in the system. This is of vital interest when precipitation of metals such as carbonates or hydroxides, which precipitate at high pH values, is not desired [13]. If a selective precipitation is desired among the metallic sulfides present, the precipitation is carried out according to the pH value from which each metallic sulfide precipitates. For example, at a pH below 2 only copper sulfide precipitates, between 2 and 4 zinc sulfide precipitates, above pH 5 iron sulfide precipitates, and around pH 8 lead sulfide precipitates [4, 14–16].

The existence of these microorganisms at a wide range of conditions does not necessarily indicate an efficient sulfate-reduction system. Suitable conditions are normally found in a mesophilic environment near neutrality [14, 17]. Many investigations have been focused on evaluating the optimal conditions for SRM performance. However, in some instances, resource consumption is analyzed in order to reach these optimal conditions. For example, to find out if a process is making efficient use of energy, thermodynamic analysis is a suitable tool. There, it is possible to determine if there is a correspondence between the used resources and the results obtained. If this situation does not occur, it is necessary to identify the operations in the process that need improvement [18].

To do a thermodynamic analysis, thermodynamic information of the process is required, such as the thermodynamic properties of the process streams involved in the operation. One of the methods used is called process simulation. The objective of process simulation is to create a simulation model that represents a physical or chemical process from the calculation of mass and energy balances, considering reaction kinetics, transport phenomena, and also phase equilibrium [19, 20]. Once developed and validated, the simulation model can be the source of the thermodynamic information of the process; it can also be used to predict the performance of the process under any perturbation to which it may be exposed and, thus, increase its performance and efficiency, optimizing the use of energy resources control capacity, and minimizing operating costs [19–21].

The objective for this part of the research was to assess the operation of a laboratory scale upflow anaerobic sludge blanket reactor (UASB) with SRM. The UASB reactor received process water from a cooperating mine enterprise from its flotation units.

20.2. METHODOLOGY

20.2.1. UASB Reactor's Experimental Data

Experimental data for sulfates, sulfides, dissolved metals, and pH of the lab scale UASB reactor's influent and effluent were taken in the pseudo steady-state phase of 28 weeks [5]. The UASB reactor had a total volume of 2.2 L and a working volume of 1.9 L [22, 23]. The influent's rate was 149 mL of the mine's flotation unit aqueous effluent with 1 mL of lactic acid added as carbon source and 0.4 g of sodium bicarbonate as a source of alkalinity. The reactor's temperature was 28 °C using a heating system with warm water [5].

For the gaseous product, a three-stage capture set-up was carried out. The first stage consisted of an alkaline solution of calcium carbonate at saturation, which had the objective of absorbing carbon dioxide. The second stage used a saturated solution of zinc acetate to react with hydrogen sulfide, leaving the reactor in the gaseous phase and forming zinc sulfide that could be quantified. The last stage had a sodium hydroxide solution, from which the displaced volume of methane formed in the UASB reactor was measured [24].

20.2.2. Laboratory Scale UASB Reactor Modeling

The collected experimental data were used for creating a model to simulate the lab scale UASB reactor operation using the Aspen Plus® software. The laboratory scale UASB reactor was theoretically modeled by dividing its height into 10 sections (Figure 20.1).

Each of the sections included two stages, the biochemical one represented as *RBIO*, and the physical–chemical one represented by a separator *B* and a reactor where the metals precipitated, *RMETAL*. The tenth physical–chemical stage included an additional separator, *B11*, that was necessary to reach the correct distribution of sulfides among the liquid and gaseous phases. The mixer, MIX, merged the gas flows coming out of each section to have a unique gaseous effluent coming out of the reactor, which might simulate the experimental one. The flow diagram also showed the heat exchangers IC1, IC2, and IC3, for adapting the UASB reactor inflow and outflow temperatures, considering both the laboratory temperature and the operating one. Laboratory temperature used in the simulation model was 18 °C, corresponding to the average temperature for the lapse of the experiments published by the Mexico's National Water Commission [25] (*Comisión Nacional del Agua, CONAGUA*, in Spanish) [13].

The biotransformation of sulfates into sulfides was modeled using Monod kinetics with the parameters proposed by Mattei et al. [26] (2015, kinetic model 1), by Torner-Morales and Buitrón [27] (2010, kinetic model 2), and by Tang et al. [28] (2007, kinetic model 3). These kinetic models were added with an inhibition parameter for pH proposed by Batstone et al. [29] for anaerobic degradation, previously employed by Kvarnström and Lönntoft [30] for SRM. Also, an adjustment parameter, Parameter P, was considered according to the effluent experimental results. Further, the hydrogen sulfide fraction that reacted with each metal on the physical–chemical stage was manually adjusted to match the experimentally characterized concentrations of dissolved metal ions in the liquid effluent [24].

20.2.3. Thermodynamic Properties and Analysis

Enthalpy, entropy, and exergy from all process streams were calculated following the procedure described in Leal-Gutiérrez et al. [24]. For the thermodynamic analysis, the energy efficiency of the system was defined as the relationship between its environmental benefit (useful work that is not lost when sent to the ambient) and the energy consumption required to obtain this environmental benefit [5, 24].

20.2.4. Sensitivity Analysis

A sensitivity analysis was performed using the Aspen Plus software to evaluate the process response to the variation of pH in the lab scale UASB reactor's influent. The pH variable was selected since it has a very significant impact on the performance of the SRM and it has the ability to be adjusted by adding or reducing sodium bicarbonate [5].

20.3. RESULTS AND DISCUSSION

The concentration of sulfates during the experimental lapse is presented in Figures 20.2a,b. It reaches an average biotransformation to sulfides by SRM of 84.9% inside the lab scale UASB reactor according to Kaksonen and Puhakka [4]. In the same figure, an average concentration of 1.24 mg sulfides/L in the influent increased to 104.85 mg sulfides/L in the effluent.

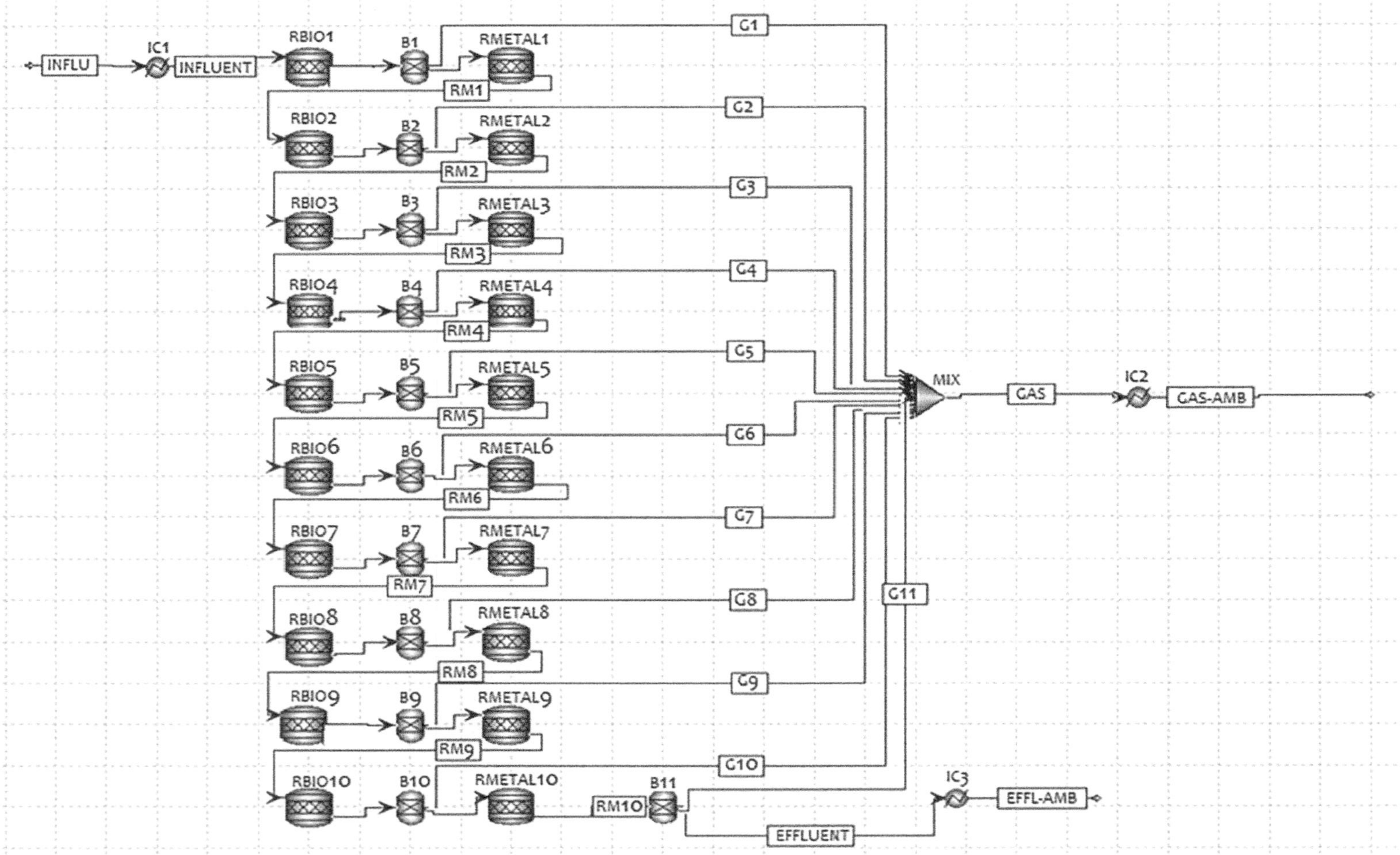

Figure 20.1 Flow diagram of the operation of the lab scale UASB reactor simulated using Aspen Plus.

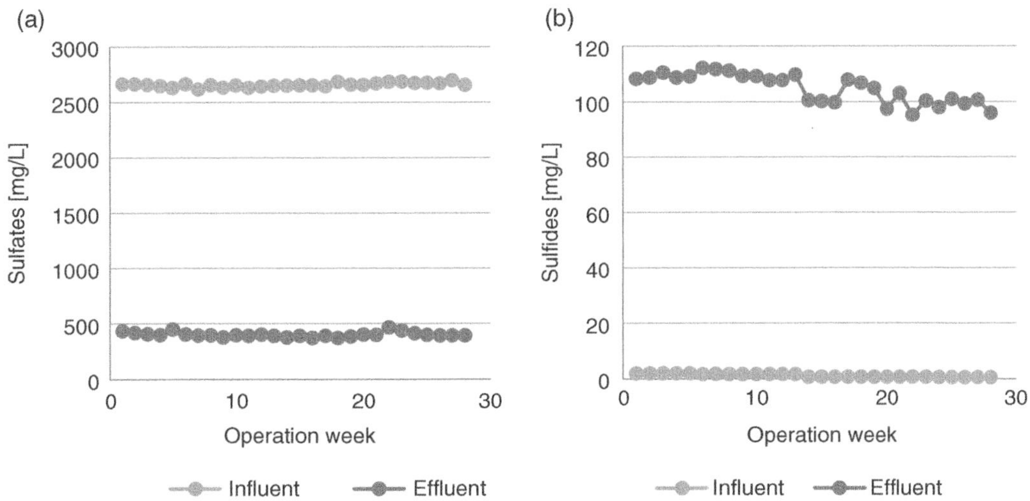

Figure 20.2 (a) Sulfate concentration behavior in the liquid phase of the lab scale UASB reactor. (b) Sulfide concentration behavior in the liquid phase of the lab scale UASB reactor.

The sulfide concentration does not account for the total amount of sulfides produced by the SRM, since some of them reacted with the metal ions present in the media. In this study, the influent had a concentration of 0.187 mg Cu/L, 8.881 mg Zn/L, 1.226 mg Pb/L, and 4.445 mg Fe/L, which are important metals for the cooperating mine industry [5]. In Figure 20.3a, the performance of each metal ion in the effluent during the experiment's lapse is observed, reaching an average removal of 84.5% for Zn, 62.3% for Fe, 52.8% for Cu, and 46.4% for Pb during this phase of the study. Additionally, the microbiological sulfate reduction also produced bicarbonate ions that in a direct form increased pH values inside the lab scale UASB reactor. During these 28 weeks, an average pH value of 3.01 in the influent, and an average pH value of 6.83 in effluent were measured (Figure 20.3b).

Table 20.1 presents the values for the Parameter P of each kinetic model evaluated. This parameter adjusted the experimental data to the simulation model. For the three kinetic models, the parameter was smaller than one, indicating that the rate of reaction of the kinetic models overestimated the lab scale UASB reactor performance.

With the simulation model developed, a thermodynamic analysis was carried out [24]. According to this analysis, it was found that the process efficiency of the experimental system corresponded to 5.34%. Therefore, three options for the operation for the use of energy in the process were studied in order to increase its efficiency.

The first option takes into account the effluent streams that leave the process at ambient temperature (18 °C) and not at the operating lab scale UASB reactor temperature (28 °C). For that, a heat exchanger might be proposed to transfer the thermal load from the effluent to the influent, increasing the process efficiency to 7.47%.

During the thermodynamic analysis, it was found that the highest energy consumption corresponded to the heating system to maintain the lab scale UASB reactor at a constant temperature. To eliminate this consumption, a suitable option would be to isolate the lab scale UASB reactor by covering its external surface, considerably reducing the heating system needs. This second option would increase the process efficiency up to 38.40%.

The third option to eliminate the heating system of the reactor would be to operate it in a geographical location with a higher ambient temperature. This location might be the mine

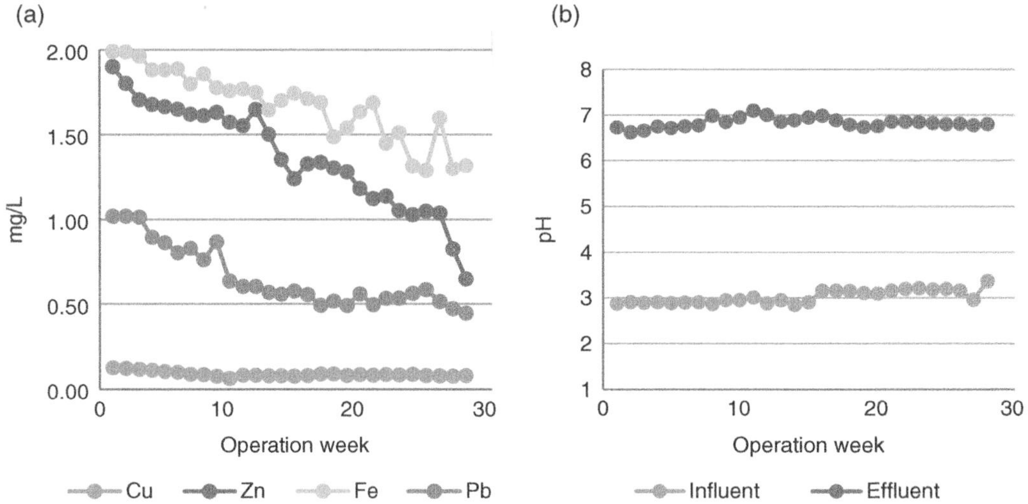

Figure 20.3 (a) Performance of the dissolved metal ions in the liquid phase of the lab scale UASB reactor. (b) Performance of the hydrogen ions (pH) dissolved in the liquid phase of the lab scale UASB reactor.

Table 20.1 Parameter P for the kinetic models evaluated.

Kinetic model	Reference	Parameter P
1	Mattei et al. [26]	0.59487
2	Torner-Morales and Buitrón [27]	0.61345
3	Tang et al. [28]	0.95032

installations. In this scenario, the heating system would not be necessary. Also, the heating of the influent would not be required. The process efficiency for this option would be 48.72%.

These three options might improve the performance of SRM within the reactor concomitantly with the energy use. To evaluate how SRM performance affected the thermodynamic study, a sensitivity analysis was done considering the pH changes of the influent. Results of this analysis are presented in Figure 20.4a,b, where expected concentrations of sulfates and metals dissolved in the lab scale UASB reactor's effluent are shown.

Studying these results, it is evident that the increase of the pH values in the influent augments the microbial sulfate reduction to such an extent that in the theoretical scenario of feeding the UASB reactor with a pH higher than 6 in the influent, a 99.9% removal of sulfates from the system is achieved.

These results are consistent with those reported by Jong and Parry [31], who calculated that the sulfate reduction process can be increased by up to 300% by increasing the pH of the experiments from 3.5 to 6 when lactate is used as a carbon source, and thus, the precipitation of the metals. This trend was also observed for the energy efficiency of each proposed option. From them, at an influent with a pH value of 6.2, the efficiency values would be 8.27, 41, and 51.43%, respectively.

20.4. CONCLUSIONS

Considering the several thermodynamic theoretical analyses based on data from the experiments at laboratory scale level, it was possible to identify improvement points that could significantly increase the energy efficiency of the process. The operation of the UASB

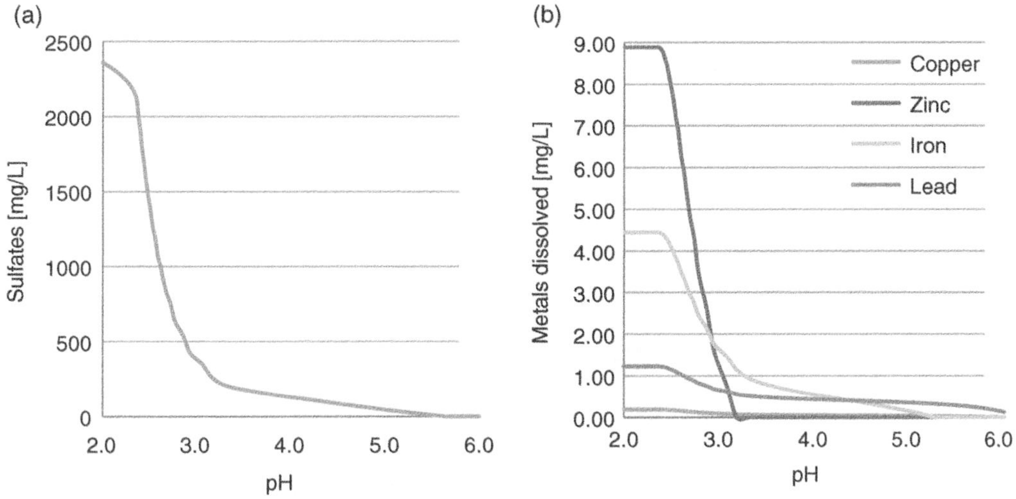

Figure 20.4 (a) Behavior of the concentration of dissolved sulfates as a function of the pH of the influent. (b) Behavior of the concentration of dissolved metal ions as a function of the pH of the influent.

reactor with a feed stream with a pH near to neutrality, and carrying out thermal insulation of the UASB reactor, could increase the efficiency from the current 5.34 to 41%. At the same time, the implementation of this technology in an area with an ambient temperature in the mesophilic range, such as the cooperating mining company, would allow an energy efficiency of over 50%, which contrasts drastically with the current efficiency.

20.5. FURTHER TRENDS

If the pandemic is controlled and access to the laboratories at full-time conditions is allowed (since presently the access is only to maintain the ongoing biosystems in a stable manner), the next steps would be to study the microbial communities proliferating in the bioreactor in the same way as has been done in this research group some years ago for similar bioreactors operating with effluents of the sugarcane-ethyl alcohol industry [32, 33].

Also, new experiments are planned to collect data at different stages of the bioreactor in order to corroborate the effects of a better isolation of the UASB reactor for improving energy efficiency (Figure 20.5a,b).

ACKNOWLEDGEMENTS

This research was partially supported by the National Autonomous University of Mexico through its General Directorate for Academic Personnel Affairs (*Universidad Nacional Autónoma de México, Dirección General de Asuntos del Personal Académico, UNAM-DGAPA*, in Spanish), to acquire materials, reagents, and small equipments (Project PAPIIT IN115118) in charge of Prof. Dr. Enrique Rodolfo Bazúa-Rueda and Prof. Dr. Ángel Enrique Chávez-Castellanos. Also, funds were provided by the Academic Secretary for Research and Graduate Studies of the Faculty of Chemistry (*Secretaría Académica de Investigación y Posgrado, SAIP, de la Facultad de Química, UNAM*, in Spanish) for other expenses with the annual funds allocated to professors María-del-Carmen Durán-Domínguez-de-Bazúa (5000-9067) and Enrique Rodolfo Bazúa-Rueda (5000-9065). The first author acknowledges the graduate

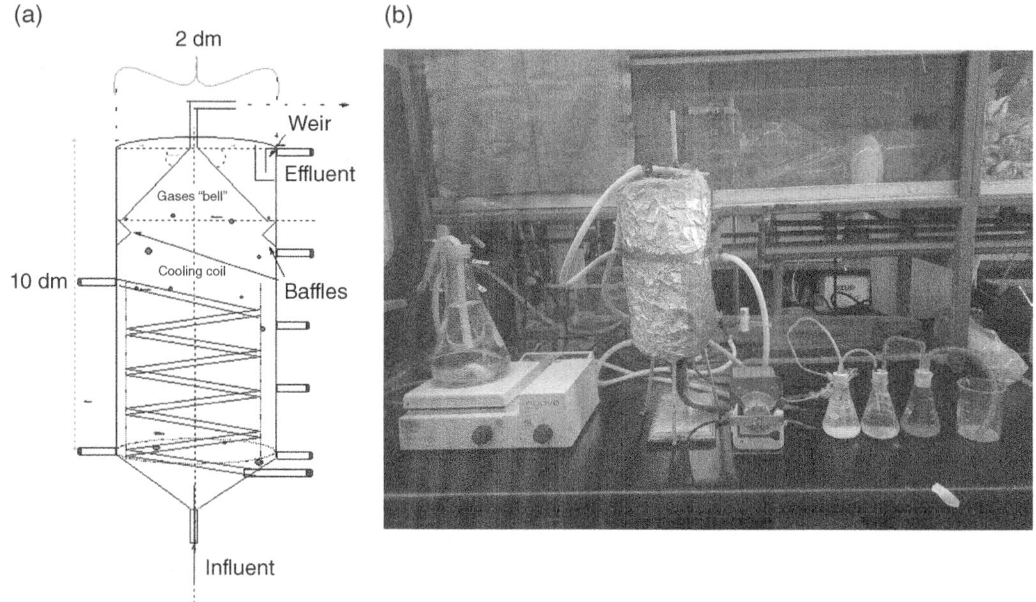

Figure 20.5 (a) Laboratory scale UASB reactor built for sampling at several stages. (b) Laboratory scale UASB reactor built for sampling at several stages and isolated to improve energy efficiency.

studies scholarship (*CVU*: 1003833) granted by the Mexico's National Council of Science and Technology (*Consejo Nacional de Ciencia y Tecnología, Conacyt*, in Spanish). Finally, authors acknowledge the mine's personnel logistic support (process information and technical data for the flotation units) as well as the sending of water effluents to the UNAM's laboratories to carry out this experimental research.

REFERENCES

1. Tang, K., Baskaran, V., and Nemati, M. (2009). Bacteria of the sulphur cycle: An overview of microbiology, biokinetics and their role in petroleum and mining industries. *Biochemical Engineering Journal 44* (1): 73–94.
2. Muyzer, G. and Stams, A.J.M. (2008). The ecology and biotechnology of sulphate-reducing bacteria. *Nature Reviews. Microbiology 6* (6): 441–454.
3. Rabus, R., Hansen, T.A., and Widdel, F. (2006). Dissimilatory sulfate- and sulfur-reducing prokaryotes. In: *The Prokaryotes. A Handbook on the Biology of Bacteria*, 3e (ed. S. Falkow, E. Rosenberg, K.-H. Schleifer and E. Stackebrandt), 659–768. Berlin, Germany: Springer https://doi.org/10.1007/0-387-3074207_22.
4. Kaksonen, A.H. and Puhakka, J.A. (2007). Sulfate reduction based bioprocesses for the treatment of acid mine drainage and the recovery of metals. *Engineering in Life Sciences 7* (6): 541–564.
5. Leal-Gutiérrez, M.J., Cuéllar-Briseño, R., Castillo-Garduño, A.M. et al. (2021). Precipitation of heavy metal ions (Cu, Fe, Zn, and Pb) from mining flotation effluents using a laboratory-scale upflow anaerobic sludge blanket reactor. *Water, Air, and Soil Pollution 232* (5): 197. https://doi.org/10.1007/s11270-021-05042-1.
6. Leal-Gutiérrez, M.J., Bazúa-Rueda, E.R., Bernal-González, M. et al. (2021). Efficient use of energy in nature: Sulfate-reducing microorganisms in the transformation of ions in aqueous effluents from the mining industry (*Uso eficiente de la energía en la naturaleza: Microorganismos sulfato-reductores en la transformación de iones en efluentes acuosos de la industria minera*, in Spanish). In Proceedings

of the 7th International Seminar of Experts on the Treatment of Industrial Effluents and Residues in Honor of Mrs. Dr. Luisa Tan Molina (October 1–2). Mexico City, Mexico. Virtual event.

7. Pacheco-Gutiérrez, L.A. (2006). Proposal for holistic reuse of process water of a mining enterprise using anaerobic biological processes. Master's thesis in Chemical Engineering (Processes). Program of Master's and Doctoral Studies in Engineering, Faaculty of Chemistry, National Autonomous University of Mexico. Mexico City, Mexico (*Propuesta de aprovechamiento integral del agua de proceso de una empresa minera empleando sistemas biológicos anaerobios. Tesis de maestría en Ingeniería Química (Procesos). Programa de Maestría y Doctorado en Ingeniería, Facultad de Química, UNAM. Ciudad de México, México*, in Spanish).
8. Tsukamoto, T.K., Killion, H.A., and Miller, G.C. (2004). Column experiments for microbiological treatment of acid mine drainage: low-temperature, low-pH and matrix investigations. *Water Research 38* (6): 1405–1418.
9. Foti, M., Sorokin, D.Y., Lomans, B. et al. (2007). Diversity, activity, and abundance of sulfate-reducing bacteria in saline and hypersaline soda lakes. *Applied and Environmental Microbiology 73* (7): 2093–2100.
10. Paulo, L.M., Stams, A.J.M., and Sousa, D.Z. (2015). Methanogens, sulphate and heavy metals: a complex system. *Reviews in Environmental Science and Biotechnology 14* (4): 537–553.
11. Utgikar, V.P., Tabak, H.H., Haines, J.R. et al. (2003). Quantification of toxic and inhibitory impact of copper and zinc on mixed cultures of sulfate-reducing bacteria. *Biotechnology and Bioengineering 82* (3): 306–312.
12. Choi, J.Y., Kim, D.S., and Lim, J.Y. (2006). Fundamental features of copper ion precipitation using sulfide as a precipitant in a wastewater system. *Journal of Environmental Science and Health – Part A 41* (6): 1155–1172.
13. Gallegos-García, M., Celis, L.B., Rangel-Méndez, R. et al. (2009). Precipitation and recovery of metal sulfides from metal containing acidic wastewater in a sulfidogenic down-flow fluidized bed reactor. *Biotechnology and Bioengineering 102* (1): 91–99.
14. Sahinkaya, E., Özkaya, B., Kaksonen, A.H. et al. (2007). Sulfidogenic fluidized-bed treatment of metal-containing wastewater at low and high temperatures. *Biotechnology and Bioengineering 96* (6): 1064–1072.
15. Tabak, H.H., Scharp, R., Burckle, J. et al. (2003). Advances in biotreatment of acid mine drainage and biorecovery of metals: 1. Metal precipitation for recovery and recycle. *Biodegradation 14* (6): 423–436.
16. Tokuda, H., Kuchar, D., Mihara, N. et al. (2008). Study on reaction kinetics and selective precipitation of Cu, Zn, Ni and Sn with H_2S in single-metal and multi-metal systems. *Chemosphere 73* (9): 1448–1452.
17. Willow, M.A. and Cohen, R.R.H. (2003). pH, dissolved oxygen, and adsorption effects on metal removal in anaerobic bioreactors. *Journal of Environmental Quality 32* (4): 1212–1222.
18. Álvarez-Hincapie, C.F. and Velasquez-Arredondo, H.I. (2013). Exergy in biological systems: Holistic approach for the study of ecosystems and environmental management (*Exergía en sistemas biológicos: Aproximación holística para el estudio de ecosistemas y el manejo ambiental*, in Spanish). *Producción + Limpia 8* (2): 106–127.
19. Eden, M.R. (2012). Introduction to Aspen Plus simulation. *Chemical Engineering Department, Auburn University, CHEN 4460*: 1–11.
20. Gil-Chaves, I.D., Guevara-López, J.R., García-Zapata, J.L. et al. (2015). *Process Analysis and Simulation in Chemical Engineering*. Berlín, Germany: Springer Verlag.
21. Espínola-Lozano, F. (2017). *Tutorial de Aspen Plus Introducción y modelos simples de operaciones unitarias*. Spain: Universidad de Jaen. Jaen.
22. Ojeda-Berra, L. (2008). Modelo matemático de un reactor anaerobio con bacterias sulfatorreductoras para el tratamiento del efluente del proceso de flotación de una planta minera. Tesis de maestría en Ingeniería Química (Procesos). Programa de Maestría y Doctorado en Ingeniería, Facultad de Química, UNAM. Ciudad de México.
23. Ríos-Vázquez, J.L. (2009). Diseño, construcción y arranque de un reactor anaerobio de lecho de lodos de flujo ascendente para el estudio de la precipitación de metales de efluentes de la industria minera. Tesis profesional de Ingeniería Química, Facultad de Química, UNAM. Ciudad de México.

24. Leal-Gutiérrez, M.J., Bernal-González, M., Durán-Domínguez-de-Bazúa, M.d.C. et al. (2021c). Exergy analysis of a laboratory-scale upflow anaerobic sludge blanket reactor with sulfate-reducing microorganisms to precipitate heavy metal ions (Cu, Fe, Zn, and Pb) using Aspen plus®. *Brazilian Journal of Chemical Engineering* in press.
25. CONAGUA. Monthly abstracts for ambient temperature and rain / *Resúmenes mensuales de temperaturas y lluvia*. https://smn.conagua.gob.mx/es/climatologia/temperaturas-y-lluvias/resumenes-mensuales-de-temperaturas-y-lluvias (in Spanish)
26. Mattei, M.R., D'Acunto, B., Esposito, G. et al. (2015). Mathematical modeling of competition and coexistence of sulfate-reducing bacteria, acetogens, and methanogens in multispecies biofilms. *Desalination and Water Treatment 55* (3): 740–748.
27. Torner-Morales, F.J. and Buitrón, G. (2010). Kinetic characterization and modeling simplification of an anaerobic sulfate reducing batch process. *Journal of Chemical Technology and Biotechnology 85* (4): 453–459.
28. Tang, Y., Pingitore, F., Mukhopadhyay, A. et al. (2007). Pathway confirmation and flux analysis of central metabolic pathways in *Desulfovibrio vulgaris* Hildenborough using gas chromatography-mass spectrometry and Fourier transform-ion cyclotron resonance mass spectrometry. *Journal of Bacteriology 189* (3): 940–949.
29. Batstone, D.J., Keller, J., Angelidaki, I. et al. (2002). The IWA anaerobic digestion model no 1 (ADM1). *Water Science and Technology 45* (10): 65–73.
30. Kvarnström, H. and Lönntoft, E. (2013). Modelling of metal recovery using sulfate reducing bacteria, simulation in MATLAB and Aspen Plus. Master's thesis. Lund University. Sweden.
31. Jong, T. and Parry, D.L. (2006). Microbial sulfate reduction under sequentially acidic conditions in an upflow anaerobic packed bed bioreactor. *Water Research 40* (13): 2561–2571.
32. Bernal-González, M., Vergara-Salgado, S., and Durán-Domínguez-de-Bazúa, M.C. (2012). Identification of methanogenic and sulfato-reducing bacteria from three upflow anaerobic sludge blanket reactors operating at 45, 55, and 65°C with sugarcane-based vinasses diluted with tap water (50:50). *Journal of Microbial & Biochemical Technology 4*: 111–120. https://www.cabdirect.org/cabdirect/abstract/20133016649, http://dx.doi.org/10.4172/1948-5948.1000080.
33. Bernal-González, M, Toscano-Pérez, D, Falcón, L.I. (2016). Archaea, a useful group for unconventional energy production: Methane production from sugarcane secondary distillation effluents using thermotolerant strains. In: *The Handbook of Microbial Bioresources* (eds. V.K. Gupta, G.D. Sharma, M.G. Tuohy and R. Gaur), 502–527. Oxfordshire, United Kingdom: CABI. ISBN: 9781780645216, https://www.cabi.org/ISC/ebook/20163199943.

21

The Human Microbiome: An Imminent Therapy for Mankind – A Review

Immanuel Suresh, Iswareya Lakshimi, and Abinaya Lakshmi

21.1. INTRODUCTION

How is a sound microbiome characterized? From an ecological standpoint, the strength of a local area (bacterial or otherwise) can be considered as a utilitarian property apart from the wellbeing of that local area. Strength alludes to the capacity of a local area to oppose change in the setting from an ecological pressure (opposition), or to get back to a harmonious state following a pressure-related irritation (versatility). These ideas of obstruction and strength as key highlights of sound microbiomes are consistent with set perspectives on microbial environments. Nonetheless, the idea that a sound microbiome can be characterized by some glorified local area made up of characterized populaces of explicit organisms is oversimplified, considering the steady interindividual contrasts seen in numerous examinations [1].

An elective conceptualization is a glorified assortment of qualities and pathways instead of explicit populaces; albeit, a "sound" set of metabolic capacities still needs to be characterized. Certain microbial appropriations may make an individual more powerless to contamination or sickness. For instance, modification of the native gut microbiota by anti-infection agents can put a person in danger of becoming contaminated from an astute microorganism, for example, *Clostridium difficile*. Diverse microbial populaces can drastically influence vulnerability to constant irritation. The presence of microorganisms that convert luminal compounds into potential cancer-causing agents likewise puts one at expanded danger for malignant growth and can prompt antagonistic reactions to chemotherapeutic specialists [1].

The absence of adequate variety or equality in a bacterial local area structure seems to decrease its capacity to withstand irritation. Hosts with such bacterial networks may not show clear illness under many circumstances. Be that as it may, their bacterial networks may still not be ideal for forestalling illness, and these individuals might be more powerless when encountering various sicknesses. An increasingly common worldview is that conditions like being overweight and internal illnesses like inflammatory bowel diseases (IBD) are related to decreased variety in the intestinal microbiome, which may constitute proof of an imperfect microbiome [1].

PG Department of Microbiology, The American College, Madurai, Tamil Nadu, India

Microbial Biotechnology: Role in Ecological Sustainability and Research, First Edition.
Edited by Pankaj Chowdhary, Sujata Mani, and Preeti Chaturvedi.

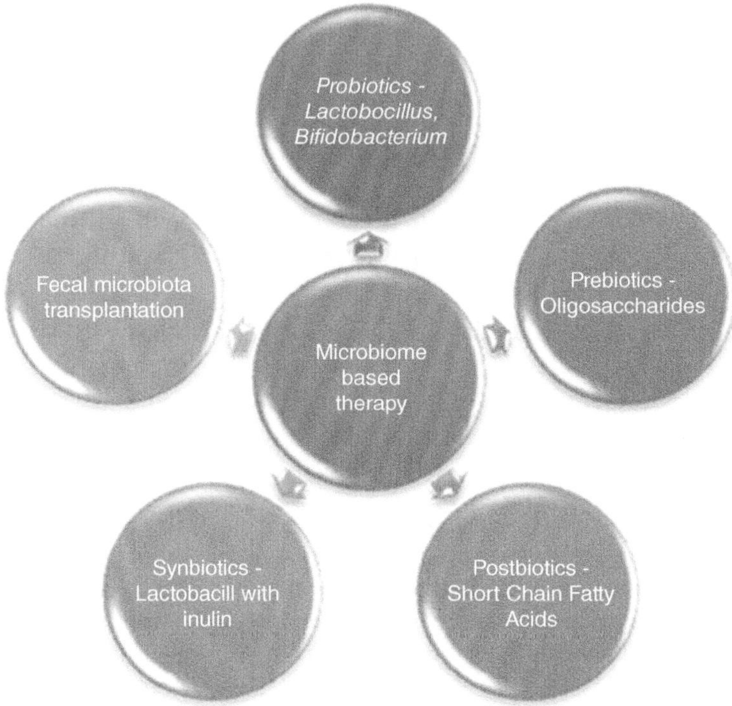

Figure 21.1 Various methods of microbiome-based therapy.

The human microbiome includes microorganisms, growths, Archaea, infections, and parasites, shaping dynamic consortia, all cooperating with themselves and human cells, and being linked to wellbeing and illness status [2]. Because of its extensive role in different diseases, the microbiome has become a significantly attractive platform for potential therapeutics. The microbiome maintains the homeostasis of the body and can be used to prevent and treat various diseases. Microbiome-based therapies include various mechanisms like probiotics, prebiotics, postbiotics, synbiotics, and fecal microbiota transplantation (FMT) (Figure 21.1). Probiotics is the feeding of viable microorganisms that can confer beneficial effects by altering the host microflora. Prebiotics are a kind of supplement to probiotics that are indigestible and promote the development and activity of probiotics and microflora. Postbiotics are bacterial metabolites or byproducts that are non-viable and are derived from probiotic bacteria, which confer beneficial effects on the host. Synbiotics is the combination of probiotics and prebiotics to confer stability and enhanced activity to probiotics that are administered. FMT is a widely studied microbiome-based therapy in which the fecal matter of the healthy donor is transplanted to the recipient to alter their gut microbiome and treat diseases [3].

21.1.1. Human Gastrointestinal Tract

The human gastrointestinal (GI) tract is home to various bacterial species, which have the capacity to support, process, and abet healthy systems, working with the development of the colonic epithelium and assurance from microorganisms. The human gut microbiome differs in each person and is usually steady and tough in the long term; however, natural components, including prebiotics, probiotics, diet, infections, and medications, especially anti-infection agents, can interfere with the GI tract. Various sicknesses are associated with the gut microbiota, including irresistible conditions (irresistible gastroenteritis, *C. difficile* contamination [CDI]), immune system illnesses (hypersensitive infection, diabetes, incendiary gut illness [IBD]), some broad conditions (being overweight, practical GI problems), and conduct diseases [4]. Currently, a wide assortment of clinical restorative procedures are utilized to address gut dysbiosis, yet an extraordinarily large number of them don't present good clinical impacts, aside from FMT. FMT, especially stool transplantation, is a procedure which transplants stool from a healthy provider into another individual's GI tract to directly change the recipient's stomach microbiota and normalize the environment, getting therapeutic benefit. Currently, various assessments have shown FMT to be a productive treatment for discontinuous and persistent CDI, even in patients who have comorbid conditions or are immunosuppressed. Beyond its application in CDI, more expansive use of FMT has been discovered recently [5].

21.1.2. Metabolic and Immunologic Functions of Indigenous Gut Microbiota

The stomach microbiota can create biochemical activities that could alternate over luminal blends to discretionary metabolites. These extra reactions (fashions recorded in red) can detoxify ingested harmful substances but in numerous instances can reap the manufacturing of blends that may be poisonous. The specific synthesis of the intestine microbiota might for that reason be capable of determining the balance between helpful and risky substance transformation responses inside the intestinal lumen. The gut microbiota can move toward the host invulnerable framework and gut epithelium to establish the sense of mucosal insusceptibility. Considering the stomach microbiota, the host will produce a grouping of cytokines and monitor effect or particles that can in this way shape the local microbiota prime host responses to regular lifts [1].

21.2. FECAL MICROBIOTA TRANSPLANTATION

Fecal microbiota transplantation (FMT) is a way to immediately extrude the recipient's stomach microbiota to normalize the mixture and it also has a remedial benefit. FMT has been recognized since the fourth century but became common beginning in 2013, when the American Food and Drug Administration upheld FMT for treating discontinuous and unshakable *C. difficile* illness. Since then, the volume of FMT packages has widened unexpectedly and considerably for GI issues, as well as for extra-GI ailments. Many tests, including oral, blood, and stool tests must be undertaken before FMT to reduce the chance of adverse effects. Determining fecal readiness for individuals by properly selecting appropriate conveyance strategies through male- and female-centric delivery methods are key aspects of the FMT cycle. Although the current evidence proves FMT is a mostly safe remedial strategy with few negative impacts, the long-term results of FMT have not been completely clarified. Thus, developing periodicity and length of standard advancement after FMT to screen for clinical sufficiency and long-term adversarial events are important issues. Eventually, we will expect modified FMT for different patients and conditions depending on individual hosts and diseases [5].

21.2.1. History of Fecal Microbiota Transplantation

FMT may have been first observed in fourth-century China, when human waste cloth become known as yellow soup and was used by sufferers of outrageous detachment of the entrails. Until the Chinese Ming Dynasty in the 16th century, there had been depictions of recent or evolved waste suspensions implemented in sufferers with GI conditions, inclusive of the diarrhea, blockages, and belly pain. Eiseman and his accomplices credibly worked with patients with FMT for pseudomembranous colitis in 1958, an important milestone in the scientific work surrounding FMT [5].

As early as the 1980s, the gut microbiota had been related to being overweight in humans and rodents through the use of society subordinate strategies, which just look at bits of microorganisms in the gut. During the 1990s, the use of culture-free sub-atomic strategies dependent on 16S ribosomal RNA (rRNA) qualities extended our insight to the human gut microbiota, overwhelmed by microscopic organisms of Bacteroidetes and Firmicutes. Additionally, relocating gut microbiome from patients after Roux-en-Y gastric detour led to a decreased fat statement in sans germ mice. These outcomes on the whole stressed that weight-related aggregates, for example, expanded weight and diminished adiposity, are contagious from individuals to mice utilizing the customized microbiome [6].

21.2.2. Etiology of Fecal Microbiota Transplantation

The study of weight gain and obesity has become more popular as they have become more prevalent in our society. After examining well-known threat elements like dietary regimen, manner of existence, and monetary status, intestine microbiota arises as a typical new issue that plays a non-public role in weight gain and loss. In spite of the truth that endeavors to sum up the many-sided connections among intestine microbiota and weight with the aid of using compositional records remains comprehensive, what we eat (i.e. food plan and anti-toxins) has been unequivocally demonstrated to have an effect on our intestinal microbiota, which finally ends up influencing host digestion and irritation. A new study has introduced those prebiotics that have been helpful to obese and larger children in bringing down their frame weight, body fat percentage, and serum stage of interleukin and fatty oil, along elevated Bifidobacterium spp., and further, decreased *Bacteroides vulgatus*. Accordingly, adjusting intestine microbiota via a weight-reduction plan and dietary supplements is not only a hypothetically effective manner to cope with mitigating the manifestations associated with corpulence, but moreover addresses a possible useful avenue to heaviness[6].

21.2.3. Application of Fecal Microbiota Transplantation

The most dependable record of FMT applied in a non-aggressive disease was disseminated in 1989 as "an exchange of stomach plant life" on a 45-year old male with unmanageable ulcerative colitis (UC), showing full and complete clinical recovery after treatment. After the clinical usage of FMT moved from aggressive to non-communicable issues, the extent of FMT applications expanded rapidly. Moreover, new encounters partner stomach microbiota to extraintestinal diseases that would further grow the clinical prowess of FMT [5].

21.2.4. Fecal Microbiota Transplantation-Based Therapy

FMT is an exceptionally viable remedy for RCDI, however expanded information on the intestinal microbiota in wellness support, as well as controlled preliminary studies of FMT for a wide scope of issues, are required before FMT can be acknowledged and applied clinically, for issues such as *C. difficile* contamination, FMT, fiery gut infection, peevish gut condition, and weight [7].

21.2.4.1. Irritable Bowel Disease

The altered intestinal microbial piece (dysbiosis) and metabolic items actuate forceful mucosal insusceptible reactions that intercede provocative gut illnesses (IBD). This dysbiosis disables the capacity of administrative insusceptible cells, which ordinarily advance mucosal homeostasis. Normalizing and keeping up with administrative invulnerable cell work by amending dysbiosis provides a promising way to treat IBD patients. Nonetheless, existing organism designated treatments, including anti-toxins, prebiotics, probiotics, and FMT, give variable results that are not ideal for current clinical applications. This study talks about ongoing advancement in understanding the dysbiosis of IBD and the reason for remedial rebuilding of homeostatic insusceptible capacity by controlling an individual patient's microbiota arrangement and capacity. We accept that distinguishing more exact remedial targets and creating proper and quick indicative devices will direct more viable and more secure microorganism-based enlistment and upkeep medicines for IBD patients that can be applied in a customized way [8].

The initiation, movement, multiplication, separation, and upkeep of an assortment of mucosal invulnerable cells are controlled by inhabitant microbiota. These enacted resistant cells participate to keep up with intestinal homeostasis in ordinary hosts. Incendiary insusceptible cells assist by attacking microorganisms by exceptionally powerful excess inborn and versatile invulnerable systems. Microbiota support the natural insusceptible reaction against microorganisms by animating discharge of antimicrobial peptides and cytokines, for example, TNFa, IL-22, and IL-17, and actuating the inflammasome to be hostile to microbe protection. In contrast, managerial resistant cells, including regulatory T cells, B cells, dendritic cells (DC), macrophages, and inborn lymphoid cells (ILCs), balance unreasonable provocative reactions. The recurrence and elements of these administrative cells are debilitated in IBD, however they can possibly be invigorated by microbial control to re-establish invulnerable homeostasis which will turn around and standardize dysregulated insusceptible capacity and improve mucosal irritation. In this way, designated acceptance and support of administrative safe cells by controlling microbial profiles and capacities offer a promising way to treat IBD patients [8].

21.2.4.1.1. A Personalized Method of Treatments Human IBD incorporates hereditarily and clinically heterozygous patient subpopulations with extremely special intestinal bacterial arrangements and capacities that assist with determining resistant reactions and infection results. Accordingly, we accept that it will be possible to assess the microorganism/invulnerable profiles by quick, demonstrative trials of microbiota utilitarian and mucosal safe profiles to coordinate profoundly successful and safe medicines in a customized way. Re-establishing weakened administrative safe cell action by rectifying dysbiosis and imperfect microbial metabolic capacities is a novel and profoundly encouraging way to deal with oversight of IBD in a more physiological, secure, and supported way. Revealing the ways fundamentally flawed microscopic organisms have associations in each IBD patient will allow for precise altering of microbiota and their capacity effectively [8].

Dysbiosis-related mucosal invulnerable malfunctions with IBD. Enteric contamination, medications including anti-toxins, immunosuppressive, and NSAIDs medications, diet, smoking and alcohol, and mental stress in vulnerable hereditary people cause microbial dysbiosis and metabolic changes. Delayed dysbiotic conditions described by expanded forceful bacterial strains and diminished administrative species lead to malfunction of the mucosal safe reaction. Intense microbial assemblies begin provocative response by activating Th1/Th17-effector cells, while lessened regulatory species impair the selection and limit of authoritative cells that incorporate administrative T cells (Treg), B cells (Breg), macrophages (MU), DC, and intrinsic lymphoid cells (ILCs). This awkwardness of mucosal cytokine profiles in combination with flawed obstruction work supports mucosal irritation and can conceivably prompt IBD in defenseless people [8].

21.2.4.2. Fecal Microbiota Transplantation in the Treatment of Sepsis

The impact of FMT is by all accounts free of the balance of the intestinal vegetation with probiotics. To affirm this theory, we produced another model of irritation. Zymosan-prompted peritonitis is a typical model to consider foundational provocative reaction and different organs' malfunctioning condition. In this model, three markers of oxidative harm –TBARS, protein carbonyls, and nitrite-nitrate – were of concern; FMT, autonomous of supplementation with *Lactobacillus*, diminished the levels of these markers. A comparative example was noted for incendiary markers. MPO action and IL-1b, IL-6, and TNF-a levels expanded after zymosan organization, as clear as the decline in gut aggravation in the LPS + FMT treatment group. Histological examination was completed on the digestive tracts of individuals submitting to zymosan treatment followed by FMT. The intestinal villi were preserved, but extensive infiltration of lymphocites into the submucosa with lymphoid follicular hyperplasia was seen in the zymosan-treated group. In the control migrate group, the perienteric fat tissue had minor lymphoplasmacytic blazing attacks. Some lymphocytic attacks were found in *Lactobacillus casei* and *Lactobacillus rhamnosus* collective events [4].

21.2.5. Regulation of Fecal Microbiota Transplantation

The guidelines for FMT use are a critical necessity due to the quickly generated interest in this field. Guidelines for FMT vary around the world. The US FDA has attested that FMT is made of an organic item and is a medication applied to analyze, treat, or forestall illness or impact the construction or capacity of the body. Wellbeing Canada views FMT as "another biologic drug" and proclaims that all clinical investigations should pass the cycle of a clinical preliminary application to guarantee they satisfy quality and security guidelines. In the United Kingdom, FMT is additionally endorsed for treating CDI, and it is viewed as protected and compelling. From a clinical and examination viewpoint, any FMT application beyond treating CDI is considered "off-label"; consequently, the advantages and dangers of clinical practice should be painstakingly considered [5].

Notwithstanding these legitimate guidelines, another fundamental point is to clarify the benefits, dangers, measurement, and follow-up for FMT to all patients before they give informed consent to the procedure. If difficulties and future considerations ignore the clinically apparent adequacy and security of FMT, clinicians and specialists are required to discover substitutes for FMT taking into account the danger of illness transmission between the contributor and beneficiary, patients' acknowledgment, undesirable results, and the dubious effects on the beneficiary's unsusceptible framework [5].

Beside normalization of benefactor screening and clear conventions for checking for adverse events, a FMT library ought to be set up to gather long-term information and follow-up results and complications. We have acquired a lot of information about the bacterial populace in the human digestive tract throughout recent years; however, little is known about viral or contagious creation in the gut and the capacity of intestinal microorganisms [5].

Further, another vulnerability of FMT is the profoundly powerful structure of live microbiota, which are susceptible to factors like eating regimen and drugs. Accordingly, future study should zero in on distinguishing the gut microbiota, characterizing their capacity, and further controlling the gut microbiota. In the years to come, we anticipate customized FMT for various patients and conditions as per individual host and infection genotypes/aggregates [5].

21.2.6. Recent Findings

It is understood that an imbalanced intestinal microbiota is inclined to *C. difficile* infection, IBD, and IBS. The unpredictable aspect of intestinal microbiota to promote wellness, however, is a more current idea that is being increasingly examined. The microbiome assumes a significant part in cell insusceptibility and energy digestion and has been embroiled in the pathogenesis of non-GI immune system illnesses, chronic fatigue conditions, obesity, and, surprisingly, some neuropsychiatric problems [7]. Significant results were reported from FMT therapy studies in patients with irritable bowel syndrome, metabolic syndrome, chronic constipation, colonization by antibiotic-resistant bacteria, and hepatic encephalopathy. Further studies using FMT-based therapy for a variety of diseases such as psoriasis, neurological diseases like Parkinson's disease and multiple sclerosis, and cancers are also under research [9].

21.3. PROBIOTICS

For many years, people have had the false belief that inoculation or ingestion of microbes causes diseases. But some beneficial microbes, called probiotics, have emerged as a source of promotion of human health and elimination of diseases. Probiotics refers to the use of viable microorganisms to promote health by altering the microflora of the host. The microorganisms used as probiotics are derived from normal human microflora, such as those obtained from the alimentary canal [3, 10]. In short, probiotics can be defined as ingestion of beneficial microbes to improve health [11]. Probiotics are not narrowed down to only one strain of bacteria; it also includes combinations of many strains, in which each microbe has its own beneficial effect [12]. Commonly administered probiotics include Lactobacilli like *L. rhamnosus, L. reuteri, L. casei, Lactobacillus acidophilus,* and other organisms like *Bifidobacterium* sp., non-pathogenic strains of *E. coli* like *E. coli Nissile 1917,* and yeasts like *Saccharomyces boulardii* [10, 13]. Probiotics have a wide range of applications in fermented and non-fermented food products, nutraceuticals, and dietary supplements [14].

There are several guidelines that must be followed to select a microorganism as a probiotic organism. They must: be of human origin; should not elicit immune response or cause pathogenicity leading to diseases; must have a set of particular beneficial effects; should not cause ill effects even in immunocompromised individuals; should be resistant to stomach acids, bile juice, and other enzymes in the gastric system; should be able to adhere to the epithelial tissue of gut; and must have a long shelf life to be stored for a lengthy periods of time [10, 13].

Probiotics confer great health benefits to individuals through a variety of mechanisms. The major principle behind the mechanism of probiotics is the interaction of probiotic microbes with the gut microbiota. Probiotic microbes adhere to and colonize the gut tissues, which causes competition between pathogens and the probiotic microorganism for adhesion receptors. Thus, pathogens will be deprived of access to adhesion receptors and hence there will be less chance for the pathogens to cause disease [15]. Antimicrobial substances like lactic acid, acetic acid, β-hydroxy propionaldehyde, bacteriocin, and hydrogen peroxide are produced. Probiotic microbes act against pathogens by suppressing the intrusion of pathogens into epithelial tissues. Similarly, they also increase the deterrent work of the intestine by improving intestinal barrier selectivity with the production of mucin, IgA, and defensins [16, 17]. Short chain fatty acids (SCFAs), minerals, vitamins and growth regulators, and digestive enzymes like carbohydrase, peptidase, sucrase, and tripeptidase are also observed.

In *Clostridium butyricum*, production of n-butyric acid is also studied [15]. Probiotic microbes like *Lactobacilli* also inhibit bacterial translocation of *E. coli* proportionally with dose [18]. They have also been studied for improving the immune activity of individuals such as promoting cytokines, antioxidants, anti-inflammatory activity, and enhancing humoral and cell-mediated immunity. Administration of probiotics also improves and stimulates the immune activity of mucosa and host system. They also prevent development of cancer by regulating apoptosis and cell differentiation [19, 20].

21.3.1. Probiotics-Based Therapy

Probiotics administration has been studied for the treatment and prevention of a variety of diseases including GI, metabolic, neurodegenerative, and many other diseases.

21.3.1.1. Diarrhea

One of the major GI diseases is diarrhea, which is comprised of three major types: infection-associated diarrhea, antibiotic-associated diarrhea (AAD), and traveler's diarrhea.

Many bacteria are used to treat diarrhea and *S. boulardii* is the only yeast that is studied in this category [13]. Infection-based diarrhea may be caused due to rotavirus or *C. difficile, among others*. Probiotic microbes like *Lactobacilli* have been studied for their efficacy against infection-based diarrhea and the major mechanism associated is the attachment of probiotic microbes to erythrocytes to colonize the crypts [10]. *L. rhamnosus GG* (LGG) is the most widely probiotic in diarrhoeal disease caused due to infection. In two studies, LGG was studied for its potential activity against rotavirus in children, and results showed that probiotic administration significantly reduced both hospitalization and duration of diarrhea [21].

Similarly, probiotic VSL#3, a probiotic mixture of eight microorganisms, improved recovery rates and stool consistency, and reduced stool frequency and ORS requirement in rotavirus diarrhea-affected children [22]. *S. boulardii* was shown to reduce the onset of *C. difficile*, another causative agent of infection-based diarrhea, by producing proteases that inhibit the *C. difficile* toxin and activating antitoxin A, an immunoglobin that plays a major role in inhibiting the factors associated with incidences of diarrhea [23].

The second major type of diarrhea is associated with inappropriate use of antibiotics. Many bacteria like *L. acidophilus, Lactobacillus delbrueckii, LGG, Leuconostoc, Bifidobacterium,* and *Bacillus* have been studied for their efficacy against antibody-associated diarrhea [10]. Generally, inappropriate use of antibiotics may disturb the normal gut microflora, causing dysbiosis leading to diarrhea. Administration of probiotics may balance the normal microflora and improve health. In 240 children affected by AAD, LGG administration in 10^{10} colony-forming units twice a day during antibiotic treatment prevented the incidence ofAAD [24]. L. casei-based drinks have preventive effects against both *C. difficile*-based diarrhea and AAD [25]. Three species of *Lactobacillus,* namely *L. acidophilus, L. bulgaricus* and *LGG,* along with *Bifidobacterium* and *Enterococcus faecalis,* were shown to have a preventative effect against AAD in a study of combined and individual activity of *Leuconostoccremoris, Lactobacillus, Lactococcus, Saccharomyces, Streptococcus*, and *Bacillus* species [15, 25].

Traveler's diarrhea is another common type, especially when people travel from urban to remote or tropical to non-tropical regions. Its etiology can generally be traced 60–85% due to *E. coli* and *Campylobacter jejuni* and 10 and 5%, respectively, due to s*almonella* and *shigella.* Many probiotics like *Lactobacilli, Bifidobacteria, Streptococci*, and *Enterococci* are generally studied for therapeutic and preventive effects against traveler's diarrhea [13]. *S. boulardii* and LGG have also shown preventive effects against traveler's diarrhea [26]. *Lactobacillus plantarum*-based probiotics were effective against HIV-associated diarrhea in children and no pathogenesis was observed, proving that probiotics-based therapy is also safe in immunocompromised individuals [27]. Nosocomial acquired diarrhea was prevented by the administration of *Streptococcus, thermophilus,* and *Bifidobacterium lactis* [28].

21.3.1.2. Inflammatory Bowel Disease

IBD is a degenerative disease associated with the GI tract that leads to bloody and watery diarrhea. Many probiotic strains have been associated with prevention and treatment of IBD, which includes Crohn's disease, UC, and pouchitis. *Bifidobacterium infantis* alleviated symptoms like straining, bloating, bowel dysfunction, and incomplete evacuation in IBD patients [13]. Similarly, *Lactobacillus salivarius* and *Lactobacillus gasseri BNR17* also helped in treating IBD symptoms [10]. *E. coli Nissile1917*, another probiotic bacteria, along with VSL3#, prevented IBD [25]. *Bifidobacterium bifidum, L. casei, L. acidophilus, B. breve,* and *S. boulardii* was studied to be effective against UC [29]. Probiotics like LGG, VSL#3, *S. boulardii,* and *E. coli Nissile 1917* were helpful in treating Crohn's disease by mechanisms like competition with pathogen and improving the immune system [30]. VSL3 also showed reduction of disease in pouchitis [31].

Probiotics like *Bifidobacterium, lactobacillus,* and its mixture Ecologic® Relief played an effective therapeutical role in reducing ill effects of constipation [32]; *S. boulardii* and *Lactobacillus johnsonii La1* have been studied against *Helicobacter pylori,* in which *S. boulardii* showed positive effects in reducing symptoms caused by *H. pylori*, which causes ulcers, gastric cancer, and other illnesses. [33].

21.3.1.3. Cancer

Probiotics also play a major role in combating various types of cancer. *L. casei Shirota* and its associated fermented foods were shown to prevent the onset of breast cancer [34]. LGG is also a major probiotic strain exhibited anticancer activity against various cancers like breast, ovary, hepatic, and colorectal cancer by their antiproliferative, antimetastatic effect and immunomodulation [25]. LGG has also shown anticancer activity against colon cancer by a similar mechanism *in vitro* [35]. Other mechanisms of probiotic anticancer activity have also been studied. Probiotics have the capability to reduce the conversion of pre-carcinogens into carcinogens by enteric flora, regulating apoptosis and cell differentiation, inhibiting activity of enzymes produced by pathogens, modulation of microbiota, and enhancement of gut barrier functions [13, 35].

21.3.1.4. Renal Disorders

Many disorders related to the renal system have been treated and prevented using probiotic intervention. Urinary tract infections (UTI) are a major problem caused by *E. coli, Salmonella, Klebsiella,* etc. Probiotic bacteria colonization has been shown to reduce infection rate [10]. *L. acidophilus, LGG,* and *Bifidobacterium* sp., have prevented the recurrence of UTI [36]. In women, improper balance of *Lactobacilli* causing bacterial vaginosis poses a risk factor for contracting sexually transmitted diseases like HIV. In a study, administration of the *Lactobacillus reuteri RC-14* probiotic strain reduced the rate of transmission of HIV [13]. Positive regulation of serum endotoxins, proinflammatory cytokinin, and interleukins were observed upon administration of *Lactobacillus* and *Bifidobacterium* in peritoneal dialysis patients with chronic kidney diseases [37]. Probiotics like *Bifidobacterium, Lactobacillus,* and *Saccharomyces* sp. have been studied for their effective role in combating bacterial vaginalis. In particular, *Saccharomyces cerevisiae* inhibited *Gardnerella vaginalis* by reducing the adhering capacity of the pathogen to epithelial cells [25].

21.3.1.5. Lactose Intolerance

Lactose intolerance is a major problem causing difficulty in the digestion of lactose in many individuals. Many probiotics are known to produce the lactase enzyme, which helps in the digestion of lactose [38]. *St. thermophilus* and *Lactobacillus bulgaricus* have shown themselves to be effective in treating symptoms related to lactose intolerance through the production of enzymes like beta-galactosidase [39].

21.3.1.6. Lipid Disorders

Many metabolic disorders are also treated using probiotics. Improper levels of lipid/cholesterol are a major problem in many individuals despite their age. Many probiotics-based studies have shown positive reductions in cholesterol levels. Probiotics like *Lactobacilli* and *Bifidobacterial* are widely studied for their therapeutic value in treating hypercholesterolemia, especially through the production of the bile salt hydrolase enzyme, and rat model studies have shown reductions in cholesterol and triacylglycerol levels and elevated high density lipoprotein [25]. *L. bulgaricus* has shown significant results in reducing high cholesterol levels; they have also been studied in managing obesity. Another probiotic strain of future importance in obesity is *Akkermansia muciniphila,* which was reduced in obese individuals [40].

21.3.1.7. Diabetes Mellitus Type 2

Diabetes mellitus is the most common metabolic disorder. *Lactococcus, Bifidobacterium,* and *Lactobacillus* are shown to modulate the gut microbiota function and regulate energy metabolism, so they are useful in treating diabetes mellitus type 2. In particular, *L. acidophilus KLDS1.0901* and *S. boulardii* were shown to balance increased levels of blood glucose and also slowed insulin resistance [41, 42].

21.3.1.8. Immune Modulation

Probiotics are shown to enhance immune function by increasing immunoglobulin production and improving the gut immunological barrier [10]. Probiotics also a play a major role in managing autoimmune diseases. Probiotics of *L. acidopilus, B. bifidum,* and *L. casei,* and in another study *L. fermentum* in addition to the above three strains, were studied in multiple sclerosis patients and they reduced the level of C reactive proteins, which are an inflammatory marker [43].

21.3.1.9. Neurodegenerative Disorders

Probiotics also play a major therapeutic role in the treatment of neurodegenerative disorders like Alzheimer's disease and Parkinson's disease. *Lactobacillus* and *Bifidobacterium* reduced oxidative stress and inflammation in peripheral blood mononuclear cells isolated from Parkinson's disease. Another probiotics mixture, SLAB51, containing eight different probiotic strains including *Lactobacillus* and *Bifidobacterium* reduced the harmful effect of 6-hydroxydopamine in Parkinson's disease. Similarly, *Lactobacillus, Bifidobacterium,* and *C. butyricum* improved the cognitive deficits and slowed the neuroinflammation observed in Alzheimer's disease [25].

21.3.1.10. Other Disorders

Dental caries can be a major problem, especially among children. Probiotics like *LGG, L. rhamnosus SP1,* and *Streptococcus salivarius M18* have been shown to be effective in reducing tooth decay in children; the possible mechanism is the production of bacteriocins that suppress pathogens causing decay [44]. Atopic diseases in high-risk infants were prevented by administration of *LGG* six months prenatally and postnatally to mother and infant, respectively [10]. *L. reuteri* administration to infants for 28 days resulted in a reduction in duration of infant crying. *LGG* has also reduced the enteric candida colonization in premature babies that can cause mortality [33]. *LGG* has shown a therapeutic effect against androgenic alopecia by exerting a growth-promoting effect on hairs; against respiratory tract infections; and against osteoporosis by inhibiting inflammation of mesenchymal stem cells derived from the mandible. Golden Bifid probiotic formulation and *Saccharomyces pastorianus* and *Acetobacter xylinum* probiotics in kombucha was found effective against hand, foot, and mouth disease [25]. *Lactobacillus* and *Bifidobacterium* showed a combined effect against necrotizing enterocolitis [35], *Lactobacillus pentosus GMNL-77* against psoriasis [45] VSL#3 against hepatic diseases, and *S. cerevisiae* and *Lactobacillus helveticus* for reducing hypertension [46]. *Bifidobacterium longum* and *Lactococcus lactis* have been studied for treatment of the influenza virus using mouse models. Recently, probiotics like *St. thermophilus DSM 32345, L. paracasei DSM 32241, L. plantarum, L. helveticus brevis,* and *B. lactis* have been studied for treating Covid 19 through the possible mechanism of reduction of oxidative stress [47].

21.4. PREBIOTICS

Prebiotics are a subcategory of healthy fibers that aren't always affected by gastric causticity and the compounds present in the gut of vertebrates that are crucial for our wellbeing. The principal benefit of prebiotics is their improvement of intestinal microorganisms related to well-being and comfort [13].

Many trillions of microorganisms, along with microscopic organisms, infection, growths, and archaea, remain in our distal digestion tracts and normally connect to co-advanced host secure cells in a beneficial way. They are impacted through hereditary traits and herbal elements, along with dietary routine. The microbiota grew to colonize unique, organic areas of the human GI tract and apply variable weight manipulation plans, at the same time as the human mucosal insusceptible framework improved to guard the host from risky microbial microorganism openings, and prevent regular intestinal irritation. Enteric inhabitant microbiota exists as a consortium that carries putative proinflammatory and protective strains [8].

The term "prebiotics" refers to particularly perennial ingredients that lead to specific shape-modifying or GI-improving actions, with scientific benefits for the host. This definition encompasses to a degree that of nutritional fiber, but provides the selectivity of prebiotics for a few precise microorganisms (e.g. the admission of fructo-oligosaccharides [FOS] and inulin especially favors bifidobacterial development). "Dietary fibre" is a greater category that alludes to exclusive sugars and lignin, which oppose hydrolysis with the aid of human stomach-associated proteins, but is probably aged with the aid of colonic microflora and moreover incompletely discharged in excrement. This definition carries the concept of fiber non-starch polysaccharides (e.g. celluloses, hemicelluloses, gelatins, gums, and adhesives), inulin, FOS, galacto-oligosaccharides (GOS), and safe starch (and starch corruption items, which aren't processed inside the small intestine of healthy people) [2].

A component of those elements of fiber carefully meet the trends important for them to be taken into consideration as prebiotics (e.g. inulin, FOS, GOS, soy-decided oligosaccharides, xylooligosaccharides, pyrodextrins, and isomaltooligosaccharides). Different segments of fiber are difficult to order. For example, guar gum, a form of fermentable dissolvable fiber, rather improves probiotic microorganisms, but is thought of as a general (vague) substrate for colonic microscopic organisms ("fermentable colonic food"), and cannot, consequently, be characterized as prebiotic inside the exacting definition of the word [2]. Prebiotics definition is happy through secure starch. In the colon prebiotic form of getting old is given through it and has hundreds of metabolic benefits, with inside the bile salt advent and laxation, brings down the danger of GI tract malignancies, and brings down the post-meal increase in glucose and blood lipid levels. Further, it enables the growth and proliferation of epithelial cells through extended butyrate fixation through the intestine microbiota [13].

21.4.1. Mode of Action of Prebiotics

- They are indigestible carbohydrates that are fermented by the resident microbes through the upper GI tract and transported into the ileum and colon.
- Prebiotics causes homeostasis of the intestine.
- Host surface receptors are coated by prebiotics.
- They produce bacteriocin.
- Microscopic organisms that are favorable produce short chain unsaturated fats with the assistance of non-absorbable starches.
- The energy source of epithelial cells are the short-chain fatty acids.
- They modulate metabolic function and regulate the immune system [13].

21.4.2. The Sway of Prebiotics on the Small Intestine

Dietary fiber, which can be important for prebiotics, can impact the serenity of the small intestine. Some research suggests both probiotics and prebiotics (wherein inedible meal ingredients are extra useful to invigorate the improvement of colonic microscopic organisms) may stifle tumor and pre-neoplastic accidents in the colons of individuals treated with cancer-inflicting synthetics. The presence of beneficial microbes in intestines of toddlers who are nourished with breastmilk is upheld via the digestion of the puzzling mixture of oligosaccharides present in the milk, where many of them expand intestine microbiota located within the milk. Inappropriate intestine microbiota (dysbiosis) can cause immune system issues, contaminations, and hypersensitive responses in old age [13].

21.4.3. Prebiotics in Corpulence as Dietary Modulators of Gut Microbiota

It is proposed that intestine microbiota are a widespread factor in stoutness and corpulence-associated comorbidities. Intestine microbiota can be a treatment option for bloatedness inclusions. It is predicted that microbial dysbiosis is associated with bloatedness, and it's presumed likely that reorganizing the useful interplay among the intestine microbiota and prebiotics holds extremely good ability. Prebiotics are used in order that intestine microbiota may be adjusted. Prebiotics aren't certified to be processed via catalysts despite the fact that that they circulate in the inner organ and are particularly grown in the GI tract where they assist in improvement of microorganisms, particularly Bifidobacterium and Lactobacillus, which have been linked with scientific advantages [13].

The current most well-known focuses for prebiotic use are Lactobacilli and Bifidobacteria. This is to a great extent dependent on their achievement in the probiotic region. However, as our insight into the gut vegetation improves (through utilizing the atomic systems depicted earlier), it might become evident that different microorganisms ought to be used. One model might be the *Clostridium coccoides–Eubacterium rectale* group that incorporates microorganisms creating butyric corrosive, a metabolite seen as valuable for gut usefulness and could conceivably defend against gut malignant growth. The probability of different microbes (counting at this point unclear genera) additionally being focuses for a prebiotic impact should be placed in context with our expanding understanding (because of new sub-atomic philosophies) of the bacterial variety in the gut microflora. In fact, the more we recognize and portray the bacterial genera, species, and even strains that form the intestinal microflora, the more we will be in a situation to depict, both subjectively and quantitatively, changes in that organization and, subsequently, to see how the hordes of bacterial cells in the digestive tract communicate and how they add to and adjust intestinal (particularly colonic) physiology. Prebiotics will then become novel devices to make, both in test subjects and in patients, colonic microflora with "controlled" arrangements that will then correspond with explicit physiological conditions. In any case, information is still too scarce to even consider speculating at this time [48].

21.4.4. Prebiotics Tweak of Safe Capacity

Immunological capacities are tweaked by our feeding routine and influence obstruction of hosts in completely different ways. Aboard elementary food segments uncalled-for supplements like non-processed sugars likewise assume a major part in dominating medicinal reactions, particularly in tissues of the gut. Studies have in the past investigated the impacts of prebiotics in increasing the resistance of the host and saw that there is adequate proof to show that prebiotics, for example, polysaccharide, help to tweak immunological capacities. In one study, the patient was counseled to require prebiotics for the regulation of immunological barriers in gut-connected lymphoid tissues, nonobligatory lymphoid tissues, and fringe course. These include actual barriers like skin and mucose layers, platelets and tissue like phagocytes, regular killer cells, and solvent middle groups, like enhanced proteins and cytokines [13].

21.4.5. Prebiotics-Based Therapy

An examination was directed to check the effect of prebiotics consumption comprising xylo-oligosaccharides, fructo-oligosaccharides, safe dextrin, and polydextrose on invulnerability and the design of gut microbiota in perioperative colorectal malignancy patients. A twofold blind, randomized medical study was conducted on perioperative colorectal malignancy sufferers, both women and men, between 40 and 75 years. Patients had been separated into gatherings comprising of an intercession group who were given 30 g/day of prebiotic supplementation throughout seven days and a manipulated group who failed to get supplementation. The immunological and health documents of the two groups were assessed and had been contrasted with their benchmark levels. Likewise, the fecal examples of an abnormal 40 sufferers from the two gatherings had been taken to study the intestine microbiota. There was a large decrease in intestinal microbiota in the benchmark group. Admission of prebiotics is prescribed to patients based on immunological statistics in patients having colorectal malignant growth seven days earlier. Prebiotics upgraded the convergence of four commensal microorganism-containing deft microbes in sufferers having colorectal disease. Careful strain faded the wide variety of populations in intestinal microbiota but accelerated the wide variety of populations in commensal microbiota and smart microorganisms [13].

21.4.5.1. Treatment of Sepsis

Late-beginning sepsis is a significant reason for mortality in preterm newborn children from industrialized and agricultural nations. At the point when led a meta-analysis of probiotic use against late-beginning sepsis in preterm babies including RCTs (randomized controlled trials) enlisting 9400 newborn children. The creators showed a huge decrease of late-beginning sepsis from 16.3% in placebo treatment to 13.9% in probiotic beneficiaries. The distinction stayed large whenever investigated in babies treated with *Lactobacilli, Bifidobacteria*, or single or numerous probiotics. A critical decrease of late-beginning sepsis and demise was additionally found in probiotic-treated preterm newborn children from non-industrial nations [49].

21.4.5.2. Diabetes and Cardiovascular Infections

Prediabetes is connected to drawn-out irritation that is related to the more serious danger of developing type 2 diabetes and cardiovascular illnesses. Expanded convergence of lipopolysaccharides relates to dysbiosis of the regular microbiota that is engaged with the advancement of type 2 diabetes and cardiovascular illnesses. Prebiotics are associated with the working of regular miniature biota, for example, inulin diminishes the centralization of endotoxin, decreases porousness of the digestive tract, and cuts off metabolic malfunction in rodents. The impact of prebiotics is still unclear on the cardiovascular regions in patients who are at the danger of type 2 diabetes. Prebiotic supplementation alongside inulin could be used as a preventive system for lowering the risk of cardiovascular illnesses in the patients in danger of type 2 diabetes. This system can affect clinical practices by setting and tolerating the dietary proposal of prebiotics on clinical and academic local areas [13].

21.4.5.3. Renal Disorders

It has been shown that prebiotics, probiotics, and harmonious supplementation has brought about the improvement of renal capacity. Analysts contemplated the renal profile of a meta-investigation of a clinical preliminary. They called attention to the degree of glomerular filtration rate that was diminished, though the convergence of creatinine was expanded in the mediation group in contrast with the placebo group and the outcomes were non-critical. The consolidated impact on the blood urea nitrogen showed that the level declined when contrasted with the placebo group, while the degrees of uric corrosive expanded in the mediation group when compared to the placebo group. The supplementation of prebiotics, probiotics, and supplements ought to be restricted to those whose renal capacity is undermined until larger randomized controlled preliminaries demonstrate the proficiency and security of prebiotics, probiotics, and cooperative supplementation for improving renal capacity [13].

21.4.5.4. Colon Disease Control

Aberrant crypt foci (ACF) is an unusual development of cells that shows up as sores in the digestive system. To distinguish ACF injuries, chromoscopic colonoscopy of high amplification is utilized, and by and large it shows up as colonic mucosa. They contain crypts that are raised on the upper side of ordinary mucosa. ACF comprises incredibly consolidated epithelia and has changed luminal openings noticeably bound from the ordinary adjoining crypts. The progress of the ACF to polyp, adenoma, and adenocarcinoma reciprocals development from various hereditary and biochemical adjustments. A small measure of ACF can be a cause of colon disease. As of now, it is unknown which crypts are responsible for the advancement of tumors. However, different studies support the possibility of advancement of colon cancer from ACF. Studies showed a few fibers that invigorate the arrangement of stable butyrate-creating colonic biological system. This sort of colonic climate reduced the pace of ACF. Along these lines, obviously a colonic environment that produces stable butyrate diminishes the dangers of developing colon cancer [13].

21.4.6. Prebiotics as Food Varieties

Probiotics and prebiotics are utilitarian meals sorts which have massive biotechnological advantages with a first-rate capacity for creation. Despite the reality that probiotics and prebiotics are in all likelihood going to be applied in treating several transferable and non-transmittable problems, it's miles quite difficult to attain inferences from diverse studies that demonstrate the useful functionality of probiotics and prebiotics due to the fact that researchers applied diverse lines and treatments for specific problems. Further exploration making use of reliable suggestions will permit greater definitive results of the effects of probiotics and prebiotics to remedy various infections. Specific devices with probiotic, prebiotics, and harmonious supplements are usually available in meals and complementary designs; now it's possible that those tools will become known for treating specific infections and will be available as medication [13].

21.4.7. Focus for Prebiotics

Fermentation of safe starches and dietary fiber is brought about by microorganisms in colon discharge short chain unsaturated fat metabolites (SCFA). SCFA are notable for gut wellbeing just as their part in the arrangement of valuable energy to the host in size is in banter. In contrast, 10% of the energy in people is given by SCFA. As indicated by energy yield theory, dysbiotic microbiomes have an amplified ability to eliminate energy from food, so the bacterial transformation of non-absorbable carbs and dietary fiber to SCFAs could give extra energy to the host and result in size of the individual over the long run. In like manner, G-protein coupled receptors (GPR) can detect SCFAs that attach them with lipid and glucose breakdown. SCFAs initiate the two significant proteins, GPR41 and GPR43, which are communicated on adipocytes and enter endocrine L cells. Peptide YY is delivered by incitement of intestinal GPR41, which improves gut section time and incremental satiety. Aggravation decreases by enactment of intestinal GPR43 just as it re-enacts glucagon-like peptide (GLP), a chemical which contributes to guidelines of insulin discharge. Enteroendocrine L cells express GLP-1 just as they discharge the gut-trophic chemical GLP-2. Superior upgrade for GLP-2 discharge is supplement utilization. SCFAs keep up with the arrival of GLP-2, which is significant in regulation of the gut barrier framework and diminishes lipopolysaccharide movement [13].

21.5. POSTBIOTICS

After the development of probiotics as therapy and as a result of deep research in probiotics, a new field developed that is using probiotic-derived metabolites termed postbiotics. Microbiota provides a variety of benefits and physiological effects due to the activity of the metabolites produced by them [50]. Postbiotics are the non-viable metabolic byproducts derived from probiotic bacteria. They may also be represented as non-viable probiotic activity due to the metabolite activity to enhance biological activity in host and prevent and treat various diseases. [51] Postbiotics components can be of various classes like lipids, carbohydrates, vitamins, organic acids, vitamins, and intricate molecules [52].

Since in some rare cases probiotics can cause inflammatory conditions in high risk individuals like immunocompromised individuals, postbiotics can be administered to avoid such adverse effects and improve the safety degree of microbiome-based therapy [53]. This is advantageous especially in preterm babies with immature intestinal barrier and weaker immune defense [3]. Since postbiotics cannot involve replication and elicit immune response like probiotics, they are much safer [54]. They have other advantages like stability and being abundant at high concentration [50]. Since postbiotics are the metabolites derived from probiotic organisms, they possess inherent stability both during the industrial process and storage. Postbiotics have advantages over probiotics (live organisms) since they can be stored at room temperature for a long time without loss in quantity, whereas probiotics requires cold temperatures, and the number of viable cells may decrease with an increase in shelf life [54].

Postbiotics like proteins, peptides, bacteriocins, and organic acids can be obtained by methods like centrifugation and ultrafiltration [55]. Postbiotics products generally include probiotic bacteria's products like vitamins, SCFA, vitamins, bacteriocins, and cell free supernatant (CFS), flavonoid-derived postbiotics like equol daidzein, terpenoid-derived postbiotics like genipin, phenolic-derived postbiotics like uolithins, etc. [56]. Microbiome associated metabolites can act either by integration into the host intracellular metabolism and provide endocrine energy source for tissues, or act by receptor-mediated signaling, which is directly recognized by the cells. Postbiotics therapy, especially SCFA, had been studied initially for the treatment of diseases like IBD, metabolic diseases, and intestinal inflammation [50].

Similar to probiotics, there are few criteria that must be studied for selection of postbiotics. Complete molecular characterization of the probiotic from which the metabolite is derived is required so that it helps in genetic screening and safety studies. Composition and health benefit of the postbiotic must be clearly studied and the method for inactivation of postbiotic along with confirmation of inactivation must be identified. Its safety must also be analyzed so that it should be pathogenic even in immunocompromised patients [54].

21.5.1. Postbiotics-Based Therapy

Postbiotics are emerging therapeutic agents that have been studied in a variety of diseases like IBD, cancer, liver-related diseases, and immunological modulation. Postbiotics like RG14, TL1, and RS5 metabolites from *L. plantarum* have been studied in piglets for reducing fatality due to diarrhea [55]. SCFA is an important postbiotic studied for the treatment of IBD and colorectal cancer by the mechanism of reduction of inflammation and anti-proliferative effects. Postbiotic SCFAs like butyrate and acetate were increased due to the administration of *Lactobacillus paracasei. CNCM I-1572* reduced inflammation in IBD patients [56]. Inflammatory mucosal response post infection of IBD was reduced by *L.casei* soluble factors [55].

Many postbiotics have been studied to reduce and regulate the inflammatory effect. Culture supernatant of *L. paracasei B21060* prevented the inflammatory effect of Salmonella in the colon [53]. S layer protein A obtained from *L. acidophilus NCFM* modulated dendritic and T cell function [55]. Pathogens like *E. coli, Salmonella enterica* and *Listeria monocytogenes* were inhibited by CFS of various strains of *L. plantarum*. Studies have reported that exopolysaccharides (EPS) from *Bifidobacterium animalis* and crude culture extract and purified EPS from *L. helveticus MB2-1* exhibited radical scavenging activity in addition to lipid peroxidation inhibition and ferrous ion chelating capacity, respectively [53].

SCFA postbiotics have shown beneficial effects in various metabolic process like regulation of lipid metabolism, glucose homeostasis, plasma cholesterol homeostasis, and insulin sensitivity by mechanisms like activation of G protein coupled receptor. Postbiotics produced by *Streptococcus* and *Faecali bacterium* have been associated with a reduction in hypertension [55]. Wound healing capacity in *in vitro* models was improved by activation of $\alpha2\beta1$ integrin collagen receptors through the administration of a culture supernatant of *S. boulardii*. Acetaminophen-induced hepatoxicity was reduced by cell lysate suspension of

Lactobacillus fermentum BGHV110 and intracellular content from *Enterococcus lactis IITRHR1* and *L. acidophilus MTCC447* [53].

21.5.2. Postbiotics in the Treatment of Cancer

Several postbiotics have been studied for their anticancer activity using various cell line models. Postbiotics produced from *L. plantarum* of six different strains have been reported to exhibit cytotoxic effect on various cancer lines like MCF-7, HT29, HeLa, Hep G2, and HL [25]. Postbiotics based on *L. casei* prevented the relapse of colorectal and bladder cancer [55]. Similarly, *L. fermentum*-based postbiotics, EPS from *L. plantarum, LGG, L. brevis, L. delbrueckii*, CWP postbiotics from *L. paracasei* sp., and SCFA from *C. butyricum* have shown effective activity against colon and colorectal cancer cell lines by various mechanisms like upregulation of Bax, Caspase 3, etc., and induction of apoptosis and anti-proliferative activity [57, 58]. *LGG* and *L. plantarum*-based postbiotics have exhibited anticancer activity against cervical and breast cancer, respectively, through various mechanisms like downregulation of BCL-2 gene and upregulation of BAX, caspase 3, which in turn causes apoptosis. CFS of *L. casei* and *Bifidobacterium sp.* have shown apoptotic and anti-proliferative activity [58] and CFS from *Faecalibacterium praunitzii* showed significant regulation of anti-inflammatory and pro-inflammatory genes in lung cancer cell line A549 [57].

General mechanisms studied for anti-cancer activity associated with postbiotics include:

(i) inducing anti-proliferative and apoptotic pathways: SCFA from *Propionibacterium freudenreichii* induced apoptotic pathway and induced anticancer activity in gut cancer cell lines
(ii) preserving intestinal barrier by inhibiting bacterial translocation
(iii) immunomodulation activity
(iv) inhibiting enzymatic activity of pathogens like *L. plantarum* b240 based postbiotics can reduce the invasive effect of *Sa. enterica serovar typhimurium*
(v) inhibiting activity of carcinogens: *LGG* was able to inhibit the activity of mutagen N-methyl-N0-nitro-N-nitrosoguanidine and 2-amino-3,4-dimethylimidazo[4,5-f]-quinoxaline (MeIQx)
(vi) lowering the intestinal pH
(vii) Deterring tyrosine kinase signaling pathway [58]

Many clinical trials are carried out at present for the therapeutic activity of various postbiotics. Postbiotic TMAO is in phase 4 clinical trial for treating chronic kidney diseases through alteration of cholestrol and sterol metabolism. Flavonoid-based postbiotics are in phase 2 clinical trials for autism spectrum disorders due to their antioxidant and anti-inflammatory activity. N-acyl amines are also in phase 1 trial for treating diabetes mellitus.

Currently an important challenge linked with postbiotic therapy is the pleiotropic function of metabolites like SCFA, which have a variety of effects on administration like metabolic control, appetite regulation, immunomodulation, etc., which must be well studied for proper application. Despite these limitations, postbiotics can be developed as a promising microbiome-based therapy [50].

21.6. SYNBIOTICS

As mentioned, probiotics and prebiotics individually have their own beneficial effect against various diseases. But *in vivo*, probiotics can survive better in the presence of prebiotics because probiotics in absence of prebiotics, that is their food source, may face problems like intolerance to oxygen, pH, and temperature changes, So, the synergistic activity of probiotic and prebiotic is termed synbiotics. In synbiotics, the count and sustainability of probiotics can be increased using prebiotics. In addition to acting as food source, prebiotics help probiotics to pass through the upper digestive tract and help in intestinal colonization. Commonly used probiotics like *Lactobacillus, Bifidobacterium, S. cerevisiae, E.coli*, and *Bacillus* are administered along with common prebiotics like fructo-oligosaccharides, GOS, isomalto-oligosaccharides, inulin, etc. Common combinations include *Lactobacilli* with lactitol/inulin/FOS, *Bifidobacterium* with FOS/GOS, and *Lactobacilli* and *Bifidobacterium* with FOS/inulin [59].

21.6.1. Synbiotics-Based Therapy

Various symbiotic formulations have been studied against a variety of diseases. Several health benefits of gut microbiota regulation of Lactobacillus and Bifidobacterium, immunomodulation, and bacterial translocation inhibition have been observed upon administration of synbiotics [60]. Systemic inflammatory response syndrome (SIRS) is characterized by symptoms like abnormality in body temperature, respiratory and heart rate, and WBC count. Synbiotic therapy of *B. breve strain Yakult, L. casei strain Shirota* with galactooligosaccharides has improved SIRS abnormalities and was more effective than respective pro, pre, and postbiotic therapy. In abdominal surgery for biliary cancer, pre and post-operative administration of similar symbiotics played an effective role in inhibiting post-operative infection compared to those who took only post-operative synbiotics [61].

Significant effect was observed on administration of symbiotic combination of *B. longum* with Synergy 1 (an inulin/oligofructose mix) in Crohn's disease patients. *B. longum* with synergy 1 administration for a month showed significant reduction in inflammatory markers in UC patients [62]. *Bacillus coagulans* with inulin administration for six weeks reduced the levels of C reactive protein and enhanced glutathione levels. Similarly, in animal studies, synbiotics of *L. acidophilus* ATCC 4962, FOS, mannitol, and inulin for two months exhibited significant hypercholesterolemic activity. Azoxymethane-induced suppression of NK-cell activity in Peyer's patches was exhibited by synbiotics but not by individual pro and prebiotics. *Lactic acid bacteria* with mannanoligosaccharides and *E. faecum* with FOS significantly reduced the mortality due to necrotic enteritis caused by *Clostridium perfringes* [60].

21.7. CONCLUSION

The detailed and deep knowledge about the interaction between human microflora alteration in terms of human physiology and disease pathogenesis will pave new ways for the use of microbiome-based therapy [10]. This review provides insight into the present studies that suggest that microbiome-based therapy could provide immense immunomodulatory benefits and have potential as an approach for the therapeutic management of various diseases. Future viewpoints have extraordinary potential as specialists continue to improve or keep a healthy intestinal microflora to upgrade wellness and prosperity.

This chapter has compiled results with respect to microbiome-based therapy characterization from current studies. As these results accumulate, the image will become clearer, empowering the classification of various methods like FMT, probiotics, prebiotics, postbiotics, and synbiotics where proof of real-time wide range application is missing. Also, as better data on structure-to-work data accumulates, just as individual metabolic profiles of target microorganisms are aggregated, it could be simpler to tailor microbiome-based therapy for explicit wellness targets. Considerably more data is required on the fine construction of the progressions achieved. With the new age of molecular microbiological methods presently opening, it will be feasible to acquire authoritative data on the effects of using microbiome-based therapy. Extensive relative studies will permit insightful decisions in incorporating the microbiome into treatment and prevention of various metabolic and non-metabolic diseases.

REFERENCES

1. Bäckhed, F., Fraser, C.M., Ringel, Y. et al. (2012). Defining a healthy human gut microbiome: current concepts, future directions, and clinical applications. *Cell Host & Microbe 12* (5): 611–622.
2. Salehi, B., Dimitrijević, M., Aleksić, A. et al. (2021). Human microbiome and homeostasis: insights into the key role of prebiotics, probiotics, and symbiotics. *Critical Reviews in Food Science and Nutrition 61* (9): 1415–1428.
3. Patel, R.M. and Denning, P.W. (2013). Therapeutic use of prebiotics, probiotics, and postbiotics to prevent necrotizing enterocolitis: what is the current evidence? *Clinics in Perinatology 40* (1): 11–25.
4. Ávila, P.R.M., Michels, M., Vuolo, F. et al. (2020). Protective effects of fecal microbiota transplantation in sepsis are independent of the modulation of the intestinal flora. *Nutrition 73*: 110727.
5. Wang, J.-W., Kuo, C.-H., Kuo, F.-C. et al. (2019). Fecal microbiota transplantation: review and update. *Journal of the Formosan Medical Association 118*: S23–S31.
6. Tseng, C.-H. and Wu, C.-Y. (2019). The gut microbiome in obesity. *Journal of the Formosan Medical Association 118*: S3–S9.
7. Aroniadis, O.C. and Brandt, L.J. (2013). Fecal microbiota transplantation: past, present and future. *Current Opinion in Gastroenterology 29* (1): 79–84.
8. Mishima, Y. and Sartor, R.B. (2020). Manipulating resident microbiota to enhance regulatory immune function to treat inflammatory bowel diseases. *Journal of Gastroenterology 55* (1): 4–14.
9. Liwinski, T. and Elinav, E. (2020). Harnessing the microbiota for therapeutic purposes. *American Journal of Transplantation 20* (6): 1482–1488.
10. O'sullivan, G.C., Kelly, P., O'Halloran, S. et al. (2005). Probiotics: an emerging therapy. *Current Pharmaceutical Design 11* (1): 3–10.
11. Guarner, F. and Schaafsma, G.J. (1998). Probiotics. *International Journal of Food Microbiology 39* (3): 237–238.
12. Chapman, C.M.C., Gibson, G.R., and Rowland, I. (2011). Health benefits of probiotics: are mixtures more effective than single strains? *European Journal of Nutrition 50* (1): 1–17.
13. Gulzar, N., Saleem, I.M., Rafiq, S. et al. (2019). Therapeutic potential of probiotics and prebiotics. In: *Oral Health by Using Probiotic Products*, 2–27. IntechOpen.
14. Di Cerbo, A., Palmieri, B., Aponte, M. et al. (2016). Mechanisms and therapeutic effectiveness of lactobacilli. *Journal of Clinical Pathology 69* (3): 187–203.
15. Goel, A.K., Dilbaghi, N., Kamboj, D.V. et al. (2006). Probiotics: microbial therapy for health modulation. *Defence Science Journal 56* (4): 513.
16. Mitall, B.K. and Garg, S.K. (1995). Anticarcinogenic, hypocholesterolemic, and antagonistic activities of *Lactobacillus acidophilus*. *Critical Reviews in Microbiology 21* (3): 175–214.
17. Bogovič-Matijašić, B., Rogelj, I., Nes, I.F. et al. (1998). Isolation and characterization of two bacteriocins of *Lactobacillus acidophilus* LF221. *Applied Microbiology and Biotechnology 49* (5): 606–612.
18. Mattar, A.F., Drongowski, R.A., Coran, A.G. et al. (2001). Effect of probiotics on enterocyte bacterial translocation in vitro. *Pediatric Surgery International 17* (4): 265–268.
19. Sharma, A. (2019). Importance of probiotics in cancer prevention and treatment. In: *Recent Developments in Applied Microbiology and Biochemistry*, 33–45. Elsevier.

20. Sartor, R.B. (2004). Therapeutic manipulation of the enteric microflora in inflammatory bowel diseases: antibiotics, probiotics, and prebiotics. *Gastroenterology 126* (6): 1620–1633.
21. Isolauri, E., Rautanen, T., Juntunen, M. et al. (1991). A human Lactobacillus strain (*Lactobacillus casei* sp strain GG) promotes recovery from acute diarrhea in children. *Pediatrics 88* (1): 90–97.
22. Dubey, A.P., Rajeshwari, K., Chakravarty, A. et al. (2008). Use of VSL♯ 3 in the treatment of rotavirus diarrhea in children: preliminary results. *Journal of Clinical Gastroenterology 42*: S126–S129.
23. McFarland, L.V. (2006). Meta-analysis of probiotics for the prevention of antibiotic associated diarrhea and the treatment of *Clostridium difficile* disease. *Official Journal of the American College of Gastroenterology| ACG 101* (4): 812–822.
24. Ruszczyński, M., Radzikowski, A., and Szajewska, H. (2008). Clinical trial: effectiveness of lactobacillus rhamnosus (strains E/N, Oxy and Pen) in the prevention of antibiotic-associated diarrhoea in children. *Alimentary Pharmacology & Therapeutics 28* (1): 154–161.
25. Aponte, M., Murru, N., and Shoukat, M. (2020). Therapeutic, prophylactic, and functional use of probiotics: a current perspective. *Frontiers in Microbiology 11*: 2120.
26. Sniffen, J.C., McFarland, L.V., Evans, C.T. et al. (2018). Choosing an appropriate probiotic product for your patient: an evidence-based practical guide. *PLoS One 13* (12): e0209205.
27. Cunningham-Rundles, S., Ahrné, S., Bengmark, S. et al. (2000). Probiotics and immune response. *The American Journal of Gastroenterology 95* (1): S22–S25.
28. Saavedra, J.M., Bauman, N.A., Perman, J.A. et al. (1994). Feeding of Bifidobacterium bifidum and Streptococcus thermophilus to infants in hospital for prevention of diarrhoea and shedding of rotavirus. *The Lancet 344* (8929): 1046–1049.
29. Méndez, I., Audivert, S., Farran-Codina, A. et al. (2019). The efficacy of probiotics, prebiotic inulin-type fructans, and synbiotics in human ulcerative colitis: a systematic review and meta-analysis. *Nutrients 11* (2): 293.
30. Jonkers, D., Penders, J., Masclee, A. et al. (2012). Probiotics in the management of inflammatory bowel disease. *Drugs 72* (6): 803–823.
31. Veerappan, G.R., Betteridge, J., and Young, P.E. (2012). Probiotics for the treatment of inflammatory bowel disease. *Current Gastroenterology Reports 14* (4): 324–333.
32. Martínez-Martínez, M.I., Calabuig-Tolsá, R., and Cauli, O. (2017). The effect of probiotics as a treatment for constipation in elderly people: a systematic review. *Archives of Gerontology and Geriatrics 71*: 142–149.
33. Salvatore, S. and Vandenplas, Y. (2010). Prebiotics and probiotics in therapy and prevention of gastrointestinal diseases in children. In: *Bioactive Foods in Promoting Health*, 181–203. Elsevier.
34. Ranjbar, S., Seyednejad, S.A., Azimi, H. et al. (2019). Emerging roles of probiotics in prevention and treatment of breast cancer: a comprehensive review of their therapeutic potential. *Nutrition and Cancer 71* (1): 1–12.
35. Nazir, Y., Hussain, S.A., Abdul Hamid, A. et al. (2018). Probiotics and their potential preventive and therapeutic role for cancer, high serum cholesterol, and allergic and HIV diseases. *BioMed Research International 2018*: 4–7.
36. Koradia, P., Kapadia, S., Trivedi, Y. et al. (2019). Probiotic and cranberry supplementation for preventing recurrent uncomplicated urinary tract infections in premenopausal women: a controlled pilot study. *Expert Review of Anti-Infective Therapy 17* (9): 733–740.
37. Wang, I.-K., Wu, Y.-Y., Yang, Y.-F. et al. (2015). The effect of probiotics on serum levels of cytokine and endotoxin in peritoneal dialysis patients: a randomised, double-blind, placebo-controlled trial. *Beneficial Microbes 6* (4): 423–430.
38. Oak, S.J. and Jha, R. (2019). The effects of probiotics in lactose intolerance: a systematic review. *Critical Reviews in Food Science and Nutrition 59* (11): 1675–1683.
39. Kechagia, M., Basoulis, D., Konstantopoulou, S. et al. (2013). Health benefits of probiotics: a review. *International Scholarly Research Notices 2013*: 4.
40. Mazloom, K., Siddiqi, I., and Covasa, M. (2019). Probiotics: how effective are they in the fight against obesity? *Nutrients 11* (2): 258.

41. Yan, F., Li, N., Yue, Y. et al. (2020). Screening for potential novel probiotics with dipeptidyl peptidase IV-inhibiting activity for type 2 diabetes attenuation in vitro and in vivo. *Frontiers in Microbiology 10*: 2855.
42. Sun, Z., Sun, X., Li, J. et al. (2020). Using probiotics for type 2 diabetes mellitus intervention: advances, questions, and potential. *Critical Reviews in Food Science and Nutrition 60* (4): 670–683.
43. Zamani, B., Golkar, H.R., Farshbaf, S. et al. (2016). Clinical and metabolic response to probiotic supplementation in patients with rheumatoid arthritis: a randomized, double-blind, placebo-controlled trial. *International Journal of Rheumatic Diseases 19* (9): 869–879.
44. Di Pierro, F., Zanvit, A., Nobili, P. et al. (2015). Cariogram outcome after 90 days of oral treatment with *Streptococcus salivarius* M18 in children at high risk for dental caries: results of a randomized, controlled study. *Clinical, Cosmetic and Investigational Dentistry 7*: 107.
45. Alesa, D.I., Alshamrani, H.M., Alzahrani, Y.A. et al. (2019). The role of gut microbiome in the pathogenesis of psoriasis and the therapeutic effects of probiotics. *Journal of Family Medicine and Primary Care 8* (11): 3496.
46. Brown, A.C. and Valiere, A. (2004). Probiotics and medical nutrition therapy. *Nutrition in Clinical Care: An Official Publication of Tufts University 7* (2): 56.
47. Sadiq, F.A. (2021). Is it time for microbiome-based therapies in viral infections? *Virus Research 291*: 198203.
48. Roberfroid, M. (2007). Prebiotics: the concept revisited. *The Journal of Nutrition 137* (3): 830S–837S.
49. Brüssow, H. (2019). Probiotics and prebiotics in clinical tests: an update. *F1000Research 8*: 5–6.
50. Wong, A.C. and Levy, M. (2019). New approaches to microbiome-based therapies. *MSystems 4* (3): e00122–e00119.
51. Vieira, A.T., Fukumori, C., and Ferreira, C.M. (2016). New insights into therapeutic strategies for gut microbiota modulation in inflammatory diseases. *Clinical & Translational Immunology 5* (6): e87.
52. Tsilingiri, K. and Rescigno, M. (2013). Postbiotics: what else? *Beneficial Microbes 4* (1): 101–107.
53. Aguilar-Toalá, J.E., Garcia-Varela, R., Garcia, H.S. et al. (2018). Postbiotics: an evolving term within the functional foods field. *Trends in Food Science and Technology 75*: 105–114.
54. Salminen, S., Collado, M.C., Endo, A. et al. (2021). The international scientific Association of Probiotics and Prebiotics (ISAPP) consensus statement on the definition and scope of postbiotics. *Nature Reviews. Gastroenterology & Hepatology* 1–19.
55. Foo, H.L., Loh, T.C., Abdul Mutalib, N.E. et al. (2019). The myth and therapeutic potentials of postbiotics. In: *Microbiome and Metabolome in Diagnosis, Therapy, and Other Strategic Applications*, 201–208. Academic Press.
56. Nataraj, B.H., Ali, S.A., Behare, P.V. et al. (2020). Postbiotics-parabiotics: the new horizons in microbial biotherapy and functional foods. *Microbial Cell Factories 19* (1): 1–22.
57. Rad, A.H., Maleki, L.A., Kafil, H.S. et al. (2021). Postbiotics as promising tools for cancer adjuvant therapy. *Advanced Pharmaceutical Bulletin 11* (1): 1.
58. Rad, A.H., Aghebati-Maleki, L., Kafil, H.S. et al. (2021). Molecular mechanisms of postbiotics in colorectal cancer prevention and treatment. *Critical Reviews in Food Science and Nutrition 61* (11): 1787–1803.
59. Malik, J.K., Raina, R., Ahmad, A.H. et al. (2021). Synbiotics: safety and toxicity considerations. In: *Nutraceuticals*, 1107–1123. Elsevier.
60. Pandey, K.R., Naik, S.R., and Vakil, B.V. (2015). Probiotics, prebiotics and synbiotics – a review. *Journal of Food Science and Technology 52* (12): 7577–7587.
61. Shimizu, K., Ogura, H., Asahara, T. et al. (2013). Probiotic/synbiotic therapy for treating critically ill patients from a gut microbiota perspective. *Digestive Diseases and Sciences 58* (1): 23–32.
62. Ghouri, Y.A., Richards, D.M., Rahimi, E.F. et al. (2014). Systematic review of randomized controlled trials of probiotics, prebiotics, and synbiotics in inflammatory bowel disease. *Clinical and Experimental Gastroenterology 7*: 473.

22

Insight into Soil Organic Pollutants: Microbial Bioremediation as a Sustainable Approach Toward Restoration of Agriculture Soil

Amarnath Mishra[1], Shrutika Singla[2], Sourabh Kumar Singh[2], and Akanksha Behl[2]

22.1. INTRODUCTION

POPs (persistent organic pollutants) are synthetic (man-made) chemicals – they are all engineered chemicals, either intentionally or non-intentionally produced/released. A few are pesticides; others are mechanical items or unintended byproducts resulting from reactions or combustions. These pollutants are persistent in the environment, their fortitude within the environment is surprising and it may take centuries for them to be degraded. Long-range transport leads to worldwide contamination. As is the case with many environmental pollutants, it is most troublesome when ailments or illnesses are found due to the introduction of a specific, persistent natural toxin in the environment [1].

Organic pollutants (OPs) are generally lipophilic: they have a propensity to stay in fat-rich tissues. This affinity for fat tissues implies that POPs are likely build up, endure, and bioconcentrate and may, inevitably, achieve toxicologically relevant concentrations – in spite of the fact that presentation may be limited. They accumulate in the environmental food chain by amassing within larger creatures as they eat smaller ones. The highest levels are found in highly evolved marine creatures – these levels are considered to conceivably cause increased mortality among highly evolved sea creatures [2]. It is hypothesized that the prevalence of OPs may lead to vitamin and thyroid insufficiencies and cause expanded susceptibility to microbial infections and regenerative disorders. The acute, high-level poisonous quality of OPs is well characterized – intense impacts after introduction of large quantities have been described for some organochlorine pesticides (e.g. aldrin, dieldrin and toxaphene). The pedosphere, the soil mantle of the Earth, covers an essential portion of the Earth's lithosphere, hydrosphere, environment, and biosphere. What happens in the soil ought to have a significant effect not only on soil quality and rural generation, but moreover on total wellbeing

[1]*Amity Institute of Forensic Sciences, Amity University Uttar Pradesh, Noida, Uttar Pradesh, India*
[2]*Research Scholar, Amity Institute of Forensic Sciences, Amity University, Noida, Uttar Pradesh, India*

Microbial Biotechnology: Role in Ecological Sustainability and Research, First Edition.
Edited by Pankaj Chowdhary, Sujata Mani, and Preeti Chaturvedi.

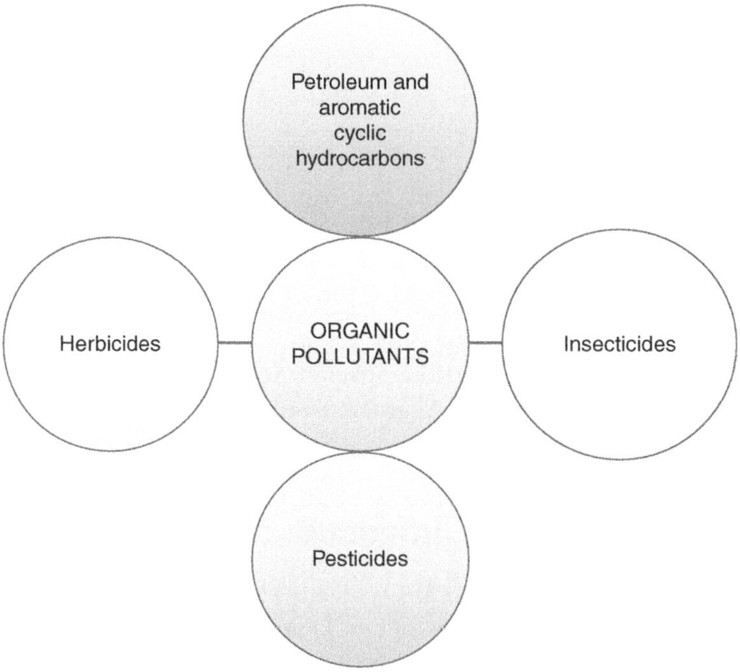

Figure 22.1 Types of organic pollutants.

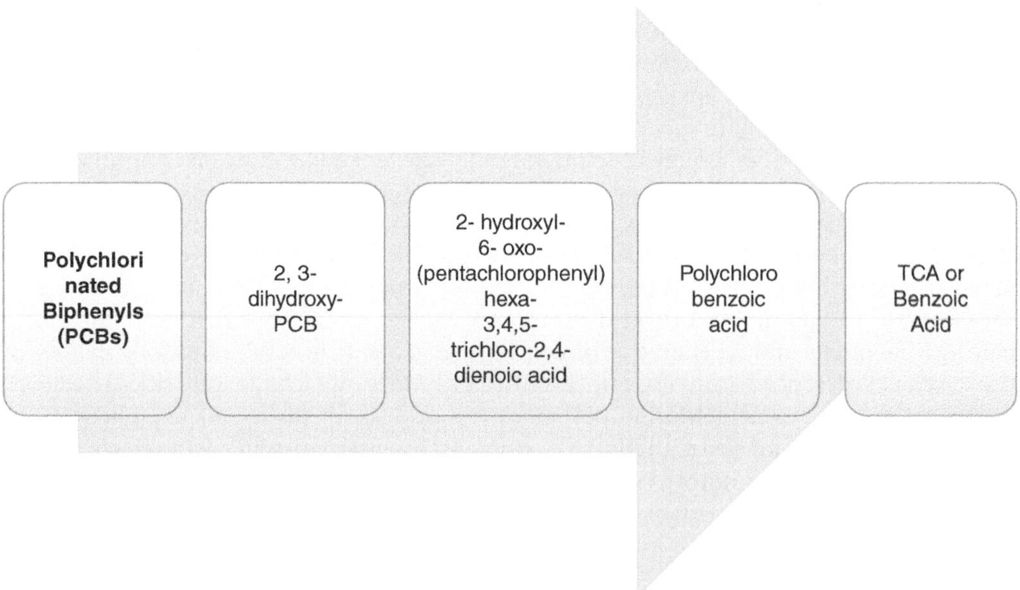

Figure 22.2 Steps in the degradation of PCBs.

of the environment, which is characterized by ecosystem maintainability in terms of action, organization, and resilience. Soil may be a central point of the environment [3]. Soil components, be they minerals, organic matter, or microorganisms, significantly influence the physical, chemical, and natural processes in the soil. Over the past decades, achievements in the subdisciplines of material science, chemistry, and science of the soil have been amazing. However, data on interactions of soil minerals with organic components and microorganisms is fragmentary and scattered within various writings of soil and natural sciences. Mineral colloids can impact microbial movement through direct and indirect impacts [4]. Direct impacts are those impacts that include a surface interaction between mineral colloids and microorganisms. Indirect impacts are characterized as those impacts of mineral colloids that modify the environment in which the microorganisms are dwelling. The refinement between direct and indirect impacts of mineral colloids on microorganisms in soil isn't always clear because many microbial effects are likely influenced at the same time, both directly and indirectly. Mineral colloids can advance the action of microorganisms in their region by keeping the pH of microhabitats inside the ideal physiological levels and by sorbing microbial metabolites that would otherwise be impeding development [5].

Knowledge of soil minerals with natural components and microorganisms exert enormous impacts on the change and elements of soil natural matter, nutrient cycling, supplement bioavailability, viability and poisonous quality of pesticides, microbial metabolic forms, development, attachment, biological enzymatic movement, and soil physical properties. In this manner, knowledge of soil minerals with organic components and microorganisms ought to have significant impacts on plant nourishment and biological productivity of soils. Xenobiotics can moreover be debased by oxidative coupling responses, which are catalyzed either naturally by polyphenol oxidases and peroxides or abiotically by metal oxides and clay minerals. These naturally occurring forms are known to result in the detoxification of xenobiotics and have been proposed as a way of purification. While indigenous proteins are as a rule not likely to supply a full cleaning of polluted locales, revising soil with chemicals inferred from particular microbial societies or plant material may fortify detoxification forms. Abiotic and biotic catalysts coexist in soil environments. Abiotic catalysts can impact microbial arrangement of proteins and enzymatic activity.

Defilement of soils with known natural poisons (POPs) has become of great concerns in recent years. The POPs of specific concern include organochlorine pesticide, polybrominated diphenyl ethers (PBDEs), halohydrocarbon, and polycyclic aromatic hydrocarbons (PAHs). Most of those compounds are also harmful to human health at low concentrations. Microbial debasement, as an imperative mechanism for the evacuation of POPs, is confined by the microorganism's bioavailability. In this manner, expanding the bioavailability of POPs to degrader or available microorganisms is vital in soil bioremediation [6].

22.2. BIOAVAILABILITY OF ORGANIC POLLUTANTS IN SOIL

It has been shown that the bioavailability of POPs was diminished after adsorption by soil. With time, soils form more grounded binding with POPs, thus causing a decay in bioavailability, a process termed as maturing. The quick desorption fractions are surface-adsorbed and conversely with the porewater, thus promptly bioavailable. Be that as it may, "slow" and "very-slow" fractions are unequivocally bounded or micropore-adsorbed, making them unavailable to microorganisms. Moderate or irreversible desorption of POPs is the most important factor limiting their bioavailability. Although microorganisms can assault adsorbed contaminants, they prefer dissolved divisions. In this manner, sorption and desorption in soil are crucial methods of controlling susceptibility of POPs to microbial debasement. Sorption of POPs is generally affected by the substance [7].

In any case, soils with the same natural matter substance may have different sorption opportunities for POPs. SOM (soil organic matter) may be a heterogeneous mixture, including recently stored biopolymers (e.g. proteins, polysaccharides, and lipids), tolerably matured humic substances (HSs), well matured kerogen, and dark carbon, each of which are potential sorbents for POPs. It was recently found that the desorbing distribution and microbial degradation ratio of PBDEs were weakly related to add up to natural carbon, but emphatically related to the substance of dark carbon. It has been detailed that structure characteristics (aromaticity and aliphaticity), spatial course of action, physical conformation, and extremity of SOM may influence its sorption capacity [8].

Whether aliphatic or aromatic, the composition of SOM is more important to sorption of POPs and has been much examined. It was demonstrated that phenanthrene would be adsorbed preferentially by fragrant carbon-rich lignin compared to paraffinic carbon-rich wax, hence bringing down its bioavailability. Be that as it may, neither aliphatic nor aromatic spaces alone can be utilized to anticipate the sorption capacity of the nitroaromatic as it is closely connected with aliphatic carbon instead of aromatic carbon. Usually reliable was the observation that aliphatic carbon of common natural matter was capable of large sorption capacity for benzene and phenanthrene.

The special structure of minerals empowers it to be associated with various compounds, particularly profoundly polar compounds. Sorption of nonpolar 1,3,5-trichlorobenzene diminished by around 86% during removal of the total SOM, whereas it was 34–54% for exceedingly polar 1,3,5-trinitrobenzene. This demonstrated higher partiality of soil minerals with polar compounds. The surface hydroxyl group, adsorbed water atomic, hydrophobic Si-O bridged surfaces, hydrated cations, and surficial metal of minerals give conceivable sorption sites for POPs. Phyllosilicate clay mineral and metal oxides are widely examined minerals, both of which are effectively sorbent in soil. In the meantime, considering that soil may be a complicated integration, natural and mineral stages interact together to make combined instead of separated mixtures. In this case, the sorption behavior of the mineral-SOM complex is also important.

Microorganisms release surface active atoms (e.g. biosurfactants) to extend bioavailability. On the one hand, biosurfactants diminish interfacial pressures of contaminants and advance their solubility and portability. This can be realized by micellar solubilization of POPs or direct modification of the contaminant lattice (i.e. decrease of soil–water interfacial pressure) In morphology, microbes create multidimensional structures to increase the contact zone with POPs, a common case of which is fungal mycelia [9].

The arrangement of mycelia empowers efficient mobilization of chemicals in soil, with the advantage of hyphal long distance transport conjointly the plausibility for hypha crossing air-filled pores confirmed dynamic transport of vesicles-associated PAH within hyphal pipelines with the assistance of cytoplasmic spilling, driving to a more efficient distribution than dissemination instruments. In physiology, microorganisms encourage POP exchange either by decreasing transportation to POPs (e.g. microbial connection), or by steepening the chemical angle of substrate sources and quickening diffusive transfer/desorption. Arrangement of biofilm on contaminant sources is critical to urge mass-transfer restrictions in low bioavailability lattice (ineffectively dissolvable and strongly adsorbed PAH).

The secretion of extracellular polymeric substances, nearness of glycolipid or glycopeptido lipid on the mycobacterial cell wall, and a more hydrophobic cell surface may encourage this process. The broken-down divisions of OPs are promptly accessible to microorganisms. However, due to microbial adjustments, OPs that are absorbed in soil can also be utilized by microorganisms. Coordinate adhesion to adsorbed OPs and generation of extracellular chemical and biosurfactant are possible ways for microorganisms to get into the adsorbed OPs. The behaviors of microbial connection and extracellular protein emission lead to coordinate degradation of OPs [10].

22.3. IMPACT OF ORGANIC POLLUTANTS ON SOIL MICROBES AND QUALITY

The general OPs that affect public health include incorporated petroleum hydrocarbons from the petroleum industry, as well as numerous insecticides and herbicides that have been utilized in agribusiness and bug control. Other natural poisons are byproducts of fabricating industries. For example, phthalates are plasticizers utilized in bottles, toys, and individual care items. PBDEs are fire retardants included in an expansive assortment of consumer items that filter into encompassing materials and can now be identified in numerous populations. The presence of hydrocarbons in the environment, whether coincidental or due to human exercises, is the biggest cause of water and soil contamination. Petroleum hydrocarbon toxins are known to have antagonistic impacts on sea-going and earthbound life, as well as soil efficiency. The extensive use of pesticides, antimicrobials, and other chemicals results in different situations that are of extraordinary concern due to their ecotoxicological impacts on various living beings [11]. The expanding awareness of the dangers to humans related to organic toxins has driven the advancement of viable procedures to detoxify or clean up these pollutants. Microbial debasement or biodegradation is the use of microorganisms to break down or corrupt, detoxify, or change natural toxins. Natural decay of different substrates is performed by a huge number of microorganisms including microscopic organisms, parasites, protozoa, etc. Organic toxins or pollutants may experience distinctive changes once they enter the environment, including change or debasement, sorption–desorption, volatilization, take-up by plants, run off into surface waters, and transport into groundwater.

Change or degradation is one of the key methods that oversee environmental processes and transport of natural pollutants comprises distinctive forms based on abiotic and biodegradation. Amid these methods, organic poisons are changed into degradation items or are totally mineralized to a carbon field. A variety of microbial properties are observed to impact the exchange of natural chemicals to metabolically active degraders. It is apparent that organic changeability limits the viability of creating an all-encompassing chemical mimicry of bioaccessibility. Important properties incorporate morphological, physiological, and behavioral adjustments of single cells and populaces as well as phenomena associated with the elements and environments of entire natural communities. The low bioaccessibility of a contaminant may genuinely obstruct both the advancement of degradative characteristics and, for extant catabolism, the maintenance of the catabolically dynamic biomass. Chemicals that enter the concentrations, or for the most part exchange at low rates, are less likely to drive advancement of metabolic pathways [12].

Organic pollutant debasement by microorganisms can happen under both aerobic and anaerobic conditions. The foremost quick and total biodegradation of a large share of OPs is caused by oxygen-consuming conditions. The beginning intracellular assault of natural toxin, for example hydrocarbon, is an oxidative practice and the enactment, as well as joining, of oxygen is the enzymatic key response catalyzed by oxygenases and peroxidases. Fringe corruption pathways change over natural toxins step by step into intermediates of the central mediator digestion system, for example, the tricarboxylic corrosive cycle. Biosynthesis of cell biomass happens from the central forerunner metabolites, for example, acetyl-CoA, succinate, and pyruvate. Sugars required for different biosynthesis and development are synthesized by gluconeogenesis. Degradation of pesticides caused by microbes may include a three major steps or stages. In the initial stage, Step I, the primary characteristics of a main compound are changed through oxidation, reduction, or hydrolysis to create a more water-soluble and usually a less poisonous item than the parent compound. The second stage, Step II, includes conjugation of a pesticide or pesticide metabolite to a sugar or amino corrosive, which increases water solvency and decreases toxicity compared to the parent pesticide. The third stage, Step III, involves change of Stage II metabolites into secondary conjugates, which are moreover non-toxic. In these processes, organisms and microscopic organisms are included, producing intracellular or additional cellular chemicals including hydrolytic chemicals, peroxidases, oxygenases, etc. [13]

22.4. BIODEGRADATION OF ORGANIC POLLUTANTS

Biodegradation is the breakdown of natural contaminants that happens due to microbial action. As such, these organic contaminants can be considered a microbial nourishment source or substrate. Biodegradation of any natural compound can be thought of as an arrangement of natural debasement steps or a pathway that eventually comes about within the oxidation of the parent compound. Complete biodegradation or mineralization involves oxidation of the parent compound to make carbon dioxide and water, a preparation that gives both carbon and energy for development and propagation of cells.

Each degradation step within the pathway is catalyzed by a specific enzyme made by the corrupting cell. Chemicals are most often found inside a cell, but are also made and released from the cell to help start corruption reactions. Enzymes found outside to the cell are known as extracellular enzymes. Extracellular proteins are imperative in the degradation of macromolecules such as the plant polymer cellulose.

A few natural contaminants are, as it were, mostly degraded by natural microorganisms. This will result from absence of the appropriate corrupting chemical as mentioned earlier. A second sort of inadequate corruption is cometabolism, in which a fractional oxidation of the substrate occurs but the vitality inferred from the oxidation is not used to bolster microbial development. The method occurs when living beings have one or more chemicals that coincidentally can corrupt a specific contaminant in expansion. Fractional or fragmented degradation can moreover result in polymerization or amalgamation of compounds that are more complex and steady than the parent compound. Beginning debasement steps, frequently catalyzed by extracellular enzymes, make receptive middle compounds [14]. These highly responsive compounds can at that point combine with each other or with other natural matter display in the environment. Amid the biodegradation process, these incorporate formation of dimers or bigger polymers, which are quite stable within the environment. Solidness is due to microorganism bioavailability (high sorption and microorganism solvency), need of degrading enzymes and the reality that a few of these buildups become chemically bound to the soil natural matter division.

Biodegradation of alkanes occurs with a high biological oxygen demand (BOD). The more common pathway is the coordinate consolidation of one atom of oxygen onto one of the end carbons of the alkane by a monooxygenase protein coming about within the arrangement of a primary alcohol. Then again, a dioxygenase protein can incorporate both oxygen atoms into the alkane to create a hydroperoxide. The conclusion result of both pathways is the production of an essential greasy corrosive. There are reported cases of diterminal oxidation, with both ends of the alkane oxidized, and of subterminal oxidation, with an interior carbon oxidized.

According to various studies, it has been concluded that alkenes and alkanes have approximately proportionate biodegradation rates. The initial step in 1-alkene breakdown can affect the terminal or a subterminal methyl group, as explained for alkanes. Then again, the introductory step can affect the twofold bond, which can result in a primary or secondary alcohol or an epoxide. Each of these beginning degradation products is oxidized to a primary fatty acid, which is debased by β-oxidation. Pesticides are the greatest nonpoint source of chemicals added to the environment. The huge share of the currently used natural pesticides are subject to broad mineralization within the time of one growing season or less. Synthetic pesticides show a stupefying assortment of chemical structures, but most can be followed to the generally simple aliphatic, alicyclic, and fragrant base structures already discussed. These base structures bear an assortment of halogen, amino, nitro, hydroxyl, carboxyl, and phosphorus substituents [15].

The use of pesticides has allowed their collection in the environment from squander dumps and spills and as a result of pesticides fabricating forms. Despite the fact that some pesticide debasement happens, it is limited by microorganism bioavailability, by the obstinacy of profoundly chlorinated pesticides congeners beneath high-impact conditions, and by incomplete corruption beneath anaerobic conditions. The extensive study that has been performed to understand pesticide corruption has recommended a few procedures for promoting biodegradation. Of these, the most promising is the use of a consecutive anaerobic aerobic plan to begin with to allow evacuation of chlorines using halo respiration, and then allow mineralization of the less chlorinated congeners. In general, heterocyclic compounds are more difficult to debase than aromatics that contain only carbon; this is usually likely due to the higher electronegativity of the nitrogen and oxygen particles compared with the carbon iota, driving deactivation of the molecule toward electrophilic substitution. Heterocyclic compounds with five-membered rings and one heteroatom are readily biodegradable, likely since five-membered ring compounds show higher reactivity toward electrophilic agents, and subsequently are more promptly biologically hydroxylated [16].

Biodegradation of petroleum hydrocarbons is a complex action that depends on the nature and the amount of the hydrocarbon display, as well as the chemical composition of the petroleum hydrocarbons. Susceptibility of a hydrocarbon to microbial degradation varies with sort and estimate of the hydrocarbon molecule. Short chain alkanes (<C10) are solvents, and as solvents, they tend to disturb the lipid layer structures of microorganisms, which makes them harmful to many microorganisms. They are fundamentally expelled by volatilization. Alkanes of middle chain length (C10–C24) are regularly degraded quickly, whereas exceptionally long chain alkanes are progressively protected from microbial biodegradation. Anaerobic microbial debasement of petroleum hydrocarbons in characteristic situations has been shown to happen at insignificant rates and its ecological significance has been for the most part considered minor. In any case, the microbial debasement of oxidized aromatic compounds such as benzoate and halogenated aromatic compounds such as the halobenzoates, chlorophenols, and polychlorinated biphenyls (PCBs) has been shown to happen under anaerobic conditions [17].

22.5. STRATEGIES OF ORGANIC POLLUTANT DEGRADING MICROORGANISMS

There are various methods under development for the removal of OPs from the environment. Certain bacteria and microorganisms are consumed widely to remove OPs from the environment as these bacteria have high capability for turning these contaminants into simpler products such as carbon dioxide and water. These microbes can adapt the environment containing OPs due to gene rearrangement and mutation. Chemoorganotroph microbe genera such as *Bacillus, Gordonia, Microbacterium, Micrococcus, Methanoseata,* etc., can degrade persistent OPs [18]. *Pseudomonas fluorescenc* and *Pseudomonas putida* are aerobic Gram-negative bacteria having the highest tendency to degrade OPs [19]. *Mycobacterium* and *Rhodococcus* are Gram-positive bacteria referred to as biodegraders of polyaromatic hydrocarbons [20].

OPs after dissolving in soil become accessible to the microorganisms present in the soil. Not only residual OPs, but OPs dissolved in soil, would be used to microorganisms. Microorganisms attach to the OPs and with the help of extracellular enzymes and biosurfactants, OPs are degraded [21].

Microorganisms adopt different methods to degrade OPs. One of the methods used by microorganisms in situ is chloroaromatic metabolic enzymes and pathways. α-ketoglutarate-dependent dioxygenase (TfdA) enzyme breaks down the side chain of 2,4-dichlorophenoxyacetate and 4-chloro-2-methylphenoxyacetate, which give rise to 2,4-dichloromethylphenol and 4-chloro-2-methylphenol [22, 23].

In other studies, it was shown that monooxygenase enzyme present in *Burkholderiacepacia* can also initiate biodegradation of 2,4,5-trichlorophenoxyacetate [24].

Many organohalogens are degraded by the process of bacterial dehalorespiration in anaerobic conditions. This group of bacteria includes Gram-positive bacteria such as *Dehalobacter* and *Desulfomonile* [25].

Many dioygenase enzymes, such as the Rieske-type non-heme iron oxygenases enzyme, are responsible for the breakdown of carbazole and dibenzofurans [26]. These oxygenase enzymes have been isolated from *Pseudomonas resinovorans* and *Sphingomnas*strains [27].

22.6. MICROBIAL BIOREMEDIATION OF ORGANIC POLLUTANTS

The transformation of any chemical into another form using a biological catalyst for reduction purposes is the crucial step in biological bioremediation. Studies are ongoing surrounding microbial bioremediation of chemicals, as it changes more complex and harmful substances into simpler ones with more efficiency and without causing lesser effects to the surrounding areas. Microorganisms use OPs as substrates for converting them into simpler products, which help in their growth and metabolism [28].

1. **Petroleum Hydrocarbon**

Petroleum hydrocarbon is one of the most widely distributed OPs because of the use of fuels and their products all over the world. These are released from various refineries and other industries, affecting the environment. There are various algae and cyanobacteria that are used for the degradation of these hydrocarbons. For example, *Protothecazopfii* is one of the algae that can reduce petroleum [29]. Another cyanobacteria, *Oscillatoria quadripunctualata,* can be used to absorb different products such as phenols and sulphide from the petroleum products released in the atmosphere [30].

Various halophilic bacterial species including *Halomonas*, *Marinobacterssp*, *Dietzia*, *Ralstonia*, etc., also degrade these hydrocarbons efficiently [31]. The degradation is decreased if salinity of the surrounding soil is increased. In some studies, bacteria like *Rhodococcus* and *Arthrobacter* degrade hydrocarbon in high salt conditions [32].

2. **Polychlorinated Biphenyls (PCBs)**

PCB is one of the main constituents of industrial waste as it is used in insulators, pharmaceuticals, and lubricating oils. It is also present in some pesticides. PCBs have been banned in many countries due to their high toxicological effects but they are still present in the environment due to their ability to escape from degradation. The degradation of PCB results in the formation of chlorobenzoate [33].

PCBs are insoluble in water, non-polar, and non-reactive species. There are more than 100 isomers of PCBs that are used in industries. PCBs are first of all hydrolyzed to 2,3-dihydroxy-PCB. This is then converted into 2-hydroxyl-6-oxo-(pentachlorophenyl) hexa-3,4,5-trichloo-2,4-dienoic acid, which is then converted into polychlorobenzoic acid. Polychlorobenzoic acid can be either degraded into trichloroacetic acid or benzoic acid [34].

In a study conducted by Takase et al., they observed that *Pseudomonas cruciviae* can degrade PCBs using the meta cleavage pathway [35]. In other study conducted by Dmochewitz et al., *Aspergillus niger* was observed to utilize mixtures of PCBs for growth and metabolism [36].

3. **Polycyclic Aromatic Hydrocarbons (PAHs)**

PAHs are released by many industries, including coal and petroleum. They enter the surrounding areas due to their use in pharmaceutical and agrochemical industries. The structure of PAH consists of only aromatic rings without any other side molecules [37]. PAHs causes mutagenic and oncogenic effects in many life forms [38].

Many algae species are known for biodegradation of PAH such as *Chlorella minister* and *Scenedesmus platydiscus*, which remove pyrene, a carcinogenic PAH, from the atmosphere. These algae absorb and transform these PAH into lesser harmful products [39]. Various factors that affect the removal of PAH by algae include the concentration of PAH and algae, algae species, surface characteristics, and the condition of area [40].

Degradation of PAH by some bacteria requires the presence of molecular oxygen to initiate the degradation process. Many bacteria and their genomic sequences have been reported for degradation. It includes *nah* genes from *P. putida*, *pah* genes from *Pseudomonas aeruginosa* [41], AN10 of *Pseudomonas stutzeri*, etc. [42]. Another bacterial strain, *Stenotrophomans* sp. IITR87, was observed degrading PAH compounds. The degradation by this species directly depends on the concentration of PAH present in the area [43].

4. **Pesticides**

Pesticides are agrochemicals that are used in crops for the prevention of attack by insects and other microorganisms. These chemicals, when used in small quantity, do not affect anyone and attack specific microorganisms, but these chemicals are used regularly in large amounts, which can lead to various diseases and disorders in animals and humans. These chemicals accumulate in humans and animals after consumption, causing dreadful effects [44].

These chemicals also stay in the soil and degrade the quality of soil, which results in degradation of the crop [45].

Pendimethalin is a herbicide used to prevent the growth of perennial grasses and other plants such as cotton, etc. *Azetobacter vinelandii* can utilize this pesticide as a carbon source to fix N_2 [46]. Transformation of pendimethalin by *A. vinelandii* occurs by the process of N-dealkylation and it results in six major metabolites. *Fusarium solani* also breaks down pendimethalin into three major metabolites [47]. *F. solani* breaks down oxadiazon in a way that no residue of it is left in the soil. The main methods for reduction of these pesticides are acetylation and nitro reduction [48].

Various algae and cyanobacteria are also used for the biodegradation of pesticides. In a study by Mansy et al., various cyanobacteria including *Anabaena cylindrical* and *Anabaena spiroides* were used for the degradation of fluometruron [49]. These cyanobacteria result in degradation by 94%. *Anabaena* sp. are also known for the reduction of methyl parathion [50].

In one of the study conducted by Bumpus and Aust, a fungus, *Phanerochaetechrysosporium,* which can degrade lignin, was identified for the degradation of DDT using oxidation and dechlorination pathways. The action of this fungus results in formation of CO_2 as a product [51].

22.7. FUTURE CHALLENGES

One of the largest problems faced by us at this moment is pollution of the environment, as we all know the level of toxicity and lethality caused by these OPs on living organisms. With the passage of time, bioaccumulation and biomagnification of these OPs will increase their concentration in the environment. So, it very important to work for the remedies to degrade these OPs to remove them from the environment.

Remediation by microorganisms has been proven as powerful, easily sustainable, and cheap to reduce pollution. Microorganisms adapt themselves to the changing environment, which make it very important to understand the microbial population, their reaction to the changing environment, how they approach the pollutant, and through which pathway degradation of OPs occur.

The study of microorganisms remains vast and neverending. In the future, it is necessary to study each and every detail about them so that scientists can work on the advancement of methods of biodegradation. There are some specific gene sequences that are responsible for degradation of OPs. These sequences can be transplanted to other organisms using different methods. Scientist can work on these gene sequences to make them more potent for degradation purposes.

One of the major challenges that will be faced by scientists in the future is the changing environment. Microorganisms work efficiently in certain conditions of temperature, pH, humidity, and alkalinity. If these conditions keep on changing, it will be very difficult for these microorganisms to work further.

A second challenge faced by scientists is mutation on gene sequences. The gene sequence of the microorganism mutates easily, which will affect the particular sequence that is responsible for the task. In many cases, the byproducts released after the degradation are more toxic and persistent to the environment than the original pollutant. This point should be taken into consideration by the researchers.

22.8. CONCLUSION

OPs are present in the environment in various forms such as PCBs, pesticides, poly aromatic hydrocarbons (PCHs), and heavy metals, which are released from agrochemical, pharmaceuticals, transport, and other industries and accumulate in the surrounding environment. These OPs are biomagnified when transferred from soil or water to living organisms, resulting in toxicity and lethality to them. It is very important to remove these OPs from the environment. Bioremediation techniques using bacteria, fungi, and algae are used for their removal because these are efficient and advantageous to the surrounding. Many challenges are faced by scientists and researchers while working with microorganisms, which need to be resolved as soon as possible.

REFERENCES

1. Mbachu, A.E., Chukwura, E.I., and Mbachu, N.A. (2020). Role of microorganisms in the degradation of organic pollutants: a review. *Energy and Environmental Engineering* 7 (1): 1–11.
2. Gioia, R., Group, Y., Dachs, J. et al. (2011, 2013). *Persistent Pollution: Past, Present and Future.* ISBN: 9783642447259.
3. Hani, H. (1990). The analysis of inorganic and organic pollutants in soil with special regard to their bioavailability. *International Journal of Environmental Analytical Chemistry* 39 (2): 197–208.
4. World Health Organization (2008). Persistent Organic Pollutants (POPs). *Children's Health and the Environment.*

5. Le Borgne, S., Paniagua, D., and Duhalt, V.R. (2008). Biodegradation of organic pollutants by halophilic bacteria and archaea. *Journal of Molecular Microbiology and Biotechnology 15* (2–3): 74–92.
6. Maier, R.M. and Gentry, T.J. (2014). Microorganisms and organic pollutants. In: *Environmental Microbiology*, 3e. Elsevier Inc.
7. O'Sullivan, G. and Sandau, C. (ed.) (2013). *Environmental Forensic for Persistent Organic Pollutants*. Elsevier Inc.
8. Reid, B.J., Jones, K.C., and Semple, K.T. (2000). Bioavailability of persistent organic pollutants in soils and sediments. A perspective on mechanism, consequences and assessment. *Environmental Pollution 108* (1): 103–112.
9. Ren, X., Zeng, G., Tang, L. et al. (2018). Sorption, transport and biodegradation – an insight into bioavailability of persistent organic pollutants in soil. *Science of the Total Environment 610–611*: 1154–1163.
10. Ren, X., Zeng, G., Tang, L. et al. (2018). The potential impact on the biodegradation of organic pollutants from composting technology for soil remediation. *Waste Management 72*: 138–149.
11. Sandrin, T.R. and Maier, R.M. (2003). Impact of metals on the biodegradation of organic pollutants. *Environmental Health Perspectives 111* (8): 1093–1101.
12. Semple, K.T., Doick, K.J., Wick, L.Y. et al. (2007). Microbial interactions with organic contaminants in soil: definition, processes and measurement. *Environmental Pollution 150* (1): 166–176.
13. Huang, P.M. (2004). Soil mineral–organic matter–microorganism interactions: Fundamentals and impacts. *Advances in Agronomy 82*: 393–472.
14. Walker, C. (2009). *Organic Pollutants: An Ecotoxicological Approach*. ISBN: 9780367386405.
15. Welander, U. (2005). Microbial degradation of organic pollutants in soil in a cold climate. *Soil and Sediment Contamination 14* (2): 281–291.
16. Welp, G. and Brummer, G.W. (1999). Effects of organic pollutants on soil microbial activity: the influence of sorption, solubility and speciation. *Ecotoxicology and Environmental Safety 43* (1): 83–90.
17. Zhang, M., Tao, S., and Wang, X. (2020). Interaction between organic pollutants and carbon nanomaterial and the associated impact on microbial availability and degradation in soil: a review. *Environmental Science: Nano 7* (9): 2486–2508.
18. Chowdhary, A., Pradhan, S., Saha, M. et al. (2008). Impact of pesticides on soil microbiological parameters and possible bioremediation strategies. *Indian Journal of Microbiology 48*: 114–127.
19. Houghton, J.E. and Shanley, M.S. (1994). Catabolic potential of pseudomonads: a regulatory perspective. In: *Biological Degradation and Bioremediation of Toxic Chemicals* (ed. G.R. Chaudhry), 11–32. Portland: Dioscorides press.
20. Nzila, A. (2013). Update on the cometabolism of organic pollutants by bacteria. *Environmental Pollution 178*: 474–482.
21. Megharaj, M., Ramakrishnan, B., Venkateswarlu, K. et al. (2011). Bioremediation approaches for organic pollutants: a critical perspective. *Environment International 37*: 1362–1375.
22. Pieper, D.H., Martin dos Santos, V.A.P., and Golyshin, P.N. (2004). Genomic and mechanistic insights into the biodegradation of organic pollutants. *Current Opinion in Biotechnology 15*: 215–224.
23. Itoh, K., Kanda, R., Sumita, Y. et al. (2002). tfdA-like genes in 2,4-dichlorophenoxyacetic acid-degrading bacteria belonging to the Bradyrhizobium-Agromonas-Nitrobacter-Afipin cluster in a proteobacteria. *Applied and Environmental Microbiology 68*: 3449–3454.
24. Kitagawa, W., Takami, S., Miyauchi, K. et al. (2002). Novel 2,4-dichlorophenoxyacetic acid degradation genes from oligotrophic Bradyrhizobium sp. strain HW13 isolated from a pristine environment. *Journal of Bacteriology 184*: 509–518.
25. Sanford, R.A., Cole, J.R., and Tsedje, J.M. (2002). Characterization and description of Anaeromyxobacterdehalogenans gen. nov., sp. nov., an aryl-halorespiring facultative anaerobic myxobacterium. *Applied and Environmental Microbiology 68*: 893–900.

26. Nam, J.W., Nojiri, H., Noguchi, H. et al. (2002). Purification and characterization of carbazole 1,9a-dioxygenase, a three-component dioxygenase system of *Pseudomonas resinovorans* strain CA10. *Applied and Environmental Microbiology 68*: 5882–5890.
27. Kilbane, J.J., Daram, A., Abbasian, J. et al. (2002). Isolation and characterization of *Sphingomonas*sp. GTIN11 capable of carbazole metabolism in petroleum. *Biochemical and Biophysical Research Communications 297*: 242–248.
28. Alexander, M. (1994). *Biodegradation and Bioremediation.* San Diego, California, USA: Academic Press.
29. Walker, J.D., Colwell, R.R., and Petrakis, L. (1975). Degradation of petroleum by an algae, *Protothecazopfii. Applied and Environmental Microbiology 30*: 79–81.
30. Joseph, V. and Joseph, A. (2001). Microalgae in petrochemical effluent growth and biosorption of total dissolved solids. *Bulletin of Environmental Contamination and Toxicology 66*: 522–527.
31. Borgne, S.L., Paniagua, D., and Duhalt, R.V. (2008). Biodegradation of organic pollutants by halophilic bacteria and archea. *Journal of Molecular Microbiology and Biotechnology 15*: 74–92.
32. Zvyagintseva, I.S., Poglasova, M.N., Gotoeva, M.T. et al. (2001). Effect of the medium salinity on oil degradation by Nocardioform bacteria. *Microbiology 70*: 652–656.
33. Kidd, K.A., Becher, G., Bergman, A. et al. (2012). Human and wildlife exposure to EDCs. In: *State of the Science of Endocrine Disrupting Chemicals*, 1–261. ISBN: 9789280732740.
34. Chaudhry, G.R. (1991). Biodegradation of halogenated organic compounds. *Microbiological Reviews 55* (1): 59–79.
35. Takase, I., Omori, T., and Minoda, Y. (1986). Microbial degradation products from biphenyl-related compounds. *Agricultural and Biological Chemistry 50*: 681–686.
36. Dmochewitz, S. and Ballschmiter, K. (1988). Microbial transformation of technical mixtures of polychlorinated biphenyls (PCB) by the fungus *Aspergillus niger*. *Chemosphere 17*: 111–121.
37. Liu, K., Han, W., Pan, W.P. et al. (2001). Polycyclic aromatic hydrocarbon (PAH) emissions from a coal-fired pilot FBC system. *Journal of Hazardous Materials 84*: 175–188.
38. Dejmek, J., Solansky, I., Benes, I. et al. (2000). The impact of polycyclic aromatic hydrocarbons and fine particles on pregnancy outcome. *Environmental Health Perspectives 108*: 1159.
39. Lei, A., Wong, Y., and Tam, N. (2002). Removal of pyrene by different microalgal species. *Water Science Technology 46*: 195–201.
40. Todd, S.J., Cain, R.B., and Schmidt, S. (2002). Biotransformation of naphthalene and diaryl ethers by green microalgae. *Biodegradation 13*: 229–238.
41. Boronin, A.M. (2001). Purification and characterization of a salicylate hydroxylase involved in 1-hydroxy2-naphthoic acid hydroxylation from the naphthalene and phenanthrene degrading bacterial strain *Pseudomonas putida* BS202-P1. *Biodegradation 12*: 179–188.
42. Habe, H. and Omori, T. (2003). Genetic of polycyclic aromatic hydrocarbon metabolism in diverse aerobic bacteria. *Bioscience, Biotechnology, and Biochemistry 67*: 225–243.
43. Tiwari, B., Manickam, N., Kumari, S., and Tiwari, A. (2016). Biodegradation and dissolution of polyaromatic hydrocarbons by *Stenotrophomonas* sp. *Bioresource Technology 216*: 1102–1105.
44. Handa, S.K., Agnihotri, N.P., and Kulshreshtha, G. (ed.) (1999). Effect of pesticide on soil fertility. In: *Pesticide residue: Significance, Management and Analysis*, 184–198. Research Publications & Book Publishing House.
45. Zhu, G., Wu, H., Guo, J. et al. (2004). Microbial degradation of fi pronil in clay loam soil. *Water, Air, and Soil Pollution 153*: 35–44.
46. Barua, A.S., Saha, J., Chaudhari, S. et al. (1990). Degradation of Pendimethalin by soil. *Fungi Pesticide Science 29* (4): 419–425.
47. Vogt, C. and Richnow, H.H. (2013). Bioremediation via in situ microbial degradation of organic pollutants. *Advances in Biochemical Engineering/Biotechnology 143*: 123–146.

48. Chakraborty, S.K., Chowdhury, A., Bhattacharyya, A. et al. (1995). Microbial degradation of oxidization by soil fungus *Fusarium*. *Journal of Agricultural and Food Chemistry 43*: 2964–2969.
49. El-Rahman Mansy, A. and El-Bestawy, E. (2002). Toxicity and biodegradation of fluometuron by slectedcyanobacterial species. *World Journal of Microbiology & Biotechnology 18*: 125–131.
50. Barton, J.W., Kuritz, T., O'Connor, L.E. et al. (2004). Reductive transformation of methyl parathion by the cyanobacterium *Anabaena* sp. strain PCC7120. *Applied Microbiology and Biotechnology 65*: 330–335.
51. Bumpus, J.A. and Aust, S.D. (1987). Biodegradation of DDT [1,1,1,-trichloro-2,2-bis (4-chlorophenyl)ethane] by the white rot fungus *Phanerochaete chrysosporium*. *Applied and Environmental Microbiology 53*: 2001–2008.

23

Agro-Wastes for Cost Effective Production of Industrially Important Microbial Enzymes: An Overview

Shruti Dwivedi[1], Aiman Tanveer[1], Sangeeta Yadav[1], Gautam Anand[1,2], and Dinesh Yadav[1]

23.1. INTRODUCTION

Every year, agricultural-based companies produce many residues that if released into the environment without being properly treated may pollute the ecosystem and threaten human and animal health. In most cases such agro-wastes are untreated and unused, and are preferably disposed of by burning, dumping, or unplanned landfilling [1]. These methods also resulted in enhanced levels of greenhouse gas (GHG) emissions as accomplished by burning of fossil fuels. Wastes produced from the cultivation and processing of agricultural products such as fruits, vegetables, meat, poultry, dairy products, and crops are called agricultural wastes [2]. The non-products of agricultural production and processing may contain material that is useful to humans. These non-products have economic value far lower than the cost for their collection, transportation, and processing [3]. They may be liquids, slurries, or solids, and their composition will vary depending on the system and type of farming.

These agro-wastes need proper management owing to the presence of possible leftover nutritive ingredients in the form of proteins, carbohydrates, and minerals as byproducts [4, 5]. Pomegranate peels, lemon peels, and green walnut husks are examples of agricultural wastes that may act as effective natural antimicrobials. These also may be a source to produce mushrooms as well as other bio-based goods such as bio-energy and bio-fertilizers [6, 7]. Animal feed is also made from some agricultural wastes. Because of their high nutritional content, these residues serve as raw materials for production of several

[1]*Department of Biotechnology, Deen Dayal Upadhyaya Gorakhpur University, Gorakhpur, Uttar Pradesh, India*

[2]*Department of Plant Pathology and Weed Research, Agricultural Research Organization, The Volcani Center, Rishon LeZion, Israel*

Microbial Biotechnology: Role in Ecological Sustainability and Research, First Edition.
Edited by Pankaj Chowdhary, Sujata Mani, and Preeti Chaturvedi.

important products and provide ideal conditions for the growth of microbes and preferred as a solid base in solid-state fermentation (SSF). It also aids in the generation of fermentable sugars by lowering the cost of production using food crops [7].

Agricultural wastes are assumed to contribute a large share of the total waste matter. The increasing agricultural productivity has further increased the quantities of livestock wastes, agricultural crop wastes, and agro-industrial byproducts. It has been reported that agricultural wastes are projected to be produced at a rate of 998 million tonnes each year. On a wet weight basis, organic wastes can account for up to 80% of total solid wastes created on a farm, with manure production reaching 5.27 kg^{-1} day^{-1} 1000 kg^{-1} live weight. The farming practices greatly influences the amount of waste generated and hence needs to be judiciously applied [8].

A substantial increase of $6.3 billion in 2021 in the global market for industrial enzymes speaks volumes about the importance of enzyme research. The group of industrially important enzymes includes several enzymes that act on polysaccharides (such as α-amylase, amyloglucosidases, cellulases, xylanases, inulinases, hemicellulases, mannanase, lactase, α-glucanases, invertases, and pectinases), proteins (proteases, transglutaminases), and other enzymes such as lipases [9, 10]. Productions of several microbial enzymes have been substantially reported both by solid state (SSF) and submerged fermentation (SmF) methods. Cost-effective production of these microbial enzymes using low-cost substrates comprising different agro-fruit wastes through SSF is a recent development. SSF is preferred over SmF based on several advantages like lower capital requirements, lower energy consumption, a simple fermentation medium, higher productivity, and less effluent creation. Further, SSF creates a habitat similar to the microorganism's natural environment, allowing it to flourish and produce enzymes efficiently [11–13]. Among different agro-wastes, lignocellulosic wastes emerging from the food industry are a cheap and widely available source of carbohydrates for valorization and value addition, particularly for the manufacture of industrially significant enzymes employing microbes. Enzymes are widely used in several industrial applications due to their strong substrate selectivity and biodegradability [14, 15]. The utility of agro-wastes for production of industrially important enzymes has been depicted in Figure 23.1.

23.2. AGRO WASTES: TYPES AND CHARACTERISTICS

Stems, leaves, seed pods, husks, seeds, roots, bagasse, peels, oil cakes, and molasses are some important examples of agricultural wastes. The majority of food and agricultural residues and wastes need to be properly managed as they are sources of carbohydrates, fat, protein, lignin, and cellulose. These wastes provide appropriate substrates for microbial growth, such as bacteria, filamentous fungus, and yeast. As a result, they could be fermented in the solid state for enzyme production. These agricultural wastes represent a type of solid wastes resulting from different farming activities [16]. Another prominent source or cause of agricultural solid wastes is food spoilage. Orange peels and banana peels, for example, are commonly dumped as agricultural solid waste in many households. Agricultural solid wastes, on the other hand, might be generated unintentionally because of food deterioration [17]. Several fruit wastes left over from industrial processing of citrus, banana, apple, and pear also contribute to agricultural wastes. Citrus fruits, such as oranges, grapefruits, lemons, limes, and mandarins, are grown abundantly throughout the world. Annually, around 100 million tonnes of citrus fruits are produced, with roughly 30 million tonnes being processed for various purposes. Citrus peel waste makes up over half of the wet fruit mass after industrial processing. Apples, bananas, and pears produce 107.1, 75.5, and 24.0 million tonnes per year, respectively, with over 40% as waste mass [18]. The waste from

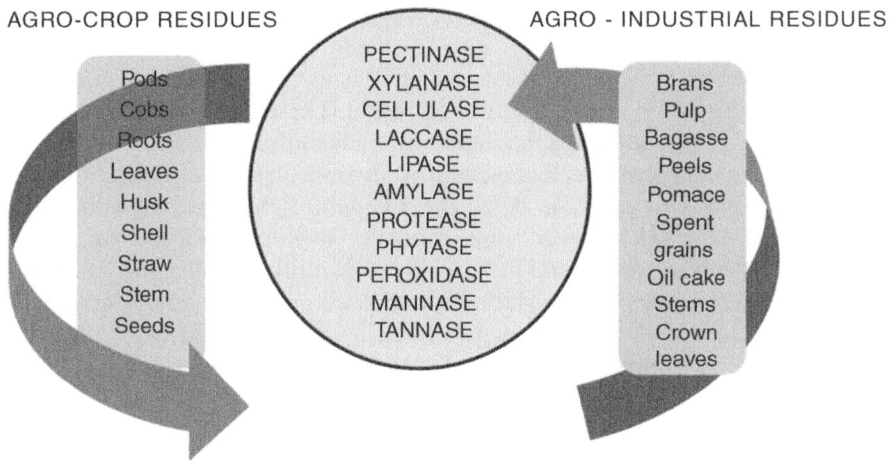

Figure 23.1 The thrust of bioconversion of agricultural wastes to produce industrially efficient enzymes has paved a path for economical and sustainable industrial implications. Enzymes obtained from different sources of plant, animal, or microbial origin can utilize agro-waste in a number of ways for efficient enzymatic production.

the fruit industry is often used as cow feed. However, due to its low protein content, this is not an ideal feedstock and most of it ends up in landfills or in the oceans. Due to the high sugar content of fruit waste, these methods may cause environmental problems. The high cost of waste disposal by landfilling, especially for fruit wastes, needs serious consideration due to several limitations like site constraints, transportation, and personnel costs. The cost of disposing apple pomace alone in the United States is $10 million per year. Fruit wastes contain abundant soluble sugars such as glucose, fructose, and sucrose, as well as structural cellulose and hemicellulose. These act as ideal substrates for fermentation. These chemical elements, together with the abundance of fruit waste, suggest that fruit waste could be an excellent waste biomass source for ethanol synthesis [19, 20].

Byproducts from agricultural and agro-industrial activities include sugarcane bagasse, citrus bagasse, fruit peel, maize straw, and corn cobs. These wastes are generally found in the form of fibrous residues and brans [21]. Fibrous residues are further categorized based on digestibility as easily digestible and difficult to digest. Moreover, there is great variability with regard to nutritional values among these wastes. Citrus pulp, corn gluten bran, soy husk, and brewing residues fall in the category of highly digestible fibrous residues, while sugarcane bagasse, cereal, corn, straws, husks, and forage grass seeds are considered less digestible fibrous wastes. These fibrous residues and brans like soy, cotton, rice, and peanut provide physical support as well as being a source of carbon and nutrients for microbial growth [22, 23]. The recent expansion of food and agricultural industries has resulted in a substantial increase in the volume of agro-industrial residues as wastes and has resulted in severe environmental and economic challenges [24]. The biotechnological management of these wastes is being explored in different sectors. Use of agro-industrial residues and wastes as raw materials to produce enzymes and value-added products is an emerging area of research into making the production process cost effective. Further, food loss caused by overproduction, harvesting damage, microbial, bug, and pest infections, postharvest handling, insufficient transport and storage, food processing, poor distribution, and inefficient consumer usage also increases the amount of wastes [25, 26].

23.3. AGRO-WASTES USED IN MICROBIAL ENZYME PRODUCTION

It has been reported that both fruit and vegetable wastes are nutritionally rich and consist of carbohydrates, proteins, vitamins, and minerals [26]. These can serve as substrates for a large number of high-value products, including biofuels and phytochemicals. Transferases, oxidoreductases, lyases, hydrolases, ligases, and isomerases are the six classes of enzymes, based on the reaction they catalyze. They are frequently employed in commercial and industrial settings [27, 28]. Microbial enzyme production by fermentation technology is being optimized for cost effectiveness. Several industrially important microbial enzymes like peroxidases, lipases, proteases, pectinases, xylanases, amylases, mannases, and glucooxidases are being produced by different agro-wastes as substrates [29, 30]. Different agro-wastes preferred for microbial enzyme production are described in the following subsections.

23.3.1. Wheat Bran

Germ, endosperm, aleurone layer, and pericarp are the different tissues that make up a wheat grain. Bioactive substances, minerals, and phytochemicals are abundant in wheat grain. The bran fractions include higher concentrations of these compounds. Wheat bran (WB) is a byproduct of the rolling milling process comprising pericarp, aleurone, and testa tissues. In WB, non-starch carbohydrates account for 55–60% of the dry matter, starch for 14–25%, protein for 13–18%, minerals for 3–8%, and fat for 3–4% [31]. WB is a low-cost agro-industrial byproduct that can assist growth of a variety of microorganisms. It is mostly used by microorganisms as a source of carbon and nitrogen. In comparison to other agricultural residuals, WB provides the highest levels of xylanase, pectinase, and amylase production [32]. One of the biggest benefits of using bran is that it contains enough nutrients on its own, thus no additional carbon or nitrogen sources are required. It contains glucose, which is required by microorganisms for proper growth and metabolism. It was found to be the most appropriate substrate for polygalacturonase (PG) synthesis [33, 34]. WB is a low-cost lignocellulosic substrate for microbial growth. It also serves as a source of feruloyl oligosaccharides, WB oil extraction, single-cell oils (SCO), and polyhydroxy butyrate synthesis as a cell immobilization carrier [35, 36].

23.3.2. Rice Bran

Rice bran (RB) is a byproduct of the milling of rice grain and is made up of the seed's external layers (pericarp, tegmen, and aleurone layer) and accounts for around 12% of the entire kernel weight. Globally, this leads to over 68 million tonnes of unmanageable material [37, 38]. The surroundings of the kernel contain a greater quantity of bioactive compounds, minerals, vitamins, dietary fiber, proteins, and lipids than the core. The inner core is characterized by simple carbohydrates and starch granules. A significant amount of lipids ($12–20\,g^{-1} 100\,g^{-1}$ RB) is present in RB. Most of these are used for oil extraction, such as oryzanol, ferulic acid, and tocopherol. Despite this, due to the quick activity of lipolytic endogenous enzymes, it is particularly susceptible to lipid oxidation, necessitating a heat stabilization step. However, RB's initial fate is in the feed formulation sector [37, 39].

23.3.3. Sugarcane Baggase

Sugarcane is a key crop grown in tropical and subtropical nations around the world. After being harvested from the fields, sugarcane is transported to mills and used for juice extraction in the sugar production process. Sugarcane bagasse, a primary byproduct of commercial sugarcane processing, is produced in vast quantities. It is commonly produced after sugarcane juice has been cleaned and extracted [40]. It is a fibrous waste made up of about 32–45% cellulose, 20–32% hemicellulose, 17–32% lignin, 1.0–9.0% ash, and other ingredients. It serves as a suitable substrate for the mass production of valuable and unique products on a large scale. Bagasse is also a feasible biomass for producing second-generation biofuels (ethanol), power, enzymes, sugars, and a range of other high-value commodities. It is also utilized as a biosorbent to remove toxic and heavy metals such as lead, zinc, copper, and cadmium. Sugarcane bagasse has been used as a sustainable biomass for biofuel generation due to its high cellulose and hemicellulose content [41]. At an industrial scale, sugarcane bagasse has been extensively used to produce numerous enzymes (cellulase, xylanase, lipase, amylase, phytase, and others) utilizing bacterial and fungal strains in SSF and SmF. Further, it has been explored for novel items such as acetoin, prebiotics, and bioplastics [42].

23.3.4. Coffee Wastes

Husks, skin, pulp, coffee mucilage, coffee parchment, coffee silver skin (CSS), and spent coffee grounds are lignocellulosic components found in coffee wastes and co-products [43]. These byproducts can be converted into highly appealing substrates suitable for bioconversion. The cellulose, hemicellulose, and lignin components present in coffee wastes can be used for bioconversion through biotechnological approaches to produce bulk chemicals, enzymes and other value-added products. Biofuel, mushroom, and fertilizer production, as well as the extraction of dietary fiber and bioactive chemicals, are now coffee co-product applications [40]. Through recycling, compound recovery, or energy valorization, coffee wastes can be a source of the most interesting and valuable products, such as metals, oils and fats, lignin, cellulose and hemicelluloses, tannins, antioxidants, caffeine, polyphenols, pigments, and flavonoids. Coffee husks are high in carbs (35%), proteins (5.2%), fibers (30.8%), and minerals when compared to other coffee byproducts (10.7%). The composition of coffee skin and pulp is comparable to that of husks, with carbs (21–32%), protein (7.5–15.0%), and fat (2.0–7.0%). Water (84.2%), protein (8.9%), sugar (4.1%), pectic compounds (0.91%), and ash are the main components of mucilage (0.7%) [44]. The main components of coffee parchment are cellulose (40–49%), hemicellulose (25–32%), lignin (33–35%), and ash (0.5–1%). CSS and wasted coffee grinds are the most common byproducts from the coffee business (spent coffee grounds, or SCG). CSS makes up roughly 4.2% of the total weight of the beans. They can be used as substrates for SSF since they are mostly composed of carbohydrates, proteins, and phenolic chemicals [9].

23.3.5. Tea Wastes

All tea companies produce enormous amounts of tea waste, such as discarded leaves, buds, and stems of tea. These wastes can pollute water, soil, and air if they are not properly disposed of. Tea residue is sometimes used as animal fodder in Assam. Only a small fraction of tea trash is purchased and used for caffeine extraction by some enterprises, while the remainder is discarded as garbage, causing a slew of environmental problems [45]. Researchers have highlighted the use of tea waste as an antioxidant substrate. It's also been utilized to make industrial enzymes, adsorbents for wastewater treatment, and biofuels. Tea trash is categorized as a lignocellulosic biomass containing cellulose, hemicellulose, lignin, polyphenols, proteins, and tannins. Tea contains biologically active components such as polyphenols (catechins, flavonoids, and proanthocyanidins), methylxanthines, alkaloids (caffeine, theophylline, and theobromine), vitamins, minerals, terpenoids, pigments, amino acids, and polysaccharides. Tea trash usually has nearly identical amounts of the same components as regular tea. Tea leaves, branches, and residues are high in cellulose, lignin, and bioactive substances such polyphenols, polysaccharide, and water-insoluble proteins, as well as cellulose, holocellulose, and lignin [46].

23.3.6. Apple Pomace

After oranges, bananas, and grapes, apples are the fourth most popular fruit crop worldwide. Based on global mass production of juice [46], a total of several million metric tonnes of pomace are expected to be generated each year. Only approximately 1% of apple pomace is appropriate for use as animal feed or as a dry product in India. The most common method of disposal for this byproduct is to dump it directly into the soil in a landfill. [47]. Apple pomace is a mixture of skin and flesh (95%), with small amount of seeds (2–4%) and stems (1%). Despite the differences in substance, it contains a variety of nutrients. It has a high carbohydrate content, as well as containing proteins, vitamins and minerals, and phytochemicals. The majority of the carbohydrates in apple pomace are insoluble sugars like cellulose (127.9 $g\,kg^{-1}$ DW), hemicellulose (7.2–43.6 $g\,kg^{-1}$ DW). Lignin being (15.3–23.5 $g\,kg^{-1}$ DW) with simple sugars like glucose (22.7%), fructose (23.6%), and galactose accounting for the rest (6–15%). Potassium (0.07–0.076%), Calcium (0.06–0.1%), Magnesium (0.02–0.36%), and iron (31.8–38.3 $mg\,kg^{-1}$, dry weight basis) respectively [48].

Polyphenols, such as cinnamate esters, dihydrochalcones, and flavonols, are found in substantial amounts. Furthermore, apple pomace has been demonstrated to include a variety of natural antioxidants, such as quercetin glycosides and phloridzin. Apple pomace is used in the food industry to produce light alcoholic beverages. This is supposed to improve their flavor. Its high pectin content makes it a most suitable fermentation substrate for pectinase enzymes [49–51].

23.3.7. Olive Pomace Wastes

Olive is a staple food in Mediterranean countries. It is widely grown in many countries of Africa, Europe and Asia. Olive oil accounts for 20% of total production, with olive pomace accounting for the remaining 33%. The solid residue from pressing, made up of pulp and olive stones, is known as pomace. Waste comprises 28.5% water, 41.5% hull, 21.5% pulp, and 8.5% oil. A lignocellulosic matrix (cellulose, hemicelluloses, and lignin), phenolic chemicals, uronic acids, and oily residues make up the olive pomace [52]. The main challenges in treating this solid waste are connected to their high phenolic chemical content, which makes them environmentally harmful. Because of their high acidity, these wastes are not employed as fertilizers, and they are not fed to animals because of their toxicity and high glycoside content. The pomace, on the other hand, retains a substantial amount of oil. Chemical extraction with a solvent that produces very acidic oil can be used to recover these oils. Olive pomace oil is frequently valorized in non-alimentary uses such as soap manufacture due to its strong acidity. Various enzyme production through SSF using olive pomace has been reported [52, 53].

23.3.8. Citrus Peel

The citrus processing industry is one of most economical industries in the agro-industrial sector. The orange is the most frequently cultivated fruit in the world, accounting for roughly 50–60% of total citrus production; however, other species such as lemon, lime, mandarin, and grapefruit are also important to the industry [54]. A large amount of fruit (40–60%) is inedible and is dumped as waste. Peels, pulp and pith residue, and seeds make up this waste biomass. The pulp and pith remnants have a lot of fermentable sugars, but the peels have a lot of limonene and other bioactive chemicals that prevent optimal fermentation. This emits an unpleasant odor and pollutes the land, air, and water [55]. Citrus trash contains 70% carbs and may produce 1.2 billion liters (or 300 million gallons) of ethanol on a global scale. Citrus peel wastes are high in soluble and insoluble carbohydrates, making them a suitable feedstock. They also contain D-limonene, a powerful microbial inhibitor [54, 56]. D-limonene synthesis from citrus peel is commercially viable because this byproduct has a high added value as a flavoring agent and in a variety of chemical applications [56].

23.3.9. Banana Peels

Because of their nutritional value, hygiene, cost-effectiveness, and digestibility, bananas are one of the most popular fruits in the world. It is the most popular fruit in South Asia, with over a hundred thousand hectares of farmed land [57]. A banana plant produces only one bunch in its lifespan, resulting in massive waste output both during banana bunch harvesting and after processing or eating. The roots, suckers, rhizome, pseudo stem, leaves, peduncle, rachis, and male bud of the banana plant decompose into solid garbage. The rhizome, also known as the corm or bulb, is the major subterranean structure that produces primary roots and suckers. Primary roots give rise to secondary and tertiary roots. Suckers are either employed to produce new crops or dumped as garbage after harvesting. When the peel is consumed or processed, it is dumped as solid waste [58]. These wastes are mostly made up of holocellulose, which is composed of cellulose and hemicelluloses. Glucose, galactose, arabinose, rhamnose, and xylose were found in pectin derived from banana peel. Starch (3%), crude protein (6–9%), crude fat (3.8–11%), total dietary fiber (43.2–49.7%), polyunsaturated fatty acids, particularly linoleic acid and linolenic acid, pectin, essential amino acids, and micronutrients are all abundant in banana peel. Banana peels have been described as an excellent fermentation substrate because they contain significant amounts of lignin (6–12%), pectin (10–21%), cellulose (7–10%), and hemicelluloses (6–9.4%). Several attempts have been made to produce industrially important enzymes like alpha-amylase produced by *Bacillus subtilis* and *Penicillium* species, and laccase produced by *Trametes pubescens*. It has also been used as a substrate to produce citric acid from *Aspergillus niger*, biohydrogen through two-phase anaerobic fermentation, and alcohol [59].

23.3.10. Pineapple Wastes

There has been substantial increase in the production and processing of fruits and vegetables as a result of changing dietary habits among the general public [60]. This leads to generation of wastes beginning with the harvesting of raw material through to its processing. Pineapple processing also generates wastes in the form of peel, core, pomace, and crown, which are considered to contain important bioactive chemicals [61]. Byproducts typically include higher levels of beneficial chemicals with greater nutritional and medicinal value than the final product. Pineapple wastes have been considered important owing to their ability to extract enzymes like bromelain, pectinase, xylanase, and cellulases, and as a low-cost substrate for the production of dietary fiber, organic acids, and phenolic antioxidants. Pineapple byproducts consisting primarily of the remaining pulp, peels, stems, and leaves are a matter for proper waste management [60]. Pineapple has high calcium, potassium, fiber, and vitamin C content. Pineapple waste is rich in lignocellulosic material, particularly the fruit's peel and leaves. Pineapple waste, such as peels and cores, account for 40–50% of the fresh pineapple. Pineapple waste is also known to be an enriched raw material, containing insoluble fibers, pectins, simple sugars, and proteins as the primary constituents, as well as a high quantity of micronutrients like vitamins, minerals, and phytonutrients. Enzymes such as protease, cellulase, and xylanase can be found in pineapple fruit and trash [61, 62].

23.3.11. Oil Cakes

Vegetable oils are an essential component of the human diet since they are a major source of calories and energy. Solids commonly referred to as cakes (residue collected after screw press), or meal if it has undergone any additional procedure (usually an organic solvent de-oiling step), are the principal byproducts of vegetable oil extraction [63]. These are distinguished by their high protein content and large fraction of plant cell wall structural components such as cellulose, hemicellulose, pectin, and lignin. Traditional uses for oilseed cakes (OCs) include animal feed and plant/soil compost [64]. Because of the volume and composition of OC produced each year, it is critical to investigate biotechnological and clean techniques that can add value to these byproducts. The chemical composition of OCs, in fact, creates the ideal environment for microbial growth. OCs are lignocellulosic materials that can be utilized as a SSF substrate [65, 66].

23.3.12. Others

Corn is a staple food for many people in Africa, Asia, and Latin America. The corn husk is a lignocellulosic biomass that is low in lignin and high in hemicellulose and celullose. A corn husk is the part of the plant that shields the kernels when the tree is young and it dries up when the tree ages, while corn stover is the residue of corn processing [67, 68].

Coconut fiber is made from unripe coconuts and has a global annual production of roughly 350 000 metric tonnes, with about 90% of it going to waste. It comprises the husk, which contains 30% fiber and 70% pith material. The physical–chemical properties of it make it extremely valuable in biotechnological processes and agronomy [65].

Brewers' spent grain is the most common byproduct of the brewing business. This accounts for 85% of all byproducts produced in the European Union. This material has a great nutritional value. It is a heterogeneous substance whose chemical makeup varies according to cereal variety, harvesting period, type of hops used, malting and mashing regime, and on adjuvant addition. This is a lignocellulosic material that is high in protein and fiber, accounting for 20 and 70% of its total content, respectively. The surface layer consists primarily of many fibrous tissues, with arabinoxylan, lignin, and cellulose as the primary components. It is rich in vitamins (biotin, folic acid, niacin, choline, riboflavin, thiamine, pantothenic acid, and pyridoxine), minerals (calcium, cobalt, copper, iron, magnesium, manganese, phosphorus, potassium, selenium, sodium, and sulfur), and amino acids (lysine, histidine, methionine, phenylalanine, and tryptophan) [68, 69].

Grape pomace is the residue left over from the winemaking process after extracting the juice from the grapes. This resource is underutilized and the majority is dumped in open places, causing considerable environmental issues. SSF, on the other hand, sees this waste as having potential for value-added products (high carbohydrate content with the fiber representing about 50% of the total mass). It has been utilized as a substrate for the manufacture of hydrolytic enzymes like as xylanase, exo-PG, cellulase, and others due to its composition. However, because this composition varies according to the season, grape variety, weather, and other factors, no reproducible enzymatic productivities can be established. Orange peels were chosen as a natural source of nutrients to blend with grape pomace in order to boost the optimal synthesis of some hydrolytic enzymes [70, 71].

Maize-gluten meal, a byproduct of corn wet milling that has historically been used for animal feed, has been discovered to be a viable substrate for the synthesis of amylase by *Bacillus amyloliquefaciens* due to its high protein (60%), vitamin, and other mineral content. The synthesis of alpha amylase by *Bacillus licheniformis* was investigated at 1% levels in agricultural raw starches such as pearl millet, rice, gramme, hordium, maize, and wheat starches. Beta glucans make up the majority of barley and oat bran. Pearl millet, which contains 56–65% starch, 20–22% amylose, free sugars ranging from 2.6–2.7%, and a total protein level of 8–19%, produced significantly more active biomolecules [72, 73]. Different agro-wastes preferred as substrates for the production of microbial enzymes are shown in Table 23.1.

Table 23.1 Agro-wastes used as substrates for production of different microbial enzymes.

Substrate	Enzymes	References
Wheat bran	Xylanases	[74, 75]
	Laccases	
	Mannanases	
	Proteases	
	Polygalacturonase (endo/exo)	
	Tannases	
	L-asparaginases	
	Glutaminases	
	Phytases	
	l-xylosidases	
	Ferulic acid esterases	
	Milk-clotting enzyme	
	Lipases	
Rice bran	Celluloses	[76, 77]
	Llipases	
	Xyalanses	
	Hydrolyating enzymes	
	Proteolytic enzymes	
Sugarcane baggase	Lipases	[78]
	Amylases	
	Xyalanses	
	Celluloses	
	Proteases	
Coffee wastes (basically coffee pulp)	Xyalanses	[79]
	Pectinases	
	Laccases	
	Tannases	
	Peroxidases	
	β-Glucosidases	
	β-Fructofuranosidases	
	Arbinofuranidase	
	Amylases	
Tea waste	Celluloses	[80, 81]
	Mannases	
	Xyalanses	
	Laccases	
	Pectinases	

(Continued)

Table 23.1 (Continued)

Substrate	Enzymes	References
Apple pomace	Pectinases Cellulases Xyalanses	[82, 83]
Citrus peel	Pectinases Glucooxidases Celluloses Xylanases Polygalacturonses Proteases Amylases Lipases Invertases	[84–87]
Oil seed cake	Cellulases Xyalanses β-glucosidases Lipase, l-Asperginases Phytases Inulases Phosphatases Amylases	[88]
Banana peel	Pectinases Xylanases Amylases Peroxidases	[89]
Pomogranate peel	Xyalanse Amylase	[90]
Pineapple waste	Xyalanses Bromoleian Pectinases Cellulases Proteases	[91, 92]
Grape pomace	Polygalacturanses Xylanses Peroxidases celluloses	[72]

23.4. MICROBIAL SOURCES FOR PRODUCTION OF ENZYMES FROM AGRICULTURAL WASTES

Several industrially important enzymes are produced by different microbes, and to make the production by fermentation technology cost effective, researchers have used several agro-wastes as substrates. Some of the important enzymes and utility of agro-wastes in the production process are discussed in the following subsections.

23.4.1. α-Amylase

α-amylases (endo-1, 4-β-D-glucanglucanohydrolase EC 3.2.1.1) are responsible for cleaving 1, 4 linkages present between glucose subunits in polysaccharides, thereby releasing short chain oligomers. These enzymes find application in various industrial sectors such as paper and pulp based ventures, textiles, food, detergents, and pharmaceuticals [93]. Wild as well as genetically modified *Bacillus* and *Aspergillus* species have been used at the industrial scale for production of alpha amylases both by SSF and SmF. A large number of microbial strains have been identified and characterized for the production of α-amylase enzyme. Many of these enzymes are found to be thermostable under a wide range of temperature and pH conditions. Since it is an industrially significant enzyme, efforts have been made for its cost-effective production using agro-wastes as substrates. Amylase production increased around 20% with the use of *Aspergillus oryzae* as inoculums and spent brewer's grain as substrate [94, 95].

RB has been used as substrate for amylase using *Bacillus tequilensis* TB5 through SmF [96]. The growth conditions were optimized and by utilizing agro-waste substrate, maximum yield ($39.736 \pm 0.296\,U\,ml^{-1}$) was obtained at pH 6.0 and temperature of 37 °C for an incubation period of 72 hours. Cassava peels have been used as a substrate for the production of alpha amylase from *A. niger* [97]. The culture conditions for optimal production of the enzyme were found to be pH 9 at 35 °C and gelatin as nitrogen source for 120 hours of incubation. Amylase has been produced by SmF using WB as a substrate by *A. niger* [98]. The optimal conditions were also screened by using different nitrogen sources and $326.12 \pm 3.68\,U\,l^{-1}$ of amylase have been produced after six days of fermentation.

23.4.2. Cellulase

Cellulase is a commercially important complex comprising three different enzymes: endoglucanase (E.C. 3.2.1.4), exoglucanase (E.C. 3.2.1.91), and β-glucosidase (E.C. 3.2.1.21). The hydrolysis of cellulose is initiated by enzymes endoglucanase (EG) and exoglucanase (CBG), which together transform cellulose into smaller molecules, oligosaccharides; these are then acted upon by β-glucosidase (BG), which converts cello oligosaccharides into glucose [99]. Cellulose containing raw material can be used for sustainable production of the cellulase enzyme to make it economically profitable. Lignocellulosic materials have been used as substrates for SSF for production of extracellular cellulases using *Aspergillus terreus* M11 [100]. Maximum cellulase activity was obtained at 45 °C at pH 3.The application of this enzyme was used for Avicel hydrolysis. Up to 63% hydrolysis of 5% Avicel (w/v) was reported for 72 hours with 20 UFPase g^{-1} substrate. Individual as well as mixed cultures of *A. niger* and *Trichoderma reseei* have been used for the production of cellulase and hemicellulase enzymes using different agricultural residues [101]. Higher cellulase, β-glucosidase, endoglucanase (CMCase), and xylanase activities were obtained when mixed culture of these two strains was used for production compared to the individual strains. So, this approach helps in waste management and low cost production of industrially important enzymes without using any sophisticated technology.

A bacterial isolate from compost, identified as *B. licheniformis* has been used for cellulase production using agro-wastes. Supplementation of the wheat straw and rice husk in the culture medium resulted in 0.08 UEA ml^{-1} min^{-1} of enzyme activity after six hours of incubation [102]. Another bacterial isolate, *Amycolatopsis* sp. GDS, has been reported to secrete high levels of cellulase and xylanase enzymes when supplemented with agriculture waste biomass [103]. The enzyme obtained from this fungus was found to be suitable for industrial applications due to its thermostability and stability under high NaCl concentrations. This enzyme was also found to be stable in the presence of several chemicals used as organic solvents, surfactants and oxidizing agents in the industries. Such enzymes may be useful in ethanol production and biomass feedstock preparation.

A cellulase producing microbe isolated from soil and identified as *Klebsiella sp*. PRW-1 has been screened for enzyme production using different agricultural wastes along with carboxymethylcellulose (CMC) and avicel [104]. Higher levels of reducing sugar were produced when culture media was supplemented with grass powder and sugarcane baggase. The cellulolytic enzymes were characterized, which confers its application to bioenergy production. A thermophilic bacterial isolate, *Bacillus sp*. PCH94, has been shown useful for sustainable bioenergy production utilizing lignocellulosic biomass [105]. Bioinformatics based searches have identified 106 carbohydrate utilizing genes. A wide range of temperature and pH stability was obtained when production was set using rice grain powder. Such enzymes obtained may be used in several industrial applications.

23.4.3. Xylanase

Xylan is the key component of plant hemicellulose, which accounts for 20–40% of total plant biomass. To breakdown this lignocellulosic biomass, a variety of hydrolysing enzymes are required [106]. Xylanases are a crucial group of depolymerizing enzymes used for the hydrolysis of the xylan. Xylanases are most commonly produced by different microorganisms and are industrially important for biofuel production, chemical and pharmaceutical sectors, wood pulp bioleaching, papermaking, food and beverage manufacturing, and animal nutrition [107].

Several research groups have reported production of xylanase using different agricultural wastes. *Aspergillus foetidus* MTCC 4898 has been used for xylanase production using WB and anaerobically treated distillery spent wash [108]. Optimal xylanase activity of 8450 U g^{-1} was obtained. Crude xylanase proved to be useful in enzymatic saccharification of agroresidues such as wheat straw, rice straw, and corncobs. Industrially relevant xylanase should possess attributes such as thermostability and alkali stability, and should be free of cellulase activity. Recently, xylanase with these properties has been produced extracellularly by *Bacillus pumilus* SV-205 through SSF using WB as a substrate [109]. A 14.4 fold increase in enzyme production has been reported under optimal conditions, suggesting the relevance of agro-residues for cost-effective production of xylanase. Its application in the bio-bleaching of kraft has been investigated, which could serve as an alternative to harsh chemicals used in the paper and pulp industries.

Production of xylanase enzyme from *Aspergillus fumigatus* strain JCM 10253 using ragi husk as a substrate has also been reported recently [110]. A central composite design (CCD) has been used for the optimization of the individual process variables and thereby studies the influence on xylanase activity. Under optimized conditions 156.7 IU ml^{-1} enzyme activity was obtained at optimal pH of 2.0 and at 49.9 °C when incubation was done for 120 hours. The enzyme can be efficiently produced using agri-waste, ragi husk which in turn can be utilized in the production of value-added products by several industrial. Recently various substrates like cassava peels, corn cobs, WB, and rice husk has been used for xylanase production by *A. tubingensis* using SSF [111]. The highest xylanase activity of 111.23 ± 0.31 U g^{-1} was observed when a corn cob was used as substrate under fermentation carried out at pH 6.0 and temperature 30 °C for a period of 72 hours. *A. niger* has also been used for the production of xylanase enzyme using rice husk as substrate [112]. The enzyme thus obtained was used in the treatment of palm oil mill effluent, which resulted in a decrease in the chemical oxygen demand and total solid present in the effluent.

A thermophilic xylanase produced from *B. licheniformis* DM5 has been reported to utilize agro-waste corncob for the production of prebiotic and anti-inflammatory xylobiose oligosaccharides [32]. Another bacterial isolate from soil, *B. subtilis,* has been used for the production of xylanase using sugarcane bagasse and malt extract as substrate in SSF along with xylose [113]. It exhibited maximum activity at pH 9.0 and at 40 °C after 48 hours' incubation. The recent report of xylanse from Trichoderma also highlights the efficacy of agro-waste in production of the enzyme [114].

23.4.4. Pectinases

Naturally occurring pectinases are responsible for the breakdown of pectin present in the middle lamella of plant cell wall. Commercial pectinases are mostly obtained from microbial sources and are useful in several industrial applications [115]. Cost-effective pectinase production has been attempted by using agricultural residues as substrates in the fermentation process. These lignocellulose-containing wastes comprise a huge amount of hemicellulose, cellulose, and pectin, which can be used for pectinase production.

Production of Exo-polygalacturonase (Exo-PG) using WB as substrate in SSF by *Penicillium fellutanum* has recently been reported [116]. Conditions for enzyme production were optimized as 40% moisture, pH 4.0, and temperature of 40 °C with an incubation period of 96 hours. The enzyme was then partially purified and characterized. It exhibited a wide range of pH 3.0–8.0 and temperature stability 30–70 °C. Its application has been evaluated in fruit juice clarification. Bioproduction of pectinases by endophytic fungi isolated from Thai orchids has been investigated [117]. Four of these isolates were screened for production of the pectinase enzyme using different agricultural waste as substrates. Among all the substrates screened, soybean meal proved to be most effective in production of the enzyme. This study suggests the use of soybean meal as a cost-effective and eco-friendly source for the production of pectinases, which will find application in different industries. The high production level of pectinase production (1366.30 ± 36.71 U ml^{-1}) was achieved by setting up fermentation with apple pomace incorporated in the medium inoculated with *A. parvisclerotigenus* KX928754 [118]. Maximum production was optimized under culture conditions at pH 7.0, temperature of 30 °C, and for an incubation period of 168 hours. Another microbial strain, *A. niger,* has been studied for biosynthesis of pectinase using media supplemented with pineapple peel pectin in SmF [119]. Optimum enzyme production was achieved at pH 5.0, temperature of 40 °C, and at 1% substrate concentration. The enzyme thus obtained exhibited better results compared to the commercially available pectin for coconut-oil extraction.

Production of Endo-polygalacturonase (endo-PG) has been evaluated among fungal strains, namely *A. niger* AUMC 4156, *Penicillium oxalicum* AUMC 4153, and *Penicillium variotii* AUMC 4149 in the presence of dried orange peel and sugar beet pulp as carbon sources [116]. Among these two substrates, sugar beet pulp exhibited better production of the pectinase enzyme during fermentation revealing sugar beet as a potential inducer for endo-polygalacturoase production. Based on the screening of different agricultural wastes, orange peel was found to be the best substrate for pectinase production using the fungus *Penicillium chrysogenum* MF318506 [120]. Maximum activity was reported as 0.48 U ml^{-1} under optimal growth conditions. Further optimization using response surface methodology led to 6.04 fold increase in the pectinase activity in presence of 2.5% orange peel and 4.5 g l^{-1} peptone as substrates. Fusarium genera have also been categorized for pectin lyase studies [121]. Apart from the above mentioned fungal strains, a bacterial strain, *B. amyloliquefaciens* TKU050, has been screened on various agriculture residues for the production of pectinase enzyme. WB has been reported to be the best carbon source for pectinase production with 0.76 U ml^{-1} of enzyme productivity at pH 6 and temperature of 50 °C [119, 122].

23.4.5. Lipases

Lipases are also known as triacylglycerol hydrolases that act on triglycerides to produce free fatty acids and glycerol. They also catalyze the process of alcoholysis, esterification, and transesterification of fatty acids. Several groups have attempted the production of lipase in the presence of agricultural residues as substrates. In a recent study, culture media formulated with agro-industrial residues led to better production of the lipase enzyme compared to the commercially available substrates using *Diutina rugosa* as enzyme producer [120]. Highest lipase production of 561 U l^{-1} was observed under optimal conditions. Another study used deoiled castor seed cake for lipase production from an organic solvent-tolerant lipase-producing microbe identified as *Acinetobacter sp* [123]. 243 U ml^{-1} of lipase was produced under the culture conditions containing agro-waste as a substrate [124]. Agro-industrial palm oil waste has been used for lipase production and a maximum activity of 4.9 U ml^{-1} was observed. Hence, palm oil waste serves as an important source of lipase production from non-pathogenic yeast *Y. lipolytica*. Soybean oil supplemented media of *Aspergillus brasiliensis* exhibited enzyme activity of 9.8 U g^{-1}, whereas another study reported *Moringa oleifera* seed waste for lipase enzyme production by the bacterial isolate *Bacillus* sp. SK II-5 [125].

23.4.6. Laccase

Laccases are industrially important enzymes that are useful in several industrial ventures such as the paper and pulp industry, food-based ventures, and pharma manufacturing units. The enzyme of this family comprises multi-copper oxidases that are responsible for the oxidation of one electron in a variety of phenolic compounds [126]. It is considered an eco-friendly enzyme as only water molecules are released as a byproduct. Fungi are the main producer of the laccase enzyme apart from some bacteria, insects, and plants. Although few laccases have been isolated from bacteria, a recent study has reported laccase production from a bacterial isolate identified as *Bacillus aquimaris* AKRC02 [127]. The enzyme production has been screened by setting up SmF using several agro-industrial wastes. Under optimal conditions of enzyme production, specific activity of 228.34 U mg^{-1} was achieved. Since the strain was isolated from the effluent of pulp and paper mills, this could serve as an important enzyme for environmental bioremediation. Similarly, production of the laccase enzyme on different agriculture and forest waste residues by white rot fugal strains *Pleurotus ostreatus* and *Ganoderma lingzhi* has been investigated [128, 129]. Oil palm empty fruit bunch (OPEFB) has been used for the production of the laccase enzyme by *Pycnoporus sanguineus* [130].

23.5. TECHNOLOGICAL INNOVATIONS IN ENZYME PRODUCTION FROM AGRICULTURAL WASTES

23.5.1 Solid State Fermentation and Submerged Fermentation Techniques

SSF is described as a fermentation process that takes place in the absence or near absence of free water and involves a solid matrix. The substrate, on the other hand, needs to be moist enough to support the microorganism's development and metabolism. The solid matrix could be a source of nutrients or merely a support impregnated with the necessary nutrients for microbial development. SSF's potential rests in getting grown microorganisms near to the substrate. As a result, the fermentation can reach its maximal substrate concentration. SSF mimics a microorganism's natural environment. Microorganisms prefer it to thrive, and develop useful value-added products because of this. On the one hand, the use of low-cost agricultural leftovers increases the process' economic viability. On the other hand, it eliminates the issue of disposal, which would otherwise result in pollution [131].

The fundamentals of SSF technology were employed in all ancient fermentation techniques. Due to the development of SmF technology, it is believed that it was lost in obscurity in western countries around 1940 [132]. Perhaps SSF was overlooked since the invention of the wonder medicine penicillin occurred in SmF, which was extremely important at the time. SSF research has always been carried out, albeit in small pockets. During the 1950s and 1960s, reports on steroid transformation using fungal culture appeared, and during the 1960s and 1970s, reports on mycotoxin generation using SSF surfaced. SSF was able to achieve the milestone as a result of this. SSF's manufacturing of protein-enriched bovine feed from agro-industrial leftovers was also reported on. For the development of bioprocesses, the SSF arena has been continuously expanded. Hazardous chemical bioremediation and biodegradation, as well as biological detoxification of agro-industrial wastes, are among them. For nutritional enrichment and biopulping, crops and crop residues are biotransformed. Biologically active secondary metabolites, such as antibiotics, alkaloids, plant growth factors, enzymes, organic acids, biopesticides such as mycopesticides and bioherbicides, biosurfactants, biofuel, fragrance compounds, and other value-added products are also produced [65].

The goal of developing appropriate models is to identify correlations between microbial physiology and physico-chemical variables. Temperature, pH, aeration, water activity and moisture, bed characteristics, and the type of solid substrate used are only a few of the variables to consider. The most essential component affecting SSF operations is moisture, which is determined by the nature of the solid substrate used. Moisture selection is influenced by both the microorganisms used and the substrate's composition. Fungi require less moisture, therefore 40–60% moisture may suffice. However, substrate selection is influenced by a number of factors, the most important of which are cost and availability. As a result, screening a variety of agro-industrial leftovers may be necessary [133].

23.5.2. Submerged Fermentation

Microorganisms are grown in a liquid medium during the SmF process. In comparison to SSF, it is more developed, and it is frequently utilized in the industrial production of enzymes. To do this, bioreactors allow accurate physiological conditions for carrying out fermentations [134]. Fermentation can be carried out in several modes such as batch, fed-batch, and continuous fermentation using different bioreactors. Since the nutrients are dissolved in the liquid medium it favors easy access to bacteria. So, the culture medium must contain all of the essential nutrients required for growth. The ability of microorganisms to adapt to the media, the utilization of low-cost raw materials, and the ability to synthesize the product gained in big quantities are all important factors in the fermentation process' success. Because the culture media accounts for the majority of costs in the industrial scale fermentation, it's critical to reduce them. Due to their low cost, industrial wastes are commonly utilized to make culture media for microbial enzyme synthesis. The simplicity with which the physical and chemical factors of the process may be controlled, as well as the relative ease with which large-scale cultivation can be accomplished, are both advantages of this method. The cost of aeration and agitation, as well as the media used in the fermentation preparation, is a downsid' of this procedure. Another downside is that the high volume of water utilized and foaming increase the risk of contamination [133, 135].

23.5.3. Limitations of Fermentation Techniques

Fermentation is a process in which microbes break down organic substances by anaerobic metabolism in order to get energy. This biological process may be used to convert organic waste into value-added goods in addition to low-cost processing, low-energy consumption, and lesser water wastage. Fungi, yeasts, and bacteria grow on the surface of organic components in SSF, which give physical support for their growth without the use of water. Fungi and yeasts are the microorganisms of choice for this application, which is carried out at a moisture level of 40–80% [65].

Each approach, either SmF or SSF, has their own set of benefits and drawbacks: SmF is used for manufacturing of a wide range of products. Despite its wide range of applications, SmF has several limitations in terms of process scale-up. It gives poor yield despite utilizing enormous amount of water. SSF, on the other hand, has long been used in the production of Asian fermented foods. It has low operating costs, low water usage, and does not need specialized bioreactors [133]. It faces the disadvantage of restricted control of the environment within the bioreactor in addition to high costs of end-product recovery and downstream processing. The ability to use low-cost or no-cost substrates like food waste, as well as the selection of the best media and microorganisms for the optimization and planning of downstream processing operations, are critical to the technique's success [134].

The modern world is rapidly changing, with continuous technological advancements powered by innovation: higher fermentation productivity, higher product end-concentration, higher product stability, lower catabolic repression, and cultivation of microorganisms specialized for water insoluble substrates or mixed cultivation of various fungi [135, 136]. Lower sterility demands due to the low water activity used in SSF appear to have several biotechnological advantages. The key hurdles that the researchers faced were scaling up, purification of end products, and biomass estimation. Scale up has long been a limiting element in SSF, but with the introduction of biochemical engineering, a variety of bioreactors have been developed that can overcome scale up issues as well as online monitoring of numerous parameters, as well as heat and mass transfer to some extent [137, 138]. Some of the recent approaches for cost-effective production by fermentation techniques are summarized in Table 23.2.

Table 23.2 Recent implications of fermentation techniques in microbial enzyme production.

Microbial enzyme	Agro-waste used in production	Mode of fermentation employed (SSF/SmF)	References
Pectinase	Citrus peel	SSF	[139]
	Orange peels	SSF	[140]
	Coffee pulp	SSF	[141]
	Grapefruit peel	SSF	[142]
	Banana peel	SmF	[143]
	Orange peel	SmF	[117]
	Wheat bran	SmF	[144]

Table 23.2 (Continued)

Microbial enzyme	Agro-waste used in production	Mode of fermentation employed (SSF/SmF)	References
Cellulase	Orange peels	SMF	[140]
	Rice straw	SmF	[137]
	Coffee husk	Solid state and submerged	[145]
	Grape pomace	Submerged	[146]
	Wheat bran	Solid state	[147]
Xylanase	Sugarcane bagasse	Submerged	[148]
	Banana peels	Submerged	[149]
	Palm waste	Solid state and submerged state	[150]
	Corn barn	Submerged	[151]
	Oil palm frond leaves	Solid state	[152]
	Wheat bran	Solid state	[153]
	Sugarcane bagasse and brewery spent grain	Solid state	[154]
Laccase	Orange waste	Solid state	[155]
	Tea residue	Solid state	[80]
	Corncobs and paddy straws	Submerged	[156]
	Corncobs	Submerged	[157]
Lipase	Brewery spent grain	Submerged	[158]
	Watermelon peels	Solid state	[159]
	Babul biomass waste	Solid state	[160]
	Wheat bran	Solid state	[161]
α-amylase	Edible oil cakes	Solid state	[162]
	Wheat bran	Solid state	[163]
		Solid state	[164]
	Soybean husk and flour mill waste	Solid state	[165]
	Brewery spent grain	Submerged	[166]
	Grapefruit peels	Submerged	[167]

23.6. CONCLUSIONS

The demand for microbial enzymes is increasing every day, but their usage is limited due to high cost of substrate that increases overall cost of production. Therefore, it is of high importance to explore and discover alternative cost-effective substrates to produce microbial enzymes. Several agro-wastes discussed in this chapter, including RB, WB, coffee and tea waste, sugarcane bagasse, oil cakes, and fruit peels, are abundantly present around us and could be used as low-cost substrates to produce industrially important enzymes. Use of these wastes for enzyme production will also help in tackling the health and environmental challenges associated with the increased amount of these wastes. The fermentation techniques discussed here are not only important tools for utilization of these agro-wastes to produce valuable enzymes by the help of microbes, but also provide a solution for the management of these wastes.

REFERENCES

1. Adejumo, I.O. and Adebiyi, O.A. (2020). Agricultural solid wastes: causes, effects, and effective management. In: *Strategies of Sustainable Solid Waste Management* (ed. H.M. Saleh), 15. IntechOpen.

2. Chukwuma, O.B., Rafatullah, M., Tajarudin, H.A. et al. (2020). Lignocellulolytic enzymes in biotechnological and industrial processes: a review. *Sustainability 12* (18): 7282.
3. Ravindran, R., Hassan, S.S., Williams, G.A. et al. (2018). A review on bioconversion of agro-industrial wastes to industrially important enzymes. *Bioengineering 5* (4): 93.
4. Bharathiraja, S., Suriya, J., Krishnan, M. et al. (2017). Production of enzymes from agricultural wastes and their potential industrial applications. In: *Advances in Food and Nutrition Research*, vol. *80* (ed. S.-K. Kim and F. Toldrá), 125–148. Academic Press.
5. Hassan, S.S., Williams, G.A., and Jaiswal, A.K. (2019). Lignocellulosic biorefineries in Europe: current state and prospects. *Trends in Biotechnology 37* (3): 231–234.
6. Neelkant, K.S., Shankar, K., Jayalakshmi, S.K. et al. (2019). Optimization of conditions for the production of lignocellulolytic enzymes by Sphingobacterium sp. ksn-11 utilizing agro-wastes under submerged condition. *Preparative Biochemistry & Biotechnology 49* (9): 927–934.
7. Cano, M.E., García-Martin, A., Comendador Morales, P. et al. (2020). Production of oligosaccharides from agrofood wastes. *Fermentation 6* (1): 31.
8. Šelo, G., Planinić, M., Tišma, M. et al. (2021). A comprehensive review on valorization of agro-food industrial residues by solid-state fermentation. *Foods 10* (5): 927.
9. Nayaka, A. and Bhushan, B. (2019). An overview of the recent trends on the waste valorization techniques for food waste. *Journal of Environmental Management 233*: 352–370.
10. Koul, B., Yakoob, M., and Shah, M.P. (2021). Agricultural waste management strategies for environmental sustainability. *Environmental Research 25*: 112285.
11. Cerda, A., Artola, A., Barrena, R. et al. (2019). Innovative production of bioproducts from organic waste through solid-state fermentation. *Frontiers in Sustainable Food Systems 3*: 63.
12. Naidu, Y., Siddiqui, Y., and Idris, A.S. (2020). Comprehensive studies on optimization of ligno-hemicellulolytic enzymes by indigenous white rot hymenomycetes under solid-state cultivation using agro-industrial wastes. *Journal of Environmental Management 259*: 110056.
13. Garg, N., Singh, B., Vaish, S., and Kumar, S. (2021). Enzyme production from different fruit and vegetable waste using lactic acid fermentation. *Journal of Eco-Friendly Agriculture 16*: 74–76.
14. da Silva, V.D., Cruz, I.A., Torres, N.H. et al. (2019). Agro-industrial wastes: environmental toxicology, risks, and biological treatment approaches. In: *Environmental Contaminants: Ecological Implications and Management* (ed. R.N. Bharagava), 1–23. Singapore: Springer.
15. Qureshi, U.A., Hameed, B.H., and Ahmed, M.J. (2020). Adsorption of endocrine disrupting compounds and other emerging contaminants using lignocellulosic biomass-derived porous carbons: a review. *Journal of Water Process Engineering 38*: 101380.
16. Lobo, M.G. and Dorta, E. (2019). Utilization and management of horticultural waste. In: *Postharvest Technology of Perishable Horticultural Commodities* (ed. E.M. Yahia), 639–666. Woodhead Publishing.
17. Senthilkumar, K., Kumar, M.N., Devi, V.C. et al. (2020). Agro-Industrial waste valorization to energy and value added products for environmental sustainability. In: *Biomass Valorization to Bioenergy* (ed. R.P. Kumar, B. Bharathiraja, R. Kataki and V.S. Moholkar), 1–9. Singapore: Springer.
18. Dey, T., Bhattacharjee, T., Nag, P. et al. (2021). Valorization of agro-waste into value added products for sustainable development. *Bioresource Technology Reports 24*: 100834.
19. González-García, S., Gullón, B., Rivas, S. et al. (2016). Environmental performance of biomass refining into high-added value compounds. *Journal of Cleaner Production 120*: 170–180.
20. Ali, M., Saleem, M., Khan, Z. et al. (2019). The use of crop residues for biofuel production. In: *Biomass, Biopolymer-Based Materials, and Bioenergy* (ed. D. Verma, E. Fortunati, S. Jain and X. Zhang), 369–395. Woodhead Publishing.
21. Kaur, P., Ghoshal, G., and Jain, A. (2019). Bio-utilization of fruits and vegetables waste to produce β-carotene in solid-state fermentation: characterization and antioxidant activity. *Process Biochemistry 76*: 155–164.
22. Gonzalez-Garcia, S., Gullón, B., and Moreira, M.T. (2018). Environmental assessment of biorefinery processes for the valorization of lignocellulosic wastes into oligosaccharides. *Journal of Cleaner Production 172*: 4066–4073.

23. Esteban, J. and Ladero, M. (2018). Food waste as a source of value-added chemicals and materials: a biorefinery perspective. *International Journal of Food Science and Technology 53* (5): 1095–1108.
24. Madeira, J.V. Jr., Contesini, F.J., Calzado, F. et al. (2017). Agro-industrial residues and microbial enzymes: an overview on the eco-friendly bioconversion into high value-added products. *Biotechnology of Microbial Enzymes 1*: 475–511.
25. Andreola, F., Lancellotti, I., Manfredini, T. et al. (2020). The circular economy of agro and post-consumer residues as raw materials for sustainable ceramics. *International Journal of Applied Ceramic Technology 17* (1): 22–31.
26. Coman, V., Teleky, B.E., Mitrea, L. et al. (2020). Bioactive potential of fruit and vegetable wastes. In: *Advances in Food and Nutrition Research*, vol. *91* (ed. F. Toldrá), 157–225. Academic Press.
27. Bajar, S., Singh, A., and Bishnoi, N.R. (2020). Exploration of low-cost agro-industrial waste substrate for cellulase and xylanase production using *Aspergillus heteromorphus*. *Applied Water Science 10* (6): 1–9.
28. Lopes, F.C. and Ligabue-Braun, R. (2021). Agro-industrial residues: eco-friendly and inexpensive substrates for microbial pigments production. *Frontiers in Sustainable Food Systems 5*: 65.
29. Prueckler, M., Siebenhandl-Ehn, S., Apprich, S. et al. (2014). Wheat bran-based biorefinery 1: composition of wheat bran and strategies of functionalization. *LWT- Food Science and Technology 56* (2): 211–221.
30. Hell, J., Kneifel, W., Rosenau, T. et al. (2014). Analytical techniques for the elucidation of wheat bran constituents and their structural features with emphasis on dietary fiber – a review. *Trends in Food Science & Technology 35* (2): 102–113.
31. Santos, F.A., de Carvalho-Gonçalves, L.C., de Carvalho Cardoso-Simões, A.L. et al. (2021). Evaluation of the production of cellulases by *Penicillium* sp. FSDE15 using corncob and wheat bran as substrates. *Bioresource Technology Reports 14*: 100648.
32. Amin, F., Arooj, T., Nazli, Z.I. et al. (2021). Exo-polygalacturonase production from agro-waste by *Penicillium fellutanum* and insight into thermodynamic, kinetic, and fruit juice clarification. *Biomass Conversion and Biorefinery 25*: 1–1.
33. Carrillo-Nieves, D., Saldarriaga-Hernandez, S., Gutiérrez-Soto, G. et al. (2020). Biotransformation of agro-industrial waste to produce lignocellulolytic enzymes and bioethanol with a zero waste. *Biomass Conversion and Biorefinery 9*: 1–2.
34. Singh, D.P., Prabha, R., Renu, S. et al. (2019). Agrowaste bioconversion and microbial fortification have prospects for soil health, crop productivity, and eco-enterprising. *International Journal of Recycling of Organic Waste in Agriculture 1*: 1–6.
35. Katileviciute, A., Plakys, G., Budreviciute, A. et al. (2019). A sight to wheat bran: high value-added products. *Biomolecules 9* (12): 887.
36. Babu, C.R., Ketanapalli, H., Beebi, S.K. et al. (2018). Wheat bran-composition and nutritional quality: a review. *Advances in Biotechnology and Microbiology 9* (1): 7.
37. Gul, K., Yousuf, B., Singh, A.K. et al. (2015). Rice bran: nutritional values and its emerging potential for development of functional food – a review. *Bioactive Carbohydrates and Dietary Fibre 6* (1): 24–30.
38. Huang, Y.P. and Lai, H.M. (2016). Bioactive compounds and antioxidative activity of colored rice bran. *Journal of Food and Drug Analysis 24* (3): 564–574.
39. Spaggiari, M., Dall'Asta, C., Galaverna, G. et al. (2021). Rice bran by-product: from valorization strategies to nutritional perspectives. *Foods 10* (1): 85.
40. Kovalcik, A., Obruca, S., and Marova, I. (2018). Valorization of spent coffee grounds: A review. *Food and Bioproducts Processing 110*: 104–119.
41. Agrawal, N., Verma, P., and Shahi, S.K. (2018). Degradation of polycyclic aromatic hydrocarbons (phenanthrene and pyrene) by the ligninolytic fungi Ganoderma lucidum isolated from the hardwood stump. *Bioresources and Bioprocessing 5* (1): 1–9.
42. Chambon, C.L., Mkhize, T.Y., Reddy, P. et al. (2018). Pretreatment of South African sugarcane bagasse using a low-cost protic ionic liquid: a comparison of whole, depithed, fibrous and pith bagasse fractions. *Biotechnology for Biofuels 11* (1): 1–6.

43. Ravindran, R., Williams, G.A., and Jaiswal, A.K. (2019). Spent coffee waste as a potential media component for xylanase production and potential application in juice enrichment. *Foods 8* (11): 585.
44. Saratale, G.D., Bhosale, R., Pugazendhi, A. et al. (2020). A review on valorization of spent coffee grounds (SCG) towards biopolymers and biocatalysts production. *Bioresource Technology 9*: 123800.
45. Guo, S., Awasthi, M.K., Wang, Y. et al. (2021). Current understanding in conversion and application of tea waste biomass: a review. *Bioresource Technology 10*: 125530.
46. Aghili, A.H., Toghyani, M., and Tabeidian, S.A. (2019). Effect of incremental levels of apple pomace and multi enzyme on performance, immune response, gut development and blood biochemical parameters of broiler chickens. *International Journal of Recycling of Organic Waste in Agriculture 8* (1): 321–334.
47. Nile, S.H., Nile, A., Liu, J. et al. (2019). Exploitation of apple pomace towards extraction of triterpenic acids, antioxidant potential, cytotoxic effects, and inhibition of clinically important enzymes. *Food and Chemical Toxicology 131*: 110563.
48. Awasthi, M.K., Ferreira, J.A., Sirohi, R. et al. (2021). A critical review on the development stage of biorefinery systems towards the management of apple processing-derived waste. *Renewable and Sustainable Energy Reviews 143*: 110972.
49. Okoro, O.V., Nie, L., Hobbi, P. et al. (2021). Valorization of waste apple pomace for production of platform biochemicals: a multi-objective optimization study. *Waste and Biomass Valorization 1*: 1–5.
50. Yates, M., Gomez, M.R., Martın-Luengo, M.A. et al. (2017). Multivalorization of apple pomace towards materials and chemicals. Waste to wealth. *Journal of Cleaner Production 143*: 847–853.
51. Paz, A., Karnaouri, A., Templis, C.C. et al. (2020). Valorization of exhausted olive pomace for the production of omega-3 fatty acids by *Crypthecodinium cohnii*. *Waste Management 118*: 435–444.
52. Alcazar-Ruiz, A., Garcia-Carpintero, R., Dorado, F. et al. (2021). Valorization of olive oil industry subproducts: ash and olive pomace fast pyrolysis. *Food and Bioproducts Processing 125*: 37–45.
53. Zema, D.A., Calabrò, P.S., Folino, A. et al. (2018). Valorisation of citrus processing waste: a review. *Waste Management 80*: 252–273.
54. Camargo, F.P., Sakamoto, I.K., Duarte, I.C. et al. (2019). Influence of alkaline peroxide assisted and hydrothermal pretreatment on biodegradability and bio-hydrogen formation from citrus peel waste. *International Journal of Hydrogen Energy 44* (41): 22888–22903.
55. Camargo, F.P., Sakamoto, I.K., Delforno, T.P. et al. (2021). Microbial and functional characterization of an allochthonous consortium applied to hydrogen production from Citrus Peel Waste in batch reactor in optimized conditions. *Journal of Environmental Management 291*: 112631.
56. Mahato, N., Sharma, K., Sinha, M. et al. (2021). Biotransformation of citrus waste-I: production of biofuel and valuable compounds by fermentation. *Processes 9* (2): 220.
57. Pathak, P.D., Mandavgane, S.A., and Kulkarni, B.D. (2016). Valorization of banana peel: a biorefinery approach. *Reviews in Chemical Engineering 32* (6): 651–666.
58. Zehra, M., Syed, M.N., and Sohail, M. (2020). Banana peels: a promising substrate for the coproduction of pectinase and xylanase from *Aspergillus fumigatus* MS16. *Polish Journal of Microbiology 69* (1): 19.
59. Sogi, D.S. (2020). Value-added processing and utilization of banana by-products. *Handbook of Banana Production, Postharvest Science, Processing Technology, and Nutrition 31*: 191–206.
60. Banerjee, S., Patti, A.F., Ranganathan, V. et al. (2019). Hemicellulose based biorefinery from pineapple peel waste: Xylan extraction and its conversion into xylooligosaccharides. *Food and Bioproducts Processing 117*: 38–50.
61. Sepúlveda, L., Romaní, A., Aguilar, C.N. et al. (2018). Valorization of pineapple waste for the extraction of bioactive compounds and glycosides using autohydrolysis. *Innovative Food Science & Emerging Technologies 47*: 38–45.

62. Ridzuan, N.A., Shaarani, S.M., Arshad, Z.I. et al. (2020). *Study on Enzyme Activities in Pineapple Fruit and Pineapple Waste to Be Applied as Poultry Supplement*, IOP Conference Series: Materials Science and Engineering, vol. *991*, No. 1, 012064. IOP Publishing.
63. Dahunsi, S.O., Ogunwole, J.O., Owoseni, A.A. et al. (2021). Valorization of pineapple peel and poultry manure for clean energy generation. *Food and Energy Security 5*: e228.
64. Sooch, B.S., Lugani, Y., and Singh, R.S. (2019). 2. Agro-industrial lignocellulosic residues for the production of industrial enzymes. In: *Industrial Biotechnology* (ed. M. Yadav, V. Kumar and N. Sehrawat), 31–50. De Gruyter.
65. Sabater, C., Ruiz, L., Delgado, S. et al. (2020). Valorization of vegetable food waste and by-products through fermentation processes. *Frontiers in Microbiology 11*: 2604.
66. Ancuţa, P. and Sonia, A. (2020). Oil press-cakes and meals valorization through circular economy approaches: a review. *Applied Sciences 10* (21): 7432.
67. Arrutia, F., Binner, E., Williams, P. et al. (2020). Oilseeds beyond oil: press cakes and meals supplying global protein requirements. *Trends in Food Science & Technology 100*: 88–102.
68. Szymczak, T., Cybulska, J., Podleśny, M. et al. (2021). Various perspectives on microbial lipase production using Agri-food waste and renewable products. *Agriculture 11* (6): 540.
69. Ricciardi, P., Cillari, G., Carnevale Miino, M. et al. (2020). Valorization of agro-industry residues in the building and environmental sector: a review. *Waste Management & Research 38* (5): 487–513.
70. Girelli, A.M., Astolfi, M.L., and Scuto, F.R. (2020). Agro-industrial wastes as potential carriers for enzyme immobilization: a review. *Chemosphere 244*: 125368.
71. Ravindran, R., Jaiswal, S., Abu-Ghannam, N. et al. (2018). A comparative analysis of pretreatment strategies on the properties and hydrolysis of brewers' spent grain. *Bioresource Technology 248*: 272–279.
72. Bianco, A., Budroni, M., Zara, S. et al. (2020). The role of microorganisms on biotransformation of brewers' spent grain. *Applied Microbiology and Biotechnology 2*: 1–8.
73. Sirohi, R., Tarafdar, A., Singh, S. et al. (2020). Green processing and biotechnological potential of grape pomace: current trends and opportunities for sustainable biorefinery. *Bioresource Technology 2*: 123771.
74. Leite, P., Sousa, D., Fernandes, H. et al. (2021). Recent advances in production of lignocellulolytic enzymes by solid-state fermentation of agro-industrial wastes. *Current Opinion in Green and Sustainable Chemistry 27*: 100407.
75. Nema, A., Patnala, S.H., Mandari, V. et al. (2019). Production and optimization of lipase using *Aspergillus niger* MTCC 872 by solid-state fermentation. *Bulletin of the National Research Centre 43* (1): 1–8.
76. Putri, D.N., Khootama, A., Perdani, M.S. et al. (2020). Optimization of *Aspergillus niger* lipase production by solid state fermentation of agro-industrial waste. *Energy Reports 6*: 331–335.
77. Navaneethapandian, U., Kumar, A.G., Liduja, K. et al. (2020). Biocatalyst: cellulase production in solid state fermentation (ssf) using rice bran as substrate. *Biointerface Research in Applied Chemistry 11*: 7689–7699.
78. Kumar, A., Kumar, V., and Singh, B. (2020). Cellulosic and hemicellulosic fractions of sugarcane bagasse: potential, challenges and future perspective. *International Journal of Biological Macromolecules 1* (169): 564–582.
79. Corrêa, C.L., Penha, E.M., Freitas-Silva, O. et al. (2021). Enzymatic technology application on coffee co-products: a review. *Waste and Biomass Valorization 12* (7): 3521–3540.
80. Xu, L., Sun, K., Wang, F. et al. (2020). Laccase production by Trametes versicolor in solid-state fermentation using tea residues as substrate and its application in dye decolorization. *Journal of Environmental Management 270*: 110904.
81. Bhardwaj, N. and Verma, P. (2021). Xylanases: a helping module for the enzyme biorefinery platform. In: *Bioenergy Research: Revisiting Latest Development* (ed. P. Verma), 161–180. Singapore: Springer.
82. Lyu, F., Luiz, S.F., Azeredo, D.R. et al. (2020). Apple pomace as a functional and healthy ingredient in food products: a review. *Processes 8* (3): 319.

83. Zhang, W., Wang, J.J., Gao, Y. et al. (2020). Bacterial cellulose synthesized with apple pomace enhanced by ionic liquid pretreatment. *Preparative Biochemistry & Biotechnology 50* (4): 330–340.
84. Prajapati, J., Dudhagara, P., and Patel, K. (2021). Production of thermal and acid-stable pectinase from *Bacillus subtilis* strain BK-3: optimization, characterization, and application for fruit juice clarification. *Biocatalysis and Agricultural Biotechnology 10*: 102063.
85. Sharma, D. and Mahajan, R. (2020). Development of methodology for concurrent maximum production of alkaline xylanase–pectinase enzymes in short submerged fermentation cycle. *Waste and Biomass Valorization 11* (11): 6065–6072.
86. Vama, L.A. and Cherekar, M.N. (2020). Production, extraction and uses of eco-enzyme using citrus fruit waste: wealth from waste. *Asian Journal of Microbiology, Biotechnology & Environmental Sciences 22* (2): 346–351.
87. Sousa, D., Salgado, J.M., Cambra-López, M. et al. (2021). Degradation of lignocellulosic matrix of oilseed cakes by solid-state fermentation: fungi screening for enzymes production and antioxidants release. *Journal of the Science of Food and Agriculture 102* (4): 1550–1560.
88. Sulong, M.R., Hamid, H., Sivam, A. et al. (2021). Utilization of banana peel-wastes for the production of low-cost and Shariah-compliant amylolytic enzyme. In: *Enhancing Halal Sustainability* (ed. S. NNM, N. Lateh, N.F. Zarmani, et al.), 249–258. Singapore: Springer.
89. Atalla, S.M. and El Gamal, N.G. (2020). Production and characterization of xylanase from pomegranate peel by *Chaetomium globosum* and its application on bean under greenhouse condition. *Bulletin of the National Research Centre 44* (1): 1–1.
90. Ahmed, N.E., El Shamy, A.R., and Awad, H.M. (2020). Optimization and immobilization of amylase produced by *Aspergillus terreus* using pomegranate peel waste. *Bulletin of the National Research Centre 44* (1): 1–2.
91. Amraini SZ, Pratiwi ED, Ismala UN. Production of Cellulase and Xylanase from Eupenicillium Javanicum by Solid-State Fermentation Utilizing Pineapple Crown Leaves Waste as the Substrate. In *Journal of Physics*: Conference Series 2020 *1655*, No. 1, p. 012113. IOP Publishing.
92. Meena, L., Sengar, A.S., Neog, R. et al. (2021). Pineapple processing waste (PPW): bioactive compounds, their extraction, and utilisation: a review. *Journal of Food Science and Technology 8*: 1–3.
93. Farooq, M.A., Ali, S., Hassan, A. et al. (2021). Biosynthesis and industrial applications of α-amylase: a review. *Archives of Microbiology 22*: 1–2.
94. Paul, J.S., Beliya, E., Tiwari, S. et al. (2020). Production of biocatalyst α-amylase from agro-waste 'rice bran' by using *Bacillus tequilensis* TB5 and standardizing its production process. *Biocatalysis and Agricultural Biotechnology 26*: 101648.
95. Ce, O., Po, O., and Obasuyi, C. (2019). Influence of nitrogen sources on the production of single cell protein and amylase by *Aspergillus niger* using wheat bran. *Sau Science-Tech Journal 4*: 75–84.
96. Adebare, B.S., Johnson, A., Olanrewaju, T. et al. (2021 Oct 5). Production and optimization of alpha amylase from *Aspergillus niger* using TME 419 cassava peel as substrate. *African Journal of Biological Sciences 3*: 50–59.
97. Nill, J.D. and Jeoh, T. (2020). The role of evolving interfacial substrate properties on heterogeneous cellulose hydrolysis kinetics. *ACS Sustainable Chemistry & Engineering 8* (17): 6722–6733.
98. Gao, J., Weng, H., Zhu, D. et al. (2008). Production and characterization of cellulolytic enzymes from the thermoacidophilic fungal *Aspergillus terreus* M11 under solid-state cultivation of corn stover. *Bioresource Technology 99* (16): 7623–7629.
99. Dhillon, G.S., Oberoi, H.S., Kaur, S. et al. (2011). Value-addition of agricultural wastes for augmented cellulase and xylanase production through solid-state tray fermentation employing mixed-culture of fungi. *Industrial Crops and Products 34* (1): 1160–1167.
100. Marco, É.G., Heck, K., Martos, E.T. et al. (2017). Purification and characterization of a thermostable alkaline cellulase produced by *Bacillus licheniformis* 380 isolated from compost. *An. Acad. Brasil. Ciênc 89*: 2359–2370.

101. Kshirsagar, S.D., Saratale, G.D., Saratale, R.G. et al. (2016). An isolated Amycolatopsis sp. GDS for cellulase and xylanase production using agricultural waste biomass. *Journal of Applied Microbiology 120* (1): 112–125.
102. Waghmare, P.R., Kshirsagar, S.D., Saratale, R.G. et al. (2014). Production and characterization of cellulolytic enzymes by isolated *Klebsiella* sp. PRW-1 using agricultural waste biomass. *Emirates Journal of Food and Agriculture 26* (1): 44–59.
103. Thakur, V., Kumar, V., Kumar, V. et al. (2021). Genomic insights driven statistical optimization for production of efficient cellulase by Himalayan *Thermophilic bacillus* sp. PCH94 using agricultural waste. *Waste and Biomass Valorization 5*: 1–3.
104. Chaudhary, R., Kuthiala, T., Singh, G. et al. (2021). Current status of xylanase for biofuel production: a review on classification and characterization. *Biomass Conversion and Biorefinery 16*: 1–9.
105. Basit, A., Jiang, W., and Rahim, K. (2020). *Xylanase and its Industrial Applications, Biotechnological Applications of Biomass* (ed. T.P. Basso, T.O. Basso and L.C. Basso). IntechOpen https://doi.org/10.5772/intechopen.92156.
106. Chapla, D., Divecha, J., Madamwar, D. et al. (2010). Utilization of agro-industrial waste for xylanase production by *Aspergillus foetidus* MTCC 4898 under solid state fermentation and its application in saccharification. *Biochemical Engineering Journal 49* (3): 361–369.
107. Nagar, S. and Gupta, V.K. (2021). Hyper production and eco-friendly bleaching of Kraft pulp by xylanase from *Bacillus pumilus* SV-205 using agro waste material. *Waste and Biomass Valorization 12* (7): 4019–4031.
108. Saroj, P., Manasa, P., and Narasimhulu, K. (2020). Optimization of xylanase production using ragi (*Eleusine coracana*) husk as a substrate by *Aspergillus fumigatus* JCM 10253 through response surface methodology. *Biomass Conversion and Biorefinery 28*: 1–1.
109. Kolade, I.O., Yousuf, M.B., Ajijolakewu, K.A. et al. (2021). Optimization of conditions for xylanase production using *Aspergillus tubingensis* under different carbon sources. *Fountain Journal of Natural and Applied Sciences 10* (1): 34–44.
110. Razali, S.A., Rasit, N., and Ooi, C.K. (2021). Statistical analysis of xylanase production from solid state fermentation of rice husk associated fungus *Aspergillus niger*. *Materials Today: Proceedings 39*: 1082–1087.
111. Ghosh, A., Sutradhar, S., and Baishya, D. (2019). Delineating thermophilic xylanase from *Bacillus licheniformis* DM5 towards its potential application in xylooligosaccharides production. *World Journal of Microbiology and Biotechnology 35* (2): 34.
112. Marimuthu, M., Sorimuthu, A., and Muruganantham, S. (2019). Production and optimization of xylanase enzyme from *Bacillus subtilis* using agricultural wastes by solid state fermentation. *International Journal of Pharmaceutical Investigation 9* (4): 169–173.
113. Sopalun, K. and Iamtham, S. (2020). Evaluation and optimization of pectinase production by endophytic fungi isolated from Thai orchids using agrowaste medium. *Journal of ISSAAS (International Society for Southeast Asian Agricultural Sciences) 26* (2): 86–98.
114. Anand, G., Leibman-Markus, M., Elkabetz, D. et al. (2021). Method for the production and purification of plant immuno-active xylanase from Trichoderma. *International Journal of Molecular Sciences 22* (8): 4214.
115. Satapathy, S., Soren, J.P., Mondal, K.C. et al. (2021). Industrially relevant pectinase production from *Aspergillus parvisclerotigenus* KX928754 using apple pomace as the promising substrate. *Journal of Taibah University for Science 15* (1): 347–356.
116. Almowallad, S.A., Aljobair, M.O., Alkuraieef, A.N. et al. (2021). Utilization of agro-industrial orange peel and sugar beet pulp wastes for fungal endo-polygalacturonase production. *Saudi Journal of Biological Sciences 29* (2): 963–969.
117. Ahmed, N.N. and Abd El Rahman, H.M. (2021). Optimizing the production of pectinase of orange peel waste by *Penicillium Chrysogenum* Mf318506 using response surface methodology in submerged fermentation. *Journal of Microbiology, Biotechnology and Food Sciences 7*: e3931.
118. Ajayi AA, Lawal B, Salubi AE, et al. Pectinase production by *Aspergillus niger* using pineapple peel pectin and its application in coconut oil extraction. In IOP Conference Series: *Earth and Environmental Science* 2021 *655*, No. 1, p. 012014). IOP Publishing.

119. Doan, C.T., Chen, C.L., Nguyen, V.B. et al. (2021). Conversion of pectin-containing by-products to pectinases by *Bacillus amyloliquefaciens* and its applications on hydrolyzing banana peels for prebiotics production. *Polymers 13* (9): 1483.
120. Maria de Fátima, M., Cavalcante, L.S., Gudiña, E.J. et al. (2021). Sustainable lipase production by Diutina rugosa NRRL Y-95 through a combined use of agro-industrial residues as feedstock. *Applied Biochemistry and Biotechnology 193* (2): 589–605.
121. Anand, G., Yadav, S., Gupta, R. et al. (2020). Pectinases: from microbes to industries. *Microorganisms for Sustainable Environment and Health 21*: 287.
122. Yadav, S., Anand, G., Singh, V.K. et al. (2021). Molecular cloning and structural insights into pectin lyase proteins from different strains of fusarium. *Current Proteomics 18* (3): 326–337.
123. Patel, R., Prajapati, V., Trivedi, U. et al. (2020). Optimization of organic solvent-tolerant lipase production by *Acinetobacter* sp. UBT1 using deoiled castor seed cake. *3 Biotech 10* (12): 1–3.
124. Irfansyah AR, Koentjoro MP, Ekawati I, et al. Utilization of *Moringa oleifera* seed waste as substrate in lipase production under different pretreatments. In IOP Conference Series: *Earth and Environmental Science* 2021 *649*, No. 1, p. 012002). IOP Publishing.
125. Eichler, P., Bastiani, D.C., Santos, F.A. et al. (2020). Lipase production by *Aspergillus brasiliensis* in solid-state cultivation of malt bagasse in different bioreactors configurations. *An. Acad. Brasil. Ciênc 12*: 92.
126. Mayolo-Deloisa, K., González-González, M., and Rito-Palomares, M. (2020). Laccases in food industry: bioprocessing, potential industrial and biotechnological applications. *Frontiers in Bioengineering and Biotechnology 8*: 222.
127. Kumar, A., Singh, A.K., Bilal, M. et al. (2021). Sustainable production of thermostable laccase from agro-residues waste by *Bacillus aquimaris* AKRC02. *Catalysis Letters 14*: 1–7.
128. An, Q., Li, C.S., Yang, J. et al. (2021). Evaluation of laccase production by two white-rot fungi using solid-state fermentation with different agricultural and forestry residues. *BioResources 16* (3): 5287–5300.
129. Hafid, H.S., Baharuddin, A.S., Mokhtar, M.N. et al. (2021). Enhanced laccase production for oil palm biomass delignification using biological pretreatment and its estimation at biorefinary scale. *Biomass and Bioenergy 144*: 105904.
130. Majumdar, S. and Bhowal, J. (2021). Studies on production and evaluation of biopigment and synthetic dye decolorization capacity of laccase produced by *A. oryzae* cultivated on agro-waste. *Bioprocess and Biosystems Engineering 30*: 1–6.
131. Confortin, T.C., Spannemberg, S.S., Todero, I. et al. (2019). Microbial enzymes as control agents of diseases and pests in organic agriculture. *New and Future Developments in Microbial Biotechnology and Bioengineering 1*: 321–332.
132. Arora, S., Rani, R., and Ghosh, S. (2018). Bioreactors in solid state fermentation technology: design, applications and engineering aspects. *Journal of Biotechnology 269*: 16–34.
133. Hadj Saadoun, J., Bertani, G., Levante, A. et al. (2021). Fermentation of agri-food waste: a promising route for the production of aroma compounds. *Foods 10* (4): 707.
134. Gullon, P., Gullon, B., Astray, G. et al. (2020). Valorization of byproducts from olive oil industry and added-value applications for innovative functional foods. *Food Research International 137*: 109683.
135. Chung, Y.M., Lee, S.Y., Lin, Y.K. et al. (2020). The high potential of a red-light submerged fermentation technique in the biofunctionality improvement of rambutan extracts. *Nutrition & Food Science 51* (6): 920–935.
136. Garrido-Galand, S., Asensio-Grau, A., Calvo-Lerma, J. et al. (2021). The potential of fermentation on nutritional and technological improvement of cereal and legume flours: a review. *Food Research International 11*: 110398.
137. Karimi, F., Mazaheri, D., Saei Moghaddam, M. et al. (2021). Solid-state fermentation as an alternative technology for cost-effective production of bioethanol as useful renewable energy: a review. *Biomass Conversion and Biorefinery 6*: 1–7.
138. Vassileva, M., Malusà, E., Sas-Paszt, L. et al. (2021). Fermentation strategies to improve soil bio-inoculant production and quality. *Microorganisms 9* (6): 1254.

139. Ahmed, T., Rana, M.R., Zzaman, W. et al. (2021). Optimization of substrate composition for pectinase production from Satkara (*Citrus macroptera*) peel using *Aspergillus niger*-ATCC 1640 in solid-state fermentation. *Heliyon 7* (10): e08133.

140. Labrath, Y.P. and Gaikar, V.G. (2020). Solid State Fermentation of Orange Peels for Production of Cellulase, Pectinase and Recovery of Orange Oil using Aspergillus Species NCIM 1432. Research square. PPR: PPR253938.

141. Hidayah AA, Azizah, Winarsa R, Muzakhar K. Utilization of coffee pulp as a substrate for pectinase production by *Aspergillus* sp. VTMS through solid state fermentation. In *AIP Conference Proceedings* 2020 Nov 16 (Vol. *2296*, No. 1, p. 020012). AIP Publishing LLC.

142. Adedayo, M.R., Mohammed, M.T., Ajiboye, A.E. et al. (2021). Pectinolytic activity of *Aspergillus niger* and *Aspergillus flavus* grown on grapefruit (citrus Parasidis) peel in solid state fermentation. *Global Journal of Pure and Applied Sciences 27* (2): 93–105.

143. Arekemase, O.M., Omotosho, I.O., Agbabiaka, T.O. et al. (2020). Optimization of bacteria pectinolytic enzyme production using banana peel as substrate under submerged fermentation. *Science World Journal 15* (1): 56–63.

144. Begum, G. and Munjam, S. (2021). Carbon and nitrogen sources effect on pectinase synthesis by *Aspergillus niger* under submerged fermentation. *Biosciences, Biotechnology Research Asia 18* (1): 185–195.

145. Catalán, E. and Sánchez, A. (2020). Solid-state fermentation (SSF) versus submerged fermentation (SmF) for the recovery of cellulases from coffee husks: a life cycle assessment (LCA) based comparison. *Energies 13* (11): 2685.

146. Kurt, A.S. and Cekmecelioglu, D. (2021). Bacterial cellulase production using grape pomace hydrolysate by shake-flask submerged fermentation. *Biomass Conversion and Biorefinery 2*: 1–8.

147. Lodha, A., Pawar, S., and Rathod, V. (2020). Optimised cellulase production from fungal co-culture of *Trichoderma reesei* NCIM 1186 and *Penicillium citrinum* NCIM 768 under solid state fermentation. *Journal of Environmental Chemical Engineering 8* (5): 103958.

148. Namnuch, N., Thammasittirong, A., and Thammasittirong, S.N. (2021). Lignocellulose hydrolytic enzymes production by *Aspergillus flavus* KUB2 using submerged fermentation of sugarcane bagasse waste. *Mycology 12* (2): 119–127.

149. Shakir, H.A., Anwar, A., Irfan, M. et al. (2020). Statistical optimization of xylanase from *Bacillus licheniformis* using Banana peels in submerged fermentation. *Iranian Journal of Science and Technology, Transactions A: Science 44*: 981–991.

150. Intasit, R., Cheirsilp, B., Suyotha, W. et al. (2021). Purification and characterization of a highly-stable fungal xylanase from *Aspergillus tubingensis* cultivated on palm wastes through combined solid-state and submerged fermentation. *Preparative Biochemistry & Biotechnology 27*: 1–7.

151. Stephen, A.C., Adeniyi, O.A., and Hadiza, J. (2020). Effect of optimization conditions on submerged fermentation of corn bran for the production of xylanase enzyme. *World Journal of Advanced Research and Reviews 5* (2): 019–025.

152. Ezeilo, U.R., Wahab, R.A., and Mahat, N.A. (2020). Optimization studies on cellulase and xylanase production by Rhizopus oryzae UC2 using raw oil palm frond leaves as substrate under solid state fermentation. *Renewable Energy 156*: 1301–1312.

153. Azzouz, Z., Bettache, A., Djinni, I. et al. (2020). Biotechnological production and statistical optimization of fungal xylanase by bioconversion of the lignocellulosic biomass residues in solid-state fermentation. *Biomass Conversion and Biorefinery 28*: 1–3.

154. Moran-Aguilar, M.G., Costa-Trigo, I., Calderón-Santoyo, M. et al. (2021). Production of cellulases and xylanases in solid-state fermentation by different strains of *Aspergillus niger* using sugarcane bagasse and brewery spent grain. *Biochemical Engineering Journal 172*: 108060.

155. Contato, A.G., In, F.D., Brugnari, T. et al. (2020). Solid-state fermentation with orange waste: optimization of laccase production from *Pleurotus pulmonarius* CCB-20 and decolorization of synthetic dyes. *Acta Scientiarum. Biological Sciences 42*: e52699.

156. Yuliana, T., Putri, N.Z., Komara, D.Z. et al. (2020). Study of *Ganoderma lucidum* in laccase production using corncob and paddies straw substrates on submerged fermentation system. *Pakistan Journal of Biological Sciences: PJBS 23* (8): 1060–1065.

157. Nuhu, A., Hussaini, I., Gide, S. et al. (2020). Production of laccase by fungi isolated from soil via submerged fermentation using corn cob as substrate. *FUDMA Journal of Sciences 4* (3): 224–229.
158. Casas-Godoy, L., González-Escobar, J.L., Mathis, A.G. et al. (2020). Revalorization of untreated Brewer's spent grain: novel and versatile feedstock to produce cellulases, lipases, and yeast biomass in a biorefinery approach. *Biomass Conversion and Biorefinery 18*: 1–2.
159. Sales, J.C., de Castro, A.M., Ribeiro, B.D. et al. (2020). Supplementation of watermelon peels as an enhancer of lipase and esterase production by *Yarrowia lipolytica* in solid-state fermentation and their potential use as biocatalysts in poly (ethylene terephthalate) (PET) depolymerization reactions. *Biocatalysis and Biotransformation 38* (6): 457–468.
160. Nadeem, F., Mehmood, T., Anwar, Z. et al. (2021). Optimization of bioprocess steps through response surface methodology for the production of immobilized lipase using *Chaetomium globosum* via solid-state fermentation. *Biomass Conversion and Biorefinery 21*: 1–2.
161. Martins, P.A., Pacheco, T.F., de Camargo, B.R. et al. (2021). Solid-state fermentation production and characterization of an alkaline lipase from a newly isolated Burkholderia gladioli strain. *Preparative Biochemistry & Biotechnology 14*: 1–10.
162. Balakrishnan, M., Jeevarathinam, G., Kumar, S.K. et al. (2021). Optimization and scale-up of α-amylase production by *Aspergillus oryzae* using solid-state fermentation of edible oil cakes. *BMC Biotechnology 21* (1): 1–1.
163. Masoodi, N., Mobini-Dehkordii, M., and Saffar, B. (2021). Optimization of alpha-amylase production using *Bacillus subtilis* PTCC 1720 under solid-state fermentation and partial purification of the enzyme. *Biological Journal of Microorganism 10* (39): 1–1.
164. Fadel, M., AbdEl-Halim, S., Sharada, H. et al. (2020). Production of glucoamylase, α-amylase and cellulase by *Aspergillus oryzae* F-923 cultivated on wheat bran under solid state fermentation. *Journal of Advances in Biology & Biotechnology 30*: 8–22.
165. Melnichuk, N., Braia, M.J., Anselmi, P.A. et al. (2020). Valorization of two agroindustrial wastes to produce alpha-amylase enzyme from *Aspergillus oryzae* by solid-state fermentation. *Waste Management 106*: 155–161.
166. Tan, Y.X., Mok, W.K., and Chen, W.N. (2020). Potential novel nutritional beverage using submerged fermentation with *Bacillus subtilis* WX-17 on brewers' spent grains. *Heliyon 6* (6): e04155.
167. Iram, N., Shakir, H.A., Irfan, M. et al. (2021). Statistical optimization of amylase production using grape fruit peels in submerged fermentation. *Acta Scientiarum. Technology 43*: e50538.
168. Vashisht, A., Thakur, K., Kauldhar, B.S. et al. (2019). Waste valorization: identification of an ethanol tolerant bacterium *Acetobacter pasteurianus* SKYAA25 for acetic acid production from apple pomace. *The Science of the Total Environment 690*: 956–964.
169. Fraga, J.L., Souza, C.P., Pereira, A.D. et al. (2021). Palm oil wastes as feedstock for lipase production by *Yarrowia lipolytica* and biocatalyst application/reuse. *3 Biotech 11* (4): 1–9.
170. Naher, L., Fatin, S.N., Sheikh, M.A. et al. (2021). Cellulase enzyme production from filamentous fungi *Trichoderma reesei* and *Aspergillus awamori* in submerged fermentation with rice straw. *Journal of Fungi 7* (10): 868.

INDEX

Microbial Biotechnology: Role in Ecological Sustainability and Research, First Edition.
Edited by Pankaj Chowdhary, Sujata Mani, and Preeti Chaturvedi.